D1294375

THE PANNONIAN BASIN

A STUDY IN BASIN EVOLUTION

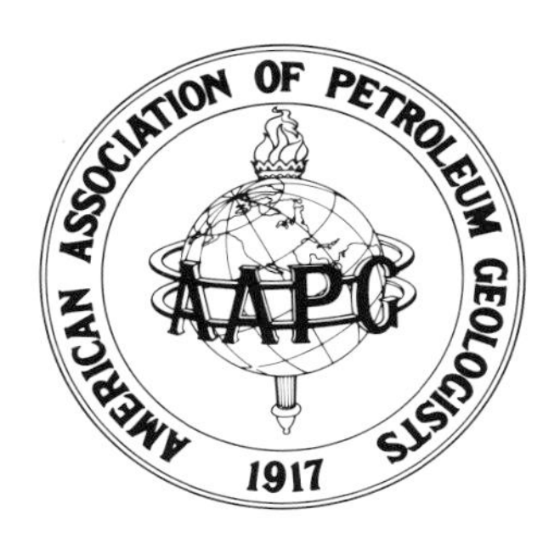

AAPG MEMOIR 45

The Pannonian Basin

A Study in Basin Evolution

AAPG Memoir 45

Edited by

Leigh H. Royden

Massachusetts Institute of Technology

and

Ferenc Horváth

Eötvös University

Published by
The American Association of Petroleum Geologists
Tulsa, Oklahoma 74101, U.S.A.
and
The Hungarian Geological Society
Budapest, Hungary

Published May, 1988

ISBN: 0-89181-322-5
ISSN: 0065-731X

For the Association:

Elected editor: James Helwig
Science director: Ronald L. Hart
Science editor: Victor V. Van Beuren
Project editor: Anne H. Thomas
Special editor and production supervisor: Kathy A. Walker

For the Hungarian Geological Society:

Chairman: G. Hámor
General Secretary: I. Bérczi

Special thanks for invaluable technical assistance to Marika Horváth and Tina Freudenberger.

Foreword

This volume is one of many scientific works published during the seven-year cooperative scientific effort by Hungarian and American earth scientists. It is designed to present an overview of the Pannonian basin and its evolution in the framework of the Alpine–Carpathian mountain system. The joint Hungarian–American project was conceived and organized through the discussions and consultations of L. Stegena and F. Horváth of Eötvös University and J. Sclater, B. C. Burchfiel, and L. Royden of the Massachusetts Institute of Technology. Support for the initiation of joint discussions and the establishment of a cooperative agreement was provided by Shell Oil Company. A three-year study of the Pannonian basin was developed and funded by the Hungarian Academy of Sciences and the International Division of the U.S. National Science Foundation. Later, because the scope of the project required additional work, continuation of the study was further supported by the Hungarian Academy of Sciences and the Earth Science Division of the U.S. National Science Foundation. As the cooperation proceeded, scientists from other organizations, because of mutual scientific interests, contributed their efforts to this study, and the cooperative effort grew to include scientists from the Hungarian Geological Survey, the Hungarian Oil and Gas Trust, and the Geochemical Institute of the Hungarian Academy of Sciences.

The present volume can also be considered as an outgrowth of a five-day symposium held in Veszprém, Hungary in 1982 entitled "Evolution of Extensional Basins within Regions of Compression with Emphasis on the Intra-Carpathian Region." The symposium was sponsored jointly by the Hungarian Academy of Sciences, the Veszprém Academy of Sciences, the Hungarian Oil and Gas Trust, the Hungarian Geological Survey, and the U.S. National Science Foundation. At this meeting the preliminary scientific results from the cooperative study were presented to an international group of 80 scientists from Europe and North America. Papers were also presented by other scientists who had interests in the Pannonian basin, other extensional basins, and general processes of lithospheric extension. It was realized that there were many scientists with diverse interests working on the Pannonian basin and related problems. Because the papers that were presented contained a wide range of data and excellent ideas and concepts, the organizers decided to gather together and present a sampling of this broad scientific effort in a single volume. They felt that a geological summary of the Pannonian basin and related extensional processes would provide a way to make these excellent scientific studies available, and they hoped that this study might serve as a good example of a basin study.

Preparation, organization, and editing of the volume has been the task of two editors, one American, L. Royden, and one Hungarian, F. Horváth. Both editors have been involved in the cooperative scientific effort from its beginning to its closing phases. Great praise should be given to the editors for the major effort of editing papers presented by authors from many different countries into a coherent and comprehensive overview of the Pannonian basin. The papers have been written and edited so that they are not just a collection of isolated papers but are cross-referenced to provide continuity throughout the volume. Although they form part of a comprehensive overview, each paper is also a complete scientific work that can stand on its own merits. The editors and the authors are to be congratulated for working together to produce what we feel is an excellent scientific presentation.

Special thanks are also due to Tina Freudenberger for typing, Pat McDowell and Laci Becker for drafting, and Marika Horváth for typing and drafting.

From the spirit of cooperation that has existed between so many scientists throughout the cooperative effort, it is only fitting that this volume is published jointly by the American Association of Petroleum Geologists and the Geological Society of Hungary.

We feel we can speak for all the participants who have been associated with this joint Hungarian–American study when we say that it has been an exciting and rewarding effort, not only scientifically, but personally as well. It has been a splendid example of how an international cooperative scientific effort can work well, and we are indebted to all the Hungarian institutions who gave us generous access to and use of their data for this cooperative project.

B. C. Burchfiel
L. Stegena

Table of Contents

Introduction to the Pannonian Region

The Pannonian basin system is an integral part of the Alpine mountain belts of east-central Europe. It is completely encircled by the Carpathian Mountains to the north and east, the Dinaric Alps to the south, and the Southern and Eastern Alps to the west (see Map 2 in the accompanying packet). In 1912, Kober defined the Pannonian basin as one of the type "Zwischengebirge," a relatively undeformed region characterized by block faulting and situated between externally vergent thrust belts. More recent studies using subsurface data have shown that the Pannonian area was extensively deformed by Mesozoic thrusting and subsequently disrupted by a complex system of Cenozoic normal and wrench faults. Thus, the Pannonian "massif" has undergone several types of deformation, which are partly hidden by a thick sequence of sedimentary rocks of Neogene–Quaternary age.

The Pannonian basin is actually a system of small, deep basins separated by relatively shallow basement blocks (Map 8). The Neogene–Quaternary sedimentary rocks exceed 7 km in thickness in some areas (Map 8), and the basin system (including the Transylvanian basin) is about 400 km from north to south and 800 km from east to west. It is currently interpreted by most workers as a Mediterranean back arc extensional basin of middle Miocene age (Horváth, this volume). Together with the surrounding mountain belts, it lies within parts of seven different countries with different languages.

The Carpathians, Eastern Alps, and Dinarides, which surround the Pannonian basin, are the result of Mesozoic and Cenozoic continental collision between Europe and several continental fragments to the south, including Africa. Thrusting was directed outward from the present Pannonian basin toward the European platform and the Adriatic region. In all the orogenic belts, the interior parts of the thrust belts were deformed in Mesozoic time, while the outer parts were deformed in Tertiary time. A Benioff zone is still present beneath the southeastern Carpathian bend.

The Pannonian basin overlies the most internal Mesozoic thrust sheets of the Carpathian Mountains (and probably of the Dinarides as well), and the existence of some large-scale overthrusts has been demonstrated in the basement of the Pannonian basement (Rumpler and Horváth, this volume). However, much of the Mesozoic structure of the Mesozoic, Paleozoic, and crystalline basement rocks has been strongly overprinted by Cenozoic tectonic events. Older zones of weakness may have localized some of the Cenozoic faulting along major tectonic lines. Paleogene sedimentary rocks occur beneath the Neogene Pannonian basin only in some places (Royden and Báldi, this volume).

Neogene sedimentation and extension in the Pannonian basin system occurred during the final stages of thrusting and folding in the outer part of the Carpathians. Sedimentation is of different ages in different parts of the basin system. Rapid sedimentation in the northernmost basins (Vienna and Transcarpathian basins) occurred mainly in early to middle Miocene time. In middle Miocene time, rapid subsidence began in the basins farther south (in the Danube, Zala, and Drava basins and in the Derecke, Makó, and Békés depressions of the Pannonian basin) (Map 1). During late Miocene time, rapid sedimentation was mainly confined to the Little Hungarian and Great Hungarian plains and to the southeastern part of the Drava region.

Large synsedimentary normal faults of Miocene age are clearly associated with rapid Miocene basin subsidence in most of the basins within the Pannonian system. In the Little Hungarian and Great Hungarian plains, however, much of the thick (more than 5 km) upper Miocene–Quaternary sedimentary sequence contains few obvious normal faults with large displacements (Rumpler and Horváth, this volume). Instead, middle Miocene subsidence within these two areas generated deep water basins that were rapidly filled by large prograding delta systems (Bérczi, this volume; Mattick et al., this volume). The complex system of faults observed throughout the Pannonian basin is thought to be related to late Miocene strike-slip and extensional deformation (Wessely, this volume; Royden, this volume; Tomek and Thon, this volume). Locally, some compressional deformation exists along some of the strike-slip faults.

Subsidence within the Pannonian basin and underthrusting in the Carpathian mountains were accompanied by eruption of Miocene calcalkaline volcanic rocks, mainly in the northern part of the Pannonian basin and in the East Carpathians (Póka, this volume; Salters et al., this volume). Alkali basalts were erupted and intruded in the Pannonian basin in Pliocene and Quaternary time. Both calcalkaline and alkaline lavas appear to be derived partly from the mantle and are probably partly related to the subduction of the European plate southward and eastward beneath the Carpathian Mountains.

Classically, the Pannonian basin has played an important role in paleontological studies of faunal migration and evolution. In 1924, Laskerev defined the Paratethys, consisting of the Pannonian basin, the Alpine foredeep areas, the Carpathians, etc. as a separate bioprovince, distinct from the Mediterranean bioprovince, and beginning at about the end of the Oligocene. The Paratethys contains biostratigraphic stages different from those in the Mediterranean (Steininger et al., this volume). In the central Paratethys, which includes the Pannonian basin system, the biostratigraphic stages can be correlated easily with the Mediterranean stages until late Miocene time when the Pannonian basin became totally and finally isolated from the Mediterranean. The late Miocene through Quaternary endemic faunas in the Pannonian area have caused difficulties in correlation of the Paratethyan biostratigraphic stages, both among individual basins in the Pannonian basin system and with the Mediterranean biostratigraphic stages (Nagymarosy and Müller, this volume). In particular, this problem has hindered correlation of the Pannonian (late Miocene) and younger stages throughout the basin areas. Current controversies revolve mainly around the absolute ages of the Paratethyan stages and their correlation with the worldwide biostratigraphic time scale (Horváth and Pogácsás, this volume). Some of the index fossils used to identify the Pannonian boundaries are probably controlled partly by facies changes, giving slightly diachronous boundaries across the basin areas.

Additional confusion has been created by the usage of the term "Pannonian" both for a late Miocene biostratigraphic stage (defined by Papp, 1953) and, in Hungary, for a stratigraphic sequence that begins at the base of Papp's Pannonian biostratigraphic stage and includes rocks up to latest Pliocene age. In this volume we have attempted to distinguish between the two usages by referring to the former as "Pannonian *sensu stricto*" (Pannonian *s. str.*) and to the latter as "Pannonian *sensu lato*" (Pannonian *s. l.*). The Pannonian *s. l.* can be divided into two parts. The lower Pannonian *s. l.* corresponds reasonably well to the Pannonian *s. str.* The upper Pannonian *s. l.* includes younger Miocene and Pliocene (Pontian, Dacian and Romanian stages). In most papers by Hungarian authors the term Pannonian is used in the sense of Pannonian *s. l.*, while most other authors use it in the sense of Pannonian *s. str.*

Abundant subsurface data is available from almost all of the Pannonian basin area. Within Hungary alone, more than 11,000 km of wildcat and production wells have been drilled, and more than 30,000 km of multifold stacked seismic profiles have been generated (Dank, this volume; see also Map 4). In addition to aiding in the primary objective of locating and recovering hydrocarbon deposits from sediments in the Pannonian basin and the Carpathian foredeep, the wealth of geologic information provided by this extensive subsurface control makes the Pannonian basin system one of the best understood basin systems in the world. The knowledge of regional and local structures gained through analysis of these and other data will, in turn, provide new strategies for locating hydrocarbon plays within the Pannonian basin.

REFERENCES

Kober, L., 1912, Über Bau und Entstehung der Ostalpen: Mitt. Geol. Ges. Wien, v. 5, p. 368–481.

Laskerev, V. N., 1924, Sur les équivalents du Sarmatien supérieur en Serbie, *in* Recueil de travaux offert a M. Jovan Cvijíc par ses amis et collaborateurs, 13 p.

Papp, A., 1953, Die Molluskenfauna des Pannon im Wiener Becken: Mitteilungen der Geologischen Gesellschaft in Wien, v. 44, p. 86–222.

CHAPTER 1

Early Cenozoic Tectonics and Paleogeography of the Pannonian and Surrounding Regions

L. H. Royden
Massachusetts Institute of Technology
Department of Earth, Atmospheric and Planetary Sciences
Cambridge, MA 02139

T. Báldi
Loránd Eötvös University
Department of Geology
H-1088, Budapest
Múzeum krt. 4/a, Hungary

A series of palinspastic and paleogeographic reconstructions has been made for the Pannonian and surrounding regions for five time periods: (1) Coniacian-Paleocene, (2) early–middle Eocene, (3) late Eocene-early Oligocene, (4) late Oligocene–early Miocene, and (5) late Miocene. These maps were constructed by grouping together various crustal blocks that underwent similar phases of deformation or sedimentation into tectonostratigraphic units. We show how the present complex distribution of Mesozoic tectonostratigraphic units could have developed from a simple initial configuration during Cenozoic deformation of the Carpathian–Pannonian region, and that the formation, duration, and disruption of various Paleogene paleogeographic elements can be directly related to contemporaneous tectonic events.

In this analysis we interpret the elongate Hungarian Paleogene basin as a wrench related basin that formed along an east-northeast-trending zone of dextral shear in Eocene–Oligocene time. This shear zone was probably responsible for the Paleogene dislocation of the Apuseni Mountains from the inner West Carpathians. We further interpret the Pieniny Klippen belt as the result of convergence and sinistral shear active in part during Eocene time. This analysis suggests that the inner flysch zones (Podhale flysch, Szolnok-Maramures flysch, and Transcarpathian flysch) originally constituted a continuous flysch basin that was subsequently disrupted by roughly east-northeast-trending dextral shear zones.

Large shear zones such as those postulated in this paper are required partly because of the diachronous nature of the convergent boundary extending from the Eastern Alps to the East Carpathians and partly because of the different directions of thrusting around the belt. These shear zones separate areas of active shortening in the outer Carpathian orogenic belt from inactive parts of the belt and also act as transform type boundaries that connect areas of shortening in the Carpathians to areas of shortening in the Dinaric Alps. The existence of such shear zones can thus be deduced almost directly from analysis of the varying rates and directions of convergence across the Carpathian belt.

INTRODUCTION

The present morphology and structure of the Carpathian–Pannonian region (Figure 1) is the result of Cretaceous to Miocene convergence and collision of the European plate with several small continental fragments to the south. Collision occurred following southward and westward subduction of an oceanic terrane beneath the Carpathian–Pannonian region, so that the Carpathian Mountains and the Pannonian basin now fill a recess in the combined European–Moesian fragment. At least several hundred kilometers of Cenozoic convergence occurred across the West and East Carpathians (Andrusov, 1968; Sandulescu, 1975, 1980; Burchfiel, 1976, 1980). The present arcuate shape of the Carpathian mountain belt clearly indicates that this Cenozoic convergence could not have occurred without considerable deformation within the interior of the Carpathian–Pannonian region.

Detailed reconstructions of the internal deformation within the Carpathian–Pannonian region during late Neogene time indicate that most of the late Neogene deformation internal to the active thrust belt was associated with major shear zones with

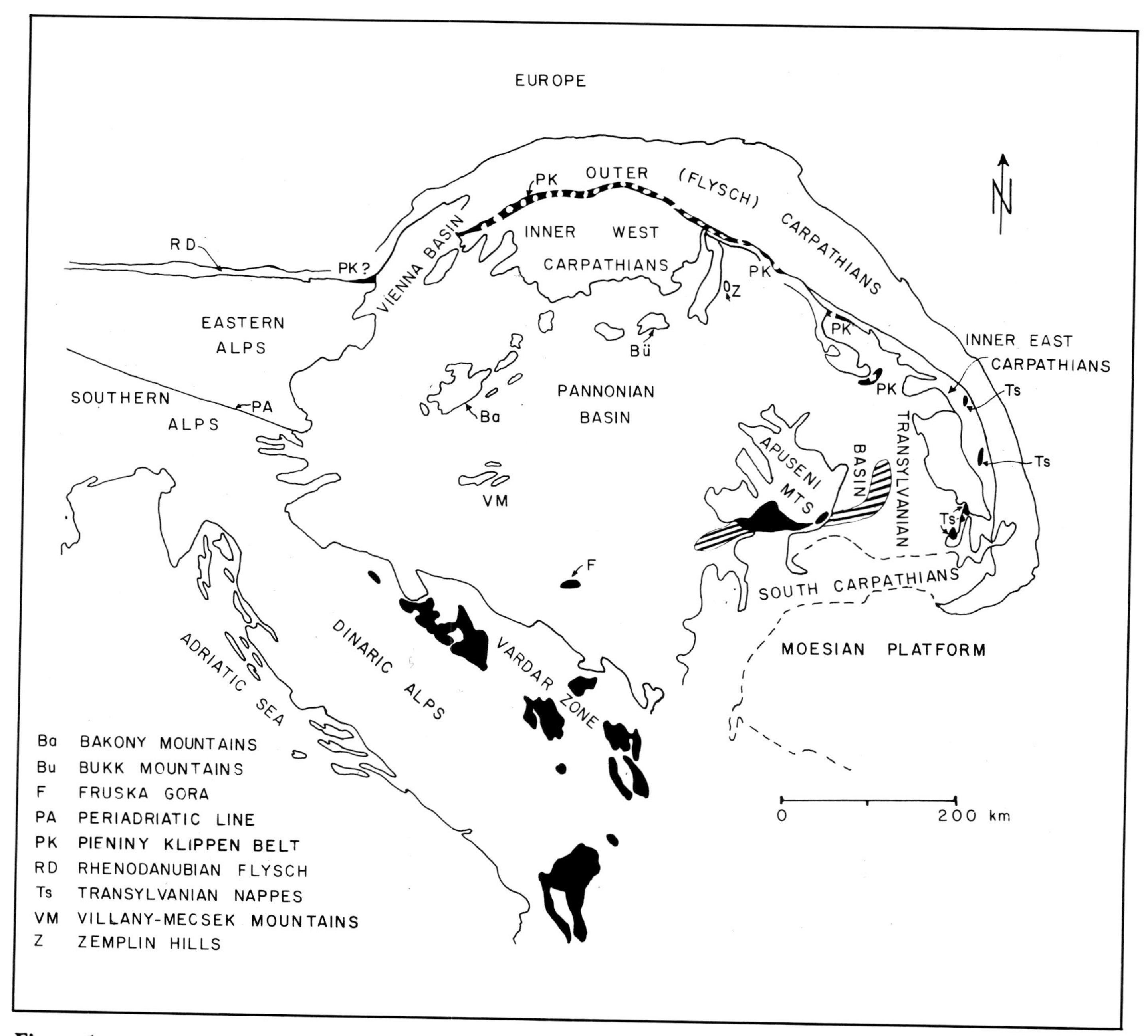

Figure 1. Location map of Carpathian–Pannonian region. Black indicates ophiolitic rocks, and black and white stripes near the Apuseni Mountains indicate subsurface ophiolites. Black with white dots indicates Pieniny Klippen belt rocks, including the Wildflysch and Botiza nappes in the east and the Hauptklippenzone in the Eastern Alps.

large (tens of kilometers) horizontal displacements (Royden et al., 1982, 1983, and this volume). In general, these strike-slip zones served to accommodate different rates and directions of convergence across the outer Carpathian thrust belt. The analysis of Royden et al. (1983) was based primarily on data from the shear zones themselves (and associated extensional sedimentary basins), but the need for such strike-slip zones as well as their sense of displacement, cumulative offset, timing, and approximate location can be deduced directly from analysis of the varying rates and directions of convergence across the Carpathians (see also Laubscher, 1971). Although less precise data are available for Paleogene tectonic events and although Paleogene events were probably of smaller magnitude than later Miocene events, a similar analysis of the internal deformation of the Carpathian–Pannonian region can be applied for Paleogene and early Miocene time.

The present distribution of Mesozoic and early Cenozoic tectonic and paleogeographic elements within the Carpathian–Pannonian region has led several authors to postulate the former existence of several different oceanic terranes within this region, as well as complex initial distributions of facies and tectonic elements (e.g., Channell and Horváth, 1976). In this paper, we hope to show that the present complex distribution of these various tectonic and paleogeographic elements could have developed from a very simple initial configuration during Cenozoic deformation in the interior of the Carpathian–Pannonian region. Moreover, we will try to relate, in a general way, the formation, duration, and disruption of various Paleogene paleogeographic elements to contemporaneous tectonic events in the Carpathian–Pannonian region. Finally, we hope to show that this Cenozoic tectonic activity within the internal part of the Carpathian–Pannonian region is not only consistent with the timing and magnitude of Cenozoic convergence across the Eastern Alps, Carpathians, and Dinaric Alps, but is in fact required to accommodate diachronous convergence along this arcuate mountain chain.

TECTONOSTRATIGRAPHIC ELEMENTS

The Cenozoic evolution of the Carpathian–Pannonian region can be described, at any given time, by a number of coherent crustal fragments bounded by broad zones of fairly intense deformation. However, these zones of deformation cannot be regarded as permanent features because old fault zones often become inactive and new fault zones are created that cut across the older fault zones and disrupt formerly coherent crustal fragments. Thus, crustal fragments that at one time move together may, at another time, move in entirely different directions, be subject to different tectonic events, and be the site of different depositional environments.

To best understand the disruption and rearrangement of crustal fragments throughout Cenozoic time, one can group together various crustal blocks that underwent similar phases of deformation or sedimentation into tectonostratigraphic units. Because of the temporary and shifting nature of the fault zones that bound coherent crustal blocks, these tectonostratigraphic units are also temporary features. For example, two crustal blocks that belonged to the same tectonostratigraphic unit in Cretaceous time may have belonged to very different tectonostratigraphic units in Eocene time. By defining such tectonostratigraphic units for various time intervals, it is possible to infer a link between each element within a given tectonostratigraphic unit over the appropriate time interval. This approach makes it possible to constrain partially the ages of formation and disruption of various crustal blocks. Combining such data with information on the timing and magnitude of displacements along major zones of deformation, such as along the outer Carpathians, and with paleogeographic data, such as the presence of faunal connections, makes it possible to restore partially the complex Cenozoic history of this region.

In this paper we concentrate primarily on tectonostratigraphic elements within the Carpathian–Pannonian region, although some discussion of adjacent areas (Eastern and Dinaric Alps) is necessary. The main tectonic and stratigraphic units discussed in this paper are summarized in the following sections. Wherever detailed correlations have been made previously, those arguments are summarized briefly and the reader is referred to those works for further details. The time scale used is shown in Figure 2.

The Inner West Carpathians

The inner West Carpathians are generally accepted to be the eastward continuation of the Eastern Alps (Tollmann, 1980). North-vergent nappes in the West Carpathians consist primarily of Mesozoic carbonate rocks, sometimes resting on low-grade Paleozoic and crystalline basement rocks. The last major phase of thrusting and folding can be narrowly constrained as intra-Turonian (pre-Gosau) because Turonian age rocks are involved in thrusting and the thrust contacts are in places overlapped by late Turonian Gosau deposits. South of the inner West Carpathians, the Mesozoic carbonate rocks of the Hungarian Mid-Mountains (Bakony and Buda) appear to have Southern Alpine affinities.

North-vergent nappes of the northern Apuseni Mountains can be closely correlated with those of the inner West Carpathians (Sandulescu, 1975). Not only are the early Mesozoic and Paleozoic facies identical to those found in the inner West Carpathians, but the changes in the stratigraphy from the lowest structural units in the north (Tatrides and Bihor "autochthon") to higher structural units in the south (Gemerides and Codru–Arieseni nappe series) are strikingly similar. Moreover, the last major phase of north-directed thrusting in the northern Apuseni Mountains is synchronous with that in the inner West Carpathians because Turonian rocks of the Bihor "autochthon" are involved in the most external thrust, which are in turn overlain by early Senonian Gosau facies.

Several other outcrops of Mesozoic, Paleozoic, and crystalline rocks in the Pannonian region have also been correlated with some inner West Carpathian units including the Villany and Mecsek Mountains (correlated with the Apuseni Mountains), crystalline and Paleozoic rocks of the Zemplin hills, and a few outcrops with Tatric affinity northeast of the Apuseni Mountains (Sandulescu, 1975).

Other interpretations of the stratigraphic affinity of these rocks within the Carpathian–Pannonian region have been based on Jurassic ammonites (Géczy, 1972) and Jurassic brachiopods (Vörös, 1977; see also Channell and Horváth, 1976) and group the Apuseni Mountains with the Eastern Carpathians in Jurassic time. However, we feel that the striking similarity in

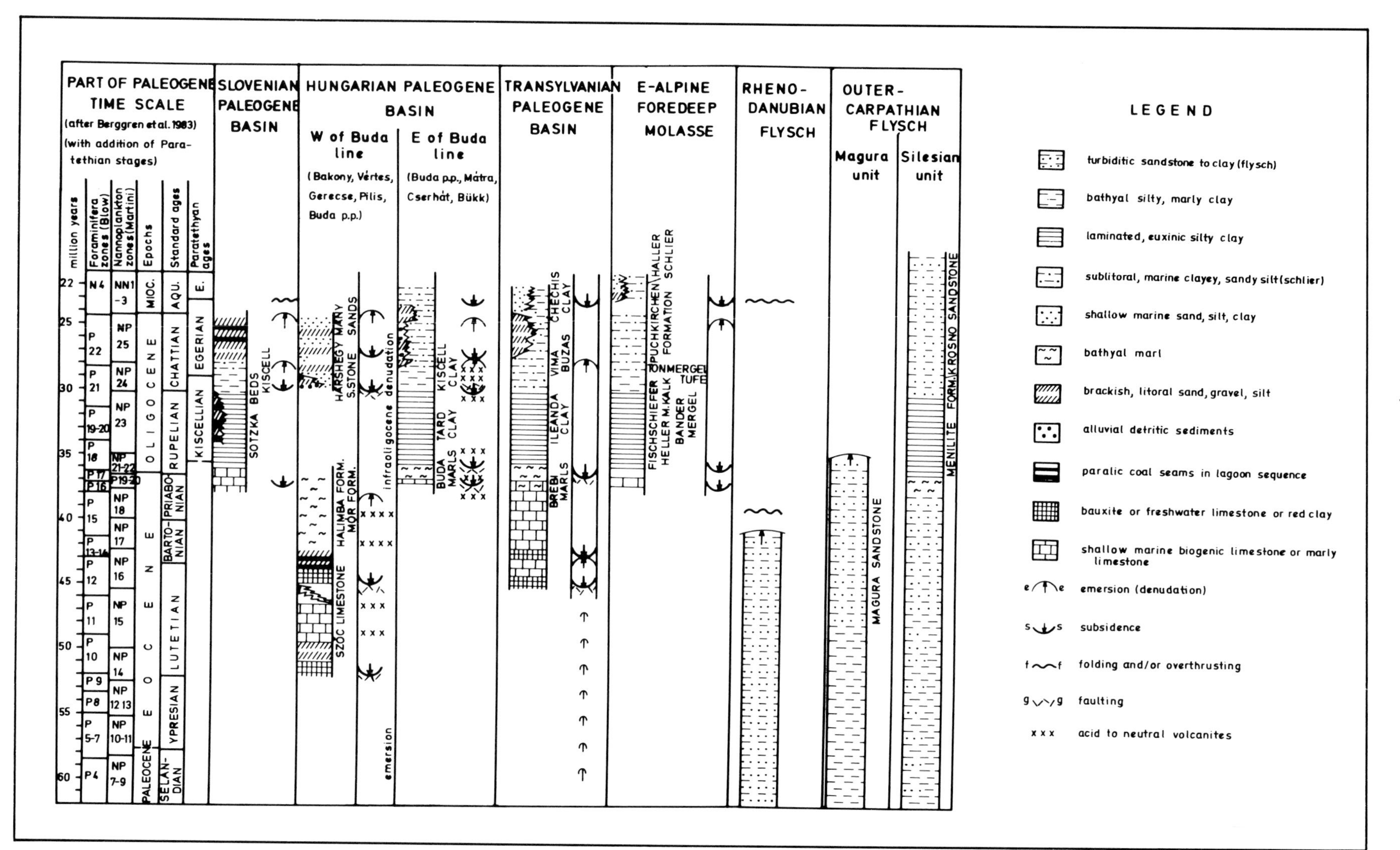

Figure 2. Timescale and correlation chart of various facies of early Tertiary age.

sedimentologic evolution, tectonic style, timing of nappe emplacement, and changes in facies with structural level in the inner West Carpathians and Apuseni Mountains provide overwhelming evidence that they were part of the same tectonostratigraphic unit until post-Senonian time.

Pieniny Klippen Belt

The inner West Carpathians are separated from the flysch nappes of the outer West Carpathians by the Pieniny Klippen belt. The Pieniny Klippen belt crops out continuously for about 800 km along strike, yet in most places it is no more than a few kilometers wide. This unit was deformed with the inner West Carpathians in Cretaceous time and probably again with the outer west Carpathians in Eocene and Miocene time. The belt contains exotic blocks (klippen) consisting primarily of Mesozoic carbonate rocks and cherts. A few of these may have been derived from the inner West Carpathians, but most of the large variety of facies found in blocks within the belt are thought to have been derived from a terrane that has largely disappeared (Andrusov, 1968). The wide variety of facies present in the klippen belt has led some authors to hypothesize the existence of several cordillera and basins as source regions for the blocks. Such an interpretation has given rise to estimates of the width of the source terrane of as much as 600 km (Unrug, 1979).

Other authors (Sandulescu, 1980; Birkenmajer, 1981) have interpreted the structure of the Pienniny Klippen belt as the result of both convergent and strike-slip displacements along the belt. We believe that large-scale shear associated with strike-slip displacement during oblique subduction and convergence can best explain the narrow but continuous nature of the belt, the wide variety of facies and rock types present in tectonic slices, and the internal imbricate structure of the belt.

In its eastern part, the Pieniny Klippen belt trends southwest of the inner structural units of the East Carpathians. An isolated outcrop of similar rocks occurs to the northeast of the Apuseni Mountains (the Botiza and Wildflysch nappes) (Sandulescu et al., 1981b). The Pieniny Klippen belt may continue westward into the Eastern Alps (the Hauptklippenzone), but the correlation is not certain (Prey, 1980).

Inner East and South Carpathians

The structure and sedimentary facies of the internal units of the East and South Carpathians are distinct from those of the West Carpathians, at least for post-Early Jurassic rocks, as some workers have correlated Triassic and Early Jurassic stratigraphic units between the East and West Carpathians (e.g., Kovács, 1982). Both the East and South Carpathian belts consist of inner crystalline unit(s), with remnants of Mesozoic (pre-Vraconian) cover, that were imbricated and thrust eastward during Early Cretaceous time. Cretaceous cover rocks are distinguished from those of West Carpathian affinity by the presence of Early Cretaceous flysch and wildflysch, and an Early–Late Cretaceous posttectonic cover (Sandulescu, 1975, 1980; Sandulescu et al., 1981a).

The lowest structural unit(s) of the inner East and South Carpathians consist of Upper Jurassic to Albian–Aptian flysch thrust eastward in latest Cretaceous–Paleocene time (e.g., the Ceahlau nappe in the East Carpathians and the Severin unit in the South Carpathians). Thus, the inner East and South Carpathians may be considered as part of the same tectonostratigraphic unit until latest Cretaceous time, and their structure and sedimentary facies are clearly different from those of the West Carpathians and Apuseni Mountains.

Ophiolitic Rocks

The highest structural unit of the East Carpathians is the east-vergent Transylvanian nappe, which contains ophiolitic rocks. The age of overthrusting of the Transylvanian nappes is roughly contemporaneous with the Early Cretaceous thrusting of the inner East Carpathian units (Sandulescu, 1980). Another ophiolitic body is present between the South Carpathians and the northern Apuseni Mountains. This ophiolitic assemblage of the southern Apuseni Mountains can be traced toward the northeast by gravity and magnetic data under the Neogene rocks of the Transylvanian basin where it is thought to comprise the root zone for the east-vergent Transylvanian nappes (Sandulescu, 1980). In the Apuseni Mountains the tectonic boundaries of the ophiolitic body probably have a major strike-slip component. Implacement of the ophiolitic rocks in the southern Apuseni Mountains (by strike-slip faulting?) is probably of latest Cretaceous–Paleocene age, as indicated by an overlap assemblage of that age (Bleahu et al., 1981).

This ophiolite zone probably continues to the west and south into the well-developed Vardar ophiolite belt of the Dinaric Alps. The entire ophiolite zone is thought to represent a once continuous ocean of Middle and Late Jurassic age that originally separated rocks with West Carpathian affinity from those of East Carpathian affinity (Sandulescu, 1975; Burchfiel, 1980). The relationship of this oceanic terrane to the Pennine ocean to the north and west is still unclear. In our opinion, the Vardar and Pennine oceans were connected through the Transylvanian basin (see Figure 4Aa). Other ultramafic rocks that have been interpreted as fragments of Jurassic ocean floor(?) are found far to the north in the Bükk Mountains.

Bükk–Igal Zone

Mesozoic rocks present in the Bükk Mountains contain Mesozoic facies of a more southerly origin than those of the West Carpathian units (Balogh, 1964; Kovács, 1982). These rocks appear to correlate with rocks far to the south in the Serbian and Bosnian zones of the Dinaric Alps. Rock types present in the Bükk Mountains are also present in the subsurface south of the Bakony Mountains. In both places they occur as narrow fault-bounded elements that probably correspond to tectonic slices with Serbian zone affinities rather than to an original configuration of a paleogeographic domain (Kovács, 1982; Báldi, 1983). These tectonic slices will be referred to as the Bükk–Igal zone.

Peri-Klippen Paleogene Deposits

Locally, a Paleocene–Eocene wildflysch is present south of the Pieniny Klippen belt along almost its entire length (Marschalko, 1980). This wildflysch may be more than 1 km thick and contains blocks of exotic Paleocene reefal limestone as much as 100 m in diameter and occasional clasts of Mesozoic carbonate rocks. The wildflysch grades upward into Eocene flysch and is separated from rocks of the klippen belt to the north and from rocks of the inner West Carpathians to the south by high-angle faults. This wildflysch–flysch unit becomes younger toward the east, being of Paleocene–early Eocene age in the west and Paleocene–Lutetian age in the east. A similar sequence of wildflysch and flysch ("wildflysch nappe") occurs south of Pieniny Klippen type rocks (Botiza nappe) and northeast of the Apuseni Mountains where the wildflysch is of Lutetian and Priabonian age. Here, the wildflysch contains exotic clasts of Lutetian or Priabonian reefal limestone (Sandulescu et al., 1981b).

These wildflysch deposits clearly indicate a Paleocene–Eocene tectonic event along the Pieniny Klippen belt, during which time the source areas for exotic blocks in the wildflysch were totally obliterated (Marschalko, 1980; Sandulescu, 1980). The narrow fault-bounded and rapidly subsiding troughs in which these rocks were deposited suggest a major component of horizontal shear during Paleogene time. Moreover, the similarity of the stratigraphy found within both the klippen belt itself and the Peri-Klippen Paleogene deposits for more than 600 km along strike strongly suggests that this narrow zone was originally a more coherent terrane that was dismembered, imbricated, and stretched into its present configuration during Late Cretaceous, Paleogene, and Neogene strike-slip and thrust faulting along the Pieniny Klippen zone.

Rhenodanubian–Carpathian Flysch Belt

The lowest structural units in the Eastern Alps and the West and East Carpathians comprise a continuous belt of Upper Cretaceous to Miocene flysch that were deformed into north- and east-vergent thrust and fold nappes in Tertiary time. Much of this flysch was deposited below the carbonate compensation depth (Van Couvering et al., 1981), and an oceanic or highly attenuated continental basement may be inferred for the inner part of the flysch belt. Flysch sedimentation ended diachronously from west to east, ending in middle Eocene time in the Eastern Alps, late Oligocene time in the West Carpathians and early Miocene time in the East Carpathians. Sedimentation in the form of molasse facies continued until middle Miocene time in the Eastern Alps and western West Carpathians, until late Miocene time in the eastern West Carpathians, and until Pliocene time in the East Carpathians. This diachroneity of deposition reflects a west to east migration of convergent tectonic activity and related arc volcanism (e.g., Lexa and Konecny, 1974; Jiricek, 1979; Royden et al., 1982).

Deformation of the flysch belt occurred from Eocene to early Miocene time in the Eastern Alps and from late Oligocene to middle and late Miocene time in the West and East Carpathians, respectively (Sandulescu, 1980; Prey, 1980). Thus, deformation of the flysch belt also ended diachronously from west to east. A poorly understood Eocene–Oligocene(?) phase of deformation also occurred within the West Carpathians (Ksiazkiewicz and Lesko, 1959; Ślaczka, 1969; Oszczypkzo, 1973; Marshalko, 1980), probably representing the eastward continuation of Eocene convergence across the Eastern Alps. Evidence for this Eocene deformation comes partly from the Pieniny Klippen belt (see above), partly from the innermost flysch nappes and partly from the presence of acidic magmatism derived from Eocene–Oligocene subduction present in the southern part of and south of the inner West Carpathians as far east as Recsk and Eger (north-central Hungary) (Figure 3). This magmatic belt is the eastward continuation of the tonalite belt of the Alps (Bauer and Schönlaub, 1980), suggesting some Eocene convergence across the West Carpathians. However, most of the direct evidence for Eocene convergent tectonic activity across the West Carpathians has been obliterated by the extensive Miocene shortening and subduction. Eocene–Oligocene convergence across the Eastern Alps was probably more than 100 km, while Miocene convergence across the Carpathian flysch belt has been estimated at several hundred kilometers (Sandulescu, 1975).

Inner Carpathian Flysch Units

A Paleogene flysch sequence now occurs in three major zones south of the Pieniny Klippen belt: the Podhale flysch, the Szolnok–Maramures flysch, and the Transcarpathian flysch (Figure 3). This inner flysch zone was probably the southward extension of the Carpathian flysch belt. The Cretaceous to Early Kiscellian Podhale flysch overlies transgressively the inner West Carpathian nappes and consists of terrigenous turbidites that were partly folded in Oligocene time (Samuel and Salaj, 1968). In the central part of the Podhale flysch, sedimentation did not begin until Eocene time.

The Eocene (or Cretaceous?) to uppermost Oligocene Szolnok–Maramures flysch (which includes molasselike deposits) is present only in the subsurface of the Neogene Pannonian basin (Báldi-Beke et al., 1981). It can be traced eastward by drill hole data into the Transcarpathian flysch, which is of Eocene–early Miocene age (Sandulescu, 1980; Sandulescu et al., 1981b). The similarity and continuity of the flysch and molasse facies present in each of these zones suggests that originally there may have been a single continuous area of deposition for the inner flysch. If so, this zone was probably disrupted during Oligocene–early Miocene time. Alternatively, these flysch basins may have evolved independently at approximately the same time. In either case, the Szolnok–Maramures flysch and the Transcarpathian flysch were subsequently deformed by early(?) Miocene convergent or transpressional (transform motion plus compression) tectonic activity.

Dinaric Flysch

The outer (western) side of the northern Dinaric Alps consists of a belt of Eocene–early Oligocene(?) flysch that has been deformed into a west-vergent series of imbricate folds and thrusts. This Dinaric flysch was deposited in a deep marine environment and shows a clear faunal association with the Ligurian ocean in the west. This southern marine realm was distinct and separated from the northern marine realm in which the Rhenodanubian and Carpathian flysch was deposited. From time to time marine connections formed between the two oceans, such as through the Hungarian Paleogene basin (see next section).

From Eocene until early Miocene time, the Dinaric flysch and some of the internal carbonate units of the Dinaric Alps were imbricated and thrust westward as the Adriatic foreland was

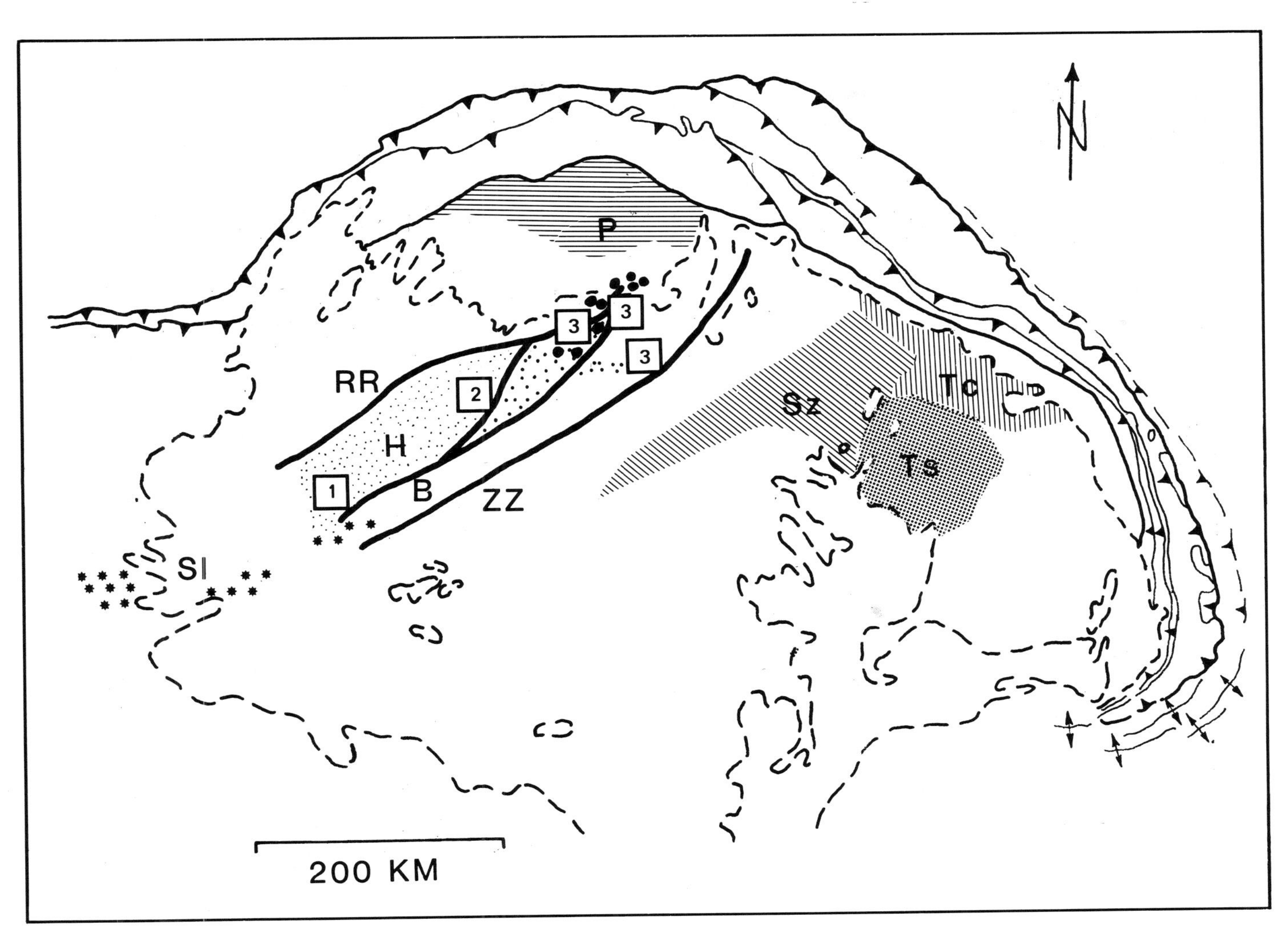

Figure 3. Schematic figure showing deposits of Hungarian Paleogene basin of Eocene (small dots), early Oligocene (medium dots), and late Oligocene (big dots) age. Associated volcanic necks or dikes are shown by numbers in boxes: (1) Eocene, (2) lower Oligocene, (3) upper Oligocene. The Transylvanian Paleogene basin is indicated by shading and the Slovenian basin by stars. P, Podhale flysch; H, Hungarian Paleogene Basin; SZ, Szolnok flysch; Tc, Transcarpathian flysch; Ts, Transylvanian Paleogene Basin; RR, Raba–Rosnava line; ZZ, Zagreb–Zemplin line; B, Buda line.

subducted eastward under the Dinarides. The total amount of shortening across the Dinarides has been estimated at several tens of kilometers in the northernmost part and several hundred kilometers in the south (Burchfiel, 1980). The inner Mesozoic carbonate units of the Dinarides that lie west of the Vardar zone have some affinities with the Southern Alps and inner West Carpathians, but are thought to represent a more southerly facies region that existed in Mesozoic time.

The Hungarian Paleogene Basin

South of the inner West Carpathians a narrow, east–west elongate Paleogene basin formed in three distinct stages, proceeding from west to east (Figure 3). First, in Eocene (Lutetian–early Priabonian) time initial subsidence of the westernmost part of the basin was accompanied by deposition of bauxites and conglomerates, which was followed by deep marine deposition in late Lutetian time (Baldi-Beke, 1984). Its basement consists of subsided Mesozoic carbonate rocks that probably represent a structurally more internal part of the West Carpathians than is now observed at the surface. Second, in late Eocene through early Oligocene time the western part of the Hungarian Paleogene basin was uplifted and partially eroded (Telegdi-Roth, 1927; Báldi-Beke, 1977), while to the east a new section of the basin began to subside. Sedimentation here also proceeded rapidly from a shallow water environment to deep water conditions. The basement for this part of the basin is composed mainly of rocks of the Bükk–Igal fragment. Third, in early Miocene (Eggenburgian) time the area of basinal sedimentation extended farther to the northeast.

The Hungarian Paleogene basin exhibits many features typical of extensional basins that form along major strike-slip faults or fault zones (see also Crowell, 1974). It is a narrow, elongate basin bounded entirely by high-angle synsedimentary faults (the Rába-Roznava, Zagreb–Zemplin, and Buda lines). It exhibits migration of subsidence along the trace of the boundary faults, rapid changes in elevation of the basin floor (such as from a deep marine to an erosional environment and vice versa), and the presence of very close source areas as indicated by diamictite and pebbly mudstone within the basin. Thus, we infer that the high-angle faults bounding this basin are part of a large strike-slip fault system that was active in Paleogene time and that trends roughly parallel to the axis of the basin.

From Eocene through Oligocene time, an andesitic volcanic arc erupted along the southern margin of the basin (see previous sections). Although these volcanic rocks are thought to be derived from a subducted slab extending beneath the inner West Carpathians, it is clear from Figure 3 that the location of these volcanic rocks was strongly controlled by the active faults associated with subsidence of the Paleogene basin, or vice versa.

TECTONIC AND PALEOGEOGRAPHIC RECONSTRUCTIONS

The tectonic and paleogeographic reconstructions given here are represented by a series of time-stage maps constructed by working backward from the most recent events to the earliest events (Figure 4). These reconstructions, however, are presented in reverse order to better illustrate the logical progression of tectonic events and depositional environments through time. The size, shape, and precise location of individual crustal fragments shown in these reconstructions are mostly uncertain, so we have shown fragments with roughly their present size and shape wherever applicable. Moreover, we have attempted to preserve the relative location of these fragments based on the biogeographic composition of faunas, as well as their present relative locations. The widths of various terranes that have been largely or entirely subducted, such as the Rhenodanubian–Carpathian flysch basin, have been estimated based on palinspastic reconstructions of thrust belts, facies, and basins. Thus, we attempt to show only the association or proximity and approximate locations of various crustal fragments through time and their positions relative to one another because exact reconstructions of position and other details are impossible.

The five time-stages shown in Figure 4 were chosen to represent convenient tectonic and stratigraphic intervals during the Cenozoic evolution of the Carpathian–Pannonian region, thus major tectonic events and depositional environments remained mostly unchanged throughout each time interval. However, because convenient tectonic and stratigraphic boundaries often do not coincide, and because tectonic and stratigraphic phases often begin and end diachronously throughout the region, such time intervals are difficult to define exactly. Therefore, for each time-stage map, we consider the tectonic or stratigraphic events that dominate the corresponding time interval; these events may extend beyond the upper or lower boundaries of the time interval or may not last throughout the entire interval. Nevertheless, the events discussed will be the dominant events during each time interval, and exact ages will be given in the text.

Each tectonic or paleogeographic map shows the reconstructed fragment locations at the beginning of the appropriate time interval. The tectonic events and paleogeographic environments shown correspond to those that occurred throughout that time stage.

Coniacian–Paleocene

For convenience, the tectonic events shown for this time stage extend from Coniacian through Paleocene time (Figure 4Ab and 4Ac), while paleogeographic reconstructions correspond only to Paleocene events (Figure 4b). The fragment locations are supposed to reflect approximate positions at about the beginning of Coniacian time (see above). During Coniacian–Paleocene time, the Eastern Alps, West Carpathians, Apuseni Mountains, and several smaller fragments (shown by small dots) are inferred to make up a coherent terrane as evidenced by similar Turonian–Senonian thrusting events. The more southerly facies of the Bükk–Igal zone and the Serbian zone are inferred to have belonged to a more southerly domain such as the Serbian and Bosnian zones of the Dinaric Alps, but they need not be separated geographically from the Eastern Alps and West Carpathians. Two other domains or coherent crustal fragments are distinguished: the East to South Carpathians and the ophiolitic root zone for the formerly oceanic terrane of the Vardar zone, Apuseni Mountains, and Transylvanian ophiolites. Note that in this reconstruction, rocks of the Bükk–Igal zone are present only a short distance from ophiolites within the Vardar zone (Figure 4Ab). Thus, the ophiolitic rocks(?) that are now associated with the Bükk Mountains may have become welded to the Bükk–Igal fragment during Cretaceous strike-slip faulting or westward-directed thrusting within the Dinaric Alps (see Burchfiel, 1980). The source terrane for the Pieniny Klippen belt is shown schematically north of the inner West Carpathians.

Coniacian–Paleocene tectonic events are dominated by north–south convergence of Europe and the Eastern Alps during southward subduction of a probable oceanic terrane, and by east–west convergence and shortening across the East to South Carpathians and the Dinaric Alps (Figure 4Ac). However, the opposite vergence of thrusting and subduction within the East Carpathians and Dinarides requires a major east-west–trending dextral transform system to connect these zones of coeval shortening (Burchfiel, 1980). This transform system probably disrupted the once continuous ophiolite belt, resulting in dextral offset of the belt between the Apuseni Mountains and the South Carpathians and in the suturing of the southern Apuseni ophiolite body (Metalliferous Mountains) to the rest of the Apuseni Mountains. However, the initial disruption of the ophiolite belt may have occurred earlier in Cretaceous time. The continuation of this dextral shear zone to the west is not required by our reconstructions, but we cannot exclude the possibility that the major strike-slip zones present in the Carpathian–Pannonian region throughout Paleogene time did extend farther west (see next section).

Paleocene paleogeography indicates two separate areas of deep marine environments: the northern area where the Rhendanubian–Carpathian flysch was deposited, and the southern area where the Dinaric flysch was deposited (Figure 4Aa). The northern marine area was connected to the Tethys through a seaway that opened into the Alpine flysch belt in the west. The southern marine area was directly connected to warmer waters in the Ligurian ocean. Some inner parts of the Alps, West Carpathians, and Dinarides were apparently elevated above sea level because the Paleocene is missing in Hungary and Transylvania (e.g., Oberhauser, 1980). The most southerly part of this Paleocene flysch belt was deposited on the northern part of the calcareous nappes of the Eastern Alps and West Carpathians, forming a part of the inner flysch belt.

Early-Middle Eocene

By the beginning of Eocene time, the Vardar–Transylvanian ophiolite belt had probably been partly disrupted by dextral strike-slip faulting, but the Eastern Alps–West Carpathians–Apuseni Mountains block may still have been a fairly coherent domain (Figure 4Bb). Tectonic activity was dominated by rapid north–south convergence of Europe and the inner Eastern Alps, probably extending eastward into the West Carpathians, and by east–west convergence and shortening across the Dinaric Alps, while the East Carpathians and eastern West Carpathians remained tectonically inactive (Figure 4Bc). Such a geometry indicates a northeastward displacement of the Eastern Alps and West Carpathians relative to the East Carpathian chain and requires a zone of dextral shear separating the Eastern Alps and West Carpathians from the East Carpathians. We suggest that this zone of dextral shear coincided partly with the east-northeast-trending Eocene–Oligocene fault system that bounded the Hungarian Paleogene basin (e.g., the Rába–Roznava line, the Buda line, and the Zagreb–Zemplin line) (Figure 3). The evolution and configuration of the Hungarian basin is typical for basins that form along zones of major transcurrent displacement, although the direction of displacement is not evident from analysis of the basin.

This interpretation suggests an Eocene–Oligocene age for the dislocation of the Apuseni Mountains from the inner West Carpathians because the fault system described above passes between them. Moreover, at least one of the faults of this fault system (Zagreb–Zemplin line) passes south of the Bükk–Igal zone so that severing of the Bükk–Igal fragment from the Serbian zone and its relative northeastward displacement can also be partly accommodated by this fault system and thus can also be of Eocene age (Figure 4Bb). Note that this inferred zone of dextral shear also acts as a transform boundary, connecting areas of coeval convergence in the West Carpathians and Dinarides. In this sense its function is much the same as the dextral transform fault inferred to connect the Dinaric and East Carpathian thrust belts in Coniacian–Paleocene time (compare Figures 4Ac and 4Bc).

We have inferred a Paleocene to late Eocene left-slip or transpressional boundary north of the inner West Carpathians along the Pieniny Klippen zone because its configuration and internal structure appear to be the result of major horizontal shear. The inferred age of this left-slip displacement is indicated by the Paleocene–Eocene wildflysch present locally along the entire Pieniny Klippen belt and adjacent to Pieniny type rocks north of the Apuseni Mountains (Marschalko, 1980, 1982; Sandulescu, 1980). The apparent continuity of this wildflysch sequence between the area of the West Carpathians and the area north of the Apuseni Mountains suggests that the Klippen belt was a continuous feature until post-late Eocene. In this interpretation the Eastern Alps, West Carpathians, Bakony and Bükk areas form part of an eastward-moving wedge bounded to the north and south by large strike-slip or shear zones. The northern shear boundary appears to have been associated with oblique convergence, while the southern boundary suggests a small amount of oblique extension.

During early–middle Eocene time, the faunas of the two deep marine areas north and south of the Alpine–Carpathian chain remained partly separated as in Paleocene time (Figure 4Ba) and both areas maintained their direct connections to the Ligurian and Dinaric oceans. However, in Eocene time the area of flysch deposition extended southward from the outer Carpathian flysch belt, covering large parts of the inner west Carpathians (Podhale flysch). The Podhale flysch contains upper Eocene to Oligocene facies similar to those present in the Szolnok trough and the Transcarpathian region. In the reconstruction shown in Figures 4Ba, 4Bb, and 4Bc, all three areas of inner flysch deposition (Podhale, Szolnok, and Transcarpathian flysches) are shown to form a continuous inner flysch belt south of the outer Carpathian flysch belt, thus explaining the similarity of facies in these areas. However, these flysch facies may have been deposited in different basins. The disruption and deformation of this perhaps continuous inner flysch zone can be explained by Eocene–Miocene shear along the major dextral fault zone described above.

In addition to the large areas of Eocene flysch deposition, two distinct basins characterized by nonflysch sedimentation developed in Eocene time. The oldest part of the elongate Hungarian Paleogene basin (Lutetian–early Priabonian) formed along the western part of this large dextral shear zone in the Bakony–Buda area and was bounded in part by smaller synsedimentary faults that may also have had dextral displacements (see Figure 3).

Detailed stratigraphic and faunal studies have shown that the transgression came from the southwest into the Bakony area (Báldi-Beke, 1984). Nannoplankton clearly indicate the proximity of an ocean (Báldi-Beke, 1984) which must have been the Dinaric sea. The coeval northern Italian *Nummulite* faunas are far more similar to the *Nummulite* fauna of the Bakony region than are those of Transylvania (Kecskeméti, 1980). Gastropod fauna of this age also exhibit a northern Italian affinity (Körmendy, 1980). Alveolina limestone found in the Hungarian Paleogene basin also occurs in the Eocene of the Dinaric, northern Italian and Karinthian region. It has not been found in the Helvetic region or in the Alpine–Carpathian flysch zone (Kahler and Papp, 1968). Thus, this basin formed an embayment of the Dinaric ocean, probably created as a result of the dextral transform boundary described above. Moreover, the partial separation of the northern and southern marine faunas further indicates that no major direct connection existed at this time.

Dating from nannoplankton (Báldi-Beke, 1984), foraminifera (Kecskeméti, 1980), and molluscs (Szöts, 1953; Strausz, 1966; Körmendy, 1980) indicates that the transgression occurred first in the southwestern Bakony in calcareous nannoplankton Zone NP14 (lowermost Lutetian) (after Berggren et al., 1985). The rest of the Bakony region became a marine basin in Zone NP16 (42–44 Ma, Figure 2). Freshwater limestone or red clays and bauxites were deposited first and were overlain by paralic coal seams grading upward into brackish and shallow marine marls. This sequence is overlain by the Szöc Limestone, a shallow marine, biogenic limestone with *Nummulites* and reef-building fossils. The Lutetian transgression at first deposited only shallow marine facies. At about the beginning of the Priabonian (40 Ma) bathyal *Globigerina* Piszke Marl was deposited in the east, and in the west, bathyal late Lutetian–early Priabonian Halimba Marl was deposited. This sudden subsidence of the basin at about 40–44 Ma, followed shortly thereafter by uplift and erosion (see below), is also a characteristic feature of wrench related basins.

In contrast, the Transylvanian Paleogene basin formed a southward embayment of the northern Carpathian ocean (Figure 3), and the marine faunas differ significantly from coeval faunas of the Hungarian Paleogene basin; thus it appears no direct connection existed between them at this time. For example, *Gryphaea eszterhazyi* has been recognized in Lutetian strata in both Transylvania and central Asia, but is not known in the Hungarian Paleogene basin (Koch, 1911). Note that the recon-

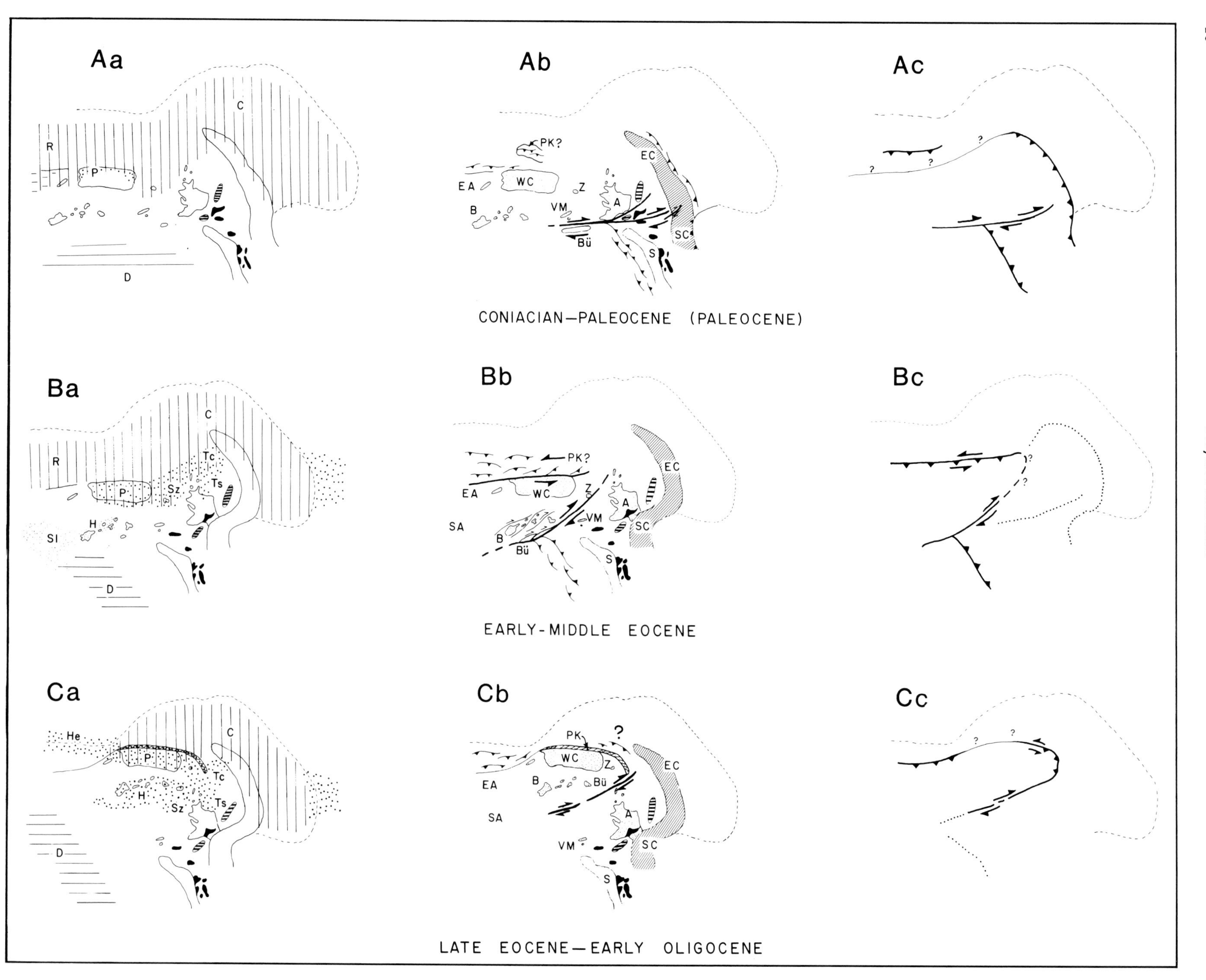
Aa
C
R
P
D
Ab
PK?
EC
EA
WC
Z
B
VM
A
SC
Bü
S
Ac
CONIACIAN—PALEOCENE (PALEOCENE)
Ba
C
R
Tc
Ts
P
Sz
H
Sl
D
Bb
PK?
EC
EA
WC
Z
A
SA
VM
SC
B
Bü
S
Bc
EARLY-MIDDLE EOCENE
Ca
He
C
P
Tc
H
Sz
Ts
D
Cb
PK
?
WC
Z
EC
EA
B
Bü
SA
A
VM
SC
S
Cc
LATE EOCENE—EARLY OLIGOCENE

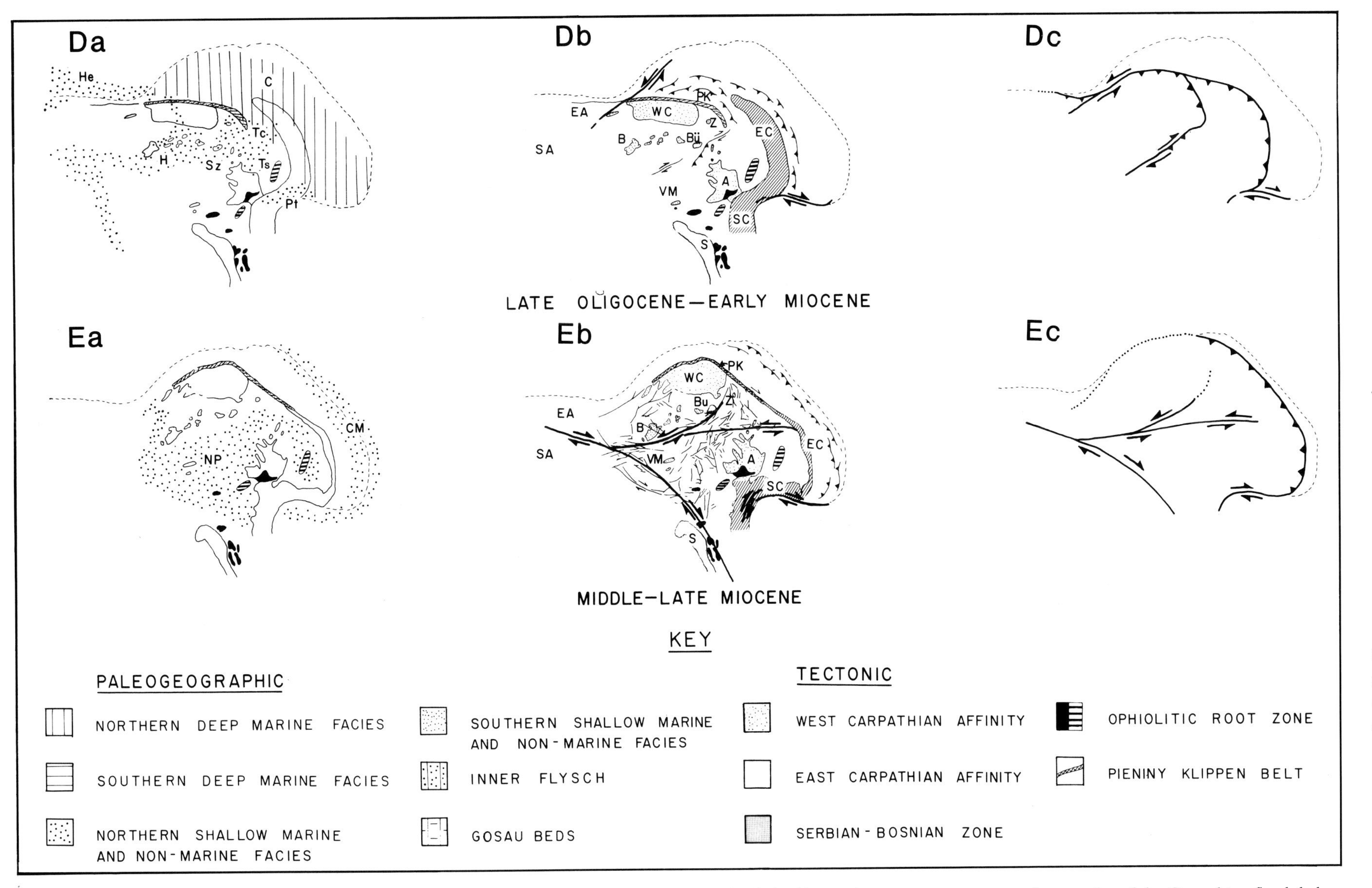

Figure 4. Series of qualitative palinspastic reconstructions for five time stages in the Tertiary. Light dashed line indicates present position of outer edge of the Carpathian flysch belt. Position of fragments is reconstructed for the beginning of each time interval, and tectonic events (middle column, Ab–Eb) and the paleogeographic environments (left column Aa–Ea) shown are those which dominate that time interval. Our interpretation of the major crustal fragments and the type of deformation that occurred along their boundaries is indicated in the right column (Ac–Ec). The dotted line indicates the position of major tectonic boundaries from the preceding time stage in Bc, Cc and Ec. The dashed line shows the present position of outermost Carpathian thrust faults for reference only. Black areas indicate outcrops of ophiolitic rocks, and horizontal black and white stripes indicate subsurface ophiolites beneath the Transylvanian basin and west of the Apuseni mountains. EA, Eastern Alps; SA, Southern Alps; PK, Pieniny Klippen belt; Z, Zemplin Hills; VM, Villany–Mecsek; B, Bakony Mountains; Bu, Bukk Mountains; WC, West Carpathians; EC, East Carpathians; SC, South Carpathians; A, Apuseni Mountains; S, Serbian-Bosnian zone; R, Rhenodanubian flysch; P, Podhale flysch; C, (outer) Carpathian flysch; D, Dinaric flysch; Sl, Slovenian basin; H, Hungarian Paleogene basin; Sz, Szolnok-Maramures flysch; Tc, Transcarpathian flysch; Ts, Transylvanian Paleogene Basin; He (NA?), Helvetic Molasse (North Alpine Molasse); NP, Neogene Pannonian basin fill; CM, Carpathian molasse.

struction in Figure 4Ba indicates that in Eocene time these two basins were separated by a wide zone of emergent rocks, so that the lack of a direct faunal connection between the two basins is not surprising despite their proximity.

Late Eocene–Early Oligocene

Throughout late Eocene–early Oligocene time, northward convergence continued across the Eastern Alps as evidenced by the Rhenodanubian flysch being overridden by north-directed nappes of the Eastern Alps in Late Eocene time (Prey, 1978; Tollmann, 1980). North of the Rhenodanubian flysch, a late Eocene–early Oligocene molasse basin formed (North Alpine Molasse), thus displaying the typical transition from flysch deposition in more internal structural areas to molasse deposition in more external areas as continental collision proceeds. The magnitude of contemporaneous shortening across the outer West Carpathians is not known, but along the entire length of the West Carpathians Eocene flysch of the inner Magura unit is steeply dipping to overturned and is overlapped unconformably by relatively flat-lying upper Eocene rocks (Ksiazkiewicz and Lesko, 1959; Oszczypkzo, 1973).

The Eocene Hungarian Paleogene basin, which was connected to the southern Dinaric ocean, was uplifted in late Eocene through early Oligocene time (Figure 4Ca). Priabonian bathyal sedimentary rocks in this basin are overlain unconformably by Oligocene alluvial or brackish sediments and were later exposed to the "infra Oligocene denudation" (Telegdi-Roth, 1927). A new Hungarian Paleogene basin developed east of the Eocene basin, also bounded by many of the same east–northeast trending (dextral) faults (e.g., Rába–Roznava line and Zagreb–Zemplin line). This narrow, elongate basin also shows many of the characteristics of a wrench basin and is interpreted here as the result of continuing dextral shear along a transformlike fault zone.

The geometry shown in Figures 4Cb and 4Cc suggests that some convergent activity may have occurred north of the easternmost part of the Pieniny Klippen belt to accommodate continued northeastward motion of the West Carpathians relative to the East Carpathians and Apuseni Mountains. This interpretation is consistent with the presence of late Eocene–early Oligocene volcanic rocks that are thought to originate from subduction south and southwest of the Buda Hills and southwest of the Bükk Mountains. However, there is little evidence for such an event in the flysch belt itself; presumably, if late Eocene–early Oligocene convergence occurred here, most of the resulting structures within the flysch belt were overridden and subducted during Miocene time.

In the inner flysch belt, deposition of the Podhale flysch ended in early Kiscellian time, while deposition continued uninterrupted in the Szolnok–Maramures and Transcarpathian parts of the inner flysch belt through the end of Oligocene time. It may be that the interruption of deposition of the Podhale flysch represents uplift associated with continued subduction beneath the inner West Carpathians. The later cessation of deposition within the other parts of the inner flysch belt may reflect that these areas were already partly disrupted from the inner West Carpathian flysch or may only reflect a more gradual change in facies from north to south.

In contrast to the earlier Hungarian Paleogene basin, it is apparent that the latest Eocene–Oligocene part of the basin opened toward the northeast into the outer Carpathian flysch ocean because the final connection with the southern Dinaric ocean through the Slovenian basin was closed at about the Eocene–Oligocene boundary (Figure 4Ca). Indeed, in early Oligocene time, the entire Carpathian flysch belt along with the inner flysch and the nonflysch Paleogene basins became anoxic as the marine connections closed or became restricted to the north.

The sedimentary succession within this late Eocene to Oligocene part of the Hungarian Paleogene basin begins with shallow marine, biogenic limestones containing occasional depositional hiatuses. It is overlain by bryozoa and *Discocyclina* marls and lies transgressively on the Mesozoic rocks. The presence of *Nummulites fabianii, Discoyclinidiae,* and nannoplankton faunas (Báldi-Beke, 1972) indicate a late Priabonian age for this transgression 37–38 Ma (after Berggren et al., 1985). The late Priabonian faunal and sedimentary affinities with northern Italy are striking, indicating that a southern marine connection still existed at this time. However, a northward connection to the Carpathian flysch trough and the Transylvanian basin most have already opened by late Priabonian time, because coeval formations within the Transylvanian region are similar (Figure 4Ca).

In latest Priabonian time, this new part of the Hungarian Paleogene basin abruptly subsided to bathyal depths, as indicated by deposition of the Buda Marl, a *Globigerina* marl with calcareous turbiditic intercalations (Nagymarosy, 1981; Horváth, 1983, Márton, 1981).

The first connection with Boreal water came from the North Sea, probably through the Rhine graben, and a cool water, monospecific *Spiratella* fauna evolved around 36 Ma (Báldi, 1984). The *Spiratella* zone and the overlying *Cardium–Ergenica* assemblage zone (35–33 Ma) are contained in the lower part of the Tard Clay. The Tard Clay is a euxinic facies, built up mainly of laminated clays; it overlies the Buda Marl and underlies the Kiscell Clay.

The first anoxic events occurred at about the beginning of the Spiratella zone (36 Ma). Between 35 and 33 Ma, the basin was almost entirely isolated from the world oceans and endemic faunas evolved (*Cardium lipoldi* assemblage). (Dating is based on magnetostratigraphy and nannozones, from Báldi, 1984). Reoxygenation occurred at about 30 Ma, the beginning of deposition of the Kiscell Clay.

The sequence of *Globigerina* marls and laminated clays in latest Eocene–early Oligocene time is strikingly similar over the entire northern Alpine molasse basin (Fischschiefer Formation) (Steininger et al., 1979), the Carpathian flysch belt (Menelitic Formation), the Hungarian Paleogene basin (Tard Clay), the Transylvanian basin (Ileanda Clay), the Slovenian basin, and as far east as the Aral Sea in the southern Soviet Union (Báldi, 1980).

The entire Alpine–Carpathian–Pannonian region in early Oligocene time can be characterized by similar sedimentation histories; endemic faunas are present at some levels of the early Oligocene sequence. Báldi (1983) has proposed that this distinct paleogeographic domain be distinguished by the term *Eoparatethys*, because faunas and facies are reasonably uniform throughout the area and are in sharp contrast with those of the North Sea and Mediterranean area.

Closure of the marine seaways to the northwest (through the Alpine foredeep) and to the southwest (through the Slovenian basin) in early Oligocene time were probably the direct result of continuing continental collision across the Central, Eastern, and Southern Alps and associated uplift. Thus, the dramatic change in paleogeography at this time (the first Paratethyian isolation) can be understood in light of the synchronous tectonic events.

Late Oligocene–Early Miocene

By this time, convergence was mostly completed across the Eastern Alps, but rapid convergence was initiated across the East and West Carpathians in late Oligocene time (Figures 4Db and 4Dc). For this convergence to be accommodated, we infer the presence of a major east-west-trending zone of dextral shear south of the South Carpathians and a northeast trending zone of sinistral shear northeast of the West Carpathians. Evidence for the former is lacking because the inferred fault zone south of the South Carpathians is now buried by thick upper Miocene molasse. However, the Egerian Formation of the brackish to marine Petrosani basin in this region may be related to such a fault zone. Evidence for the second fault zone proposed is given by Roth (1980), but the age of this fault (or fault system) is not certain and may be partly middle Miocene. Other faults within the Moesian platform may have been active at this time (Sandulescu, 1980) and may have helped to accommodate convergence across the bend between the East and South Carpathians.

The final overthrusting of the Transcarpathian flysch unit and several other allochthonous units in the same vicinity (e.g., Botiza nappe and Wildflysch nappe) are of early Miocene age (Eggenburgian), which indicates shortening by an unknown amount and suggests that the final suturing of the internal parts of the West and East Carpathians is of early Miocene age (Sandulescu, 1980). Sedimentation (in molasse facies) within the Szolnok and Carpathian flysch troughs ended in latest Oligocene time. Subsequent deformation and north-directed thrusting of the Szolnok flysch occurred during early or early middle Miocene time. Some workers have interpreted the deformed Szolnok flysch as a major suture zone marking a former ocean, but we do not believe that paleogeographic or tectonic evidence supports this conclusion. Local compression (or transpression) along a shear zone such as that proposed above would be sufficient to account for the structures present in the Szolnok flysch trough, as well as for the tectonostratigraphic units present to the north and south of it.

In late Kiscellian–Egerian time, the Hungarian Paleogene basin subsided along its entire length by late Oligocene–early Miocene block faulting. This block faulting may have been accompanied by horizontal shear, but we have no strong indication of the sense of motion along such a shear zone at this time.

The faunal compositions found in the Eoparatethys indicate the reopening of a marine connection toward the Ligurian sea in the south, as northern Italian faunal elements, such as larger foraminifera (*Lepidocyclina, Nummulites vascus*) and warm subtropical molluscs reached the central Eoparatethys (the Harshegy Sandstone) (Figure 4Da). The full marine Slovenian sedimentary sequence of late Kiscellian age suggests the reopening of a marine passage through Slovenia. Hence it appears that the late Oligocene subsidence of the Hungarian Paleogene basin was probably responsible for the reopening of the Eoparatethys to the southern marine elements. This provides another example of how tectonic events directly control the creation and destruction of specific paleogeographic environments.

At the same time as a southern marine connection was established, boreal molluscan taxa migrated southward from the North Sea through the Rhine graben and the Russian epicontinental sea (Báldi, 1980, 1983). Thus, a passageway between the northern and southern oceans was opened through the Alpine chain and Pannonian region in late Kiscellian time (30 Ma). Egerian faunas in both shallow and deep marine facies in the Alpine–Carpathian–Pannonian region show the same mixture of boreal and southern provinces. However, because of the cool Oligocene climate (relative to Eocene time), the dominant direction of migration of faunas was from north to south (Cavelier, 1979).

Middle–Late Miocene

Detailed discussions of Neogene to Recent tectonics (Sandulescu, 1975; Royden et al., 1982, 1983; Royden, this volume) and paleogeography (Steininger and Rögl, 1979; Rögl and Steininger, 1983) have been published elsewhere, but we include a brief discussion of these events here for the sake of completeness.

From middle to late Miocene time, north- and east-directed thrusting and convergence continued across the outer Carpathian flysch belt (Figures 4Eb and 4Ec). Southward and westward subduction of the European plate beneath the Pannonian region was accompanied by eruption of a Miocene volcanic arc that extended from the Graz basin eastward to the junction between the East and South Carpathians. These primarily andesitic rocks are geochemically similar to those erupted along the southern margin of the Hungarian Paleogene basin in Eocene–Oligocene time and are thought to be derived from subduction (Gy, Pantó, personal communication, 1984). The eruption of volcanic rocks and the cessation of thrusting in the outer flysch Carpathians show a clear west to east migration through time (Póka, this volume). By latest Miocene time convergence was essentially finished everywhere across the outer flysch belt from the central Alps to the East Carpathians.

The final phase of convergence in the outer Carpathians was accompanied by considerable extension and subsidence of the Pannonian basin area associated with large strike-slip fault zones (Royden et al., 1982, 1983; Royden, this volume). Some of these major shear zones have been generalized by heavy lines, but a slightly more detailed picture of the late Miocene fault system is shown by the lighter lines. In a gross sense, these Miocene fault systems functioned as transformlike boundaries that accommodated zones of diachronous shortening and different magnitudes and directions of convergence in the outer flysch belt.

Similarly complex fault patterns most have also existed for earlier Paleogene deformation of the Pannonian region, but the configuration of such fault patterns has been largely destroyed by later tectonic events. For this reason it is only possible to generalize about the locations and sense of offset of the Paleogene fault systems. Moreover, it is clearly impossible to reconstruct the exact or relative locations of various crustal blocks for pre-Miocene events because complex fault patterns such as those shown in Figure 4Eb can easily change the distance between and relative locations of any two small crustal fragments.

Marine deposition ended in the Carpathian foredeep in late Miocene (upper Badenian) time. As in the Eastern Alps, this transition from deep marine to shallow marine and finally non-marine sedimentation clearly reflects continued continental collision, convergence, and uplift across an active thrust belt (Figure 4Ea). In the areas of the Pannonian and Transylvanian basins, marine conditions prevailed until the end of the middle Miocene (Badenian). By the end of Miocene time, the Pannonian basin was covered by a freshwater lake. Fluvial sedimentation continues today throughout the Pannonian basin area, and as much as 7 km of late Miocene to Recent sedimentary rocks have accumulated in parts of the basin.

DISCUSSION AND CONCLUSIONS

It seems that present arrangement of small tectonostratigraphic elements throughout the Carpathian–Pannonian region can be explained as the result of Cenozoic slivering of larger, originally contiguous tectonostratigraphic units. Although we know that disruption and juxtaposition of crustal fragments must have occurred along rather complex fault systems (e.g., see Figure 4Eb), these detailed fault systems cannot be completely restored. Nevertheless, they can be reconstructed in a simplistic fashion from analysis of tectonostratigraphic and paleogeographic elements and from geometric constraints imposed by better known zones of tectonic activity. Yet even the simplified fault patterns so constructed are capable of accounting for most of the complex distribution of tectonostratigraphic elements observed in the Carpathian–Pannonian area.

These fault systems form a coherent pattern whose basic character remained largely unchanged throughout Cenozoic time. Deformation occurred along the boundaries of eastward-moving fragments or wedges whose northern boundaries were usually dominated by convergent or convergent plus strike-slip tectonic activity and eastern boundaries by convergent activity (Figures 4Ac, 4Bc, 4Dc, and 4Ec). The southern boundaries of these fragments generally were zones of dextral shear and were probably responsible for most of the slivering within the internal part of the Carpathian–Pannonian region. For example, such zones of dextral shear were probably responsible for (1) the severing of the Bükk–Igal zone from the Serbian zone and its subsequent northeastward translation, (2) for the dislocation of the Apuseni Mountains from the inner West Carpathians, (3) for the eastward displacement and suturing of the Apuseni and Transylvanian ophiolites, and (4) for the deformation of the inner Carpathian flysch zone (Podhale, Szolnok, and Transcarpathian flysch zones), (see Figures 4Ab, 4Bb, 4Cb, 4Db and 4Eb.) Clearly, the effects of Tertiary dextral shear within the Carpathian region most be restored before the early Tertiary and Mesozoic tectonostratigraphic elements can be understood in relation to one another. Conversely, realigning elements that originally belonged to the same tectonostratigraphic unit into coherent zones yields valuable information about the nature of tectonic events that subsequently disrupted the region.

The large shear zones within the Carpathian region discussed in this paper are required partly because of the diachronous nature of the convergent boundary extending through the Eastern Alps and West and East Carpathians and partly because of the different directions of thrusting around the belt. These shear zones separate areas of active shortening in the outer Carpathians from inactive parts of the outer Carpathian belt, as well as connecting areas of shortening to one another. For example, the zone of dextral shear shown in Figures 4Bb and Bc separates areas of convergence in the Eastern Alps and West Carpathians from the inactive East Carpathians, and also acts as a transform boundary that connects this convergent zone to another convergent zone in the Dinaric Alps. Thus, the presence of a zone of dextral shear within the Carpathian region can be inferred from reconstruction of relative fragment motions based on examination of the locations of active and inactive convergent boundaries. In some cases, however, shear may have been accomplished by bending of crustal elements rather than by the development of discrete fault zones.

Old strike-slip zones may now appear as inactive high-angle fault zones sometimes associated with coeval rift basins or en echelon folds. In places they may be totally obscured by posttectonic sedimentation. Although it is often difficult to document the direction of motion on these faults directly from field studies, fault-bounded, elongate basins with rapid changes in depositional depths, both up and down, are indicators of possible strike-slip activity.

Tertiary deformation within the Carpathian–Pannonian region seems to have occurred along the boundaries of eastward-moving crustal fragments. Many of the deformational events that occur along the boundaries of these fragments show a west to east migration through time, such as the following: wildflysch deposition along the Pieniny Klippen belt, subsidence of the Hungarian Paleogene basin, the eruption of Paleogene (subduction-derived) volcanic rocks, the end of Neogene thrusting in the outer Carpathians, the eruption of Neogene volcanic rocks within the Carpathian region, and subsidence of the Neogene Pannonian and other intra-Carpathian basins. This persistent pattern suggests that many of these events may have been initiated to the west in the area of the Eastern Alps. We suggest that this observation may be explained as follows. Tertiary north–south convergence across the central and Eastern Alps was partly accommodated by eastward displacement of continental fragments (or wedges) into a recess in the combined European–Moesian plate (Burchfiel, 1980). Thus, the Tertiary evolution of the Carpathian fragment is one of repeated episodes of "continental escape" (Burke and Sengör, 1986) as material is transported laterally away from a collision zone toward a free interface (that is, a zone containing lithosphere with easily subducted oceanic or attenuated continental crust). Most of the slivering and slicing that occurred in the Pannonian region can be explained by the temporary nature of the fragment boundaries. As the major fault or faults that make up the fragment boundaries shift with time, originally contiguous tectonostratigraphic units were disrupted and small pieces of different tectonostratigraphic units were sutured together in a complex fashion.

ACKNOWLEDGMENTS

This work was supported by NSF grants #INT-7910275 and # EAR-8115863, the Hungarian Academy of Sciences, and Shell Research and Development Co. We are grateful to M. Sandulescu, B. C. Burchfiel, A. M. C. Sengör, F. F. Steininger, and F. Rögl for helpful reviews and discussions.

REFERENCES

Andrusov, D., 1968, Grundriss der Tektonik der Nordlichen Karpaten: Verlag der Slowakischen Akademie der Wissenschaften, Bratislava, Czechoslovakia, 187 p.

Báldi-Beke, M., 1972, The nannoplankton of the upper Eocene bryozoan and Buda marls: Acta Geol. Acad. Sci. Hung., v. 16, p. 211–228.

Báldi-Beke, M., 1977, Stratigraphical and faciological subdivision of the Buda Oligocene as based on nannoplankton: Földt. Közl., v. 107, p. 59–89.

Báldi-Beke, M., 1984 A dunántuli paleogén képzödmények nannoplanktonja [Nannoplankton of the Transdanubian Paleogene (Hungary)]: Geol. Hung. Ser. Pal., v. 43, 307 p.

Báldi-Beke, M., M. Horváth, and A. Nagymarosy, 1981, Biosztratigráfiai vizsgálatok az alfoldi flisképzödményekröl [Biostratigraphic investigation of Flysch formations in the Great Hungarian Plain]: Annual Rept. Hungarian Geol. Survey from the Year 1979, Budapest, 143–158 p.

Báldi, T., 1980, A korai Paratethys története [The early history of the Paratethys]: Földt. Közl. v. 110, p. 456–472.

Báldi, T., 1983, Mid-Tertiary stratigraphy and paleogeographic evolution of Hungary (late Eocene through early Miocene): Budapest, Akademiai Kiado, 293 p.

Báldi, T., 1984, The terminal Eocene and early Oligocene events in Hungary and the separation of an anoxic, cold Paratethys: Eclogae Geologicae Helvetiae, v. 77, p. 1–27.

Balogh, K., 1964, A Bükkhegység földtani képzödményei [Die geologischen Bildungen das Bükk-Gebirges], MÁFI Évk., 48, 719 p.

Bauer, F. K., and H. P. Schönlaub, 1980, Der Drauzug (Gailtaler Alpen–Nordkarawanken), *in* R. Oberhauser, ed., Die Geologische Auf bau Österreich: New York, Springer-Verlag, p. 405–425.

Berggren, W. A., P. V. Kent, J. J. Flynn and J. A. Van Couvering, 1985, Cenozoic geochronology: Geol. Soc. Am. Bull., v. 96, p. 1407–1418.

Birkenmajer, K., 1981, Strike-slip faulting in the Pieniny Klippen belt of Poland: Paper presented at the 12th Congress, Carpatho-Balkan Geol. Assoc., Bucharest, Romania, Sept. 8-12, 1981.

Bleahu, M., M. Lupu, D. Patrulius, S. Bordea, A. Stefan, and S. Panin, 1981, The structure of the Apuseni Mountains, *in* Guide to Excursion B-3, XII Congress of the Carpatho-Balkan Assoc, Bucharest, 103 p.

Burchfiel, B. C., 1976, Geology of Romania: Geol. Soc. Am. Special Paper 158, 82 p.

Burchfiel, B. C., 1980, Eastern European Alpine System and the Carpathian orocline as an example of collision tectonics: Tectonophysics, v. 63, p. 36–61.

Burke, K., and A. M. C. Sengör, 1986, Tectonic escape in the evolution of the continental crust, *in* M. Barazangi and L. Brown, eds., Reflection Seismology: a global perspective: Geodynamics Series v. 14, p. 41–53.

Cavelier, C., 1979, La limite Éocéne–Oligocéne en Europe occidentale: Sci. Géol. Mém. 54, CNRS, Strasbourg, 280 p.

Channell, J. E. T., and F. Horváth, 1976, The African/Adriatic promontory as a paleogeographical premise for the Alpine orogeny and plate movements in the Carpatho-Balkan region: Tectonophysics, v. 53, n. 1-3, p. 7–102.

Crowell, J. C., 1974, Origin of late Cenozoic basins in southern California, *in* W. R. Dickinson, ed., Tectonics and sedimentation: SEPM Special Publication 22, p. 190–220.

Géczy, B., 1972, The origins of the Jurassic faunal provinces and the Mediterranean plate tectonics: Annals, Univ. Sci. R. Eotvos Sect. Geol., v. 16, p. 99–114.

Horváth, M., 1983, Eocene/Oligocene boundary and the terminal Eocene events on the basis of planktonic foraminifera: TEE Meeting Abstracts, Budapest-Visegrád (with unpublished manuscript).

Jiricek, R., 1979, Tectonic development of the Carpathian arc in the Oligocene and Neogene: *in* M. Mahel, ed., Tectonic profiles through the West Carpathians: Geol. Ustav Dionyza Stura, Bratislava, Czechoslovakia, p. 205–214.

Kahler, F., and A. Papp, 1968, Uber die bisher in Karnten gefunden Eozängerölle: Carinthia II, v. 78, p. 80–98.

Kecskeméti, T., 1980, A Bakony hegysegi *Nummulites* fauna paleobiogeografiai attekintese [An outline of the paleobiogeography of the *Nummulites* fauna of the Bakony Mts. in Hungary]: Földt. Közl., v. 110, p. 432–449.

Koch, A., 1911, Uj adalékok a Gryphaea Eszterházy elterjedéséhez és geológiai jelentöségéhez [New data to the distribution and geologic significance of the Gryphaea Eszterházy]: Földt. Közl, v. 41, p. 42–45.

Körmenady, A., 1980, Az északkeleti Bakony eocén medence facies ének puhatestü faunája [Mollusc fauna of the basin-facies in the NE-Bakony Eocene]: MÁFI Évk., v. 63, 227 p.

Kovács, S., 1982, Problems of the "Pannonian Median Massif" and the plate tectonic concept, Contributions based on the distribution of Late Paleozoic–Early Mesozoic isotopic zones: Geol. Rundschau, v. 71, p. 617–640.

Ksiazkiewicz, M., and B. Lesko, 1959, On the relation between the Krosno and Magura flysch: Bull. de Acad. Polonaise Sci.: Serie Des Sciences Chimiques, Géol. et Geograph., v. 7, n. 10, p. 773–780.

Laubscher, H. P., 1971, Dss Alpen-Dinsriden problem und die pslinspsstik der sidlichen Tethys: Geol. Rundschau, v. 70, p. 813-832.

Lexa, J., and V. Konecny, 1974, The Carpathian volcanic arc: a discussion: Acta Geol. Acad. Sci. Hung., v. 18, n. 3-4, p. 279–293.

Marschalko, R., 1982, Paleotektonickíj význam flysvvých sedimentov pri analýze kory alpínskej geosynklinály [Paleotectonic importance of flysch during analysis of Alpine geosyncline]: Geol. Práce, Zprávy, v. 77, p. 125–133.

Marschalko, R., 1980, Evolution of Paleocene–Lower Eocene Pieniny Klippen belt and central West Carpathian block (an example of Sulovske Vrchy Hills): Geol. Zborn. Geol. Carpath., v. 31, n. 4, p. 513–521.

Márton, P., 1981, Report of Paleomagnetic studies on Oligocene cores, borehole Kl-1: Budapest, TEE Meeting Abstracts, Budapest-Visegrádo.

Nagymarosy, A., 1981, Chrono- and biostratigraphy of the Pannonian basin: a review based mainly on data from Hungary: Earth Evol. Sci., v. 1. n. 3, p. 183–194.

Oberhauser, R., ed., 1980, Der geologische Aufbau Österreichs: New York, Springer Verlag, 699 p.

Oszczypkzo, N., 1973, Geology of the Nowy Sacz Basin (the Bull. Geol. Institute, Warsaw, p. 101–197.

Prey, S., 1978, Rekonstruktionsversuch der alpidischen Entwicklung der Ostalpen: Mitt. Österr. Geol. Ges., v. 69, p. 1–25.

Prey, S., 1980, Helvetikum, flysche und Klippenzonen von Salzburg bis Wien: *in* R. Oberhauser, ed., Der geologische aufbau Österreichs: New York, Springer Verlag, p. 189–217.

Rögl, F., and F. F. Steininger, 1983, Von zerfall der Tethys zu Mediterran und Paratethys. Die Neogene Palaeogeographie und Palinspastik der zirkum-mediterranen Raumes: Annal. Naturhist. Museum Wien, v. 85A, p. 135–163.

Roth, Z., 1980, Západni Karpaty-tercíerní struktura stredni Evropy [West Carpathians, Tertiary structure of central Europe]: Ustred. Ustav Geol., v. 55, 128 p.

Royden, L., F. Horváth, and B. C. Burchfiel, 1982, Transform faulting, extension and subduction in the Carpathian–Pannonian region: Geol. Soc. Am. Bull., v. 73, p. 717–725.

Royden, L., F. Horváth, and J. Rumpler, 1983, Evolution of the Pannonian basin system, 1. Tectonics: Tectonics, v. 2, p. 63–90.

Samuel, O., and J. Salaj, 1968, Microbiostratigraphy and foraminifera of the Slovak Carpathian Paleogene: Geol. Ust. d. Stura, 232 p.

Sandulescu, M., 1975, Essai de synthese structurale des Carpathes: Bull. Geol. Soc. Fr., v. 17, p. 299–358.

Sandulescu, M., 1980, Analyse geotectonique des chaines alpines situees au tour de las mer noire occidentale: Ann. Inst. Geol. Geophys., v. 56, p. 5–54.
Sandulescu, M., M. Stefanescu, A. Butac, I. Patrut and P. Zaherescu, 1981a, Genetical and structural relations between flysch and molasse (the East Carpathian model), *in* Guide to Excursion A5: XII Congress of the Carpatho-Balkan Geological Association, Bucharest, 95 p.
Sandulescu, M., H. G. Kräutner, I. Balintoni, D. Russo-Sandulescu, and M. Micu, 1981b, The structure of the East Carpathians, *in* Guide to Excursion Bl: XII Congress of the Carpatho-Balkan Geological Association, Bucharest, 92 p.
Ślaczka, A., 1969, Final stages of geosynclinal development in the SE part of the Polish Carpathians: Acta Geol. Acad. Sci. Hung., v. 13, p. 331–335.
Steininger, F., and F. Rögl, 1979, The Paratethys history—a contribution toward the Neogene geodynamics of the Alpine orogen (abs.): Ann. Geol. Pays Hellen., VII Congr. RCMNS, v. 3, p. 1153–1165.
Strausz, L., 1966, Dudari eocén csigák [The Eocene gastropods of Dudar]: Geol. Hung. Ser. Psl., v. 33, 199 p.
Szöts, E., 1953, Magyarország eocén puhatestüi. I. Gánt környéki puhatetstüek [Eocene molluscs of Hungary. I. Gánt]: Geol. Hung. Ser. Pal., v. 22, 245 p.
Telegdi-Roth, K., 1927, Ingraoligocén denudáció nyomai a Dunántuli Középhegység északnyugati peremén [Traces of an Infra-Oligocene denudation in the NE margin of the Transdanubian Mid-Mts]: Földt. Közl. v. 57, p. 32–41.
Tollmann, A., 1980, Grosstektonische Ergebnisse an den Ostalpen in sinne der Plattentektonik: Mitt. Österr. Geol. Ges., v. 71–72, p. 37–44.
Unrug, R., 1979, Palinspastic reconstruction of the Carpathian arc before the Neogene tectonogenesis: Rocznik Polsk. Towarz. Geol., v. 49, n. 1-2, p. 3–21.
Van Couvering, J. A., M. P. Aubrey, W. A. Berggren, J. P. Bujak, C. W. Naeser, and T. Wieser, 1981, The terminal Eocene event and the Polish connection: Palaeogeogr., Palaeoclim., Palaeoecol., v. 36, p. 321–362.
Vörös, A., 1977, Provinciality of the Mediterranean Lower Jurassic brachiopod fauna: causes and plate tectonic implications: Palaeogeogr. Palaeoclim., Palaeoecol., v. 21, p. 1–16.

CHAPTER 2

Cenozoic Tectonic History of the Carpathians

Mircea Sandulescu
Institute of Geology and Geophysics,
str. Caransebes 1
78344 Bucharest, Romania

The Carpathian Mountains formed during closure of the Tethys ocean during Cretaceous and Miocene convergent events. Three main shortening events occurred in the outer flysch Carpathians during Miocene time: (1) early Miocene events (old Styrian "phase," 18–20 Ma), (2) middle Miocene events (young Styrian "phase," ~15.5 Ma), and (3) late Miocene events (Moldavian "phase," 11–12 Ma). Early Miocene events include deformation and thrusting of the Pienides (Pieniny Klippen belt and Magura group nappes), en bloc overthrusting of the inner Dacides over the Pieniny Klippen belt, and thrusting within the most internal Moldavidian nappes (Audia–Czernahora, Macla, Convolute flysch, and Dukla(?) units). Middle Miocene events include thrusting within the Moldavidian nappes (Tarcau, Marginal Folds, Silesian, Skole, Subsilesian, and Waschberg units). Late Miocene events include thrusting within the outer Moldavidian nappes (Marginal Folds, Tarcau, Skole, Subsilesian, and Silesian units) and the Subcarpathian nappe. Minor Pliocene–Pleistocene folding occurred in the Subcarpathian nappe in the area of the southeast Carpathian bend. All of these Pienide and Moldavidian nappes consist only of sedimentary rocks, without their original crystalline basement. The westernmost limit of shortening migrated progressively eastward from one event to the next; in addition, shortening within each event may have occurred diachronously, being older in the west than in the east.

INTRODUCTION

The Carpathian Mountains lie between the Alps to the west and the Balkan and Rhodope mountains to the southeast. The Carpathian Mountains and the Dinaric Alps form two opposite branches of the Alpine chain east of the Eastern and Southern Alps (Figures 1 and 2). The Carpathians underwent several episodes (or "phases") of compressional deformation during Cretaceous and Cenozoic time. Two main periods of compressional deformation can be recognized in the Carpathians: (1) the Dacidian period, which includes middle Cretaceous (100–120 Ma), middle Late Cretaceous (~90 Ma), and latest Cretaceous (63–68 Ma) events; and (2) the Moldavidian period, which includes early Miocene (early Styrian, ~18–20 Ma), middle Miocene (late Styrian, ~15.5 Ma), and late Miocene (Moldavidian, ~11–12 Ma) events. The Cretaceous Dacidian events were mainly responsible for the deformation and present geometry of the inner Carpathians (or Dacides), while the Miocene Moldavidian events were restricted to the outer Carpathians (or Moldivides) (Figure 3). The Pienide zone recorded both the Late Cretaceous Dacidian events and early Miocene Moldavidian events to the same extent.

This paper concentrates primarily on the Cenozoic deformation within the Carpathian chain, and only a brief review of the Mesozoic events is presented. Mesozoic events are of interest in this paper in so far as earlier structures influenced the later tectonic and sedimentary evolution of this region. Specific references are not included in the text, but a general bibliography on the subject is given at the end of the paper.

MAIN PRE-CENOZOIC TECTONIC EVENTS IN THE CARPATHIAN AREA

The Alpine history of the Carpathian area can be divided in two parts: an extensional period lasting from Triassic until Neo-

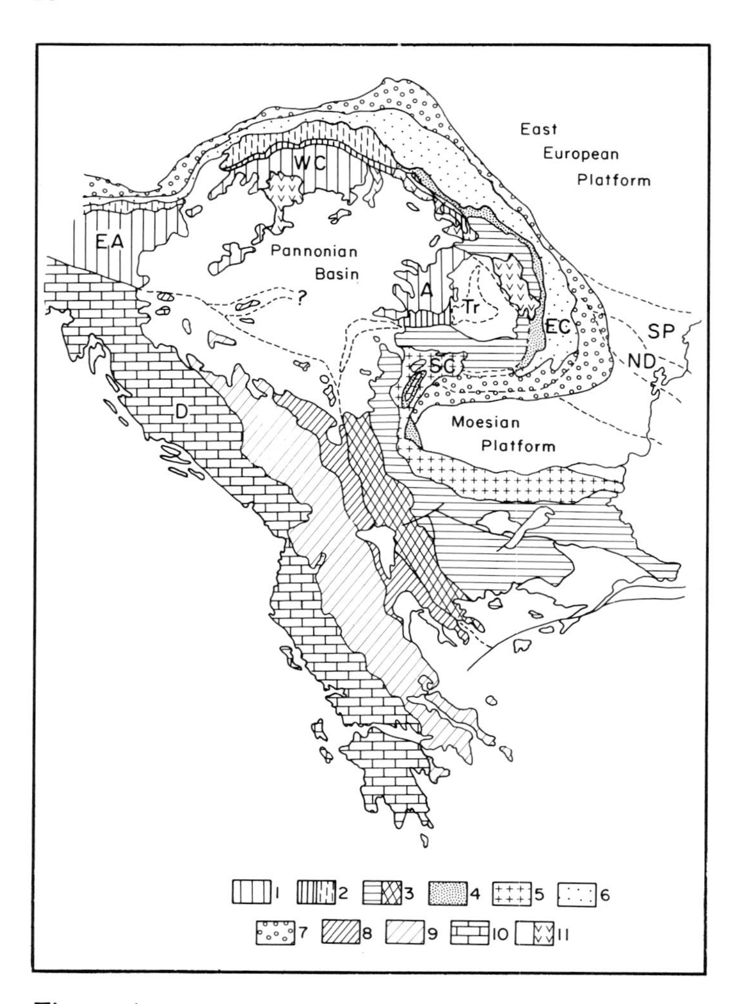

Figure 1. Tectonic sketch of the Alpine fold belts in central and southeastern Europe. Symbols shown are as follows: Carpathians: (1) inner Dacides; (2a) Transylvanides and Pienides, (2b) Magura group; (3a) middle Dacides, (3b) Serbo-Macedonian massif; (4) external Dacides; (5) marginal Dacides (Danubicum and outer Balkanides); (6) Moldavides; (7) foredeep. Dinarides and Hellenides: (8) Vardar zone; (9) inner Dinarides and Hellenides; (10) outer Dinarides and Hellenides; and (11a) basins, (11b) Neogene volcanics. Tr, Transylvanian basin; EEP, eastern European platform; SP, Scythian platform; MP, Moesian platform; ND, North Dobrogea orogen; EA, Eastern Alps; WC, West Carpathians; EC, East Carpathians; SC, South Carpathians; D, Dinarides; A, Apuseni Mountains.

comian time, and a compressional or convergent period lasting from Neocomian through Miocene time (Figure 3).

The extensional history of the Carpathian area led to the opening and spreading of the oceanic Tethys. (The Tethys sea was underlain by both continental and oceanic crust, thus it is necessary to distinguish between an oceanic and a continental part of the Tethys.) Extension began in Middle Triassic time and led to generation of oceanic floor now preserved as ophiolites in the main Tethyan suture. The suture can be followed through the inner part of the Carpathians (Figure 4), as well as through the Vardar zone and Asia Minor Ophiolitic zone(s). Ophiolitic complexes of Middle and Late Triassic, and Jurassic age are known in the Transylvanian nappes and Metaliferi Mountains (southern Apuseni Mountains). The youngest ophiolites known in the Carpathians are thought to be Tithonian, or even Neocomian, in age.

The Carpathian nappes represent not only oceanic Tethys (the main Tethyan suture) and its European continental margin, but also the opposite margin, the Austro-Bihorean margin. South of the Danube, the oceanic suture separates rocks belonging to the European and Apulian continental margins. An oceanic trough, the South Pannonian sphenochasm, which is represented by a specific suture (Figures 4 and 5), separates the Austro-Bihorean block from the Apulian one. It has been suggested that the basic rocks that crop out in the Büuk Mountains are ophiolitic fragments of the South Pannonian suture. Nevertheless, it is possible that these rocks are the remnants of an intracontinental rift that developed at the same time as the outer Dacidic and/or Pindus rifts.

Tethys rift valley structures comparable with the Afar–Red Sea were generated within the continental margins of the oceanic Tethys during this extensional period. These paleorifts were floored by thinned continental or even oceanic crust that generally conformed to the shape of the continental margins. In the Carpathians, one such paleorift is represented by the outer Dacidic units (in the East and South Carpathians) (Figure 5A). These units have the same tectonic position and significance as the Valais units in the Central Alps of western Europe. On the other side of the main Tethyan (oceanic) suture, along the Apulian continental margin, a similar paleorift is represented by the Pindus–Serbian zone.

Compressional events within the Carpathians began during Early Cretaceous time in (late) Barremian or Aptian time and represent the beginning of middle Cretaceous tectogenetic events. Until Albian time, sediments and their continental basement were sheared and emplaced as thrust sheets and some of the oceanic crust was obducted as a thin sheet (Figure 3). Thus, the Transylvanian nappes are obducted ophiolite sheets, the middle Dacidic nappes represent continental margin sediments and their basement (shearing nappes), and the outer Dacidic nappes originated from the marginal troughs within the continental margin. It should be emphasized that middle Cretaceous crustal shortening mainly involved the oceanic part of Tethys and its European continental margin. The Austro-Bihorean and/or the Apulian margins were not deformed until later.

Disruption of the Austro-Bihorean continental block occurred during or at the end of Turonian time (Mediterranean or pre-Gosau tectogenetic events). The Austro-Alpine nappes of the Eastern Alps and their direct continuation as the inner Dacidic nappes of the Carpathians (comprising the central West Carpathians, the northern Apuseni mountains, and the Bakony, Mecsek, and Villany mountains) consist of sedimentary rocks sheared along with their continental basement and emplaced during this compressional period.

The Pienides consist of the Pieniny Klippen belt and the nappes of the Magura group, and they correspond to the Tethyan suture. Their Cretaceous tectonic history is partly similar to that of the Transylvanide nappes. Evidence for middle Cretaceous compression and corresponding crustal shortening is recorded in the structural and sedimentary evolution of the Pienides. Later, during pre-Maestrichtian and/or intra-Maestrichtian deformation, north-vergent cover nappes (with no basement) were emplaced in the Pieniny belt and northwest-vergent ophiolitic (Transylvanide) nappes were emplace in the

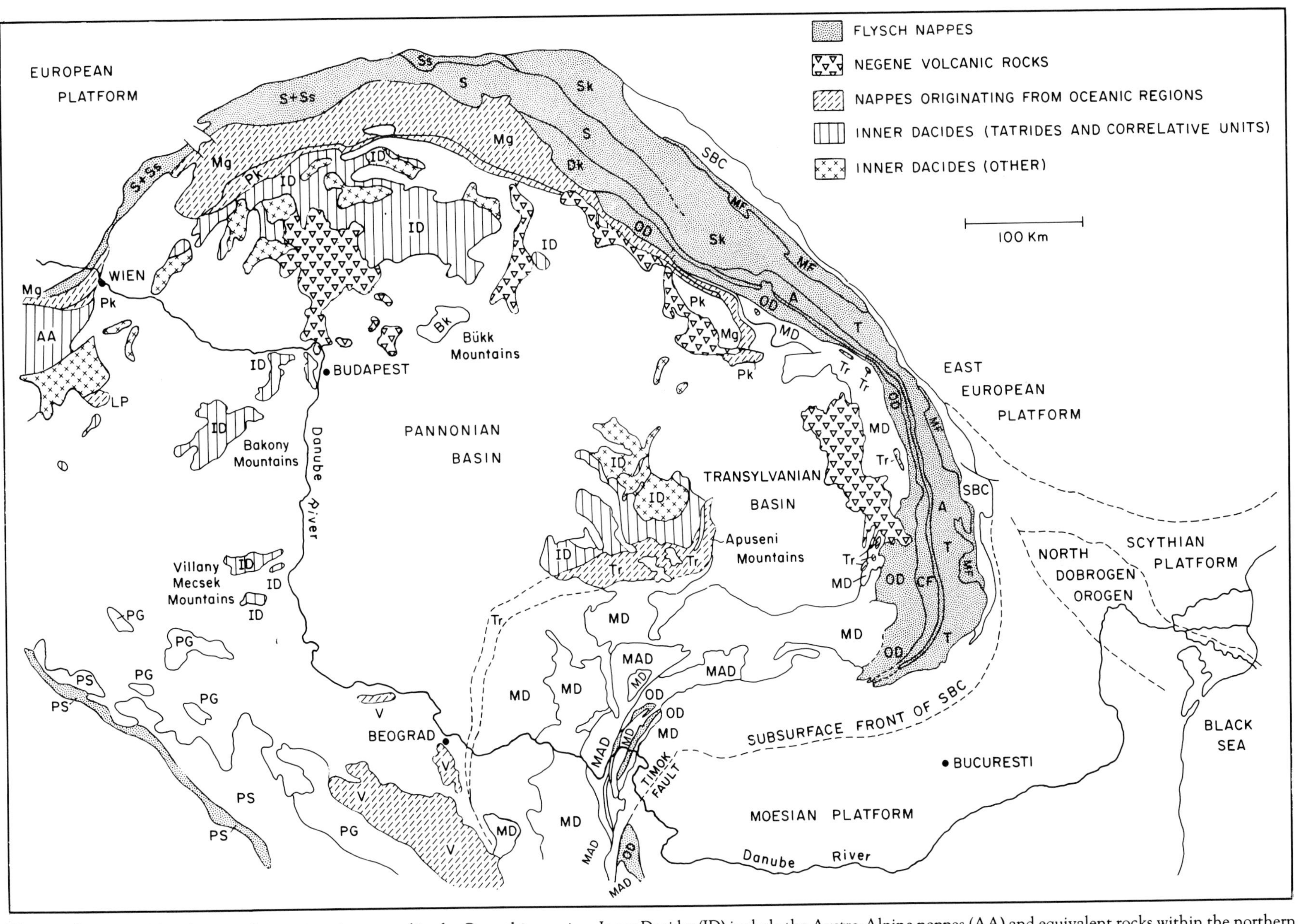

Figure 2. General distribution of tectonic units exposed in the Carpathian region. Inner Dacides (ID) include the Austro-Alpine nappes (AA) and equivalent rocks within the northern Apuseni, central West Carpathian, Villany–Mecsek, and Bakony mountains. These rocks are separated from the middle Dacides (MD) and the marginal Dacides (MAD) by rocks that were originally deposited on oceanic crust (Pieniny Klippen belt, PK; Magura group, Mg; and Transylvanides, Tr) as well as by ophiolitic assemblages that include oceanic crust (Transylvanides; lower Pennine zone, LP; Vardar zone, V). The middle Dacides include nappes of the central East Carpathians and the Getic and Supragetic nappes. The outer Dacides include flysch nappes deformed in Mesozoic and Paleocene time (Black Flysch, Ceahlau, and Severin nappes). The Moldavidian nappes, deformed during Miocene time, include the Convolute Flysch (CF), Dukla (Dk), Audia (A), Silesian (S), Subsilesian (Ss), Skole (Sk), Tarcau (T), Marginal Folds (MF), and Subcarpathian (SBC) nappes. These nappes and the Magura nappe consist mainly of flysch (except for the Subcarpathian nappe, which is mainly molasse). Other units are the Bükk (Bk), Pelagonian and equivalent rocks (PG), and Pindus-Serbian and equivalent rocks (PS).

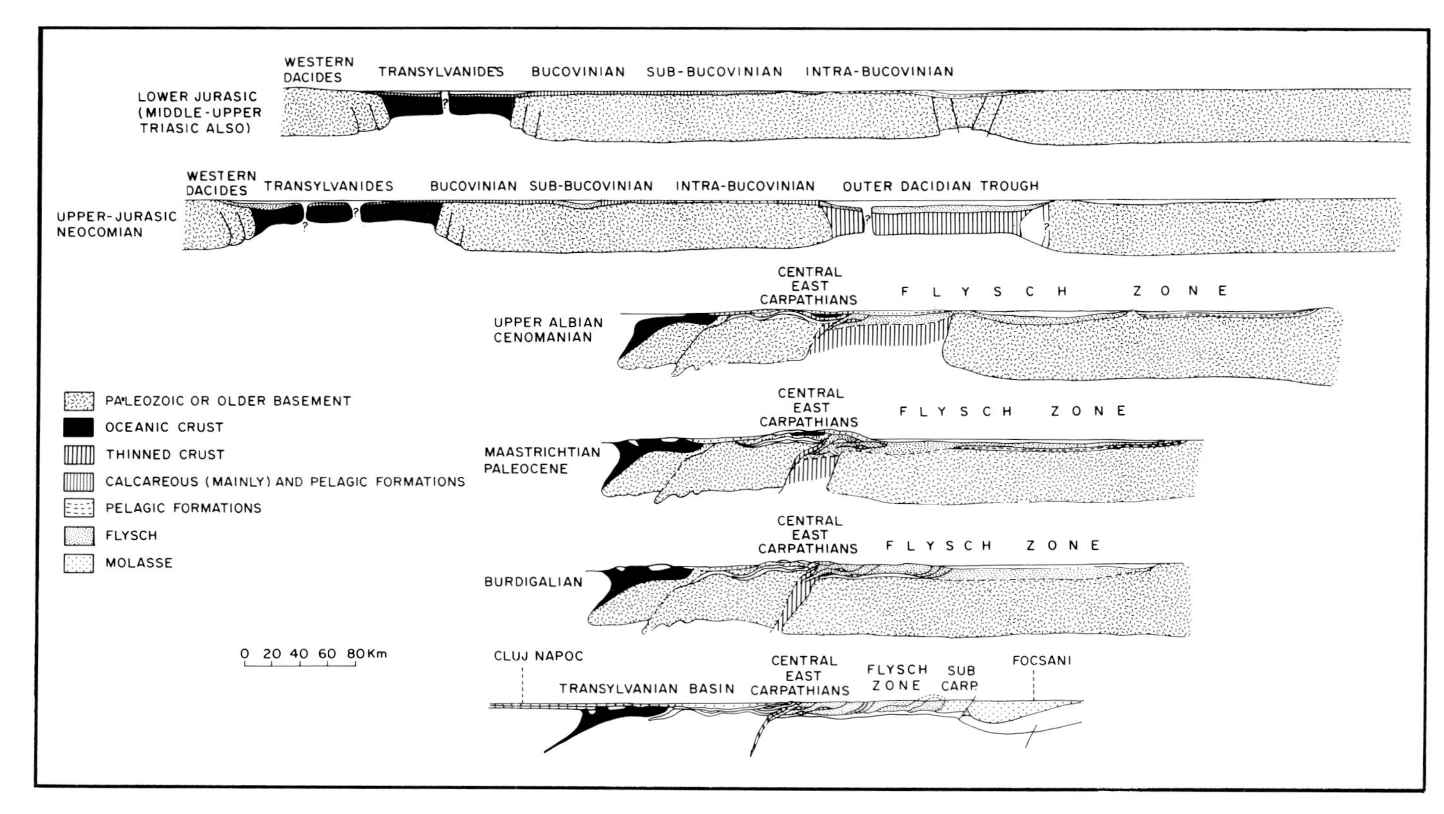

Figure 3. Palinspastic evolution of the East Carpathian–Transylvanian geotraverse.

Metaliferi Mountains (southern Apuseni Mountains). These latest Cretaceous compressional tectogenetic events also involved the European continental margin (middle Dacides in the South Carpathians and outer Dacides along the whole belt). In contrast to the Transylvanide and the Dacidic nappes, the Pienides were the site of renewed sedimentation followed by deformation in Cenozoic time.

Except for the Pienides, the whole inner Carpathian realm attained its present structure and morphology by the end of Cretaceous time. After the Cretaceous deformation, these areas behaved roughly as rigid blocks and were overlapped by a posttectonic (postthrusting) sedimentary cover, which was locally slightly deformed. These posttectonic covers were deposited in different depositional environments and on top of a differentially subsiding basement. This resulted in highly differentiated lithologies and thicknesses in the Paleogene posttectonic cover, which includes coarse-grained (molasselike) epicontinental flysch and pelagic deposits.

Cretaceous deformation of the inner Carpathians was also responsible for several arcuate bends in the chain, including the South Carpathian reentrant west of Bucharest, the Transylvanian double bend, and the southeast Carpathian salient north of Bucharest (Figures 2 and 5). All of these arcuate structures of Cretaceous age are covered discordantly by mainly undeformed Paleogene sedimentary rocks. Thus, Cenozoic deformation and/or differential rotation did not create these arcuate Cretaceous bends. In Cenozoic time, deformation involved large blocks separated by strike-slip faults, and Cenozoic rotations recorded in the inner Carpathians most be understood as a result of this process. Differences in the amount and direction of rotation within the Carpathians is determined by the offset between blocks along the strike-slip faults and not by Tertiary bending.

It is important to emphasize that rocks (mainly in large nappes) situated below the young Pannonian and Transylvanian molasse basins were deformed during Dacidian (Cretaceous) tectogenetic events.

PALEOGENE AND EARLIEST MIOCENE SEDIMENTARY MOBILE AREAS

The most important zones of subsidence, characterized mainly by flysch type sedimentation in Paleogene and earliest Miocene time, were in the Pienide and Moldavidian domains (Figure 5B). In the Pienide zone (mainly in the Pieniny Klippen Belt and the inner Magura nappes), Paleogene sedimentary rocks overlapped the units deformed in Cretaceous time (see previous section). The Magura zone is characterized by practically uninterrupted sedimentation (mostly flysch) from Late Cretaceous through early Miocene time (the local unconformities and gaps in the section are pre-Paleogene).

The Magura nappes were formed from an important trough of Paleogene flysch in the Pienide zone. This trough continues toward the southeast as the Petrova nappe of the East Carpathians. The Pieniny Klippen Zone formed from another trough of Paleogene flysch located south of the Magura zone. The northern part of the Pienides is thought to have been a relatively elevated ridge underlain by middle and outer Dacidic units deformed in Cretaceous time. It was partly overlain by posttectonic cover and formed source areas for the flysch (Figure 6A).

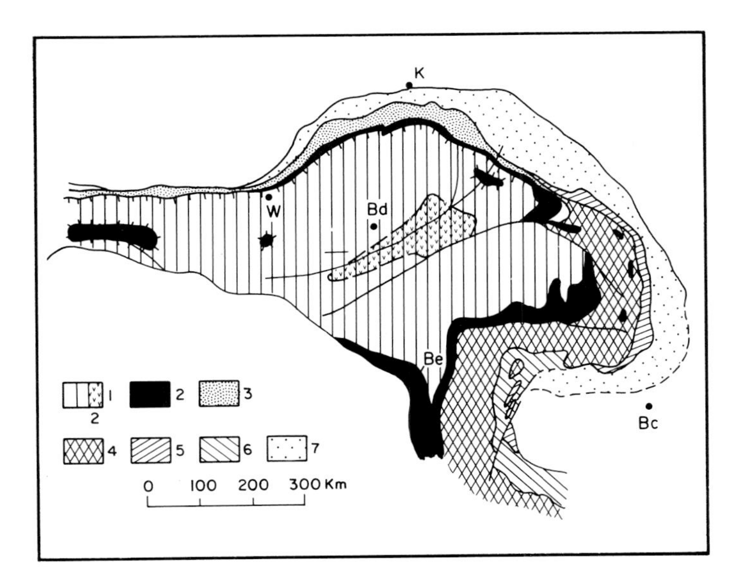

Figure 4. The major Tethyan suture in the Carpathian realm (posttectonic cover is not shown). (1a) Inner Dacidian (Austro-Alpine) units, (ab) Bükk Mountains and correlative units; (2) major Tethyan sutures (Vardar, South Pannonian suture, Transylvanides, Pieniny Klippen belt, Tauern and Rechnitz Windows); (3) Magura group flysch (belonging to the suture); (4) middle Dacides (central East Carpathian nappes, Getic, and Supragetic nappes); (5) outer Dacides (outer Dacidic flysch); (6) marginal Dacides (Danubicum); and (7) Moldavides. Bc, Bucharest; Be, Beograd; Bd, Budapest; K, Krakow; W, Wien.

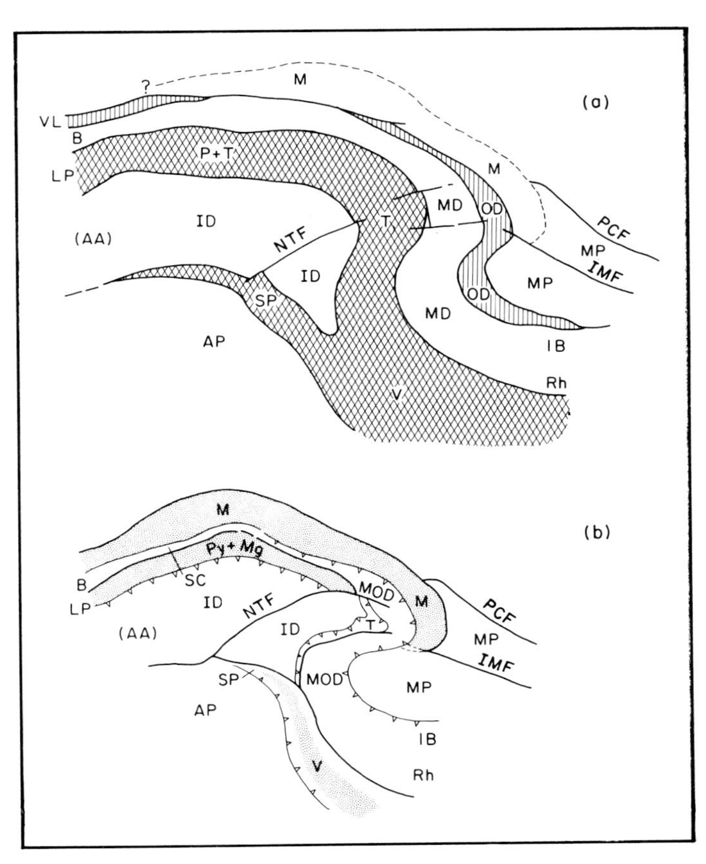

Figure 5. Palinspastic sketch of the Carpathians (A) during the earliest Cretaceous (at the end of spreading) and (B) during the Paleogene (after Cretaceous tectonism but before Miocene tectonism). AA, Austro-Alpine; AP, Apulian; B, Brianconais; ID, inner Dacides; IB, inner Balkanides; M, Moldavides; MD, middle Dacides; MOD, middle and outer Dacides (deformed); MP, Moesian platform; OD, outer Dacides; PG + Mg, Pieniny and Magura nappes; P + T, Pienides and Transylvanides; Rh, Rhodopian units; SC, Silesian cordillera; T, Transylvanides; V, Vardar zone; VL, Valais zone; LP, Liguro-Piemont zone; SP, South Pannonian rift; PCF, Peceneaga-Camena fault; IMF, Intramoesian fault; NTF, North Transylvanian fault. Vertical lines indicate oceanic Tethys; criss-crossed area, thinned crust, locally oceanic type; and dotted area, the main Paleogene flysch troughs.

South of the Pienide flysch troughs the central Carpathian Eocene–Oligocene posttectonic flysch (Podhole flysch) was deposited over the Dacidic nappes deformed in Cretaceous time. The Podhole flysch is flat-lying to slightly deformed. The southernmost Paleogene posttectonic basins (in the Pannonian area) contain pelagic epicontinental or paralic lithofacies.

At the exterior margin of the Silesian cordillera, a large flysch basin developed corresponding to the Moldavidian nappe system. This flysch basin continues into the East Carpathians. It lies outside of the deformed units of the middle and outer Dacides, which occupy the same position as the Silesian cordillera and are a source for part of the flysch arenites. These sediment sources are inside of the subsiding area; another external source for the flysch is located in the foreland outside of the subsiding area (Figure 6B).

The Moldavidian or outer Carpathian flysch trough ends roughly west of the Romanian (or southeast) Carpathian reentrant against an important strike-slip fault, the Intramoesian fault, which can be followed from the foreland northwestward into the interior of the mobile belt (Figure 5B).

The Pienide flysch troughs end toward the southeast against another important strike-slip fault, the North Transylvanian fault, which continues northeastward from the southern border of the Szolnok graben (Figure 5B). (The Szolnok graben is located internal to the Pieniny Klippen belt and the Magura or Petrova flysch.) South of the North Transylvanian fault, epicontinental lithofacies were deposited during Eocene, Oligocene, and early Miocene time.

The subsiding Paleogene to earliest Miocene flysch troughs described above became the site of important Neogene compressional events during major Neogene crustal shortening and crustal subduction. In the outermost part of the East Carpathian (Moldavidian) area, early Miocene molasse formations are also involved in the Neogene overthrusting.

The Paleogene to earliest Miocene flysch troughs developed on continental crust (generally thought to be thinned continental crust, but this has not yet been well documented) and above the structures deformed in Cretaceous time. In places where the flysch overlies oceanic crust, the basement is thought to be Mesozoic oceanic crust that was not subducted or deformed during Cretaceous time, such as below the Magura and Pieniny Paleogene troughs (Figure 6A).

A tentative correlation between the Paleogene flysch troughs in the Carpathians and in the Alps depends on the origin of the Rhenodanubian flysch of the Eastern Alps. Within the flysch zone of the Eastern Alps, rocks equivalent to the Moldavidian and Magura flysch can be recognized. If the whole Rhenodanubian flysch belongs to the Liguro-Piemont (oceanic) suture, it is correlative with only the Magura and Pieniny flysch. If so, then the Subalpine units are the only possible correlatives of the

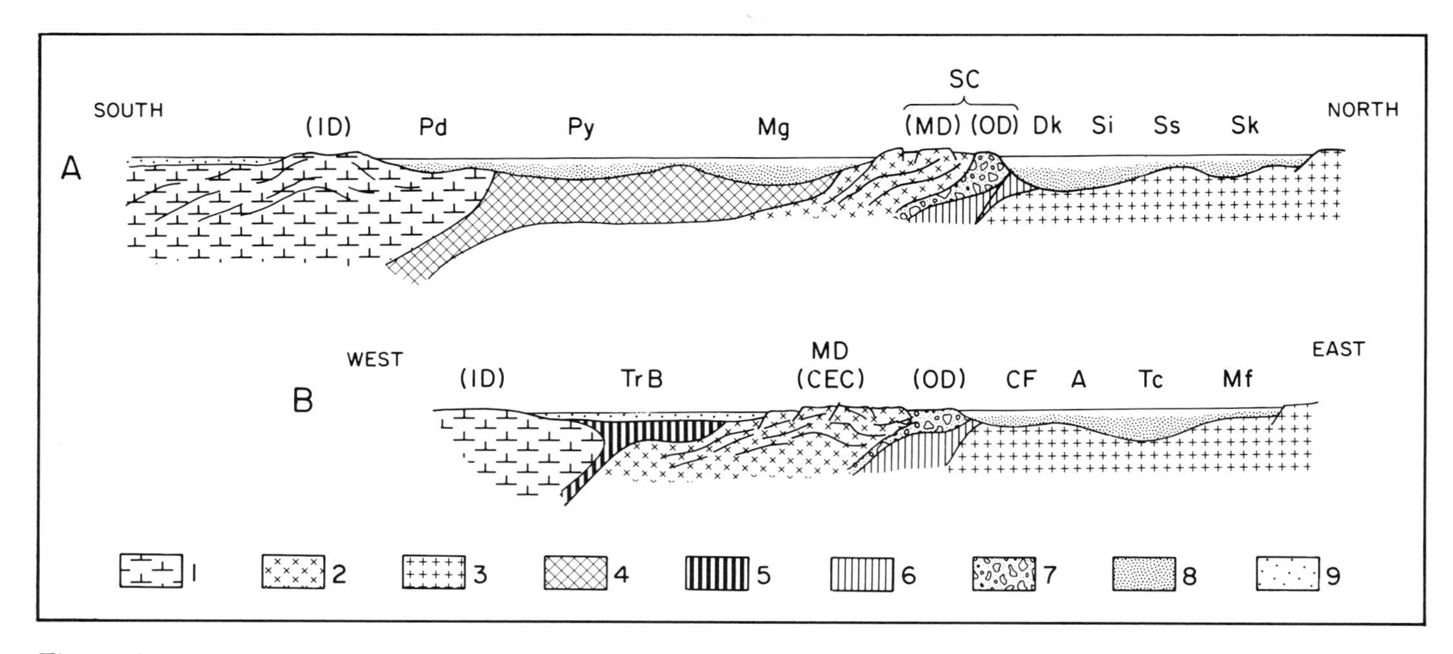

Figure 6. Palinspastic cross sections showing the main Paleogene flysch troughs in the West Carpathians (A) and East Carpathians (B). Symbols are as follows: (1) inner Dacidic nappes (ID), tectonized during Mediterranean deformation; (2) middle Dacidian nappes (MD), tectonized during middle Cretaceous deformation; (3) Moldavidian flysch basement, mainly continental; (4) Pieniny and Magura basement, mostly oceanic; (5) Transylvanides, oceanic, deformed during Cretaceous tectonism; (6) outer Dacidian thinned continental crust; (7) outer Dacidian flysch, deformed; (8) Paleogene flysch; and (9) epicontinental Paleogene deposits. Pd, Podhale flysch; Py, Pieniny units (in general); Mg, Magura unit; SC, Silesian Cordillera; MD, middle Dacides; OD, outer Dacides; Dk, Dukla unit; Si, Silesian unit; Ss, Subsilesian unit; Sk, Skole unit; TrB, Transylvanian basin; CF, Convolute flysch; A, Audia unit; Tc, Tarcau unit; Mf, Marginal Folds unit.

Moldavidian flysch. If the outer part of the Rhenodanubian flysch originated from a zone external to the Liguro-Piemont zone (e.g., if it is the prolongation of the Valais trough), then it may be correlated with the Moldavidian units and the Magura flysch is correlative with units within the Alpine suture.

The Paleogene flysch troughs of the Carpathians do not have a direct connection with those developed in the Balkans. The age of the main deformation of these two areas is also different (Neogene for the former, latest Eocene for the latter).

EARLY MIOCENE TECTOGENETIC EVENTS

The first important Cenozoic compressional deformation of the Carpathians occurred during early Miocene time and involved the Pienides and a part of the Moldavides. For a long time this was considered to be the equivalent of the Savian "phase" and thus thought to occur at the Oligocene–Miocene boundary. In fact, these events occurred after the Savian phase and can be dated as late Aquitanian or Burdgalian. (The Burdigalian chronostratigraphic stage is younger than the Aquitanian and older than the early Badenian or Langhian. Its lower boundary is dated at about 22 Ma and its upper boundary at about 17 Ma.)

On the northern border of the Transylvanian basin, deformed units containing rocks as young as Burdigalian are covered by uppermost Burdigalian deposits (the Hida Formation of Karpatian age). These folded Burdigalian rocks are overthrust by Pienide units, indicating that part of the thrusting of the Pienides was of Burdigalian age. The same situation is known below the inner part of the Vienna basin where both St. Veit Klippen (equivalent of the Pieniny Klippen belt) and flysch nappes (equivalent of the Magura group of nappes?) are unconformably overlapped by the upper Burdigalian (Karpatian) Laa Formation.

Along the East Carpathian flysch zone, the inner nappes of the Moldavides (Audia-Czernahora, Macla and Convolute Flysch nappes), and perhaps also the Dukla nappe of the East and North Carpathians, were overthrust during early Miocene time (Figure 3). The youngest rocks overthrust by these nappes are of Aquitanian or even lower Burdigalian age (Vinetisu flysch equivalent to the upper Krosno flysch, Cornu Formation). Along the East Carpathian bend these inner nappes of the Moldavides and their eroded frontal thrust faults are overlapped by uppermost Burdigalian deposits (Dofteana molasse).

The nappes just discussed consist entirely of sedimentary rocks detached from their basement (cover nappes). Their structural and depositional relationships, which are well developed in the Carpathian area, demonstrate that the first important tectogenetic events of Cenozoic (Neogene) age occurred in the early Miocene. It is possible that along the chain the tectogenetic events were not strictly synchronous. For example, near the Alps in the western part of the outer Carpathians, these events are probably of latest Aquitanian age, while in the eastern part of the outer Carpathians and in the Maramures area (Pienides), they are intra-Burdigalian.

The early Miocene events can be correlated with the en bloc thrusting of the inner Dacides over the Pienides. The inner Dacides are the continuation of the Austro-Alpine nappe system, which was thrust over the Penninic nappes and the Rhenodanubian flysch after Eocene time. The diachroneity of en bloc overthrusting of the inner Alpine–Carpathian nappes (which were thrust in the Cretaceous) over the Pienide zone is certainly important. Conversely, en bloc overthrusting of the Austro-Alpine units in the Eastern Alps along with the Penninic elements may be younger, specifically, of early Miocene age, which is in agreement with the early Miocene overthrusting in the Carpathians.

Early Miocene deformation involving the inner Dacides occurred only north of the North Transylvanian fault. This fault is an important zone of right-slip that allowed for differential movement of the inner Dacidic blocks on either side of the fault (Figure 5). Thus, Cenozoic overthrusting of Austro-Alpine and inner Dacide units occurred north of this fault and ended against it.

Early Miocene tectogenetic events in the Carpathians can be summarized as follows:

1. The Pienides (Pieniny Klippen belt and Magura group of nappes) were strongly deformed and overthrust and developed their present structure.
2. The inner Dacides, situated at the interior of the Pienides, overthrust the Pienides en bloc during early Miocene time or even earlier (depending on correlation with Eastern Alpine units).
3. The inner part of the Moldavidian nappes were thrust during early Miocene tectonic events.

The early Miocene tectogenetic events occurred at about the same time as continent–continent collision. The northward overthrusting of the Pienides in front of the inner Dacides determined the tectonic overlap of the middle and outer Dacides (and of the corresponding Silesian cordillera) west of the East Carpathians. This is similar to the situation in the Alps where the Brianconais (which has a tectonic position similar to the middle Dacides) is overlapped toward the east by nappes of the Austro-Alpine and Liguro-Piemont zone.

Crustal shortening associated with early Miocene tectogenetic events involved mainly old oceanic crust below the Pienides and thinned continental crust below the outer Dacides that had not been subducted during the Cretaceous. Andesitic volcanic rocks were erupted during these events and are subduction related.

Early Miocene compressional events were closely followed by extensional faulting that occurred prior to the end of Langhian time. One example is the Satmarean graben: its southern boundary is the North Transylvanian fault which extends westward to the southern border of the Szolnok graben (Figure 7). These grabens cut across older units of different tectonic ages (Cretaceous and early Miocene). Even Pienide nappes are preserved within the Satmarean graben.

Similar extensional structures that formed shortly after compressional events are also known for older times. For example, a graben trending roughly north–south occurs along the western margin of the Apuseni Mountains and below the eastern margin of the Pannonian basin (Figure 7). This is a Senonian graben that formed shortly after the compressional tectogenetic events that formed the inner Dacidian nappes (during or at the end of Turonian time).

These grabens have been reactivated periodically. The western part of the Satmarean graben was reactivated in Sarmatian and Pannonian time (about 8–13 Ma). Deposits of this age accumulated within the graben are several times thicker than the equivalent formations situated to the north and south. A cross-cutting relationship exists between younger Sarmatian–Pannonian grabens and the Senonian north-south-trending graben. These younger grabens cut perpendicularly or obliquely across the Senonian graben (Figure 7).

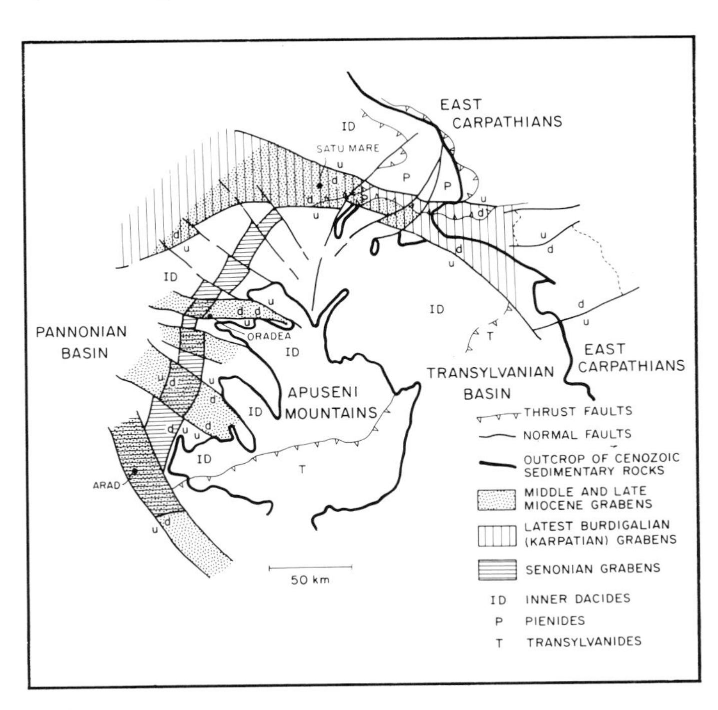

Figure 7. Senonian, latest Burdigalian (Karpatian), and middle–late Miocene grabens in the area of the Apuseni Mountains and Transylvanian basin. Middle–late Miocene normal faulting reactivated a late Burdigalian graben (the Satmarean graben) north of the Apuseni Mountains, while other Middle–late Miocene grabens cross-cut the Senonian graben west of the Apuseni Mountains.

MIDDLE MIOCENE TECTOGENETIC EVENTS

The second period of Neogene compression, which involved only the Moldavides, occurred during Badenian time. The Tarcau and Marginal Folds nappes in the flysch nappes of the East Carpathians (and the equivalent Skibas and Boryslaw units); the Silesian, Skole, and Subsilesian (Zdanice) nappes in the flysch nappes of the West Carpathians; and the Waschberg unit in the Eastern Alps all were overthrust during this period. All of these nappes consist only of sedimentary rocks.

In the Romanian part of the East Carpathians, the youngest deformed rocks in the Tarcau and Marginal Folds nappes are of earliest Langhian age (Praeorbulina zone). The youngest deformed rocks within the Subcarpathian nappe and overthrust by the Tarcau and Marginal Folds nappes are of Langhian age (Slanic Tuff and/or Salt Formation). Nowhere are upper Badenian (Kossovian or Serravalian) rocks overthrust by the Tarcau and/or Marginal Folds nappes. In the Ukrainian part of the East Carpathians, a similar situation is present.

In the Polish and Czechoslovakian parts of the Flysch Carpathians, evidence for intra-Badenian tectogenetic events is also present. Locally, the frontal part of the Magura nappes overthrust lower Badenian rocks. The frontal parts of the Silesian, Subsilesian and Skole nappes (e.g., Bochnia and Rzerzow) contain complex imbricate structures in which lower Badenian rocks are involved.

The intensity of middle Miocene tectogenetic events is difficult to establish. It is clear that these events are compressional but the previous (early Miocene) and subsequent (late Miocene) events appear to have been more important. Late Miocene events clearly reactivated nappes created in middle Miocene time and greatly increased the amount of overthrusting. Therefore, the middle Miocene events are more difficult to recognize than the late Miocene ones.

Middle Miocene tectogenetic events can be thought of as equivalent to the young Styrian "phase." Crustal shortening at this time was probably in proportion to the intensity of deformation of the nappes, and andesitic volcanic rocks were erupted behind the thrust belt.

Middle Miocene evaporites known in the Transylvanian and Transcarpathian depressions and in the Subcarpathians appear to be roughly contemporaneous with middle Miocene tectogenetic events. This temporary cessation of subsidence over a large area that was not greatly affected by compressional deformation may be related to development of a compressional strain field. A similar relationship between early Miocene compressional events and deposition of evaporites (the Burdigalian Salt Formation in the Subcarpathians and Marginal Folds) can also be documented, but the areal extent of the evaporites was less than during middle Miocene time.

It is important to note that after the deposition of the Badenian evaporites, important changes occurred in the biofacies. At this time the final faunal separation of the Neogene Paratethys and the Tethys began.

Middle Miocene tectogenetic events can be summarized as follows:

1. Overthrusting of part of the Moldavidian nappes in the East and West Carpathians occurred during middle Miocene time.
2. The front of the Pienides (Magura Group) was locally reactivated but without significant horizontal displacement.
3. Important paleogeographic changes occurred over the whole Carpathian realm.

The subcircular shape of the Transylvanian basin was established in early Badenian (early Langhian) time before the middle Miocene compressional events but after the extensional fracturing that followed the early Miocene events (see above). The basal unit of the Transylvanian depression is a tuff layer of regional extent (Dej Tuff) below which are preserved several subsided troughs. They formed contemporaneously with the extensional fracturing of latest Burdigalian (Karpatian) age and are filled with mainly coarse-grained deposits. These troughs may also be extensional, but this has not yet been well documented.

LATE MIOCENE TECTOGENETIC EVENTS

The last important period of compressional events involved only the outer zones of the Carpathians (the Moldsvides) and took place during Bessarabian (late Sarmatian) time. To correlate this last tectonic period with other events, the precise chronostratigraphy within the Carpathian region and adjacent areas most be examined. The most important point is that the term Sarmatian can have two different meanings: a narrow one (the Sarmatian *s. str.*) in the inner Carpathian zones and an expanded one (Sarmatian *s. l.*) in the external zones and on the foreland.

The last compressional events in the outer Carpathians (the Moldavian "phase") are intra-Sarmatian (*s. l.*). This corresponds to the stratigraphic boundary between the Sarmatian (*s. str.*) and the Pannonian. When correlated with events in the Mediterranean, the Moldavian "phase" corresponds to Tortonian tectogenetic events known in North Africa, Sicily, Calabria, and elsewhere.

The age of the youngest tectogenetic events of late Miocene age are best documented in the eastern Subcarpathians. There, the Subcarpathian nappe is thrust over lower Sarmatian (*s. l.*) formations. Chersonian deposits, locally with upper Bessarabian rocks at their base, unconformably overlap the Subcarpathian nappe and its erosional frontal contact (Trotus Valley area). Lower Sarmatian (*s. l.*) formations have been found in deep boreholes, which penetrated the Moldavidian nappes in many places (from Moldavia in the Romanian Carpathians, throughout the Ukrainian Carpathians, and up to the Eastern Polish Flysch Carpathians). In addition to the Subcarpathian nappe, which was thrust above the lower Sarmatian (*s. l.*) cover of the foreland, several Moldavidian nappes, including the Marginal Folds, Tarcau, Skole, Subsilesian, and Silesian nappes were also thrust. These nappes were thrust together as a homogeneous unit and were moved from the position that they had occupied after the middle Miocene tectogenesis.

In the Czechoslovakian flysch Carpathians (Moravia), the youngest sedimentary rocks penetrated by drilling below the Silesian and Subsilesian (Zdanice) nappes are of late Badenian age. Two reasons for this can be proposed: (1) in the westernmost outer Carpathians the youngest overthrusting event is of latest Badenian age, or (2) along the whole outer Carpathians the last tectogenetic event is isochronous and is intra-Sarmatian (*s. l.*). The absence of the lower Sarmatian below the nappes in Moravia would thus be due to erosion or nondeposition of Sarmatian rocks.

The last two tectogenetic events known in the outer Carpathians, the middle and the late Miocene events, decrease in magnitude westward toward the Alps. The magnitude decreases rapidly west of Vienna. There, the only unit that may have been affected by these events is the Waschberg unit.

The late Miocene tectonic events extended farther along the Carpathian chain, into the southern Subcarpathians, than did the early and middle Miocene events. The Subcarpathian nappe frontal contact (known also as the Pericarpathian fault) has been traced in the subsurface by boreholes west of the South Carpathian salient in Romania to the Jiu Valley. (The dashed line north of the Moesian platform on Figure 2 shows the southern extent of this nappe in the subsurface.) It is the only Moldavidian nappe that is known south of the South Carpathians. The Subcarpathian nappe does not follow the bend of the South Carpathians, which is mainly Mesozoic, but rather ends against the northern prolongation of the Timok fault which is an important strike-slip fault with dextral displacement (Figure 2). The strike-slip displacement of this fault is gradually transferred into thrusting and, farther to the east, into overthrusting of the Subcarpathian nappe.

PLIOCENE–PLEISTOCENE

Locally, folding occurred during early Pleistocene time and involved the Subcarpathian zone in the southeast Carpathian bend area. This tectonic event is known as the Wallachian "phase." It is not comparable with the other Neogene events (which generated large nappes and affected a large area), but is important because it occurred in an area having high seismicity (Vrancea). The Wallachian deformation consists of tilting, folding, and thrusting with a maximum horizontal displacement of not more than 1–2 km. This deformation involved young sedimentary rocks of the Subcarpathian nappe, postnappe cover of the Subcarpathian nappe, and the inner part of the Carpathian foredeep.

The areal extent of the Wallachian deformation is controlled by two sinistral strike-slip faults, the Peceneaga–Camena and the Intramoesian faults (Figure 5). These faults, which are seismically active, have allowed the displacement of a specific block of the foreland toward the Carpathian bend area and thus accommodated the Pleistocene deformation.

CONCLUSIONS AND GENERAL REMARKS

Cenozoic tectonic events within the Carpathian Mountains followed a major Cretaceous compressional deformation that affected the internal parts of the belt. During Cenozoic time, these areas of Cretaceous deformation acted roughly as rigid blocks, and their differential movement was accommodated by large strike-slip faults.

Areas of active Paleogene and early Miocene subsidence formed troughs with flysch sedimentation and include (1) the Paleogene Pienide flysch troughs, corresponding to the Pieniny Klippen belt and Magura nappes (including the Petrova nappe in the Maramures area); and (2) the Moldavidian flysch troughs which have been the site of practically uninterrupted (mainly flysch) sedimentation since Early Cretaceous time.

These two groups of troughs were separated by a morphologic high formed by (1) the middle and outer Dacides (nappes deformed in Cretaceous time), which partly formed a source area for the flysch arenites in the East Carpathians, and (2) the Silesian cordillera in the West Carpathians, which was an extension of the Dacidic morphologic high. This morphologic high had a tectonic position similar to the Brianconais zone in the Western Alps. The Pienide flysch was deposited on top of oceanic crust that was partly deformed and represented the main Tethyan suture (oceanic Tethys). The Moldavidian flysch was deposited on top of continental or thinned continental crust.

These Cenozoic troughs were deformed during three compressional tectogenetic events or "phases":

Event	Age
1. Early Miocene tectogenesis (old Styrian "phase")	18–20 Ma
2. Middle Miocene tectogenesis (young Styrian "phase")	15.5 Ma
3. Late Miocene tectogenesis (Moldavian "phase")	11–12 Ma

Each of these tectogenetic events corresponded to an episode of crustal shortening and subduction, as evidenced by andesitic volcanic arcs erupted in the inner part of the Carpathians.

Except for the early Miocene en bloc thrusting of the inner Dacides aver the Pienides, all the nappes generated during these three Neogene tectogenetic episodes are cover nappes consisting only of sedimentary rocks.

Along the Carpathian chain all of these events may be diachronous. If they are diachronous, then within each event, the oldest deformation occurred in the west (Eastern Alps and West Carpathians) and the youngest events in the southeast (Romanian Carpathians). Extensional graben structures formed mainly in the inner Carpathians shortly after the compressional tectogenetic events. They cut older units of different ages. These grabens may have been reactivated periodically or were cut by younger grabens with different trends.

The subcircular Transylvanian basin was established during early Langhian time after the early Miocene compressional events but before the middle Miocene ones. Its basal lithostratigraphic unit (Dej Tuff) is slightly faulted and is roughly dish-shaped.

BIBLIOGRAPHY

Andrusov, D., J. Bystritzky, and O. Fusán, 1973, Outline of the structure of the West Carpathians: Carp.-Balk. Geol. Assoc. 10th Congr., Dionz Stúr. Geol. Inst. Publ., Bratislava, 45 p.

Birkenmajer, K., 1977, Jurassic and Cretaceous lithostratigraphic units of the Pieniny Klippen Belt, Carpathians, Poland: Stud. Geol. Pol. XLV, Warsaw, 158 p.

Buday, T., and M. Mahel, eds., 1968, Regional geology of Czechoslovakia, Part II, The West Carpathians: Ed. Academia, Praha, 723 p.

Danys, V. V., J. O., Kulcicid, V. A. Sakin, and O. S. Vialov, 1974, Soviet Carpathians—The Flysch Carpathians, *in* Tectonics of the Carpathians Balkan regions: Dionz. Stúr Geol. Inst. Publ., Bratislava, p. 210–220.

Koszarsky, L., W. Síkora, and S. Wdowiaŕz, 1974, Polish Carpathians—The Flysch Carpathians, *in* Tectonics of the Carpathians Balkan regions: Dionz. Stur Geol. Inst. Publ., Bratislava, p. 180–187.

Kruglov, S. S., 1974, Soviet Carpathians—The Pieniny Klippen Zone, *in* Tectonics of the Carpathians Balkan regions: Dionz. Stúr Geol. Inst. Publ., Bratislava, p. 205–209.

Ksiazkiewicz, M., J. Oberec, and W. Pozariysky, 1977, Geology of Poland, Geol. Publ. House, Warsaw, v. IV, 718 p.

Mahel, M., 1974, The inner West Carpathians, *in* Tectonics of the Carpathians Balkan regions: Dionz. Stúr Geol. Inst. Publ., Brstislava, p. 91–134.

Prey, S., 1974, Austrian Eastern Alps—External zones, *in* Tectonics of the Carpathians Balkan Regions: Dionz. Stur Geol. Inst. Publ., Bratislava, p. 75–85.

Roth, Z., 1974, The western sector of the outer Carpathians in Czechoslovakia, *in* Tectonics of the Carpathians Balkan regions: Dionz. Stúr Geol. Inst. Publ., Bratislava, p. 163–172.

Sandulescu, M., 1975, Essai de synthese structurale des Carpathes: Bull. Soc. Géol. Fr., Paris, v. XVII, n. 3, p. 299–358.

Sandulescu, M., 1980, Analyse géotectonique des chaines alpines situées autour de la Mer Noire occidentale: Ann. Inst. Geol. Geophys., Bucuresti, v. LVI, p. 5–54.

Sandulescu, M., H. G. Kräutner, I. Balintoni, D. Russo-Sandulescu, and M. Micu, 1981, The structure of the East Carpathians (Moldavia–Maramures area): Carp.-Balk. Geol. Assoc. 12th Congr. (Guide Book B1), Inst. Geol. Geophys. Publ., Bucharest, 92 p.

Senes, J., ed., 1976, Chronostratigraphie et Neostratotypen, Miozän, M4, Badenien: Ed. Acad. Slov. Sci., Bratislava, 599 p.

CHAPTER 3

Late Cenozoic Tectonics of the Pannonian Basin System

L. H. Royden
Massachusetts Institute of Technology
Department of Earth, Atmospheric and Planetary Science
Cambridge, MA 02139

The Miocene evolution of the Carpathian–Pannonian system appears to have been controlled by events within the adjacent Alpine mountain belts. Early Miocene initiation of northward to eastward thrusting in the outer Carpathians is best ascribed to the eastward escape of the Pannonian continental lithospheric fragment(s) away from the zone of collision in the Eastern Alps. The subsequent "back arc" extension in the Pannonian basin system in middle–late Miocene time was coeval with the late stages of thrusting in the adjacent Carpathian belt. Net east–west extension within the basin system can be related both to the arrangement of continental lithospheric fragment boundaries outside of the Pannonian area, which prohibited continued convergence of the Pannonian fragment with Europe, and to the continued subduction and shortening beneath the East Carpathians at the same time.

Basin extension was heterogeneous and diachronous throughout the Pannonian basin system. Variations in basin development were intimately related to contemporaneous thrust belt activity in the Carpathian Mountains. Extension occurred along a conjugate system of strike-slip faults that connected areas of coeval extension to one another and to coeval areas of shortening within the Carpathian thrust belt, thus providing a mechanical link between basin extension and thrusting. The style of extension at depth was controlled by the geometry of the thrust belt at depth and the distance from the thrust front.

The style of sedimentation within each basin was also influenced by the proximity of each basin to the thrust front. Basins located near the thrust belt contain thick synextensional fault-bounded sedimentary rocks overlain by thin postextensional sediments. The normal faults reach nearly to the surface. Basins located far from the thrust belt contain thin sequences of synextensional fault-bounded sedimentary rock sequences overlain by thick sequences of postextensional, unfaulted, flat-lying sedimentary rocks. These differences can be explained by differences in the thermal subsidence rate of the basement after extension and by the proximity of each basin to the sediment sources in the Carpathians.

INTRODUCTION

Several examples of late Cenozoic back arc type extensional basins exist adjacent to zones of coeval subduction and convergence within the Mediterranean region (Figure 1). The similarities between these systems are striking. The basins, of roughly comparable size, contain abundant evidence for extensional tectonism. Basin development began by stretching of continental crust, which in some systems developed later into oceanic spreading (the Tyrrhenian sea). These basins developed adjacent and internal to coeval thrust belts and are often superimposed on older, internal structural units of the same belts. The thrusting within these convex belts is outwardly directed and appears to be synthetic with the direction of subduction, so that the basins lie on the overriding plate of the corresponding subduction zone. These and other parallels between these basin–thrust belt pairs strongly suggest that the space–time relationships between extension and convergence are not inci-

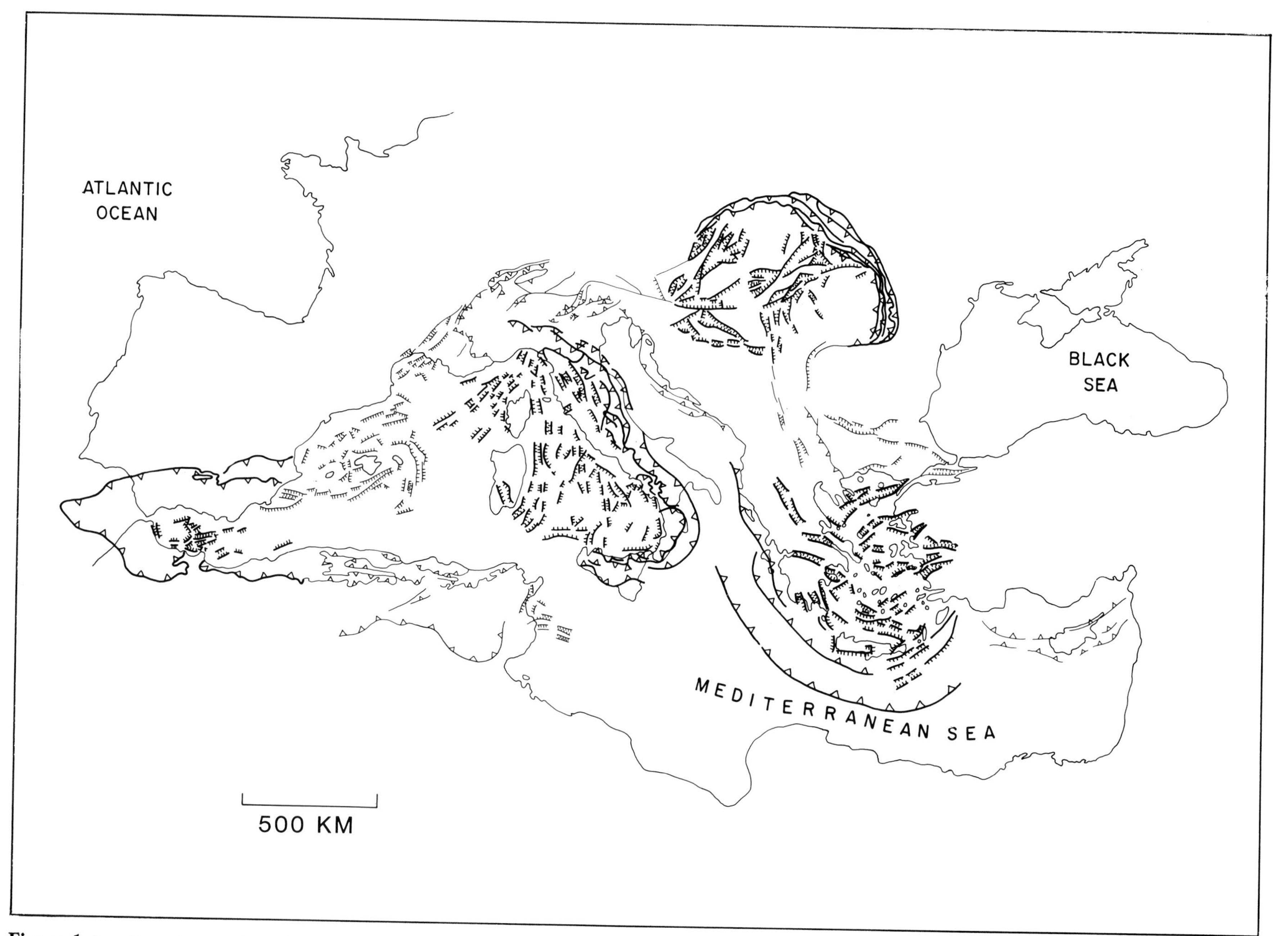

Figure 1. Late Cenozoic examples of thrust belts (with heavy barbed lines indicating late Cenozoic thrust faults) associated with coeval back arc type extensional systems (with heavy ticked lines indicating late Cenozoic normal faults) in the Mediterranean region. Some other late Cenozoic faults are also shown (light barbed and ticked lines). From west to east, the thrust belt–basin pairs are (1) the Alboran–Betic Cordillera–North African system, (2) the Tyrrhenian (Ligurian)–Apennine system, (3) the Pannonian–Carpathian system, and (4) the Aegean system.

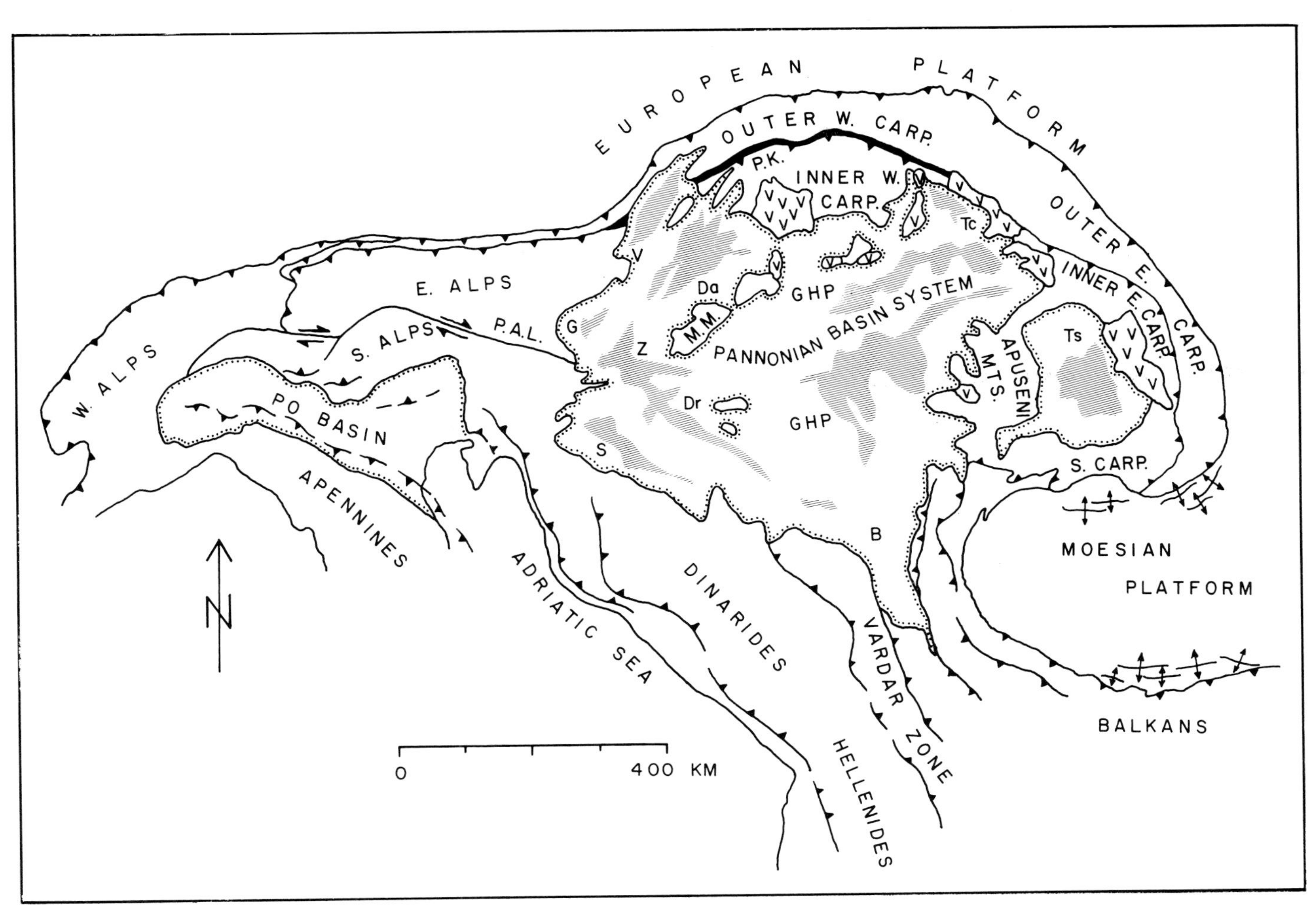

Figure 2. Tectonic sketch map showing the position of the Carpathian Mountains and the Pannonian basin within the Alpine belts of central and eastern Europe. Stippled area indicates parts of the Pannonian basin where the depth to base of Miocene exceeds 3 km. Subbasins are: V, Vienna; Da, Danube; G, Graz; Z, Zala; Dr, Drava; S, Sava; Tc, Transcarpathian; Ts, Transylvanian; GHP, Great Hungarian Plain; B, Banat. Other abbreviations: P.A.L., Peri-Adriatic Line; P.K., Pieniny Klippen Belt; M.M., Hungarian Mid (or Central) Mountains.

dental but rather reflect similar dynamic processes operating in each system.

This paper examines the heterogeneous tectonic development of one such extensional basin system, the Pannonian system, and the temporal, spatial, and kinematic connections between the basins and the adjacent Carpathian thrust belt. The diachronous nature of both extension and convergence in this area affords an excellent opportunity to study the relationship between extension and convergence through time and to evaluate the role of the strike-slip faults that connect them. The scope of this paper encompasses mainly the timing and geometry of events within the upper crust. Associated events within the lower crust and mantle, as inferred from subsidence and heat flow data, are described in a complimentary paper (Royden and Dövényi, this volume).

GENERAL TECTONIC SETTING

The Pannonian basin system is a large, topographically low area that lies between the Carpathian Mountains and the Dinarides (Figure 2), but it is underlain by several small, deep basins. The basin system formed by rapid Miocene subsidence of pre-Miocene basement and in places contains as much as 7 km of Miocene to Quaternary sedimentary rocks (see Maps 2 and 8 and Figure 1 in Royden and Dövényi, this volume). The basins are underlain mainly by nappes of Mesozoic age of the inner Carpathian and Dinaric orogenic belts. Initial subsidence of the basins was synchronous with the Miocene thrusting of the outer Carpathians toward the European foreland.

The basin system is a result of middle to late Miocene lithospheric extension (e.g., see Sclater et al., 1980; Horváth and Royden, 1981; Royden et al., 1982) and occupies an area with thin continental crust, high heat flow, and high temperatures within the lithosphere (e.g., see, Dövényi and Horváth, this volume; Royden and Dövényi, this volume). It has been classified as a Mediterranean backarc or interarc type basin formed during westward and southward subduction of the European plate beneath the inner Carpathians (e.g., see Stegena et al., 1975). The basins and their basement are intruded by Miocene magmatic rocks that are mainly calcalkaline.

The individual basins or troughs within the basin system are areas of local crustal extension related to strike-slip faults or fault zones (Figure 3). In general, northeast-trending left-slip faults and northwest-trending right-slip faults comprise a set of conjugate shears and correspond to roughly east–west extension across the Pannonian region. These strike-slip faults not only accommodate local extension within the deep Miocene troughs, but also connect these areas of Miocene extension to one another and to zones of shortening and convergence in the outer Carpathian thrust belt. They thus function in a general way as transform faults.

To best understand the role of this late Cenozoic extensional strike-slip and convergent deformation within the Carpathian–Pannonian system, it is necessary to analyze concurrent events at several different scales. In the following sections, I describe, first, how large-scale regional events of the Alpine mountain belt controlled the overall timing and geometry of deformation in the Carpathian–Pannonian system and, second, how local events within different parts of the Carpathian–Pannonian system, although heterogeneous and diachronous, together constitute a consistent, coherent pattern of basin–thrust belt evolution.

REGIONAL CENOZOIC EVENTS

The eastern European Alpine mountain belt is the result of Jurassic to Recent suturing of Europe and several smaller continental fragments that have been subject to considerable internal deformation. At any time, the boundaries of these fragments can be defined as broad zones of deformation that contain the most prominent faults in a continuously deforming continental crust. At various times, old fragment boundaries became inactive and new fragment boundaries formed that cut across the older boundaries. The development of the Pannonian basin system can be related to such a rearrangement of fragment boundaries in the Eastern Alps, Carpathians, and Dinarides in late Oligocene–early Miocene time. Therefore, it is necessary to examine both early Cenozoic events that predate the formation of the Pannonian basin system and late Cenozoic events that are contemporaneous with basin formation.

This section contains only a short summary of Cenozoic events within the eastern European Alpine mountain belt. A more complete discussion and analysis of the Cenozoic evolution of this mountain belt is given by Burchfiel (1980), Royden et al. (1983), and Royden and Báldi (this volume). No discussion of Miocene tectonics within the Carpathian region would be possible without reference to central Paratethyan biostratigraphic time stages. I have attempted to keep the use of these stage names to a minimum and to accompany the occurrence of each stage name with its approximate isotopic age, based on the time scale given in Steininger et al. (this volume).

Early Cenozoic Events

Throughout early Cenozoic time, southward-directed subduction of Europe beneath the Apulian continental fragment (consisting mainly of Italy, the Adriatic area, and the western Pannonian area) can be related to roughly north–south convergence between Europe and Apulia (Figure 4). Eocene(?) continental collision across the Eastern Alps was followed by continued shortening and imbrication of the continental crust until late Oligocene or early Miocene time (Burchfiel, 1980). To the east, north–south convergence was probably accommodated by Eocene consumption of oceanic lithosphere, subducted southward beneath the West Carpathians (Ksiazkiewicz and Lesko, 1959; Oszczypko, 1973). To the southeast, early Cenozoic intracratonic shortening occurred within the Dinarides, probably related to east-dipping subduction of Apulia beneath the Dinarides (Burchfiel, 1980). The magnitude of shortening within the Dinarides becomes significantly less northward, and some Eocene convergence was probably transferred to the West Carpathians via a north-northeast-trending right-slip transform boundary. A zone of dextral shear may have begun to develop parallel to the Dinarides at this time. No deformation occurred from Eocene to Oligocene time within the East or South Carpathians. Thus, the main fragment boundaries in early Cenozoic time were a convergent boundary across the Eastern Alps and a convergent plus right-slip(?) boundary across the outer Dinarides, which were connected through a poorly defined transform zone across the eastern end of the Eastern Alps and, in Eocene time, through the Pannonian area. Early Cenozoic tectonics can thus be described as a two-fragment system, consisting of Europe and Apulia.

Early Miocene Events

By late Oligocene time, major shortening had ceased in the Eastern Alps. Continued motion between Europe and Apulia seems to have been accommodated by right-slip along the peri-Adriatic-Vardar zone (Laubscher, 1971), by shortening in the Western Alps, and perhaps by southward-directed overthrusting in the Southern Alps (Figure 4) (Laubscher, 1985). In addition, continued convergence between Apulia and Europe in Miocene time seems to have been accommodated partly by eastward and northeastward displacement of the Pannonian fragment relative to Europe and Apulia, thus initiating the main northward and eastward overthrusting along the outer Carpathian flysch belt in early Miocene time (see below). This would appear to represent the lateral escape of material away from a zone of collision and thick crust (Eastern Alps) toward an area where oceanic or thin continental crust was easily subducted (Carpathian flysch basin, later deformed into the outer flysch Carpathians). In this way, continued north–south convergence across the Eastern Alps could be accommodated without continued thickening of the crust beneath the Eastern Alps ("continental escape" of Burke and Sengör, 1986; see also McKenzie, 1972; Molnar and Tapponier, 1975).

The main fragment boundaries present in early Miocene time can be defined as (1) a right-slip transformlike boundary along the peri-Adriatic-Vardar zone and coupled to a zone of east–west shortening in the Western Alps, and (2) a convergent boundary along the outer Carpathian flysch belt, probably linked to a zone of right-slip in the South Carpathians (now buried by late Miocene–Pliocene molasse) (Figure 4). Early Miocene tectonics thus can be described as a three-fragment system, consisting of Europe, Apulia, and an eastward-moving Pannonian fragment. Note that in early Miocene time the Pannonian fragment, as shown in Figure 4, consisted partly of material formerly belonging to the Apulian fragment and partly of material formerly attached to Europe during early Cenozoic time.

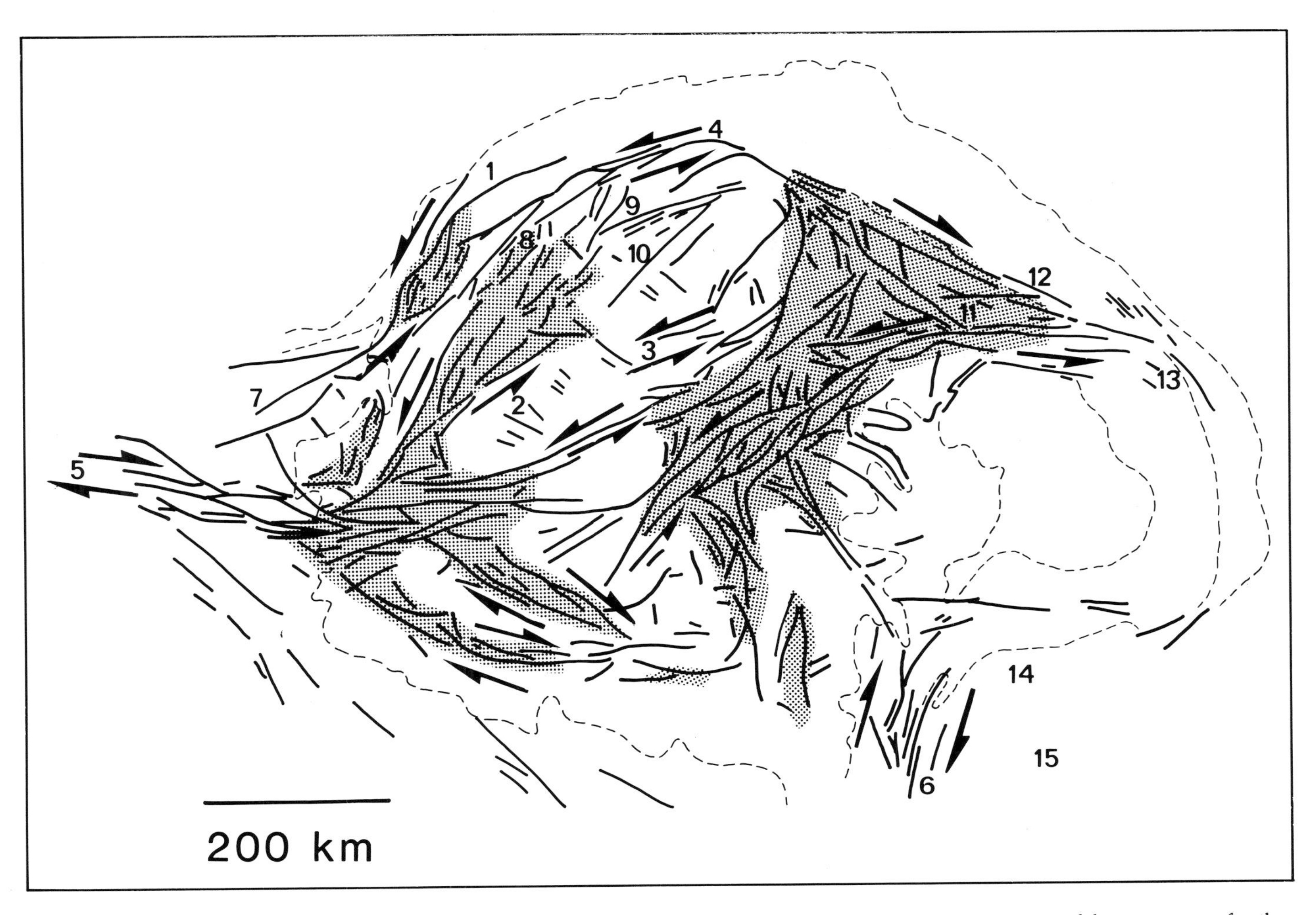

Figure 3. Generalized map of Neogene strike-slip faults in the Carpathian–Pannonian region. Arrows indicate sense of shear on zones of strike-slip displacement. Numbers 1–15 refer to faults, fault zones or locations discussed in the text. Compiled from Landsat photos, isopach maps, and seismic data for subsurface faults (e.g., Map 2). Other sources include the Institut Géologique (1967–1970), Fusán et al. (1967), and Mahel (1973).

Late Miocene Events

Throughout Miocene time, the broad right-slip boundary along the peri-Adriatic-Vardar zone appears to have separated the Pannonian fragment from Apulia, so that during Miocene time these fragments moved and evolved independently from one another. Throughout middle and late Miocene time, thrusting and subduction ceased progressively from west to east along the Carpathian thrust belt (see later sections for details). By middle Miocene (Badenian–Sarmatian) time, active thrusting had ceased in the outer West Carpathians and continued only in the outer East Carpathians. At about the same time, middle Miocene east–west extension occurred within the Pannonian basin system (Figures 3 and 4). It appears that this extension occurred to accommodate continued westward subduction of the European crust beneath the East Carpathians after eastward translation of the entire Pannonian fragment became incompatible with the motion of the Apulian block relative to Europe in middle Miocene time. Roughly comparable magnitudes of middle–late Miocene east–west shortening in the flysch nappes and east–west extension beneath the Pannonian basin system (~ 100 km) tend to support this interpretation (see next section). Extension within the Pannonian basin system and the middle–late Miocene thrusting and subduction within the outer Carpathian thrust belt can thus be interpreted as related phenomena that developed progressively from earlier events within the Eastern Alps and Dinarides.

PANNONIAN-CARPATHIAN SYSTEM

The relationship between Miocene convergence in the outer Carpathians and extension within the Pannonian basin system is examined here in greater detail. The contemporaneous evolution of these two zones of deformation is examined by evaluating the spatial and timing relationships of convergence and extension and how these two processes are linked both at the surface and at depth.

Outer Carpathian Thrust Belt

The Carpathian thrust belt consists of externally directed thrust and fold nappes that can be divided into two major belts:

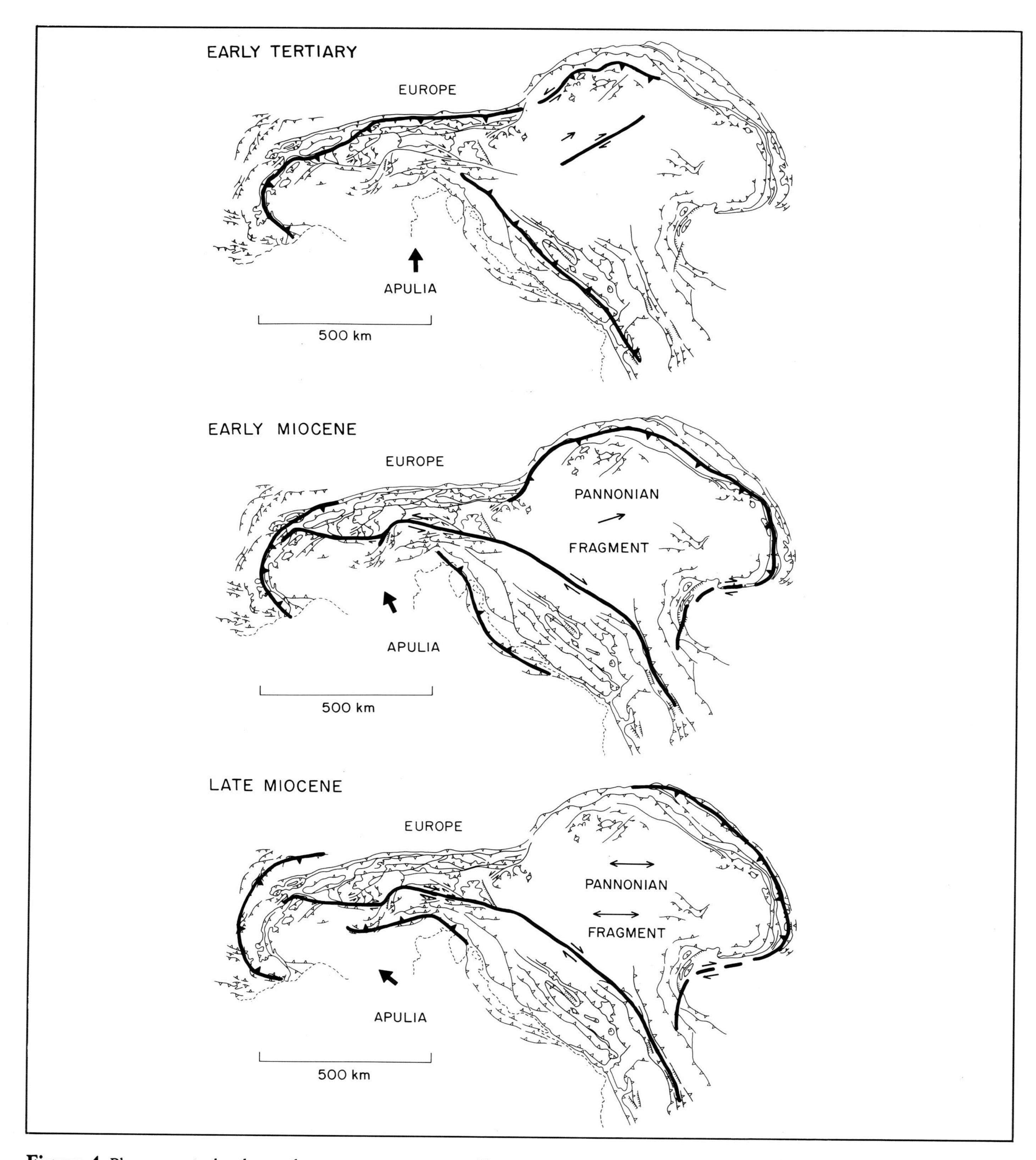

Figure 4. Plate tectonic sketch map showing present positions of fragment boundaries active in early Tertiary, early Miocene, and late Miocene time. The boundaries between continental fragments are shown simplistically by single heavy lines, but in reality are broad zones of deformation. The sequence of events is inferred to have been as follows: (1) Eocene(?) collision in the Eastern Alps, (2) eastward escape of the Pannonian fragment away from the collision zone, (3) continental collision along the Carpathian belt not later than early Miocene time, and (4) back arc or intraarc extension of the Pannonian basin during continued late convergence across the Carpathian mountain belt in late Miocene time.

an outer morphologically and structurally continuous belt and an inner discontinuous belt. Deformation within the inner belt occurred in Cretaceous to early Paleocene time. On the basis of lithology and ages of deformation the inner belt can be divided into three distinct parts: the West, East, and South Carpathians (Figure 2). A discussion of this inner belt is beyond the scope of this paper (see Burchfiel, 1980; Sandulescu, 1980, and this volume; Royden and Báldi, this volume). The outer belt consists mainly of Cretaceous to early Tertiary flysch and molasse. Except for Eocene deformation of the innermost flysch units (inner Magura nappe) in the West Carpathians, flysch deposition was continuous from early Paleocene time until the main thrusting within the outer (flysch) nappes in early Miocene time.

Deformation within the outer Carpathian belt involves only the sedimentary cover originally deposited on the European plate (including the passive continental margin of Europe and adjacent oceanic[?] crust). Deformation of the crystalline rocks that formed the original basement for the nappes is generally not observed, and the outer Carpathians form a thin-skinned foreland fold and thrust belt that overlies the autochthonous European basement above a major detachment surface. This detachment can be traced using drilling and seismic reflection data for about 50 km toward the Pannonian basin from the thrust front. Along the inner part of the thrust belt the detachment is observed at about 10–15 km depth (Paraschiv, 1979; Tomek and Ibrmajer, 1982). The thin-skinned nature of this outer thrust belt has important implications for the style of extension within the Pannonian basin system (see next section). To the west, the outer Carpathian flysch belt can be correlated directly with the early Tertiary flysch nappes of the Eastern Alps. At the southeastern end of the Carpathian flysch belt, the flysch nappes disappear below the thick upper Miocene molasse deposited on top of the inner South Carpathians and the Moesian platform. Here the nappes can be traced westward in the subsurface for about 50 km by drilling.

The outer flysch Carpathians are separated from the inner West Carpathians (deformed in Cretaceous time) by the Pieniny Klippen belt, a narrow, continuous belt of tectonic "lozenges" that extends for about 800 km along strike. The Pieniny Klippen belt was deformed in both Mesozoic and Tertiary time and has been interpreted as a major convergent boundary with significant left-slip displacements in both early Tertiary and Miocene time (Sandulescu, 1980; Birkenmajer, 1981; Royden and Báldi, this volume). The Pieniny Klippen belt and the innermost flysch units of the outer Carpathians (Magura group) pass inside (west) of the inner East Carpathians and thus separate the inner East Carpathians from the inner West Carpathians. Final juxtaposition of the inner West and East Carpathians occurred in early Miocene time when shortening and probable strike-slip displacements occurred between them (Sandulescu et al., 1981).

The minimum amount of Miocene shortening required across the outer Carpathians has been determined to be about 100 km across the outer East Carpathians (Burchfiel, 1976) and 60 to 80 km across the outer West Carpathians (e.g., Oszczypko and Tomaś, 1985). Because much sedimentary material has been lost to erosion or subduction, these are minimum estimates; a generally accepted figure for the total Miocene shortening across the outer flysch Carpathians is a few hundred kilometers (Burchfiel, 1980).

The timing of nappe transport within the outer Carpathians is well known and varies along the strike of the belt (e.g., see Jiricek, 1979; Oszczypko, 1982; Oszczypko and Tomaś, 1985; Sandulescu, this volume). In general, thrusting becomes younger from more internal to more external parts of the flysch belt and from west to east (Figure 5). Sandulescu (this volume) distinguishes three main "phases" of overthrusting (Figure 5): an early Miocene (Eggenburgian–Karpatian) phase; a middle Miocene (intra-Badenian) phase; and a late Miocene (Sarmatian–Pannonian) phase. Within each phase, thrusting appears to be older in the west and younger toward the east. Indeed, it is not clear that distinct phases of thrusting can be identified for the whole Carpathian chain. The youngest events within each "phase" in the eastern part of the belt appear to overlap in time with the oldest events of the subsequent "phase" in the western part of the belt. There is also a clear eastward migration of the final convergent events along the most external part of the Carpathian thrust belt (Figure 5).

Folding and small-scale overthrusting (<2 km total shortening) have occurred in the southeastern part of the Carpathian foredeep in Pliocene–Quaternary time. Intermediate depth earthquakes beneath the southern East Carpathians occurring in modern times may represent the very final stage of subduction of Europe beneath the Carpathians and may result from a remnant slab subducted in Miocene time (Roman, 1970; Fuchs et al., 1979). These earthquakes at 70–140 km depth yield focal mechanisms that indicate down-dip extension on steeply dipping nodal planes. Shallow earthquakes (as deep as 30 km) yield focal mechanisms that suggest eastward overthrusting along a surface dipping gently to the west.

Evidence for major Miocene strike-slip displacement parallel to the trend of the Carpathians exists in several places. Such strike-slip zones are important for understanding the kinematic link between extension in the Pannonian basin system and coeval thrusting in the outer Carpathians. These faults are discussed in the section entitled "Strike-Slip Faults (Outside of Basins)."

The Pannonian Basin System

The pre-Cenozoic basement of the Pannonian basin system consists of the internal structural units of the inner Carpathians, Apuseni Mountains, and Dinarides, which were imbricated during Mesozoic thrusting (Figure 2). Seismic reflection profiles and deep drilling indicate that thrust sheets exist in the subsurface below the Neogene sedimentary cover of the Pannonian basin system (Ciupagea et al., 1970; Brix and Schultz, 1980; Rudinec et al., 1981; Horváth and Rumpler, this volume). These Paleozoic and Meoszoic rocks are in places overlain by early Cenozoic marine sedimentary rocks (Szepesházy, 1973; Báldi, 1983; Royden and Báldi, this volume).

Normal faulting and extension within the Neogene Pannonian basin system began in Ottnangian–Karpatian time (early Miocene, about 16.5–24 Ma), or perhaps even earlier in a few places (Figure 6). It continued through Sarmatian time (late Miocene, 12–13 Ma), and into early Pannonian time (late Miocene, 10–12 Ma) in parts of the Great Hungarian Plain. In some places, particularly in parts of the Vienna basin, some extension and strike-slip faulting continues at present. Weak seismic events (magnitude ≤ 5.5) have been detected within the Pan-

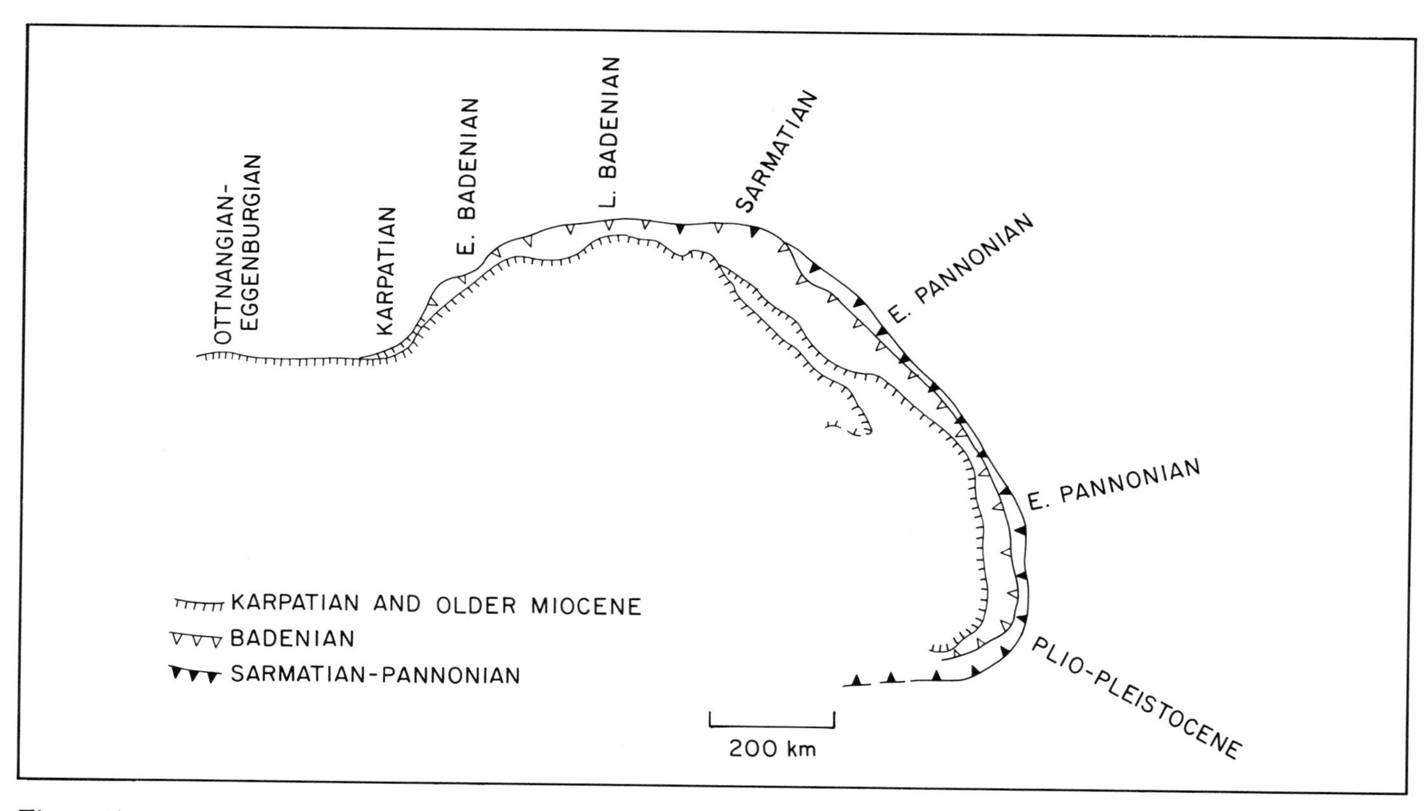

Figure 5. Approximate distribution and ages of thrusting in the outer Carpathians showing areas of Karpatian (16.5–17.5 Ma) and older Miocene thrusting, Badenian (13–16.5 Ma) thrusting, and Sarmatian (12–13 Ma) to Pannonian (8–12 Ma) thrusting. The age of the last significant shortening within the outer Carpathians is also given at various positions along the length of the belt. Note the inner to outer and west to east progression of events through time. Compiled from Mahel et al. (1968), Jiricek (1979), Oszczypko and Slaczka (1979), Fuchs (1980), Oszczypko (1982), Oszczypko and Tomaś (1985), Sandulescu (personal communication, 1985); Sandulescu (this volume).

nonian area, and the sense of motion and local strain pattern inferred from fault plane solutions is compatible with that inferred for Miocene events (Gutdeutsch and Aric, this volume). However, throughout most of the Pannonian system, extension terminated during late Miocene time.

The older sedimentary rocks within each basin were deposited mainly during active faulting and generally lie within well-defined fault-bounded troughs that form the deepest part of each basin (e.g., see Brix and Schultz, 1980; Rudinec et al., 1981; Szalay and Szentgyörgyi, this volume; Rumpler and Horváth, this volume; Tomek and Thon, this volume; Wessely, this volume). This older basin fill, which was deposited during active faulting, is overlain, often unconformably, by younger, flat-lying, posttectonic deposits. Those sedimentary rocks are generally unfaulted, except for compaction features, and usually onlap onto the pre-Miocene basement.

Style of Sedimentation

The detailed configuration of both syn- and posttectonic deposits varies greatly from basin to basin within the Pannonian system and depends on the interaction of deep lithospheric processes (producing thermal subsidence) with the rate of sediment supply. Both processes appear to vary systematically from basin to basin and can be related to the distance of each basin from the outer Carpathian thrust belt. Thus, the style of sedimentation and configuration of sediments throughout the Pannonian basin system varies with distance from the thrust belt.

The posttectonic (thermal) subsidence of the basement below each basin can be related directly to the temperature of the lithosphere following extension: a hot lithosphere cools quickly, producing rapid thermal subsidence, while a cold lithosphere cools slowly (or not at all), producing slow thermal subsidence. Within the Pannonian basin system, extension of those basins located close to the outer Carpathian thrust front was accompanied by little heating of the lithosphere and they thus show almost no thermal subsidence (Royden and Dövényi, this volume). Extension of those basins located far from the outer Carpathian thrust front was accompanied by considerable heating of the lithosphere and rapid thermal subsidence of the basin after extension (equivalent to as much as 2 km of sediment accumulation during 10 m.y.).

The availability of sediments to fill the subsiding basins also depends on the distance of each basin from the outer Carpathian flysch belt because the flysch belt, together with the Eastern Alps, is thought to have been the major source for sediments supplied to the Pannonian basin system. Sediment transport directions within the more internal basins indicate transport from the northwest and northeast, and the lithology of the basin sediments is similar to that of the flysch belt (Pogácsás, 1984; Mattick et al., this volume; Bérczi, this volume).

Basins that formed adjacent to the outer flysch belt thus formed adjacent to a major source of sediment supply, and sedimentation generally kept pace with basement subsidence. Deep water facies are rarely observed in these external basins and do not persist for long periods of time. In contrast, basins that formed far from the outer flysch belt were distant from the main source of sediment supply. Sediments reaching these internal basins were transported for as many as several hundred kilometers across more external parts of the basin system. In these basins, sedimentation lagged behind basement subsidence, especially within the deep troughs. Deep water, abyssal facies are common in these deep troughs, and they persisted for several million years following extension. Filling of these basins after extension occurred when thick deltaic or turbiditic sequences prograded into the central basin area from north to south (in the western part of the system, transport is generally from the northwest, while in the eastern part it is from the northeast).

The interaction of these two processes results in very different styles of sediment configuration within the individual basins. The basins located closer to the outer Carpathian thrust belt are filled mainly with synextensional, shallow water, fault-bounded sedimentary rocks. Posttectonic deposits are thin. Basins located farther from the thrust belt contain only a relatively thin sequence of synextensional, largely deep water sedimentary rocks. These thin sequences are overlain by a thick sequence of posttectonic sediments. The lower part of this posttectonic sequence consists of turbiditic or deltaic units that represent filling of deep water areas created during extension. The upper part consists of a maximum of 2 km of shallow water to terrestrial deposits that represent rapid posttectonic subsidence of the basement during lithospheric cooling.

For example, the Vienna basin formed partly on top of the flysch nappes of the outer Carpathians and is closer to the outer Carpathian thrust front than any of the other basins of the Pannonian system. Thus, the Vienna basin is located immediately adjacent to the source area for its sedimentary fill. Although some parts of the Vienna basin underwent as much as 5 km of syntectonic subsidence in roughly 5 m.y., the rate of sediment supply was sufficient to keep pace with basement subsidence, and deep water facies within this basin are relatively rare. Moreover, because the Vienna basin is located so close to the Carpathian thrust front, there was little or no heating of the lithosphere during extension and little or no thermal subsidence of the basement after extension. Thus, even though a large sediment supply was always available to fill this basin, there are almost no posttectonic deposits present in the Vienna basin. Instead, there is a thick sequence of Miocene syntectonic, fault-bounded sedimentary rocks that were deposited in mainly shallow water to fluviatile conditions and reaching nearly to the surface (Figure 7A).

In contrast to the Vienna basin, the Great Hungarian Plain is located far from the Carpathian thrust belt and thus far from the sediment source in the outer flysch Carpathians. Within this area, the rate of sediment supply was clearly insufficient to keep pace with subsidence of the basement during extension. Within the deep troughs of the Great Hungarian Plain, only a thin (< 2 km, commonly < 500 m) sequence of syntectonic sediments are present. These are mainly in abyssal facies in the deeper parts of the basin. The deltaic units that subsequently filled the basin are posttectonic and mainly unfaulted, but represent filling of deep basin areas created during the extension. Moreover, extension within the area of the Great Hungarian Plain involved considerable heating of the lithosphere and resulted in rapid posttectonic subsidence of the basement in this region. Thus, the deltaic sequence is overlain by a 1–2-km-thick section of young sedimentary rocks that reflect mainly rapid posttectonic subsidence of the basement. A profile through one of the deep troughs of the Great Hungarian Plain shows this thin synextensional, fault-bounded, largely deep water sequence (Figure 7B) overlain by a relatively thick deltaic sequence (as much as 2 km thick). This is overlain in turn by a thick (as much as 2 km), flat-lying, unfaulted sequence that represents posttectonic basement subsidence.

The Vienna basin and the Great Hungarian Plain, described above, represent two extremes of extensional basin development within the Pannonian basin system. Except for the Transylvanian basin, the other basins are intermediate in style between the Vienna basin and the Great Hungarian Plain. Basins that are more similar to the Vienna basin include the Transcarpathian basin and the northernmost part of the Danube basin (where it separates into several north-northeast-trending grabens). Basins that are more similar to the Great Hungarian Plain include most of the Danube basin and parts of the Drava trough. Intermediate basins include parts of the Zala and Drava basins.

Fault Geometry

The geometry of both extensional and strike-slip faults also appears to vary from basin to basin in the Pannonian system. The external basins located close to the thrust belt, such as the Vienna and Transcarpathian basins, appear to have both strike-slip and dip-slip displacement on the same set of faults. These faults are relatively steep, with average dips around 45°.

Within the more internal basins, such as the Danube basin or Great Hungarian Plain, strike-slip and dip-slip displacements may be accommodated along two different sets of faults. The most obvious is a set of steep faults that, on reflection seismic lines, show features characteristic of strike-slip faults. A second set of gently dipping Miocene normal faults also exists within the basin. Tomek (1985) presented a deep seismic reflection profile through the central Danube basin that showed such a fault clearly displacing Badenian–Sarmatian (middle–late Miocene, 12–16.5 Ma) sedimentary rocks. This fault dips roughly 10–15°, northwest and is clearly present within the pre-Miocene basement for a distance of 40 km. Unfortunately, such faults are difficult to observe in the subsurface.

Within the Great Hungarian Plain, normal faults can be observed to displace and rotate Karpatian–Badenian–Sarmatian age rocks (middle Miocene, 12–17.5 Ma) (Kőrössy, 1981; Rumpler and Horváth, this volume). Related faults may also displace some of the Pannonian (late Miocene, 8–12 Ma) deltaic sequence, which locally dips into the adjacent basement rocks. If the extensional style within the Great Hungarian Plain is similar to that of the Basin and Range Province of the United States (Figure 8), then some of the gently dipping contacts between Pannonian age sedimentary rocks and the underlying basement may be tectonic (e.g., see Wernicke and Burchfiel, 1982). This will be difficult to demonstrate without extensive coring of the contact zone.

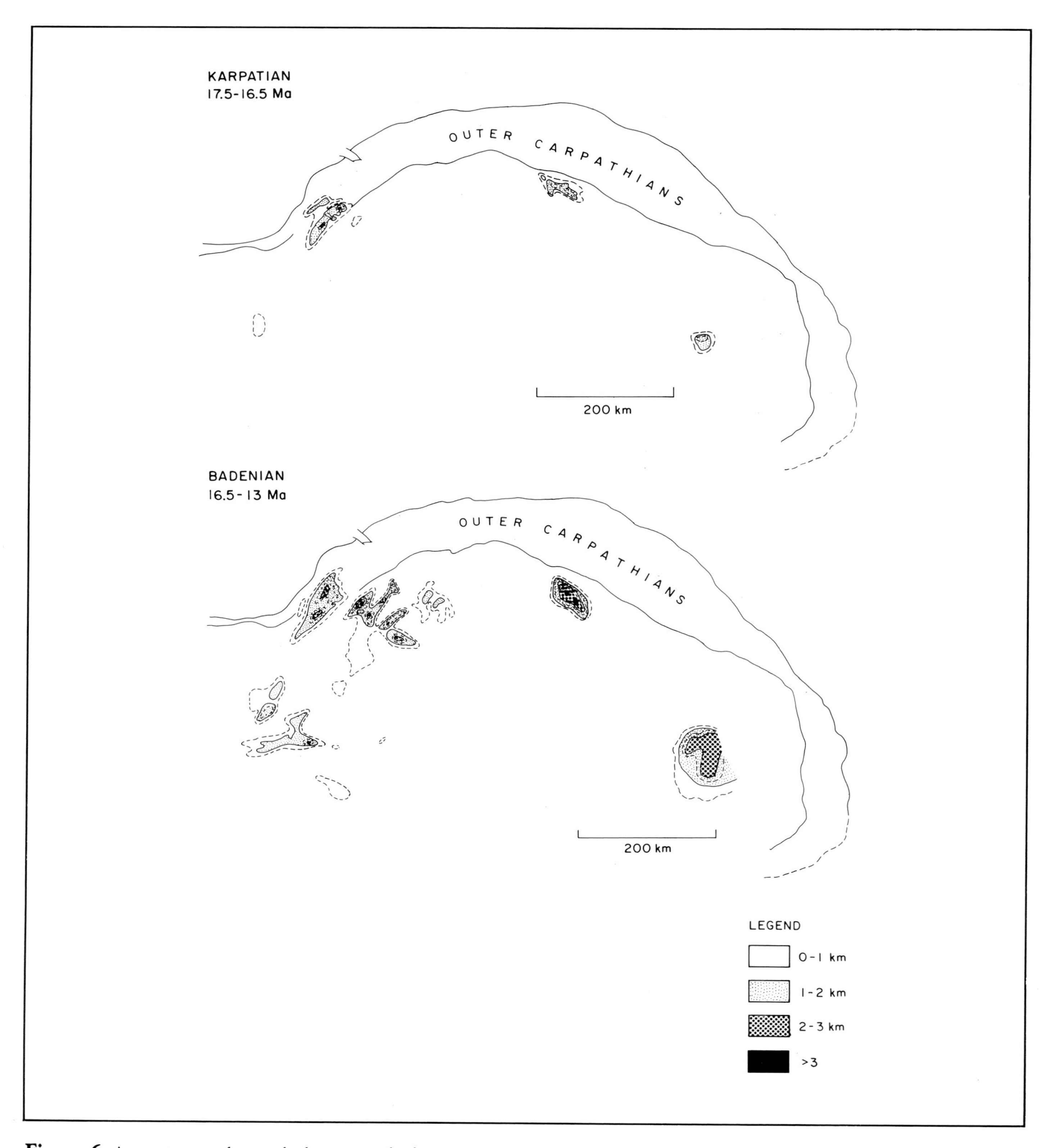

Figure 6. Approximate sediment thickness maps for five time periods: Karpatian (16.5–17.5 Ma), Badenian (13–16.5 Ma), Sarmatian (12–13 Ma), Pannonian (8–12 Ma), and Pannonian–Quaternary. Contour interval is 500 m. Sources: Southern Vienna basin, Pannonian–Quaternary, from Köves and Krobot (1980) in Brix and Schultz (1980), and Sarmatian–Karpatian, estimated from cross sections in Wessely (this volume) and from total Neogene thickness (index Map 2). Northern Vienna and Danube basins, all maps, from Spicka (1972). Transcarpathian basin, all maps, from Rudinec (1978). Transylvanian basin, all maps, from Ciupagea et al. (1970). Pannonian basin in Romania, all maps, estimated from total thickness of Neogene sediments and cross sections in Paraschiv (1979). Hungary, Badenian and Sarmatian, from Kőrössy (1980); Pannonian–Quaternary, modified after Kőrössy (1970); and Pannonian (s. str.), modified after Kőrössy (1964). Yugoslavia, all maps, from Filjak et al. (1969).

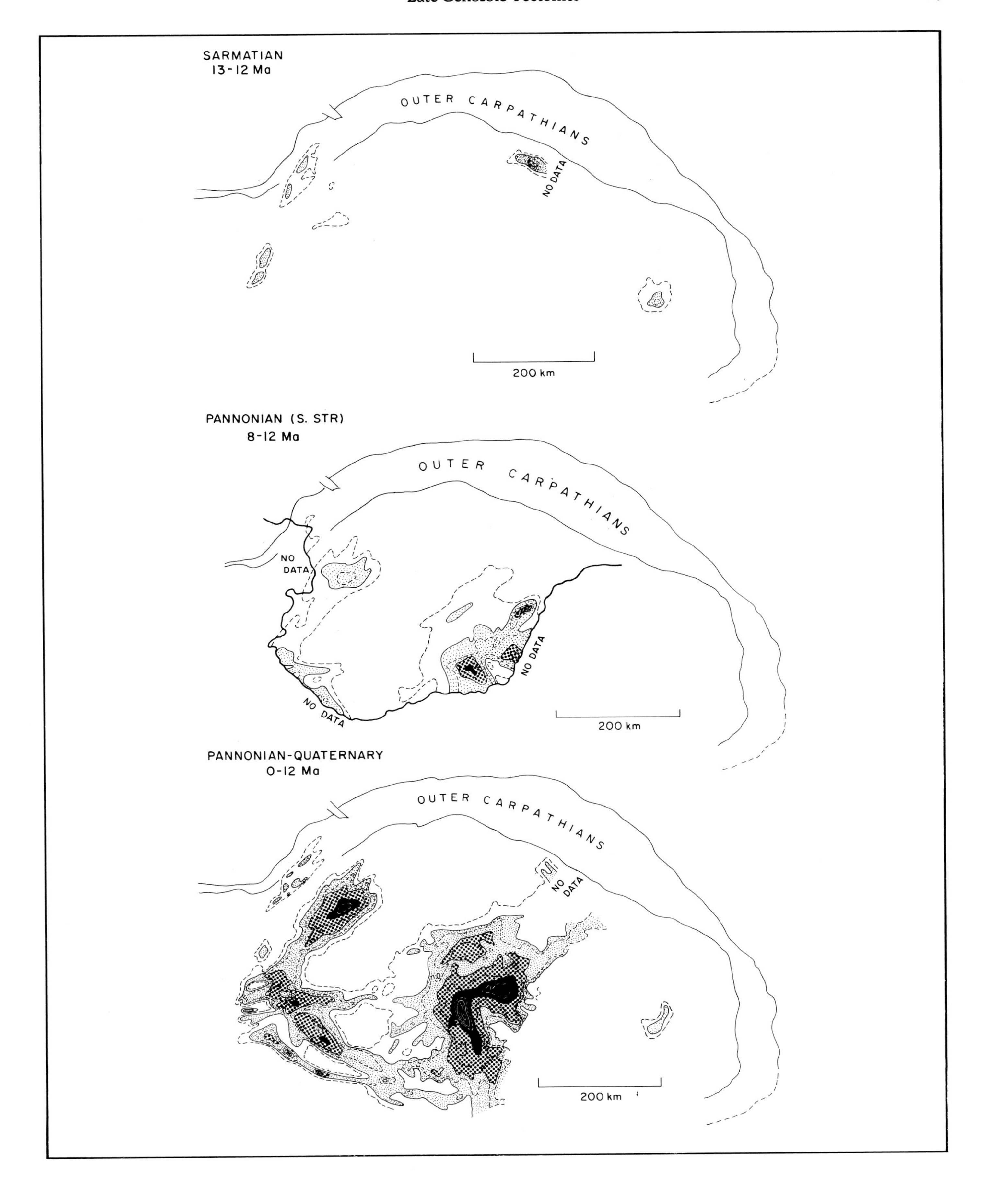
SARMATIAN
13-12 Ma
OUTER CARPATHIANS
NO DATA
200 km
PANNONIAN (S. STR)
8-12 Ma
OUTER CARPATHIANS
NO DATA
NO DATA
NO DATA
200 km
PANNONIAN-QUATERNARY
0-12 Ma
OUTER CARPATHIANS
NO DATA
200 km

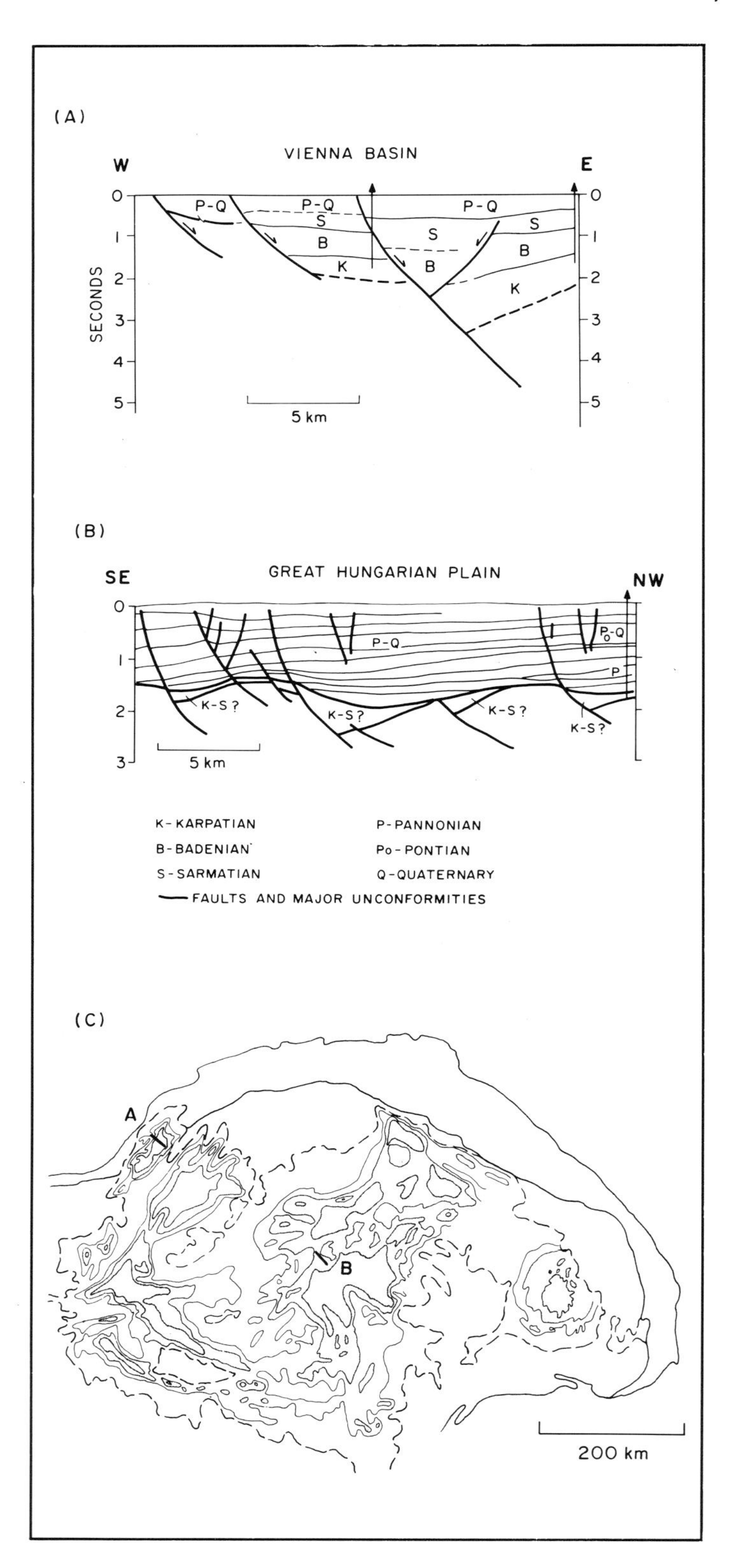

Figure 7. Line drawings from seismic reflection lines across the Vienna basin (A) and Great Hungarian Plain (B) illustrate their different styles of sedimentation and faulting. (A) is after Tomek and Thon (this volume), (B) is after Pogacsas (1984). (C) Locations of profiles: 1-, 2-, and 3-km contours of Neogene–Quaternary sediments are shown; shading represents areas with more than 2 km of Neogene–Quaternary sediments.

The Neogene Transylvanian basin is structurally different from the other Neogene basins of the Pannonian system. Except for a few small grabens of earliest Miocene age, the Neogene Transylvanian basin is a 3–4-km-deep saucer-shaped depression without significant faults (Maps 2 and 8) (Ciupagea et al., 1970). Its present surface elevation is high (~600 m above sea level), although it contains mainly marine sediments. It is probably not extensional in origin (see next section), and has very low surface heat flow (Map 5).

Timing of Extension

Faulting began and ended at different times in different basins, so that extension of the Pannonian system occurred diachronously. The ages of deposition of syn- and posttectonic sedimentary rocks are also different. In basins where the sedimentation rate kept pace with basement subsidence and where extensional features are clear, the age of extension is well defined. However, where syntectonic subsidence occurred in sediment-starved basins and where extensional features are more subtly developed, the age of extension is more difficult to determine. In the latter case, the ages of faulting in areas adjacent to the basins can be a useful addition to subsurface constraints.

The location of active extensional troughs at various times is reflected by the thickness and distribution of the sedimentary rocks that were deposited in the extending and subsiding troughs (Figure 6). The sources used to compile the stage isopach maps shown in Figure 6 were highly variable in accuracy and in conventions. Some of the maps in this figure were constructed by adding or subtracting data from two or more contour maps. The 0 m contour is not shown because different authors use different criteria for defining this boundary. Because of the paucity of information about the thicknesses of the Pannonian (*s. str.*) (available only from Hungary and Czechoslovakia), a map for the Pannonian–Quaternary was also constructed. Because of the problems in preparing this figure, the contours are only estimates for Yugoslavia (Karpatian–Sarmatian), Austria (except for the Pannonian–Quaternary in the Vienna basin), and the Romanian part of the Pannonian basin (Karpatian–Sarmatian).

Within the Vienna basin, normal growth faults as old as Ottnangian–Karpatian age (16.5–19 Ma) indicate that extension of the Vienna basin began at this time, although Karpatian basin subsidence probably occurred within a larger molasse basin. The main age of extension within much of the Vienna basin is of Karpatian–Badenian age (13–17.5 Ma), but in parts of the basin as much as 2 km of Sarmatian and younger sedimentary rocks are related to younger extensional and strike-slip deformation that has continued to the present (Gutdeutsch and Aric, this volume). Extension within the Transcarpathian basin began at roughly the same time, but probably ended mostly before or during Sarmatian (12–13 Ma) time (Figure 6).

Badenian extension (13–16.5 Ma) can be demonstrated within the northernmost part of the Danube basin (where it separates into several north-northeast-trending grabens), the deep central part of the Danube basin, the Zala and Drava basins, and the deep parts of the Great Hungarian Plain. For example, faulted and tilted strata in the western part of the Drava trough

Figure 8. Gently west-dipping normal detachment on the east side of Death Valley in the Basin and Range Province, western United States. Quaternary(?) conglomerate above the dramatically planar fault (dipping to the right in the lower part of the photo) is only slightly discordant with the sheared Precambrian gneiss exposed below the fault contact. The youngest conglomerate units in this area dip gently away from the hillside, while older units have been rotated along the fault and now dip into the hillside (not shown). In gross scale, and at the level of seismic resolution, the overlying beds would appear to onlap the basement rocks above a depositional unconformity. In fact, the contact is an extensional fault that may have large displacement. (Similar features mapped in the area may have more than 10 km of displacement in Panamint Valley; B. C. Burchfiel, et al., unpublished data). Note also the small normal faults within the hanging wall that end against the main detachment fault. The purpose of this photo is to illustrate the difficulty in recognizing such features from seismic or drilling data.

(Figure 6) are of Badenian (13–16.5 Ma) and possible Karpatian age (16.5–17.5 Ma) and are overlain unconformably by flat-lying, mainly undisturbed rocks of Pannonian age (8–12 Ma). This relationship indicates Badenian age extension and indicates no extension after deposition of Pannonian age rocks above the unconformity.

The age of the major extension within the central part of the Danube basin and the Great Hungarian Plain is ambiguous. Badenian–Sarmatian rocks are displaced on normal faults within the Danube basin (Tomek, 1985), and Badenian rocks are similarly displaced within the Great Hungarian Plain. Unfortunately, the Sarmatian is commonly unfossiliferous and generally thin in the deep basin areas of the Great Hungarian Plain, making it difficult to demonstrate displacements of Sarmatian age. Faults exposed at the surface near the central Danube basin and Great Hungarian Plain clearly displace Sarmatian rocks and are thought to be intra-Sarmatian (or younger) (e.g., Paraschiv, 1979; Mészáros, 1983; Némedi-Varga, 1983). In early Pannonian time, muddy conglomerates were shed into the deep basin areas of the Great Hungarian Plain (Dorozsma Formation), perhaps indicating faulting and creation of topographic relief in Sarmatian to early Pannonian time (Bérczi, this volume). The deltaic sequence within the basins lies directly above these conglomerate beds. The age of major extension within these two basins is inferred to be of intra-Sarmatian to earliest Pannonian age, with some extension occurring earlier (in Badenian time) (Rumpler and Horváth, this volume). (See following sections for additional constraints from structures adjacent to the basins.)

Basin Extension along Strike-Slip Faults

Miocene extension within the Pannonian basin system appears to have been accommodated by several major shear zones or strike-slip zones. These strike-slip zones form a conjugate set of faults, with northeast-trending fault zones being left-slip and northwest-trending fault zones being right-slip (Figure 3). Such a conjugate system produces net east–west extension of the Pannonian region. These faults serve to connect areas of local crustal extension to one another and to the thrust belt. Some basins appear to have extended as simple pull-apart basins along strike-slip zones (Vienna and Transcarpathian basins). Others appear to have extended along broad shear zones (northern Great Hungarian Plain) or at the intersection of two or more strike-slip zones (Zala basin). Still others may have opened as simple grabens that end against strike-slip zones (southern part of the Great Hungarian Plain).

Strike-slip displacements are difficult to prove with only subsurface data from beneath the basins themselves. Indirect morphologic and structural data for strike-slip displacement within the Neogene basins themselves include characteristic en echelon patterns of normal faults, folds (Sava folds), and small en echelon rhombohedral horsts and troughs. The sense of motion implied by these en echelon features is consistent with the senses of displacement shown in Figure 3. Additionally, many of the faults shown on seismic reflection profiles have characteristic features commonly associated with wrench tectonics, such as "flower structures" (Rumpler and Horváth, this volume).

Many faults exposed adjacent to the young basins can be traced into the basins and clearly belong to the fault systems controlling basin extension. Some of these faults can be shown to have significant Miocene strike-slip displacements. The field evidence for offsets on faults exposed outside of the basins is discussed in the next section.

STRIKE-SLIP FAULTS (OUTSIDE OF THE BASINS)

Some of the strike-slip faults that underlie the basins are also exposed outside of the basins. All of the direct evidence for the sense of horizontal displacement comes from field observations outside of the basin areas, as indicated by the numbered localities on Figure 3. These data (keyed to Figure 3) include the following: (1) 80 km of Tertiary left-slip displacement along a fault trending northeast from the northern end of the Vienna basin (Schrattenburg-Bulhary line; Roth, 1980); (2) 20 km cumulative Sarmatian (late Miocene, 12–13 Ma) right-slip along northwest-trending faults of the Hungarian Mid-Mountains (Mészáros, 1983); and (3) 25–30 km post-Badenian (13–16.5 Ma) and probably Sarmatian (12–13 Ma) left-slip along an east-northeast-trending fault in the northern Pannonian basin (e.g., in the Matra Mountains; Z. Balla, personal communication, 1980). Other large zones of strike-slip displacements include the following: (4) the contact between the inner and outer West Carpathians (Pieniny Klippen belt), thought to be a major zone of Miocene convergence and left-slip, at least in the central and western part of the belt (Sandulescu, 1980; Birkenmajer, 1981); (5) the peri-Adriatic-Vardar fault system, with perhaps one hundred to several hundred kilometers of Tertiary right-slip, some of which is middle Miocene (Laubscher, 1971; Schönlaub, 1980; Grubic, 1980); (6) the north-south-to east-west-striking high-angle faults in the South Carpathians (Timok fault and faults along the Cerna-Petrosani grabens), considered to have 50–100(?) km of Tertiary right-slip (Sandulescu, 1984); and (7) the high-angle fault trending southeast from the southern end of the Vienna basin having focal mechanisms consistent with pure left-slip.

Other faults exposed at the surface within the Carpathian region can be interpreted as major strike-slip faults, but horizontal displacements on these faults have not been documented from field data. One area in which this interpretation could be tested with the proper field observations is in the inner West Carpathians north and east of the Danube basin. In my interpretation, the north-northeast-trending grabens at the northern end of the Danube basin are Badenian pull-apart basins along northeast-striking left-slip faults (e.g., faults 8, 9, and 10 on Figure 3) that continue northward and merge with the Pieniny Klippen belt (fault 4). Although young strike-slip displacements have not been suggested for these faults, pre-Miocene structures or units cannot be tied directly across the projected traces of these steeply dipping faults, which are covered by post-Badenian molasse in many places.

Another area in which circumstantial evidence exists for strike-slip displacements along steep to vertical faults is within the East Carpathians and north of the Transylvanian basin (e.g., faults 11, 12, and 13 in Figure 3). Fault 11 (North Transylvanian fault) is exposed at the surface as a steeply dipping to vertical fault that forms the eastward continuation of a major east-northeast-trending shear zone within the Pannonian basin. This shear zone is inferred to have had middle to late Miocene left-slip displacement, although older displacements along this fault zone may have been right-slip (Sandulescu, 1980, and this volume). This fault may cut a young 3–13 Ma) volcanic complex north of the Transylvania basin (near Baia Mare Mountains) and displace the southern end of the complex about 25 km to the east. This may be demonstrable from field data, but volcanic rocks that are clearly younger than the faulting probably cover the fault trace in many places. If such displacement could be demonstrated, it would also prove a Sarmatian–Pannonian age for major strike-slip displacements in the Pannonian basin. Sandulescu et al. (1981) showed this fault splaying eastward into a set of high-angle faults each with 1–5 km of left-slip separation and showed that it appears to merge south into the contact zone of the inner and outer Carpathians (Ceahlau and Audia nappes).

Strike-slip displacement along the contact zone between the inner East Carpathians and the outer flysch Carpathians (fault 12 in Figure 3) is suggested by the pronounced straightness of the entire zone. The expected sense of displacement is not clear, although it is tentatively interpreted here as right-slip. The fault zone may have had a different sense of displacement at different times depending on the relative movements of various crustal fragments. It extends from the Transcarpathian basin southward to the latitude of fault 11 and probably lies along a zone near the contact of the inner and outer Carpathians (Pienniny Klippen belt in the north and the Convolute Flysch nappe and the outermost part of the Ceahlau nappe in the south).

Where the projections of faults 11 and 12 (Figure 3) intersect one another, the outer Carpathian belt departs from its northwest–southeast linear trend and toward the south, trends roughly north–south. This is also the narrowest part of the outer flysch belt in the East Carpathians. I suggest that north of the area of intersection, the straight and narrow Ceahleau nappe and perhaps the Convolute Flysch nappe functioned as strike-slip zones in middle to late Miocene time. South of this area, strike-slip (left-slip) was accommodated along the Audia nappe, which is 1–3 km wide for several hundred kilometers along strike and consists of imbricate tectonic slices with steep to vertical contacts. The change in strike of the belt is accommodated by left-slip along the North Transylvanian fault, transferred into the Ceahlau nappe along a series of left-slip plays. This interpretation may also be testable in the field, although the outcrop within this region is poor.

The southern end of the outer Carpathian flysch belt disappears beneath young molasse of the Moesian platform, but the flysch nappes can be traced for several tens of kilometers westward by drilling. For the kinematic history for the Carpathian mountains and Pannonian basin to be consistent, large right-slip displacements (100 km?) must be inferred along this buried thrust front (fault 14 on Figure 3) or within the nappes of the South Carpathians. Such a fault system may easily connect to the large right-slip zone formed by the Timok fault or faults along the Cerna and Petrosani grabens (fault 6 on Figure 3).

SPATIAL AND TIMING RELATIONSHIPS BETWEEN BASINS AND THRUST BELT

Strike-Slip Faults as Transform Faults

The strike-slip faults that connected the extensional basins to one another also connected the extensional basins to a zone of

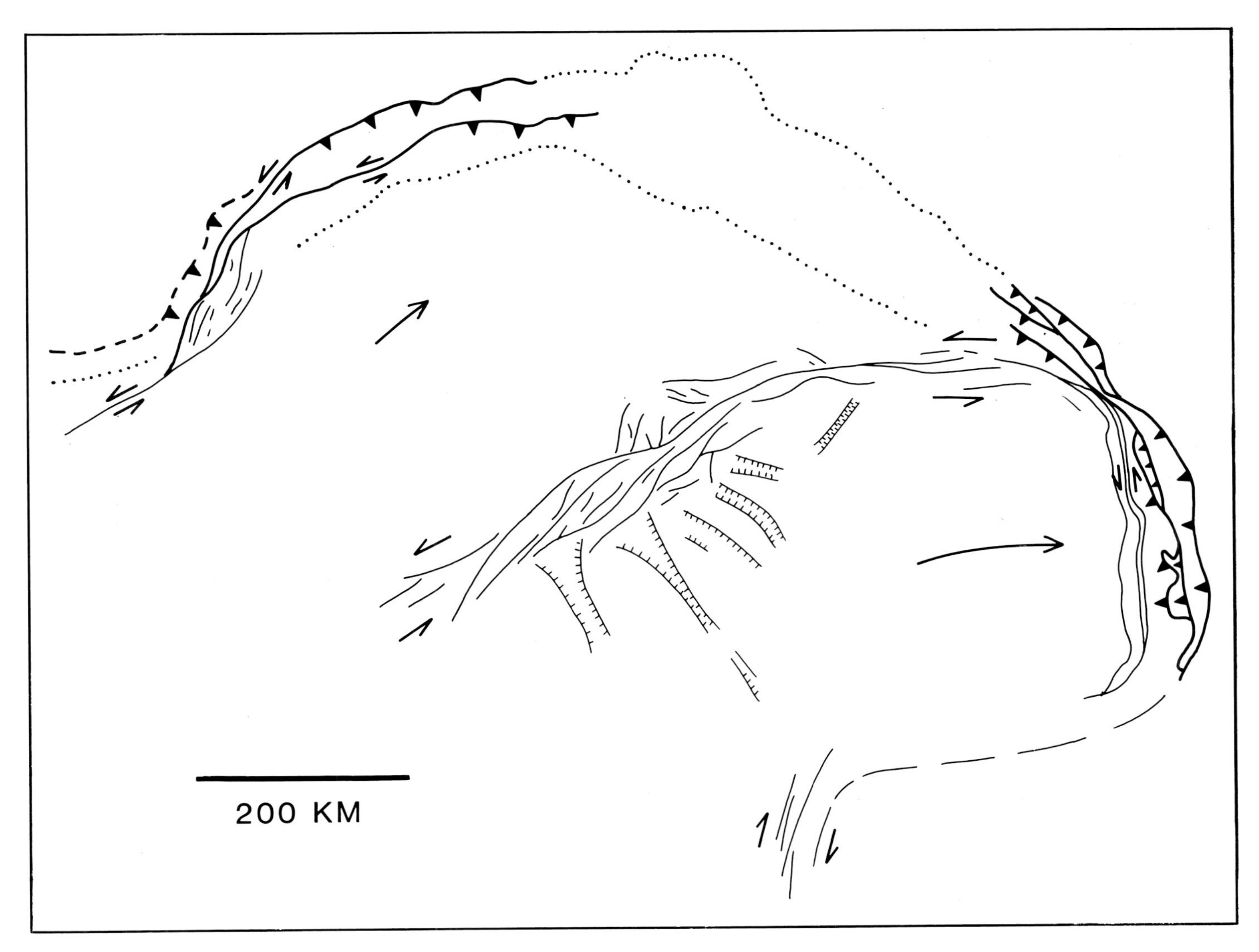

Figure 9. Schematic diagram showing the relationship of extension within the Vienna basin and part of the Great Hungarian Plain to shortening in the thrust belt. Large arrows show the movement of crustal fragments relative to Europe. In both examples, the area of extension is connected to the area of shortening by strike-slip faults. These strike-slip faults accommodate and separate parts of the thrust belt that have different directions or rates of convergence. Note that within the Great Hungarian Plain, basins formed both as pull-apart features within the strike-slip zone and as graben structures south of the strike-slip zone.

shortening within the outer Carpathian thrust belt. In general, the extensional basins opened as pull-apart basins or accommodation structures along zones of contemporaneous strike-slip (Figure 3). These same strike-slip zones also separated adjacent parts of the outer Carpathian thrust belt in which shortening occurred at different times or at different rates. Because shortening within the thrust belt ended diachronously from west to east in middle–late Miocene time (Figure 5), these strike-slip faults usually separated or accommodated greater shortening south or east of the strike-slip zones from areas of lesser shortening north or west of the fault.

The kinematic relationship between basin extension, strike-slip displacement, and thrust belt activity is nowhere more clearly illustrated than in the vicinity of the Vienna basin, where the age and direction of extension, the age of thrusting, and the location and sense of strike-slip displacement are well known (Figure 9). This basin formed as a pull-apart basin at a left-stepping discontinuity along a left-slip fault system, with a proven displacement of about 80 km of Tertiary left-slip north of the basin. The obvious rhombohedral shape of the basin, the en echelon fault pattern within the basin, and the pattern of sedimentation within the basin are characteristic of such pull-apart basins (for a more detailed discussion see Wessely, this volume; Royden, 1985). Basin extension began in about Ottnangian–Karpatian time (16.5–19.0 Ma) and was finished mostly by Sarmatian–Pannonian time (8–13.0 Ma). The main age of extension was Badenian–Sarmatian (12–16.5 Ma).

An examination of contemporaneous events within the thrust belt shows that within the easternmost Eastern Alps, thrusting ended during Ottnangian time (17.5–19.0 Ma) (Figure 5). In the westernmost Carpathians (west of the Vienna basin), thrusting ended in Karpatian time (16.5–17.5 Ma), and farther to the northeast (west of Krakow), thrusting ended during Badenian time (13.0–16.5 Ma). This geometry implies that during Karpatian and Badenian time the part of the Carpathians that was northeast and east of the Vienna basin was translated generally northward relative both to the European plate and to the inactive nappes in the external part of the Alps west of the Vienna basin (Figure 9). Relative motion occurred partly by left-slip along the northeast-trending fault zone described above. The timing and direction of displacement along this fault are in good agreement with those required to accommodate continued thrusting to the northeast. (Some of the relative motion was probably also related to left-slip in the Pieniny Klippen belt along faults trending northeastward from the northeastern corner of the Vienna basin.) Thus, extension of the Vienna basin, which resulted from strike-slip motion along this boundary fault, is closely related to diachronous crustal shortening in the outer Carpathians (Figure 9).

Many of the other extensional basins of the Pannonian system appear to be connected to the Carpathian thrust belt by strike-slip faults in much the same way as the Vienna basin. An analogous situation occurs in the northern Danube basin, where the basin separates into several northeast-trending troughs. I have interpreted these troughs as small pull-apart basins along a series of northeast-striking left-slip faults (faults 8, 9, and 10 on Figure 3) that merge northeastward with the Pieniny Klippen belt. As discussed above, left-slip displacements along these faults have not been documented, but the field relationships do not exclude that possibility. This interpretation is testable.

Less clear is the relationship of the Transcarpathian basin to thrust belt activity in the outer Carpathians. This basin lies adjacent and internal to the outer Carpathian thrust belt, and extension is thought to have occurred as a pull-apart feature during right-slip(?) along the contact between the inner and outer Carpathians. The sense of middle-late Miocene displacement along this northwest-striking strike-slip zone and the relationship to diachronous events within the thrust belt are not well constrained.

Extension within the Great Hungarian Plain also appears to be connected directly to the thrust belt. Left-slip along the large east-west-trending shear zone within the Great Hungarian Plain feeds into the Carpathian thrust belt along a high-angle fault north of the Transylvanian basin (Figure 9 and faults 11 and 13 on Figure 3). The continuation of this fault within the thrust belt is not known, but it may partly coincide with the contact of the inner East Carpathians deformed in Mesozoic–Paleocene time (Ceahlau nappe) with the outer Carpathian flysch belt deformed in Miocene time (Convolute Flysch and Audia nappes) (Figure 9). One set of en echelon pull-apart basins opened along the main axis of the shear zone (Figures 3 and 9). A second set of north-south-trending extensional troughs opened south of this left-slip shear zone.

Shortening within the southern part of the outer Carpathian thrust belt was probably accommodated by right-slip south of the Transylvanian basin and north of the Moesian platform (Figure 9). Much of this displacement (more than 100–200 km?) may have occurred at the frontal (southern) edge of the flysch nappes that are now buried by younger molasse (fault 14 on Figure 3). This displacement may be transferred to the Timok and related faults (fault 6 on Figure 3). Figure 9 suggests that Miocene thrusting within the outer East Carpathians may have been accommodated by clockwise rotation of the mainly undeformed Apuseni-Transylvanian block about a pole in the western Moesian platform (pole located at about point 15 on Figure 3). Note that both the Timok and related faults (fault 6 on Figure 3) and the large shear zone in the Great Hungarian Plain (fault 11 on Figure 3 and its southwestward continuation) form roughly concentric circular arcs. The orientation of the extensional grabens in the Great Hungarian Plain and the Apuseni Mountains is also consistent with this interpretation. In addition to field observations, paleomagnetic measurements of rotations of the young volcanic rocks in the Apuseni Mountains may provide a useful test of this hypothesis.

Two of the strike-slip zones within the Pannonian basin system are probably not related directly to thrust belt activity in the Carpathians, but rather to the large peri-Adriatic-Vardar right-slip fault system that extends through the Dinarides into the Eastern and Western Alps. Two splays of this fault system extend through the Pannonian region and lie along the Sava and Drava troughs in the southwestern corner of the Pannonian basin system. However, the middle Miocene age of extension of these basins (Badenian–Sarmatian) (Vucković et al., 1959) is probably a reflection of the thrust belt activity because extension is roughly the same age as that of the other basins in the Pannonian system.

Geometry of Strike-Slip Faults, Extensional Domains, and Thrust Belt at Depth

Seismic reflection profiles across the outer Carpathian thrust belt show that the strike-slip faults within the thrust belt (including the Schrattenburg–Bulhary fault northeast of the Vienna basin, the Pieniny Klippen belt, and perhaps parts of the Ceahlau and Audia nappes) do not extend downward into the autochthonous cover of the European plate (e.g., Tomek, 1985). Rather these faults appear to function as tear faults that flatten into gently dipping detachment horizons at depth and end against or merge with thrust faults at the surface (Figure 10). These strike-slip faults can best be described as thin-skinned features restricted to the allochthonous nappes of the overriding plate.

More generally, all of the strike-slip faults within the Pannonian region (Figure 3) can be interpreted as tear faults that are restricted to the upper "plate" of the subduction system. Near and within the thrust belt, where the Pannonian fragment is thin, these faults extend through the Pannonian fragment and flatten into or merge with the thrust décollement at depth (Figure 11). Far from the thrust belt, where the Pannonian fragment is thick, these faults would accommodate strike-slip or wrench deformation throughout the Pannonian lithosphere to great depths. The strike-slip fault system within the Pannonian area would thus be a thin-skinned system near the thrust belt, but would accommodate wrench deformation at greater depths far from the thrust belt (Figure 11). In this interpretation, the depth of penetration of the strike-slip faults is controlled both by the geometry of the thrust belt (or subduction zone) at depth and by the position of each fault relative to the thrust belt.

A similar analysis can be constructed for the depth to which extensional deformation penetrates into the lithosphere. For example, extension of the Vienna basin occurred as a pull-apart feature along a thin-skinned strike-slip fault zone. If these northeast-trending faults merge into a gently dipping detachment horizon at depth, then it seems likely that extension of the basin also was restricted to shallow crustal levels above that detachment horizon (Figure 11). In this interpretation, the large

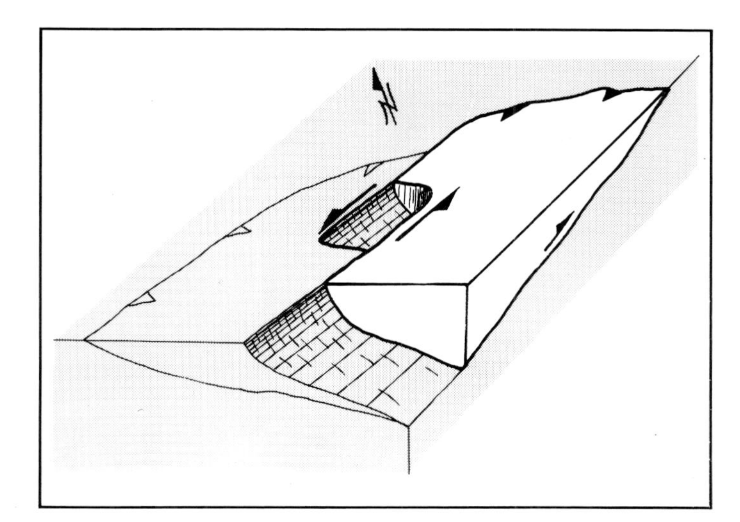

Figure 10. Block diagram to illustrate proposed style of extension within the Vienna basin. Shaded area shows European platform and inactive nappes of the West Carpathians, which may be considered as a part of Europe at this time. Unshaded area shows active nappes east and north of the Vienna basin that are being transported northeastward. Active and inactive nappes are separated by transcurrent faults that flatten into a shallow-dipping detachment surface. Vienna basin pulls apart by superficial extension above the detachment surface. Figure from Royden et al. (1983).

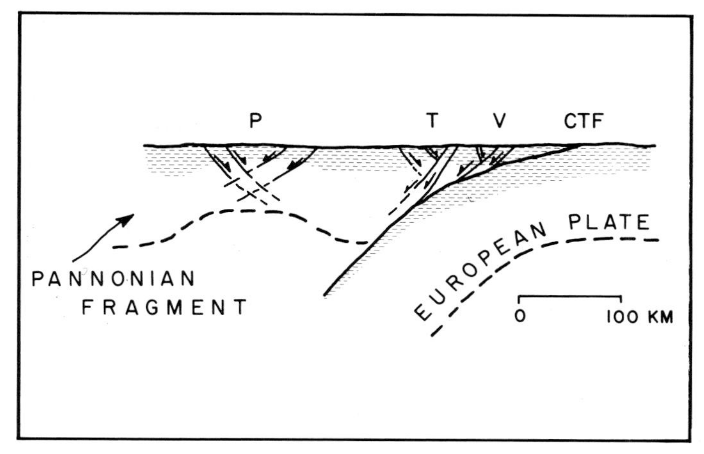

Figure 11. Schematic diagram illustrating how the thickness of the Pannonian fragment is related to the distance from the Carpathian thrust front and showing the relative distances of the Vienna (V) and Transcarpathian (T) basins and the Great Hungarian Plain (P) from the Carpathian thrust front (CTF). Extension and strike-slip faulting are thought to have been mainly confined to the Pannonian fragment, so that the depth to which extension penetrated should increase with distance from the thrust front. Shaded area represents crustal rocks. Areas of extension are indicated schematically by normal faults; the fault geometry shown here is not meant to be realistic. Figure modified after Royden et al. (1983).

normal faults that bound the Vienna basin on the west would flatten at depth toward the southeast and have both a normal and left-slip component of displacement (for details, see Royden, 1985).

If the depth to which extension penetrates into lithosphere is related to the depth of penetration of the strike-slip fault system, then extension would also be restricted to the overriding (Pannonian) fragment above the main thrust decollement. In this interpretation, basins such as the Vienna basin that are located close to the thrust front would have formed by thin-skinned extension of upper crustal rocks only. Extension beneath those basins that are located farther from the thrust front would have involved rocks at deep crustal levels or rocks within the uppermost mantle of the overriding plate. The most internally situated basins would have formed by extensional processes that involved the entire lithosphere of the Pannonian fragment (Figure 11).

Quantitative analysis of subsidence and heat flow data supports this interpretation and shows that both surface heat flow and thermal (posttectonic) subsidence rates increase systematically with increasing distance from the thrust front. This implies that the amount of mantle heating that occurs during basin extension also increases systematically with increasing distance from the thrust front (Royden and Dövényi, this volume). These results are consistent with thin-skinned extension beneath the Vienna basin and with a gradual transition to whole lithosphere extension beneath the more internal basins (Great Hungarian Plain). The transition from thin-skinned to whole lithosphere extension appears to occur within about 200 km of the Carpathian thrust front.

This interpretation also explains the widely different thicknesses of posttectonic sediments that accumulated within the different basins of the Pannonian system (see previous section on structure and sedimentation). The thin-skinned extension that occurred beneath the external basins located near the thrust belt produced no thermal anomaly at depth, and hence no thermal (posttectonic) subsidence occurred in these basins. Whole lithosphere extension beneath the internal basins located far from the thrust belt produced elevated temperatures throughout the lithosphere and resulted in rapid thermal subsidence in these basins. These differences in basement subsidence rates are exaggerated by the relative availability of sediments in the external basins, resulting in the external basins containing relatively thick syntectonic deposits and relatively thin posttectonic deposits, and the internal basins containing thin syntectonic deposits and very thick posttectonic ones.

Palinspastic Reconstruction

Extension within the basins of the Pannonian system was diachronous, beginning first in the most external basins (Vienna and Transcarpathian) in Ottnangian–Karpatian (early Miocene) time (Figure 6). During middle–late Miocene time, the main locus of extension shifted southward into the area of the Danube basin and Great Hungarian Plain. Over the same time interval, the locus of thrusting migrated eastward around the Carpathian thrust belt (Figure 5 and previous sections).

The temporal and spatial relationships between thrust belt activity and extension within individual basins are described in the previous sections and illustrated in Figure 9. The relationship between thrust belt activity and extensional deformation of the whole Pannonian region involves reconstruction of extension within many individual basins and displacement along many strike-slip zones at the same time and is geometrically more complex. This relationship is thus best investigated through a series of palinspastic reconstructions that examine basin and thrust belt evolution in a series of time stages. This approach takes advantage of the diachronous nature of both extension and thrusting throughout the Carpathian–Pannonian system and allows an examination of how changes in thrust belt activity are mirrored by changes within the extensional system.

Palinspastic reconstructions (Figure 12) were constructed for three time stages during the Miocene: (1) Karpatian (early–middle Miocene, 16.5–17.5 Ma), (2) Badenian (middle Miocene, 13–16.5 Ma), and (3) Sarmatian–Pannonian (middle–late Miocene, 8–13 Ma). These palinspastic reconstructions involve only relative motions of upper crustal rocks, thus the behavior of the lithosphere at depth is irrelevant except in so far as it effects estimates of crustal extension. Total crustal extension for most of the areas examined by Royden and Dövényi (this volume) and Horváth et al. (this volume) was between about 50 and 200%. For simplicity, we assumed that each basin (except the Transylvanian basin) extended by 100%. The uncertainty in surface area occupied by each basin prior to extension is less than about ±30% due to this approximation. A more rigorous approach is probably unwarranted because uncertainties in estimating the present size of the extended areas and the direction of extension are also fairly large.

During each time interval, extension was accommodated by strike-slip motion along the major fault zones shown in Figure 3 and ultimately by compression and shortening across the outer Carpathian flysch belt. The magnitudes of displacement are those required to close the relevant basins during the appropriate time intervals. The direction of displacement was determined by the trend of the strike-slip faults that connect the basins to one another and to the outer Carpathian thrust belt. Reconstructions for each phase of extension were developed by working backward from the present. Although presented in chronologic order from oldest to youngest, reconstructions at each time are dependent on those of each younger phase.

The reconstructions shown in Figure 12 are based solely on data from the extensional basins and not on any data from the thrust belt itself. In each reconstruction, the thrust belt was simply pulled back without allowing for shortening within the thrust belt. Only the areas of extension, compression, and strike-slip faulting that can be identified and assigned to specific time periods are shown in Figure 12. Other structures are required to complete the fragment boundaries, but such structures have not yet been identified from field studies.

Karpatian (16.5–17.5 Ma) extension occurred primarily in the northern Vienna and Transcarpathian basins and beneath some of the deepest areas in the Great Hungarian Plain (Figures 6 and 12). During Karpatian time, extension in the Vienna basin was accommodated by left-slip along a northeast-trending fault system north of the Vienna basin and along the Pieniny Klippen belt. Extension within the Transcarpathian basin was accommodated by right(?)-slip along the Pieniny Klippen belt.

Reconstruction of the basins at the beginning of Karpatian time shows that the Karpatian basin extension requires little shortening within the Carpathian thrust belt (Figure 12). These small, early basins probably reflect only local extensional structures along strike-slip faults related to diachronous shortening along the thrust belt and do not represent regional extension of the Pannonian fragment. In Karpatian time, considerably more shortening must have occurred within the Capathian thrust belt than can be accounted for simply by closing up the basins.

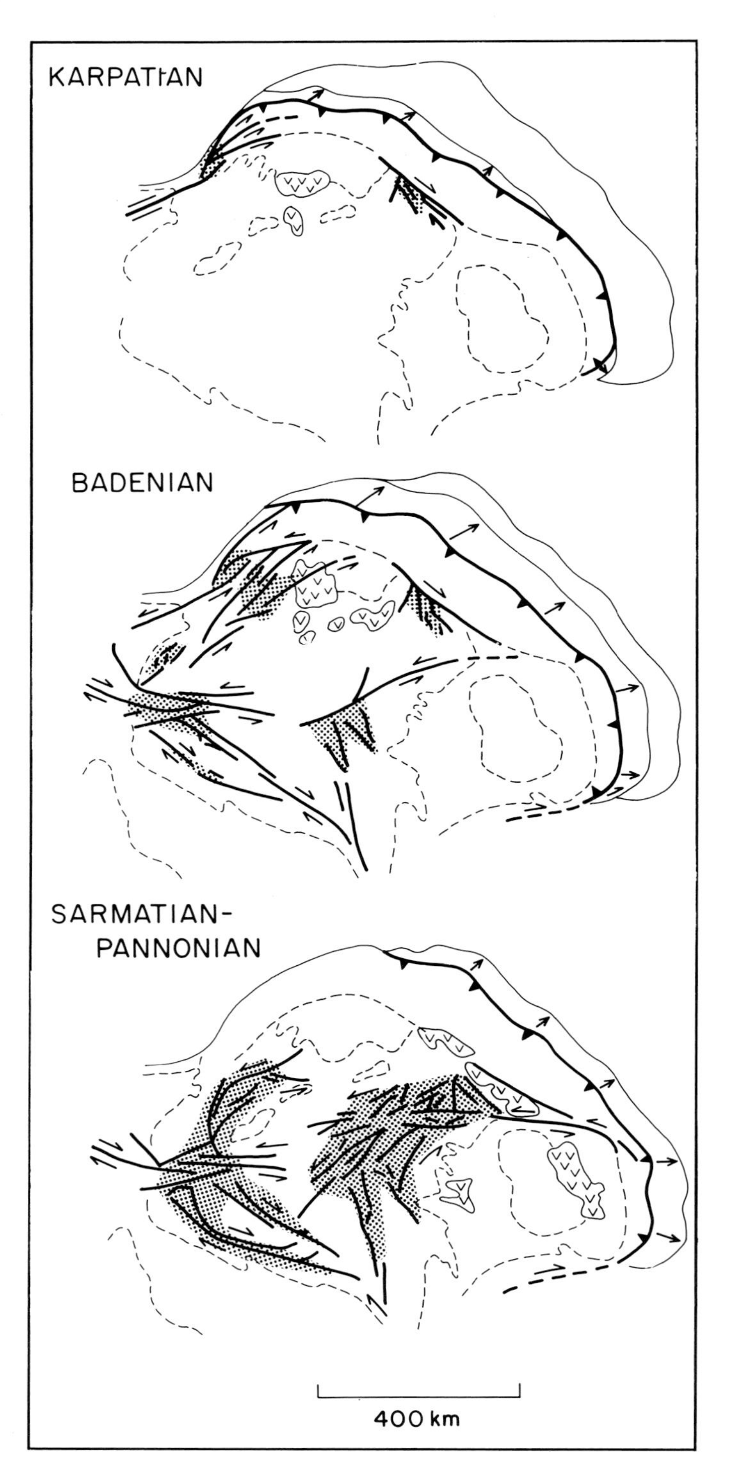

Figure 12. Palinspastic reconstruction of the Carpathian–Pannonian region during Karpatian time (~16.5–17.5 Ma), Badenian time (~13–16.5 Ma), and Sarmatian–Pannonian time (~8–13 Ma). Heavy lines with barbs indicate the restored position of the Carpathian thrust front at the beginning of each time stage. Arrows show the amount of closure in the thrust belt for each time stage. Lighter lines show the reconstructed position of the thrust front at the end of each time stage. The present position of the outermost thrust fault is shown in each diagram for reference. Areas of active extension are indicated by shading and heavy lines indicating the major shear zones inferred to link basin extension to the thrust belt. V's show the locations of the andesitic magma erupted during each time stage.

Nevertheless, note that the western limit of thrusting at the beginning of Karpatian time as constrained by data from the thrust belt is consistent with the western limit of the area of shortening in the thrust belt as calculated from closing up the basins.

During Badenian time (13–16.5 Ma) the locus of the extension moved southward to the southern Vienna and Transcarpathian basins, the west Danube lowland, the Zala basin, the central (and now deepest) part of the Great Hungarian Plain, and along the Sava and Drava troughs (Figures 6 and 12). According to this reconstruction, Badenian shortening occurred from the central part of the West Carpathians eastward to the East Carpathian bend. The western limit of the area of active thrusting based on thrust belt data is similar to that obtained by calculated closure in the thrust belt due to basin extension. The total shortening calculated across the thrust belt in the East Carpathians was about 50 km in Badenian time.

By Sarmatian–Pannonian time (8–18 Ma), the areas of major extension had moved still farther south into the main part of the Danube lowland, the Zala basin, and the Pannonian basin. The Sava and Drava troughs, located along the continuation of the peri-Adriatic line, appear to have been still active during Sarmatian time (Figures 6 and 12). The location of closure within the thrust belt as predicted from reconstruction of the basins is in good agreement with that given by data for the thrust belt itself.

The palinspastic reconstructions of the Pannonian basin system allow one to compare thrust belt activity, as determined from data within the thrust belts, with the history of basin opening in the Pannonian system. Figures 5 and 12 show that the western limit of active thrusting in the West Carpathians through time was consistent with the coeval changes in the location of active basin extension. Additionally, the amount of Badenian–early Pannonian shortening (roughly 120 ± 60 km) across the East Carpathian thrust belt (as reconstructed in Figure 12) is consistent with direct reconstructions of shortening in the thrust belt, where Badenian and younger ages of deformation are known for the Tarcau, Marginal Folds, and Sub-Carpathian nappes (Sandulescu, this volume; see also Burchfiel, 1976). (A sequence of palinspastic reconstructions for the Carpathian thrust belt is shown by Oszczypko and Ślaczka (1985). Their reconstructions are for the same time periods as shown in Figure 12, but are constructed using only data from the thrust belt. They show a distribution of shortening in space and time along the thrust belt that is almost identical to that shown in Figure 12, although their reconstructions were made independently and were unknown to me at the time that Figure 12 was constructed.)

These observations suggest that pre-Karpatian convergence along the Carpathian mountains was probably associated with wholesale eastward or northward displacement of the Pannonian fragment, although considerable internal deformation of the fragment must have occurred by strike-slip faulting. By Badenian time, convergence began to be accommodated primarily by extension within the Pannonian fragment, creating deep extensional basins over much of the region. During this time period and later, basin extension and thrust belt activity were intimately related.

In each of the three reconstructions in Figure 12, significant extension (locally 50%) was required within the outer Carpathian thrust belt parallel to the strike of the belt. This occurred as units were thrust radially from a more internal to a more external position, so that the total length of the thrust belt increased along strike. Such lengthening of the belt may have been accommodated partly by steep strike-slip faults, such as along the Pienniny Klippen belt in the West Carpathians and the Audia(?) and Ceahlau nappes(?) in the East Carpathians. In other cases, thrust displacements on gently dipping detachment surfaces may have had a large component of strike-slip displacement, such as northeast of the Vienna basin and beneath the outer Carpathian nappes north of the Moesian platform.

DISCUSSION

The spatial and timing relationships between crustal shortening in the outer Carpathians and extension within the Pannonian basin system suggest that the two phenomena are related and that they are probably the result of a single dynamic system. One possible explanation is that both thrusting and extension were the result of continued subduction of the European plate, perhaps driven by negative buoyancy of the subducted slab. Because the position of the peri-Adriatic-Vardar fault system was roughly fixed with respect to Europe, subduction of European crust and lower lithosphere required corresponding extension of the crust and lower lithosphere of the Pannonian fragment (Figure 13). Thus, extension of the intra-Carpathian basins occurred as Pannonian lithosphere moved eastward to fill the space originally occupied by the European plate, while the western edge of the Pannonian fragment remained fixed with respect to the peri-Adriatic-Vardar fault system and the European foreland (see Figure 4). This geometry implies migration of the subduction zone toward the European foreland, possibly accompanied by steepening of the subducted slab, because seismic data suggest that it is now in a nearly vertical position (Fuchs et al., 1979).

Analysis of the Carpathian foredeep in the southeastern Carpathian area provides results consistent with this interpretation. Royden and Karner (1984) showed that the depth and geometry of this Miocene–Quaternary molasse basin, which reflects flexural loading of the European plate during nappe emplacement, are inconsistent with the load present within the Carpathian thrust sheets and the present low elevation of the Carpathian mountains. They calculate that an extra vertical load is required to generate the present foredeep basin geometry and that the magnitude of this load is equivalent to the negative buoyancy of a subducted slab with a density contrast of $+0.1$ g/cm^3 and a cross-sectional area perpendicular to the thrust belt of 3000 km^2 (incorrectly given as 1500 km^2 in their original paper). Thus, negative buoyancy of the subducted slab (or dynamic effects) can explain the continued subduction of the European plate in middle–late Miocene time, the generation of coeval back arc extension, and the anomalously deep foredeep basin of the southeast Carpathians (Figure 13).

This interpretation also offers a mechanism for subsidence of the nonextensional Transylvanian basin, which has been interpreted as the result of temporary dynamic loading from below during subduction in late Miocene time. Loading may have occurred as this basin was "dragged down" by the sinking of the European plate into the asthenosphere beneath the basin, and thus may also be the indirect result of gravity acting on the downgoing slab (Figure 13). Pliocene and Quaternary uplift of the Transylvanian basin and adjacent mountains can be inter-

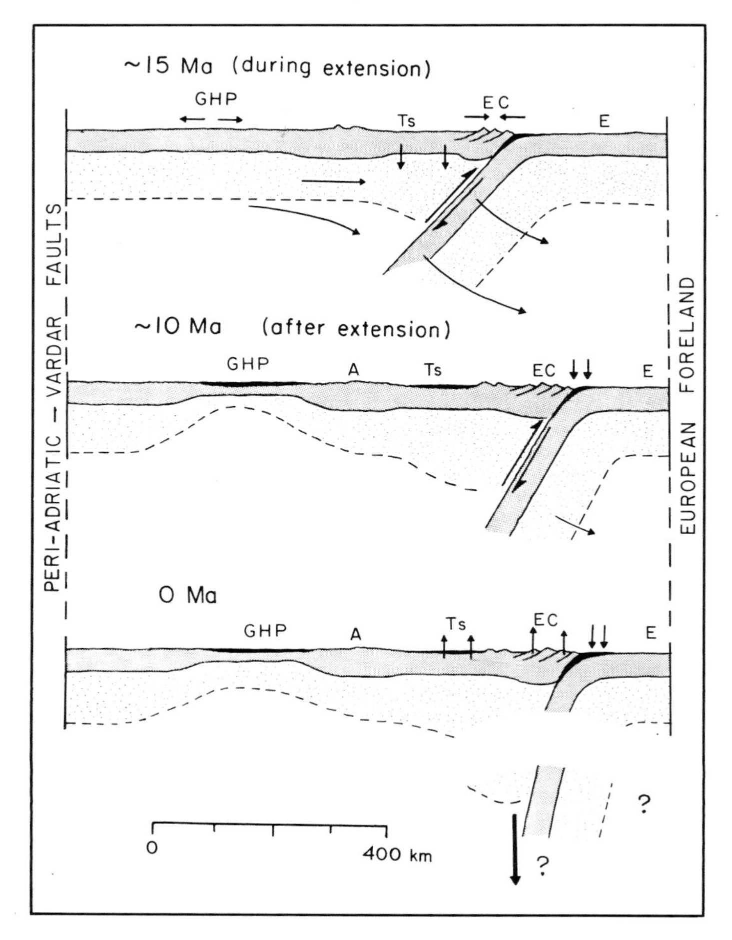

Figure 13. Schematic east–west section across the Pannonian fragment showing how continued subduction of the European plate could have resulted in extension of the Pannonian fragment. The European foreland and the peri-Adriatic-Vardar fault system are considered fixed relative to one another. During subduction, eastward movement is induced in the overriding plate to fill the space formerly occupied by the subducted plate. Note that subduction also implies migration of the subduction zone toward the European foreland. Some steepening of the subducted slab has also been shown. Big arrows show sense of shear between European plate and Pannonian fragment. E, Europe; EC, East Carpathians; Ts, Transylvanian basin; A, Apuseni Mountains; GHP, Great Hungarian Plain. Shaded area represents crustal material, dotted area represents subcrustal lithosphere, and black area represents sedimentary cover within the Neogene basins and in the outer flysch Carpathians.

preted as the result of unloading as the subducted slab (and possibly thickened lithosphere beneath the basin) became detached or decoupled from the overlying plate. Perhaps this scenario can explain why the age of early basin subsidence within the Pannonian and Transylvanian basins are similar, although the mechanisms are very different.

CONCLUSIONS

The Pannonian basin system can be most simply described as a middle Miocene back arc type extensional basin that opened behind the coeval Carpathian thrust belt. This basin–thrust belt pair can also be viewed as the result of contemporaneous events within adjacent parts of the Alpine chain. In particular, eastward displacement of the Pannonian fragment(s) in early Miocene time occurred as a product of continental escape of the Pannonian fragment(s) eastward away from the collision zone in the Eastern Alps, thus initiating Miocene thrusting and subduction along the outer Carpathian chain. By middle Miocene time, wholesale eastward displacement of the Pannonian fragment was apparently inhibited by changes in the directions and location of deformation outside the Pannonian region. From this time onward, continued convergence across the East Carpathians was accommodated by east–west extension of the Pannonian fragment.

This simple large-scale description of the Carpathian–Pannonian system can easily encompass many of the detailed space and time relationships between thrust belt activity and basin evolution without fundamental modification. Indeed, understanding of such large-scale processes is enhanced by many of these more detailed relationships. For example, east-west middle–late Miocene extension of the Pannonian fragment does approximately compensate for the synchronous east–west convergence in the Carpathian thrust belt, but, in detail, extension occurred in small basins connected by a roughly conjugate set of strike-slip faults. These same strike-slip faults also connected the extending basins to the thrust belt and transformed the extension within the basins into shortening within the thrust belt. Moreover, extension was diachronous across the basin system, and this diachroniety appears to be directly linked to the eastward migration of thrusting (and the termination of thrusting) around the outer Carpathians. At depth, the style of extension beneath the basins is also a function of thrust belt geometry and of the location of each basin relative to the thrust belt. Extension appears to have involved the entire thickness of the overriding plate above the subduction boundary. Extension of basins located near the thrust belt appears to have involved only shallow upper crustal rocks above the thrust decollement, while extension of basins located far from the thrust belt involved extension of a thick overriding plate and involved rocks at great depths.

Several points emerge from the detailed nature of this basin-thrust belt interaction. First, the initiation of thrust belt activity was the result of events outside the Carpathian–Pannonian system. Second, the initiation of basin extension was the result of a combination of events, including (1) events outside the basin–thrust belt system that precluded continued wholesale convergence between the Pannonian fragment and Europe, and (2) events within the basin–thrust belt system that caused continued subduction of Europe below the Pannonian fragment even after wholesale convergence became impossible. Third, local events, such as extension of individual basins and their connection to particular parts of the thrust belt along a few strike-slip zones, must reflect local zones of weakness and the geometries of crustal blocks as well as larger scale processes acting at depth.

ACKNOWLEDGMENTS

This work was supported by NSF Grants #INT-7910275 and EAR-8115863, the Hungarian Academy of Sciences, and Shell Research and Development Co. Additional support was provided by PYI funds from NSF, with matching funds from Shell Companies Foundation, Inc.; Texaco, U.S.A.; Chevron Oil Fields Research Co.; Kerr-McGee Corp.; and Intel Corp. Particular thanks are due to Frank Horváth, Tina Freudenberger, and Clark Burchfiel. Pat McDowell drafted most of the figures.

REFERENCES CITED

Báldi, T., 1983, Mid-Tertiary stratigraphy and paleogeographic evolution of Hungary (Late Eocene through Early Miocene): Akademiai Kiado, Budapest, 293 p.

Birkenmajer, K., 1981, Strike-slip faulting in the Pieniny Klippen belt of Poland: Paper presented at 12th Congress, Carpatho-Balkan Geol. Assoc., Bucharest, Romania, Sept. 8–12, 1981.

Brix, F. and O. Schultz, eds., 1980, Erdöl und Erdgas in Österreich: Naturhistorisches Museum Wien and F. Berger Horn, Vienna, 312 p.

Burchfiel, B. C., 1976, Geology of Romania: Geol. Soc. Am. Spec. Publ. 158, 82 p.

Burchfiel, B. C., 1980, Eastern Alpine system and the Carpathian orocline as an example of collision tectonics: Tectonophys., v. 63, p. 31–62.

Burke, K., and A. M. C. Sengör, 1986, Tectonic escape in the evolution of the continental crust, *in* M. Barazangi and L. Brown, eds., Reflection Seismology: a global perpective: Geodynamics Series, v. 14, p. 41–53.

Ciupagea, D., M. Pauca, and Tr. Ichem, 1970, Geologia Depresiunii Transylvaniei: Editura Academiei Republicii Socialiste Romania, Bucharest, 256 p.

Filjak, R., Z. Pletikapić, D. Nikolić, and V. Askin, 1969, Geology of petroleum and natural gas from the Neocene complex and its basement in the southern part of the Pannonian basin, Yugoslavia, *in* P. Hepple, ed., The exploration of petroleum in Europe and North Africa: Institute of Petroleum, London, p. 113–130.

Fuchs, W., 1980, Die Molasse und ihr nichthelvetischer Vorlandantiel am Untergrund eisschiesslich der sedimente auf der Bohmischen Masse, *in* R. Oberhauser, ed., Der Geoligische aufbau Österreichs: New York, Springer-Verlag, p. 144–176.

Fuchs, K., P. Bonjer, G. Bock, D. Cornea, C. Radu, D. Enescu, D. Jiame, G. Nouescu, G. Merkler, T. Moldoveanu, and G. Tudorache, 1979, The Romanian earthquake of March 4, 1977: aftershocks and migration of seismic activity: Tectonophys., v. 53, p. 225–247.

Fusán, O., Kodym, O., Matejka, A., and L. Urbánek, 1967, The geological map of Czechoslovakia: Ustredni ustav geologický, Prague, scale 1:500,000, 2 sheets.

Grubić, A., 1980, Yugoslavia: an outline of Yugoslavian Geology: 26th International Geological Congress, Guidebook to Excursions 201A–202C, Paris, 97 p.

Horváth, F. and L. Royden, 1981, Mechanism for formation of the intra-Carpathian basins: a review: Earth Evol. Sci., v. 1, n. 3–4, p. 307–316.

Institut Géologique, 1967–1970, Geological map of Romania: Comité d'Etat pour la Géologie, Institut Géologique, Bucharest, scale 1:200,000, 50 sheets.

Jiricek, R., 1979, Tectogenetic development of the Carpathian arc in the Oligocene and Neogene, *in* M. Mahel, ed., Tectonic profiles through the West Carpathians: Geologický ústav D. Stúra, Bratislava, p. 205–214.

Kőrössy, L., 1964, Magyar köolaj-és földgáz elöfordulások [Oil and gas fields in Hungary]: Acta Geol. Acad. Sci. Hungaricae, v. 14, p. 421–429.

Kőrössy, L., 1970, Entwicklungsgeschichte der Neogenen Becken in Ungarn: Acta Geol. Acad. Sci. Hung., v. 14, p. 42–49.

Kőrössy, L., 1980, Neogén ösföldvajzi viszgálatok a Kárpátmedencében [Investigations into Neogene paleogeography in the Karpathian basin]: Földtani Közlöny: Bull. Hung. Geol. Soc., v. 110, p. 473–484.

Kőrössy, L., 1981, Regional geological profiles in the Pannonian basin: Earth Evol. Sci., v. 1, n. 3–4, p. 223–231.

Köves, S., and W. Krobot, 1980, Wiener Becken-Structurkarte oberkante Sarmat (map), *in* F. Brix and O. Schultz, eds., Erdöl und Erdgas in Österreich: Naturhistorisches Museum Wien und F. Berger, Horn, Vienna.

Ksiazkiewicz, M., and B. Lesko, 1959, In the relation between the Krosno- and Magura flysch: Bull. Acad. Polonaise Sci., Serie des Sciences Chimiques Géologie et Geographigue, v. 7, p. 773–780.

Laubscher, H., 1971, Das Alpen–Dinariden problem und die palinspastik der südlichen Tethys: Geol. Rundsch., v. 60, p. 813–833.

Laubscher, H., 1985, Large-scale, thin-skinned thrusting in the southern Alps: Kinematic models: Geol. Soc. Am. Bull., v. 96, p. 710–718.

Mahel, M., ed., 1973, Tectonic map of the Carpathian–Balkan mountain system and adjacent areas: D. Stúr Geological Institute, Bratislava and UNESCO, Vienna, scale 1:1,000,000.

Mahel, M., T. Buday, I. Cicha, O. Fusán, E. Hanzlíková, F. Chmelík, J. Kamenický, T. Koráb, M. Kuthan, A. Matejka, et al., 1968, Regional Geology of Czechoslovakia, Vol. II, The West Carpathians: Stuttgart, E. Schweizerbart'sche Verlagsbuchhandlung, 723 p.

McKenzie, D., 1972, Active tectonics of the Mediterranean region, Geophys. J. Royal Astron. Soc., v. 30, p. 109–185.

Mészáros, J., 1983, A Bakonyi vízsintes eltolódások szerkezeti és gazdaságföldtani jelentösége [Structural and economic–geologic significance of strike-slip faults in the Bakony Mountains]: Yearbook of the Hungarian Geological Survey for 1981, p. 485–502.

Molnar, P., and P. Tapponier, 1975, Effects of a continental collision: Science, v. 189, p. 419–426.

Némedi-Varga, Z., 1983, A Mecsek Hegység szerkezetalakvlása az alpi hegység képzödési ciklushan [Structural history of the Mecsek Mountains in the Alpine orogenic cycle]: Yearbook of the Rungarian Geological Survey for 1981, p. 485–502.

Oszczypko, N., 1973, Geology of the Nowy Sacz basin (the middle Carpathians): Bull. Geol. Inst., Warsaw, p. 101–197.

Oszczypko, N., 1982, Explanatory notes to lithotectonic molasse profiles of the Carpathian foredeep and in the Polish part of the Western Carpathians: Veröff. Zentralinst. Phys. Erde AdW DDR, Potsdam, v. 66, p. 95–115.

Oszczypko, N., and A. Ślaczka, 1979, Relations between flysch and molasse deposits in the northern Carpathians: Veröff. Zentralinst. Physik Erde AdW DDR, Potsdam, v. 58, p. 209–219.

Oszczypko, N. and A. Ślaczka, 1985, An attempt to palinspastic reconstruction of Neogene basins in the Carpathian foredeep: Ann. Soc. Geol. Pol., v. 55, p. 55–75.

Oszczypko, N., and A. Tomaś, 1985, Tectonic evolution of marginal part of the Polish flysch Carpathians in the middle Miocene: Kwartalnik Geologiczny, v. 29, p. 109–128.

Paraschiv, D., 1979, Oil and gas in Romania: Institutul de Geologie si Geofizica, Studii Tehnice si Economice, Series A, Bucharest, 382 p.

Pogácsás, Gy., 1984, Seismic stratigraphic features of Neogene sediments in the Pannonian Basin: Geophysical Transactions, Eötvös Loránd Geophysical Institute of Rungary, v. 30, p. 373–410.

Roman, C., 1970, Seismicity in Romania—evidence for the sinking lithosphere: Nature, v. 288, p. 1176–1178.

Roth, Z., 1980, Západni Karpaty-tercíerní struktura Stredni Evropy [West Carpathians—Tertiary structure of Central Europe]: Ustred. Ustav. Geol., v. 55, 128 p.

Royden, L. H., 1985, The Vienna basin: a thin-skinned pull-apart basin, *in* K. Biddle and N. Christie-Blick, eds.: SEPM Spec. Publ. No. 37, p. 319–338.

Royden, L., and G. D. Karner, 1984, Flexure of lithosphere beneath Apennine and Carpathian foredeep basins: evidence for an insufficient topographic load: AAPG Bull., v. 68, p. 704–712.

Royden, L., F. Horváth and B. C. Burchfiel, 1982, Transform faulting, extension and subduction in the Carpathian–Pannonian region: Geol. Soc. Am. Bull., v. 73, p. 717–725.
Royden, L., F. Horváth, and J. Rumpler, 1983, Evolution of the Pannonian basin system, 1. Tectonics: Tectonics, v. 2, p. 63–90.
Rudinec, R., 1978, Paleogeographical, lithofacial and tectogenetic development of the Neogene in eastern Slovakia and its relaion to volcanism and deep tectonics: Geol. Zborn.-Geol. Carpath., v. 29, p. 225–240.
Rudinec, R., C. Tomek and R. Jiricek, 1981, Sedimentary and structural evolution of the Transcarpathian depression, Earth Evol. Sci., v. 1, n. 3–4, p. 205–211.
Sandulescu, M., 1980, Analyse geotectonique des chaines alpines situées au tour de la mer noire occidentale: Ann. Inst. Geol. Geophys., v. 56, p. 5–54.
Sandulescu, M., 1984, Geotectonica Romaniei: Editura Tehni a, Bucharest, 336 p.
Sandulescu, M., H. G. Kräutner, I. Balintoni, D. Russo-Sandulescu, and M. Micu, 1981, The structure of the East Carpathians (Moldavia–Maramures area): 12th Carpatho-Balkan Congress, Guidebook to Excursion B1, Bucharest, Romania, 92 p.
Schönlaub, H. P., 1980, Das Palaozoikum der Karnischen Alpen, der Westkarawank-en und des Seeberger Aufbruchs, *in* R. Oberhauser, ed., Der Geologische aufbau Österreich: New York, Springer-Verlag, p. 429-447.
Sclater, J. G., L. Royden, F. Horváth, B. C. Burchfiel, S. Semken, and L. Stegena, 1980, The formation of the intra-Carpathian basins as determined from subsidence data: Earth Planet. Sci. Lett., v. 51, p. 139–162.
Spicka, V., 1972, Paleogeographie neogénu ceskoslovenských Západních Karpat [Neogene paleogeography of the Czechoslovakian West Carpathians]: Sborn. Geol. Ved Geol., v. 22, p. 65–108.
Stegena, L., B. Géczy, and F. Horváth, 1975, Late Cenozoic evolution of the Pannonian basin: Tectonophys., v. 26, p. 71–90.
Szepesházy, K., 1973, A Tiszántul északnyugati részenek felsökrétá es paleogén koru képzödményei [Upper Cretaceous and Paleogene formations of the NW Tiszantul, Hungary]: Akadémiae Kiadó, Budapest, 96 p.
Tomek, C., 1985, Deep seismic reflection lines (to 12 seconds) in Czechoslovakia: Presented at the NATO Advanced Study Institute on Tectonic Evolution of the Tethyan Regions, Istanbul, Turkey, Sept. 23–Oct. 2.
Tomek, C., and I. Ibrmajer, 1982, New seismic reflection profiles in the West Carpathians and peri-Pannonian basins (abs.), *in* Evolution of extensional basins within regions of compression, with emphasis on the intra-Carpathians: Workshop and Discussion Meeting, Veszprém, Hungary, June 20-26, p. 31.
Vucković, J., R. Filjak, and V. Aksin, 1959, Survey of exploration and production of oil in Yugoslavia: Fifth World Petroleum Congress, Section 1, New York, p. 1003–1021.
Wernicke, B. and B. C. Burchfiel, 1982, Modes of extensional tectonics: J. Struct. Geol., v. 4, p. 105–114.

CHAPTER 4

Neotectonic Behavior of the Alpine-Mediterranean Region

F. Horváth
Geophysical Department, Eötvös University,
1083 Budapest. Kun B. tér 2, Hungary

A review of seismicity, focal mechanisms, and neotectonic data for the Alpine–Mediterranean region shows that convergence occurs mainly along narrow, reasonably well-defined zones dominated by thrust and strike-slip deformation, while extension occurs over broad areas dominated by normal and strike-slip faulting. Three generalized tectonic settings can account for many aspects of thrust belts in the Mediterranean region, and thrust belts can be classed accordingly into three types. Type 1 belts are those that result directly from north–south convergence of Africa and Europe. Type 2 belts form during lateral escape of crustal material away from a type 1 thrust belt. Type 3 belts evolve in response to local processes at a preexisting subduction zone, usually of type 2. Type 3 belts are those that are associated with zones of extension and lithospheric thinning behind the thrust belt. In the progression from type 1 to type 3 settings, development may stop at any point. Thrust belts in each of these different settings are controlled by the overall convergence of Europe and Africa and by the arrangement of small continental fragments between the converging continents.

INTRODUCTION

Seismicity, focal mechanism solutions, and geologic data for recent deformation have been used to explain the neotectonics of the Pannonian basin and the surrounding mountain belts (Horváth, 1984; Gutdeutsch and Aric, this volume). These studies suggest that convergence of the Adriatic and European plates is a major factor controlling the interaction between crustal blocks in this region. Many areas in the Alpine–Mediterranean region offer examples of active continent–continent collision, related extension, and arc deformation within and along the edges of crustal fragments (Figure 1). The aim of this chapter is to examine this broader area in order to understand the relationship between plate convergence, regional compression, and development of extensional terranes.

WESTERN ALPS

A remarkably regular pattern of seismic stress in the Western Alps is well documented by a number of reliable fault plane solutions (Pavoni, 1975; Mayer-Rosa and Mueller, 1979; Godefroy, 1980). As in the Eastern Alps (Gutdeutsch and Aric, this volume), horizontal compressional stresses predominate and they are normal to the main structural trend of the Alps. P-axes are generally horizontal and change orientation progressively from east to west in concert with the sharp bending of the mountain chain (Figure 2). Extensional stresses are also usually horizontal, and therefore many of the active faults are strike-slip faults. The coincidence between P-axes and the orientation of postmolassic Neogene and Quaternary horizontal crustal shortening indicates that similar stress fields have prevailed during the last 5–10 m.y. (Pavoni and Mayer-Rosa, 1978). In a recent review of Alpine geodynamics, Miller et al. (1982) concluded that this stress field must be attributed to the push of the African plate against the Eurasian plate through the Adriatic "promontory."

APENNINES, CALABRIAN ARC, AND SICILY

The Alpine compressional regime terminates in the northernmost Apennines and in the Ligurian Sea, and the remainder of peninsular Italy is characterized by extension (Cagnetti et al., 1978; Gasparini et al., 1980). In the northern and central

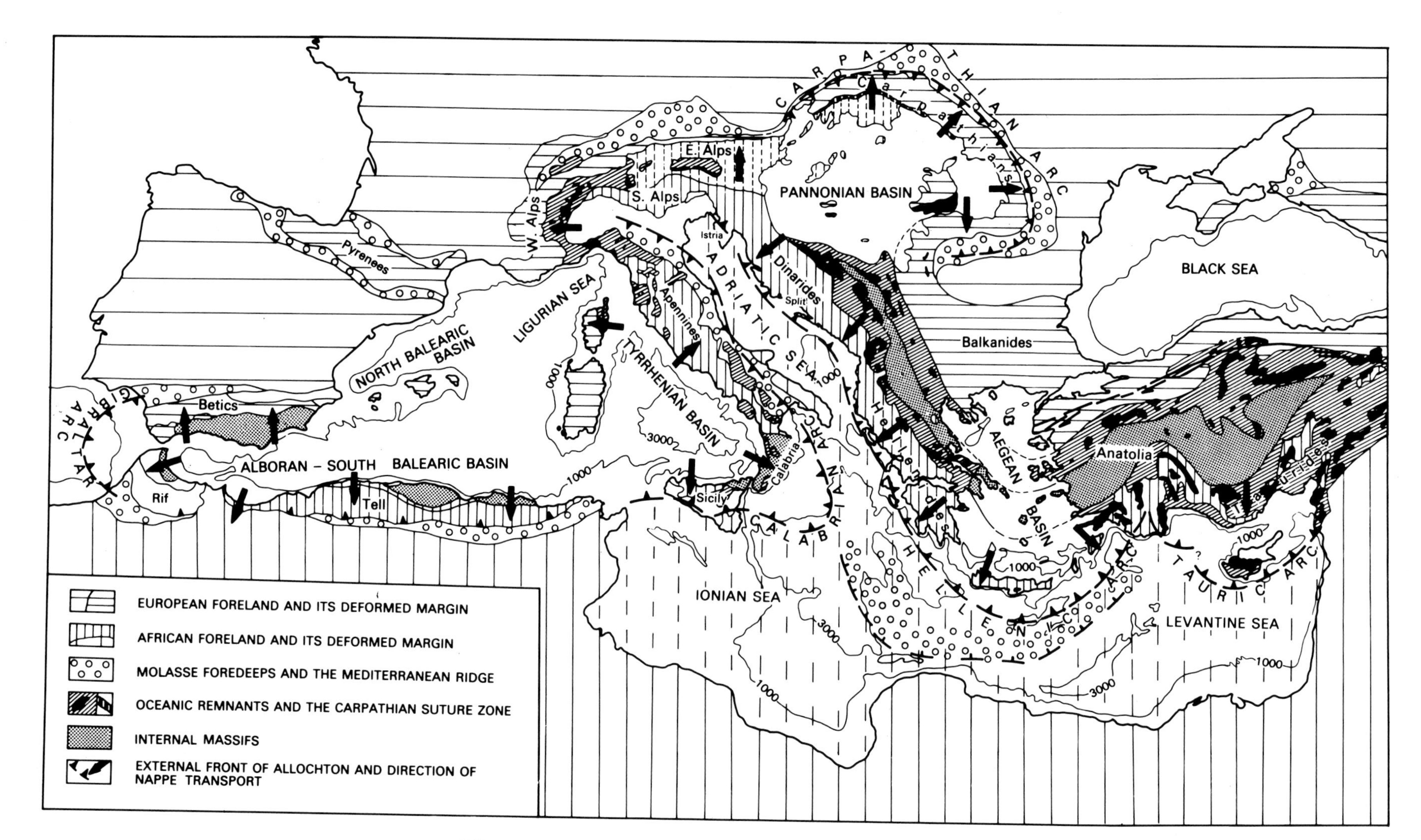

Figure 1. Tectonic scheme of the Alpine–Mediterranean region.

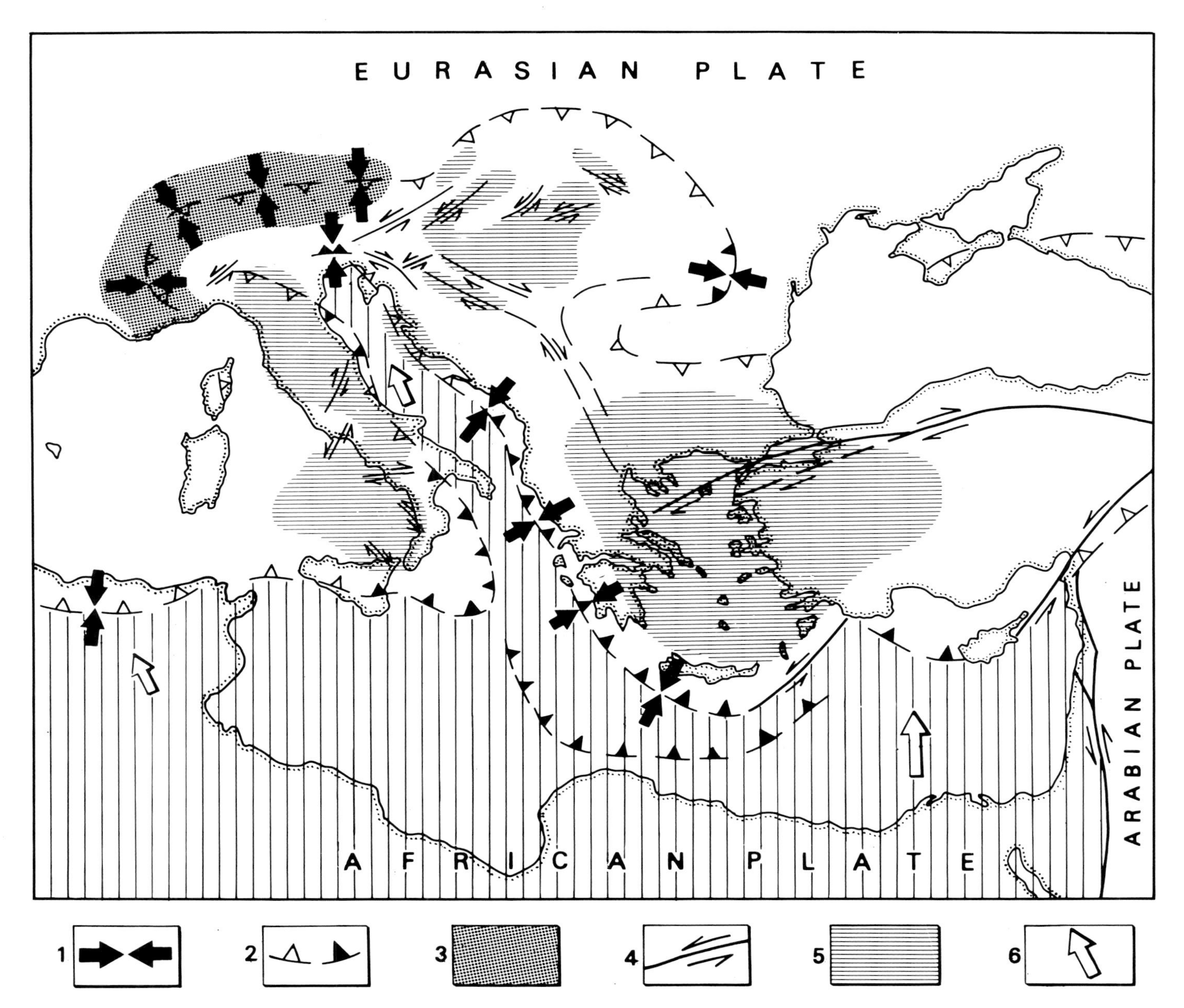

Figure 2. Neotectonic features and regional pattern of seismic compressional stresses in the Alpine–Mediterranean region. Compiled mostly after McKenzie (1972), Ritsema (1974), Pavoni and Mayer-Rosa (1978), Cagnetti et al. (1978), Mercier (1981), Horváth and Royden (1981), Fourniguet et al. (1981), Gasparini et al. (1982), and Udias (1982). Explanation: 1. Average direction of predominant horizontal compression. 2. External front of allochthon, inactive (open barb) and active (closed barb). 3. Compressional regime characterized by a complex system of transcurrent faults and folds. 4. Major transcurrent fault. 5. Areas of active extension and/or subsidence. 6. Motion vector of the African plate relative to Europe.

Apennines, P-axes are commonly subhorizontal and focal mechanisms vary from pure normal to pure strike-slip faulting. In the southern Apennines, earthquake mechanisms show predominantly normal faulting with subordinate strike-slip displacements and the T-axes are perpendicular to the main structural trend of the chain (Gasparini et al., 1982). The pattern of faulting is variable in the Calabrian arc and in Sicily. Strike-slip mechanisms predominate, but in the Messina strait dip-slip mechanisms are more significant. Thrust mechanisms in western Sicily indicate predominantly north–south compression, similar to the pattern observed in the Tellian Atlas of northern Africa (Udias, 1982). This compressional regime continues toward the east in the Ionian sea along the external Calabrian arc, as shown by seismic data (Barone et al., 1982).

Reflection seismic data demonstrate that active overriding of southern Calabria onto the Ionian bathyal plain has taken place from Tortonian time to the present (Barone et al., 1982) and is synchronous with the opening of the Tyrrhenian basin. Rotation of thrust sheets has resulted in progressive bending of the Calabrian arc (D'Argenio et al., 1980). Studies of the recent stress field and neotectonic features support this interpretation and suggest a crustal shear zone along which the Apennine and Calabrian nappe systems move toward the Ionian and Adriatic seas (Ghisetti et al., 1982).

TYRRHENIAN SEA

Recent relocation data and new fault plane solutions of the intermediate and deep earthquakes below the Tyrrhenian area have led to a better understanding of the geometry and mechanism of this seismic activity (Gasparini et al., 1982). The Benioff zone has a maximum length of 650 km and is sharply arcuate, discontinuous and changes dip markedly with depth. It is almost vertical down to 200 km. Most of the earthquakes occur in the slab where it dips about 50° between 230 and 340 km. A few shocks have been recorded from 400 to 480 km depth in a shallowly dipping part of the slab. Fault plane solutions indicate the predominance of down-dip compression in the central part of the slab and of strike-slip motion at its northern and southern ends.

The Tyrrhenian basin was the site of major extension during late Miocene and Pliocene time (Fabbri and Curzi, 1979). Extension appears to have migrated eastward from the center of the basin and is still active in mainland Italy. It is now generally accepted that the Tyrrhenian Sea has been formed by extension of an older collisional belt (Scandone, 1979; Horváth et al., 1981). Extension of the basin has been associated with successive outward migration of a compressional front.

DINARIC AND HELLENIC ALPS

There is a well-defined belt of shallow seismic activity along the Adriatic coast of the Dinarides (Figure 2). The frequency of earthquakes is particularly high to the south, where strong shocks (greater than magnitude 6 on the Richter scale) may also occur. The inferred stress pattern is remarkably regular: horizontal compression predominates with an orientation normal to the coastline and perpendicular to the main strike of the Dinarides and Hellenides (Ritsema, 1974; Cvijanovic and Prelogovic, 1977). Extensional stresses are locally also nearly horizontal, thus giving rise to some strike-slip faulting. Steeply dipping T-axes are more characteristic, and the consequent thrust faults have resulted in emplacement of the external Dinarides onto the Adriatic foreland, as shown by reflection seismic data (Dragasevic, 1975–1976). A recent seismological example is given by the fault plane solution of the destructive Montenegro earthquake on April 15, 1979 (Boore et al., 1981). The northern segment of the zone between the city of Split and the Istria peninsula may be tectonically more complex. Ritsema (1974) suggested that horizontal extensional stresses tend to prevail along the coast and offshore, in other words, T-axes plunge at a smaller angle than the P-axes. Skoko et al. (1977) argued, however, that the dominant focal mechanism is thrusting along faults steeply dipping toward the northeast.

AEGEAN SEA

The Aegean area is seismically the most active part of the Alpine–Mediterranean realm. Shallow to intermediate depth events occur, with the deeper earthquakes located toward the central Aegean. The earthquakes delineate an arcuate Benioff zone dipping from the Hellenic trench toward the center of the Aegean at an angle of about 35° and reaching a maximum depth of 180 km. Fault plane solutions of shallow earthquakes (Ritsema, 1974; Papazachos and Comninakis, 1977; McKenzie, 1978) show that the northeast-southwest-oriented horizontal compressive stress field of the Dinarides continues southward along the external Hellenides and the Hellenic trench. At the Pliny and Strabo trenches to the south of Crete, P-axes tend to change azimuth to north–south and then to northwest–southeast and the T-axes become horizontal. Both the focal mechanisms and morphological features suggest that this is a sinistral transform-dominated system (Le Pichon and Angelier, 1981). The rest of the Aegean region, including most of mainland Greece, western Anatolia, and the southern Balkans is dominated by a broad extensional regime characterized by predominant north–south horizontal extension (Figure 2).

A detailed analysis of focal mechanism data and the geometry of Neogene–Quaternary graben systems led Le Pichon and Angelier (1981) to propose that the present extensional regime has been much the same since late middle Miocene time (13 Ma). Some workers, however, have interpreted field relationships to indicate that two compressional events recurred within this extensional period (e.g., Mercier, 1981). Jackson et al. (1982) argued that these apparently compressional features are not regional in extent and may not be truly compressional in origin. In places, for example, high-angle normal faults change orientation to appear as reverse faults during tilting and rotation of blocks along listric normal faults.

TURKEY

East–west horizontal extension in Turkey is accompanied by north–south horizontal compression, which results in right lateral strike-slip with some thrust component along the North Anatolian fault (Figure 2). Westward escape of Anatolia from the Arabia–Eurasia collisional zone started sometime between Burdigalian (early Miocene) and Pliocene time, and the total displacement is about 90 km in the eastern sector of the North Anatolian fault (Sengör and Canitez, 1982). In the west, no boundaries can be defined for Anatolia and its westward motion is apparently relieved by north–south extension over the broad Aegean region. Using several kinematic constraints, Le Pichon and Angelier (1981) showed that the motion of the Hellenic arc toward Africa since Serravallian (middle Miocene) time can be described (as a first approximation) by a rotation of 30° about a pole situated near 40°N lat., 18°E long.

This results in 21 to 62% extension of the Aegean and the overriding of the originally less arcuate and shorter Hellenic arc onto the Ionian lithosphere with a velocity of about 4 cm/year.

IONIAN AND LEVONTINE SEAS

There is no seismic evidence for an active plate boundary between the African plate and the Adriatic plate (or fragment). Recent surface wave studies suggest that there is continental crust under the Ionian and Levontine seas (Cloetingh et al., 1980; Farrugia and Panza, 1981).

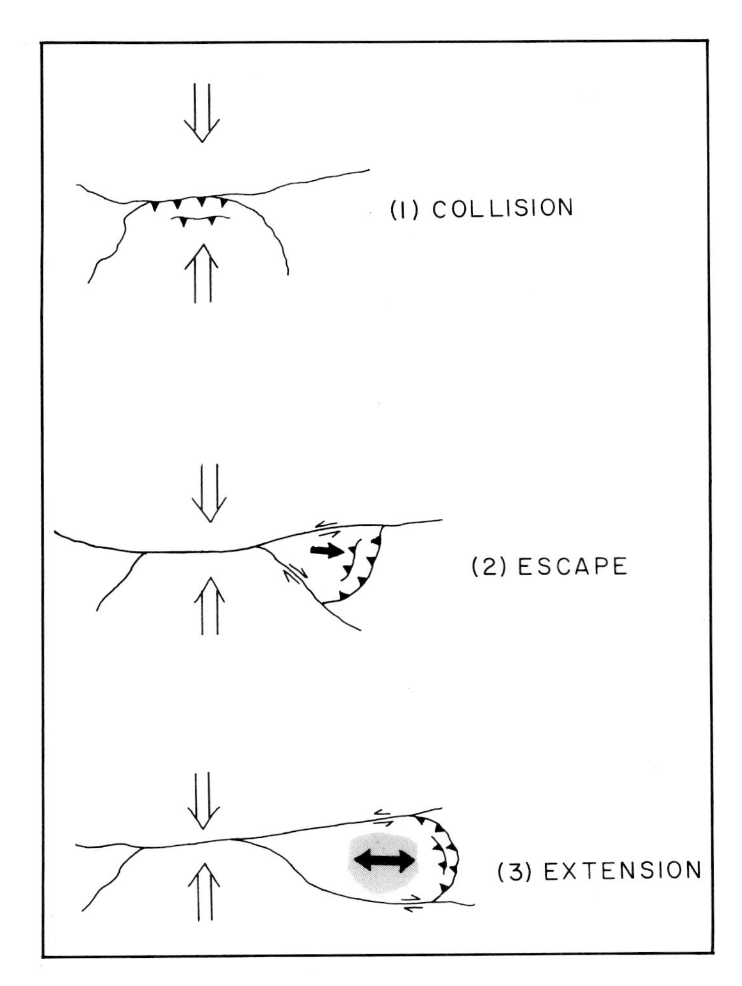

Figure 3. Cartoon showing the three types or stages of thrust belt development outlined in the text. 1. Fragment collision (or convergence) that directly accommodates convergence between Europe and Africa. 2. Thrust belt development due to lateral escape of crustal material away from a type 1 convergent zone. 3. Development of an extensional zone behind the advancing thrust belt following (or partly during) lateral crustal escape. This developmental sequence may end at any of the stages shown.

DISCUSSION

Seismic focal mechanism solutions combined with neotectonic data suggest a coherent pattern for the distribution of convergent zones and related zones of extension. Convergence seems to occur along narrow, reasonably well-defined zones dominated by thrust and strike-slip deformation, while extension occurs over broad areas dominated by normal and strike-slip faulting. This pattern of deformation can be explained partly as the direct result of northward movement of Africa, Arabia, and the Adriatic plate relative to Europe and partly as the result of deformation along the edges of small crustal fragments caught between the converging continents. In this chapter I propose three different tectonic settings for zones of convergence and crustal shortening in the Mediterranean region.

Type 1

The first type of convergent belt is represented by those belts that form in direct response to north–south convergence between Europe and Africa (Figure 3). These convergent belts may be thought of as zones of major plate boundary interaction, although smaller crustal fragments between the two major plates may be involved in the deformation. Active examples include the southern Dinaric and Hellenic Alps, the Southern Alps, and the thrust belts of eastern Turkey. Inactive (or semiactive) examples include the Central to Western Alps and northern Dinaric Alps.

Type 2

The second type of convergent belt is represented by belts that form as the result of lateral escape of crustal material away from the type 1 collision belts just described above (Figure 3) (see also Burke and Sengör, 1986). These are distinguished from type 1 belts because they do not accommodate north–south convergence of Africa and Europe directly. Instead, shortening in these belts facilitates convergence across the type 1 belts, and thus they are only indirectly controlled by the convergence of Europe and Africa. There is no active example in the Mediterranean region. The submarine fold and thrust belt of the Hellenic arc, which is thought to have formed to allow westward escape of crustal material away from the collision of Arabia and Europe along the Bitlis suture (McKenzie, 1972), initiated as a type 2 convergent belt. Inactive examples include early stages in development of the outer (flysch) Carpathian thrust belt and probably early stages in development of the Apennine–Sicilian belt.

Type 3

The third type of convergent belt is associated with major extension and crustal thinning behind the active thrust belt (Figure 3). These belts appear to evolve commonly from type 2 thrust belts and only rarely, if at all, from type 1 belts. Thrusting within these belts does not seem to occur in response to convergence of Africa and Europe, but rather seems to be largely the result of local processes acting near the preexisting subduction zone (see below). The submarine thrust and fold belt of the Hellenic arc is an active example (probably transitional between type 2 and 3), while the late stages of the outer Carpathian belt and the Apennine–Sicilian belt are inactive examples.

Conclusions

In the preceding paragraphs, I have outlined three different settings for thrust belts within the Mediterranean region. I do not mean to imply that these belts are necessarily different from one another structurally, but only that they may be classed according to their tectonic setting. These belts form a developmental sequence (from type 1 to type 2 to type 3), although development can stop at any point in the progression. Conceptually, it is easiest to understand the development of type 1 and type 2 belts because in these belts crustal shortening and subduction accommodate convergence of plates or crustal fragments across the belts. Although type 3 belts may also serve to accommodate some convergence, the presence of an extending zone behind the belt implies that plate convergence is not driving the shortening process within these belts.

In agreement with Le Pichon and Angelier (1981), I propose that thrusting in type 3 belts and the related extension are controlled by progressive downward bending and retreat of the subducted slab (see also Horváth et al., 1981; Horváth and Berckhemer, 1982; Royden et al., 1983). This tectonic setting, however, does not preclude flow in the asthenosphere (i.e., active rifting) (Keen, 1985), which may have contributed to initiating and maintaining lithospheric thinning in the Alpine–Mediterranean region. Furthermore, downward bending of the subducted slab does not necessarily imply that the dip of the slab increases; if trench retreat and sinking of the slab occur at appropriate rates, the dip of the slab can easily be maintained as constant. For example, the steep dip of the Benioff zones beneath the Tyrhennian Sea and the East Carpathians suggests that the slab dip increased during subduction (Roman, 1970; Gasparini et al., 1982). In contrast, the shallow (35 °) dip of the Benioff zone that dips under the Aegean from the Hellenic trench suggests that trench retreat has probably kept pace with sinking of the slab, thus the slab dip need never have been less than currently observed.

Extension behind these type 3 thrust belts is an areal expansion of the lithosphere and occurs to accommodate subduction along the thrust belt or to accommodate part of the subduction. (Some subduction may be accommodated partly by fragment convergence as well.) For example, the net present-day convergence between Europe and Africa is smaller than the velocity of the Hellenic and Calabrian arcs relative to Africa (Minster and Jordan, 1978; Ritsema, 1979; Le Pichon and Angelier, 1981). The subducted slab is overridden by the front of the extending terrane, giving rise to a narrow external belt of thrusting and folding.

REFERENCES

Barone, A., A. Fabbri, S. Rossi, and R. Sartori, 1982, Geological structure and evolution of the marine areas adjacent to the Calabrian arc: Earth Evol. Sci., v. 2, n. 3, p. 207–221.

Boore, D. M., J. D. Sims, H. Kanamori, and S. Harding, 1981, The Montenegro, Yugoslavia earthquake of April 15, 1979: source orientation and strength: Phys. Earth Planet. Int., v. 27, p. 133–142.

Burke, K., and A. M. C. Sengör, 1986, Tectonic escape in the evolution of the continental crust, *in* M. Barazangi, and L. Brown, eds., Reflection seismology: a global perspective: Geodynamics Series, v. 14, p. 41–53.

Cagnetti, V., V. Pasquale, and S. Polinari, 1978, Fault-plane solutions and stress regime in Italy and adjacent regions: Tectonophysics, v. 46, p. 239–250.

Cloetingh, S., G. Nolet, and R. Wortel, 1980, Crustal structure of the eastern Mediterranean inferred from Rayleigh wave dispersion: Earth Planet. Sci. Let., v. 51, p. 336–342.

Cvijanovic, D., and E. Prelogovic, 1977, Seismicity and neotectonic movements of the Croatian region (SFR Yugoslavia): Publ. Inst. Geophys. Polish Acad. Sci., v. A-5, n. 116, p. 281–290.

D'Argenio, B., F. Horváth, and J. E. T. Channell, 1980, Paleotectonic evolution of Adria, the African promontory, *in* Geology of Alpine chains, born of the Tethys: Publ. 26th Intern. Geol. Congress, Paris, C5, BRGM Mem. v. 115, p. 331–351.

Dragasevic, T., 1975–1976, Results of geophysical exploration and tectonic structure of the southeastern part of the Adriatic Sea: Vesnik. Geofizika, Beograd, v. 16/17, p. 37–59.

Fabbri, A., and P. Curzi, 1979, The Messinian of the Tyrrhenian Sea: seismic evidence and dynamic implication, Giorn. Geol., v. 43, n. 1, p. 215–248.

Farrugia, P., and G. F. Panza, 1981, Continental character of the lithosphere beneath the Ionian sea, *in* R. Cassinis, ed., The solution of the inverse problem in geophysical interpretation: New York, Plenum, p. 327–334.

Fourniguet, J., J. Vogt, and C. Weber, 1981, Seismicity and recent crustal movements in France: Tectonophysics, v. 71, p. 195–216.

Gasparini, C., G. Iannaccone, and R. Scarpa, 1980, On the focal mechanism of Italian earthquakes: Rock Mech. Suppl., v. 9, p. 85–91.

Gasparini, C., G. Iannaccone, P. Scandone, and R. Scarpa, 1982, Seismotectonics of the Calabrian arc: Tectonophysics, v. 84, p. 267–286.

Ghisetti, F., R. Scarpa, and L. Vezzani, 1982, Seismic activity, deep structures and deformation processes in the Calabrian arc, southern Italy, Earth Evol. Sci., v. 2, n. 3, p. 248–260.

Godefroy, P., 1980, Apport des mecanismes au foyer a l'étude sismotectonique de la France: un exemple de distribution des contraintes en domain intraplaque: BRGM Service Geol. Nat. 032 GEG, Orlean, p. 1–62.

Horváth, F., 1984, Neotectonics of the Pannonian basin and the surrounding mountain belts: Alps, Carpathians and Dinarides: Annal. Geophys., v. 2, n. 2, p. 147–154.

Horváth, F. and H. Berckhemer, 1982, Mediterranean backarc basins, *in* H. Berckhemer and K. Hsü, eds., Alpine–Mediterranean geodynamics: Amer. Geophys. Union and Geol. Soc. Amer. Geodyn. Ser. v. 7, p. 141–173.

Horváth, F. and L. Royden, 1981, Mechanism for the formation of the intra-Carpathian basins: a review: Earth Evol. Sci., v. 1, n. 3, p. 307–316.

Horváth, F., H. Berckhemer, and L. Stegena, 1981, Models of Mediterranean back-arc basin formation, Philos. Trans. R. Soc. London Ser. A, v. 300, p. 383–402.

Jackson, J. A., G. King, and C. Vita-Finzi, 1982, The neotectonics of the Aegean: an alternative view: Earth Planet. Sci. Lett., v. 61, p. 303–318.

Keen, C. E., 1985, The dynamics of rifting: deformation of the lithosphere by active and passive driving forces: Geophys. J. R. Astron. Soc., v. 80, p. 95–120.

Le Pichon, X., and J. Angelier, 1981, The Aegean sea: Philos. Trans. R. Soc. London Ser. A, v. 300, p. 357–372.

Mayer-Rosa, D., and St. Mueller, 1979, Studies of seismicity and selected focal mechanism in Switzerland: Schweiz. Mineral. Petrogr. Mitt., v. 59, p. 127–132.

McKenzie, D., 1972, Active tectonics of the Mediterranean region: Geophys. J. R. Astron. Soc., v. 30, p. 109–185.

McKenzie, D., 1978, Active tectonics of the Alpine–Himalayan belt: the Aegean sea and surrounding regions: Geophys. J. R. Astron., Soc., v. 55, p. 217–254.

Mercier, J. L., 1981, Extensional–compressional tectonics associated with the Aegean arc: comparison with the Andean Cordillera of south Peru–north Bolivia. Philos. Trans. R. Soc. London Ser. A, v. 300, p. 337–355.

Miller, H., St. Mueller, and G. Perrier, 1982, Structure and dynamics of the Alps: a geophysical inventory, *in* H. Berckhemer and K. Hsü, eds., Alpine–Mediterranean Geodynamics: Amer. Geophys. Union and Geol. Soc. Amer. Geodyn. Ser. v. 7, p. 175–203.

Minster, J. B., and T. H. Jordan, 1978, Present-day plate motions: J. Geophys. Res., v. 83, p. 5331–5354.
Papazachos, B. C., and P. E. Comninakis, 1977, Modes of lithospheric interaction in the Aegean area, *in* B. Biju-Duval and L. Montadert, eds., Structural history of the Mediterranean basins: Edit. Technip. Paris, p. 319–332.
Pavoni, N., 1975, Zur Seismotektonik des Westalpenbogens: Vermess. Photogramm. Kulturtechn, v. 3-4, p. 185–187.
Pavoni, N. and D. Mayer-Rosa, 1978, Seismotektonische Karte der Schweiz 1:750,000: Ecol. Geol. Helv., v. 71, n. 2, p. 293–295.
Ritsema, A. R., 1974, The earthquake mechanisms of the Balkan region, Survey of the seismicity of the Balkan region: UNDP Project, Roy. Neth. Met. Inst. De Bilt Sci. Rep. 74-4, p. 1–36.
Ritsema, A. R., 1979, Active or passive subduction at the Calabrian arc: Geol. Mijnb., v. 58, n. 2, p. 127–134.
Roman, C., 1970, Seismicity in Romania—evidence for the sinking lithosphere: Nature, v. 228, p. 1176–1178.
Royden, L., F. Horváth, and J. Rumpler, 1983, Evolution of the Pannonian basin system: 1. Tectonics: Tectonics, v. 2, n. 1, p. 63–90.
Scandone, P., 1979, Origin of the Tyrrhenian sea and Calabrian arc: Boll. Soc. Geol. It., v. 98, p. 27–34.
Sengör, A. M. C., and N. Canitez, 1982, The North Anatolian fault, *in* H. Berckhemer and K. Hsü, eds., Alpine–Mediterranean Geodynamics: Amer. Geophys. Union and Geol. Soc. Amer. Geodyn. Ser., v. 7, p. 205–216.
Skoko, D., D. Cvijanovic, and E. Prelogovic, 1977, Seismic activity based on neotectonics, *in* Proc. Symp. Analysis of Seismicity and Seismic Risk, Liblice, 17–22 Oct., 1977: Publ. Czechoslov. Acad. Sci., p. 81–91.
Udias, A., 1982, Seismicity and seismotectonic stress field in the Alpine–Mediterranean region, *in* H. Berckhemer and K. Hsü, eds., Alpine–Mediterranean Geodynamics: Amer. Geophys. Union and Geol. Soc. Amer. Geodyn. Ser., v. 7, p. 75–82.

CHAPTER 5

Neogene Sedimentation in Hungary

I. Bérczi,[1] G. Hámor,[2] Á. Jámbor,[2] and K. Szentgyörgyi[1]
[1]Hungarian Hydrocarbon Institute, 2443 Százhalombatta, POB 32, Hungary
[2]Hungarian Geological Institute, 1142 Budapest, Nepstadion út 14., Hungary

The last major cycle of sedimentation in the Pannonian basin began in early Miocene time and has continued to the present. Stratigraphically, the Miocene section is most complete in northern and southern Hungary where the section is mainly continuous. In the deep basin areas, sedimentation began with deposition of variegated terrigenous deposits (up to about 30 m thick) overlain by clastic and calcareous marine sedimentary rocks. In early Badenian time and later, creation of topographic relief led to the development of small emergent islands, where no deposition occurred, surrounded by areas receiving clastic sediments. These islands were gradually covered by transgressive sedimentary sequences, with sediments being deposited on the last island in early Pannonian time. In the deep basins, sedimentation was mainly continuous from Badenian to Quaternary time. In places a gap is present in the sedimentary record because of extensive redeposition by turbidity currents in Sarmatian time. In the deep Neogene basins, five major depositional environments can be distinguished through time.

INTRODUCTION

The late Tertiary to Quaternary sedimentary basins of Hungary are situated in the central part of the Carpathian–Pannonian region (Maps 1 and 2). Miocene to Quaternary sedimentary rocks locally attain a total thickness of 7000 to 8000 m. The thickness of pre-Pannonian Miocene age rocks varies from 1000 to 3000 m, including volcaniclastic complexes of considerable thickness. These sediments were deposited in deeply subsided zones separated by horsts of the Paleozoic–Mesozoic basement complex. In northern Hungary, uninterrupted sedimentation can be observed between the Oligocene and Miocene, while elsewhere the Mesozoic, Paleozoic, and Precambrian basement complex is directly overlain by middle Miocene and younger sediments.

In the Central Paratethys region, the Oligocene–Miocene boundary is spanned by the Egerian biostratigraphic stage (Steininger et al., this volume). The Egerian sediments of the Pannonian basin should be grouped sedimentologically with the Oligocene; they form the last regressive phase of Oligocene sedimentation.

Localities cited in this chapter are shown in Figure 1 and Map 1.

EGGENBURGIAN

One transgressive–regressive sedimentary cycle occurred during Eggenburgian time (19–23 Ma) (Nagymarosy, 1981). Early Miocene (Savian) folding and faulting in southwestern Hungary (Sava folds) caused a massive influx of terrigenous sediments into the Eggenburgian basins. These rocks now comprise the Szászvár Formation and the upper part of the Csatka Formation in southwestern Hungary (Figures 1 and 2) and consist of coarse-grained fluvial sediments of continental derivation and variegated shales (Figures 3 and 4). In northern Hungary, a marine transgression from the north or north-northwest followed a late Egerian regression. Nearshore facies consist of conglomerate,

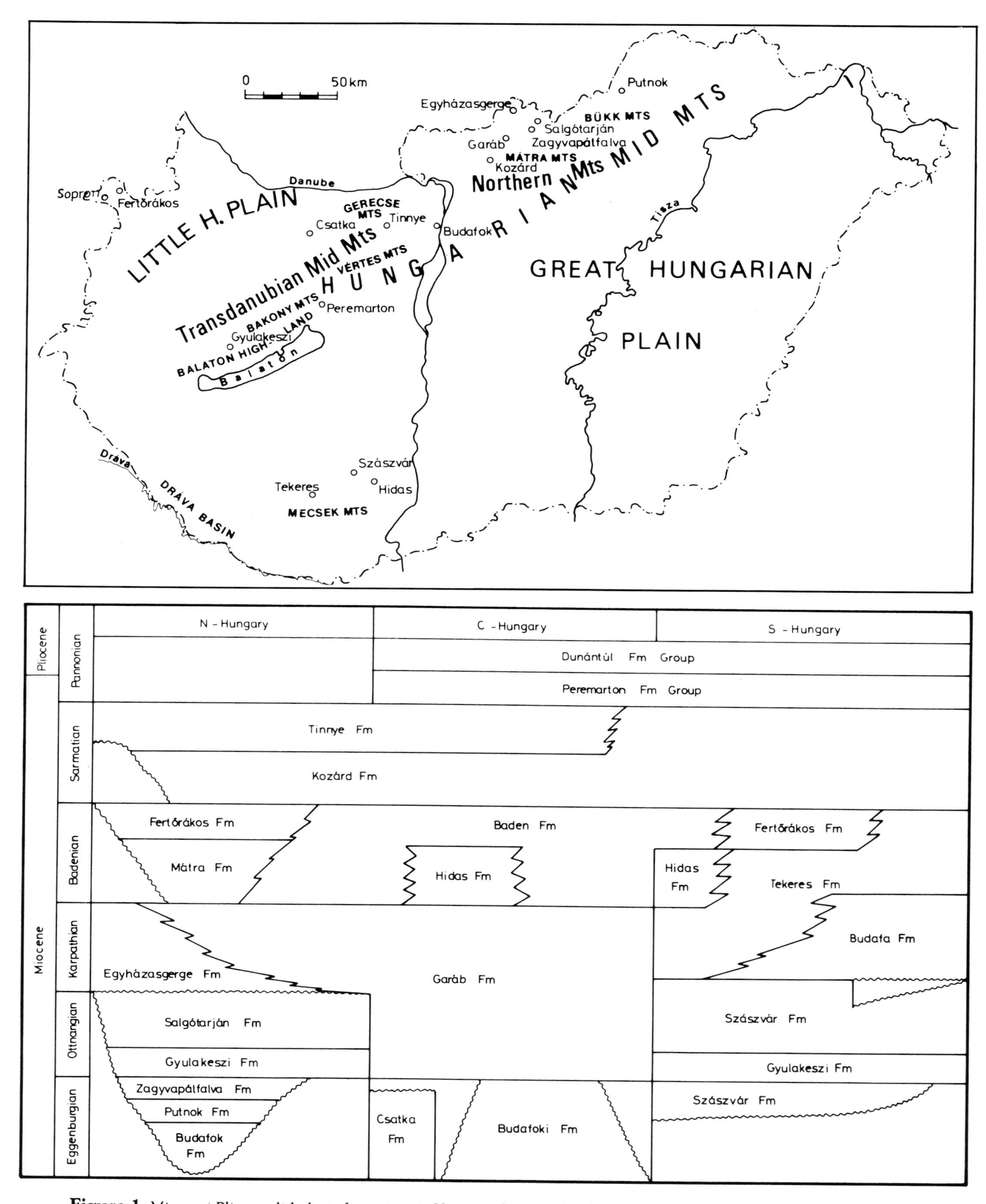

Figure 1. Miocene–Pliocene lithologic formations in Hungary, their type localities, and other geographic names used in the text.

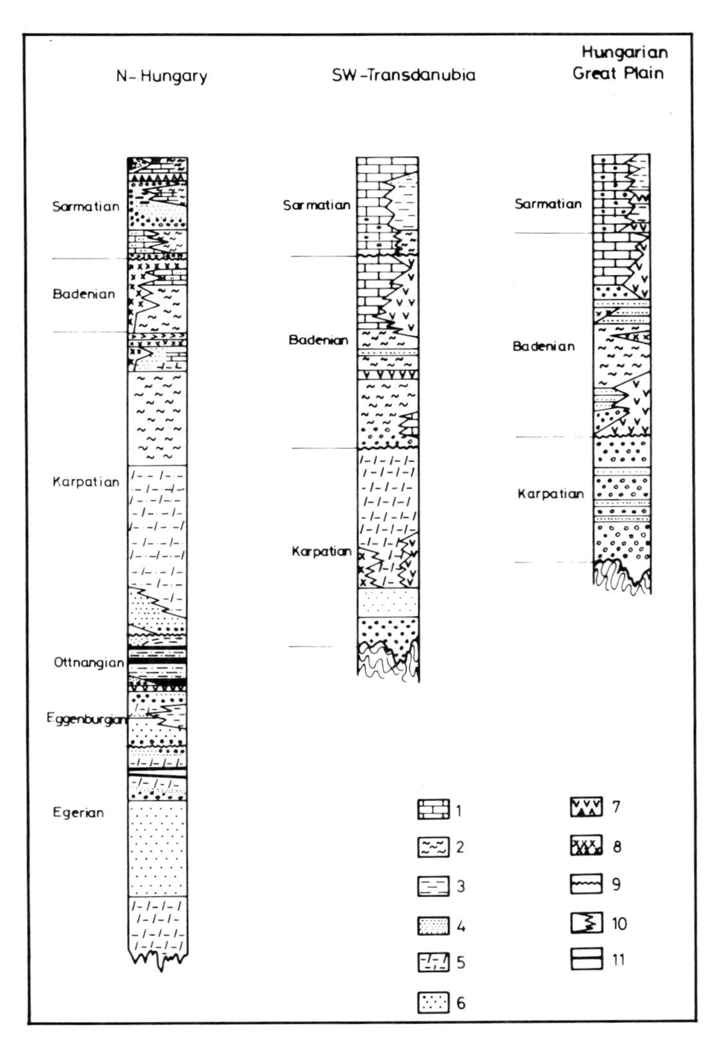

Figure 2. Principal types of pre-Pannonian Miocene Formations. Symbols: (1) limestone, (2) marl, (3) clay and silt, (4) sandstone, (5) schlieren, (6) pebble (conglomerate), (7) tuff, and, (8) andesite, rhyolite, (9) discordant bedding, (10) facies pinch-out, and (11) coal.

limestone (the Bretka beds), and sand and sandstone containing "big pectinids" (the Budafok Formation) (Figure 1). Shallow marine glauconitic sandstone and amusium-bearing schlieren (the Putnok Formation, Zone NN2) were deposited in the center of these northern basins, and lignite was deposited in locally developed lagoons. At the top of the Eggenburgian, the marine sequence is overlain by regressive deltaic–fluvial pebbly and sandy beds, which are in turn overlain by variegated clays (the Zagyvapálfalva Formation).

OTTNANGIAN

Sedimentary rocks deposited during the Ottnangian transgressive–regressive cycle have a geographic distribution similar to that of the Eggenburgian sedimentary rocks (Figure 5). Early Miocene folding and faulting in southwestern Hungary produced a series of uplifted and subsided blocks. A rhyolite tuff horizon about 100–200 m thick and dated at 19.6 ± 1.5 Ma by K-Ar was deposited during Ottnangian volcanic activity (Hámor et al., 1979). This tuffaceous unit is known as the Gyulakeszi Formation or the "lower rhyolite tuff" (Figure 1).

The oldest Ottnangian beds in north Hungary are pebbly variegated terrigenous rocks overlain by sandstone and a shale sequence with limnic and paralic limnic lignite seams (the Salgótarján Lignite Formation).

At this time the area of sedimentation increased greatly in western Hungary (Transdanubia). The lower rhyolite tuff is overlain by a 500–600-m thick fining-upward sequence with fluvial beds in its uppermost part. North of the Mecsek Mountains (southern Hungary) coarse-grained clastic sediments with intermittent lignite seams are exposed. This coarse-grained clastic sedimentary unit can be traced eastward (without the lignite seams) to the southwesternmost corner of the Great Hungarian Plain. On the southeastern rim of the Transdanubian Central Mountains (Várpalota–Bántapuszta) an isolated patch of marine rocks without any known paleogeographic connection is locally present. In the westernmost part of the country (Sopron, Brennbergbanya) there are limnic lignite beds.

In general, at the basin margins there is a well-defined unconformity at the Ottnangian–Karpatian boundary. Within the basins, Ottnangian coarse-grained clastic rocks are conformably overlain by Karpatian age sedimentary rocks.

The paleogeographic pattern of the Eggenburgian and Ottnangian is dominated by northwest-trending structures and paleogeographic connections. The faunal assemblages are characterized by both Atlantic and Indo-Pacific faunas. Paleogeographic connections existed between the Molasse foreland of the Northern Alps, the Vienna basin, northeastern Hungary, Transylvania, and the Caucasus.

KARPATIAN

The Karpatian sequence is also composed of a transgressive–regressive cycle, but the paleogeography changed dramatically from Ottnangian time. As a result of the sudden subsidence of the Dinaric foreland in the Adriatic (associated with the Styrian orogeny), a direct connection was established to the Mediterranean (Hámor and Szentgyörgyi, 1981). Evaluation of macro- and microfossils indicates that the direction of the Karpatian regional transgression was from the southwest instead of from the north (Figure 6). Marine conditions progressed from the depressions of the Pannonian basin to the Carpathians, which were being uplifted at that time.

The oldest beds of the Karpatian are of brackish water origin and contain *Congeria* and *Oncophora* (*Rzehakia*). In littoral environments, 400 m of sandstone and conglomerate were deposited (Budafa and Egyházasgerge formations). In the neritic zones, schlieren (clay and siltstone forming the Tekeres and Garáb formations) was deposited. West of the Danube (in Transdanubia), the basal Karpatian is a transgressive coarse-grained series with intercalations of schlieren and variegated, brackish, lignitiferous, and fish scale–bearing lagoonal rocks (Figure 6). The total thickness of the sequence is 600–1000 m. Lagoonal environments became dominant by the end of Karpatian time.

In southern Hungary (Mecsek Mountains and the southern part of the Great Hungarian Plain), Ottnangian rocks are overlain by onlapping clastic rocks that were deposited in brackish water. In the inner parts of the basin, Ottnangian rocks are over-

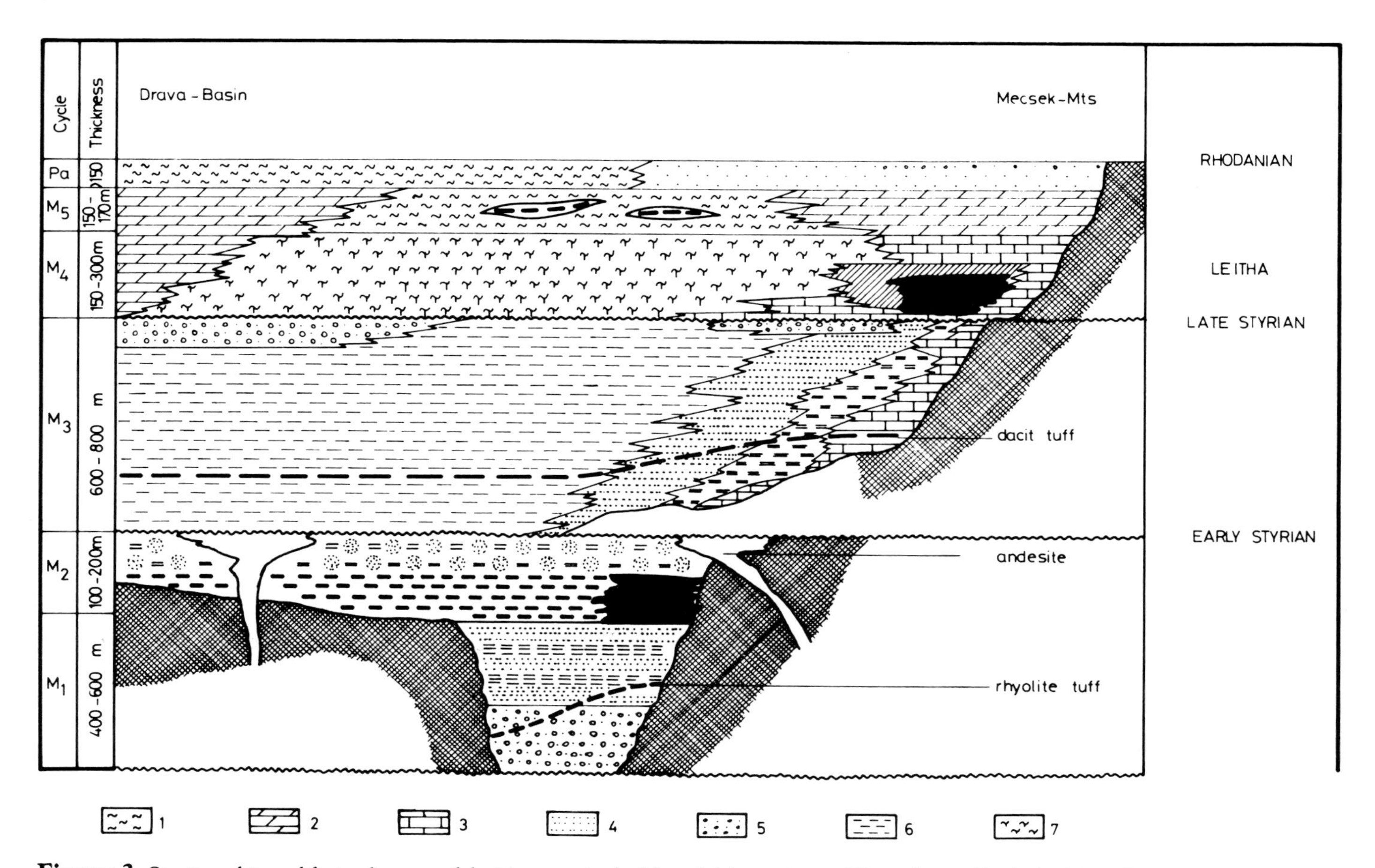

Figure 3. Stratigraphic and facies diagram of the Neogene in the Mecsek Mountains and Drava basin. Symbols: (1) marl, (2) dolomite, (3) limestone, (4) sand and sandstone, (5) sandy conglomerate and pebbly sandstone, (6) shale, and (7) calcareous marl. M_1, Eggenburgian; M_2, Ottnangian; M_3, Karpatian; M_4, Badenian; M_5, Sarmatian; Pa, Pannonian (s. l.).

lain by rhythmic molassic beds with a gradual transition to schlieren in the central part of the basin. Recently recognized Karpatian sedimentary rocks in the southeastern corner of the Great Hungarian Plain are mostly schlieren with a few interbeds of littoral and lagoonal rocks.

Southeast of the Transdanubian Central Mountains, Karpatian sediments were mostly deposited in shallow water (with barnacle- and bryozoa-bearing pebbly sandstone, and *Congeria* limestone deposited along the basin margins). Northward the Karpatian transgressive sequence begins with *Congeria*- and *Oncophora*-bearing beds that are overlain by *Chlamys*-bearing sandstone and its pelagic counterparts.

EARLY BADENIAN

The sedimentary formations of the early Badenian are products of one transgressive–regressive cycle. They are onlapping and have a completely different faunal content from the older sedimentary rocks. Early Badenian sedimentation was controlled by the differential subsidence of the basement, which is reflected by considerable lateral variation in facies (Figure 7). The basement in some areas subsided drastically, greatly increasing the areas and depth of marine deposition. The "Leitha limestone," a coarsely crystalline, sandy–pebbly, biogenic limestone of littoral and shallow water origin, was deposited along the margins of the subsided area. The "Baden clay" was deposited in the hemipelagic zones (Zone NN5), and interfingers with the limestone facies toward the basin margins. Because of a restricted terrigenous influx, the deposits that make up the oldest part of the transgressive sequence are conglomerates along the basin margin but sandstone in the central basin area.

Subsidence and uplift of the basement resulted in formation of brackish water lagoons where lignite accumulated along the basin margin (the Hidas Lignite Formation in the Mecsek and Bakony mountains). In the central parts of the sedimentary basins, the end of the regression was accompanied by deposition of a lithothamnium-bearing sandstone that blankets the older formations.

LATE BADENIAN AND SARMATIAN

The late Badenian and Sarmatian sequence also represents one transgressive–regressive sedimentary cycle. Uplift of the

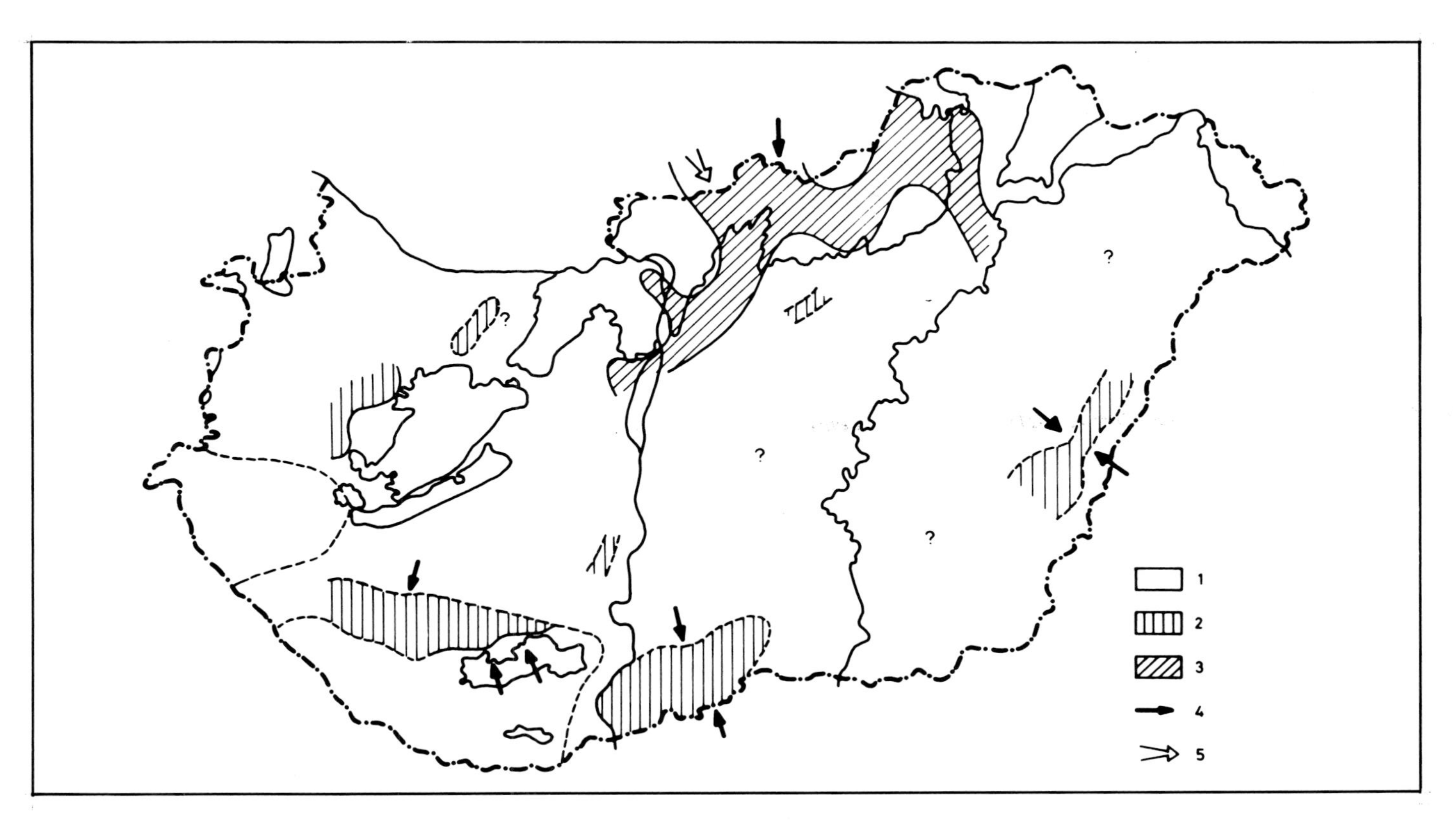

Figure 4. Paleogeographic sketch map of the Eggenburgian sequence. Symbols: (1) dry land, (2) continental formations, (3) marine formations, (4) direction of the transport of detritus, and (5) direction of transgression.

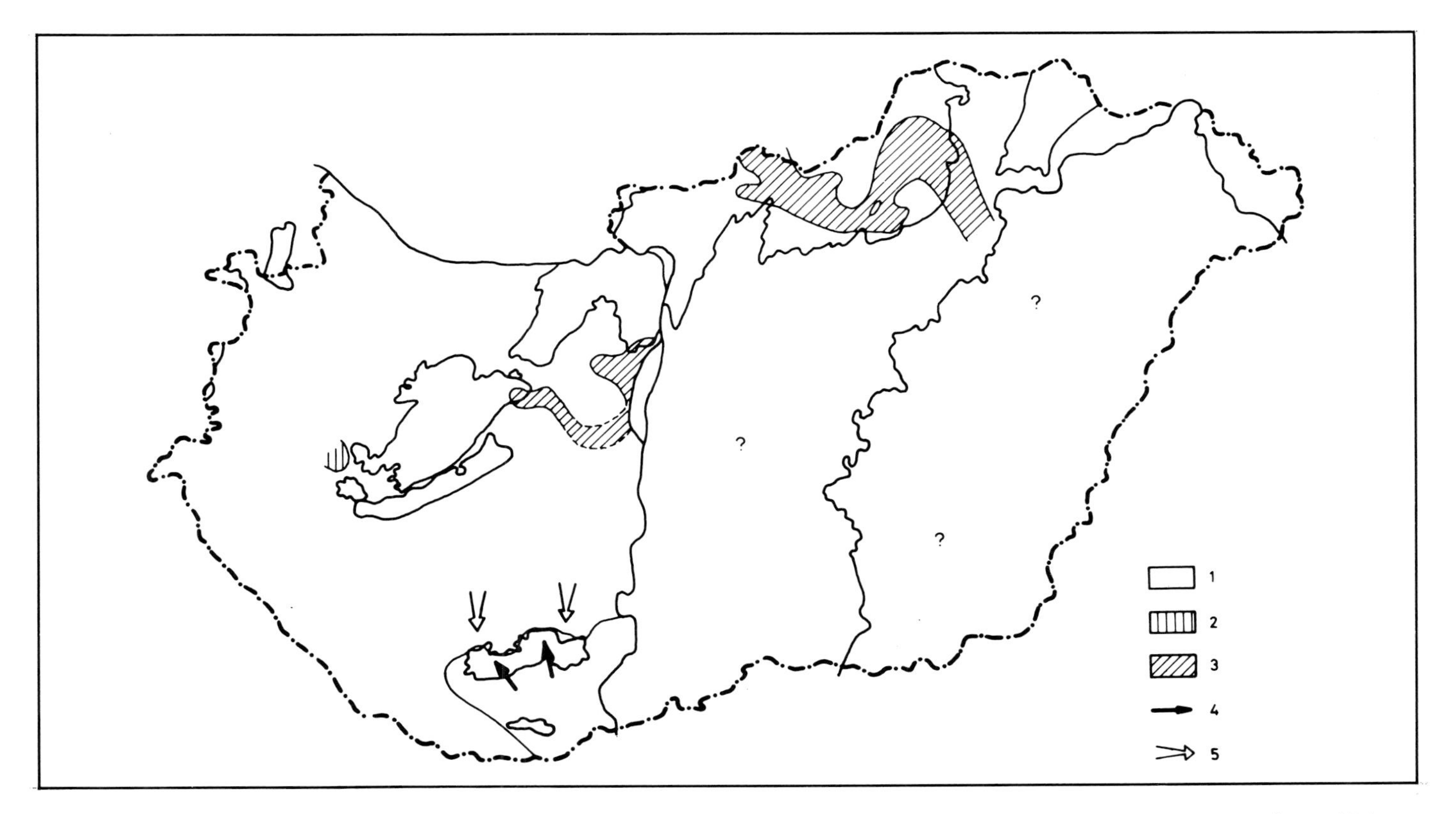

Figure 5. Paleogeographic sketch map of the Ottnangian sequence. Symbols: (1) dry land, (2) continental formations, (3) marine facies, (4) direction of the transport of detritus, and (5) direction of transgression.

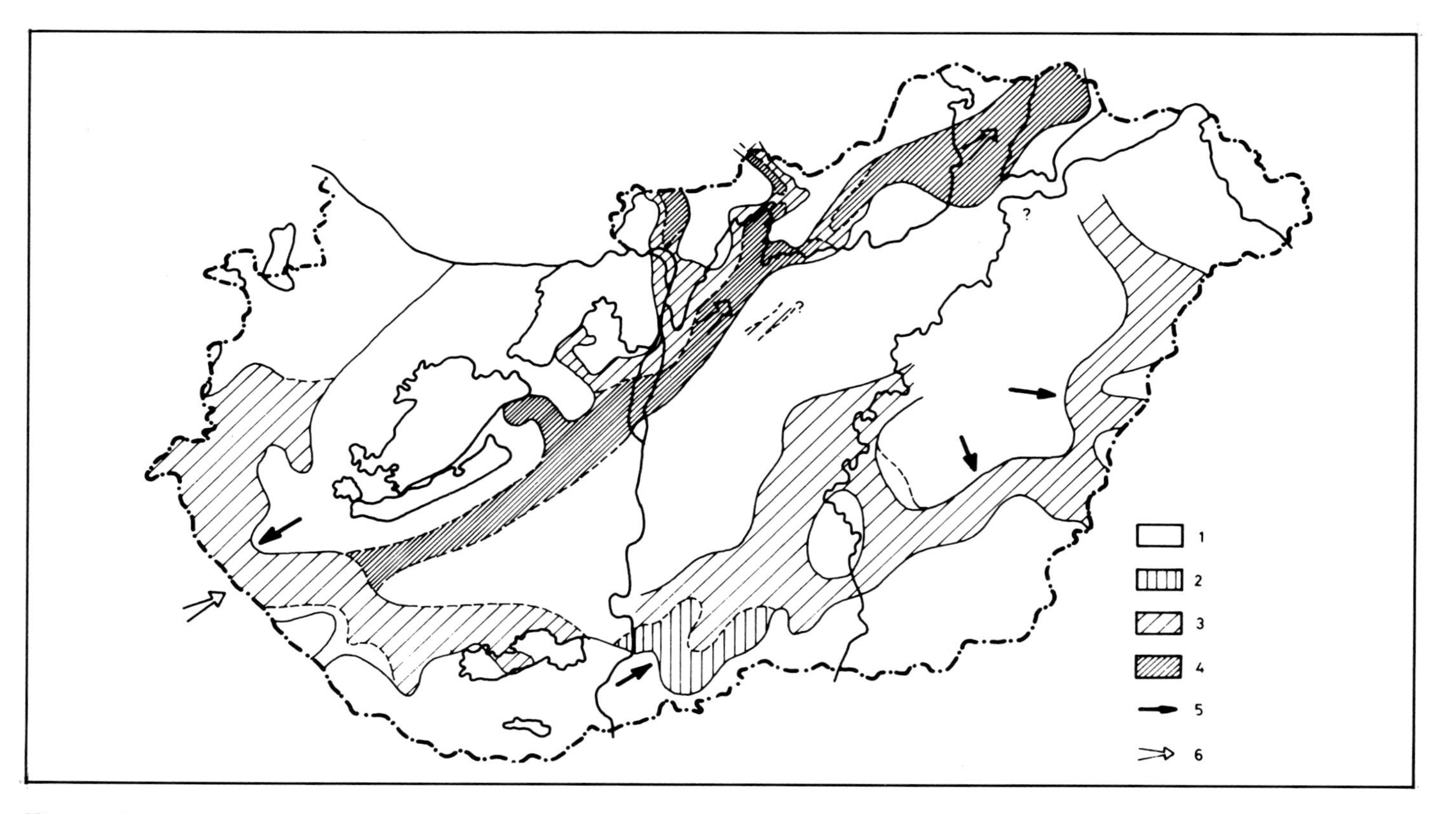

Figure 6. Paleogeographic sketch map of the Karpatian sequence. Symbols: (1) dry land, (2) continental formations, (3) near shore facies, (4) hemipelagic facies, (5) direction of the transport of detritus, and (6) direction of transgression.

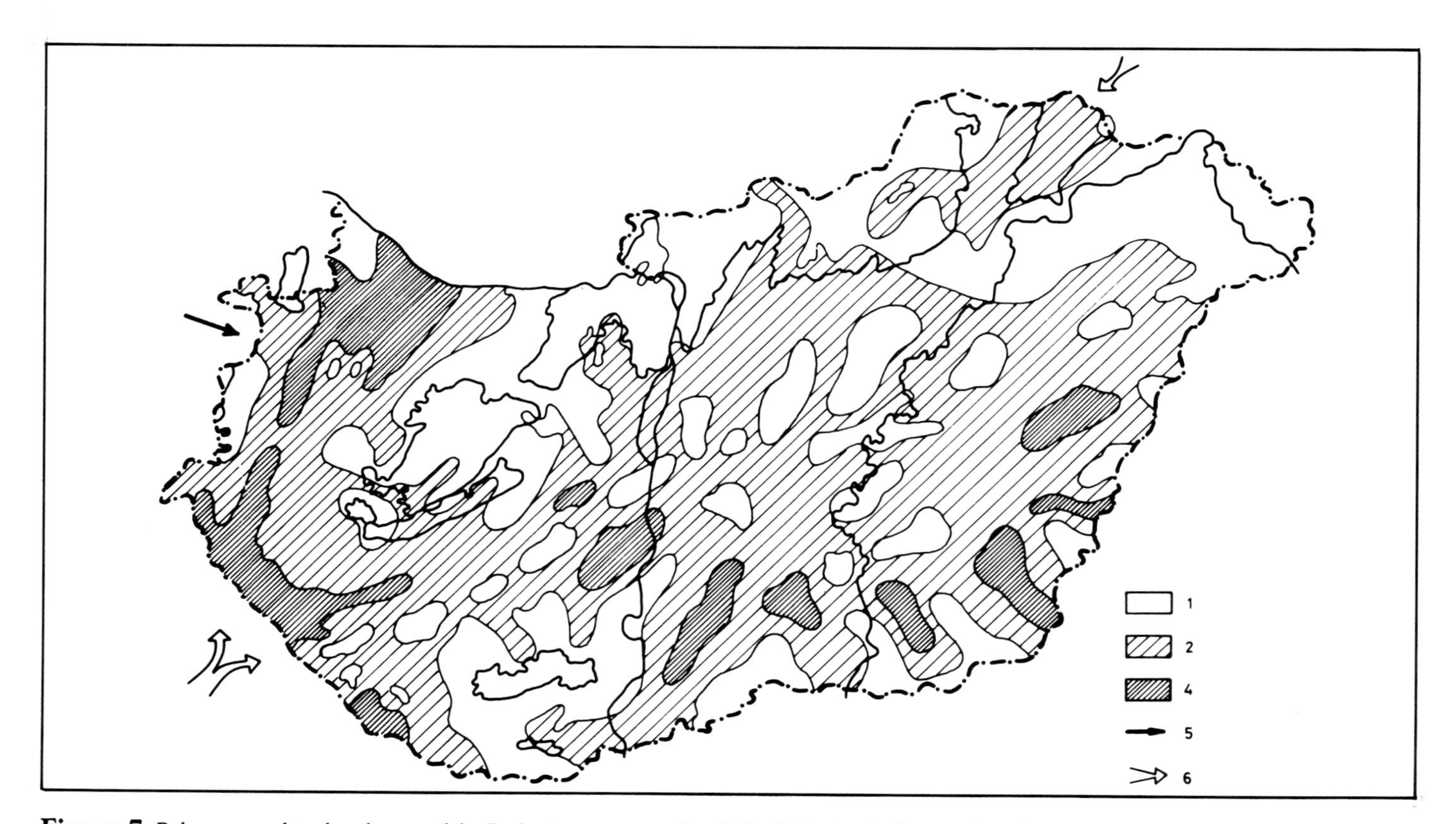

Figure 7. Paleogeographic sketch map of the Badenian sequence. Symbols: (1) dry land, (2) near shore facies, (3) hemipelagic facies, (4) direction of the transport of detritus, and (5) direction of transgression.

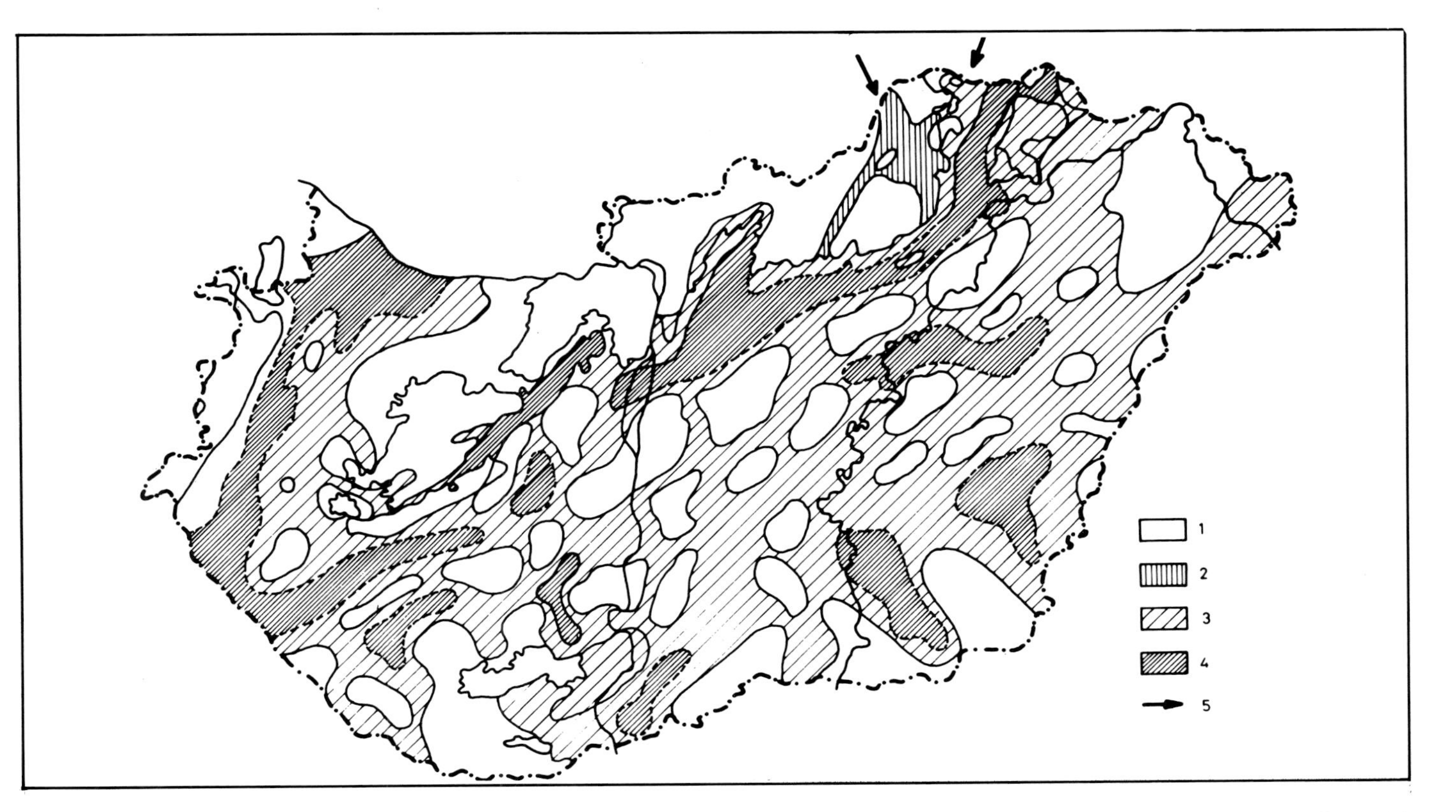

Figure 8. Paleogeographic sketch map of the Sarmatian sequence. Symbols: (1) dry land, (2) continental formations, (3) shallow water facies, (4) neritic facies, (5) direction of the transport of detritus.

Alps and Carpathian Mountains occurred in middle to late Badenian time, and the Mediterranean marine connection was closed permanently to the southwest. At the same time, a new marine connection opened toward the Aralo–Caspian basin in the southeast.

An increase in the rate of subsidence in the Pannonian basin resulted in deposition of a transgressive sequence (the Fertörákos Formation) that unconformably overlies the lower Badenian beds at the edges of the basin. After a slight regression, deposition resumed in the depression areas with deposition of a foraminifer-bearing argillaceous marl sequence 300–500 m thick containing upwardly increasing reef intercalations. This paleogeography and pattern of sedimentation persisted into early Sarmatian time. In several parts of southeastern Hungary (Ásotthalom, Csanádalberti, and Sarkadkeresztur) there are onlapping deposits (Figure 8).

The Sarmatian regression is suggested by a decrease in water depth, the predominance of littoral facies rocks, and the appearance of lagoonal and halite- and gypsum-bearing deposits. A characteristic sedimentary deposit is brackish water, coarsely crystalline limestone that shows a gradual transition to silty and shaly rocks in the areas of greater water depth. The thickness of Sarmatian beds exceeds 300 m only in areas where the rate of subsidence accelerated dramatically in late Badenian time (the maximum thickness of the Sarmatian sequence in Hungary is 900 m). This occurred in a northwest-trending trough of supposed Sarmatian age south of the Hungarian Mid-Mountains. The pelitic sequence deposited in this trough is unfossiliferous and was assigned to the Sarmatian on sedimentological grounds. This trough is bounded by both normal and reverse faults.

The Sarmatian marine regression was the final regression in the central Paratethys, and marine (Mediterranean) faunal assemblages gave way to brackish Aralo–Caspian fauna in Sarmatian time. During Pannonian time, the faunas became endemic (congeria and melanopsis).

PANNONIAN (*S. L.*)

The Pannonian (*s. l.*) formations represent the next to last transgressive–regressive cycle of the Pannonian basin. These sediments were deposited from the end of Sarmatian time to Quaternary time (11.5–2.4 Ma). The stratigraphic correlation of this huge mass of sedimentary rocks has been made on the basis of lithostratigraphy and the presence of mollusks, ostracodes, and microflorae (Balázs et al., 1981).

In general, the lower Pannonian (*s. l.*) (the Peremarton Formation) has a fairly homogeneous lithological composition and is composed of sediments deposited in deep, isolated, brackish water sedimentary basins. The upper Pannonian (*s. l.*) (the Dunántúl Formation) shows a more variable lithological composition, especially in the upper part of the sequence where paludinal, fluvial and lacustrine interbeds become more common. The total thickness of the Pannonian (*s. l.*) sequence is highly variable. At the basin margins, its thickness is 100–600 m, while in areas of greater subsidence its thickness is 600–4500 m (Figure 9).

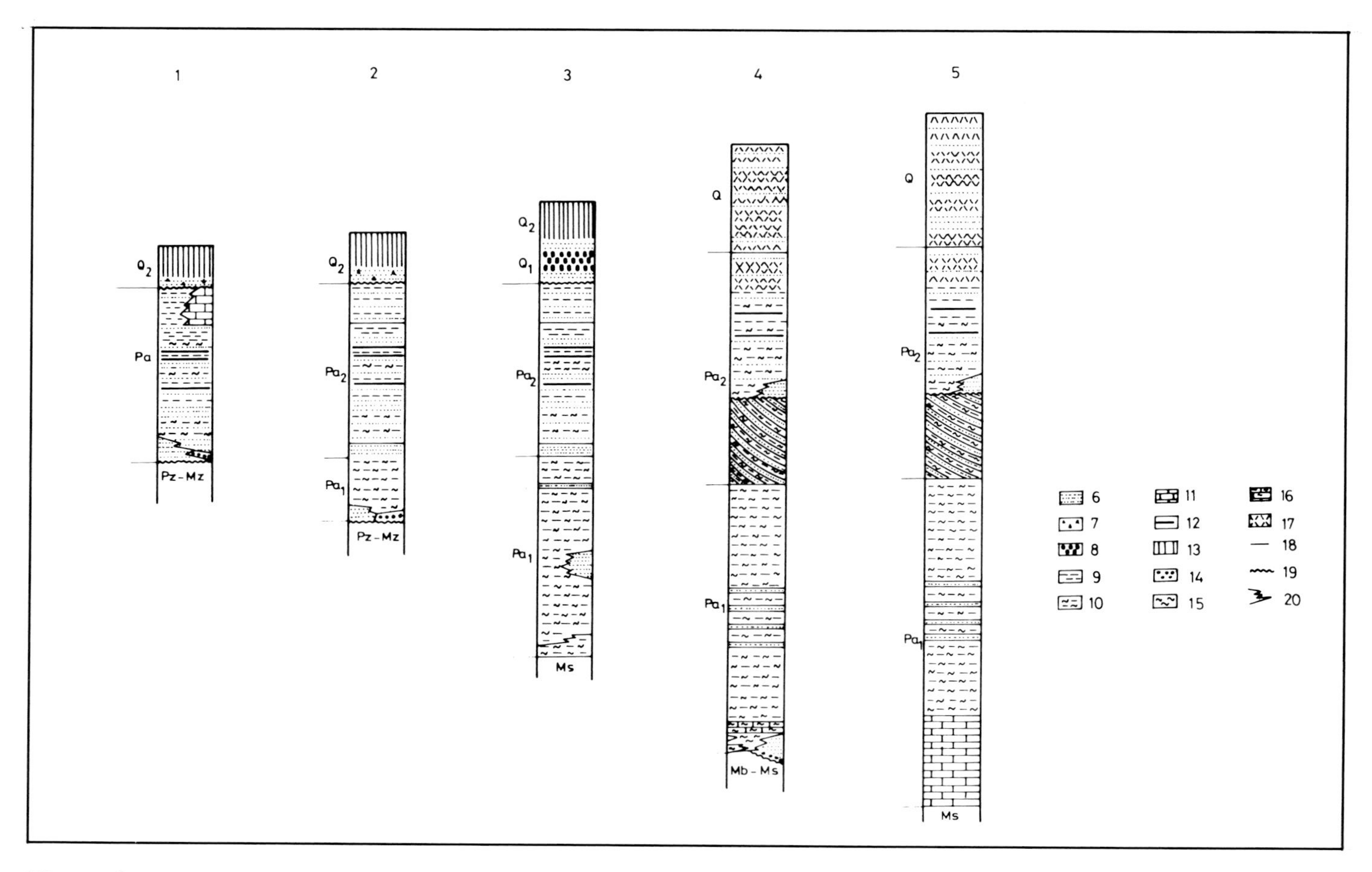

Figure 9. Principal types of the Pannonian (*s. l.*) sequence. Symbols: (1) piedmont (50–100 m), (2) above basin fringe elevations (150–300 m), (3) basin fringe subsidence (150–600 m), (4) above intrabasin elevations (500–2000 m), (5) filling intrabasin subsidence (2000–4500 m), (6) silt and siltstone, (7) breccia, (8) conglomerate, (9) shale, (10) argillaceous marl, (11) limestone, (12) coal, (13) loess, (14) sand and sandstone, (15) marl, (16) calcareous marl, (17) marshy deposits, (18) formation boundaries, (19) unconformity, and (20) pinchout.

In general, two main facies types can be distinguished within the Pannonian (*s. l.*) series in the basin areas: deep basin and basin margin facies that can be traced into one another through a series of transitional facies (Figure 9). In the deep basin areas, the oldest beds of the Pannonian are transitional from the Sarmatian and consist of platy marl, marl, and calcareous marl. At the margins of the basins, the oldest beds of Pannonian (*s. l.*) age consist of conglomerate and pebbly sandstone. These rocks are commonly cross-bedded or graded and coarse grained and are overlain by onlapping calcareous marl. In the areas where the Pannonian (*s. l.*) is relatively thin, the calcareous marl is poorly consolidated, and has a cream-yellow color. As the calcareous marl is traced toward the deeper basin areas, it thickens, becomes darker, and contains both pyritic plant remnants and conglomerate intercalations, indicating deposition by turbidity currents.

The calcareous marl horizon is overlain by a cyclic alternation of dark gray argillaceous marl, gray siltstone, and sandstone. The average sandstone content is about 30%. Sedimentary structures indicative of turbidity currents are common. In addition, the distribution of arenitic sediments was apparently influenced by currents that followed the bottom relief. Sedimentation during turbidite-free periods was dominated by the deposition of argillaceous marl and siltstone beds (Bérczi, 1970, 1972). The rate of subsidence in the individual basins exceeded the rate of sedimentation during deposition of the lower two-thirds of the lower Pannonian (*s. l.*), while rates of subsidence and sedimentation were roughly equal during deposition of the upper third. Prograding deltaic facies produced onlapping sandstone beds of regional importance (Bérczi, this volume).

Two main types of delta accumulations are observed in the lower Pannonian (*s. l.*): (1) those deposited during moderate basement subsidence whose individual beds have a large horizontal extent and are associated with minor thin rhythmic beds and (2) those deposited during faster basement subsidence and which have greater sediment thicknesses. The deltaic beds are gradually replaced upward by fluviolacustrine sedimentary rocks at about the boundary of the upper Pannonian (*s. l.*) and lower Pannonian (*s. l.*).

Surrounding the margins of the Pannonian basin, and in small intramontane basins, lower Pannonian (*s. l.*) sedimentary rocks vary in thickness between 50 and 500 m. In addition to the three main rock types mentioned above (gray clay-marl, sandstone, and brownish-gray calcareous marl), there is here an abundance of white calcareous marl, variegated clay, lignite-

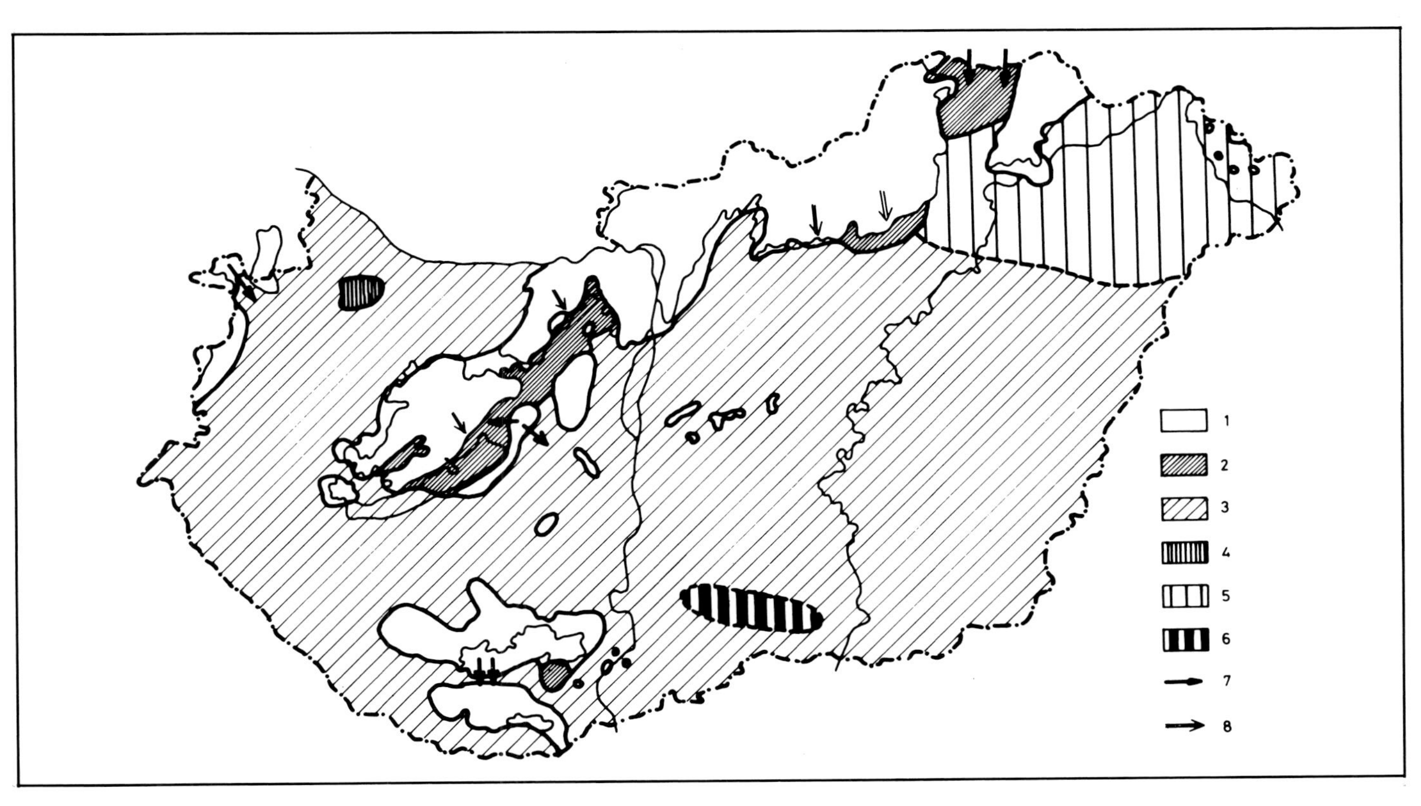

Figure 10. Paleogeographic sketch map of the lower Pannonian (*s. l.*) sequence. Symbols: (1) dry land, (2) lagoonal facies, (3) brackish-water inland sea facies, (4) brackish water inland facies with acid pyroclasts, (5) brackish water inland facies with basalt volcanites, (6) brackish water inland facies with trachyte volcanites, (7) direction of the transport of detritus, and (8) direction of transgression.

bearing beds, plankton and algae-bearing layers, pebbly quartz sand, freshwater limestone, oil shale, volcanic rocks, and volcanigenic sediments (Figure 10).

The lower Pannonian (*s. l.*) sedimentary rocks near the mountain areas of Hungary (Peremarton Formation) are mainly composed of gray pelites. In some places, such as in the northeastern and southwestern parts of the Transdanubian Central Mountains, the middle part of the Peremarton Formation shows transgressive sedimentary features. In the former coastal areas, the Peremarton Formation comprises pebbly quartz sand facies. The lower part of the Peremarton Formation contains thin layers of rhyolite tuff and bentonite. The lower and middle parts of the Peremarton Formation are replaced by white marl (in the vicinity of the Mecsek Mountains); by gray marl rich in organic walled microplankton, diatoms, or sponge spicules (southeast of the Transdanubian Central Mountains, southeast of the Bükk Mountains, and in the Nagyvázsony basin near Gyulakeszi); and by variegated clay, and sand layers with lignite interbeds (in Borsod basin near Peremarton and in Varpalota basin near Salgótarján). The upper part of the Peremarton Formation is transgressive. In the Sopron area, in the northwestern and sometimes the southeastern parts of the Transdanubian Central Mountains, and at the southern edge of the Mecsek Mountains, this horizon rests directly on rocks older than Pannonian age. In these areas, the base of the Peremarton Formation consists of thin layers of well-rounded quartz pebbles 0.5–1.0 cm in diameter and quartz sand, which is generally overlain by a homogeneous, mollusk-bearing, gray clay-marl (Figure 8). The upper two mollusk horizons of the Peremarton Formation are characterized by abundant dinofagellates. From a stratigraphic point of view, the most important species is *Spiniferites bentori*. The upper and middle horizons are distinguished on the basis of the abundance of *Pontianidium* spp.

In the basinal areas, the lower boundary of the upper Pannonian (*s. l.*) is accompanied by an increased sand content upward, as observed in well logs. The proportion of sandstone beds increases suddenly to 40–50% at the boundary between the upper and lower Pannonian (*s. l.*). The most important rock types in the upper Pannonian (*s. l.*) are sandstone, siltstone, and clay-marl. Well-compacted sandstone with calcareous cement is also common, along with sparse occurrences of calcareous marl, marl lignite, and lignite-bearing clay interbeds. Quartz pebble beds are also present. The rhythmic succession is characterized by coarsening-upward units. The thickness of these units is 10–30 m (Magyar and Révész, 1976).

The appearance of quartz pebbles in the fine-grained sediments indicates periodically strong currents. The succession is characterized by the presence of vertical root remnants, leaf impressions, and coalified fossil plant detritus. These coal-bearing beds are separated in each rhythmic sedimentary bed, which are characterized by coarsening-upward sediments. These observations, together with the settling of fossils and the alternating thicknesses of sandstone in a rhythmic fashion, indicate a deltaic environment of deposition (Mucsi and Révész, 1975).

The upper Pannonian (*s. l.*) near the mountain areas of Hun-

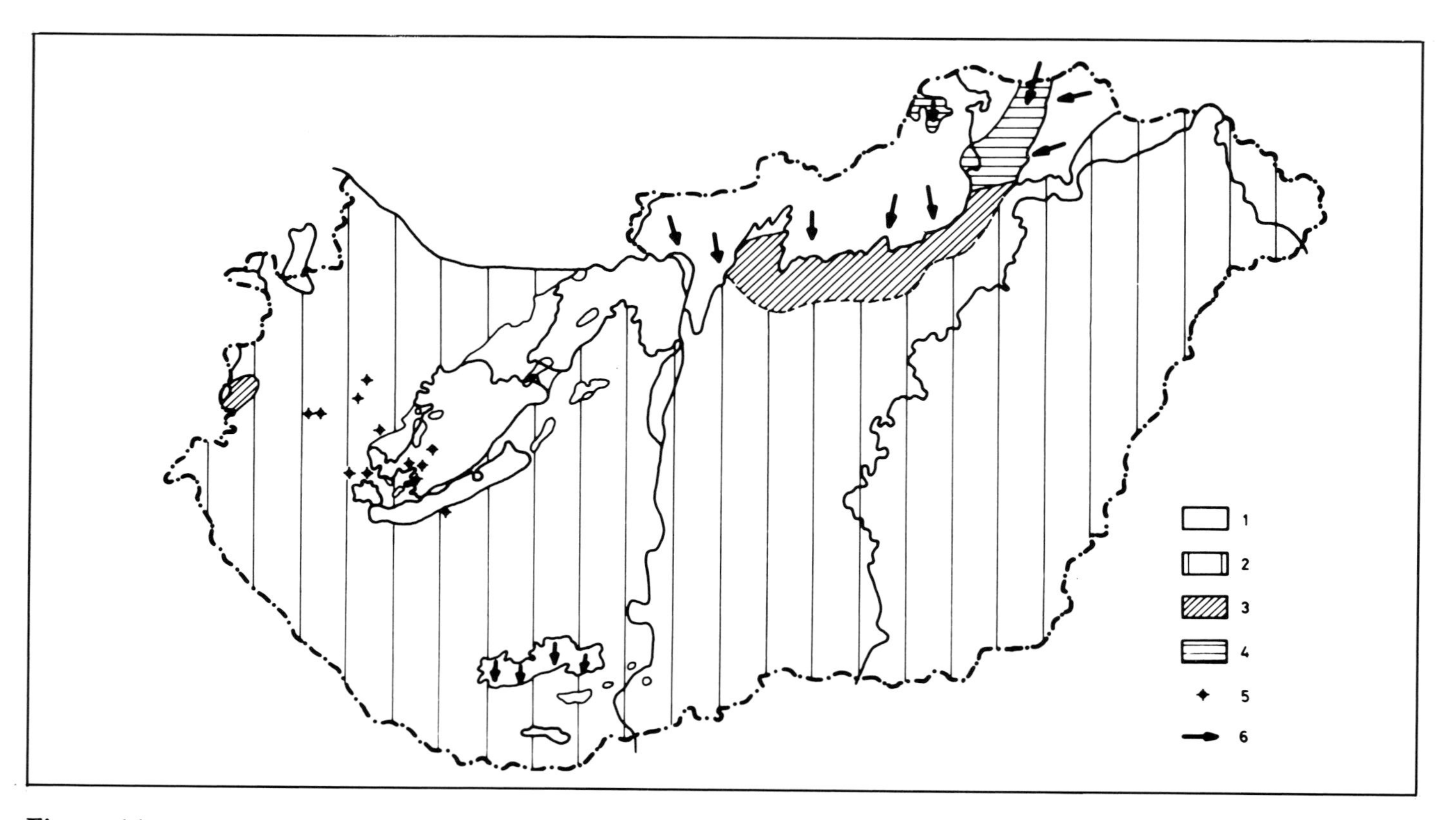

Figure 11. Paleogeographic sketch map of the upper Pannonian (s. l.) sequence. Symbols: (1) dry land, (2) fluviolacustrine, (3) paludal facies, (4) fluvial facies, (5) basalt volcanites, and (6) direction of the transport of detritus.

gary consists of alternating layers of gray argillaceous marl, silt, and sand; younger beds are characterized by the presence of coal-bearing clays, lignite, and variegated clay. In intramontane basins, the appearance of freshwater limestone is typical.

Near the mountains, the lower part of the Transdanubian Formation is transgressive (Figure 11). Coastal facies are present in the Balaton Highland, southeastern Vértés, northern and western Gerecse, southern and western Mecsek, and southern Mátra. The formation is characterized mainly by pebble conglomerate, but coarse conglomerate (Balaton Highland, southern Mecsek, and Velence Mountains), quartz sand, and rare sandstone-quartzite are also present. In the deep basins, the presence of gray argillaceous marl and sand and mollusks (great limnocardium with thick shells and congeria) attest to a deeper water depositional environment. Sedimentary rocks deposited in swamps occur in the Transdanubian Central Mountains (Csákvár Basin).

The middle and upper parts of the Dunántúl Formation comprise similar lithologies. They can be distinguished from the lower part by the absence of pebble layers and an upward increase in variegated swampy (coal-bearing clay and lignite) and desiccated lagoonal clay and sand layers. To the south and southeast, extensive swampy areas were developed at the southern margins of the North Hungarian Mountains. As a result, brackish water, limnic, and fluvial faunal assemblages appear in the middle horizon and limnic and terrigenous assemblages appear in the upper horizon.

At the beginning of the Quaternary, uplift of some marginal parts of the basins resulted in erosion of Pannonian (*s. l.*) rocks. As a consequence, the Pannonian (*s. l.*) is unconformably overlain by Quaternary (Würm) sediments in these areas. In contrast, there was a continuous transition from Pannonian (*s. l.*) to earliest Quaternary rocks in the depression areas.

VOLCANOGENIC SEDIMENTS

The volcanogenic sediments of the Pannonian region are important from both a stratigraphic and a tectonic point of view. Three main volcanogenic sedimentary layers can be distinguished. The oldest of these is composed of rhyolite flood tuffs and ignimbrites. This layer occurs stratigraphically at about the Eggenburgian–Ottnangian boundary and has been dated at 18 ± 3.8 Ma (Hámor et al., 1979). This unit is about 100–200 m thick and is called the "lower rhyolite tuff." It can be found in north Hungary, in the northern rim of the Mecsek Mountains, and in the subsurface of the Drava basin.

The second phase of volcanic activity occurred at the end of Karpatian time. Volcanogenic sediments deposited during this phase have the greatest areal extent. Andesites, dacites, and rhyodacites were erupted, partly as pyroclastic material (Póka, this volume). These late Karpatian pyroclastics and lava flows have been dated at 16.4 ± 0.8 Ma by K-Ar method (Hámor et al., 1979). This pyroclastic horizon is called the "middle rhyolite tuff."

The youngest layer of volcanogenic sediments in Hungary is of late Badenian age. Pyroclastics of andesite, rhyodacite, and rhyolite were produced and form the "upper rhyolite tuff." The greatest thickness of the tuffaceous volcanogenic sediments of the upper rhyolite tuff occurs in northeastern Hungary, where it reaches more than 2 km in thickness. The upper rhyolite tuff has been dated at 14 Ma (Hámor et al., 1979). Related subsurface volcanic rocks of northern Hungary were erupted along northeast- and northwest-trending fault zones (such as the Mátra Volcanic Formation) dated at 14.5 ±0.4 Ma. Older volcanic centers were reactivated during some of these eruptions.

According to new radiometric data (Árva-Sós, 1983), the production of andesite, basaltic andesite, and their tuffs, as well as rhyolite tuffs, continued into the early part of the early Pannonian (*s. l.*) in northeastern Hungary. In other areas, the post-Sarmatian volcanic activity is more basaltic. In the south-central part of the Great Hungarian Plain basalt lava and tuffs are present (Kecel Formation, 8–10 Ma) (Balogh et al., 1983). These basalts belong stratigraphically to the lower Pannonian (*s. l.*) or they immediately underlie it. The thickness of the basalts and basaltic tuffs reaches 600 m.

In the Little Hungarian Plain, the main volcanic activity started at the Sarmatian–Pannonian boundary, producing trachites and younger basalts and dolerites from separate eruption centers. The thickest volcanic series that has been penetrated in drill holes is 1800 m thick. The trachitic volcanism is characterized by widespread dispersal of volcaniclastics sediments over the entire basin area. The trachites are overlain by basalts or intruded by basalt dikes. These volcanic rocks in the Little Hungarian Plain are associated with a large ring-shaped area of subsidence (cauldron).

During the late Pannonian (*s. l.*), significant basalt volcanism took place in the southwestern part of the Transdanubian Central Mountains and in the Little Hungarian Plain (2.7–5.0 Ma) (Balogh et al., 1983). Some of the 70 eruption centers are ring-shaped, while stratovolcanos formed in the Balaton Highland.

REFERENCES

Árva-Sós, E., K. Balogh, G. Hámor, A. Jámbor, and L. Ravasz-Baranyai, 1983, Chronology of Miocene pyroclastics and lavas of Hungary: Am. Inst. Geol. Geofiz, Bucuresti, v. 61, p. 353–358.

Balázs, E., I. Bérczi, I. Gajdos, Á. Jámbor, M. Korpás-Hódi, L. Mészáros, G. Németh, A. Nusszer, S. Pap, I. Révész, A. Somfai, Á. Szalay, K. Szentgyörgyi, M. Széles, and L. Völgyi, 1981, Outlines of geological structure and evolution of the Pannonian, *in* Excursion guide of molasse formations in Hungary: Hung. Geol. Inst. Publ., p. 56-77.

Balogh, K., A. Jámbor, Z. Partényi, K. Ravasz-Baranyai, G. Solti, and A. Nusszer, 1983, Petrography and K-Ar dating of Tertiary and Quaternary basaltic rocks in Hungary: Am. Inst. Geol. Geofiz. Bucuresti, v. 61, p. 365–373.

Bérczi, I., 1970, Sedimentological investigation of the coarse-grained clastic sequence of the Algyö hydrocarbon-holding structure: Acta Geol. Acad. Sci. Hung., v. 14, p. 287–300.

Bérczi, I., 1972, Sedimentological investigation of pre-Pannonian sedimentary formations in the Szeged-Basin, SE-Hungary: Acta Geol. Acad. Sci. Hung., v. 16, p. 229–250.

Hámor, G., and K. Szentgyörgyi, 1981, Outlines of geological structure and evolution of the Miocene, *in* Excursion guide of molasse formations in Hungary: Hung. Geol. Inst. Publ., p. 42–55.

Hámor, G., L. Ravasz-Baranyai, K. Balogh, and E. Árva-Sós, 1979, K/Ar dating of Miocene pyroclastic rocks in Hungary: Ann. Geol. Pays Helleniques, Tome Hors Sér. II., p. 491–501.

Magyar, L., and I. Révész, 1976, Data on the classification of Pannonian sediments of the Algyö-area, Acta Min. Petr., Szeged, v. 22, n. 1, p. 267–283.

Mucsi, M., and I. Révész, 1975, Neogene evolution of the southwestern part of the Great Hungarian Plain on the basis of sedimentological investigations: Acta Min. Petr., Szeged, v. 22, n. 1, p. 29–49.

Nagymarosy, A., 1981, Chrono- and biostratigraphy of the Pannonian basin: a review based mainly on data from Hungary: Earth Evol. Sci., v. 1, n. 3-4, p. 183–194.

CHAPTER 6

Some Aspects of Neogene Biostratigraphy in the Pannonian Basin

A. Nagymarosy
Department of Geology,
Eötvös University
H-1088 Budapest Múzeum krt. 4/a,
Budapest, Hungary

P. Müller
Hungarian Geological Institute
H-1142 Budapest, Népstadion út 14,
Budapest, Hungary

The Pannonian basin was a part of the Central Paratethys sea, which existed from Oligocene to Pliocene time. The Paratethys depositional sequence can be divided into three distinct phases: the Eoparatethys, (early Oligocene to early Miocene time), the Mesoparatethys (early Miocene to middle Miocene time), and the Neoparatethys (middle Miocene to Pliocene time). Both the Eo- and Mesoparatethys seas were characterized by only short periods of endemism, but the Neoparatethys was completely isolated from the world ocean system. Thus, the Central Paratethyan biostratigraphic stages can be well correlated with the Mediterranean stages for the Eo- and Mesoparatethys, but considerable difficulties exist in correlating the stages of the Neoparatethys with the contemporaneous Mediterranean stages and in correlating stages from one part of the Neoparatethys to another.

In this chapter, we review the key data used to correlate the Central Paratethyan stages with the Mediterranean and worldwide stratigraphic stages. In particular, we concentrate on stratigraphic data from the Pannonian basin in Hungary and on some of the special problems inherent in dating sediments within the Pannonian basin of Oligocene to Pliocene age.

INTRODUCTION

The Committee of Mediterranean Neogene Stratigraphy (1971), Steininger and Nevesskaya (1975), Steininger and Rögl (1979), Nagymarosy (1981), Rögl and Steininger (1983), and Steininger et al. (this volume) have given a comprehensive description of the concept of the Paratethys and the definitions of the Paratethyan Neogene regional stages. The gradual separation of the Paratethyan basins from the world ocean system is reflected in the increasing endemism of the Paratethyan aquatic biota. In the period of the early Paratethys Sea (Eoparatethys), the regional stages can be correlated easily with Mediterranean and Atlantic stages, although a short period of endemism in early Oligocene time indicates that the Paratethys was easily separated from the oceans, even prior to Miocene time. In late

Miocene and Pliocene time, the endemic faunal development makes inter- and intrabasinal correlations within the Paratethys uncertain and highly debatable. This periodic endemism has made correlation of many of the Paratethyan stages with Mediterranean Neogene stages difficult and has made it necessary to create separate biostratigraphic stages for the Paratethys realm.

In this chapter, we summarize the important "handholds" of biostratigraphy, which enable us to fit the Paratethyan history into a global chronological framework, despite the difficulties mentioned above. All of the stages are based on molluscan assemblages but it is not possible to use the molluscan assemblages to correlate over large areas. Therefore planktonic organisms are the best means of correlation of the worldwide stages. Figure 1 shows the Neogene regional stages, the Mediterranean stages, the nannoplankton zonation, and some important geological events in the Pannonian basin. For further chronostratigraphic information on the Paratethyan Neogene time scale, see the correlation chart of Steininger et al. (this volume). For further information on lithostratigraphy in Hungary see Bérczi et al. (this volume) and Hámor (1985).

THE EOPARATETHYS

The Eoparatethys sea (Figure 2) began in early Oligocene time with a short period of endemism (Báldi, 1984). The subsequent stages (Egerian, Eggenburgian, and early Ottnangian) are well correlated with the standard global stages (Figure 1). The first widespread Neogene endemic event in the Paratethys sea occurred only in late Ottnangian time.

The area of sediment deposition in the Pannonian basin during Eoparatethyan time extended along a southwest-northeast-trending zone where the Transdanubian and North Central Mountains are now located (Figure 2). The composition of the macrofaunas, which include Mediterranean, Boreal, and local elements, indicates an extensive faunal exchange. The paleogeographic position of the Transylvanian basin in Oligocene and early Miocene time, as compared to the central basins of the Pannonian system, is not yet clear. However, these two areas were lithologically and biostratigraphically similar until Ottnangian time.

Egerian (Formerly "Chattian" or "Aquitanian")

The stratotype of the stage Egerian in the town of Eger represents the early, Oligocene part of the Egerian stage (*Globorotalia opima opima* planktonic foraminifera zone and nannoplankton Zones NP24 and NP25). The presence of Zone NN1 in the stratotype is dubious (Lehotayová and Báldi-Beke, 1975). The lower part of the Egerian is coeval with the Chattian and its upper part with the Aquitanian. Accordingly, the upper Egerian formations contain Miocene molluscs, such as *Chlamys incomparabilis*, *Ch. carryensis*, *Ch. rotundata*, and assemblages from the nannoplankton Zone NN1. The upper Egerian is represented in the Pannonian basin by a sublittoral schlier facies in northern Hungary and southern Slovakia, and elsewhere it is represented by a littoral facies, the Bretka Limestone. The early Miocene age of the latter is supported by the occurrence of larger foraminifera (*Miogypsina gunteri*) (Vanova, 1975). Brackish and freshwater sedimentation occurred over large areas of Transdanubia (western Hungary).

Eggenburgian (Formerly "Untere Mediterranenstufe" or "Burdigalian")

After a partial regression at the end of Egerian time, an early Eggenburgian marine transgression produced a new molluscan fauna, "loibersdorftype," whose appearance defines the beginning of the Eggenburgian stage. The Indian–West Pacific character of this faunal type has been described by Rögl et al. (1978). This fauna occurs together with assemblages of the planktonic foraminifera Zone N5, with typical *Globigerina woodi s. l.*, *G. scalena*, and *Globorotalia* cf. *kugleri* in the Szécsény and Putnok schlieren of northern Hungary (Horváth and Nagymarosy, 1979). The Eggenburgian stage spans the NN2 nannoplankton zone and the lower part of Zone NN3.

In the rapidly subsiding basins of northern Hungary and southern Slovakia, the marginal Budafok sand, the Petervására glauconitic sandstone, and intrabasinal schlieren were deposited (Figure 3). The sand and sandstone show strong lithological and faunal affinities to the Loibersdorf beds in Austria and to the Corus beds in Transylvania, demonstrating the faunal connection of the Pannonian basin with areas to the northwest and east.

Ottnangian (Formerly "Early Helvetian")

The base of the Ottnangian stage is defined by the first appearance of euhaline Ottnangian-type mollusc fauna, including *Chlamys albina*, *Pecten (F.) hermannseni*, and *P. fotensis*. Nannoplankton assemblages from Zones NN3 and NN4 have been described from the stratotype of the Ottnangian (Martini and Müller, 1975).

The youngest part of the lower Miocene transgressive sequence in northern Hungary (Zagyvapalfalva Formation) may be Ottnangian in age, but its terrestrial character prevents accurate dating. The last marine datum (nannoplankton Zone NN3) occurs a few meters below this terrestrial formation within a continuous depositional sequence, but typical Ottnangian molluscs are missing. The Ipolytarnóc footprinted sandstone occurs immediately above this last marine datum.

The transgressive sedimentary sequence ends with a thick layer of rhyolite tuff (the "lower" or Gyulakeszi rhyolite tuff). Its average radiometric age is 19.6 ± 1.4 Ma (Hámor et al., 1980), and its deposition was followed by a relatively short period of erosion. This radiometric age is slightly inconsistent with the Zone NN3 age of the rocks below the tuff. Similar inconsistencies are observed between radiometrically (and magnetically) determined ages of oceanic planktonic zones and epicontinental planktonic zones for the whole early Miocene (Haq, 1983; Hsü et al., 1984).

In the upper part of the Ottnangian, but still in nannoplankton Zone NN3, a younger transgressive sequence begins. The only known euhaline Ottnangian molluscan assemblage from the central part of the Pannonian basin was described in upper Ottnangian rocks (Bántapuszta beds) in Varpalota (in the Bakony Mountains, west-central Hungary) by Kokay (1971).

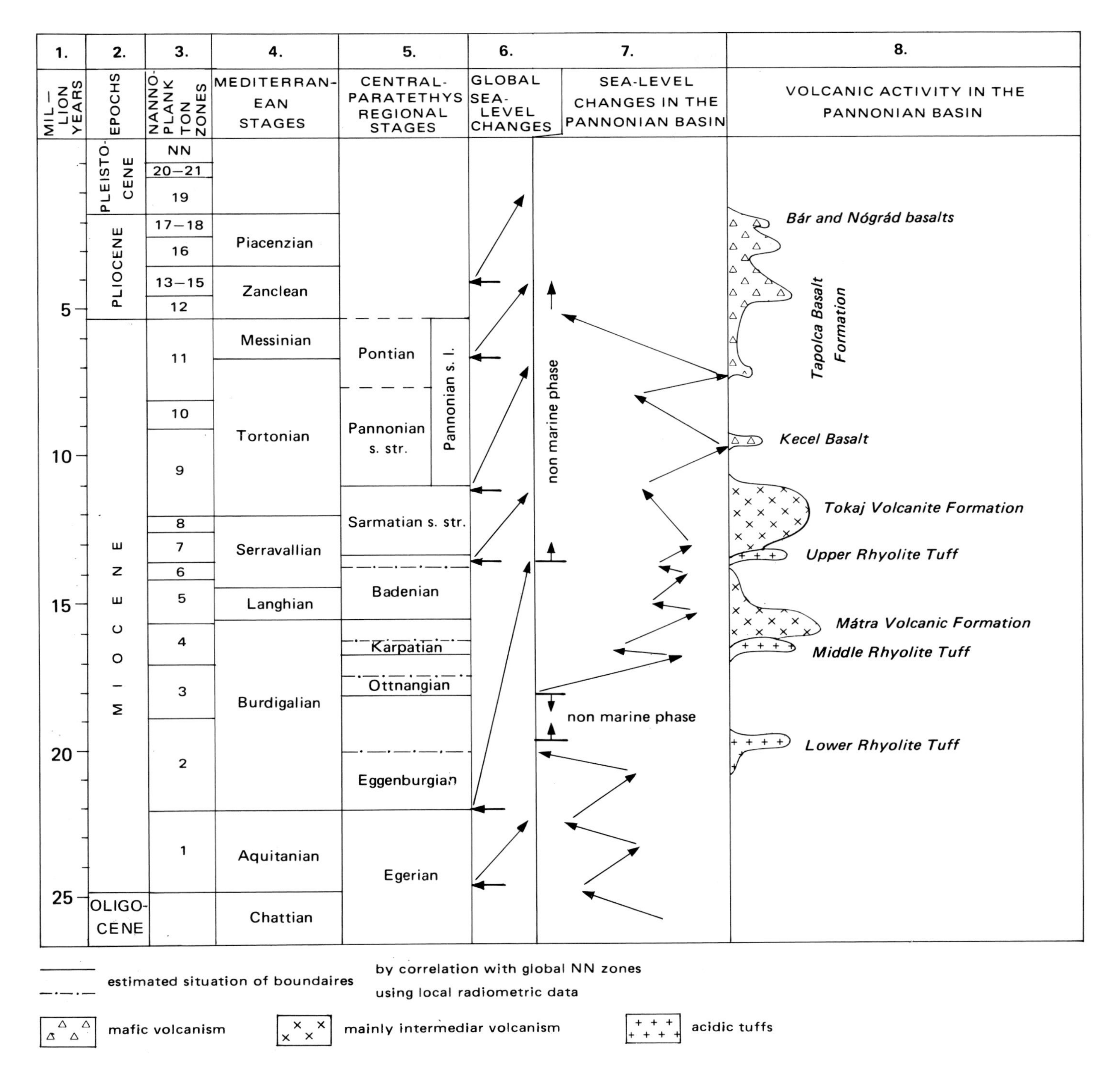

Figure 1. Neogene chronostratigraphy of the Central Paratethys and some important geological events in the Pannonian basin. Columns: (1) radiometric ages in Ma; (2) geological epochs; (3) Neogene standard nannoplankton zones of Martini (1971) and after Haq (1983); (4) Mediterranean stages correlated with the nannoplankton zones; (5) Central Paratethyan regional stages; (6) global sea level changes; (7) sea level changes in the Pannonian basin; arrows pointing to the left indicate regressions, and to the right, transgressions; and (8) volcanic activity in the Pannonian basin. (After Vail et al., 1977; Hámor et al., 1980; Haq, 1983; Rögl and Steininger, 1983; Balogh et al., in press). For more detailed chronostratigraphy see Steininger et al. (this volume); for Eoparatethyan events in the Pannonian basin see Nagymarosy (1981).

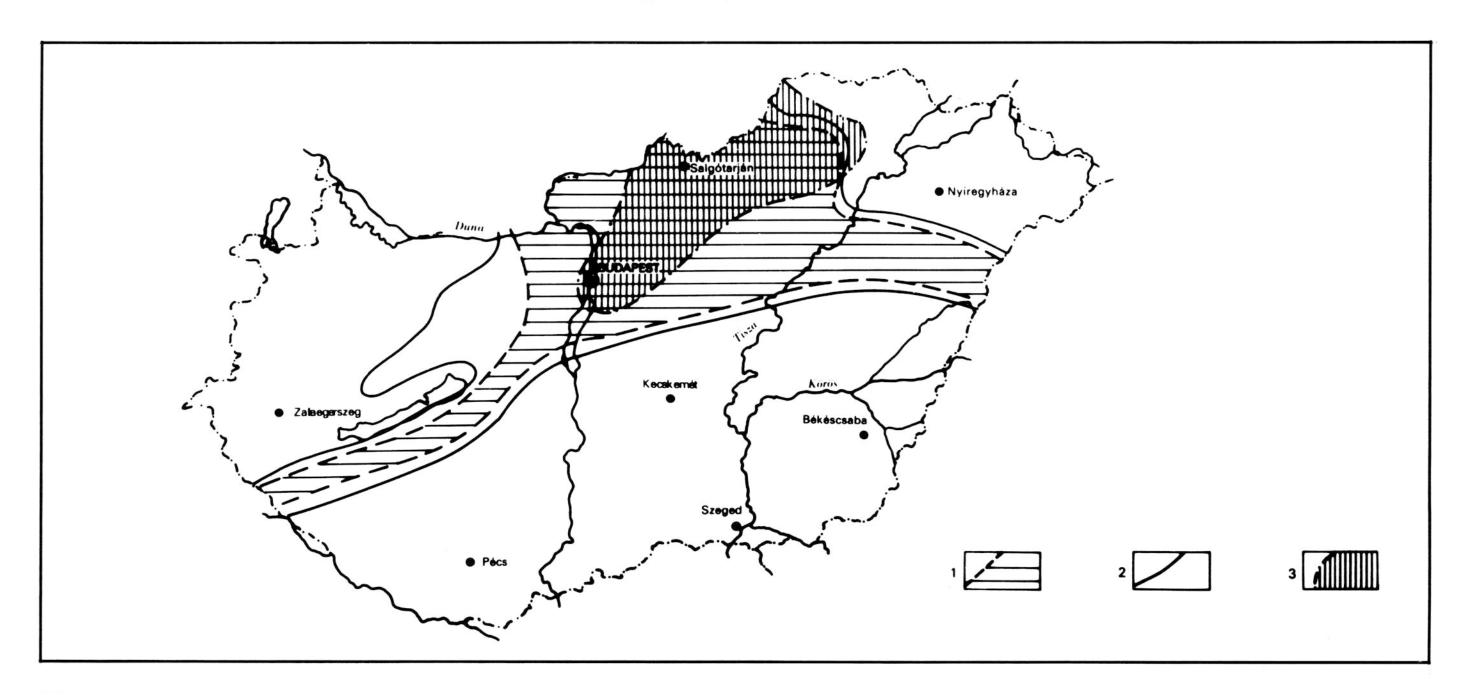

Figure 2. Sketch map of the Eoparatethys (early Oligocene–early Miocene) in Hungary. Symbols: (1) maximum extent of the formations of Kiscellian age, (2) maximum extent of strata of Egerian age, and (3) maximum extent of strata of Eggenburgian age.

The depositional site of these sediments was apparently isolated from that of the other Ottnangian sediments. Their age may be constrained to be within the short interval of overlap of Okada and Bukry's (1980) Zone CN3 with Martini's (1971) Zone NN3 and the lower part of Zone NN4. The mollusc fauna has Mediterranean affinities.

Thick coal beds (Salgotarjan brown coal formation) were deposited in northern Hungary in late Ottnangian time and form the lower member of a younger transgressive sequence. The overlying *Rzehakia (Oncophora)* beds represent the first widespread Neogene endemic event, specifically, the evolution and migration of an endemic, Paratethys-dwelling molluscan assemblage. Apparently isolated basins with *Rzehakia* faunas extend as far east as Georgia and as far west as Bavaria, but the seaways that connected these distant brackish *Rzehakia* basins with each other are not known. Coeval marine sedimentation is not known in the Pannonian area, but a number of shark teeth, planktonic foraminifera, and Zone NN4 nannoplankton (Báldi-Beke and Nagymarosy, 1979) prove an open connection with the sea in late Ottnangian time.

THE MESOPARATETHYS

The area of marine sedimentation in Hungary during the Paleogene and the early part of the early Miocene was restricted to the narrow zone now occupied by the Central Mountains (Figure 2). Beginning at the Ottnangian–Karpatian time boundary, larger and larger areas became inundated by the sea as the basement of the Pannonian basin subsided (Figure 4). While sedimentation in the Central Mountains of Hungary continued, new areas of sedimentation formed in western Hungary and along a northeast-southwest zone extending from the Mecsek Mountains (near Pécs) to the Great Hungarian Plain (to Kecskemét and Szeged). Marine connections existed to the Carpathian foredeep and the Mediterranean.

Karpatian (Formerly "Upper Helvetian")

The Karpatian stage is characterized by the appearance of new mollusc (*Chlamys scabrella, and Ch. macrotis*) and benthic foraminifera species (*Cyclammina karpatica* and a number of *Uvigerinas*). The nannofloras at the base of the Karpatian belong to Zone NN4, while the higher part of the Karpatian sequences contains Zone NN5 nannofloras (Figure 3). The planktonic foraminifera *Globigerina sicanus* (or *bisphaericus*) also had its first appearance in the late Karpatian.

The first euhaline member of this transgressive sedimentary sequence, which began in late Ottnangian time, is the Karpatian Egyházasgerge formation. A deep water, contemporaneous facies, the thick Garab schlieren, was deposited in the deeper parts of the basin.

Badenian (Formerly "Obere Mediterranenstufe" or "Torton")

The Badenian stage yields extremely rich marine micro- and macrofaunas. Its lower boundary is defined in continuous sections by the first appearance of the *Praeorbulina* planktonic foraminifera above which are found *Orbulinas* in lower Badenian strata. Planktonic foraminifera assemblages from Zones N8 to N12 allow correlation with the Mediterranean Langhian and lower Serravallian stages, although these zones can be detected only in pelitic intrabasinal sequences in the Pannonian basin.

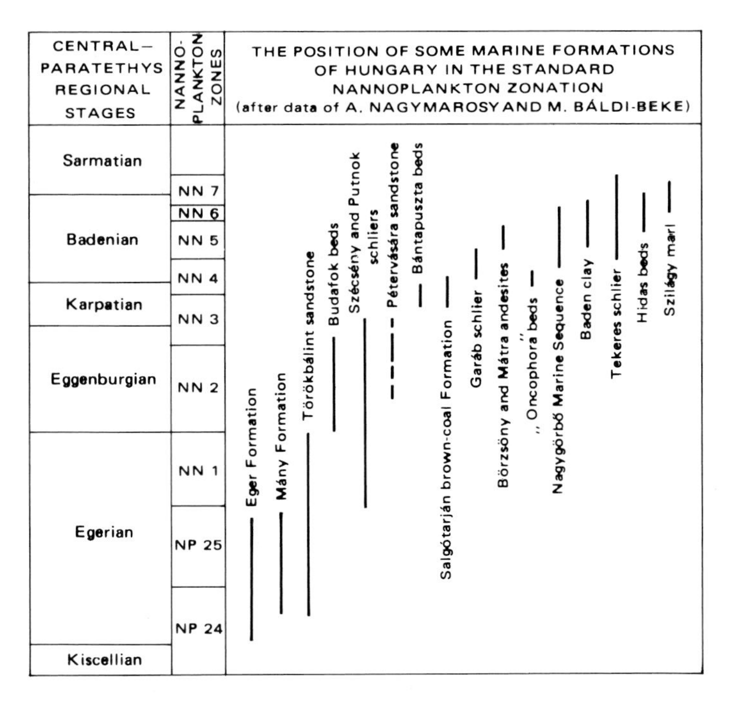

Figure 3. The position of some marine formations of Hungary in the Standard Nannoplankton Zonation. (From data of A. Nagymarosy and M. Báldi-Beke).

Other groups, such as the evolutional lineages of *Uvigerinas* and *Heterosteginas* give additional constraints on dating of the Badenian stage.

The most commonly used criterion for Badenian microbiostratigraphy is Grill's (1943) zonation, which distinguishes four benthic foraminifera zones: the lower and upper *Lagenid* zones, the *Spiroplectammina* or Sandschaler zone, and the *Bulimina–Bolivina* zone. A fifth zone, the "Verarmungszone," can also be included and has an impoverished foraminifera fauna at the top of the stage. These zones are not evolutionary. They strongly reflect environmental conditions and should be regarded as ecostratigraphic units. However, they can be used for dating with rather good results throughout the whole Pannonian basin.

The nannoplankton Zone NN5, which in our opinion begins in upper Karpatian strata (in contrast to the interpretation of Rögl and Steininger [1983], who indicate the base of the zone within the Badenian stage), ends at about the top of the *Lagenid* zone. The *Spiroplectammina* zone corresponds to Zone NN6, and the *Bulimina–Bolivina* zone to the lower part of Zone NN7. The existence of the latter zone in the Paratethys basin is widely disputed because the index fossil *Discoaster kugleri* has been so far described only from the Mecsek Mountains (near Pécs in Figure 4) (see also Nagymarosy, 1980, 1985). We infer that this tropical species was restricted to the southern part of the Pannonian area.

Although the Badenian stage is defined by the appearance of a specific molluscan fauna with Mediterranean affinities, molluscs do not play a leading role in the subdivision of the stage. These and other macroinvertebrate faunas require more investigation before they can be used as important criteria for detailed stratigraphy.

In earliest Badenian time, a large volume of volcanic rocks was erupted in the area of the Pannonian basin. The deposition of the "middle" or Tar rhyolite tuff (16.4 ± 0.8 Ma) occurred at about the Karpatian–Badenian boundary. The Mátra, Börzsöny, Cserhát, Stiavnicky, and Ostrovsky volcanic mountains were erupted mainly in early Badenian time in northern Hungary. (See Póka, [this volume] for locations; radiometric dates for eruptions from Börzsöny Mountains are 16.8–17.8 ± 1.3 Ma [Hámor et al., 1980].) These volcanic events occurred entirely within the nannoplankton Zone NN5. In other places the volcanic activity persisted until the end of Badenian time (for example, a tuff dated at 13.6 Ma occurs at Rákos, Budapest [Müller, 1984]).

The early Badenian basins, which had been partly separated by early Badenian volcanic and tectonic events, became further separated during the middle Badenian regression. This resulted in the deposition of evaporites in Transylvania and outside the Carpathians, while coal beds were formed in the central basins in Hungary.

A large part of the Pannonian basin was covered by water during late Badenian time. Faunal connections with the Mediterranean probably persisted until the end of Badenian time. Rögl et al. (1978), however, suggested that the Pannonian basin was connected to the Indian–West Pacific ocean in early and late Badenian time and that the Mediterranean connection closed in late Badenian time (Rögl et al., 1978).

In latest Badenian time there was a strong faunal influence from the Eastern Paratethys. The decreasing salinity of water in the Pannonian area resulted in the immigration of the Konka fauna, presumably due to density stratification of the water body, such as exists at present in the Sea of Marmara (Kókay, 1985).

THE NEOPARATETHYS

In Neoparatethys time, the Central Paratethys sea was completely isolated from the world oceans, although in the Eastern Paratethys three periods of marine influence occurred during late Miocene and Pliocene time (Figure 5). Consequently, events within the Eo- and Mesoparatethys seas roughly reflected global eustatic sea level changes, but global sea level changes had no effect on events within the Neoparatethys sea. This huge and long-lasting lake system underwent a gradual decrease in salinity and change in water chemistry throughout its existence. The increasing endemism of aquatic biota in the Neoparatethys sea makes it impossible to correlate directly with marine faunas and even to correlate aquatic biota between subbasins with different salinities within the Neoparatethys. Mammal stratigraphy, magnetostratigraphy, and radiometric methods are used in making stratigraphic correlations within Neoparatethys strata.

Sarmatian (*s. str.*)

The Sarmatian (*s. str.*) is restricted to the Pannonian basin. The Sarmatian (*s. l.*) of the Eastern Paratethys includes not only the Sarmatian (*s. str.*) (Volhynian and lower Bessarabian substages) but also the Chersonian and upper Bessarabian substages.

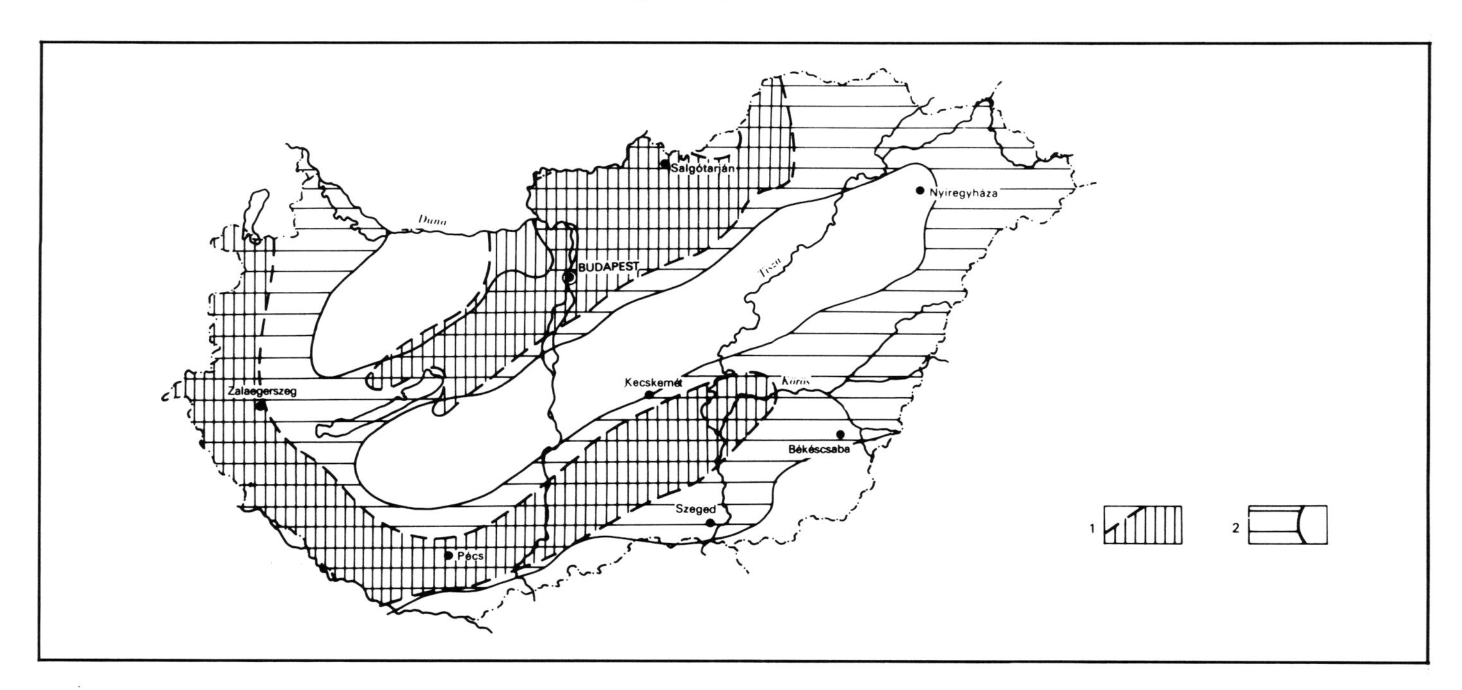

Figure 4. Sketch map of the Mesoparatethys (middle Miocene) in Hungary. Symbols: (1) maximum extent of strata of Karpatian age and (2) maximum extent of strata of Badenian age.

Typical Sarmatian molluscan assemblages consist of brachyhaline species, for example, *Mactra, Cerastoderma, Ervilia,* and *Abra.* Some benthic foraminiferas and bryozoans with high tolerance are also found in Sarmatian deposits, but these are of no assistance in correlations with the euhaline biostratigraphic zonations. Correlation of the Eastern Paratethys time scale with Mediterranean stages is rather good because of the uniformity of the faunas. Radiometric ages, rather than biostratigraphic correlations, demonstrate the synchroneity of the Sarmatian with the upper Serravallian–lowermost Tortonian (Vass and Bagdasarjan, 1978).

Over most of the Pannonian basin, Sarmatian deposits lie transgressively on rocks of Badenian age or older. The water depth apparently remained generally shallow. The "upper" or Galgavölgy rhyolite tuff marks an increase in volcanic activity near the beginning of Sarmatian time (13.7 ± 0.8 Ma, Hámor et al. [1980]). Many Sarmatian volcanos were erupted along the northeastern margin of the Pannonian basin.

Pannonian *s. str.* (Formerly "Lower Pannonian")

The beginning of the Pannonian stage is characterized by the appearance of a particular mollusc assemblage that might have been derived from (1) Sarmatian ancestors (for example, Lymnocardiidae), (2) immigrated euryhaline forms, or (3) freshwater forms that adapted to the brackish environment. Such fauna do not form a basis for extrabasinal correlation.

The stratigraphic position of the Pannonian can be defined by the first appearance of the *Hipparion,* which is placed at the lower boundary of the Pannonian. The early Pannonian Csákvár mammal fauna is well correlated with the Eastern Paratethys upper Bessarabian and Chersonian (Kretzoi, 1954).

The most widely used biostratigraphic zonation of the Pannonian is that of Papp (1953), which is based mainly on the evolutionary lineages of different molluscan families. Papp divided the Pannonian stage into five zones, designated A through E. This zonation, however, is useful mainly at the margins of the basins because the poverty of faunas within deep basin areas does not permit extension of these faunal zones into the deep basin. Other zonation systems have been proposed by other authors (for example, Korpás-Hódi, 1983).

One of the most promising new methods for zonation of the Pannonian is dating with organic-walled microplankton. Sütő-Szentai (1984, 1985) divided the Pannonian and Pontian together into eight zones, half of which belong to the Pannonian (*s. str.*). These planktonic dinoflagellates were presumably not influenced significantly by local environments on the lake bottom and therefore offer a good correlation between deeper and shallower parts of the basin.

Radiometric dating of some middle–late Pannonian basalts intercalated with lacustrine sediments give an age of 9.61 ± 0.38 Ma. This suggests that the Sarmatian–Pannonian boundary may be not younger than 11.0–10.5 Ma (Balogh et al., in press).

Another problem concerns the depth of the Pannonian lake. Although the presence of turbidite layers suggests steep paleomorphology, the paleontological record does not offer definitive evidence for great water depths. It may be assumed that no real deep lake dwelling molluscan assemblage developed within the Pannonian lake.

Pontian (Formerly "Upper Pannonian")

The Pontian stage also contains highly endemic lacustrine fauna. Its name was borrowed from the South Russian–South

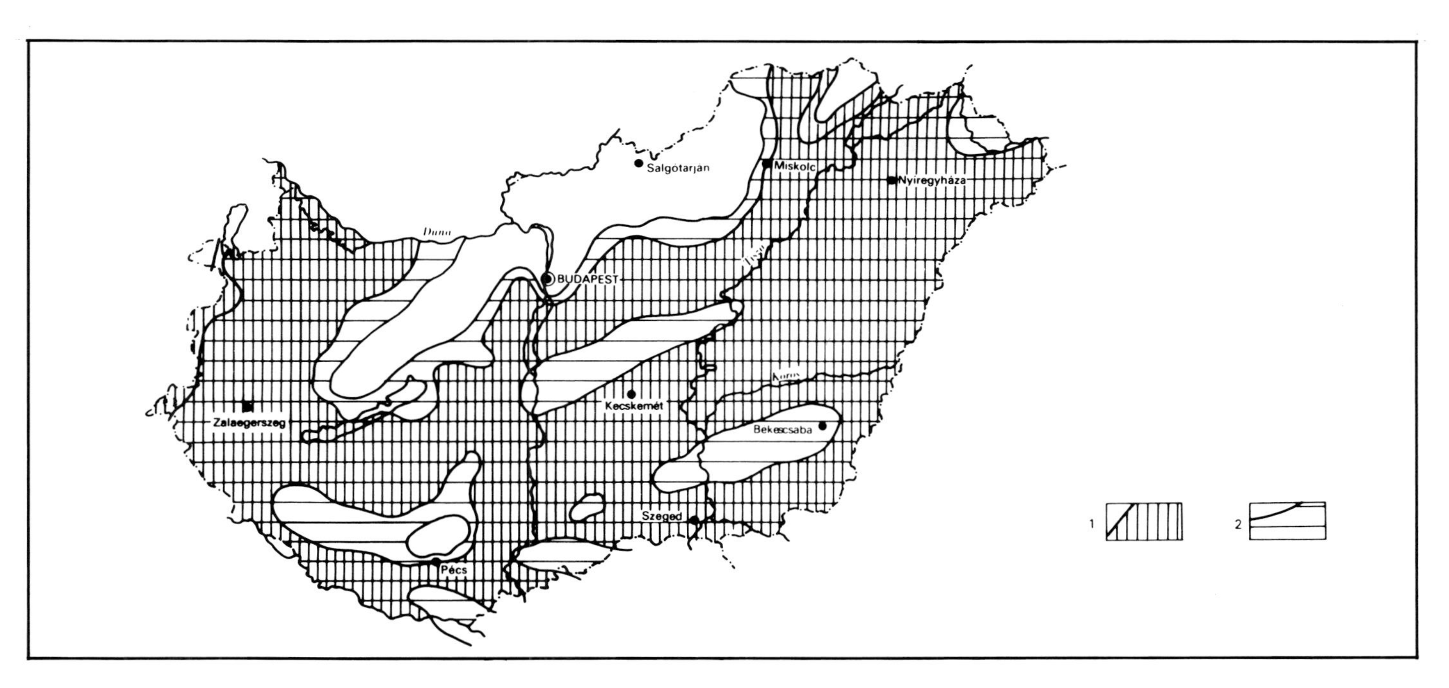

Figure 5. Sketch map of the Neoparatethys (late Miocene–Pliocene) in Hungary. Symbols: (1) maximum extent of strata of Sarmatian age and (2) maximum extent of strata of Pannonian (*s. str.*) and Pontian age.

Dacian area, using correlative molluscan faunas. Indeed, less than 10% of the Pontian molluscan species of the Pannonian basin have been found in the Eastern Paratethyan stratotype area. Common biostratigraphic events in the two areas include the evolutionary lineage leading to *Congeria rhomboidea* and such common forms as *Paradacna abichi, Dreissenomya* div., sp., and perhaps *Dreissena auricularis.* Correlation with other faunal groups has not yet been made. The stratigraphic subdivision of the Pontian in the Pannonian basin can be made using molluscs and several other organisms. Among the most promising ones are the organic-walled microplankton (Sütő-Szentai, 1984, 1985).

Immigration datums (such as from the Eastern Paratethys) may serve as good datum levels for high-tolerance organisms. At present, the immigration of the genus *Prosodacna* seems to provide a solid basis for a stratigraphic division. Evolutionary lineages widely used for subdivisions, however, may be confused with ecophenotypic responses to changing salinity, decreasing water depths or other events that may be diachronous throughout the basin.

Interbasinal endemism also occurs. A characteristic assemblage with the Lymnocardiid genus *Budmania* occurs in southern Transdanubia (west of the Danube) in northern Yugoslavia and Rumania. This form is lacking in northern Transdanubia, although coeval lacustrine sediments are clearly present. This situation suggests the existence of interbasinal barriers. It is not impossible that several lakes with different salinities may have existed in the Pannonian basin during late Pontian time.

Such difficulties in interbasinal correlation may be overcome by mammal and plankton stratigraphy. The search for micromammals in shallow water lacustrine deposits has yielded promising results in recent years (Kordos, 1987; Müller and Szónoky, in prep.). These mammal faunas all belong to the Turolian mammal stage, that is, to the late Miocene Zones MN.

At the beginning of Pontian time, the deep basin areas of the Pannonian basin were already filled with sediment. Therefore, the Pontian stage is represented predominantly by shallow water sediments often intercalated with terrestrial deposits. This filling of the basin must have been gradual, but its exact timing is still unresolved. Radiometrically dated basalts occur in the uppermost part of the Pontian sequences. These are mainly of Pliocene age. One reasonably reliable date is known from Tihany (7.56 ± 0.5 Ma) (Balogh et al., in press) on bombs belonging to a volcanosedimentary formation that is intercalated with mollusc-bearing Pontian deposits containing *Congeria balatonica.* Another data of about 4.0–4.2 Ma was recently determined (Balogh et al., in press) from a basalt interlayered with sediments, which, on the basis of lithology and position, can be correlated with upper Pontian fauna-bearing sediments in the vicinity.

PLIOCENE

The Pontian lake system disappeared by about the beginning of Pliocene time, and the lacustrine environment was replaced by a fluvial–alluvial environment. The endemic Lymnocardiid–*Congeria* fauna became extinct. From this time onward it is preferable to use European mammal zones ("Mein-zones," abbreviated MN) or mammal stages (Ruscinian, etc.) rather than stage names from other basins where these stages are characterized by endemic lacustrine faunas that are totally lacking in the Pannonian basin (such as the Dacian or Romanian stages). Introduction of new stage names for the Pliocene is unnecessary.

Several lines of evidence suggest that the disappearance of the Lymnocardiid-bearing lake was coeval with the latest Messinian. The arguments for this are as follows:

1. The Messinian correlates with the type Pontian, although the weight of this argument is reduced by the small number of the common organisms (see above).
2. The magnetostratigraphic age suggested for the Nagyalföld Variegated Clay Formation (in boreholes of Dévaványa-1 and Vësztő-1) (Rónai and Szemethy, 1979; Rónai, 1982) is about 5.4 Ma. This terrestrial clay-silt series covers the typical Pontian in the deep basins.
3. The Baltavár mammal fauna occurs in a fluvial sequence with *Margaritifera* aff. *flabellatiformis* (equivalent to *Unio wetzleri*) (Kormos, 1914). This sequence overlies the typical Pontian over much of Transdanubia. The Baltavár fauna is well correlated with upper Turolian–upper Messinian localities in Spain.

The 4.0–4.2 Ma radiometric age for Pontian sediments needs additional documentation, but if accepted, it implies that the lake system and the typical Pontian fauna may have survived the Miocene–Pliocene boundary in some parts of the Pannonian basin. Paleontological evidence for Pliocene sediments is found in sand and sandstone of terrestrial origin in Gödöllő (Mottl, 1939) and other places. These sediments overlie Pontian deposits. The Ruscinian type mammal fauna of Gödöllő needs further investigation. Some other Pliocene mammal assemblages have been found in fissure fillings and caves, and their relationship to basin sediments is unclear.

REFERENCES

Báldi, T., 1984, The terminal Eocene and early Oligocene events in Hungary and the separation of an anoxic, cold Paratethys: Eclo. Geol. Helv., v. 77, p. 1–27.

Báldi-Beke, M., and A. Nagymarosy, 1979, On the position of Ottnangian and Karpatian Neogene stages on the Tertiary nannoplankton zonation: Ann. Géol. Pays Helléniques, Tome Hors Sér., v. 1, p. 51–60.

Balogh, K., E. Árva-Soós, Z. Pécskay, and L. Ravasz-Baranyai, in press, K/Ar dating of post-Sarmatian alkali basaltic rocks in Hungary: Acta Min. Petr. Univ. Szeged.

Committee of Mediterranean Neogene Stratigraphy, ed., 1971, Stratotypes of Mediterranean Neogene stages: Giornale di Geol., v. 37, fasc. II, p. 1–266.

Grill, R., 1943, Über mikropaläontologische Gliederungsmöglichkeiten im Miozän des Wiener Beckens: Mitteilungen des Reichsanstalt für Bodenforschung, v. 2, p. 33–44.

Hámor, G., 1985, A. Nógrád-Cserhát Kutatási terület földtani viszonyai (Geology of the Nógrad-Cserhát area): Geol. Hung., Ser. Paleontol., v. 22, Budapest, 307 p.

Hámor, G., L. Ravasz-Baranyai, K. Balogh, and E. Arva-Soós, 1980, A Magyarországi miocén riolittufa-szintek radiometrikus kora, Radiometric ages of the Miocene rhyolite tuffs in Hungary: Magyar Állami Földtani Intézet Évi Jelentése az 1978. Évröl, p. 65–74.

Haq, B. U., 1983, Jurassic to Recent nannofossil biochronology: an update, *in* B. U. Haq, ed., Nannofossil biostratigraphy: Benchmark Papers in Geology, v. 78, p. 358–378.

Horváth, M., and A. Nagymarosy, 1979, On the boundaries of Oligocene/Miocene and Egerian/Eggenburgian in Hungary: Ann. Géol. Pays Helléniques, Tome Hors Sér., v. 2, p. 543–552.

Hsü, K. J., J. L. Brecque, S. F. Percival, R. C. Wright, A. M. Gombos, K. Pisciotto, P. Tucker, N. Peterson, J. A. McKenzie, H. Weissert, A. M. Karpoff, M. F. Carman, Jr., and E. Schreiber, 1984, Numerican ages of Cenozoic biostratigraphic datum levels: results of South Atlantic Leg 73 drilling: Geol. Soc. Amer. Bull., v. 95, p. 863–876.

Jámbor, Á., E. Balázs, I. Berczi, J. Bónd, I. Gajdos, J. Geiger, M. Hajós, L. Kordos, A. Korecz, I. Korecz, et al., 1985, General characteristics of Pannonian s.l. deposits in Hungary (abstract): Abstracts of VIIIth RCMNS Congress, Sept. 15–22, 1985, Budapest, Hungary.

Kókay, J., 1971, Das Miozän von Várpalota: Föld. Közl., v. 101, p. 74–90.

Kókay, J., 1985, Central and Eastern Paratethys interrelations in the light of late Badenian salinity conditions: Geol. Hung. Ser. Palaeont., v. 48, p. 3–95.

Kordos, L., 1987, Neogene vertebrate biostratigraphy in Hungary, *in* Proceedings of VIII Congress of RCMNS, Budapest, Ann. Inst. Geol. Publ. Hungary, v. 70, p. 393–396.

Kormos, T., 1914, Uber die Resultate meiner Ausgrabungen im Jahr 1913: Jahresbericht der Kóniglich, Ungarischen Geologischen Reichanstalt für 1913, p. 559–604.

Korpás-Hódi, M., 1983, A Dunántuli-középhegység északi elötere pannoniai molluszka faunájának paleoökologiai és biosztratigráfiai vizsgálata [Palaeoecology and biostratigraphy of the Pannonian mollusca fauna in the Northern foreland of the Transdanubian Central Range]: Magyar Állami Földtani Intézet Évkönyve, v. 66, p. 1–163.

Kretzoi, M., 1954, Rapport final des fuilles paléontologiques dans le grotte de Csákvár: Magyar Állami Földtani Intézet évi jelentése az 1952. Evröl, p. 37–69.

Lehotayová, R., and M. Báldi-Beke, 1975, Kalkige Nannoflora der Sedimente des Egerien, *in* T. Báldi and J. Senes, eds., Chronostratigraphie und Neostratotypen, v. 5: Oligocene–Miocene Egerian VEDA, Bratislava, p. 1–577.

Martini, E., 1971, Standard Tertiary and Quaternary calcareous nannoplankton zonation: Proc. Second Conf. Planktonic Microfossils, v. 2, p. 739–786.

Martini, E., and C. Müller, 1975, Calcareous nannoplankton and silicoflagellates from the type Ottnangian and equivalent strata in Austria: Proc. VI Congress CMNS, v. 1, p. 121–124.

Mottl, M., 1939, Die mittelpliozäne Säugetierfauna von Gödöllö bei Budapest: A Magyar Királyi Földtani Intézet Evkönyve, v. 32, p. 257–350.

Müller, P., 1984, Decapod crustacea of the Badenian: Geol. Hung. Ser. Palaeont., v. 42, p. 1–121.

Müller, P., ed., 1985, Excursion Al, Pre-Congress: Field guide of the VIII Congress RCMNS, Budapest, p. 1–68.

Müller, P., and M. Szónoky, in preparation, Faciostratotype Tihany-Fehérpart, *in* P. Stevanovic, ed., Chronostratigraphie und Neostratotypen, v. 8: Pontian.

Nagymarosy, A., 1980, A magyarországi bádenien korrelációja nannoplankton alapján [Correlation of the Badenian in Hungary on the basis of Nannoplankton]: Földt. Közl., v. 110, p. 206–245.

Nagymarosy, A., 1981, Chrono- and biostratigraphy of the Pannonian basin: a review based mainly on data from Hungary: Earth Evol. Sci., v. 1, p. 183–194.

Nagymarosy, A., 1985, The correlation of the Badenian in Hungary based on nannofloras: Ann. Univ. Sci. Budapest. Rolando Eötvös Nominatae, Ser. Geol., v. 25, p. 33–86.

Okada, M., and D. Bukry, 1980, Supplementary modification and introduction of code numbers to the low-latitude coccolith biostratigraphic zonation (Bukry, 1973, 1975): Marine Micropaleontology, v. 5, p. 321–325.

Papp, A., 1953, Die Molluskenfauna des Pannon im Wiener Becken: Mitteilungen der Geologischen Gesellschaft in Wien, v. 44, p. 86–222.

Rögl, F., and F. Steininger, 1983, Vom Zerfall der Tethys zu Mediterran und Paratethys. Die neogene Paläogeographie und Palinspastik des zirkum-mediterranen Raumes: Ann. Naturhis. Museums Wien, v. 85/A, p. 135–163.

Rögl, F., F. Steininger, and C. Müller, 1978, Middle Miocene salinity crisis and Palaeogeography of the Paratethys (middle and eastern Europe), *in* K. J. Hsü, L. Montadert, et al., Init. Rep. DSDP, v. 42, pt. 1, p. 985–990.

Rónai, A., 1982, A negyedidöszaki és felsö pliocén süllyedés menete a Körös-medencében [History of Quaternary and Upper Pliocene Subsidence of the Körös basin]: Magyar Állami Földtani Intézet Évi Jelentése az 1980. Évrôl, p. 77–80.
Rónai, A., and A. Szemethy, 1979, Az Alföld-kutatás ujabb eredményei. Paleomágneses vizsgálatok laza üledékeken [Latest results of lowland research in Hungary: Paleomagnetic measurement on unconsolidated sediments]: Magyar Állami Földtani Intézet Évi Jelentése az 1977. Évröl, p. 67–84.
Steininger, F., and L. A. Nevesskaya, eds., 1975, Stratotypes of Mediterranean Neogene stages,, v. 2: Bratislava, p. 1–364.
Steininger, F., and F. Rögl, 1979, The Paratethys history—a contribution towards the Neogene geodynamics of the Alpine orogene (abs.): Ann. Géol. Pays Helléniques, Tome Hors Sér., v. 3, p. 1153–1165.
Sütő-Szentai, M., 1984, Szervesvázú mikroplankton vizsgálatok a Mecsek-hegység környékének pannoniai rétegeiböl [Studies of organic microplankton from the Pannonien of the Mecsek Mountains]: Folia Comloensis, v. 1, p. 55–77.
Sütő-Szentai, M., 1985, Die Verbreitung organischen Mikroplankton-Vergesellschaftungen in den pannonischen Schichten Ungarns, *in* A. Papp, ed., Chronostratigraphie und Neostratotypen, v. 7: M6 Pannonian, Akad. Kiadó, Budapest, p. 1–636.
Vail, P. R., R. M. Mitchum, S. Thompson, III, 1977, Seismic stratigraphy and global changes of the sea-level, *in* Ch. E. Payton, ed., Seismic stratigraphy—applications to hydrocarbon exploration. Amer. Assoc. Pet. Geol Memoir 26, p. 83–98.
Vanova, M., 1975, Faciostratotypus der Bretkaer Formation, *in* T. Báldi and J. Senes, eds., Chronostratigraphie und Neostratotypen, v. 5: OM Egerien, VEDA, Bratislava, p. 1–577.
Vass, D., and G. P. Bagdasarjan, 1978, Radiometric time scale for the Neogene of the Paratethys region, *in* Contributions to the Geological Time Scale: AAPG Studies in Geology, v. 6, p. 179–203.

CHAPTER 7

Correlation of Central Paratethys, Eastern Paratethys, and Mediterranean Neogene Stages

IGCP-Project 73/1/25, "Stratigraphic Correlation Tethys–Paratethys Neogene"

Fritz F. Steininger
Institute for Paleontology, University of Vienna
Universitatsstr. 7, A-1010 Vienna, Austria

C. Müller
1, Rue Martignon; F-92500 Reuil-Malmaison, France

F. Rögl
Department of Geology and Paleontology, Natural History Museum Vienna
Burgring 7, A-1014 Vienna, Austria

The most important biostratigraphic datum planes and radiometric and paleomagnetic events are given for the correlation of the European Neogene stage systems of the circum-Mediterranean area.

INTRODUCTION

Accurate stratigraphic correlations of local and regional stage concepts form the basis for almost all geological and paleontological progress. The Regional Committee on Mediterranean Neogene Stratigraphy (RCMNS) and IGCP Project No. 25, titled "Stratigraphic Correlation of Tethys-Paratethys Neogene," consider such correlations to be their primary goal for the Neogene of the circum-Mediterranean area. At the 6th RCMS Congress in Bratislava in 1975, the final succession of chronostratigraphic stages for the Mediterranean, the Central Paratethys, and the Eastern Paratethys was agreed upon unanimously (Berggren et al., 1975; Menner, 1975) and a tentative correlation table of these stage concepts was given in the second volume of the proceedings (Senes, 1975). Progress since 1975 was summarized by A. Papp for the RCMNS Congress held in Athens in 1979 (Papp, 1981). Today, the Mediterranean stages are well defined, although with some minor boundary problems. Due to the progress made by the deep sea drilling program, the Mediterranean stages have been well dated by planktonic biochronologies and paleomagnetic and radiometric data. In general, they are used as worldwide standard stages, and the most recent calibrations are given by Berggren (1984) and Berggren et al. (1984a,b). In calibration of planktonic biochronologies with magnetostratigraphic units, however, there exist a number of more recent papers (especially Berggren et al., 1984a,b; Haq, 1983; Keller, 1980, 1981; Langereis and Zachariasse, 1984; Poore et al., 1983; and others) with different interpretations. The general correlation of the different stage systems is based primarily on the identification of planktonic biochronology. Absolute ages and paleomagnetic epochs assigned to these stages are those of the worldwide time scales. Local Paratethyan radiometric dates are summarized in the text, but the uncertainties in some of these measurements, because of large standard deviations, make them less reliable than the biostratigraphic correlations.

Most of the Paratethys stages have been defined and documented in a series of publications (Figure 1) (Carloni et al., 1971: Senes, 1979a; Senes, 1967-1978; Steininger and Nevesskaja, 1975). An accurate calibration and correlation between the Western–Central and Eastern Paratethys and the Mediterranean is still under discussion as demonstrated by a selected list of more recent papers: Andreescu (1981), Cicha (1975), Baldi-Beke and Nagymarosy (1979), Fahlbusch (1981), Gabunia (1981), Hochuli (1978), Horváth and Nagymarosy (1979), Jung (1982), Kojumdgieva (1979a,b), Kretzoi (1982), Nagymarosy (1981), Marinescu (1980), Paramonova et al. (1979), Pevzner and Vangeingeim (1982), Rögl et al. (1979), Semenenko (1979), Senes (1979a,b), Steininger (1979), Steininger et al. (1976), Steininger and Papp (1979), Vass and Bagdasarian (1978), and Veselov (1979).

The aim of this chapter is to synthesize the most recent data on the correlation of the Paratethys stage concepts.

THE PARATETHYS CONCEPT

The different faunal and tectonic evolution of the Neogene epicontinental seas north and south of the Alpine–Caucasian orogenic belt motivated Laskarev (1924) to separate this northern bioprovince—the Paratethys—from the Neogene Mediterranean Tethys bioprovince (Figure 2). Its geodynamic evolution makes it possible to distinguish an Eo-, Meso- and Neoparatethys (Senes and Marinescu, 1974). Changing seaways and successive transgressions and regressions throughout the Paratethys were followed by the development of different ecosystems (Figure 3), thus permitting subdivision into three biosubprovinces: the Western Paratethys (Rhone Basin to Bavaria), the Central Paratethys (the Alpine–Carpathian foredeep and intramontane basins from Austria to the Ukraine and Roumania), and the Eastern Paratethys (the Ponto–Caspian realm from the Black Sea to the Aral Sea) (Senes, 1959). The paleogeographic and geokinematic development of the Paratethys and the circum-Mediterranean region was presented recently by Rögl and Steininger (1983).

In the Eoparatethys, early Miocene marine transgression (late Eggenburgian to early Ottnangian) created a sea that extended from the Western Mediterranean across the Rhone and Molasse basins to the Ponto–Caspian region. From Ottnangian to middle Badenian time, the Yugoslavian corridor provided marine connections from the northern Italian sedimentary basins to the Central Paratethys (Mesoparatethys). The closure of this seaway brought about the isolation of the Neoparatethys. The connection from the Eastern Paratethys to the Indo-Pacific across the Mesopotamian Trough is faunally very important, but the connection changed rapidly because of the active tectonic zones east of the Paratethys. This seaway opened and closed repeatedly from the late Oligocene to the end of the middle Miocene. During late Miocene and Pliocene time, brief marine ingressions extended into the Eastern Paratethys from the Aegean Sea across the Dardanelles.

CORRELATION OF CENTRAL PARATETHYS, EASTERN PARATETHYS, AND MEDITERRANEAN NEOGENE STAGES

In calibrating marine plankton biochronologies and the Mediterranean stages, we followed, with some modification, the biostratigraphy of DSDP-Leg 42A (Berggren, 1984; Berggren et al., 1984a,b; Bizon and Müller, 1979; Cita et al., in press). Major differences exist between these publications in their assessment of (1) the duration of Martini's calcareous nannoplankton zones and their calibration with the various plankton zonations, especially in early, middle and late Miocene and (2) the geochronometric calibration and the biostratigraphic correlation between the early and middle Miocene Mediterranean stages. A brief outline follows on the most important datum levels that can be used for a better correlation of the Central and Eastern Paratethys stages with the Mediterranean stages and the European Mammal Ages.

Late Oligocene–Early Miocene

Egerian, Caucasian, and Chattian–Aquitanian

A horizon of larger foraminifera, including *Miogypsina septentrionalis, Miogypsinoides formosensis, Lepidocyclina (Neophrolepidina) morgani, Cycloclypeus, Operculina and Heterostegina* occur in the upper part of nannoplankton zone NP 25 (Adams, 1984; Drooger, 1979; McGowran, 1979; and Steininger et al., 1976).

Within this marine horizon, a vertebrate fauna known from Linz in Upper Austria can be assigned to mammal zone MN-0 (Rabeder and Steininger, 1975). Nannoplankton zone NN1 is present in the uppermost Egerian of the Austrian Molasse zone (Rögl et al., 1979).

Early Miocene

Eggenburgian, Sakaraulian, and Burdigalian

The early Miocene horizon of "giant" mollusc taxa within nannoplankton zone NN 2 correlates the early Eggenburgian and Sakaraulin with the early Burdigalian (Horváth and Nagymarosy, 1979; Martini, 1981; Rögl et al., 1979; Steininger et al., 1976).

Early Orleanian (MN 3) micromammal association appeared in Early Eggenburgian marine sediments with NN 2 nannoplankton flora (P. Mein, personal communication).

The first appearance of Proboscidea (MN 3b) in the Paratethys is dated as late Eggenburgian by nannoplankton zone NN 3 (Rögl and Steininger, 1983). *Miogypsina intermedia* appear in the late Eggenburgian within sediments dated by nannoplankton as NN 2/NN 3 (Rögl et al., 1979).

Ottnangian, Kozachurian, and Burdigalian

Ottnangian nannoplankton ages from the Central Paratethys range from zone NN 3 to zone NN 4 (Báldi-Beke and Nagy-

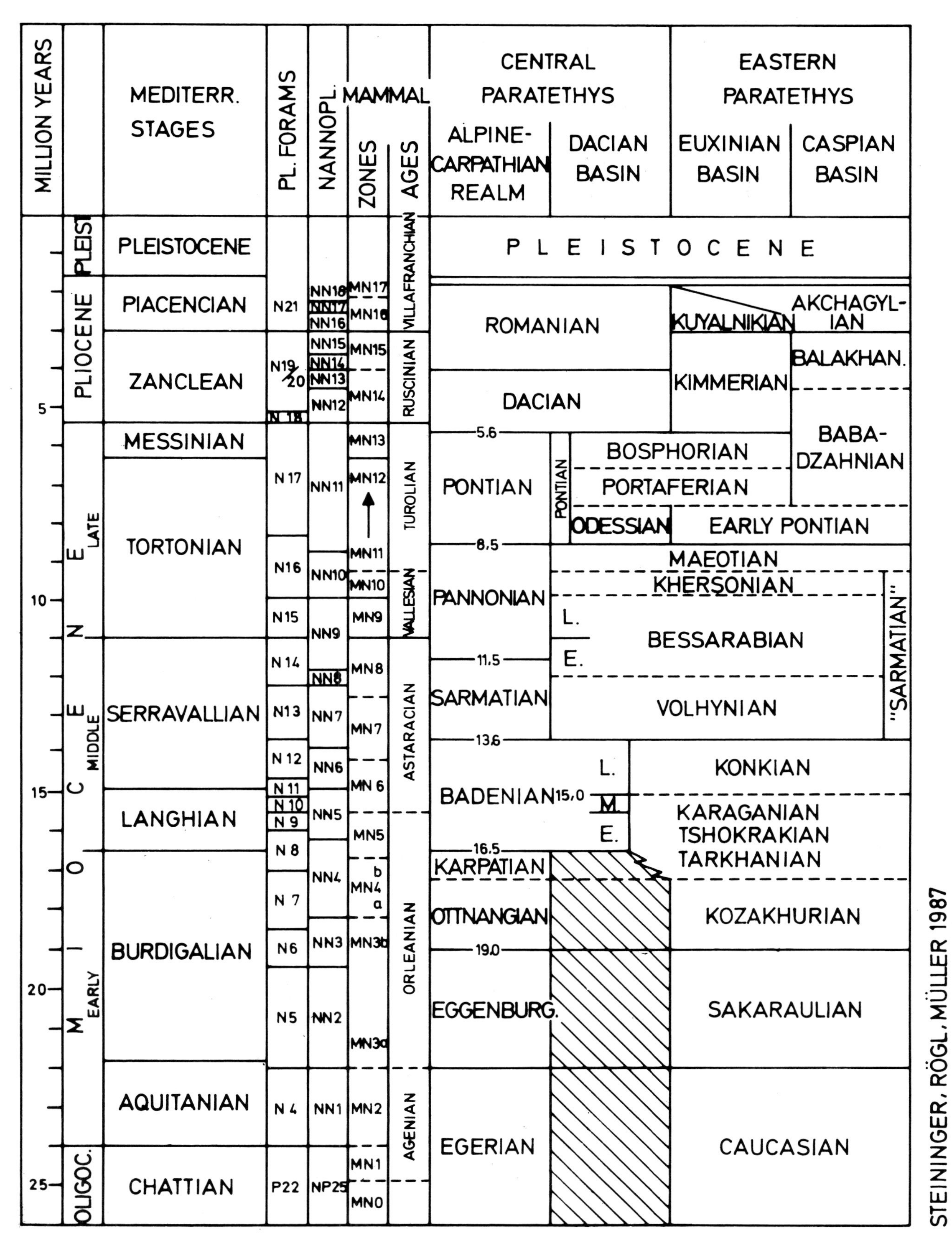

Figure 1. Correlation chart of Neogene biostratigraphic and chronostratigraphic units for the Mediterranean, the Central, and the Eastern Paratethys as well as for the European Mammal Ages. (See the Addendum for additional comments.)

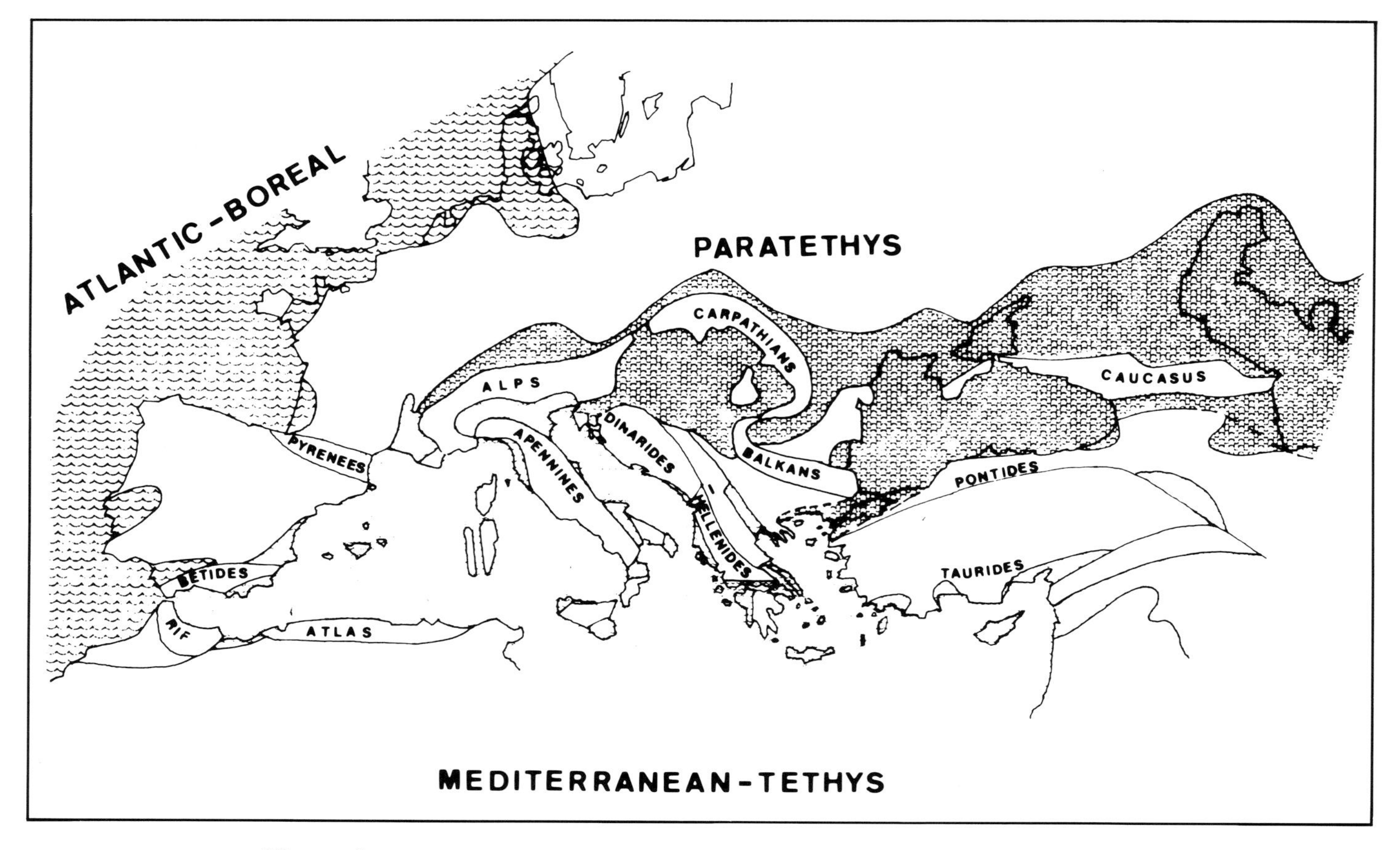

Figure 2. Neogene marine sedimentation areas and general bioprovinces throughout Europe.

marosy, 1979; Martini, 1981; Martini and Müller, 1975a; Rögl et al., 1979).

During the late Ottnangian regressive phase and during the early Karpatian(?), the "Oncophora" beds, with typical (reduced brackish salinity) to freshwater mollusc assemblages, are widespread from the Western to the Eastern Paratethys (Ctyroky et al., 1973). Nannoplankton floras from these beds belong to NN 4 zone (Báldi-Beke and Nagymarosy, 1979).

The "Oncophora" beds overlie the important mammal fauna of Orechov, CSSR (Cicha, Fahlbusch, and Fejfar, 1972), indicative for mammal zone MN 4a (Mein, 1979). For the "lower Rhyolitic Tuff," a widespread horizon of early Ottnangian age in northern Hungary, radiometric dates average 19.6 ± 1.4 Ma (Hámor et al., 1979) and range from 22.0 to 18.0 Ma (Hámor et al., 1978). These dates are in accordance with its biostratigraphic age (Rögl and Steininger, 1983).

Karpatian and Late Burdigalian

The most important datum levels are NN 4 to NN 5 nannoplankton floras and the first appearance datum (FAD) of *Globigerinoides bisphericus* (sometimes incorrectly called "sicanus") in the upper part of the Karpatian Schlier-formations (Báldi-Beke and Nagymarosy, 1979; Martini and Müller, 1975b; Steininger et al., 1976). According to Rehakova (1977), diatom assemblages out of the upper Karpatian "Schlier" formations allow a correlation with Scharder's North Pacific Diatom Zonation (NPDZ 25/24). Because the late Kozachurian of the Eastern Paratethys is characterized by the evolution of many endemic molluscs, its direct correlation with the marine sequences of the Karpatian in the Central Paratethys is not possible (Rögl and Steininger, 1983).

Middle Miocene

Badenian, Tarchanian to Konkian, and Langhian to Middle Serravallian

The worldwide planktonic datum plane with the FAD of *Praeorbulina* is known from the base of the Badenian and the Tarchanian (Cicha, 1970; Steininger et al., 1976).

The entire Badenian and the Eastern Paratethys stages range from nannoplankton zone NN 5 into zone NN 7 (Fuchs and Stradner, 1977; Lehotayova and Molcikova, 1978; and Stradner and Fuchs, 1978). According to Rehakova (1977), the diatom floras of the Badenian can be correlated to NPDZ 24 to NPDZ 22.

Biostratigraphically well-controlled radiometric ages are known throughout the Badenian. They range from 17.1 to 13.3 Ma (Hámor et al., 1978, 1979; Vass et al., 1978; and Vass and Bagdasarjan, 1978).

Important mammal faunas that can be correlated directly with the marine sequences belong to mammal zones MN 5 and MN 6 (Mein, 1979; Rabeder and Steininger, 1975).

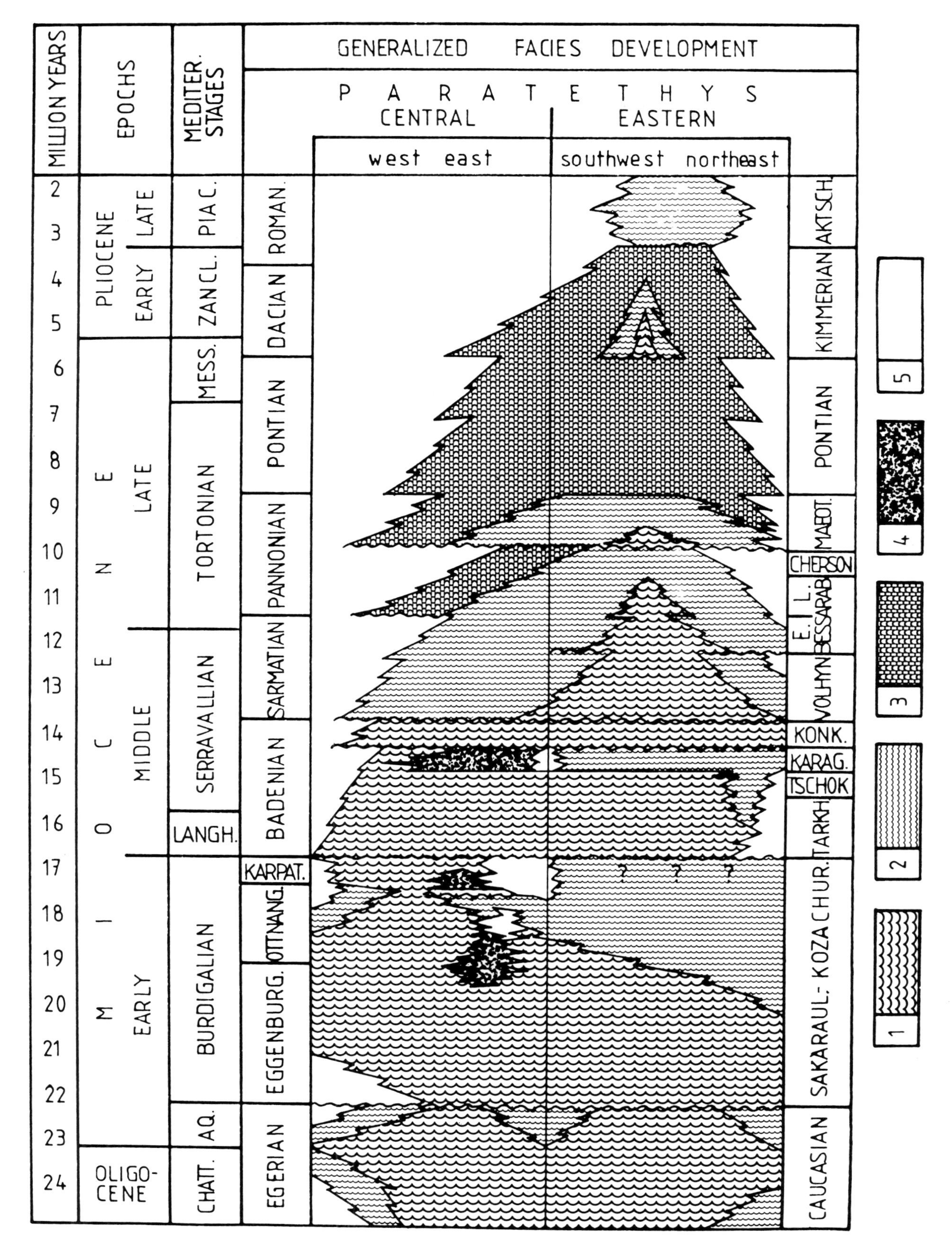

Figure 3. Generalized late Oligocene to Neogene facies distribution from west to east in Central Paratethys and Southwest to Northeast in Eastern Paratethys. (1) marine; (2) reduced marine; (3) endemic "Congeria-Melanopsis" facies in late Miocene and Pliocene; (4) evaporitic; and (5) continental.

The paleobiogeographic and geodynamic evolution of the Badenian was extensively treated by Rögl et al. (1978) and Rögl and Steininger (1983).

The more or less continuous marine sedimentation of the Paratethys is terminated at the end of the late Badenian. The latter part of the middle Miocene, the Sarmatian, is characterized by a reduced salinity facies, which prevails in the Eastern Paratethys into the late Miocene. The late Miocene of the Central Paratethys is dominated by "Congeria-Melanopsid" assemblages and finally by fluvial-limnic and continental deposits (see Figure 3). Correlation of the late–middle Miocene, late Miocene and Pliocene stages is rather well established by the evolution of endemic invertebrate faunas in the Central and Eastern Paratethys (Andreescu, 1981; Kojumdgieva, 1979a; Paramonova et al., 1979; Semenenko, 1979).

Correlations of this time span with the marine Mediterranean stages could be established as follows: (1) by mammal faunas interspersed into those sediments with endemic invertebrate faunas, (2) through short marine transgressive events into the Ponto–Caspian realm, and (3) by radiometric and paleomagnetic dating.

Sarmatian–Volhynian to Early Bessarabian-Late Serravallian

Marine diatom floras of the Early Sarmatian correlate with diatom zone NPDZ 19 (Schrader, personal communication). Late Asteracian mammal faunas of zone MN 8 are interspersed into Sarmatian sediments with endemic molluscs (Feru et al., 1980). A key feature is the lack of *Hipparion* in all Sarmatian mammal faunas found so far (Rabeder and Steininger, 1975; Steininger et al., 1976).

Biostratigraphically controlled radiometric ages from Sarmatian volcanic rocks range between 13.6 to 10.7 Ma (Vass and Bagdaserjan, 1978) and yield 13.7 ± 0.8 Ma for the "upper Rhyolite Tuff" of Hungary (Hámor et al., 1979).

Late Miocene

Pannonian, Bessarabian to Maeotian, and Early to Middle Tortonian

Vallesian mammal faunas of zone MN 9 with the FAD of *Hipparion* are known from lower Pannonian and upper Bessarabian sediments. The FAD of *Hipparion* was dated as early Tortonian in the Mediterranean (Benda and Meulenkamp, 1979; de Bruijn and Zachariasse, 1979; Steininger and Papp, 1979).

Late Vallesian mammal faunas of zone MN 10 are intercalated into upper Pannonian and Maeotian sediments (Steininger and Papp, 1979).

Because of a marine ingression into the Euxinian Basin in early Maeotian time, nannofloras of zone NN 10 occur in Pannonian sediments. Simultaneously, endemic Maeotian Paratethys mollusc faunas migrated into the Aegean Sea as far south as Athens (Papp and Steininger, 1979; Semenenko, 1979; Semenenko and Ljulieva, 1978).

Pontian and Late Tortonian to Messinian

Lower Pontian sediments yielded rich lower Turolian mammal faunas of zone MN 11. Mammal faunas of this zone are intercalated into marine upper Tortonian sediments in the Mediterranean (de Bruijn et al., 1975; Steininger and Papp, 1979).

The typical, rich Messinian mammal faunas of zone MN 13 are interspersed in the Central and Eastern Paratethys within late Pontian sediments (de Bruijn et al., 1975; van Couvering et al., 1976; Kojumdgieva, personal communication; Mein et al., 1973; and Steininger and Papp, 1979).

Biostratigraphically well-controlled radiometric dates are scarce in the Paratethys; they range from 9.4 to 7.0 Ma (Vass and Bagdasarjan, 1978; Bérczi et al., 1982).

Paleomagnetic dating of late Miocene to Pleistocene Paratethyan sections in the Dacian Basin (Roumania) and Black Sea area with endemic mollusc faunas and mammals allow for an excellent correlation to the paleomagnetic time scale (Andreescu, 1981; Andreescu et al., 1981; Semenenko, 1979; Semenenko and Pevsner, 1979).

Pliocene

Dacian, Kimmerian, and Zanclean

The Early Pliocene marine transgression extended as far as the Ponto–Caspian realm in the early Kimmerian and nannoplankton floras of zone NN 12 are known from this area (Semenenko and Ljulieva, 1978; Vail et al., 1977).

Romanian, Aktschagylian, and Piacenzian

A last marine transgression through the Dardanelles into the Black Sea provides typical late Pliocene Mediterranean ostracodes and nannoplankton in Aktschaglyian sediments (Semenenko and Ljulieva, 1978; Taner, 1982).

The correlation of the upper part of the Roumanian and Aktschagylian stages was recently supported by paleomagnetic dating of mollusc and mammal bearing sections in Romania that straddle even the Neogene–Quaternary boundary (Adreescu et al., 1981).

ADDENDUM

The calcareous nannoplankton associations of the late Badenian–Sarmatian boundary interval are being reinvestigated in typical sections of the Central Paratethys by M. P. Aubry and W. Berggren. These investigations point to a Badenian–Sarmatian boundary within calcareous nannoplankton zone NN 7. This is not in contrast to the boundary shown on Figure 1, where this boundary falls into calcareous nannoplankton zone NN 7, as was indicated by earlier studies.

Berggren et al. (1984a,b) have shown recently that the previous correlations of magnetic anomalies 5 and 5A to magnetic polarity epochs 9 and 11 are incorrect. The revised correlations between magnetostratigraphic and biostratigraphic boundaries are 1.5 to 2.0 m.y. younger in the middle Miocene than those shown in Figure 1. These new results shift the Burdigalian–Langhian (Karpatian–Badenian) boundary to 16.1 m.y. The Badenian–Sarmatian boundary in the Central Paratethys changes to 13.2 m.y. The Sarmatian–Pannonian boundary in the Central Paratethys changes to 12.0 (or 12.5) m.y. The Pannonian–Pontian boundary remains at approximately 8.5 m.y., as shown on Figure 1.

These new results point to the following possible correlations of Mediterranean and Central Paratethys Stages. Badenian: Langhian to lower Serravallian; Sarmatian: middle Serravallian; Pannonian: upper Serravallian to lower Tortonian; Pontian: upper Tortonian to Messinian. The correlation between Central and Eastern Paratethys stages is not affected by these new results. These new results should be kept in mind in using Figures 1 and 3.

REFERENCES

Adams, C.-G., 1984, 6. Neogene larger foraminifera: Evolutionary and geological events in the context of datum planes, *in* N. Ikebe and R. Tsuchi, eds., Pacific Neogene Datum Planes, p. 47–67.

Alberdi, M. T. and E. Aguirre, 1977, Round-Table on mastostratigraphy of the W. Mediterranean Noegene: Trab. Neogeno-Quaternario, Sec. Paleont. Vertebr. Humana, p. 1–47.

Andreescu, I., 1981, Middle–Upper Neogene and early Quaternary chronostratigraphy from the Dacic Basin and correlations with neighbouring areas: Ann. Géol. Pays Hellán., hors sér., fasc. 4, p. 129–138.

Andreescu, I., C. Radulescu, P. Samson, A. Tshepalyga, and V. Trubikhin, 1981, Chronologie (mollusques, mammiféres, paléomagnetisme) des formations plio–pleistocenes de la zone de Slatina (Basin dacique), Romanie: Trav. Inst. Speol. "Emile Racovitz," v. 20, p. 127–131.

Báldi-Beke, M. and A. Nagymarosi, 1979, On the position of the Ottnangian and Karpatian regional stages in the Tertiary nannoplankton zonation: Ann. Géol. Pays Hellén., tome hors sér. 1979, p. 51–59.

Benda, L. and J.-E. Meulenkamp, 1979, Biostratigraphic correlations in the Eastern Mediterranean Neogene. 5. Calibration of sporomorph associations, marine microfossil and mammal zones, marine and continental stages and the radiometric scale: Ann. Géol. Pays Hellén., tome hors sér., 1979, fasc. 1, p. 61–70.

Bérczi, I., G. Hámor, Á. Jámbor, and K. Szentgyörgyi, 1982, Characteristics of Neogene sedimentation in the Pannonian Basin, *in* F. Horváth, ed., Evolution of extensional basins within regions of compression, with emphasis on the Intracarpathians, p. 36–38.

Berggren, W.-A., M.-B. Cita, R. Selli, V. Nosovsky, and F. F. Steininger, 1975, Establishment of standard regional stages: Proc. 6 Congr. Reg. Comm. Mediterranean Neogene Strat., v. 2, p. 23–24.

Berggren, W.-A., 1981, Correlation of Atlantic, Mediterranean, and Indo-Pacific Neogene stratigraphies: Geochronology and Chronostratigraphy: Proc. IGCP-114 Intern. Workshop Pacific Neogene Biostrat., p. 29–60.

Berggren, W.-A., 1984, Correlation of Atlantic, Mediterranean, and Indo-Pacific Neogene stratigraphies: Geochronology and Chronostratigraphy, *in* N. Ikebe and R. Tsuchi, eds., Pacific Neogene Datum Planes, p. 93–110.

Berggren, W.-A., D.-V. Kent, and J.-A. Van Couvering, 1984a, Neogene geochronology and chronostratigraphy, 72 p. (preprint), *in* N. J. Snelling, ed., Geochronology and the geologic time scale, Geol. Soc. London, spec. paper.

Berggren, W.-A., D.-V. Kent, J.-J. Flynn, and J.-A. Van Couvering, 1984b, Cenozoic geochronology, 25 p. (preprint), *in* N. J. Snelling, ed., Geochronology and the geologic time scale, Geol. Soc. London, spec. paper.

Bizon, G. and C. Müller, 1979, Report on the working group on micropaleontology: Ann Géol. Pays Hellén., tome hors sér., 1979, p. 1335–1364.

Blow, W. H., 1969, Late middle Eocene to recent planktonic foraminiferal biostratigraphy: Proc. 1st Int. Conf. Plankt. Microfoss., Geneve 1967, v. 1, p. 199–422.

Bruijn, de, H., P. Mein, C. Montenat, and A. van de Weerd, 1975, Corrélations entre les gisements de rongeurs et les formations marines du Miocene terminal d'Espagne méridionale, I: Provinces d'alicante et de Murcia: K. Nederl. Akad. Wet., Proc., Ser. B, v. 78, p. 1–32.

Bruijn, de, H., and W. J. Zachariasse, 1979, The correlation of marine and continental biozones of Kastellios hill reconsidered: Ann. Géol. Pays Hellén., tome hors sér., 1979, fasc. 1, p. 219–226.

Carloni, G. C., P. Marks, R. F. Rutsch, and R. Selli, eds., 1971, Stratotypes of Mediterranean Neogene stages: Comm. Mediterranean Neogene Strat., v. 1, Gior. Geol. Ser. 2, v. 37, 226 pages.

Cicha, I., 1970, Stratigraphical problems of the Miocene in Europe: Rozpravy U.U.G., Svaz, v. 35, 134 pages.

Cicha, I., ed., 1975, Biozonal division of the upper Tertiary basins of the Eastern Alps and West Carpathians: Geol. Surv. Prague, 147 pages.

Cicha, I., V. Fahlbusch, and O. Fejfar, 1972, Biostratigraphic correlation of some Late Tertiary vertebrate faunas in Central Europe: N. Jb. Paläont. Abh., v. 140, p. 129–145.

Cita, M. B., A. Vismara-Schilling and E. Robba (in press), Correlation of the younger Neogene in the Tethys area (from middle Miocene to Pleistocene): IGCP-Project 25, final results, Elsevier.

Ctyroky, P., J. Senes, F. Strauch, A. Papp, V. Kantorová, A. Ondrejickova, D. Vass, and M. Bohn-Havas, 1973, Die Entwicklung der Rzehakia (Oncophora) Formation—M 2 c-d in der Zentrlen paratethys: Chronostrat. & Neostratotypen, v. 3, p. 89–113.

Drooger, C. W., 1979, Marine connections of the Neogene Mediterranean, deduced from the evolution and distribution of larger foraminifera: Ann. Géol. Pays Helén., tome hors sér., 1979, fasc. 1, p. 361–369.

Fahlbusch, V., 1981, Miozän und Pliozän—Was ist was? Zur Gliederung des Jungtertiärs in Süddeutschland: Mitt. Bayer. Staatsslg. Paläont. hist. Geol., v. 21, p. 121–127.

Feru, M., C. Radulescu, and P. Samson, 1980, La faune de micromammifères du Miocène de Comanesti (Dép d' Arad): Trav. Inst. Spéol, "Emile Racovitza," v. 19, p. 171–190.

Fuchs, R., and H. Stradner, 1977, Úber Nannofossilien im Badenian (Mittelmiozän) der Zentralen Paratethys: Beitr. Paläont. Ósterr., v. 2, p. 1–58.

Gabunia, L., 1981, Traits essentiels de l'évolution des faunes de Mammifères Néogénes de la région mer Noire-Caspienne: Bull. Mus. natn. Hist. nat. Paris, 4 sér., 3, 1981, sect. C, no. 2, p. 195–204.

Hámor, G., L. Ravaszné-Baranyai, K. Balogh, and E. Árvané-Soós, 1978, Radiometric age of the Miocene rhyolite tuffs in Hungary: Inst. Geol. Publ. Hungaricum, 1978, p. 65–73.

Hámor, G., L. Ravaszné-Baranyai, K. Balogh, and E. Árvané-Soós, 1979, K/Ar dating of Miocene pyroclastic rocks in Hungary: Ann. Géol. Pays Hellén., tome hors séf., fasc. 2, p. 491–500.

Haq, B.-U., 1983, Jurassic to recent nannofossil biochronology: An update, *in* B.-U. Haq, ed., Nannofossil biostratigraphy: Benchmark Pap. Geol., v. 78, p. 358–378.

Hochuli, P., 1978, Palynologische Untersuchungen im Oligozän und Untermiozän der Zentralen und Westlichen Paratethys: Beitr. Paläont. Ósterr., v. 4, p. 1–132.

Horváth, M. and A. Nagymarosy, 1979, On the boundaries of Oligocene/Miocene and Egerian/Eggenburgian in Hungary: Ann. Géol. Pays Hellén., tome hors sér., 1979, p. 543–552.

Jung, P., 1982, ed., Nouveaux résultats biostratigraphiques dans le bassin molassique, depuis le Vorarlberg jusqu'en Haute-Savoie: Docum. Lab. Géol. Lyon, H.S.7. 1982, 91 pages.

Keller, G., 1980, Early to middle Miocene planktonic foraminiferal datum levels of the equatorial and subtropical Pacific: Micropaleontol., v. 26, p. 371–391.

Keller, G., 1981, Miocene biochronology and paleoceanography of the North Pacific: Mar. Micropaleontol., v. 6, p. 535–551.

Kojumdgieva, E., 1979a, Le IXième symposium du groupe de travail

"Paratéthys" (11-18.Ix.1978-Sofia): Geol. Balcanica, v. 9, p. 112–113.

Kojumdgieva, E., 1979b, Critical notes on the stratigraphy of Black Sea boreholes (Deep Sea Drilling Project, Leg 42B): Geol. Balcanica, v. 9, p. 107–110.

Kretzoi, M., 1982, Tentative correlation of Late Cenozoic stratigraphy in the Carpathian basin: Inst. Geol. Publ. Hungaricum, p. 407–416.

Langereis, C.-G., W. J. Zachariasse, and J. D. A. Zijderveld, 1984, Late Miocene magnetobiostratigraphy of Crete: Mar. Micropaleontol., v. 8, p. 261–281.

Laskarev, V., 1924, Sur les équivalents du Sarmatien supérieur en Serbie, *in* Recueil de travaux offert à M. Jovan Cvijić par ses amis et collaborateurs, 13 pages.

Lehotayova, R. and V. Molcikova, 1978, 8. Die Nannofossilien des Badenian. 8.1. Das Nannoplankton in der Tschechoslovakei, *in* A. Papp, I. Cicha, J. Senes, and F. Steininger, M_4—Badenian (Moravien, Wielicien, Kosovien): Chronostrat. & Neostratotypen, v. 6, p. 481-486.

Marinescu, Fl., 1980, Scara cronostratigrafica a Neogenului Istoria evolutiei corelarilor: Subdiviunile Strat. Neogenului Romani, p. 9–22.

Martini, E., 1981, Nannoplankton in der Ober-Kreide, im Alttertiär und im tieferen Jungtertiär von Süddeutschland und dem angrenzenden Österreich: Geol. Bavarica, v. 82, p. 345–356.

Martini, E., 1971, Standard Tertiary and Quaternary calcareous nannoplankton zonation: Proc. II. Plankt. Conf., Roma 1970, v. 2, p. 739–785.

Martini, E., and C. Müller, 1975a, Calcareous nannoplankton and silicoflagellate from the type Ottnangian and equivalent strata in Austria (lower Miocene): Proc. VIth Congr. Reg. Comm. Mediterrn. Neog. Strat., v. 1, p. 121–123.

Martini, E., and C. Müller, 1975b, Calcareous nannoplankton from the Karpatian in Austria (middle Miocene): Proc. VIth Congr. Reg. Comm. Mediterrn. Neog. Strat., v. 1, p. 125–127.

McGowran, B., 1979, Some Miocene configurations from an Australian standpoint: Ann. Géol. Pays Hellén., tome hors sér., fasc. 2, p. 767–779.

Mein, P., 1979, Rapport d'activité du groupe de travail vertébrés mise à jour de la biostratigraphie du Néogène basée sur les mammifères: Ann. Géol. Pays Hellén., tome hors sér., 1979, fasc. III, p. 1367–1372.

Mein, P., G. Bizon, J. J. Bizon, and C. Montenat, 1973, Le gisement de mammifére: de La Alberca (Murcia, Espagne mèridionale), correlations aves les formation marines du Miocène termina L. C. R. Acad. Sci. Paris, Sér. D 276, p. 3077–3080.

Menner, V. V., 1975, 8. Worldwide correlation possibilities: Superstages and proposition of a standard global chronostratigraphic scale for the Neogene: Proc. 6. Congr. Reg. Comm. Mediterranean Neogene Strat., v. 2, p. 24–27.

Nagymarosy, A., 1981, Chrono- and biostratigraphy of the Pannonian basin: A review based mainly on data from Hungary: Earth Evol. Sc., v. 3–4, p. 182–184.

Papp, A., 1981, Calibration of Mediterranean, Paratethys and Continental Stages: Proc. 7. Internat. Congr. Mediterranean Neogene: Ann. Géol. Pays Hellén., tome hors sér., fasc. 4, p. 73–78.

Papp, A., and F.-F. Steininger, 1979, Paleogeographic implications of late Miocene deposits in the Aegean region: Ann. Géol. Pays Hellén., tome hors sér., fasc. 2, p. 955–959.

Paramonova, N. P., E. N. Ananova, A. S. Andreeva-Grigorovic, L. S. Belokrys, L. K. Gabunia, K. F. Grusinsaja, S. O. Hondkarian, G. I. Carmischina, T. F. Kozirenco, L. S. Majsuradze, et al., 1979, Paleontological characteristics of the Sarmatian s. l. and Maeotian of the Ponto-Caspian area and possibilities of correlation to the Sarmatian s. str. and Pannonian of the Central Paratethys: Ann. Géol. Pays Hellén., tome hors sér., 1979 (fasc. II), p. 961–971.

Pevzner, M. A., and E.-A. Vangeingeim, 1982, Vexed questions in the concept of volume and stratigraphic position of the Pannonian: Proc. USSR Acad. Sc., Geol. Ser., n. 11, p. 42–56.

Poore, R. Z., L. Tauxe, S. F. Percival, Jr., J. L. LaBrecque, R. Wright, N. P. Peterson, C. C. Smith, P. Tucker, and K. J. Hsu, 1983, Late Cretaceous-Cenozoic magnetostratigraphic and biostratigraphic correlations of the South Atlantic Ocean: DSDP Leg 73: Paleogeogr., Palaeoclimat., Palaeoecol., v. 42, p. 127–149.

Rabeder, G., 1981, Die Arvicoliden (rodentia, mammalia) aus dem Pliozän und dem älteren Pleistocän von Niederösterreich: Beitr. Paläont. Ósterr. v. 8, p. 1–373.

Rabeder, G., and F. Steininger, 1975, Die direkten biostratigraphischen Korrelationsmöglichkeiten von Säugetierfaunen aus dem Oligo/Miozän der zentralen Paratethys: Proc. 6th Cong. RCMNS, v. 1, p. 177–183.

Rehakova, Z., 1977, Marine planktonic diatom zones of the Central Paratethys Miocene and their correlation: Vest. Ustred. Ustav. Geol., v. 52, p. 147–157.

Rögl, F., F. Steininger, and C. Müller, 1978, Middle Miocene salinity crisis and paleogeography of the Paratethys (Middle and Eastern Europe): Init. Rept. DSDP, v. 42(1), p. 985–990.

Rögl, F., P. Hochuli, and C. Müller, 1979, Oligocene-early Miocene stratigraphic correlations in the Molasse basin of Austria: Ann. Géol. Pays Hellén., tome hors sér. 1979, p. 1045–1049.

Rögl, F., and F.-F. Steininger, 1983, Vom Zerfall der Tethys zu Mediterran und Paratethys. Die neogene Palaeogeographie und Palinspastik des zirkum-mediterranen Raumes: Ann. Naturhist. Mus. Wien, v. 85, p. 135–163.

Semenenko, V. N., 1979, Correlation of Mio-Pliocene of the Eastern Paratethys and Tethys: Ann. Géol. Pays Hellén., tome hors sér., fasc. 3, p. 1101–1111.

Semenenko, V. N., and S.-A. Ljulieva, 1978, Opyt prjamoj korreljacii Mio-Pliocene vostocnogo Paratetija i Tetija [Attempt of a direct correlation of the Mio-Pliocene of the Eastern Paratethys and Tethys]: Sb. "Stratigr. Kainozoa Severn. Pricern. Kryma", Vyp. 2, Geol. Ist. Dnjepropetrovs. Univ. (DEV), p. 95–105.

Semenenko, V. N., and M.-A. Pevsner, 1979, A correlation of Miocene and Pliocene of the Pont-Caspian on the biostratigraphic and paleomagnetic data: Proc. USSR Acad. Sci., Geol. Ser., v. 1, p. 5–9.

Senes, J., 1969, Unsere Kenntnisse úber die Paläogeographie der Zentralparatethys: Geol. práce, v. 55, p. 83–108.

Senes, J., ed., 1967–1978, Chronostratigraphie und Neostratotypen—Miozän der Zentralen Paratethys: v. 1 (Karpatien), 1967; v. 2 (Eggenburgien), 1971; v. 3 (Ottnangien), 1973; v. 4 (Sarmatien), 1974; v. 5 (Egerien), 1975; v. 6 (Badenian), 1978.

Senes, J., 1975, ed., Proceedings of the VIth Congress: Reg. Comm. Mediterr. Neogene Stratigraphy, v. 2, 69 pages.

Senes, J., 1979a, Géochronologie des stratotypes des étages du Miocène inférieur et moyen de la Paratéthys centrale utilisables pour la corrélation globale: Geol. Zbornik–Geol. Carpathica, v. 30, p. 99–108.

Senes, J., 1979b, Correlation du Néogène de la Téthys et de la Paratéthys—Base de la reconstitution de la géodynamique récente de la région de la Méditerranée: Geol. Zborn.-Geol Carpat., v. 30, p. 309–319.

Senes, J., and F. Marinescu, 1974, Cartes paléogéographiques du Néogène de la Paratéthys centrale: Mém. B.R.G.M., v. 78, p. 785–792.

Steininger, F., 1979, Integrated assemblage-zone biostratigraphy at marine-non-marine boundaries: Examples from the Neogene of Central Europe, *in* E.-G. Kauffman, and J.-E. Hazel, eds., Concepts and methods of Biostratigraphy, p. 235–255.

Steininger, F.-F., and L.-A. Nevesskaja, 1975, eds., Stratotypes of Mediterranean Neogene stages: Comm. Mediterranean Neogene Strat., v. 2, 364 pages.

Steininger, F., F. Rögl, and E. Martini, 1976, Current Oligocene/Miocene biostratigraphical concept of the Central Paratethys (Middle Europe): Newsl. Strat., v. 4, p. 174–202.

Steininger, F., and A. Papp, 1979, Current biostratigraphic and radiometric correlations of late Miocene Central Paratethys stages (Sarmatian s. str., Pannonian s. str., and Pontian) and Mediterranean stages (Tortonian and Messinian) and the Messinian Event in the Paratethys: Newsl. Stratigr., v. 8, p. 100–110.

Stradner, H., and R. Fuchs, 1978, 8. Die Narnofossilien des Badenian-8.3. Das Nannoplankton in Österreich, *in* A. Papp, I. Cicha, J. Senes, and F. Steininger, M_4—Badenian (Moravien, Wielicien, Kosovien): Chronostrat. et Neostratotypen, v. 6, p. 489–497.

Taner, G., 1982, Die Molluskenfauna und pliocene Stratigrafie der Halbinsel Gelibolou: Comm. Fac. Sci. Univ. Ankara, ser. C., Geol., v. 25, 27 pages.

Vail, P.-R., R.-M. Mitchum, Jr., and S. Thompson, III, 1977, Seismic stratigraphy and global changes of sea level, part 4: Global cycles of relative changes of sea level: Mem. AAPG, v. 26, p. 83–97.

Van Couvering, J.-A., W.-A. Berggren, R.-E. Drake, E. Aguirre, and G.-H. Curtis, 1976, The terminal Miocene event: Marine Micropaleont., v. 1, p. 263–286.

Vass, D., and G.-P. Bagdasarjan, 1978, A radiometric time scale for the Neogene of the Paratethys region, *in* G.-V. Cohee, M.-F. Glaessner, and H.-D. Hedberg, eds., Contributions to the geologic time scale: Stud. Geol. no. 6, p. 179–203.

Vass, D., G.-P. Bagdasarjan, and F. Steininger, 1978, The Badenian radiometric ages: Chronostrat. et Neostrat., v. 6, p. 35–45.

Veselov, A.-A., 1979, To the accurate definition of the stratigraphical correlation of the Oligocene–Lower Miocene border-marking horizons of the eastern and central Paratethys: Ann. Géol. Pays Hellén., tome hors sér., 1979 (III), p. 1243–1252.

CHAPTER 8

A Method for Lithogenetic Subdivision of Pannonian (s. l.) Sedimentary Rocks

Á. Szalay
Petroleum Exploration Company,
H-5001 Szolnok,
Munkásör-út 43, Hungary

K. Szentgyörgyi
Hungarian Hydrocarbon Institute
H-2443 Százhalombatta
P.O. Box 32, Hungary

A method is presented here for subdividing the Neogene sedimentary rocks of the Pannonian basin based on the fact that these Neogene sediments are made up almost exclusively of pelite and psammite. The alternation of pelitic and psammitic layers and their thicknesses are known from geophysical well logs. We construct lithologic trend diagrams for several drill holes, and for each drill hole we obtain a characteristic trend diagram. Changes in lithologic trends can be used to define lithogenetic units, which may be used as a basis for subdividing the Pannonian strata. Similar lithologic trend diagrams from neighboring boreholes allow for local correlation of lithogenetic units. This correlation has been extended to a regional system of lithogenetic units through analysis of many lithogenetic trend curves and reflection seismic sections.

The Pannonian (*s. l.*) complex of the Great Hungarian Plain was divided into (from deepest to shallowest) units Pa_1^{1a}, Pa_1^{1b}, Pa_1^2, and Pa_2. The lowermost units can be shown to occur only in the deep basin areas where sedimentation was probably uninterrupted from Sarmatian to Quaternary time.

INTRODUCTION

The evolution of a sedimentary basin can be thoroughly understood only if the subsidence of and sedimentation within the basin can be reconstructed in space and time. Toward this end, it is necessary to reconstruct the geomorphology of the subsiding basin based on the facies, depositional environments, and lithologies of the basin fill. We present here a method for subdividing the Pannonian (*s. l.*) strata on the basis of the dominant lithologies or lithologic pattern present within certain depth intervals (called the lithologic trend) and for correlating such units between deep basin areas and shallower parts of the basin.

A detailed knowledge of the sediments within the basin has been attained in the last decade as a result of extensive drilling during exploration for hydrocarbons. A lithologic and stratigraphic model for the Pannonian (*s. l.*) sedimentary rocks, which comprise a large part of the sediments within the Pannonian basin, has been summarized by Körössy (1968) and later improved by Szalay and Szentgyörgyi (1979, 1982) using lithologic trend analysis. The lithogenetic units broken out by lithologic trend analysis can be correlated across most of the Pannonian basin with the help of reflection seismic profiles.

METHODS

The Pannonian sedimentary complex is made up primarily of pelite (clay, marl, and silt) and psammite (sand and sandstone). Locally gravel, sandy gravel, and coarser conglomerate are

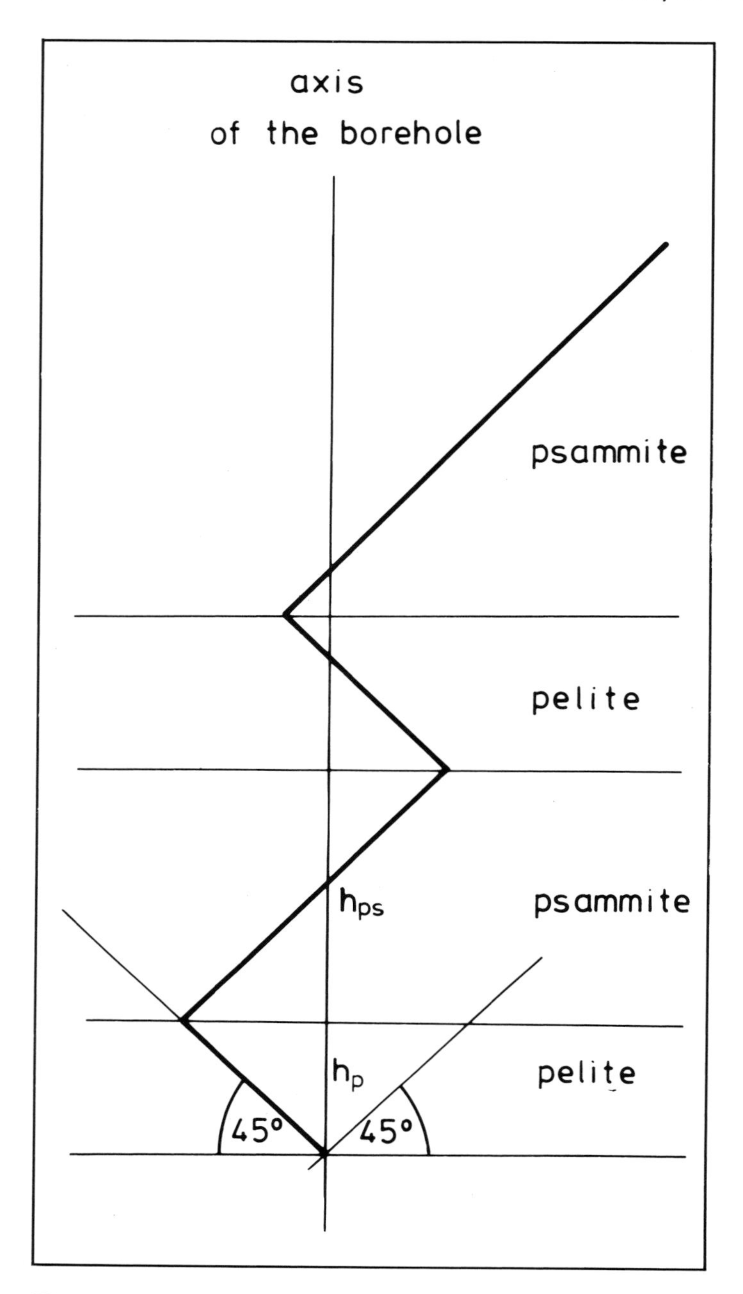

Figure 1. The method used to construct lithologic trend diagrams from the thickness of pelitic (h_p) and psammitic (h_{ps}) beds determined from lithologic logs. See text for details.

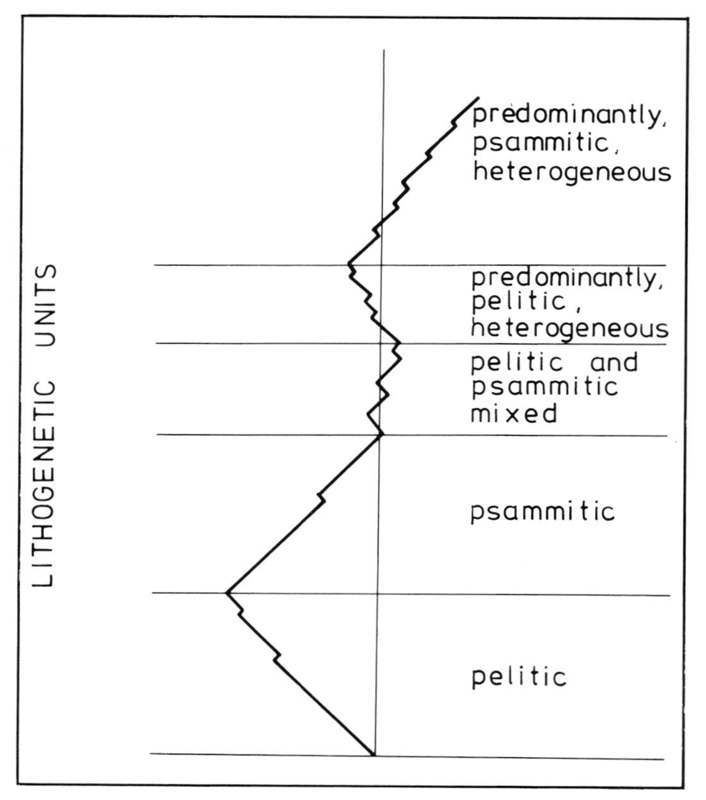

Figure 2. An example of different types of lithologic intervals on a lithologic trend diagram. When the average slope of the lithologic trend is close to 45°, the lithology is nearly 100% pelite or psammite. When the slope is between 45° and 90°, the lithology is mixed.

present. Analysis of the lithologic trend is based on interpretation of well logs that can be used to distinguish between pelitic and psammitic rocks. Normally, the resistivity and the microlaterolog are used, but sometimes other logs, such as natural radioactivity and self-potential logs, are also taken into account.

The changing lithology in a borehole is shown in the following way (Figure 1). Starting from the bottom of the sedimentary complex we draw a straight line with a 45° slope to the left if the composition of a bed is pelitic and 45° to the right if the composition is psammite. The length of the line is determined by the thickness of the individual layer (Figure 1). The zig-zag curve constructed in this way is called here the lithologic trend diagram (Figure 2). Although the individual units may be very thin, the average slope of the line indicates the dominant lithology. A line that slopes toward the right, proceeding from deeper to shallower levels, indicates a predominantly psammitic lithology, while one that slopes toward the left indicates a predominantly pelitic lithology. The average angle that the lithologic trend line forms with a vertical line is also important. A 45° slope to the right indicates 100% psammite, while a lesser slope to the right indicates predominantly psammite. A vertical line indicates equal components of psammite and pelite. The lithogenetic units are thus defined as intervals characterized by distinctive lithologic patterns or trends (Figure 2).

We suggest that these lithogenetic units constructed as above can be correlated throughout the Pannonian basin. Although the subsidence of the Pannonian basin varied in space and time, the subsidence should have followed a regular pattern over the entire region. Significant phases of basin development in the Pannonian basin can be associated with marked changes in paleogeographic conditions and with changes in the lithologic trend. Although lithogenetic units can be correlated throughout the Pannonian basin, an individual lithogenetic unit need not be the same lithology everywhere. For example, a lithogenetic unit comprised predominantly of psammite in the deep basin areas may become continuously pelitic toward the margins of the basin. For such correlations between distant localities, the use of seismic reflection profiles is necessary.

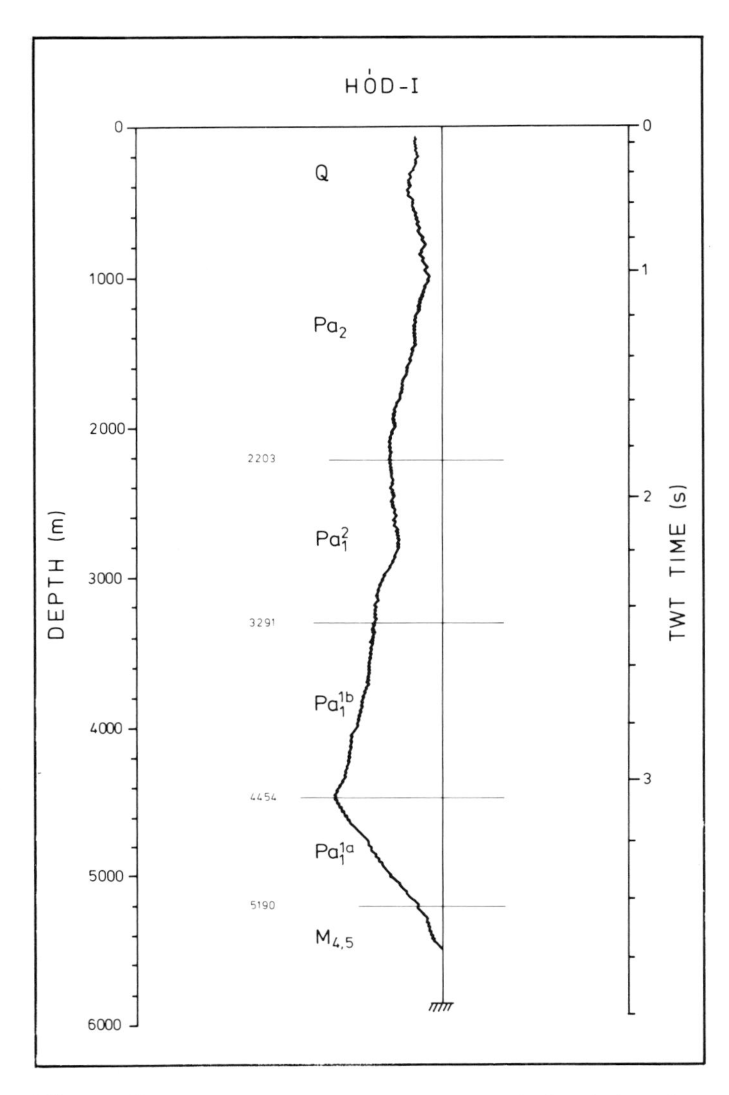

Figure 3. Lithologic trend diagram of the Hód-I borehole and its subdivision into lithogenetic units. The pelite content of the different units are as follows: Pa_2 = 46.2%, Pa_1^2 = 45.6%, Pa_1^{1b} = 39.4%, Pa_1^{1a} = 87.0%, and M_4 = 74.7%. For easy comparison with a seismic section, the vertical two-way traveltime is also shown. (See location of the borehole in Figure 5.)

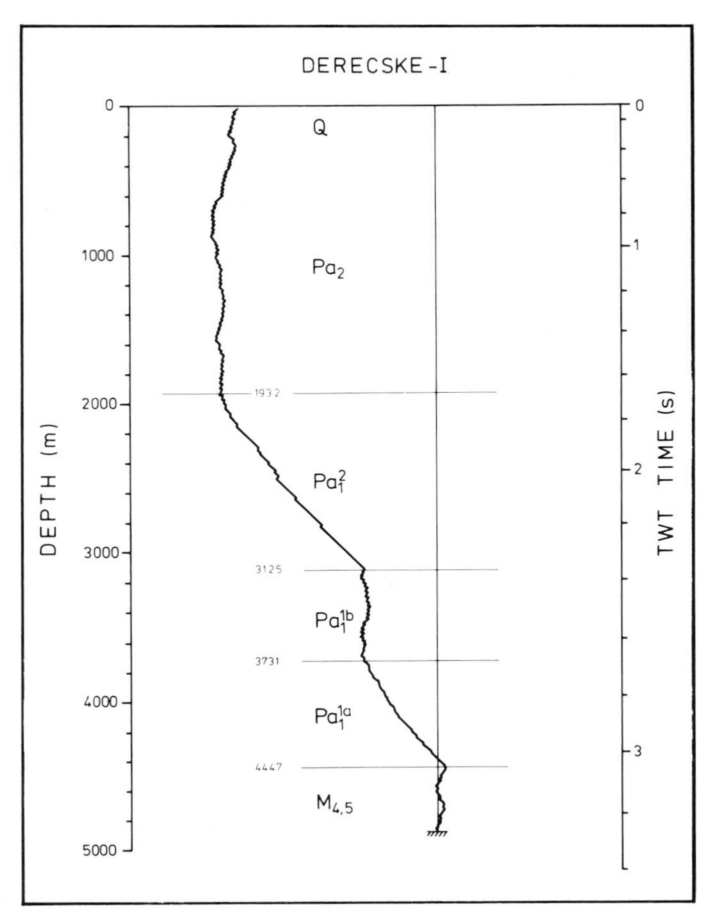

Figure 4. Lithologic trend diagram of the Derecske-I borehole and its subdivision into lithogenetic units. The pelite content of the different units are as follows: Pa_2 = 47.2%, Pa_1^2 = 88.5%, Pa_1^{1b} = 48.8%, Pa_1^{1a} = 87.6%, and M_4 = 43.5%. For easy comparison with a seismic section, the vertical two-way traveltime is also shown. (See location of the borehole in Figure 5.)

LITHOGENETIC UNITS OF THE LOWER PANNONIAN (*s.l.*)

The lower Pannonian (*s. l.*) sequence can be characterized by a distinctive lithologic trend that is similar in all areas of the Pannonian basin where the lower Pannonian (*s. l.*) is present. The consistency of the lithogenetic units and of the order in which they were deposited indicates that the evolution of the subbasins within the Pannonian basin followed a consistent pattern, even though the timing and rate of subsidence varied from place to place. Within the deepest parts of the Pannonian basin, the pre-Pannonian Miocene strata (mostly Badenian and Sarmatian rocks) usually cannot be distinguished from the lowermost Pannonian strata by lithologic trend analysis or seismic stratigraphy (Mattick et al., this volume), and deposition appears to have been continuous across the Sarmatian–Pannonian boundary. Over the basement highs, deposition began later and the subsidence of the basement was often slower than in the deep zones, so that the lowermost part of the Pannonian is not present.

In the deepest part of the Pannonian basin, the lower Pannonian (*s. l.*) unit can be divided into two subunits (Pa_1^1 and Pa_1^2), the first of which can be further subdivided (into Pa_1^{1a} and Pa_1^{1b}). Subunits Pa_1^{1a} and Pa_1^{1b} can only be shown to occur in the deepest depressions of Hungary (Figures 3 and 4). The lowermost Pannonian subunit, Pa_1^{1a}, is dominantly pelitic and consists of dark gray, strongly diagenetic marls and, subordinately, calcareous marl and clayey marl. Thin gravel intercalations are also present. The average thickness of this unit is 500 m in the deep subbasins of eastern Hungary and slightly less than this in southwestern Hungary (Drava and Zala basins). An abrupt change in lithologic trend marks the base of the overlying Pa_1^{1b} subunit, which is mainly psammitic and varies between 300 and 600 m in thickness (Figures 3 and 4; refer to Figure 5 for locations).

Unit Pa_1^1 is conformably overlain by unit Pa_1^2, which is areally more widespread and overlies transgressively the Pa_1^{1b} subunit

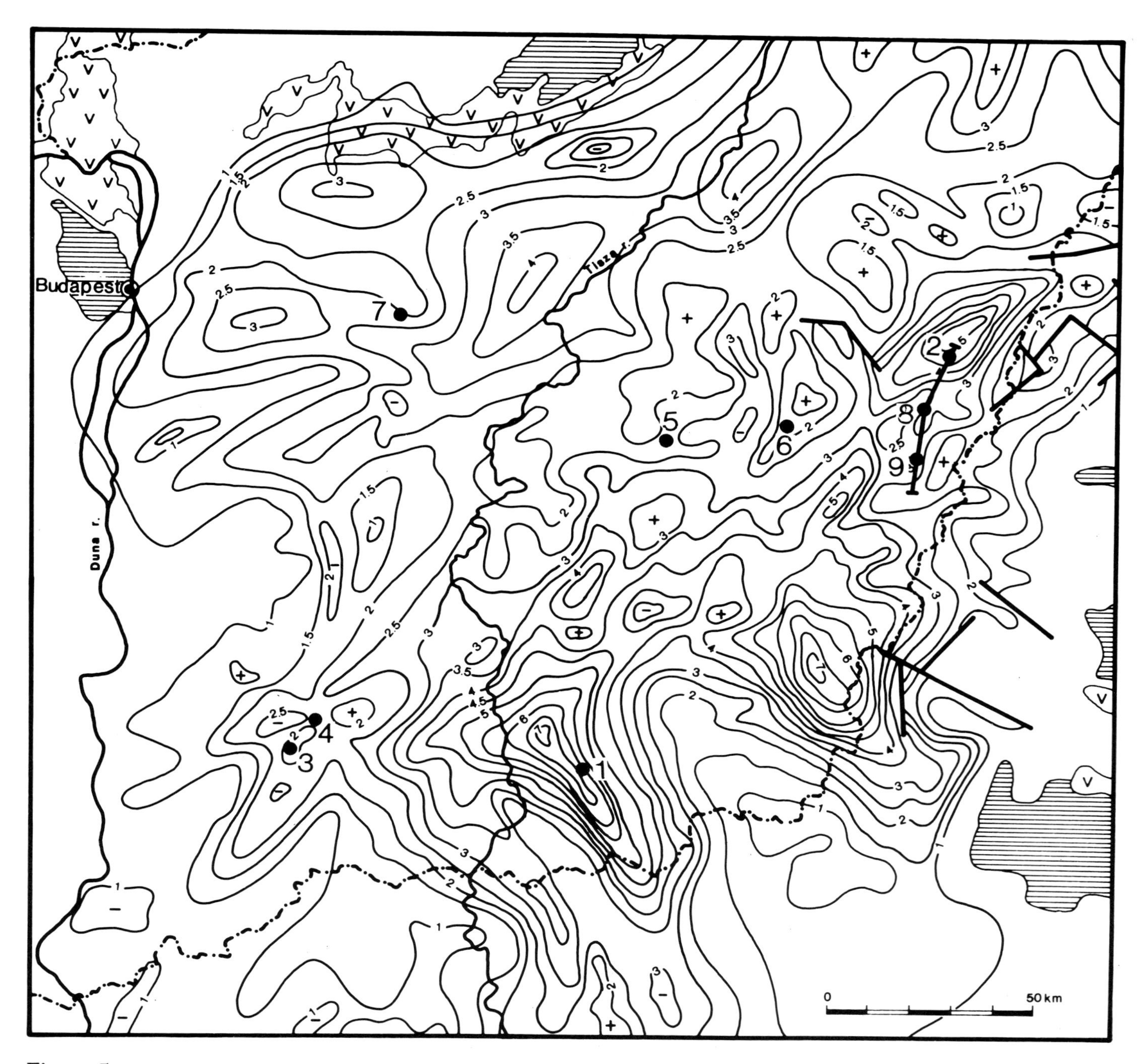

Figure 5. Index map to show the location of boreholes and seismic sections referred to in the text. The boreholes are as follows: (1) Hód-I, (2) Derecske-I, (3) Kiskunhalas-ÉK-2, (4) Tázlár-12, (5) Turkeve-1, (6) Füzesgyarmat-1, (7) Farmos-6, (8) Bem-1, and (9) Sas-K-1.

and the basement. Unit Pa_1^2 can often be subdivided into three subunits (Pa_1^{2a}, Pa_1^{2b}, and Pa_1^{2c}). Subunit Pa_1^{2a} is dominantly pelitic, while the overlying subunit Pa_1^{2b} is thick and less pelitic. Above this, subunit Pa_1^{2c} is again dominantly pelitic (Figure 6). Between the deepest parts of the basins and the basement highs, this threefold character of Pa_1^2 is more pronounced than in the deep basin areas. Above the basement highs Pa_1^2 is usually thin and consists mainly of pelitic rocks (Figure 7). The thickness of Pa_1^2 is about 800–1200 m in the deep depressions and about 400 m above the basement highs.

The lithogenetic units and subunits always occur in the same order in each subbasin so that the regional correlation of this unit is straightforward.

THE UPPER PANNONIAN (*S. L.*) LITHOGENETIC UNITS

The upper Pannonian (*s. l.*) sequence is heterogeneous and tends to be psammitic (Figure 7). In general it cannot be subdi-

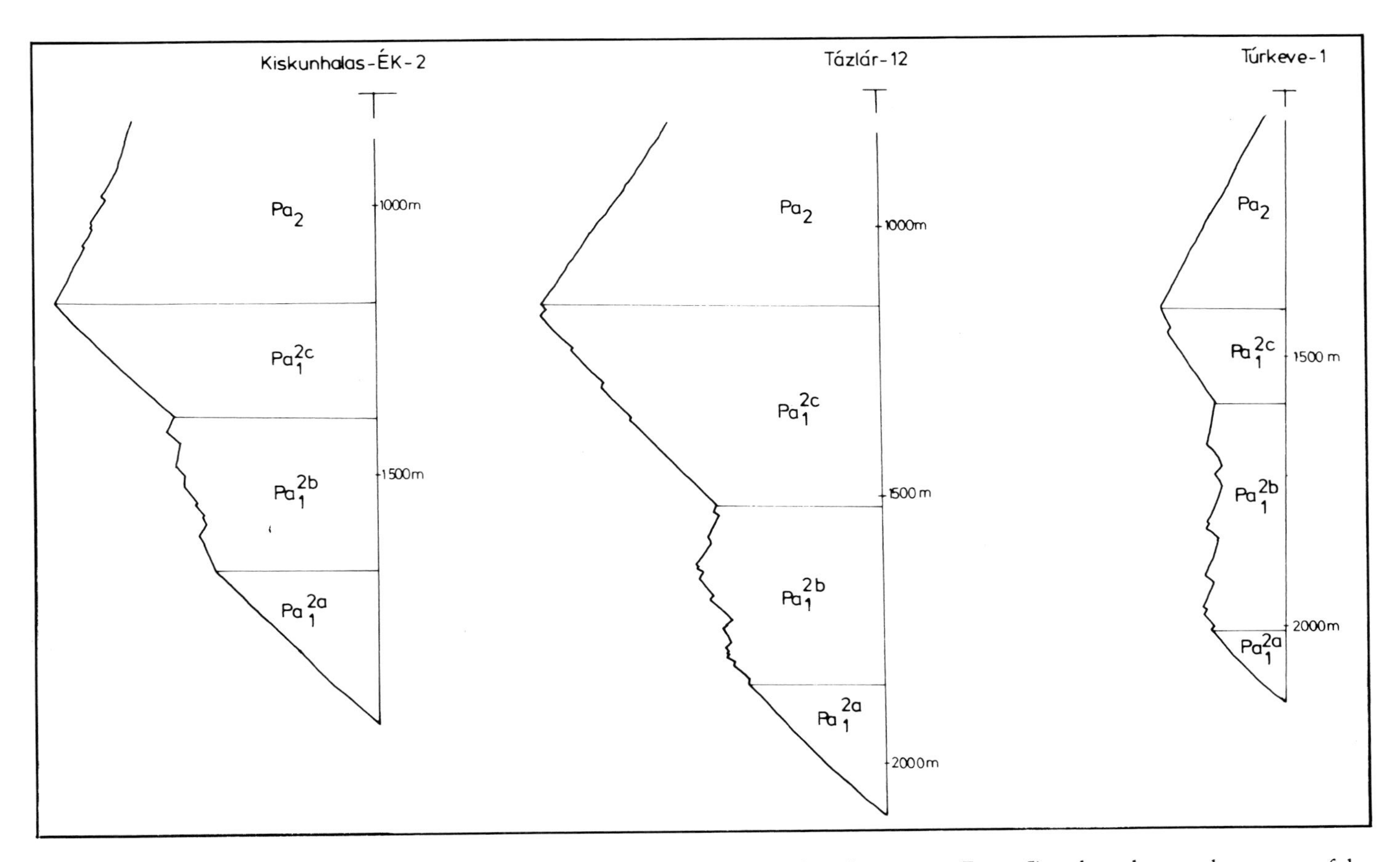

Figure 6. Lithologic trend diagram for three boreholes in the Great Hungarian Plain (locations in Figure 5) to show the complex pattern of the Pa_1^2 unit.

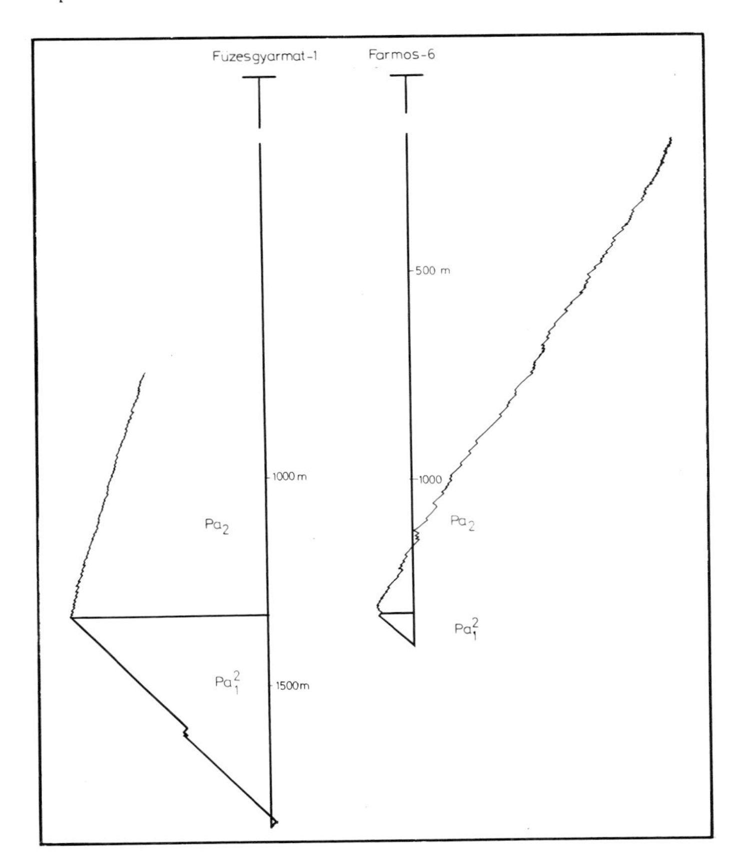

Figure 7. Lithologic trend diagram for two boreholes in the Great Hungarian Plain (locations in Figure 5) to show the pelitic nature of unit Pa_1^2 over basement highs and the very heterogeneous but dominantly psammitic unit Pa_2.

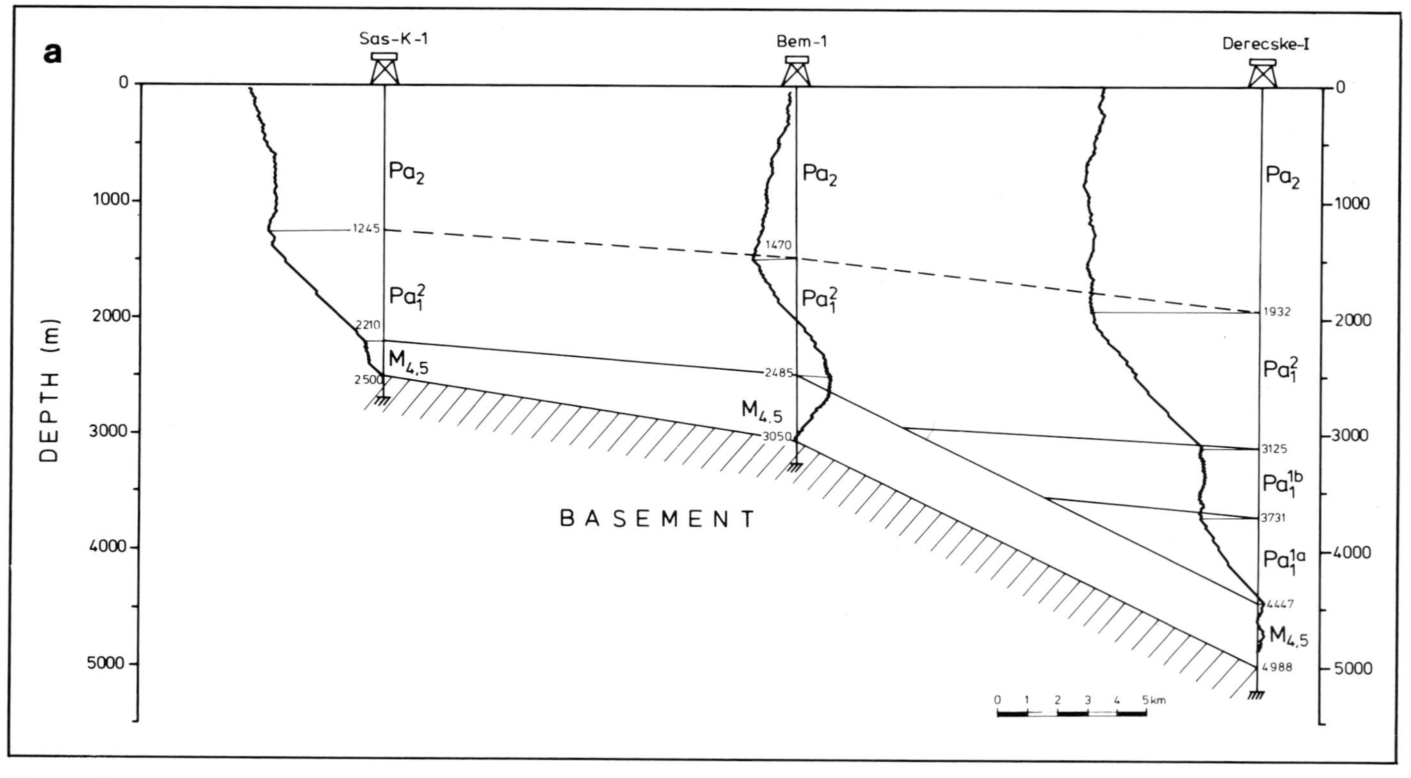

Figure 8. Characteristic example to show the correlation of lithogenetic units (a) (above) by the use of seismic reflection profiles (b) (opposite, top). A line drawing (c) (opposite, bottom) shows an interpretation of parts A and B. Depth was converted to two-way traveltime using the velocity–depth functions for the drill holes. See text for further discussion.

vided by lithologic trend analysis. This sandy unit lies conformably on the lower Pannonian (*s. l.*) strata and is a typical fluviolacustrine complex.

CORRELATION OF LITHOGENETIC UNITS

Figure 8 shows cross-sections across the Derecske depression (eastern Hungary) (Figure 5) and illustrates the configuration of lithogenetic units that is characteristic for most of the deep basin areas of Hungary. Unit M_4 is made up of middle Miocene (Badenian and Sarmatian) strata and covers a large part of the basement as a thin blanket. Determination of the age of this strata is based on fossil findings in cores. An interesting feature of the section shown in Figure 8 is that unit M_4 can also be distinguished with the help of lithologic trend analysis. In the deep Derecske-I borehole (Figure 4) the lithology of unit M_4 is heterogeneous. Marl alternates with psammite and also with psephite. On the basin flank in borehole Bem-1, unit M_4 is dominantly psammitic, while above the basement high it is mostly pelitic. The lowermost Pannonian (*s. l.*) subunits, Pa_1^{1a} and Pa_1^{1b}, are present only in the deepest part of each subbasin and onlap unit M_4 or the basement around the margins of the deep basins. The area covered by subunit Pa_1^{1b} is always larger than that covered by Pa_1^{1a}.

The lowest unit that is present everywhere in the Pannonian basin is unit Pa_1^2. It is dominated by pelite with some sandstone intercalations in the Derecske area. In Figure 8 it can be seen to overlie unconformably unit M_4 along most of the section, and only in the central part of the Derecske trough is there no hiatus in the sedimentary sequence. A distinctive change in lithologic development can be observed on all lithologic trend curves at the base of the upper Pannonian (*s. l.*) fluviolacustrine sediments (base of unit Pa_2). As is shown in Figure 8, the base of unit Pa_2 cannot be correlated along individual seismic reflectors. This change in lithology takes place at the top of a series of prograding reflectors and below a series of flatter reflectors. In the deep part of the basin (borehole Derecske-I) the transition from unit Pa_1^2 to Pa_2 is stratigraphically below the transition from Pa_1^2 to Pa_2 in boreholes Bem-1 and Sas-K-1, as can be seen by tracing seismic reflectors within unit Pa_1^2.

CONCLUSIONS

The lower Pannonian (*s. l.*) lithogenetic units discussed in this chapter may be very close to chronostratigraphic units because the correlation of these lithogenetic units between distant drill holes can be performed by following reflectors shown on seismic profiles. However, rather than defining these units by the configuration of reflectors within each unit, we have primarily defined them by the dominant lithologies present within each interval.

From a stratigraphic point of view, one of the most important results of this study is that the lowermost part of the Pannonian

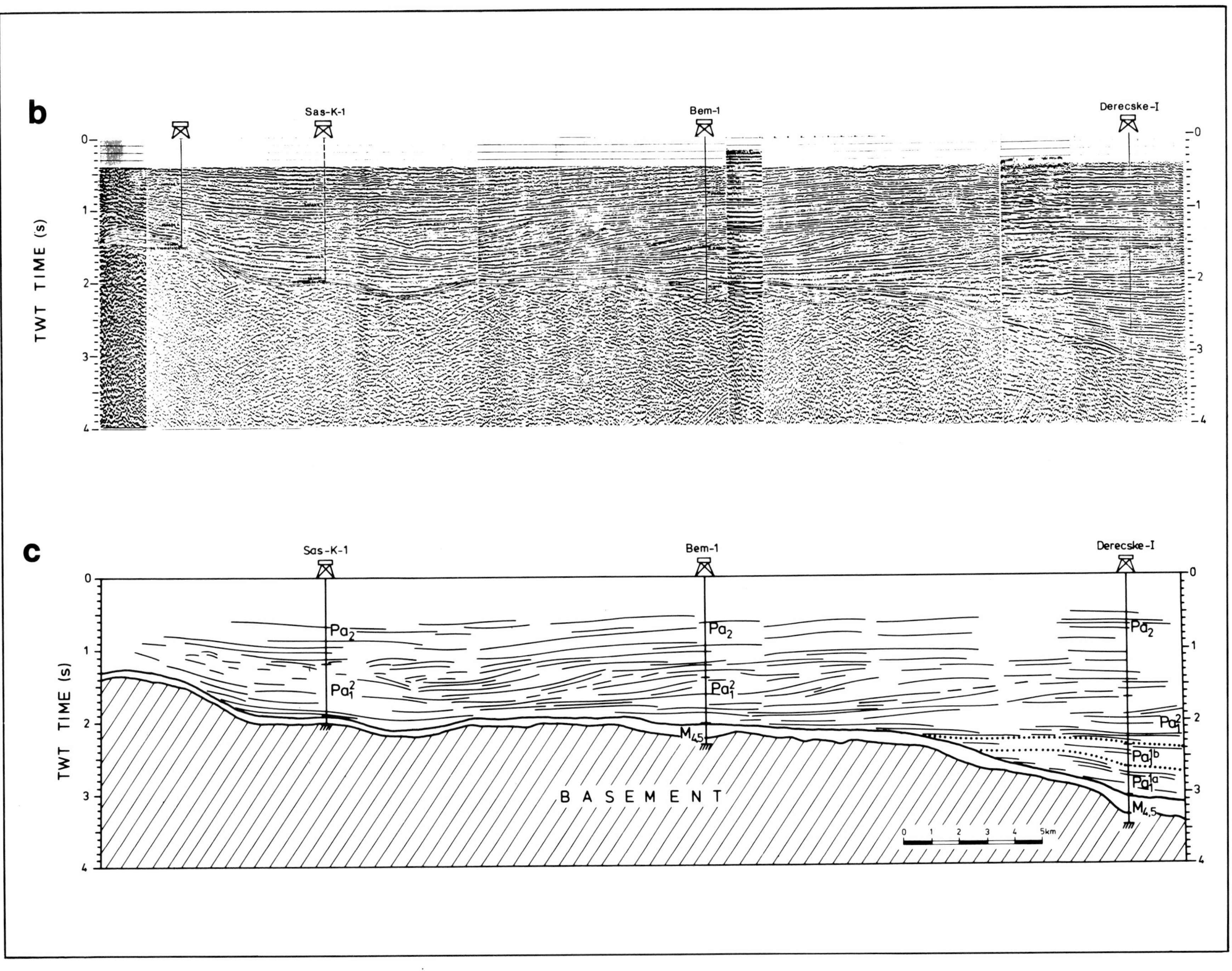

Figure 8. (continued)

(*s. l.*) sequence is not present everywhere within the Pannonian basin, but is found only in the deepest depressions. This implies that in areas of shallower deposition, Pannonian (*s. l.*) rocks lie unconformably on the underlying units. It seems that the base of the Pannonian (*s. l.*) must become younger toward the basin margins and older toward the basin centers. The entire Pannonian (*s. l.*) sequence can only be present in the deep basin areas.

The practical significance of the lithogenetic subdivision is that it makes it easy to distinguish between potential source rocks (dominantly pelitic units) and carrier beds (dominantly psammitic units). Thus, it offers a good basis for reconstruction of hydrocarbon generation and migration.

REFERENCES

Körössy, L., 1968, Entwicklungsgeschichtliche und palaogeographische Grundzüge des ungarischen Unterpannons: Acta Geol. Hung., v. 12, p. 384–417.

Szalay, A., and K. Szentgyörgyi, 1979, Contribution to the knowledge of lithologic subdivision of Pannonian basin formations explored by hydrocarbon drilling: reconstruction based on trend analysis: MTA X Oszt. Közl., Budapest, v. 12, n. 4, p. 401–423.

Szalay, A., and K. Szentgyörgyi, 1982, Subdivision of the Pannonian formations on the basis of lithological trend analysis: Paper presented at workshop/discussion meeting on the evolution of extensional basins within regions of compression, Veszprém, Hungary, June 20–26.

CHAPTER 9

Contribution of Seismic Reflection Data to Chronostratigraphy of the Pannonian Basin

F. Horváth
Geophysical Department, Eötvös University
H-1083 Budapest
Kun Bela ter. 2, Hungary

Gy. Pogácsás
Geophysical Exploration Company
H-1068 Budapest
Népköztársaság útja 59, Hungary

Available magnetostratigraphic and radiometric age data for Neogene sedimentary rocks within the Great Hungarian Plain can be shown to be consistent with each other if primary seismic reflectors are interpreted as geological time lines. Using this premise, we argue that the boundary between the lower and upper Pannonian (*s. l.*) units is diachronous and varies from 9 Ma to about 7 Ma. Lithogenetic units, determined by analyzing the trend of lithologic development in drill holes, and the correlation of these units by reflection seismic data provide a means of dividing the lower Pannonian into three subunits (Pa_1^{1a}, Pa_1^{1b}, and Pa_1^2). By combining age data with these correlations, the unconformity between synrift and postrift sediments can be shown to represent a time gap of one to several million years in many places in the Pannonian basin.

INTRODUCTION

The Pannonian basin and other depressions at its periphery, including the Alpine–Carpathian molasse foredeep, developed as part of the Central Paratethys. This marine area originated in early Oligocene time as a separate sea, which was partly connected to the Mediterranean through various seaways. At about the beginning of late Miocene time, however, these marine connections closed and an interconnected Pannonian lake system, isolated from the oceans, was formed. It eventually filled up and disappeared in latest Pliocene time.

Traditionally, time stratigraphy of the Pannonian basin has been based on biostratigraphy and lithologic data. Due to the presence of open marine or mixed fauna, the Central Paratethys regional stage system is reliable and well correlated to the Mediterranean stages for Oligocene to late Miocene time (Senes, 1979; Steininger et al., this volume). Stage boundaries are determined to an accuracy of about ± 1 m.y.

Within the peripheral basins of the Pannonian basin system, where the Pannonian through Quaternary section is thin (a few hundred meters in the Vienna and Transcarpathian basins), the dating of Pannonian and younger stages is also reasonable well constrained. However, the extension of these stages into the Little Hungarian Plain, the Zala and Dráva basins, and the Great Hungarian Plain, where the late Miocene to Holocene sediments reach up to 5 km in thickness, has been controversial. Difficulties in correlating these stages have arisen partly from the facies dependence of index fossils (e.g., molluscs and ostracods), (Korpás-Hódi, 1983) and partly from the diachronous nature of facies (and thus faunas) throughout the basin. This has hampered the proper interpretation of radiometric dates and led to the development of contradictory time scales for the late Miocene and Pliocene (Steininger and Papp, 1979; Balázs et al., 1981; Nagymarosy, 1981).

A recent breakthrough in understanding time stratigraphy in the Pannonian basin has come about through the use of new dating methods. These include the study of lithologic trends (Szalay and Szentgyörgyi, 1979, and this volume), radiometric dating of magmatic rocks (Hámor et al., 1980; Balogh et al., 1983), magnetostratigraphy (Cook et al., 1979; Rónai, 1981; Elston et al., 1985), and seismic stratigraphy (Pogácsás, 1984, 1985; Mattick et al., this volume).

In this chapter, we first review the general stratigraphy of the Pannonian basin as it is known from the reflection seismic data and lithogenetic trend diagrams. Then, we correlate radiometric age data, measured on basalts and rhyolites interbedded with Miocene sediments, and magnetostratigraphically dated sections using seismic reflection profiles. Our basic assumption throughout the paper is that primary seismic reflections are generated by bedding planes, stratal surfaces, and unconformities (Vail et al., 1977).

THE RELATIONSHIP BETWEEN BIOSTRATIGRAPHIC–LITHOGENETIC UNITS AND SEISMIC SEQUENCES

The Mediterranean and Central Paratethys stage system for Miocene to Quaternary time is shown in Figure 1. The subdivision that we prefer, and that we used in making model calculations (Horváth et al., this volume), is also shown. Our subdivision follows the working method that has been used for many years by the Hungarian Oil and Gas Trust. That is, we use the term "Pannonian" in *sensu lato* (*s. l.*), meaning all post-Sarmatian, pre-Quaternary rocks. We subdivide this 10-m.y. interval into smaller units. This subdivision is based on lithologic trend diagrams (Szalay and Szentgyörgyi, this volume).

Generalized stratigraphic features of the Pannonian basin and the method of correlation of lithogenetic units are shown in Figure 2. The base of the basin fill is a marked unconformity nearly everywhere in the Pannonian basin. Apart from localized Paleogene sedimentary basins (Royden and Báldi, this volume), Miocene beds directly overlie Mesozoic or Paleozoic basement. The basin fill starts with thick lower Miocene rocks in Transdanubia, which are absent or negligible in the Great Hungarian Plain. Middle Miocene rocks commonly occur in the Pannonian basin. These rocks are relatively thick (as much as 2000 m) in the deep troughs and thin toward the flanks, where they tend to drape the elevated basement. Lower and middle Miocene strata represent the early to synrift sedimentary succession, and their upper boundary is a marked unconformity in tectonically active areas (Rumpler and Horváth, this volume). In the center of deep basins (e.g., the Makó trough), however, the boundary is conformable and the lithologic trend and the seismic character above and below the boundary are similar. In such cases, distinction between middle Miocene (Sarmatian) strata and the overlying Pannonian strata is difficult using geophysical methods.

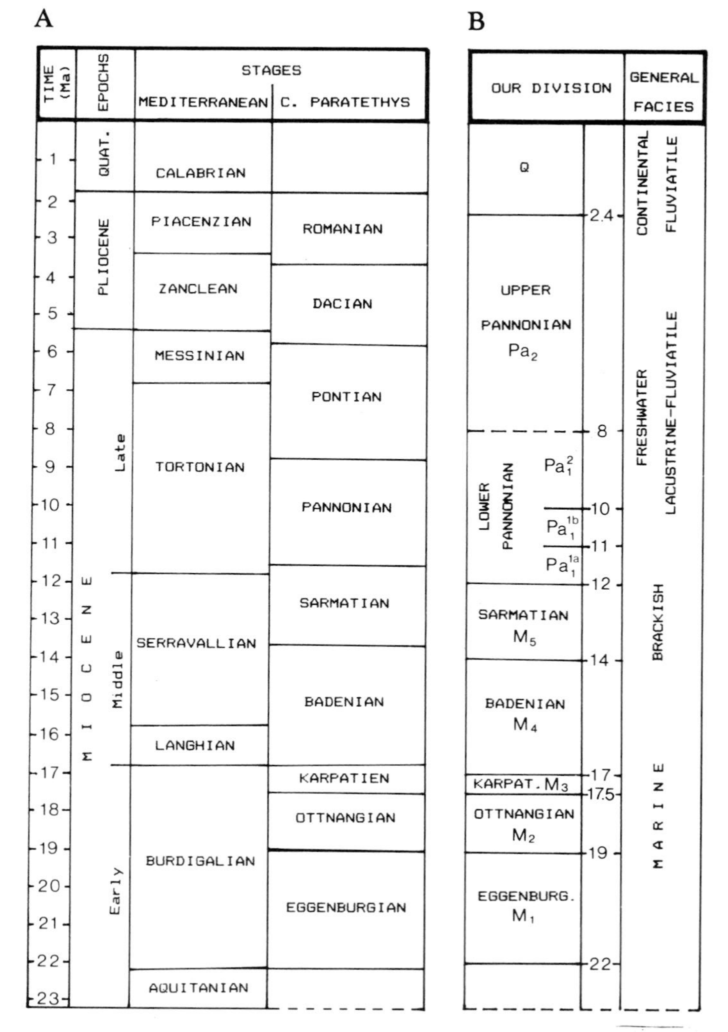

Figure 1. Late Cenozoic geologic time scale showing the Mediterranean and Central Paratethys stages (A) (Steininger et al., this volume) and the stage system used in Hungary (B). In our division, the Sarmatian and older units are the same as those defined for the whole Central Paratethys, although we use slightly different absolute ages for the stage boundaries. Moreover, the late Miocene through Pliocene time period is called Pannonian (*s. l.*) and is divided into smaller units on the basis of lithogenetic trend diagrams (Szalay and Szentgyörgyi, this volume). The last column shows the main sedimentary facies of the basin fill.

In the past, it was customary to take 10–40 core samples from lower Pannonian and older Miocene sedimentary rocks during drilling for hydrocarbons. However the number of corings has significantly decreased in recent years. Paleontologic examination of these cores makes it possible to determine the lower and middle Miocene stage boundaries. Well log markers are very useful in correlating these boundaries. We think that this method is suitable for determining lower and middle Miocene stages (e.g., Sarmatian, Badenian, and Karpatian). The dates we have used for their boundaries (Figure 1) are widely accepted.

The overlying lower Pannonian (*s. l.*) complex can be divided into the Pa_1^{1a}, Pa_1^{1b}, and Pa_1^2 subunits by lithologic trend diagrams. The lowermost Pannonian subunit (Pa_1^{1a}) is dominated by fine-grained sedimentary rocks, usually dark shale. These are overlain by subunit Pa_1^{1b}, which is mostly made up of sandstone.

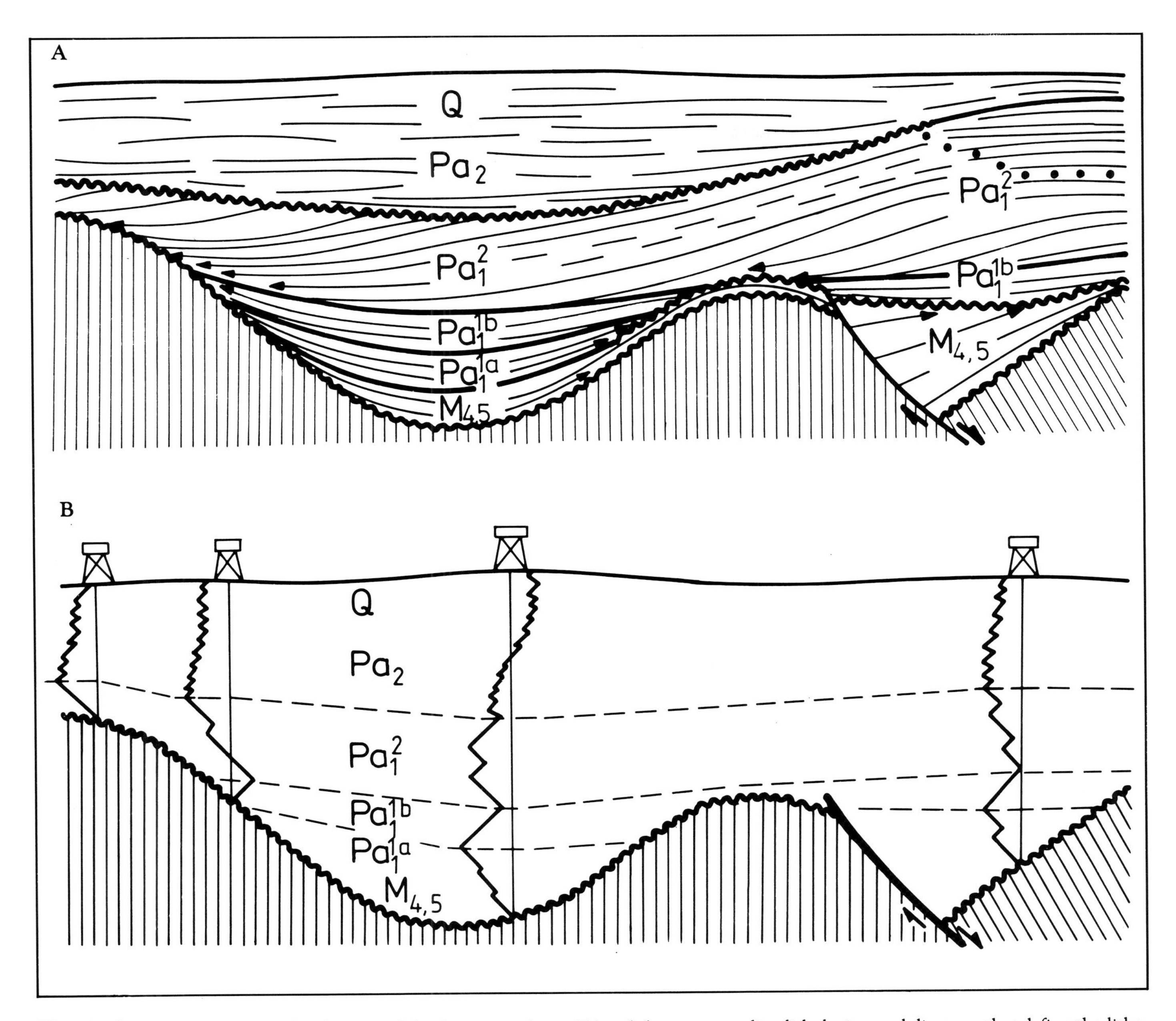

Figure 2. The main stratigraphic features of the Pannonian basin (A) and the corresponding lithologic trend diagrams that define the lithogenetic units in four boreholes (B). Correlation of these units can be made using reflection seismic sections. The boundary between units Pa_1^2 and Pa_2, however, cannot be tied together by following one single seismic reflector (see dotted line in A) suggesting that this boundary is not isochronous. There are three regional unconformities in the basin: the lower one is the top of the basement, the middle one separates the synrift ($M_{4,5}$, middle Miocene) from the postrift (Pannonian and Quaternary) sedimentary rocks, and the upper one locally coincides with the base of unit Pa_2.

Subunit Pa_1^2 consists of both pelite and psammite, and the average composition varies from place to place. Unit M_4 (Badenian and Sarmatian), together with subunit Pa_1^{1a} (and partly Pa_1^{1b}) correlate with the TM (turbidite-marl) sequence of Mattick et al., (this volume).

Correlation of the lower Pannonian subunits Pa_1^{1a}, Pa_1^{1b}, and Pa_1^2 between boreholes by reflection seismic sections has led to two important results. First, subunits Pa_1^{1a} and Pa_1^{1b} are confined to deep troughs, and individual bedding planes onlap either the basement or the middle Miocene (unit M_4) cover of the basement (Figure 2a). Second, upper and lower boundaries of subunits Pa_1^{1a} and Pa_1^{1b}, determined in distant boreholes in a single deep trough, can usually be tied together by following one continuous seismic reflector. This suggests that Pa_1^{1a} and Pa_1^{1b} can be considered to be time units.

It is surprising that the Pa_1^{1b}–Pa_1^2 boundary appears to be a time line, because unit Pa_1^2 correlates with the prodelta to delta front sequence of Mattick et al. (this volume). They argue that the delta development occurred first at the basin margin and prograded southward toward the basin center. In our interpretation, however, delta systems began to dominate the sedimentation of the Pannonian basin when rapid uplift of the

Carpathian mountains produced a sudden increase in transport of clastic material into the basin. Therefore, we think that this important regional event within the Carpathians changed the water circulation and sedimentation patterns throughout the whole Pannonian basin at about the same time.

A marked change in lithologic development defines the boundary between units Pa_1^2 and Pa_2, which is also the boundary between the upper and lower Pannonian. The upper Pannonian sequence consists of many thin beds, often with a slight predominance of psammite. The lithologic boundary between the upper and lower Pannonian usually agrees fairly well with the paleontologically determined boundary, which is roughly equivalent also to the Pannonian (*s. str.*)–Pontian boundary (see Nagymarosy and Müller, this volume). In the interior of the Great Hungarian Plain, the upper Pannonian generally corresponds to the LMT sequence of Mattick et al. (this volume), which represents shallow lake, marsh, and fluvial deposits and terrestrial soils. The sequence can easily be recognized on seismic sections because its lower boundary is a characteristic seismic unconformity (Figure 2). At the margins of the Pannonian basin, however, the lithologic change that defines the base of unit Pa_2 cuts across seismic reflectors and occurs at a deeper seismostratigraphic level (Figure 2). From this it may be inferred that the base of unit Pa_2 is diachronous in some places.

An example from the Derecske basin is shown in Figure 3. Comparison of the middle and lower diagrams shows that the base of unit Pa_2 cuts across seismic reflectors and roughly follows a facies change from delta front to delta plain sediments. Accordingly, the boundary between units Pa_1^2 and Pa_2 is not isochronous.

This result sheds light on much of the debate about the age of the upper–lower Pannonian boundary, because the upper Pannonian (defined either lithologically or paleontologically) is a heterochronous unit, which generally appeared first at the margin of the Pannonian basin and progressively shifted toward the basin interior. We think that the same is also true for paleontologically defined subunits of the upper Pannonian (Pontian, Dacian, and Rumanian) (see Steininger et al., this volume).

Upper Pannonian chronostratigraphic units cannot be accurately defined until the filling of the basin is better understood on a regional scale. We note that the LMT sequence of Mattick et al. (this volume) is also not a perfect time unit as demonstrated by the characteristic onlap pattern at the base of the sequence. Although unit Pa_2, as defined by lithology, is clearly not isochronous, the age range spanned by its lower boundary may be small (see next section).

Toward the end of the Pliocene, the Pannonian lake filled up completely and disappeared. This process was accelerated in Transdanubia by regional uplift with an amplitude of 100 to 300 m. Earlier sedimentation in a lacustrine environment was replaced by subareal erosion and deposition of terrestrial and fluvial sediments in the Pannonian basin. At about the same time, the climate changed from hot and dry to warm and humid. This event is therefore usually well documented in the sedimentary record and defines the beginning of the Quaternary period in the Pannonian basin (Rónai, 1984; Grosz et al., 1985). Magnetostratigraphic sections in two boreholes at Dévaványa and Vésztő suggest that this event can be correlated with the boundary between the Matuyama and Gauss polarity epochs. Accordingly, we take 2.4 Ma for the onset of the Quaternary in the Pannonian basin (Figure 1).

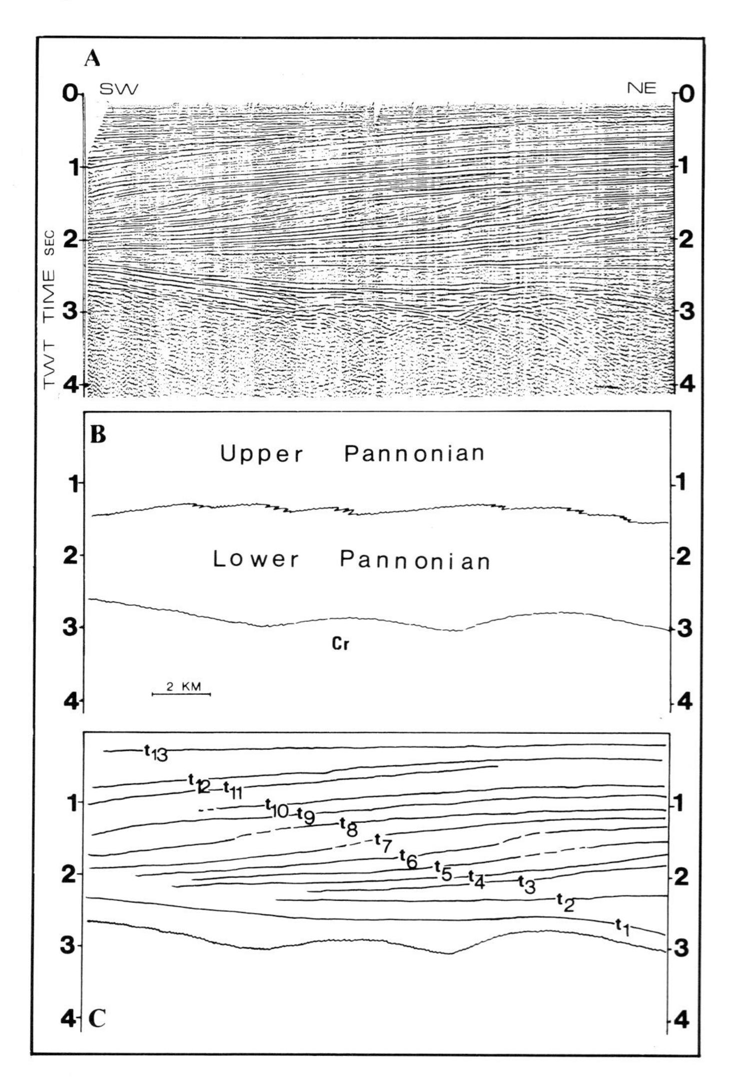

Figure 3. Seismic section (A) in the Derecske basin showing an example where the bottom of upper Pannonian lithogenetic (and biostratigraphic) unit is not an isochron. Lithologic trend analysis defines the upper Pannonian–Lower Pannonian boundary (B). Geologic time lines from t_1 to t_{13} are shown (C). The base of the upper Pannonian corresponds to about time t_9 in the southwest and t_5 in the northeast.

ABSOLUTE AGES FOR LATE MIOCENE TO QUATERNARY SEDIMENTARY ROCKS AND THEIR CORRELATION BY SEISMIC SECTIONS

Absolute age data for late Miocene to Quaternary sedimentary rocks in the Pannonian basin are provided by radiometric age determinations and magnetostratigraphic studies in boreholes (Figure 4). K-Ar ages have been measured in a few places on lava flows that interfinger with late Miocene sediments (Balogh et al., 1983; Árva-Sós et al., 1983), and magnetostratigraphic studies provide a continuous record of age versus depth in three boreholes in the Great Hungarian Plain: Dévaványa and Vésztő (Cook et al., 1979; Rónai, 1981) and Kaskantyú-2 (Elston et al., 1985). The absolute ages of the sediments in these holes were constructed by correlating the polarity zonation in

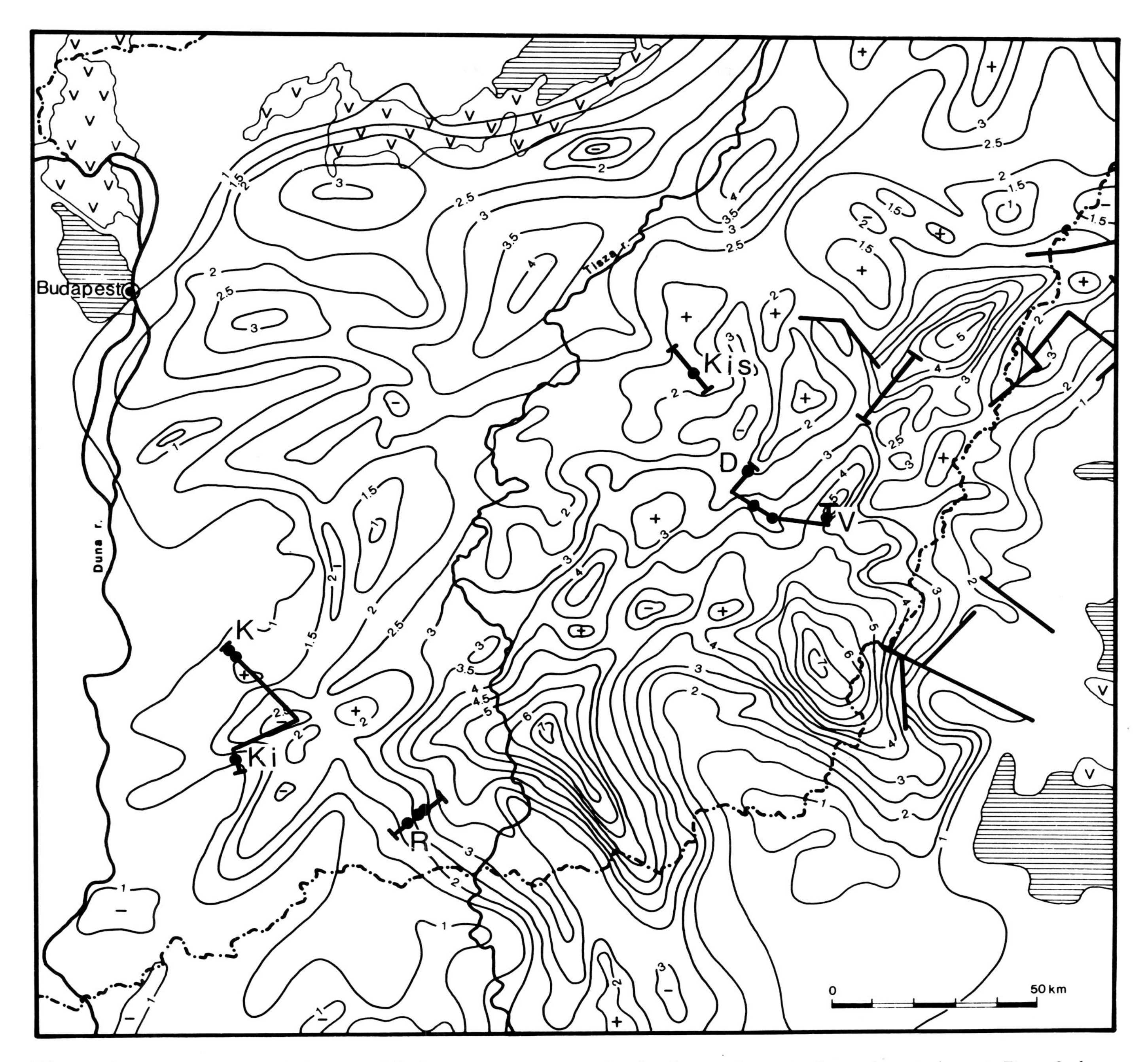

Figure 4. Index map showing the location of the five seismic sections cited in this chapter: the one in the northeast is shown in Figure 3; the one with Kis-ÉK-1 (Kis) is shown in Figure 9; the composite sections with Dévaványa (D) and Vésztő (V) and with Kaskantyú-2 (K) and Kiha-Ny-3 (Ki) are shown in Figures 6 and 7, respectively; and the section with Ruzsa-4 (R) is shown in Figure 8. The base map shows the thickness isolines of the Neogene–Quaternary sedimentary rocks (in kilometers).

these holes with the magnetic polarity time scale (Lowrie and Alvarez, 1981). The sedimentation rate was rather uniform in each hole, but the average rate of sedimentation varied significantly between holes (Figure 5). The K-Ar dates and the magnetostratigraphically determined ages can be correlated in places with seismic reflection profiles.

Figure 6 shows a composite seismic profile that connects wells Dévaványa and Vésztő. The depths of the end of the Gilbert polarity epoch in both wells (dated at 5.26 Ma and roughly equivalent to the Miocene–Pliocene boundary) were converted to two-way traveltimes and plotted on Figure 6. Although there are no long continuous reflectors in this depth range, this boundary can be tied from one well to the other by tracing a series of discontinuous reflectors but always moving parallel to seismic reflectors. This indicates that an age date measured in a borehole can be extended for some tens of kilometers by following one continuous seismic reflector or, if such reflectors are not available, moving parallel to the seismic stratification.

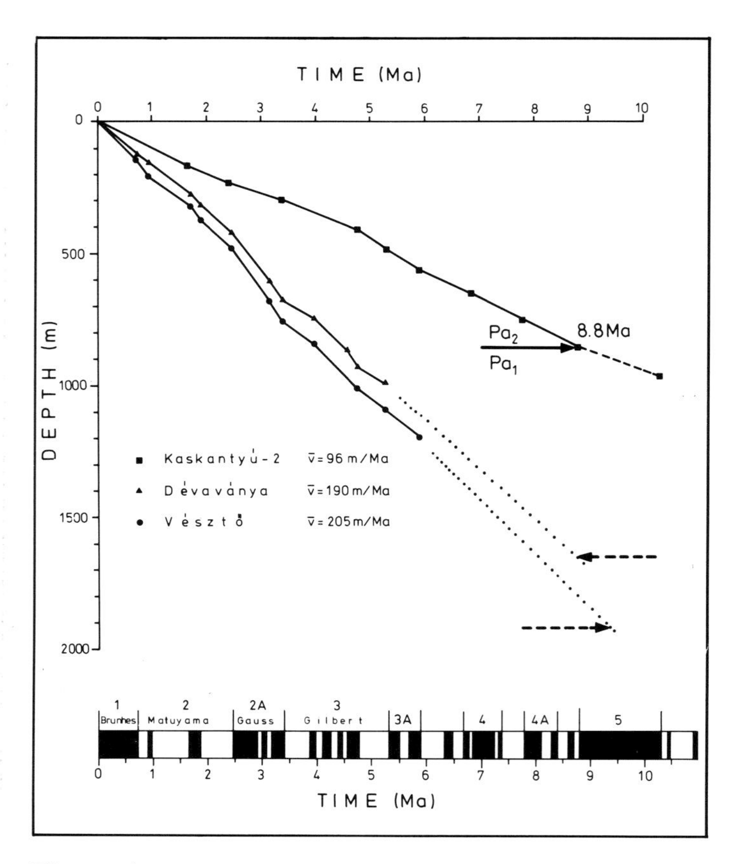

Figure 5. Accumulation rates of sediments in three wells in the Pannonian basin, derived from the correlation of magnetic polarity zonation measured in the wells with the polarity time scale of Lowrie and Alvarez (1981). Location of wells is shown in Figure 4. The solid arrow indicates the depth of the lower–upper Pannonian boundary determined in Kaskantyú-2, and the broken arrows are the same boundaries derived by seismic correlation for Dévaványa and Vésztő. Data for Kaskantyú-2 are after Elston et al. (1985) and for Dévaványa and Vésztő after Cook et al. (1979) and Rónai (1981).

Two other wells, drilled for hydrocarbon exploration, are located between Dévaványa and Vésztő and lie along the seismic profile shown in Figure 6. Lithologic trend analysis shows that these wells penetrated a thick upper Pannonian (Pa_2) sequence overlying unit Pa_1^2. Unit Pa_1^2 unconformably overlies a relatively thin middle Miocene unit (M_4) or the basement. Note that the oldest Pannonian subunits (Pa_1^{1a} and Pa_1^{1b}) are missing in these wells and near Dévaványa, although they may be present in the deep trough below Vésztő. The upper Pannonian–lower Pannonian boundary (Pa_2–Pa_1^2) can be clearly followed from these two wells to both Dévaványa and Vésztő by seismic correlation. It can be seen that the Miocene–Pliocene boundary lies well within the upper Pannonian (Pa_2) unit on Figure 6. Thus, unit Pa_2 includes the whole Pliocene and a part of the upper Miocene. A very tentative age may be assigned to the base of unit Pa_2 below Dévaványa and Vésztő by noting that the sedimentation rate was remarkably constant in both wells between about 0 and 6 Ma. Because deposition of the entire unit Pa_2 reflects the thermal subsidence following extension and because sediments of unit Pa_2 were deposited in a shallow water environment, it is probably that the sedimentation rate was roughly constant throughout deposition of unit Pa_2. Extrapolating downward to the base of Pa_2 in both Dévaványa and Vésztő suggests an age of about 9 Ma for the upper Pannonian–Lower Pannonian boundary (Figure 5).

Figure 7 shows a composite seismic section connecting the Kaskantyú-2 borehole with Kiha-Ny-3, which was drilled for hydrocarbon exploration and in which K-Ar dating gives a 9.61 ± 0.38 Ma age for basalts at a depth of 1162 to 1167 m (Balogh et al., 1983). Unlike Dévaványa and Vésztő, a complete biostratigraphic profile could be constructed for Kaskantyuú-2 and compared to the magnetostratigraphically determined ages in the well. Both paleontologic and lithologic analysis of core samples place the upper Pannonian–lower Pannonian boundary at a depth of 865 m. The measured polarity zonation suggests that the age of this boundary is about 8.8 Ma (Elston et al., 1985). If this depth is converted to two-way traveltime and plotted on Figure 7, it roughly coincides with a seismic reflector that can be followed fairly continuously for 40 km to well Kiha-Ny-3. This 8.8 Ma age agrees well with the 9.61 ma age measured on basalts at a slightly deeper level in Kiha-Ny-3.

This 8.8 Ma age for the upper Pannonian–lower Pannonian boundary agrees well with the 9 Ma age estimated for wells Dévaványa and Vésztő and with the data given by Steininger et al. (this volume). We note, however, that this boundary is diachronous and slightly younger in the central part of the Great Hungarian Plain (for example, in the Makó trough) than near Kaskantyú-2. Unfortunately, there are no data to place reliable constraints on the age of the upper Pannonian–lower Pannonian boundary in the central part of the basin. Taking into consideration, however, that the outward building of the delta was very fast, a minimum age of 7 to 8 Ma can be estimated.

The lower Pannonian section is incomplete all along the profile in Figure 7. Extrapolation of lithogenetic units onto this section from boreholes at some distance away indicates that only the upper part of unit Pa_1^2 is developed below unit Pa_2. Note that the seismic reflector dated at 9.6 Ma in Kiha-Ny-3 onlaps the unconformity at the top of middle Miocene (mostly Badenian) strata. Thus, unit Pa_1^{1a}, unit Pa_1^{1b}, part of unit Pa_1^2, and possible the Sarmatian are missing across this unconformity in the middle of the section. On the left and right sides of the section where the basin is shallow, the incomplete lower Pannonian directly overlies the basement.

Figure 8 shows a seismic reflection profile through three hydrocarbon exploration wells located 50 km southeast of the profile in Figure 7. In well Ruzsa-4, K-Ar dating yields an age of 10.4 ± 1.8 Ma for basalts at a depth of 2657 to 2666 m. Lithogenetic trend diagrams show that units Pa_2, Pa_1^2, and Pa_1^{1b} and middle Miocene (mostly Badenian) sedimentary rocks are present in these wells. The basalts occur just above the base of unit Pa_1^{1b}. These age data, combined with the K-Ar date from well Kiha-Ny-3, suggest that unit Pa_1^{1b} is older than 9.61 ± 0.38 Ma, but younger than 10.4 ± 1.8 Ma if unit Pa_1^{1b} is complete in this section. We suggest taking 10 Ma and 11 Ma, respectively, for the ages of the upper and lower boundaries of unit Pa_1^{1b} (Figure 1). The unconformity above Badenian and Sarmatian(?) sediments in this area thus represents a small stratigraphic gap. A larger gap between Pannonian and older Miocene rocks may be present near well Kis-ÉK-1 (Figure 9). Here, the basement is made up of Eocene to middle Oligocene flysch overlying about 200 m of lower to middle Eocene rocks. Rhyolites from the middle of the lower–middle Miocene sequence have yielded a K-Ar age of 18.2 ± 0.3 Ma (Árva-Sós et al., 1983). The lowermost Pannonian units (Pa_1^{1a} and Pa_1^{1b}) are not present in this area.

In general, the unconformity and stratigraphic gap between Pannonian and older Miocene rocks separates synrift sediments (below) from postrift sediments (above). As a rule of thumb, it is present in parts of the Pannonian basin where the present depth to basement is less than about 4 km. In deeper troughs the section appears to be continuous. We suggest that in places where the middle Miocene strata are thicker than about 100 m, this stratigraphic gap is probably caused by submarine erosion resulting from changes in the water circulation pattern at the end of the rifting phase. Above basement highs, where middle Miocene marine sediments are not present, it reflects nondeposition (or minor accumulation of terrestrial clastics) because the basement did not subside below the water level until Pannonian time.

CONCLUSIONS

From the data we present in this chapter, the following conclusions can be drawn:

1. Primary seismic reflectors approximate geologic time lines on a fine scale (1 m.y. or less) in the Pannonian basin.
2. Seismic correlation shows that lower Pannonian lithogenetic units (Pa_1^{1a} and Pa_1^{1b}) are very close to chronostratigraphic units. Available absolute age data suggest that the time interval they span is about 2 m.y.
3. The base of the upper Pannonian lithogenetic (and also biostratigraphic) unit is heterochronous. We suggest an average age of 8 Ma for this boundary. In places it is as old as 9 Ma and in other places it may be as young as 7 Ma.
4. The unconformity between the synrift and postrift sediments is generally associated with a stratigraphic gap. Units Pa_1^{1a} and Pa_1^{1b} and locally even the lower part of unit Pa_1^2, as well as the upper part of the Sarmatian, are missing, which indicates a stratigraphic gap corresponding to one to several million years.
5. The stage system described in this paper represents neither biostratigraphic nor chronostratigraphic units exactly. However, the uncertainties in absolute ages of the stage boundaries for Miocene and younger strata are small, so that these ages provide sufficiently accurate information for many practical purposes, such as reconstruction of basin evolution.

REFERENCES

Árva-Sós, E., K. Balogh, G. Hámor, Á. Jámbor, and L. Ravasz-Baranyai, 1983, Chronology of Miocene pyroclastics and lavas of Hungary: Ann. Inst. Géol. Géophys., Bucuresti, v. 61, p. 353–358.

Balázs, E., A. Barbás, L. Bartkó, I. Bérczi, I. Gajdos, et al., 1981, Molasse formations of Hungary. Exc, *in* Guide prepared for the meeting of WG. 3.3 of Problem Comm. IX. of socialist countries: Hugn. Geol. Inst., Budapest, p. 1–179.

Balogh, K., A. Jámbor, Z. Partényi, L. Ravasz-Baranyai, G. Solti, and A. Nusszer, 1983, Petrography and K/Ar dating of Tertiary and Quaternary basaltic rocks in Hungary: Ann. Inst. Géol. Géophys., Bucuresti, v. 61, p. 365–373.

Cook, H. B. S., J. M. Hall, and A. Rónai, 1979, Paleomagnetic, sedimentary and climatic records from boreholes of Dévaványa and Vésztő: Acta Geol. Hung., v. 22, n. 1–4, p. 89–109.

Elston, D. P., G. Hámor, A. Jámbor, M. Lantos, and A. Rónai, 1985, Magnetostratigraphy of Neogene strata penetrated in two deep core holes in the Pannonian basin: preliminary results: Geophys. Transact., v. 31, n. 1–3, p. 75–88.

Grosz, A. E., A. Rónai, and R. Lopez, 1985, Contribution to the determination of the Plio-Pleistocene boundary in sediments of the Pannonian basin: Geophys. Transact., v. 31, n. 1–3, p. 89–99.

Hámor, G., K. Balogh, and L. Ravasz-Baranyai, 1980, Radiometric age of the Tertiary formations in North Hungary: Ann. Rep. Hung. Geol. Inst. from 1976, Budapest, p. 61–76.

Korpás-Hódi, M., 1983, Palaeoecology and biostratigraphy of the Pannonian mollusca fauna in the northern foreland of the Transdanubian Central Range. Ann. Hugn. Geol. Inst., v. 64, p. 1–163.

Lowrie, W., and W. Alvarez, 1981, One hundred million years of geomagnetic polarity history: Geology, v. 9, p. 392–397.

Nagymarosy, A., 1981, Chrono- and biostratigraphy of the Pannonian basin: a review based mainly on data from Hungary: Earth Evol. Sci., v. 1, n. 3–4, p. 183–194.

Pogácsás, Gy., 1984, Results of seismic stratigraphy in Hungary: Acta Geol. Hung., v. 27, n. 1–2, p. 91–108.

Pogácsás, Gy., 1985, Seismic stratigraphic features of Neogene sediments in the Pannonian basin: Geophys. Transact., v. 30, n. 4, p. 373–410.

Rónai, A., 1981, Magnetostratigraphy of Pliocene–Quaternary sediments in the Great Hungarian Plain: Earth Evol. Sci., v. 1, n. 3–4, p. 265–268.

Rónai, A., 1984, The development of the Quaternary geology in Hungary: Acta Geol. Hung., v. 27, n. 1–2, p. 75–90.

Senes, J., 1979, Géochronologie des stratotypes des etages du miocene inférieure et moyen de la Paratethys Centrale utilisables pour la corrélation globale: Geol. Zborn.-Geol. Carp., Bratislava, v. 30, n. 1, p. 100–111.

Steininger, F., and A. Papp, 1979, Current biostratigraphic and radiometric correlation of late Miocene Central Paratethys stages (Sarmation s. str., Pannonian s. str. and Pontian) and Mediterranean stages (Tortonian and Messinian) and the Messinian event in the Paratethys: Newsl. Stratigr., v. 8, n. 2, p. 100–110.

Szalay, Á., and K. Szentgyörgyi, 1979, Contribution to the knowledge of lithogenetic subdivisions of Pannonian basin formations explored by hydrocarbon drilling: reconstructions based on trend analysis: MTA X. Oszt. Közl., Budapest, v. 12, n. 4, p. 401–423.

Vail, P. R., R. M. Mitchum, Jr., R. G. Todd, J. M. Widmier, S. Thompson, III, J. B. Sangree, J. N. Bubb, and W. G. Hattelid, 1977, Seismic stratigraphy and global changes of sea level, *in* C. E. Payton, ed., Seismic stratigraphy—application to hydrocarbon exploration: Amer. Assoc. Petrol. Geol. Memoir 26, p. 49–205.

CHAPTER 10

Preliminary Sedimentological Investigation of a Neogene Depression in the Great Hungarian Plain

I. Bérczi
Hungarian Hydrocarbon Institute
2443 Százhalombatta
P.O.B. 32, Hungary

The Makó-Hódmezovásárhély trench, situated in the southeastern part of Hungary, is filled with more than 7 km of thick Miocene to Holocene sedimentary rocks. Seismic profiles, stratification features in core samples, grain size distributions, and well logs in three boreholes clearly reveal a shoaling-upward sequence within the trough. We have distinguished five principal depositional environments that we interpret (from bottom to top) as: coarse-grained basal turbidites; deep basin fine-grained sediments; delta front turbidites; delta slope sand, silt, and marl; and shallow lake to braided stream deposits.

INTRODUCTION

The Makó-Hódmezovásárhély trough contains more than 7 km of Neogene–Quaternary sedimentary rocks, and the deepest borehole in Hungary (Hód-I) terminated in Badenian strata in this trough at a depth of 5842 m below sea level (Figure 1). This preliminary study of the basin-fill facies present in this trough is based on a detailed investigation of cores from three boreholes that yield a rough cross-section through the trough: Maroslele-1, Hódmezővásárhely-I, and Békéssámson-1 (Figure 2). Analysis of stratification features determined from grain size distribution and well log data enabled us to distinguish five main depositional facies that clearly represent a shoaling-upward sequence in a basin that at times contained steep prograding slopes. We have tentatively identified these five facies as: (1) basal facies; (2) deep basin facies; (3) prodelta facies; (4) delta front–delta slope facies; and (5) delta plain facies.

THE BASAL FACIES

The basal facies is composed of sandy to marly conglomerate and is found only in the axial part of the trench. Borehole Hód-I penetrated 373 m into this unit, but failed to reach even the stratigraphic base of the unit. The age of the upper part of this unit is early to middle Badenian. The interbedded sandstone and marl beds may dip about 7–11° due to deformation and slumping, and the sandstone layers display a coarsening-upward sequence. Most of the sandstone layers show graded bedding (A, B, and C parts of the Bouma sequence), are frequently distorted, and contain marl rip-up clasts (Plate 1). In the nongraded sandstone layers, thin amalgamated beds with small-scale cross-strata or parallel to wavy lamination with marl interbeds are common. This unit can be clearly distinguished from the basal conglomerate that occurs around the margins of the basin. This conglomerate has an obvious psefitic nature, with undistributed bedding and, in contrast to the conglomerate in the deep axial part of the trough, shows little evidence for significant sedimentary transport (Figures 3–5).

A second influx of muddy conglomerates represented by the Dorozsma Formation appeared in the Pannonian (*s. str.*) overlying the deep basin facies (Figure 2).

DEEP BASIN FACIES

The deep basin facies overlaps the basal facies in the axial part of the trench. This 900-m-thick series consists of calcareous and argillaceous silty marls of laminated and (to a lesser extent) mas-

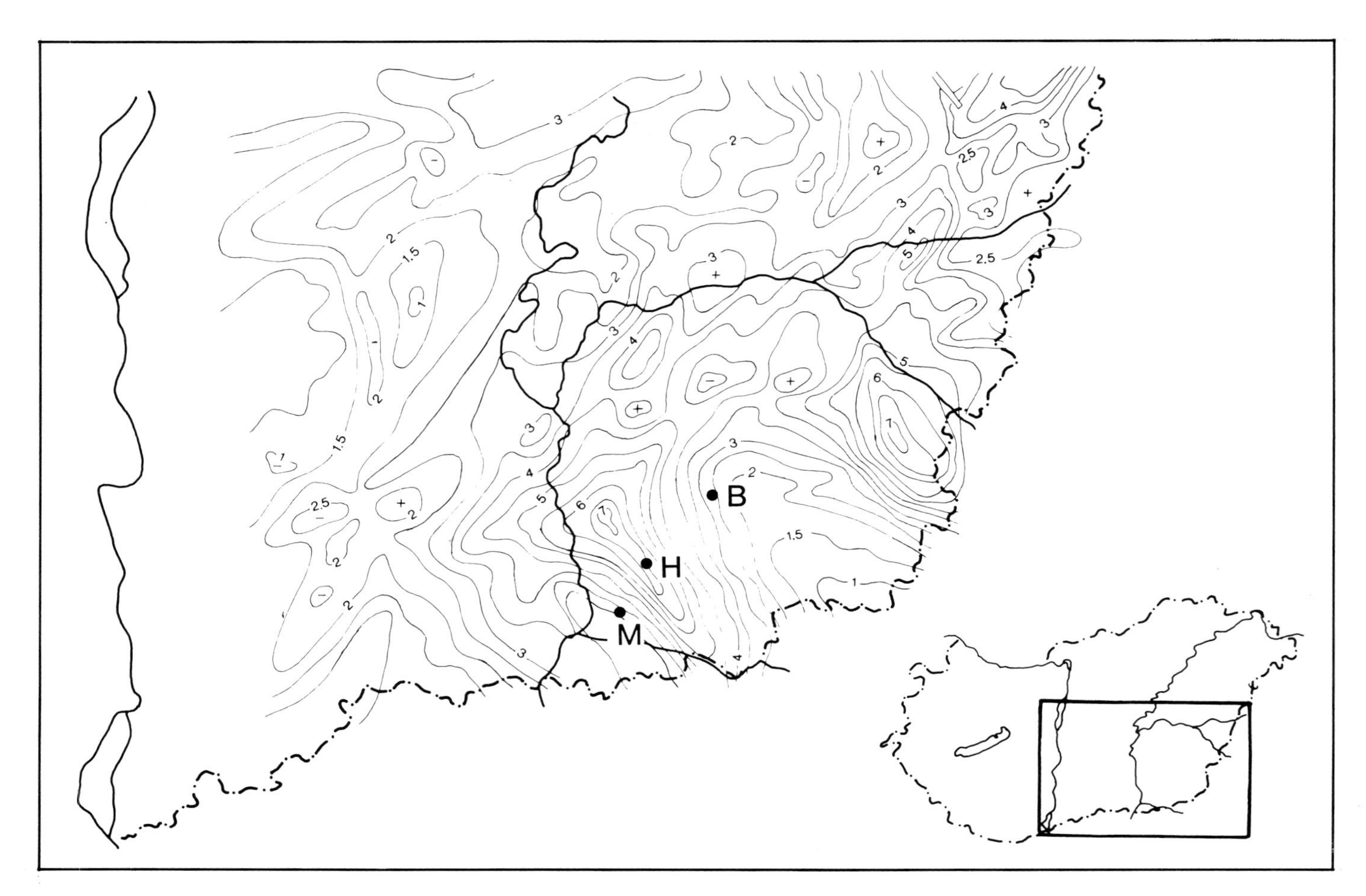

Figure 1. Location map with total thickness of the Neogene in the area studied (contours in kilometers), and the locations of the three boreholes used. The Makó-Hódmezováśárhély trough contains wells Hódmezővásárhely-I (H), Maroslele-1 (M), and Békéssámson-1 (B).

sive appearance. No bioturbation was observed (Figure 6). This unit is of middle to late Badenian to Pannonian (*s. str.*) age.

PRODELTA FACIES

The prodelta facies can be divided into two parts: the upper part is more sandy and is considered to be a product of fan deposits (proximal turbidities) represented by thick graded sandstones (A-B, B-C, and C-E parts of the Bouma sequence) (Figure 7 and Plate 2) and slumped deposits. Marl rip-up clasts are rarer than in the delta front–delta slope facies. Dish structures as well as load structures at sand–marl contacts are common (Figure 7).

The lower part (distal turbidities) are characterized by gradual decreases in the sand bed thickness, in the sand to marl ratio, and in grain size, as well as by the predominance of C-E units of the Bouma sequence. No bioturbation was observed in any part of the prodelta facies.

DELTA FRONT–DELTA SLOPE FACIES

In the delta front–delta slope facies, sandstone and marl may occur as horizontal parallel beds, but the most characteristic feature is the 4–25° dip of the bedding (the most common dip is 5–7°). Similar values can be measured as dips of foresets in regional profiles. The sandstone beds contain load casts, flame structures, and marl rip-up clasts showing well-defined graded bedding, with a predominance of A, A-B, and A-B-C units of the Bouma sequence (Figure 8 and Plate 2). In the most steeply dipping part of the sequence, soft sediment deformation (synsedimentary faulting, slumps, rotation of beds, and overturned strata) are common (Figure 7 and Plate 2). Bioturbation is abundant in the upper two thirds of this sequence.

DELTA PLAIN FACIES

The delta plain facies is composed of alternating layers of horizontally bedded sandstone, siltstone, and marl. The sandstone and siltstone layers are graded and repeated. Occurrence of small-scale cross-strata as well as bioturbation is also common (Figure 8 and Plate 2). Oxidized mudstone intercalations are common, indicating that this region was sometimes dry land (Figure 9). The increasing importance of terrestrial conditions is also indicated by the common occurrence of lignite beds (Reineck-Singh, 1980).

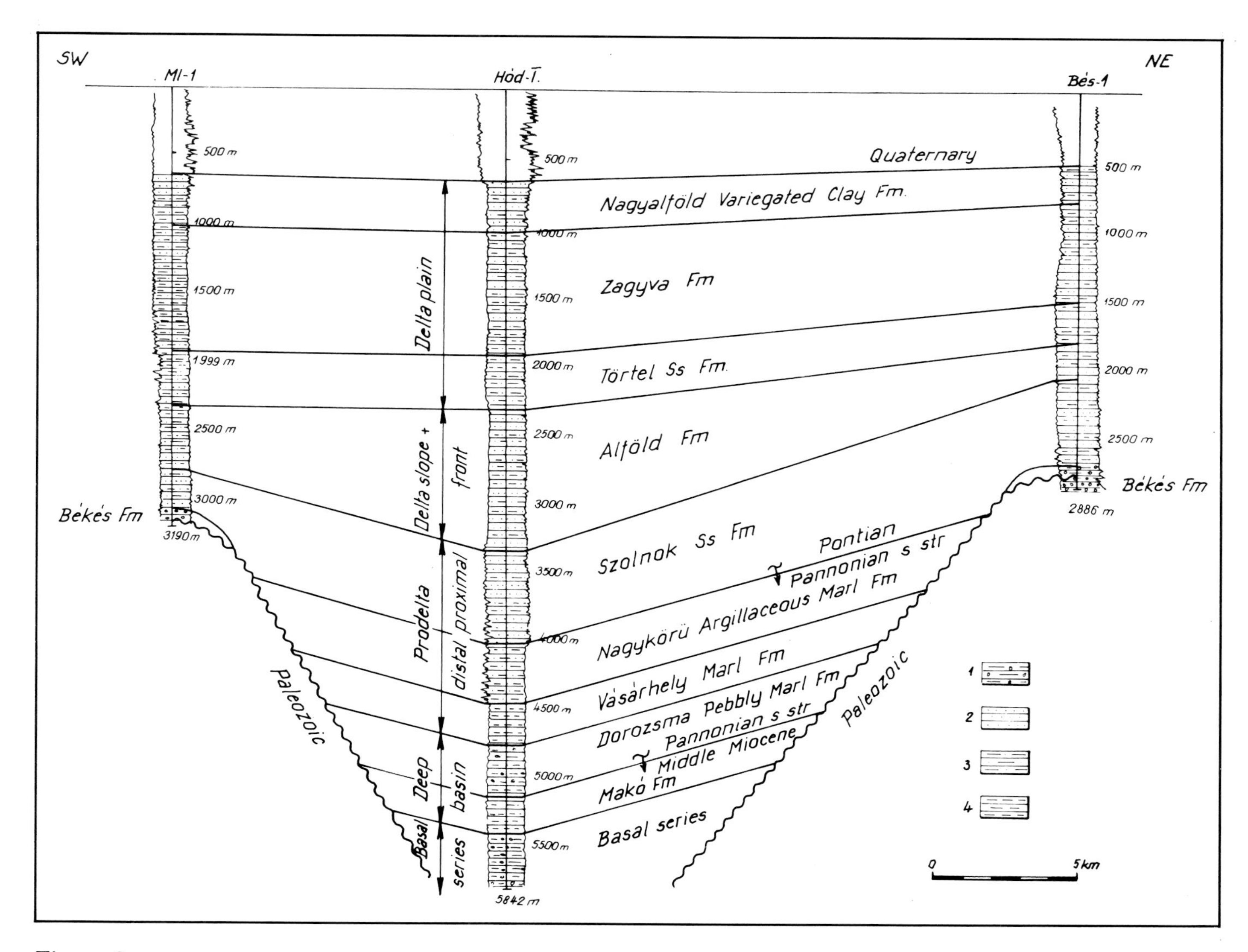

Figure 2. Stratigraphic sections in the three boreholes studied, showing depth below sea level of each formation. Symbols in legend: (1) muddy conglomerate, (2) sandstone, (3) siltstone, and (4) shale.

DISCUSSION

Two- and three-dimensional views of the facies discussed above are summarized in Figures 10 and 11. We propose that a highly constructional, fluvial-dominated delta system (Reading, 1978) existed in the Makó trench, and that it prograded from northwest to southeast. Such progradation is also indicated by the mineralogical composition of the sandstones: most of them are lithoarenites containing rock fragments and heavy minerals of metamorphic (Alpine) origin. No volcanic debris from the north and east could be detected (Bérczi, 1970, 1972).

The Neogene evolution of the trench can be tentatively summarized as follows. At the beginning of Badenian time, the rapidly subsiding areas quickly reached the stage of deep basin (pelitic) sedimentation. The steep slopes of the trench initiated the reworking of coarse-grained sediments as turbidites. Thus, the matrix of the basal facies is suspended mud and chemically precipitated $CaCO_3$, to which were added periodic influxes of coarse-grained sediments. The turbidities may have flowed either along the axis of the trench or down its flanks. Periodic occurrences of steeply dipping marl and sandstone interbeds suggest irregular bottom topography with slumping strata and/or infill of previously formed turbidite channels. (We use "turbidite" to mean predominantly sediment gravity flow deposits.)

The deep basin facies is composed of mud deposited from suspension and chemically precipitated $CaCO_3$. Fossils with $CaCO_3$ shells are encountered very scarcely, thus biogenic effects in $CaCO_3$ precipitation can certainly be excluded. The horizontal bedding shows no disturbances of any kind.

The proximal part of the overlying prodelta facies was dominated by periodic influxes of sand by turbidity flows that exhibit decreasing energy toward the center of the trough(?). The amalgamated sandstone beds (these represent the top of the prodelta series) sometimes adhere to the lower part of the slumps of fans deposited in the surrounding (sublake) channel system. The structure of the sandstone is dominated by the upper units (C-E) of the Bouma sequence and by horizontal bedding.

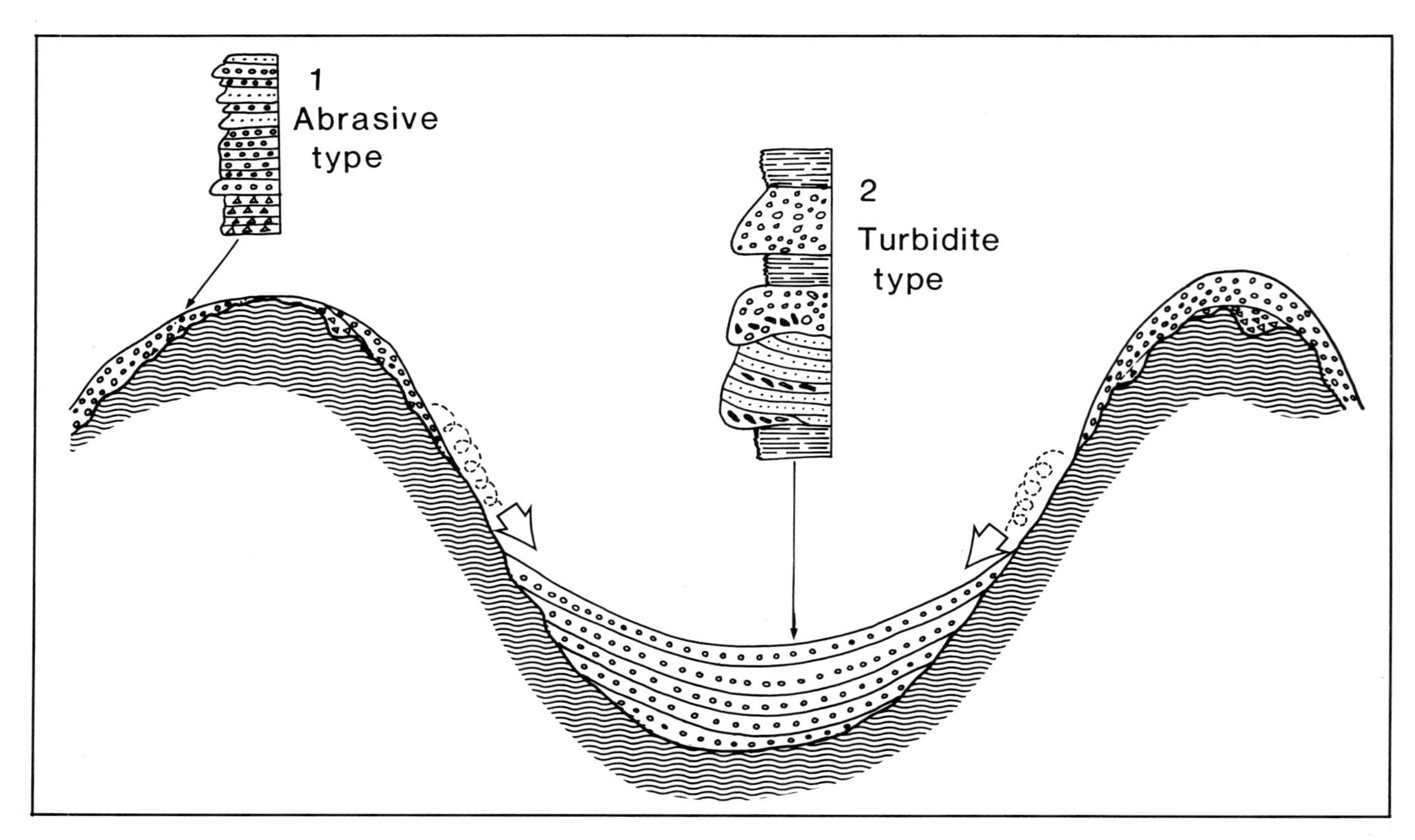

Figure 3. The genesis of the two types of basal conglomerates: (1) an abrasive type formed by more or less *in situ* erosion, and (2) a turbiditic type formed by transport of clasts into the deep basin areas.

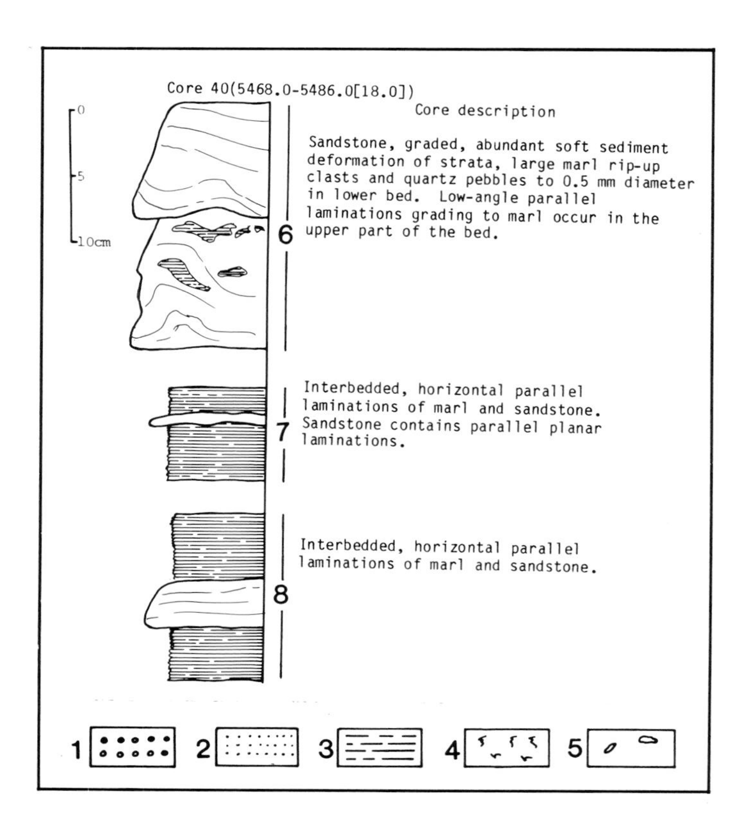

Figure 4. Characteristic core from the basal facies: Hód-I well, Core 40, 5468.0–5486.0 m. (1) Sandstone, graded, soft sediment deformation, large marl rip-up clasts and quartz pebbles to 0.5 mm diameter in lower bed. Low-angle parallel wavy laminations grading to marl occur in the upper part of the bed. (2) Interbedded, horizontal parallel laminations of marl and sandstone. Sandstone contains wavy to planar laminations. (3) Interbedded, horizontal parallel laminations of marl and sandstone.

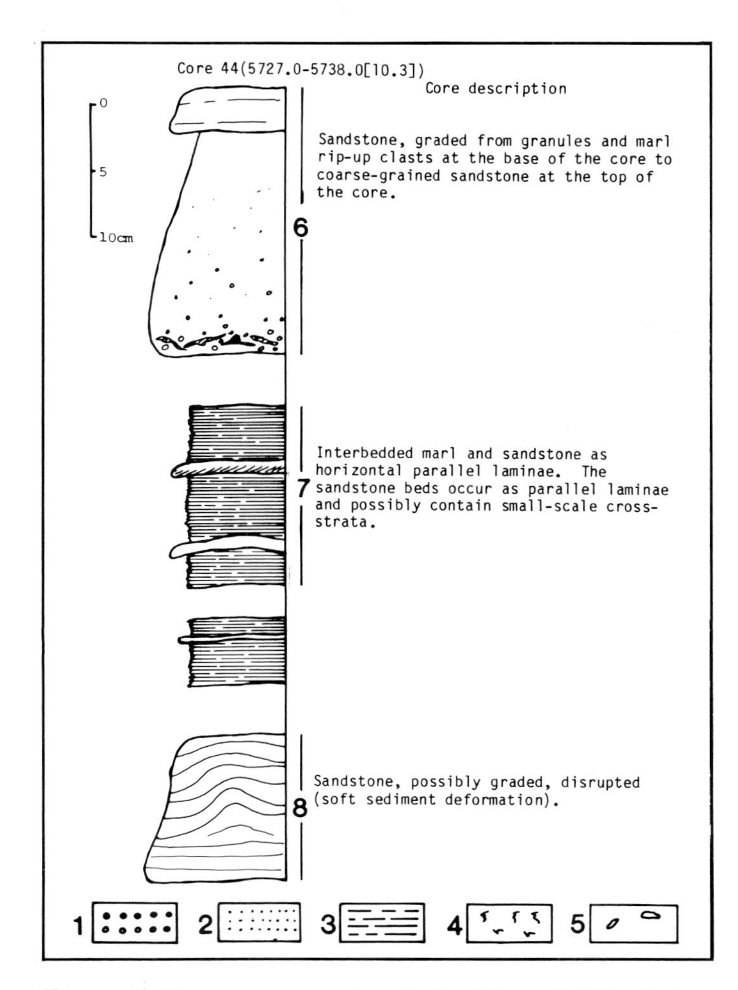

Figure 5. Characteristic core from the basal facies: Hód-I well, Core 44, 5727.0–5738.0 m. (1) Sandstone, graded from granules and marl rip-up clasts at the base of the core to coarse-grained sandstone at the top of the core. (2) Interbedded marl and sandstone in horizontal parallel laminae. The sandstone laminae possibly contain small-scale cross-strata. (3) Sandstone, possible graded, and disrupted (soft sediment deformation). (Other symbols as in Figure 4.)

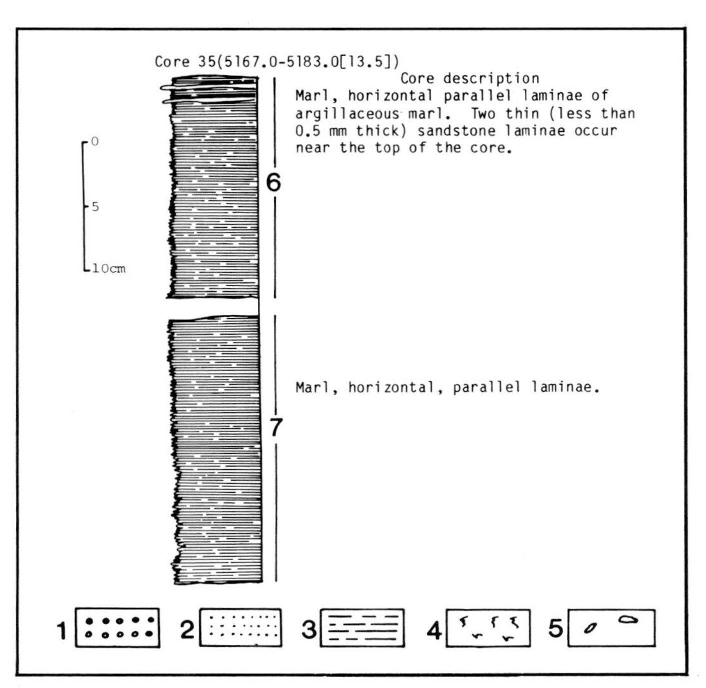

Figure 6. Characteristic core from the deep basin facies: Hód-I well, Core 35, 5167.0–5183.0 m. (1) Marl, horizontal parallel laminae of clay marl. Two thin (less than 0.5 mm thick) sandstone laminae occur near the top of the core. (2) Marl, horizontal, parallel laminae.

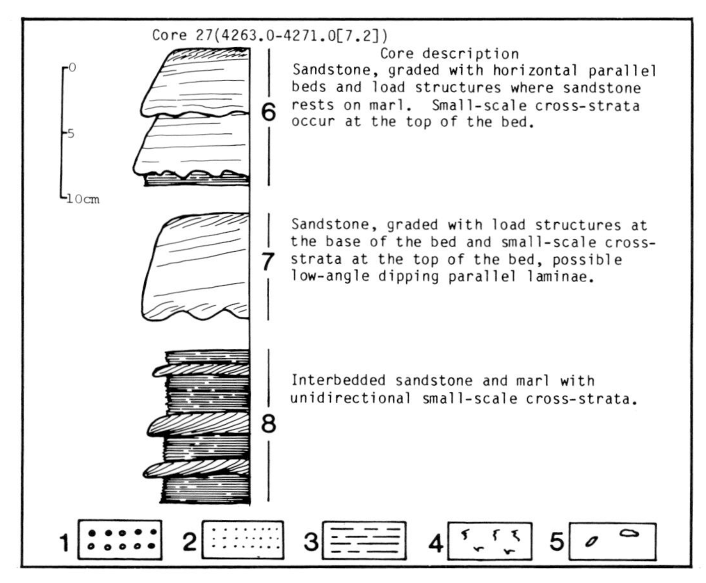

Figure 7. Characteristic core from the prodelta facies: Hód-I well, Core 27, 4263.8–4271.0 m. (1) Sandstone, graded, with horizontal parallel beds and load structures where sandstone rests on marl. Small-scale cross-strata occur at the top of the bed. (2) Sandstone, graded, with load structures at the base of the bed and small-scale cross-strata at the top of the bed, possible low-angle dipping parallel laminae.(3) Interbedded sandstone and marl with unidirectional small-scale cross-strata.

Plate 1. Hód-I well, Core 40, 5468–5486 m. Graded sandstone with large marl rip-up clasts and quartz pebbles overlain by horizontally bedded coarse-grained sandstone forming A-B units of a Bouma sequence (basal series).

Plate 2. Hód-I well, Core 40, 5468–5486 m. Extremely large, rounded marl rip-up clast surrounded by mud supported conglomerate of grain flow origin (basal series).

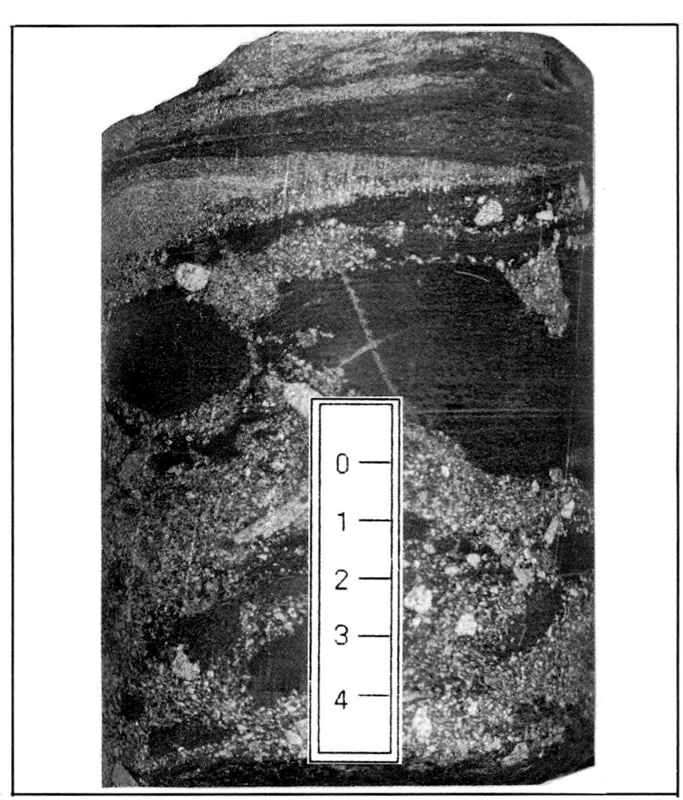

Plate 3. Hód-I well, Core 40, 5468–5486 m. Inversely(?) graded Bouma A unit (with extraordinarily large marl rip-up clasts in the upper part of the unit) overlain by wavy laminated sandstone with marl interbeds (basal series).

Plate 4. Hód-I well, Core 36, 5250–5267 m. Coarse-grained mud-supported conglomerate with faint horizontal grading (basal series).

Plate 5. Hód-I well, Core 34, 5070–5074 m. Alternation of marl and sandstone with planar horizontal graded beds. Sandstone load structure resting on marl in the lower left corner, (prodelta distal "turbidite").

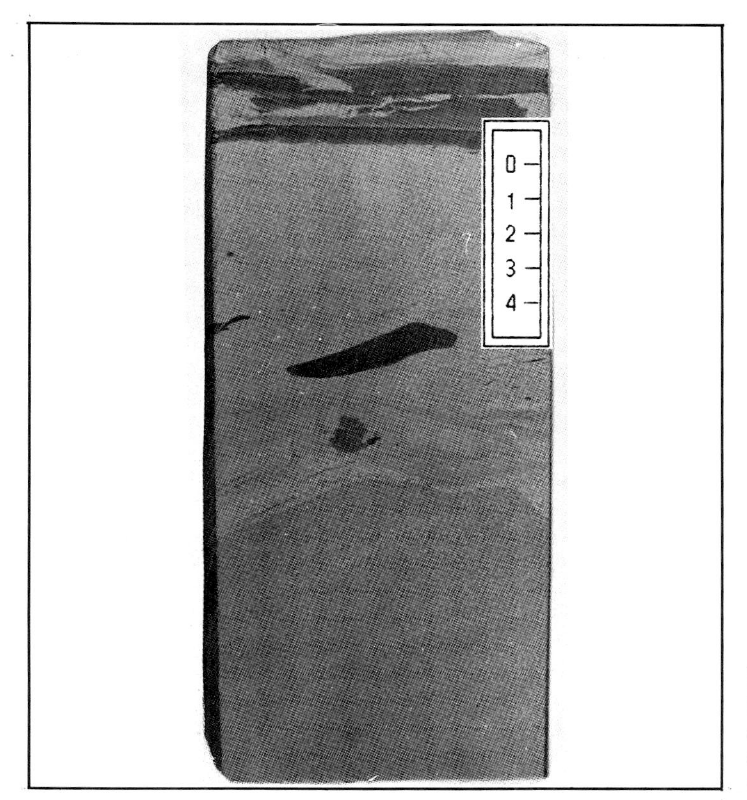

Plate 6. Hód-I well, Core 16, 3433–3442 m. Large marl rip-up clasts in sandstone with soft sediment deformation overlain by graded sandstone unit. Uppermost part represents sandstone and marl with soft sediment deformation (sand intrusion), flame structures, and rip-up clasts (delta front facies).

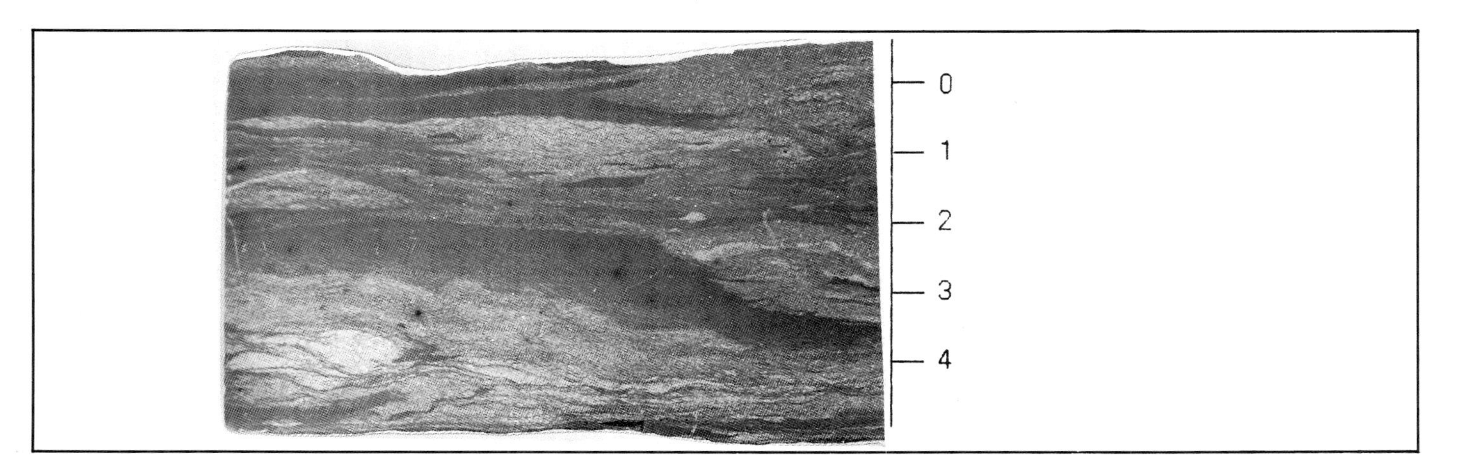

Plate 7. Hód-I well, Core 19, 3582–3590 m. Intricate interbedding of sandstone and marl. Sandstone is mostly disrupted and contains marl rip-up clasts in abundancy. The marl has a sweeping appearance and many flame structures (delta slope facies).

Plate 8. Hód-I well, Core 2, 2290–2296 m. Sandstone–siltstone rich in organic material shows low-angle cross-strata, parallel wavy lamination, and disrupted sandstone lenses. Bioturbation is frequent (delta plain facies).

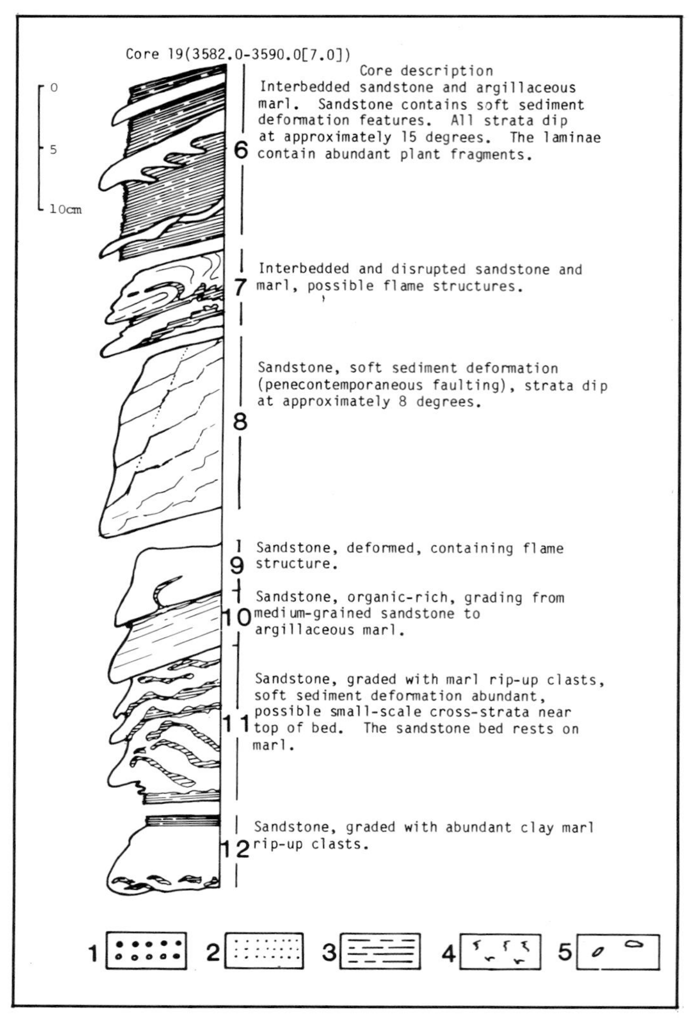

Figure 8. Characteristic core from the delta front–delta slope facies ("foreset"): Hód-I well, Core 19, 3686–3693 m. (1) Interbedded sandstone and clay marl. Sandstone contains soft sediment deformation features. All strata dip at approximately 15°. The laminae contain abundant plant fragments. (2) Interbedded and disrupted sandstone and marl, possible flame structures. (3) Sandstone, soft sediment deformation (penecontemporaneous faulting), strata dip at approximately 8°. (4) Sandstone, deformed, containing flame structure. (5) Sandstone, organic-rich grading from medium-grained sandstone to clay marl. (6) Sandstone, graded with marl rip-up clasts, soft sediment deformation abundant, possible small-scale cross-strata near top of bed. The sandstone bed rests on marl. (7) Sandstone, graded with abundant clay marl rip-up clasts. (Symbols as in Figures 4 and 5).

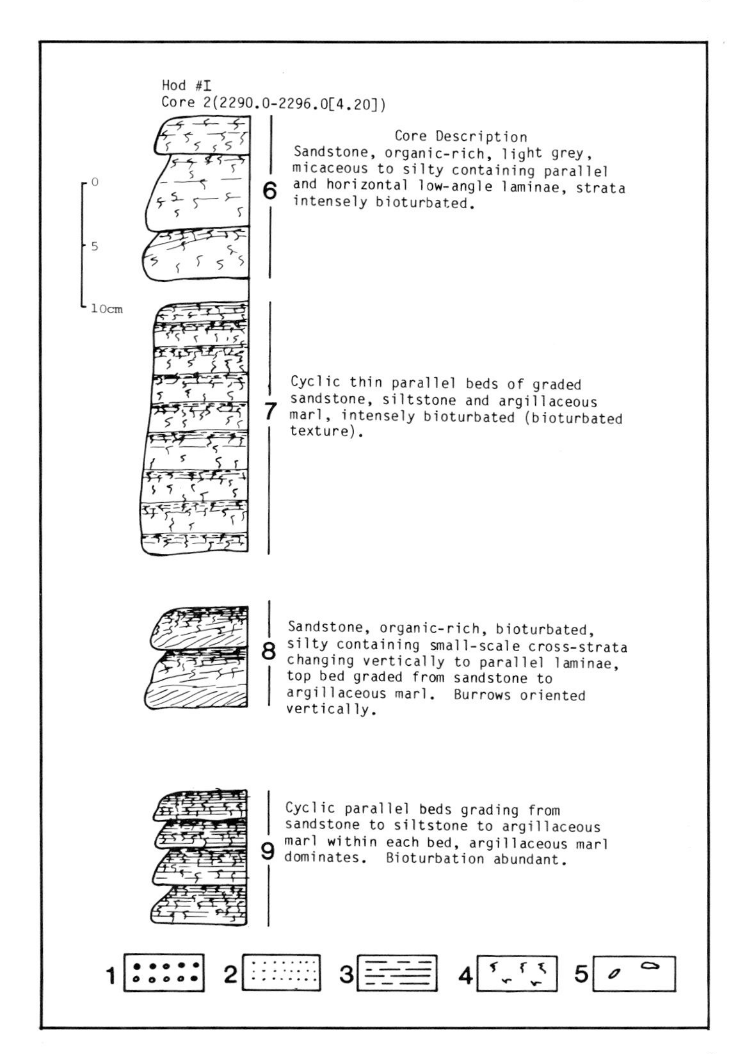

Figure 9. Characteristic core from the delta plain facies: Hód-I well, Core 2, 2290.0–2296.0 m. (1) Sandstone, organic-rich, light gray, micaceous to silty, containing horizontal to low-angle laminae, strata intensely bioturbated. (2) Cyclic, thin horizontal beds of graded sandstone, siltstone, and clay marl, intensely bioturbated. (3) Sandstone, organic-rich, bioturbated, silty, containing small-scale cross-strata changing vertically to horizontal laminae, top bed graded from sandstone to clay marl. Burrows oriented vertically. (4) Cyclic horizontal beds grading from sandstone to siltstone to clay marl within each bed, clay marl dominates. Bioturbation abundant.

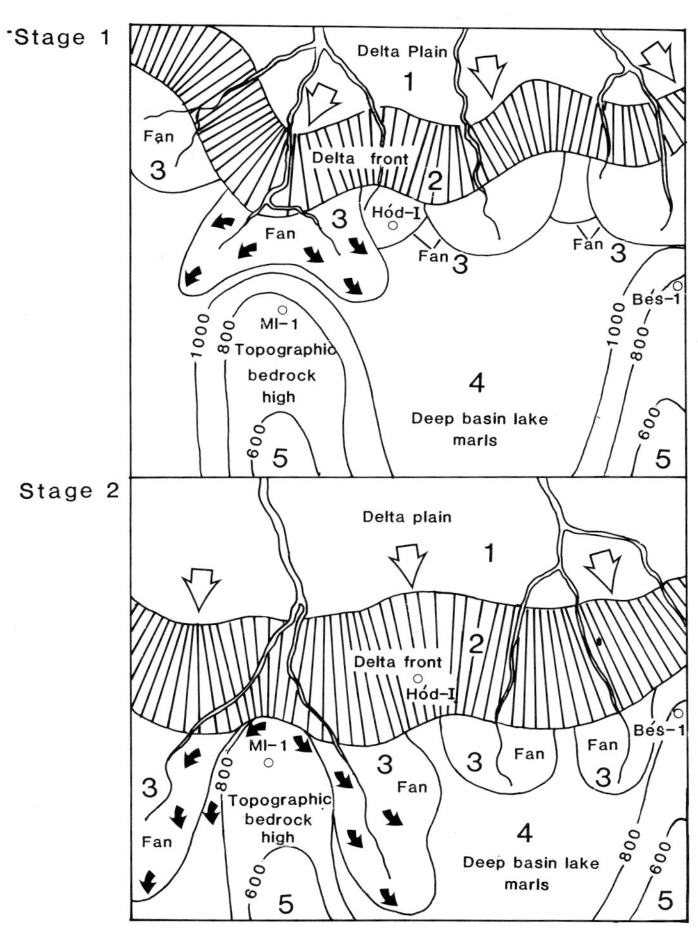

Figure 10. Two-dimensional sketch view of the progradation of the delta system in the area studied. Stage 1, late Badenian and Sarmatian; Stage 2 early Pannonian (*s. str.*). Facies: (1) delta plain, (2) delta front, (3) fan, (4) deep basins lake marls, and (5) topographic bedrock highs.

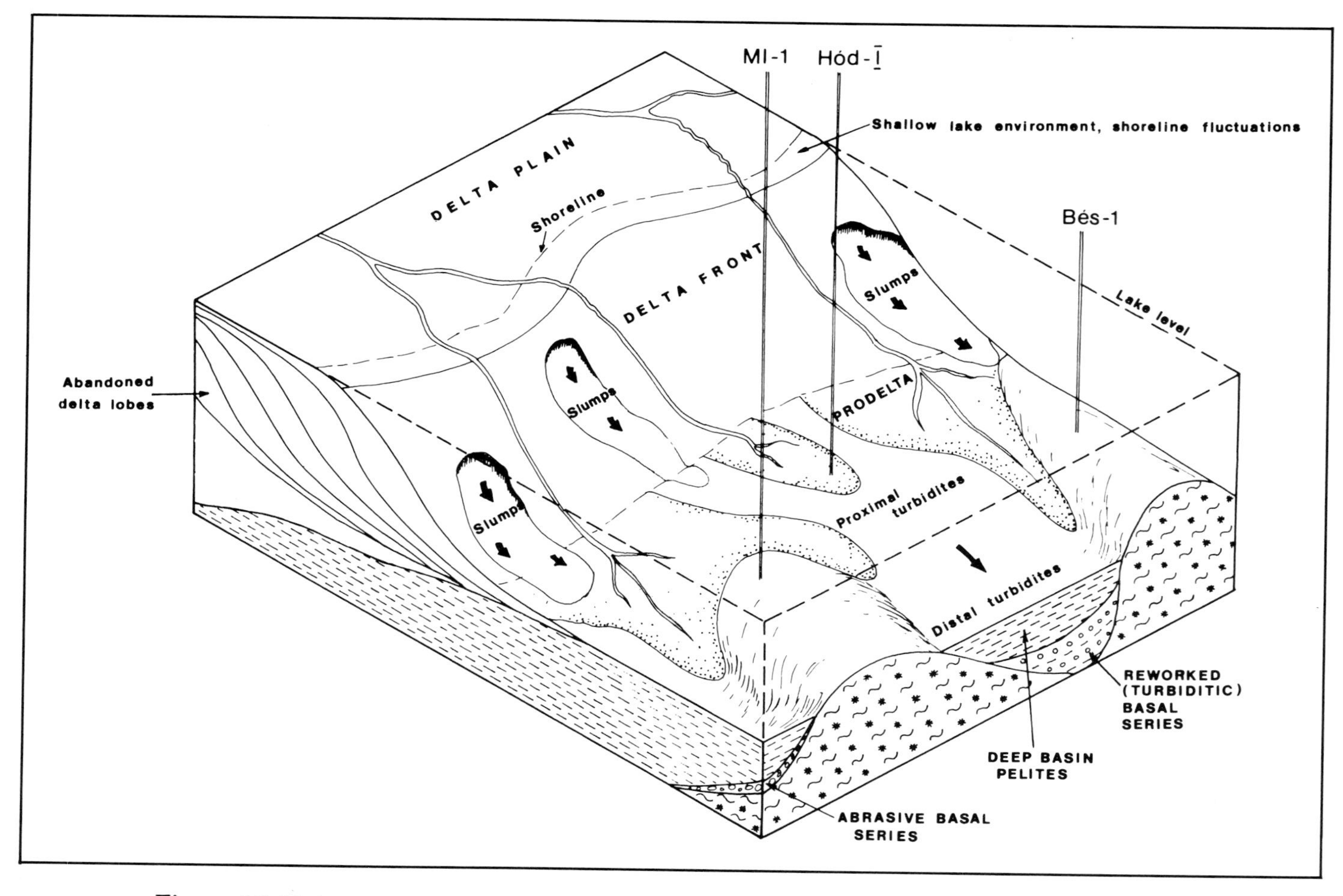

Figure 11. High-construction fluvial dominated delta system as a model for the Makó-trench, southeast Hungary.

The steeply dipping and laterally extensive beds of the delta slope–delta front facies were deposited mainly by gravity flows or grain flows of various intensity, as well as by slumping of previously deposited soft sediments. To all these deposits suspended sediments and chemical precipitation of $CaCO_3$ were added. The vertical distance between the topset and bottomset sediments indicates that a maximum of about 900 m of water depth may have existed during deposition of this unit in the deepest part of the depression (Pogácsás, this volume). This unit forms the most readily identifiable unit on seismic reflection profiles through this region. The seismic profiles show a generally southeast dip, but in the central part of the Great Hungarian Plain a southwest dip can be observed (Pogácsás and Völgyi, 1982; Horváth and Pogácsás, this volume) suggesting a multiple fluvial–deltaic advancement into the basin.

Finally the shallow lake, fluvial, and especially the marshy and terrestrial sediments in the delta plain facies signal the latest stages of sedimentation in this trench.

ACKNOWLEDGMENTS

This study has been done under the auspices of a scientific cooperative project between the Central Office of Geology (Hungary) and the USGS with the kind approval and support of the National Oil and Gas Trust of Hungary.

REFERENCES

Bérczi, I., 1970, Sedimentological investigation of the coarse-grained clastic sequence of the Algyö hydrocarbon-holding structure: Acta Geol. Acad. Scient. Hung., v. 14, p. 287–300.

Bérczi, I., 1972, Sedimentological investigation of Pre-Pannonian sedimentary formations in the Szeged-Basin, SE-Hungary: Acta Geol. Acad. Scient. Hung., v. 16, p. 229–250.

Pogácsás, Gy, and L. Völgyi, 1982, Correlation of the E. Hungarian Pannonian sedimentary facies on the basis of CH-prospering seismic and well-log sections: 27th International Geophysical Symposium, Proceedings, Bratislava, vol. A(I), p. 322-336.

Reading, H. G. (ed.), 1978, Sedimentary Environments and Facies: Oxford; Blackwell, 576 p.

Reineck, H. E., and I. B. Singh, 1980, Depositional Sedimentary Environments: Berlin/New York, Springer–Verlag, 549 p.

CHAPTER 11

Seismic Stratigraphy and Depositional Framework of Sedimentary Rocks in the Pannonian Basin in Southeastern Hungary

Robert E. Mattick
U.S. Geological Survey
Reston, Virginia 22092

R. Lawrence Phillips
U.S. Geological Survey
Menlo Park, California 94025

J. Rumpler
Geophysical Exploration Company
H-1068 Budapest VI
Nepköztársaság útja 59, Hungary

Seismic stratigraphic analyses and studies of core samples from three wells indicate that infilling of the Pannonian basin of Hungary resulted primarily from deltaic sedimentation from the northwest, north, and northeast. Infilling of the basin involved a single cycle of sedimentation which probably began in Sarmatian or earliest Pannonian time when water depths in the basin were >1,000 m. The subsequent history of the basin, during Pannonian and Quaternary time, reflects continuously shoaling waters. This shoaling resulted from sediment input rates that were generally higher than basin subsidence rates.

In general, two stages of delta construction can be recognized. In an early stage of construction, turbidite-fronted, deep water deltas were built in water depths as deep as 800–900 m. During this early constructional stage, subsidence rates and associated sediment input rates were high, and upbuilding and southward progradation of large deltaic sediment wedges filled subbasins near the source areas, overwhelmed local basement highs, and spilled sediments into subbasins in the southern part of Hungary. During a later stage of construction, prograding shallow water deltas were built in water depths of 200–400 m, and topographically low areas in the southern part of Hungary were filled by sediments discharged from river systems that advanced about 100 km southward across strata deposited during the initial stages of construction.

Seismic evidence indicates that in some areas of the Pannonian basin, the sedimentary rocks representing the two stages of delta construction are separated by a depositional unit which possibly represents a destructive phase. This unit may have been deposited during a short-lived transgressive phase or, perhaps, it was deposited following a period of accelerated lake shoaling.

The youngest and final stage of deposition is represented by delta plain facies; depositional environments varied from shallow lake, fluvial, and marsh to terrestrial soils. This unit is inferred to represent more widespread lake conditions coupled with continued shoaling and eventual disappearance of the Pannonian lake. During this period of sediment deposition, basin subsidence rates and associated sediment input rates were probably lower, and the sediment input rate is inferred to have kept pace generally with the basin subsidence rate.

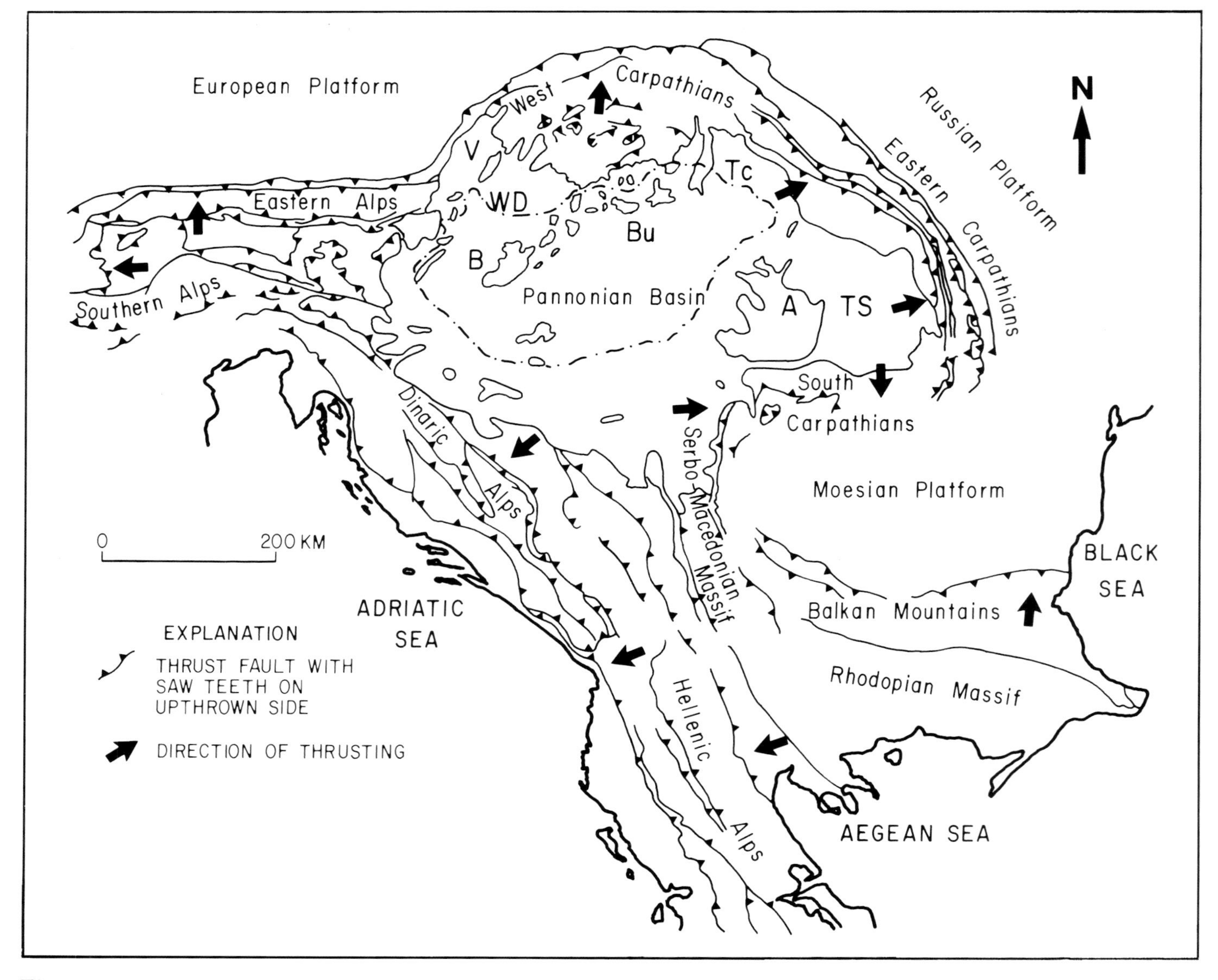

Figure 1. Location of the Pannonian basin within the intra-Carpathian region. The Pannonian, Vienna (V), West Danube (WD), Transcarpathian (TC), and Transylvania (TS) basins comprise the larger Carpathian basin. Bakony Mountains (B), Bukk Mountains (BU), Apuseni Mountains (A). Outline of Hungary shown by dash-dot line. Figure modified from Burchfiel and Royden (1982, Figure 1).

INTRODUCTION

The Pannonian basin lies within a large intramontane basin that comprises parts of five countries and is completely encircled by mountain belts of the Carpathian mountain system and the Dinaric Alps (Figure 1). The general geographic name for the region inside the mountain arc is the Carpathian basin or the intra-Carpathian region (Lerner, 1981). The intramontane region, however, is not topographically uniform, and emergent ranges divide it into several subbasins, the largest of which is the Pannonian basin. That part of the Pannonian basin that lies within Hungary generally is subdivided on the basis of topography into the Great Hungarian Plain, the Little Hungarian Plain, and the Transdanubian basin (Figure 2).

In this chapter, the principles of seismic stratigraphy (Vail et al, 1977) are applied in an attempt to reconstruct on a regional basis, the post-Mesozoic depositional history of the Pannonian basin in Hungary. Unconformities are used to bound groups of reflections into seismic sequences that can be correlated over wide areas. The seismic sequences, in turn, are correlated with depositional sequences that can be ordered in a relative, time-stratigraphic sense. The unconformities are picked from the seismic records on the basis of discordances between reflectors (or strata) at sequence boundaries. The types of discordance that occur at sequence boundaries are shown in Figure 3.

Because the main interest of the authors was the Neogene–Quaternary stratigraphy, older seismic sequences are mapped as basement rocks (basement consists chiefly of Paleozoic and Mesozoic rock units). Well data, however, has shown that, in

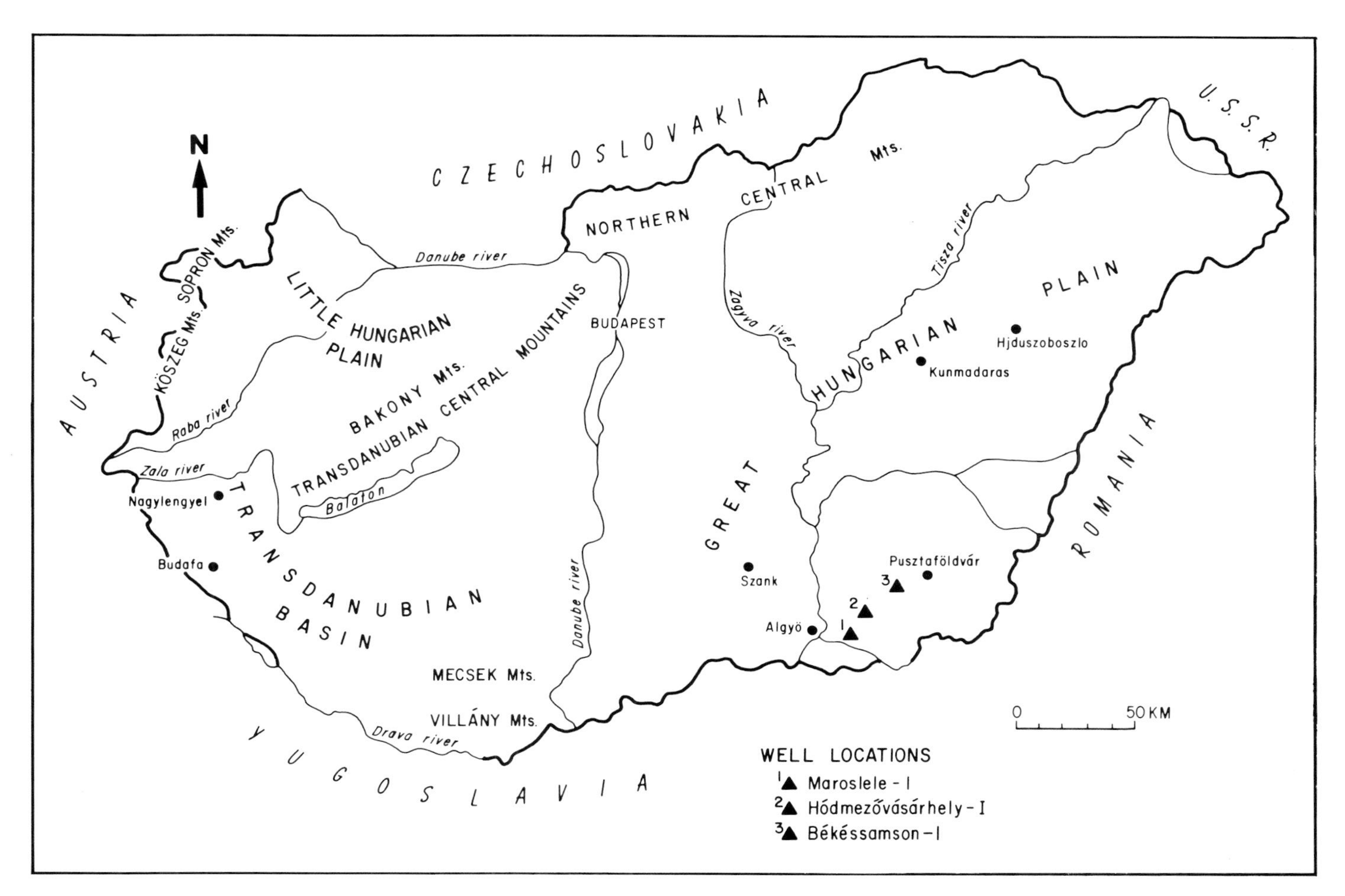

Figure 2. Geographical place names.

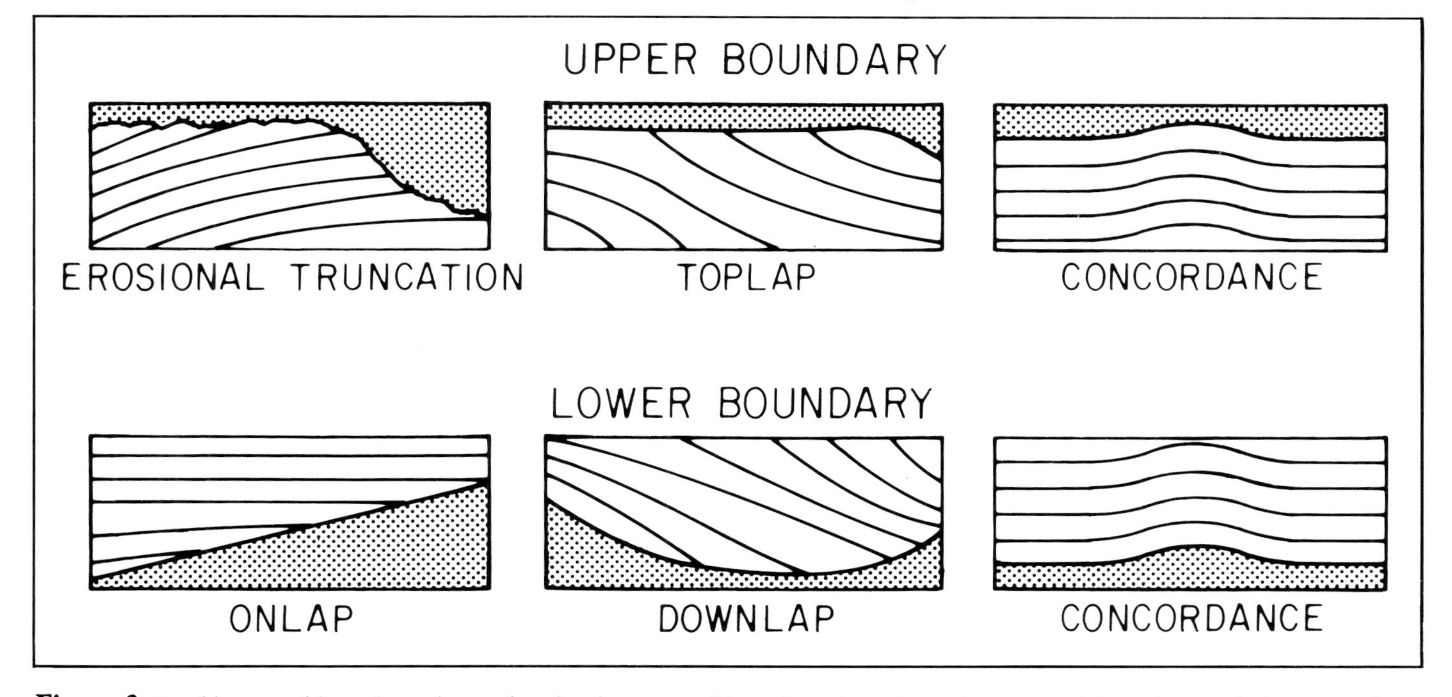

Figure 3. Possible types of discordant relations found at the upper and lower boundaries (unconformities and disconformities) of seismic or depositional sequences. Figure modified from Vail et al. (1977).

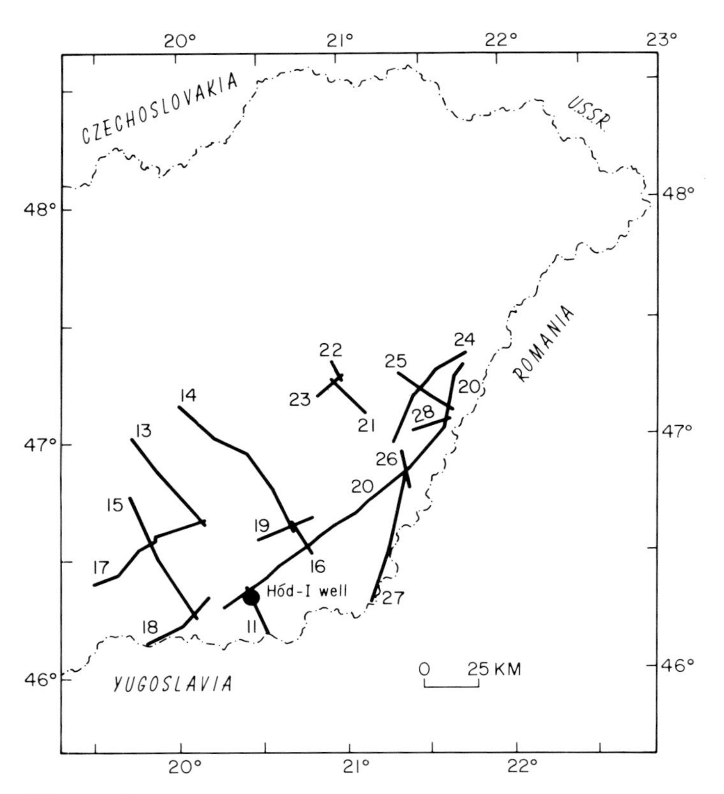

Figure 4. Location of seismic profiles interpreted in this study. Table 1 references the profile numbers used on this map to the seismic profile numbering system used by the National Oil and Gas Trust of Hungary. Profile numbers shown here correspond to figure numbers in which seismic records are displayed. Dash-dot symbols show outline of Hungary within studied area.

Table 1. Seismic Profile Numbers and Corresponding Seismic Numbering System[1]

Number in This Report	OKGT Number System
26	KO-14-7824
19	OR-24-7824
21	VFL-2-7712
11	A-11-7824
16	A-12/A-7712
13	VA-11/C-7924
15	A-10/A-7924, VA-10/B 7824
27	A-16/A/L-7612
14	A-12/B/C/D/E-7824
22	FL-20-7512
23	FL-7-7606
17	A-19/D/E/F/G-7824
18	A-16/X/Y-7824
24	KO-16-7824,KA-32-7612,KA-67-8024
25	KA-35/A-7912, KA-105-8124,KA-35I-7912
20	AL-1-7612,A-16/A/B/C/D/E/F/G-7624-8024
28	KO-44-8024

[1]Seismic profile numbers used in Figure 4 are shown in the first column. The second column shows the corresponding seismic profile numbering system used by the National Oil and Gas Trust of Hungary. The first two digits of the last number in column 2 indicate the year in which the seismic record was recorded. The second two digits indicate the fold of the final stacked record section; for example, 7824 indicates that the seismic profile was recorded in 1978 and stacked twenty-four-fold.

places, the Paleozoic and Mesozoic rocks are overlain by a thin cover of pre-Pannonian Miocene and Paleogene sedimentary rocks that may have been involved in the tectonic deformation of the basement or deposited during a marine transgression as a thin disconformable or unconformable sheet over the basement surface. In some cases, this thin unit is indistinguishable on seismic records from the underlying Paleozoic and Mesozoic units, and, therefore, the basement complex as mapped on seismic records may include pre-Pannonian Miocene and Paleogene sedimentary rock units as well as older Paleozoic and Mesozoic rock units.

The seismic profiles analyzed in this study are located in the southeastern part of the Great Hungarian Plain area (Figure 4). Table 1 references the profile numbers used in this report to the seismic profile numbering system used by the National Oil and Gas Trust of Hungary (OKGT). The depth to the basement surface (pre-Cenozoic rocks) is shown in Figure 5.

In figures and in the text, abbreviations are used in place of well names. In Table 2, well abbreviations are listed along with corresponding well names.

GEOLOGIC FRAMEWORK

The Carpathians can be divided into three tectonic elements: the inner Carpathians, the outer Carpathians, and a molasse foredeep. The inner Carpathians are the result of deformation during the Cretaceous and they consist of pre-Mesozoic crystalline and sedimentary rocks overlain by a Mesozoic sequence of chiefly carbonate rocks. In contrast, the outer Carpathians, which were deformed chiefly during the Cenozoic, consist of a sequence of thrust sheets that form a typical foreland fold and thrust belt. Rocks in this region consist of a sequence of predominantly flysch and terrigenous sandstone, shale, and locally derived conglomerates that range in age from latest Jurassic to Miocene. The rocks of the outer Carpathians grade structurally and stratigraphically into the rocks of the molasse foredeep where the rocks are dominantly Neogene in age.

The intra-Carpathian basins are believed to have formed by crustal extension as crustal shortening was taking place in the outer Carpathian belt during the last stages of thrusting. According to Royden et al. (1982, 1983) and Rumpler and Horváth (this volume), extensional (or pull-apart) tectonics behind the Carpathian arc is evidenced by drill hole data and by seismic reflection profiles that show mostly normal and growth faults bounding many of the subbasins in the region. Horváth and Royden (1981) inferred that the formation of the intra-Carpathian basins was controlled primarily by strike-slip fault movement along northeast- and northwest-trending sets of conjugate shear zones (Figure 6).

During its early history, the Pannonian Basin was a part of the central Paratethys Sea, which existed from Oligocene to Pliocene time and formed a chain of epicontinental sea basins from the Eastern Alpine foredeep to Transylvania (Nagymarosy, 1981; Nagymarosy and Müller, this volume). During this time, it was intermittently connected along temporary seaways to the Mediterranean Sea and the Indian Ocean. The closing of Mediterranean seaways to the Paratethys region was characterized by a gradual freshening of waters in the intra-Carpathian region and by the formation of a great lake system that extended from the Vienna Basin to Soviet Central Asia.

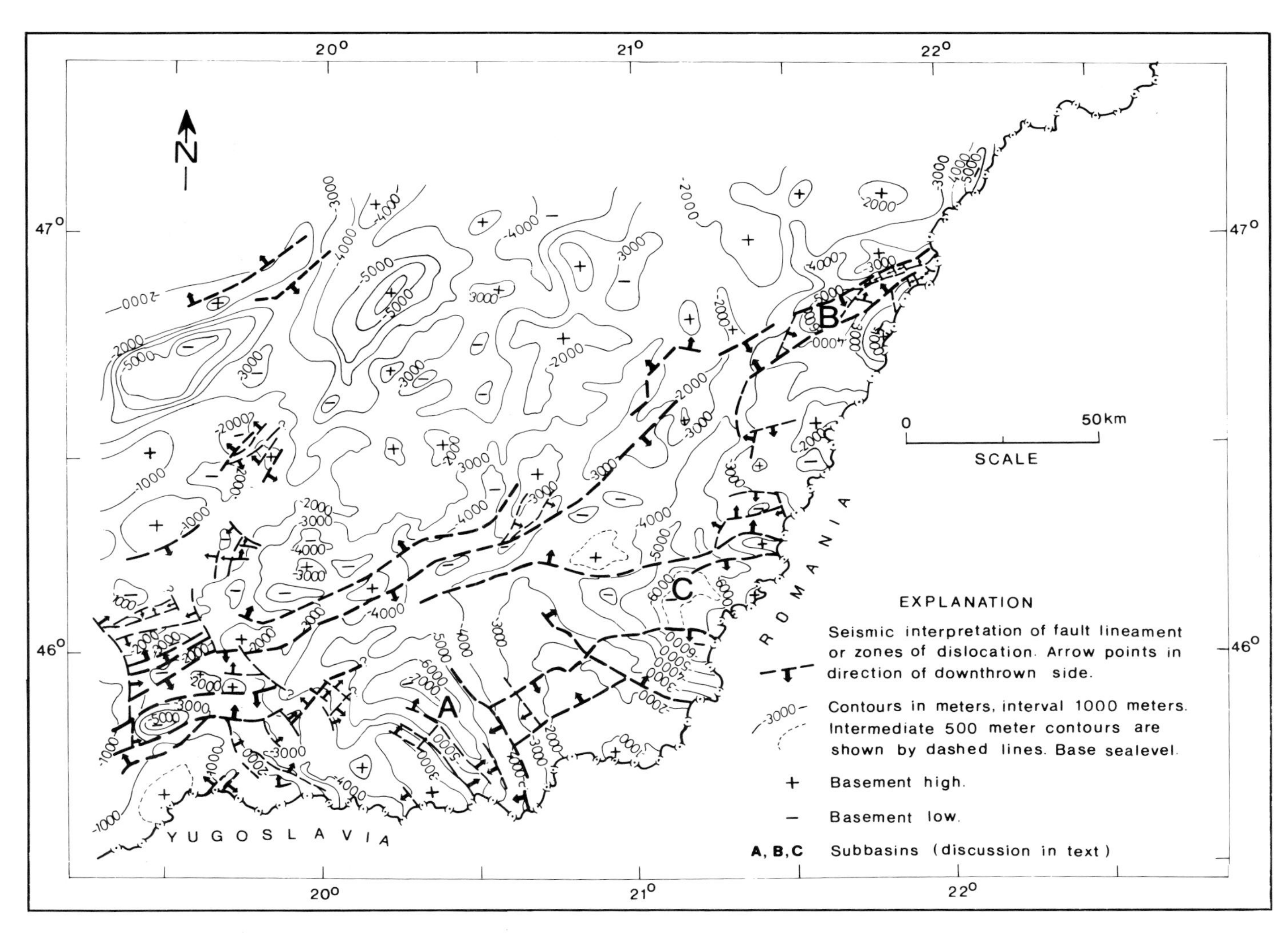

Figure 5. Structure map of pre-Cenozoic rocks in Southeastern Hungary. Dash-dot symbols show outline of Hungary within studied area. Sedimentation patterns within the three subbasins [Makó-Hódmezővásárhely trough (A), Békés basin (C) and Derecske basin (B)] are discussed in text. (Published with permission of the Hungarian Geological Institute.)

Table 2. Well Names and Corresponding Abbreviations Used in This Paper

Abbreviation	Well Name	Abbreviation	Well Name
AB-1	Abony	Kiha-D-2	Kiskunhalas
Algyő-19	Algyő	Jasz-2	Jászszentlászló
Algyő-162	Algyő	Jasz-3	Jászszentlászló
Algyő-176	Algyő	Jasz-4	Jászszentlászló
Ás-1	Ásotthalom	Makó-2	Makó
Bés-1	Békéssámson	Mar-I	Martfü
		Ml-1	Maroslele
DO-2	Dorozsma	Mora-1	Mórahalom
DO-4	Dorozsma	Mora-4	Mórahalom
DO-5	Dorozsma	NK2	Nagykörös
Der-1	Derecske	NK4	Nagykörös
ER-7	Eresztö	NK20	Nagykörös
Felgyö-1	Felgyö	NSZ-2	Nagyszénás
Felgyö-2	Felgyö	Palm-1	Pálmonstora
Felgyö-IMF	Felgyö	Palm-2	Pálmonstora
FU-1	Furta	SZK-91	Szank
FU-10	Furta	SZK-95	Szank
Hód-I	Hódmezővásárhely	SZR DNY-1	Szarvas
KB-1	Kaba		
KB-3	Kaba		

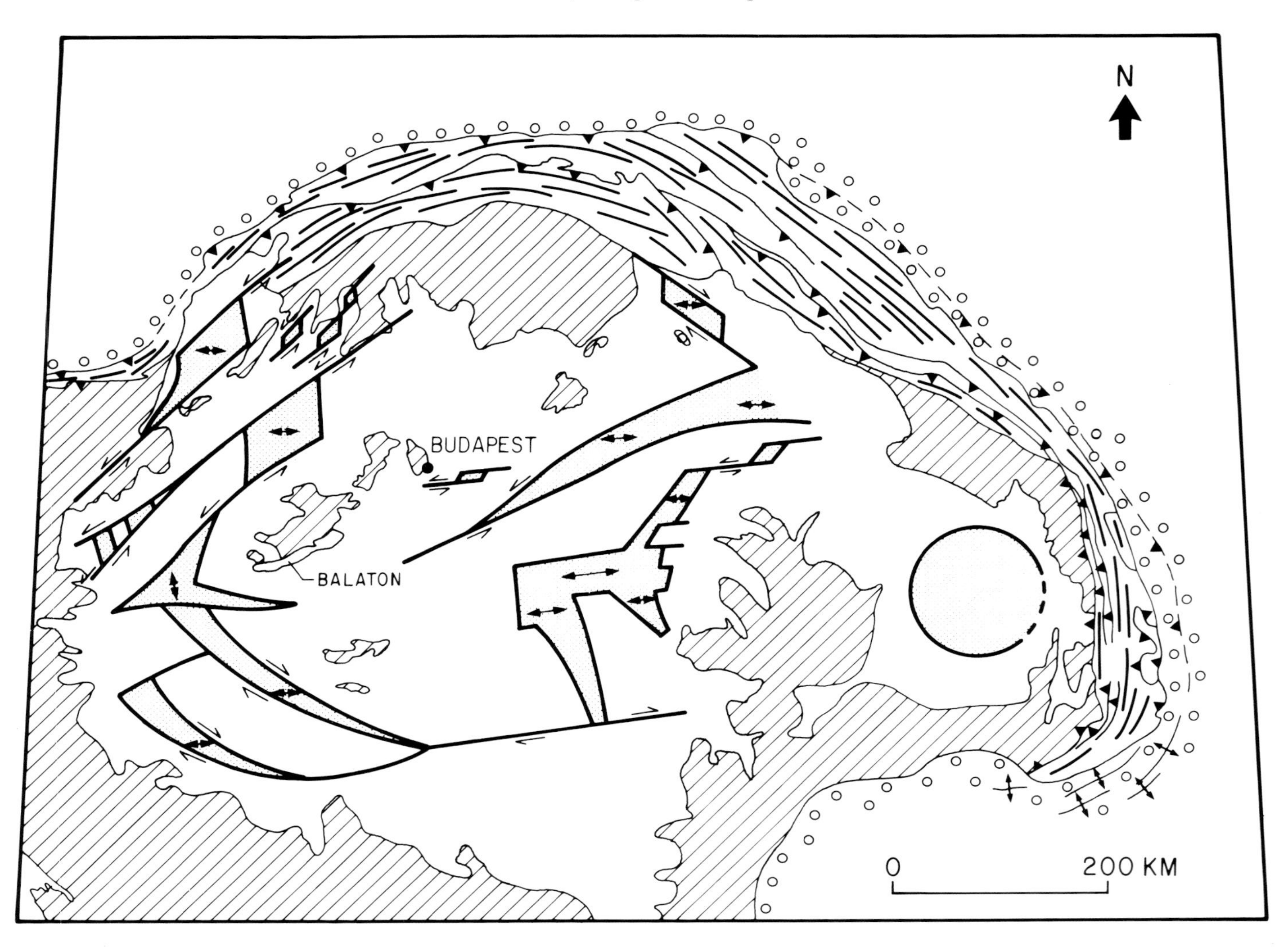

Figure 6. Generalized pattern of strike-slip fault system which could have controlled the formation of the intra-Carpathian basins. Areas of major extension are dotted; Transylvania basin (large circular area) is believed to have subsided without major extension. Figure modified from Horváth and Royden (1981).

PANNONIAN BASIN SEDIMENTARY ROCKS

The meaning of the name "Pannonian" as applied to sedimentary rocks in Hungary has not been firmly established in a chronostratigraphic sense. Despite the fact that at least 10,000 core samples are available for investigation in Hungary, where the Pannonian age section together with thin Quaternary rocks reaches 7 km in thickness (Mucsi and Révész, 1975), subdivision of Pannonian strata in a chronostratigraphic and a lithostratigraphic sense is difficult. The difficulty is caused by the scarcity of outcrops in which the succession of horizons can be observed (Dank and Kókai, 1969, p. 13B) and by the multiple lithofacies differences observed in core samples. According to Magyar and Révész (1976), faunal assemblages generally follow changes in lithology, and are, therefore, indicative of depositional environments rather than of age. In addition, some fossils from older parts of the section may have been transported and reworked into younger parts of the sedimentary section.

Earlier studies found it difficult to subdivide Pannonian strata into formations belonging to the same sedimentary cycles because most of the early geologic investigations did not recognize the extensive deltaic nature of the sedimentation pattern in the Pannonian basin. As is shown in this chapter, many of the rock-stratigraphic units become progressively younger toward the south in the Great Hungarian Plain area as a consequence of deltaic progradation from the northeast and northwest.

The stage names applied in Hungary (Central Paratethys area) along with corresponding Mediterranean usage is shown in Figure 7. Based on lithologic character, the Pannonian rock section in Hungary usually is divided into two subunits, the upper and lower Pannonian. Stefanovic (1951) states that the Pannonian (*s. str.*) corresponds to the Hungarian term "lower Pannonian" and that the term "upper Pannonian" includes the Pontian, Dacian, and Romanian stages.

Mucsi and Révész, (1975, p. 47) noted that the accepted boundary between the upper and lower Pannonian units is based on faunal and sandstone content and does not mark nec-

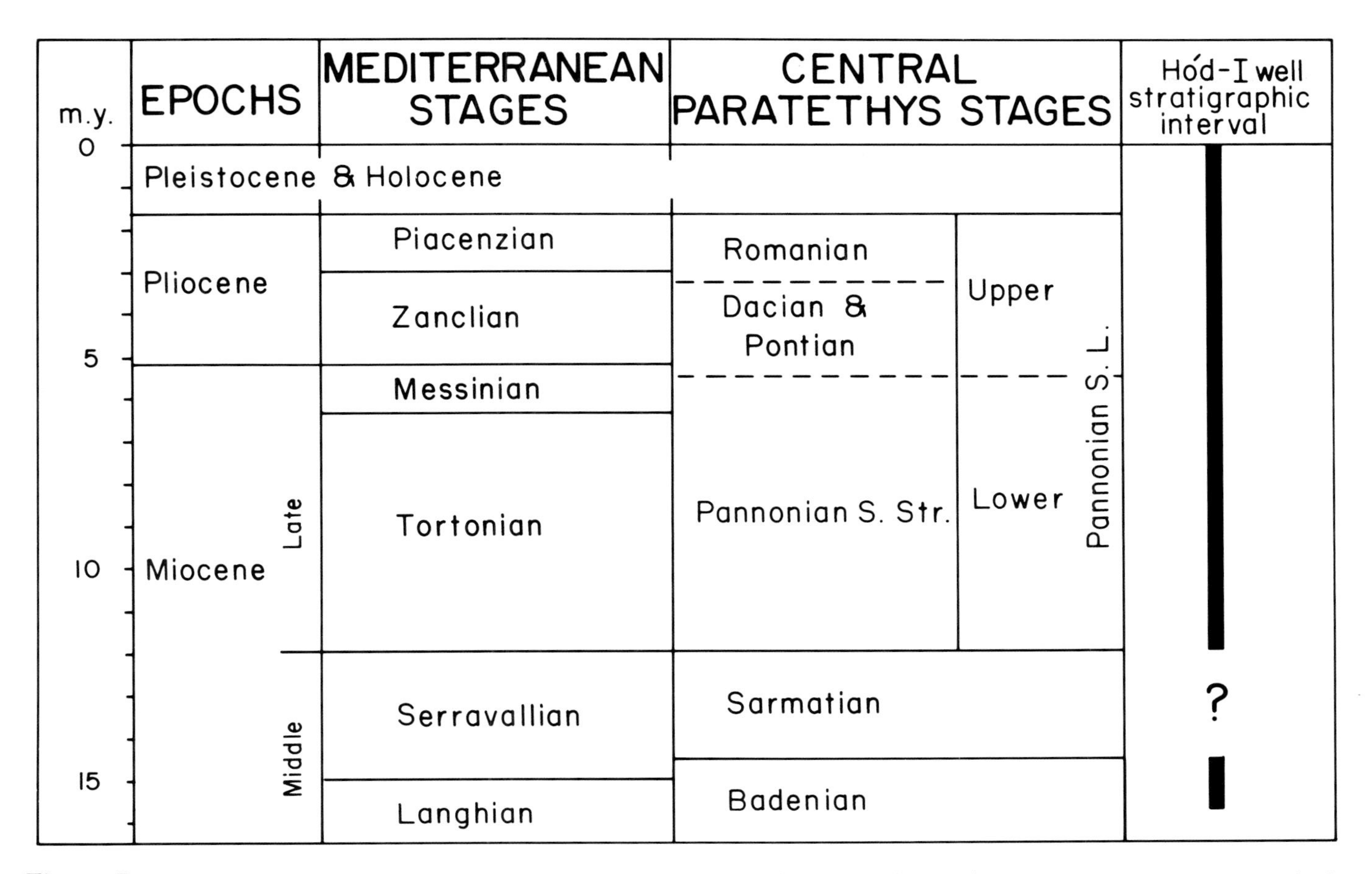

Figure 7. Stage names applied in Hungary (Central Paratethys area) along with corresponding Mediterranean usage. It is uncertain whether Sarmatian age sedimentary rocks were penetrated in the Hód-I well. Figure modified from Nagymarosy (1981).

essarily a time boundary. Preliminary analyses of seismic records by the authors of this paper appear to confirm the conclusion that the upper–lower Pannonian boundary as mapped in southeastern Hungary marks a rock-stratigraphic rather than a time-stratigraphic boundary.

CORE ANALYSES

Based on analyses of 98 cores (approximately 542 m) from the Hód-I, the Ml-1, and the Bés-1 wells, the sedimentary section of the southeastern Great Plain region of Hungary can be divided into five units, representing five different lithologic facies and two subfacies as follows: delta plain, delta front, prodelta (subfacies A and B), deep basin, and a basal facies (Figures 8 and 9) (Bérczi, this volume; Phillips and Bérczi, 1985). The locations of the three wells used in this analysis are shown in Figure 2. Brief descriptions of the lithologies, sedimentary features, and inferred depositional environments follow.

Delta Plain Facies

(Cores from 2,049 to 2,505 m in Hód-I well, cores from 801 to 2517 m in the Ml-1 well, and cores from 940 to 1905 m in the Bés-1 well.) These rocks consist chiefly of thin, horizontal, graded beds of sandstone, siltstone, and clay marl. Interbeds of siltstone and mudstone are organic rich, and lignite beds are locally common and are up to 50 cm thick. Oxidized mudstone (redbeds) containing plant fragments is abundant. All strata show evidence of intense bioturbation and possible algal mat structures. The inferred environments of deposition vary from shallow lake, fluvial, and marsh to possible terrestrial soils.

Delta Front Facies

(Cores from 2759 to 3184 m in the Hód-I well, cores from 2609 to 2981 m in the Ml-1 well, and cores from 2609 to 2756.5 m in the Bés-1 well.) These rocks consist chiefly of interbedded sandstone, siltstone, and clay marl. Beds are 1.0–1.5 cm thick and they generally are graded, inclined, and deformed. Dip angles of the strata measure 4–20°, with dip angles of 5–7°, being the most common. The inclined strata usually contain abundant soft sediment deformation features such as penecontemporaneous faults and disrupted slumped bedding. Evidence of bioturbation is widespread in the upper two-thirds of the section. Sandstone beds contain load casts, flame structures, and abundant marl rip-up clasts.

It is inferred that this section represents sediments deposited on the front of a prograding deltaic system with slopes of

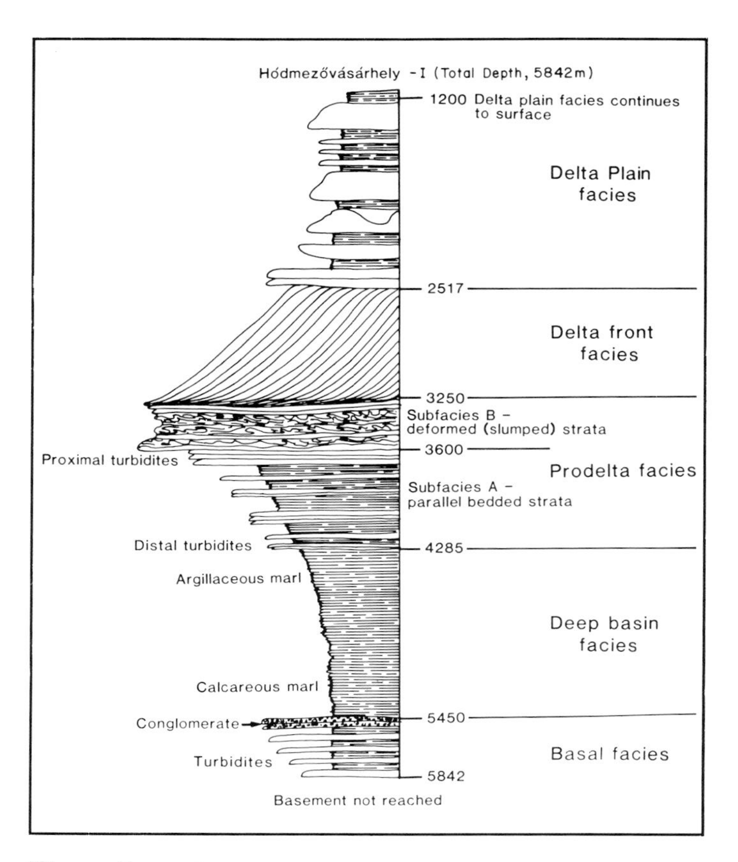

Figure 8. Five facies units penetrated in the Hód-I well. The lithologies of the various units are described in the text. Depths are shown in meters.

approximately 5 degrees. Sedimentation during deposition probably was dominated by turbidity currents, grain flow, and slumping of slope deposits as well as by fine-grained deposition from suspension and chemical precipitation of $CaCO_3$.

Prodelta Facies

(Cores from 3270 to 4271 m in the Hód-I well and possibly a core from 2806.5 to 2810.5 m in the Bés-1 well.) This section is divided into two subfacies, A and B, representing distinct depositional processes produced by an advancing delta or delta system.

Subfacies B

(Deformed strata, cores from 3270 to 3590 m in the Hód-I well.) This section consists chiefly of packages of parallel bedded sandstone (abundant) and marl as inclined and deformed strata dipping between 4 and 25° with truncated upper surfaces overlain by undeformed beds. The inclined strata contain abundant soft sediment deformation features such as penecontemporaneous faults, disrupted bedding (slumped), or completely overturned strata. Graded sandstones contain abundant marl rip-up clasts and load casts as well as dish and flame structures.

This section is inferred to represent deposition in a relatively deep basin with periodic influx of slumped deposits derived from a delta front as well as sedimentation by turbidity currents, grain flow, mud flow, and fine-grained sediments deposited from suspension along with precipitation of $CaCO_3$.

Subfacies A

(Underlying the deformed beds, cores from 3686 to 4271 m in the Hód-I well.) This section consists of parallel-bedded, graded sandstone and marl. A succession of sedimentary structures, from amalgamated sandstone beds to separate small-scale cross strata to thin horizontal sandstone laminae in marl, suggest deposition in proximal to distal regions. Increases in marl thickness toward the distal basin region can be observed in the vertical sequence in the cores. Observed sedimentary structures include load structures in which sand rests on marl and dish structures. No evidence of bioturbation is present.

This section is inferred to represent deposition in a relatively deep basin, with periodic influx of sands associated with turbidite flows and a basinward decrease in sediment transport power. The amalgamated sandstone beds near the top of the section could represent fan deposits adjacent to or within a channel system at the base of the prograding delta front. The lack of evidence for bioturbation suggests that deposition occurred in a zone of oxygen minimum.

Deep Basin Facies

(Cores from 5167 to 5418 m in the Hód-I well.) The cores from this section consist almost entirely of thinly laminated to massive, calcareous to silty marl. A decreasing $CaCO_3$ percentage, from 67% (calcareous marl) at the base of this facies to 33% (argillaceous mud) toward the top of this facies, records increasing clastic sediment input into the basin.

This section is inferred to represent deposition in a low-energy, deep-water, euxinic environment where deposition of pelagic and chemical precipitation of $CaCO_3$ and mud from suspension occurred. Time-equivalent rocks are absent in the Ml-1 well; the Ml-1 site apparently was a topographic high throughout much of early Pannonian time (Figure 10).

Basal Facies

(Cores from 5450 to 5823 m in the Hód-I well.) Cores from this section contain mainly interbedded sandstone, marl, and conglomerate with an overall upward coarsening in texture toward the top of the facies. The beds are horizontal to inclined (7–11°). The beds generally contain coarse-grained sandstone at the base and fine upward, in places, to marl. The sandstone and conglomerate beds are graded and contain large marl rip-up clasts and quartz pebbles up to 0.5 mm in diameter; some show evidence of soft sediment deformation.

These sedimentary rocks are inferred to have been deposited in a deep-water, euxinic environment characterized by precipitation of $CaCO_3$ and the deposition of muds from suspension. Periodic influx of coarse-grained sediment probably was related to turbidite flows that either moved along the axis of the basin or were derived locally from the flanks of the basin.

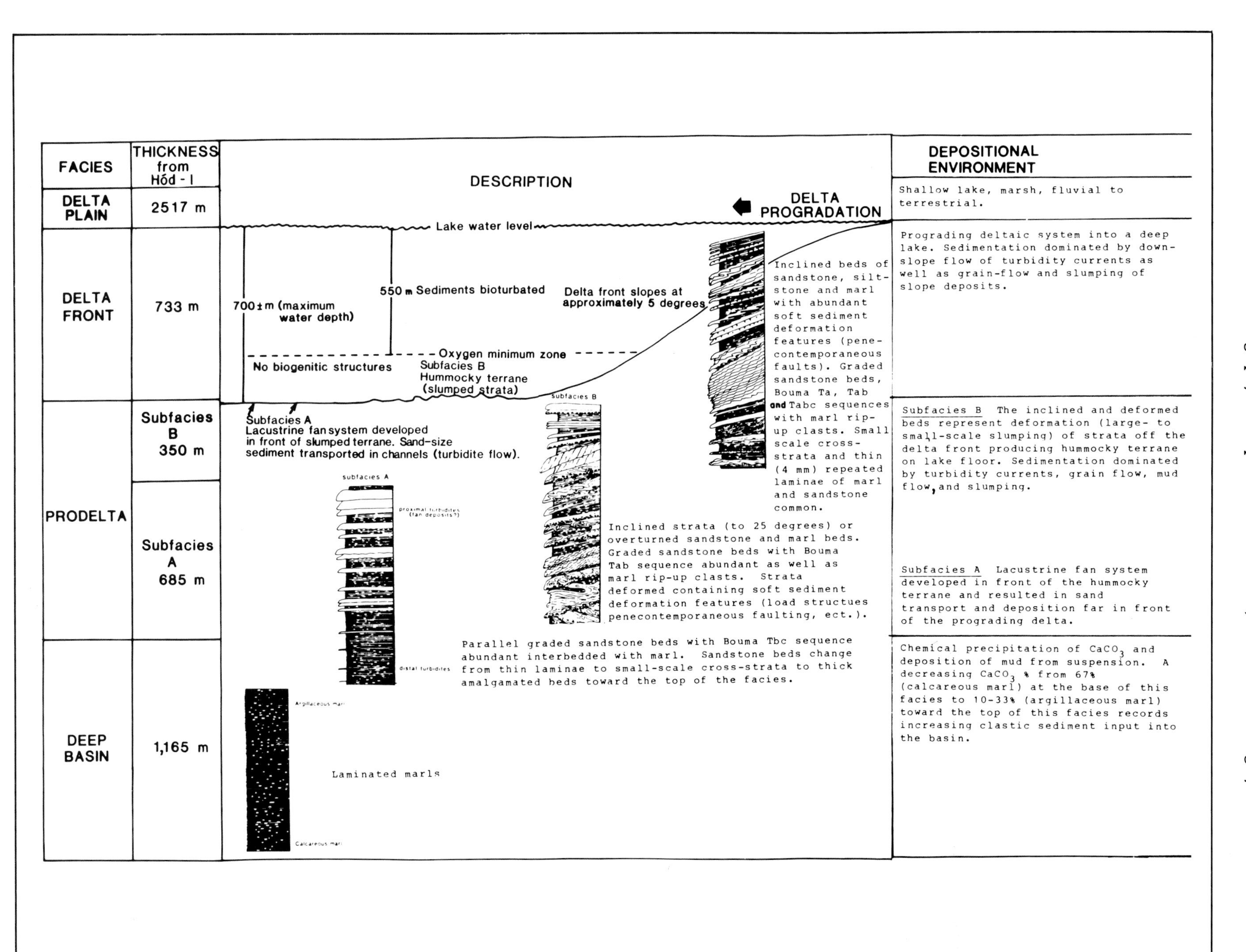

Figure 9. Summary of the facies, thicknesses, descriptions, and environments of deposition of sedimentary rock units penetrated in the Hód-I well.

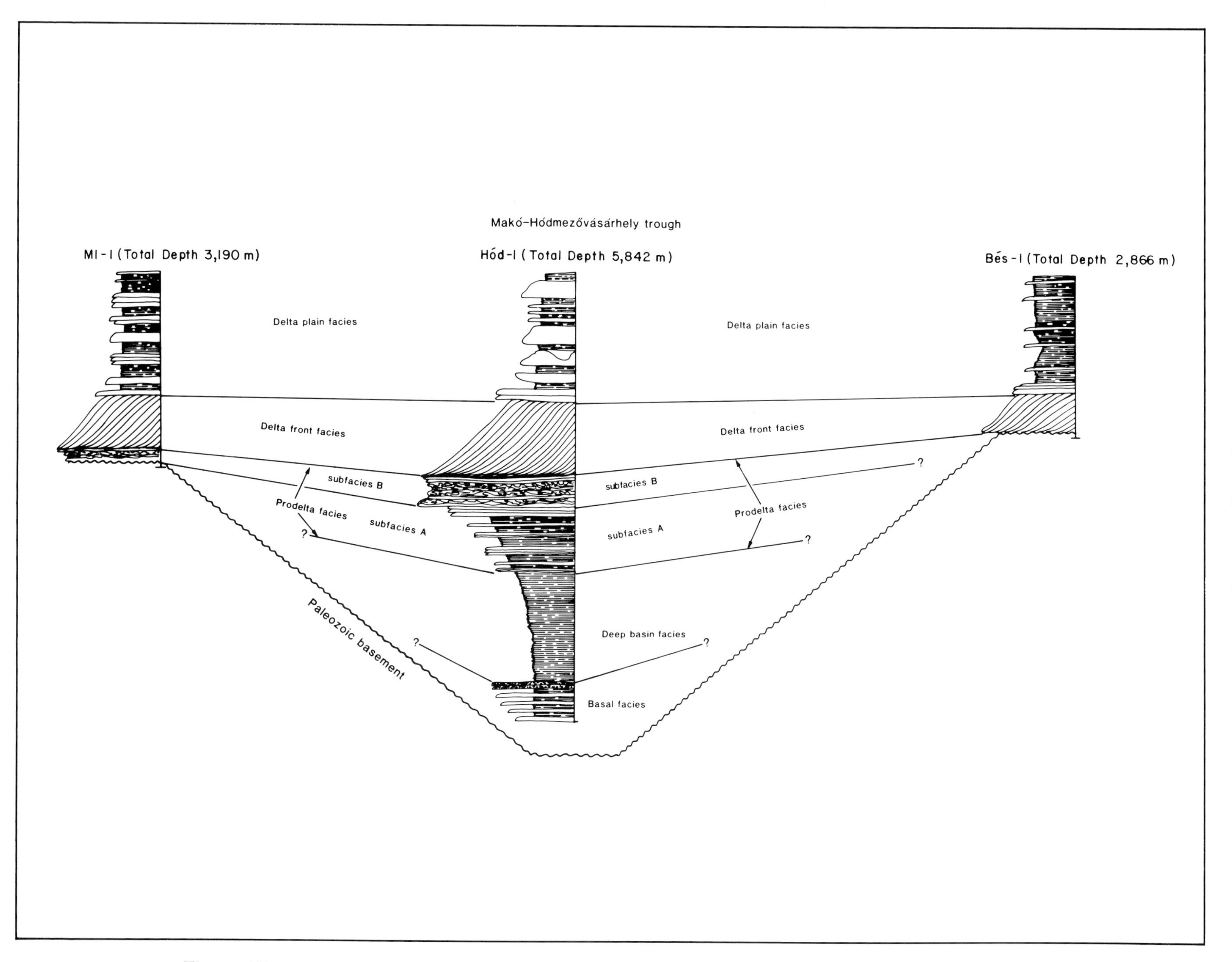

Figure 10. Summary of the correlation between the Ml-1, Hód-I, and Bés-1 wells. Lithologies of the various units are described in text.

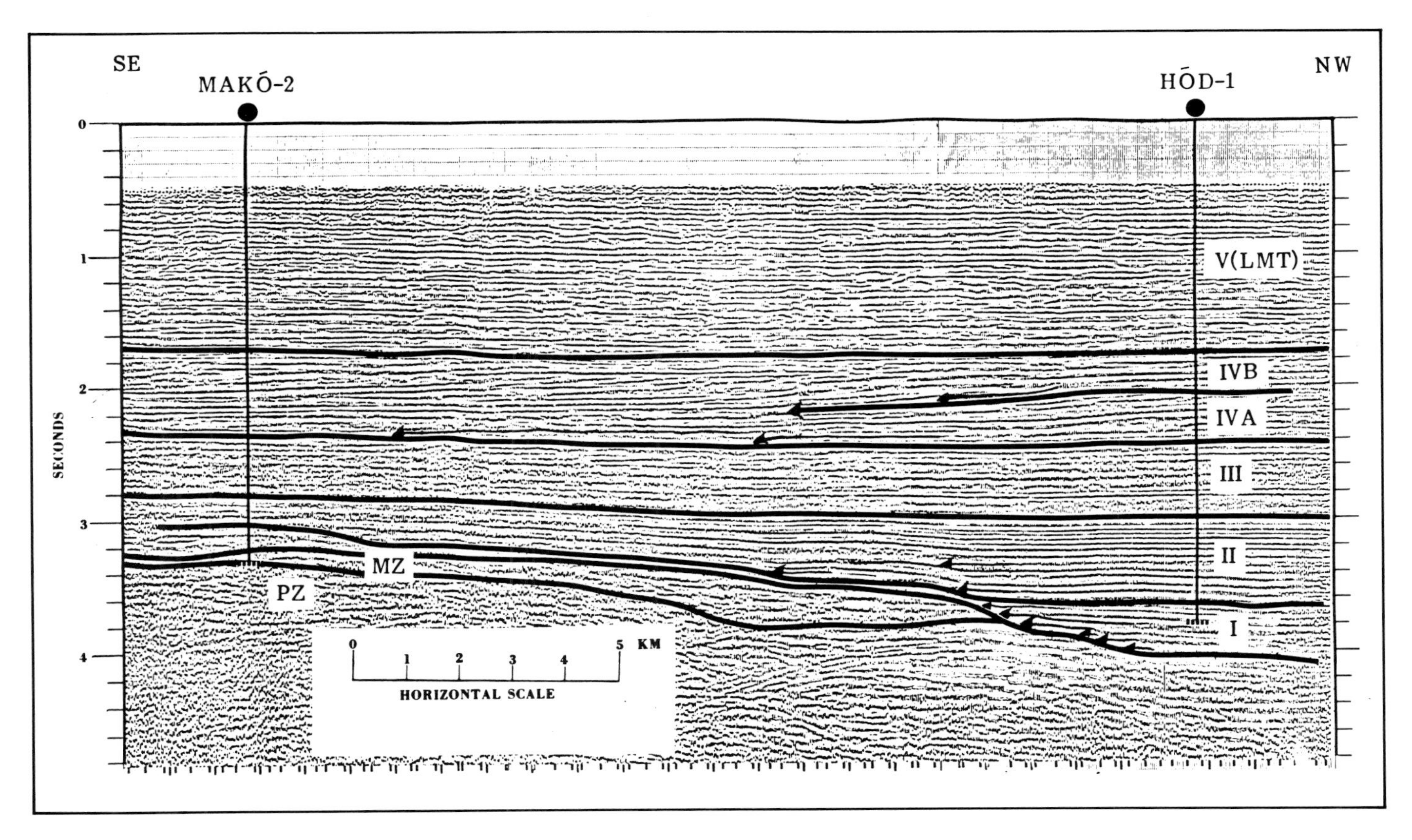

Figure 11. Interpreted seismic-reflection profile 6 recorded near the Hód-I well site showing division of the sedimentary rock section into five seismic sequences. Roman numerals identify seismic sequences. Sequence V is also referred to as LMT in other sections of this report. Geologic interpretation is shown in Figure 12. Location is shown in Figure 4. PZ and MZ are Paleozoic and Mesozoic rocks, respectively, of the basement complex. Note that vertical scale is in time; depths to the tops of seismic sequences are given in text. These depths were determined from a velocity survey conducted at the Hód-I well site.

SEISMIC STRATIGRAPHY AT HÓD-I WELL

The seismic stratigraphy in the vicinity of the Hód-I well is illustrated best on seismic profile 11 (Figure 11). The seismic profile can be divided into five seismic sequences based on the geometric relationship of reflections to sequence boundaries and on the internal configuration of reflections within sequences (Figure 11). The typical reflection pattern of each sequence is shown in Figure 12. A top to bottom description of these sequences follows:

Sequence V

(0.0–1.74 sec; 0–2003 m.) Within this sequence, reflectors are concordant to the bottom sequence boundary. Individual reflectors vary from parallel to wavy. Although reflections are strong (high amplitude), few, if any, can be traced for any distance without interruption.

Correlation with core data (described previously) indicates that seismic sequence V represents delta plain facies; depositional environments varied from shallow lake, fluvial, and marsh to possible terrestrial soils. At the Hód-I well site, the bottom boundary of sequence V is placed at 2003 m, about 514 m higher than the base of delta plain facies as determined from core analyses (Figure 8). This difference in unit boundaries is a good example of the differences that result from seismic sequence analysis in contrast to lithologic analysis. Although delta plain facies lie above and below the bottom boundary of sequence V, it will be shown later that the bottom boundary of sequence V represents an unconformity, probably related to more widespread lake conditions and shoaling of the Pannonian Lake.

Sequence IV

(1.74–2.44 sec; 2003–3250 m.) In general, the internal reflectors form an irregular sigmoid pattern. The relation between reflections and sequence boundaries is one of toplap at the upper boundary and downlap at the lower boundary.

Sequence IV can be subdivided into subsequences IVA and IVB at the Hód-I well site. The subdivision is placed at 2.06 sec (2517 m). Reflectors above 2.06 sec lie relatively flat, whereas those below 2.06 sec show a definite pattern of progradation to the southeast.

At the Hód-I well, subsequence IVB is inferred to represent delta plain facies; to the southeast, where the reflectors of subsequence IVB exhibit a strong progradational pattern with a dip

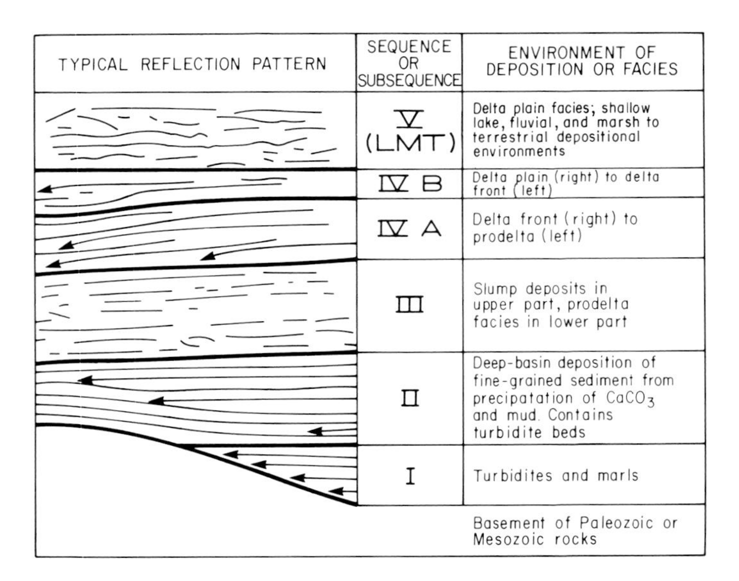

Figure 12. Typical seismic-reflection patterns of five seismic sequences at the Hód-I well site and their inferred environments of deposition or facies. The heavy lines are sequence boundaries, and the arrows show the relationship of internal reflectors to sequence boundaries (V, concordant; IV, downlap at basal boundary; III, concordant; II, concordant and onlap, I, onlap of basal boundary).

of about 4.5 °, the facies probably grade to delta front and prodelta. The base of seismic sequence IVB is placed at a depth of 2517 m, in agreement with the base of the delta plain facies as determined from lithologic analysis (Figure 8).

Subsequence IVA is inferred to represent delta front facies at the Hód-I well site. Southeast of the Hód-I well, where the reflection pattern in subsequence IVA suggests lesser dips, the facies probably grade to prodelta.

In general, seismic sequence IV represents prodelta, delta front, and delta plain facies. It will be shown, on seismic records from other parts of the basin, that sequence IV is part of a supersequence that represents a system of stacked deltas built during a late depositional stage. The term "late depositional stage" is used to distinguish this supersequence from an earlier and more widespread supersequence that represents delta construction during a period when the lake was deeper.

Sequence III

(2.44–3.00 sec; 3250–4285 m.) The boundary between sequences IV and III is placed immediately below the basalmost reflectors that exhibit the progradational pattern characteristic of sequence IV. Reflections in this sequence are concordant to the upper and lower sequence boundaries. The sequence is characterized internally by weak, highly discontinuous, wavy reflections.

The absence of strong, continuous reflections, which usually represent alternating beds with different reflection coefficients, together with the preponderance of sandstone in the cores from the Hód-I well, indicate that this sequence may represent a relatively massive sandstone unit. However, the absence of strong continuous reflections also may be caused by extensively slumped strata, and evidence of slumping was present in some of the cores (Figure 8). The vertical position of sequence III in the stratigraphic column, between a delta (sequence IV) and a deep-basin facies (sequence II), suggests that the sands could have been deposited in front of an advancing sand-prone delta or deltas.

At the Hód-I well site, seismic sequence III lies at depths of between 3250 and 4285 m. At this site, therefore, sequence III is equivalent to the lithologic unit inferred to represent prodelta facies (Figure 8). Based on core analyses, the prodelta facies is divided into subfacies B (slumped strata) and subfacies A (parallel-bedded strata). Such a division cannot be made from the seismic data, and it would appear that the lithologic data makes a finer distinction in comparison to the seismic data. As will be shown later, however, seismic evidence from other parts of the Pannonian basin suggests that updip from the Hód-I well site, at least part of seismic sequence III represents distal deposits associated with an early system of deep-water deltas and that deep-water delta construction may have been followed by a short-lived destructive phase. The lower part of seismic sequence III, therefore, apparently represents prodelta facies of the older, deep-water delta system, and the upper part of sequence III (slumped strata) may represent sands deposited in front of an advancing sand-prone delta system or, perhaps, it represents deposition during a destructive phase.

Sequence II

(3.00–3.69 sec; 4285–5576 m.) Reflections within this sequence are strong and, in general, can be traced across the entire record section or until they terminate to the southeast in an onlap configuration against underlying reflections or reflections that represent the basement surface. Onlapping reflectors alternate with concordant reflectors. The concordant reflectors are parallel with the basal sequence boundary and have the appearance of being draped over the basement surface southeast at the Hód-I well.

The alternation of concordant and onlapping reflectors suggests that layers of deep basin marl alternate with turbidite deposits. Sediments precipitated from the water column would tend to drape over basements highs. The sediments associated with turbidite deposits, however, would have been carried through canyons into the deeper parts of the basin near the water–sediment interface and, therefore, would be expected to show an onlap relation to underlying horizons.

Although seismic sequence II correlates with deep basin facies at the Hód-I well site, the sequence can be traced seismically updip (northward) where it grades to prodelta, delta front, and delta plain facies. In other sections of this chapter, it is shown that sequence II represents a deep water, lateral equivalent (in time) of a supersequence of stacked deep water deltas built during an early depositional stage.

Sequence I

(3.69–4.05 sec; 5576–6150 m.) The pattern of reflections within this sequence is similar to that noted for sequence II, except that most reflections appear to terminate against the basal boundary (basement surface) in an onlap configuration. At the Hód-I site, seismic sequence I correlates with the basal

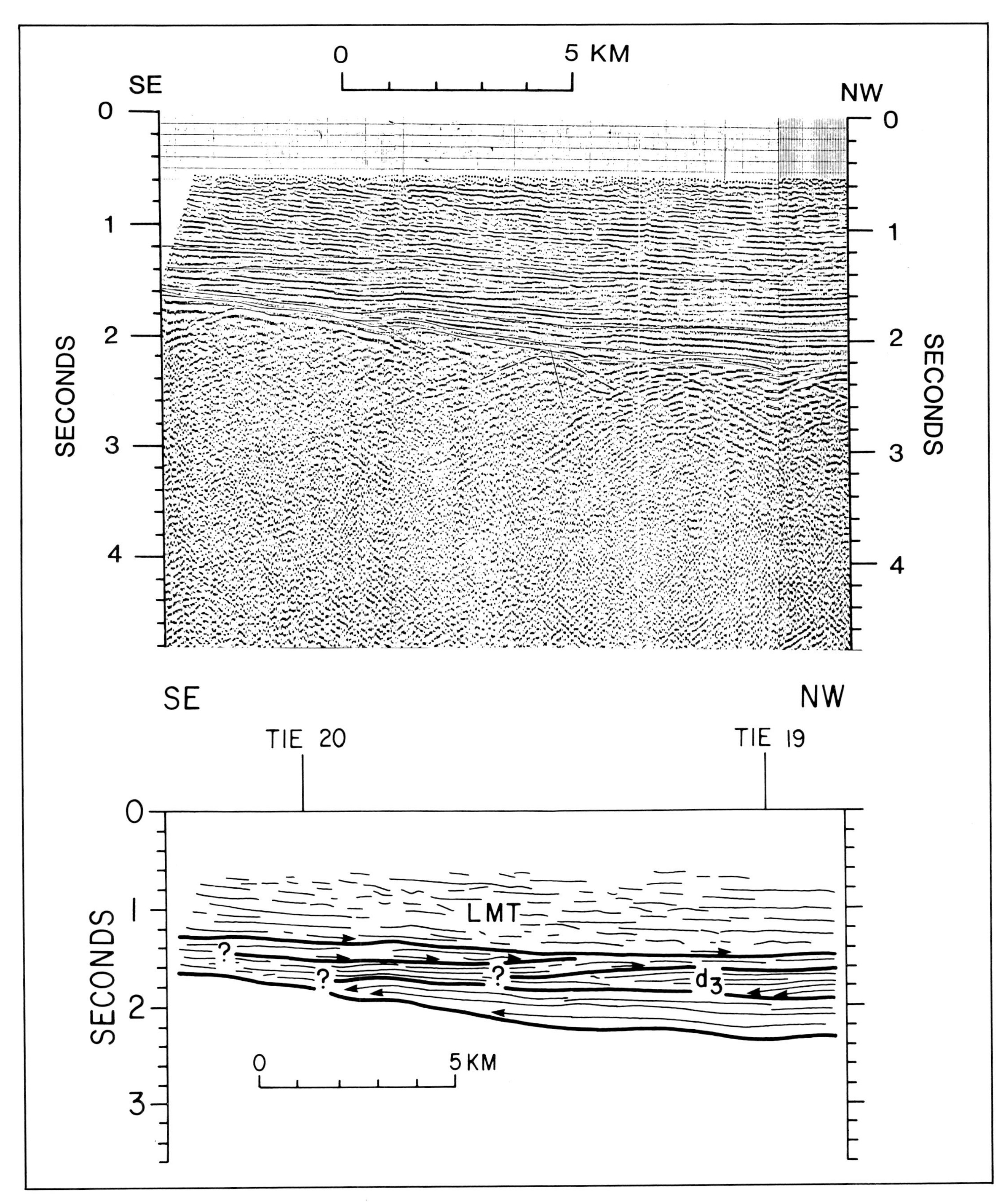

Figure 16. See Figure 13 for explanation of symbols and abbreviations.

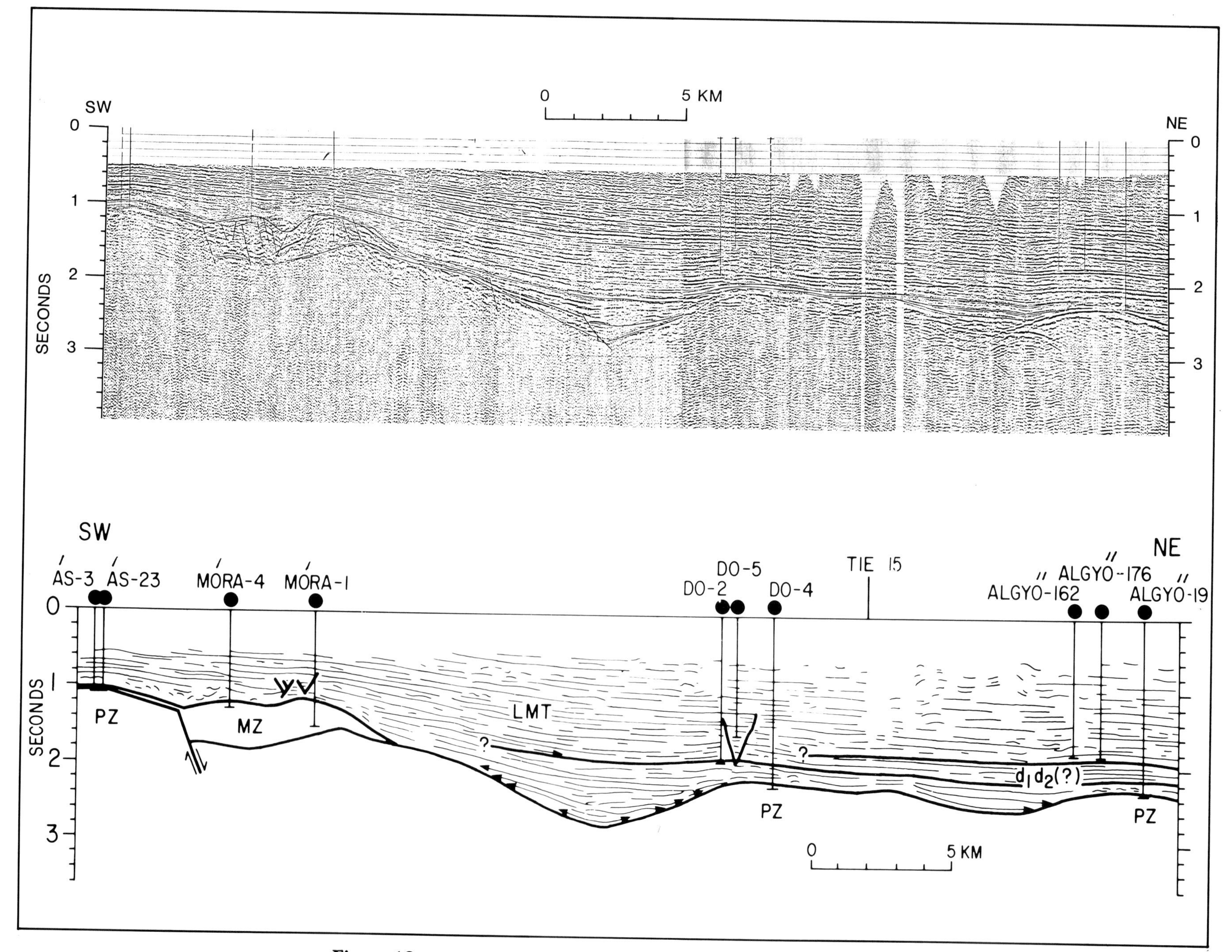

Figure 18. See Figure 13 for explanation of symbols and abbreviations.

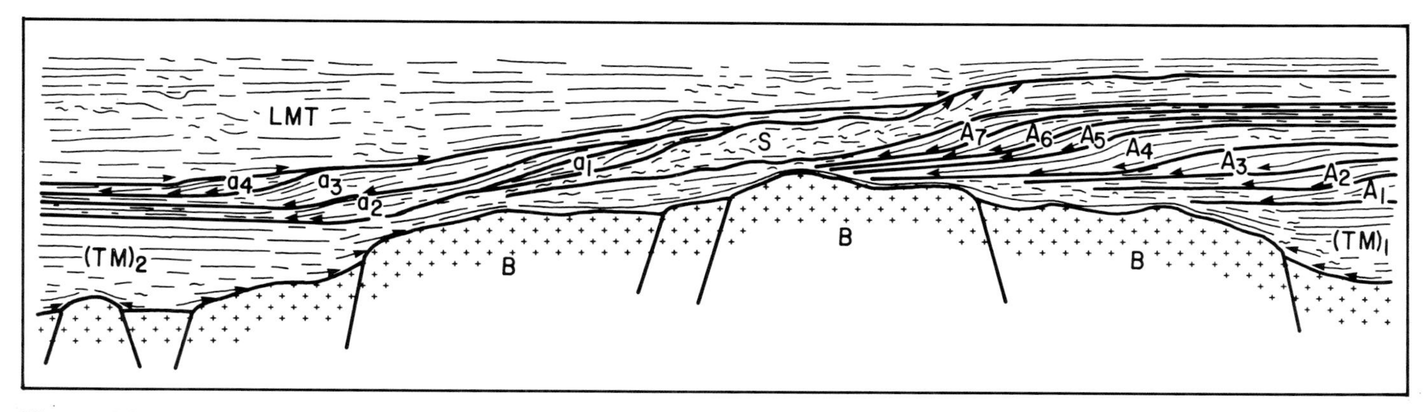

Figure 29. General configuration of seismic sequences and supersequences in the Pannonian Basin of Hungary. Heavy lines are sequence boundaries. The lowest sequence (B), which represents basement rocks (Paleozoic or Mesozoic), is shown by a cross pattern. Lines within each sequence show general seismic character of internal reflectors. Arrows show type of discordance at sequence boundaries. Figure is not to scale and represents a composite of numerous seismic records. Sequences $(TM)_1$ and $(TM)_2$ represent turbidites and marls; A_1–A_7, stacked deltaic system deposited in water depths of 800–900 m; S, slumped strata deposited when delta system prograded across basin margin and probably related to shoaling of lake; a_1–a_4, delta system deposited in water depths of 200–400 m; LMT, shallow lake, marsh, and fluvial deposits and terrestrial soils, unit is inferred to have been deposited as the lake continued to shoal and eventually to dry up.

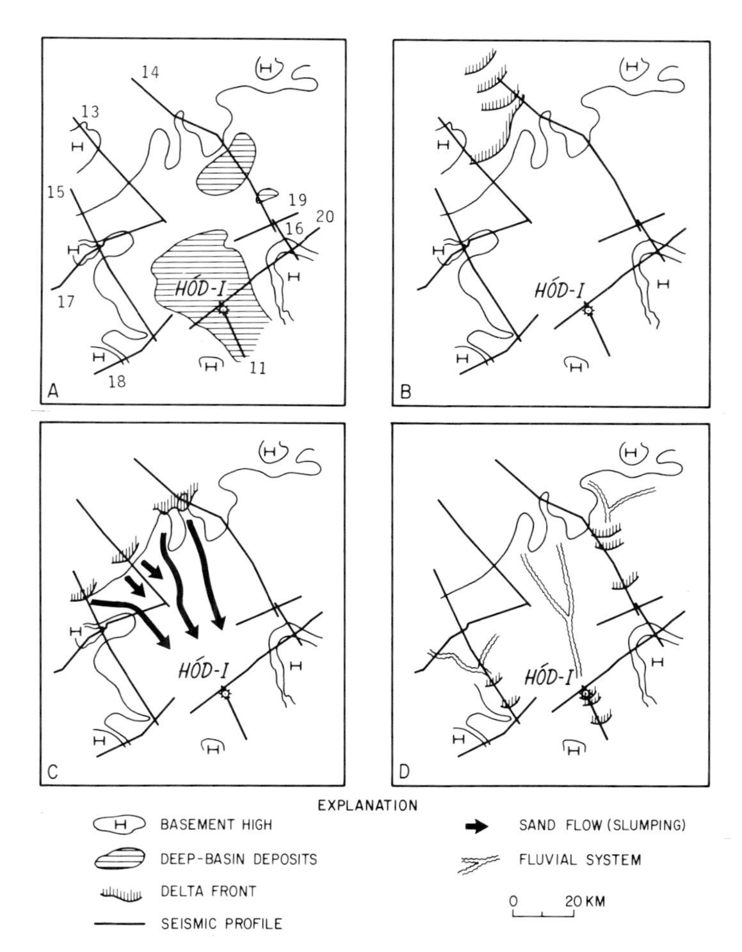

Figure 30. Four stages of deposition in the Makó-Hódmezővásárhely trough: A, deposition of early deep-basin turbidites; B, first stage of delta construction; C, deposition of slump deposits and/or turbidite flow; D, shallow water delta construction following shoaling of lake. Seismic profile numbers are shown in Figure 30A.

parts of the basin (Figure 31A). The direction of turbidite deposition is not clear but was probably from the northeast as well as the northwest. The earliest deltas are mapped on seismic profile 20 (Figure 31B). This system of deep-water (800–900 m) deltas prograded southwestward, where the delta toes are seen to pinchout against a basement high on seismic profile 20. Concurrent with delta progradation, turbidite flows funneled down canyons from a northwest direction.

Subsequently, deep water (800-900 m) delta construction shifted slightly to the west in the vicinity of seismic profile 24 (Figure 31C). Southwest progradation continued, and delta upbuilding overwhelmed the basement high shown near the bottom of Figure 31B. Massive turbidite flows continued to enter the basin from the northwest. Evidence of these flows is seen on seismic profile 24, on which numerous unconformities can be mapped, and strong onlap of the deltas by turbidite sedimentation is evident. During this cycle, minor(?) amounts of sediment entered the basin from the south and east. Seismic evidence from profile 28 suggests that these sediments were not transported any great distance but rather were derived from nearby basement highs and transported by slump and creep and downslope mass transport processes.

Békés Basin

The Békés basin differs from the two subbasins discussed above in that it is located far from the sediment source areas which lay to the northwest, north, and northeast (Figure 32). Other subbasins were located between the source areas and the Békés basin. Therefore, it received chiefly finer grained sediments until deltaic sedimentation had prograded across much of the remainder of the Pannonian basin late in its history. Deep basin deposits in the Békés basin can be expected to be much thicker and more widespread than in the areas discussed above (Figure 32B). Coarse-grained sediments probably are limited to areas in the vicinity of local basement highs where sediment input depended on weathering of local basement rocks and subsequent transport by slump and creep and downslope mass

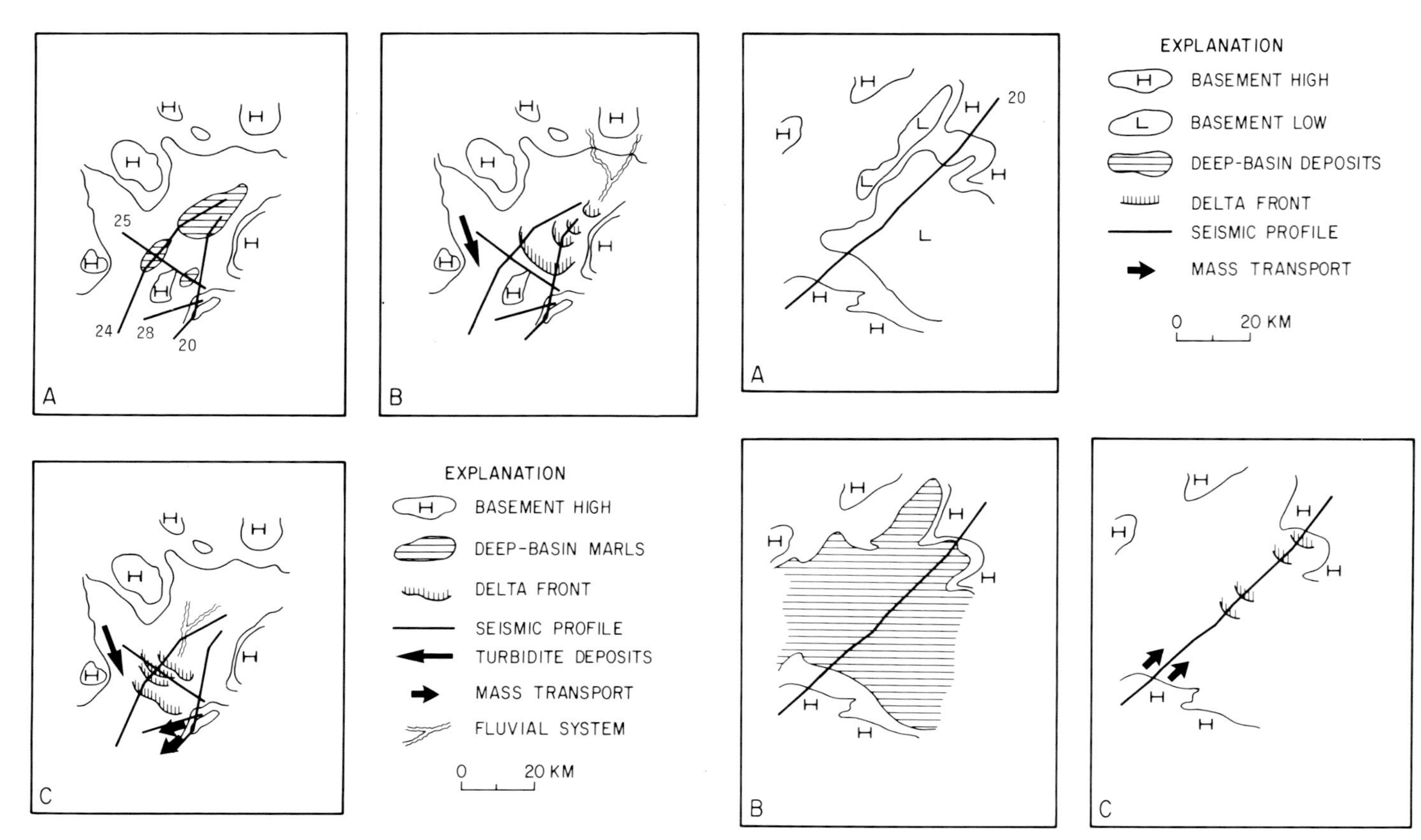

Figure 31. Three stages of deposition in the Derecske basin: A, deposition of early deep-basin turbidites; B, first stage of delta construction and turbidite flow down canyon axis; C, shift of delta construction southwestward. Seismic profile numbers are shown in Figure 31A.

Figure 32. Two stages of deposition in the Békés basin; A, location of basement highs and lows at start of deposition; B, deposition of deep basin turbidite and marl deposits; C, stage of shallow water delta construction following shoaling of lake. Seismic profile numbers are shown in Figure 32A.

transport processes (Figure 32C). Delta construction was limited to shallow water deltas (200–400 m), which are probably equivalent in time to the last stages of delta construction in the Makó-Hódmezővásárhely trough (Figure 32C).

SUMMARY OF CONCLUSIONS

The following conclusions regarding the sedimentary rocks of the Pannonian basin in Hungary are based on stratigraphic analysis of seismic records and studies of core samples from three wells:

1. Deltaic sedimentation may have started as early as Sarmatian time and continued throughout much of the Pannonian (*s. l.*).
2. Sediment input was chiefly from the northwest, north, and northeast. These directions probably were determined by prerift or early rift topography at least as early as Badenian time.
3. Two distinct stages of delta construction can be recognized: an early, deep-water stage and a later, shallower water stage.
4. In the early stage of construction, turbidite-fronted deltas were built in water depths as deep as 800–900 m. During this stage, subsidence rates and associated sediment input rates were high, and upbuilding and southward progradation of large, deltaic sediment wedges filled subbasins near the source areas, overwhelmed local basement highs, and spilled sediments into subbasins in the southern part of Hungary.
5. During a later stage of construction, deltas were built in water depths of about 200–400 m, and topographically low areas in the southern part of Hungary were infilled by sediments discharged from river systems that had advanced about 100 km southward across strata of the older constructional stage.
6. Differences, from a standpoint of petroleum exploration, may exist between the two deltaic sequences. Coarse-grained sediments in the deep-water delta system may be limited to large delta slope channels. The shallow water deltas, in contrast, may be associated with delta front sheet-sands which are related to merging and progradation of mouth bar sands.
7. In some areas, the sedimentary rocks representing the two stages of delta construction may be separated by a unit that represents a destructional phase. This unit is inferred to be related to a shoaling of the lake or, possibly, a short-lived transgressive event.
8. The final stage of sedimentation is represented by delta plain facies; depositional environments varied from shallow lake, fluvial, and marsh to terrestrial soils. During this last stage, the lake became more widespread but continued to shoal.

9. Basal turbidite-marl sections in the deeper parts of the Pannonian basin become progressively younger in the upper part of the section in a direction away from the sediment source areas. This tendency is coupled with a thickening of the overall turbidite marl section in the same general direction.

10. Distal (in reference to sediment source areas) subbasins received chiefly finer grained sediments until late in their history, when deltaic sedimentation had prograded across much of the remainder of the Pannonian basin. Older, coarse-grained sediments in these basins probably are limited to the vicinity of local basement highs.

11. The authors have interpreted the Pannonian and Quaternary history of the Pannonian basin in terms of a continuously shoaling lake due to sediment infilling.

ACKNOWLEDGMENTS

The interpretations contained in this report resulted from studies conducted under a cooperative agreement between the U.S. Geological Survey (USGS) and the Central Office of Geology of Hungary pursuant to a bilateral agreement between these organizations and as a result of an exchange agreement for culture, education, science, and technology between the United States and Hungary. The USGS authors wish to thank the many gracious Hungarian geologists and geophysicists who spent many hours introducing us to the geology of Hungary. We are especially indebted to I. Bérczi for his detailed explanations of stratigraphy and sedimentary processes, to Á. Jámbor and É. Kilényi for their descriptions of regional geology, and to G. Pogácsás for his review of seismic stratigraphic interpretations in Hungary. Finally, all the authors wish to thank F. Horváth for his critical review of, and many helpful additions to, the final manuscript.

REFERENCES CITED

Berg, O. R., 1982, Seismic detection and evaluation of delta and turbidite sequences: Their application to exploration for the subtle trap: American Association of Petroleum Geologists Bulletin, v. 66, n. 9, p. 1271–1288.

Burchfiel, B. C., and L. Royden, 1982, Carpathian foreland fold and thrust belt and its relation to Pannonian and other basins: American Association of Petroleum Geologists Bulletin, v. 66, n. 9, p. 1179–1195.

Coleman, J. M., 1976, Deltas: Processes of deposition and models for exploration: Continuing Education Publication Co., Champaign, Illinois, 102 pages.

Collinson, J. D., 1976, Deltaic evolution during basin fill-Namurian of central Pennine Basin, England: Abstracts with program, 1976 AAPG-SEPM Annual Convention, May 23–26, American Association of Petroleum Geologists Bulletin, v. 60, n. 4, p. 659.

Dank, V. E., and Kokai, J., 1969, Oil and Gas exploration in Hungary, *in* Peter Hepple, ed., The exploration for petroleum in Europe and North Africa: London, Institute of Petroleum, p. 131–145.

Horváth, F., and L. Royden, 1981, Mechanism for the formation of the intra-Carpathian basins: A review: Earth Evolution Sciences, v. I, n. 3–4, p. 307–316.

Körössy, L., 1981, Regional geological profiles in the Pannonian basin: Earth Evolution Sciences, v. 1, n. 3-4, p. 223–231.

Lerner, J., 1981, Satellite image map of the Carpatho-Pannonian region: Earth Evolution Sciences, v. I, n. 3–4, p. 180–182.

Magyar, L., and I. Révész, 1976, Data on the classification of Pannonian sediments of the Algyö area: Acta Mineralogical Petrographica, Szeged, Hungary, v. 22, p. 267–283.

Massari, S. F., 1978, High-constructive coarse-textured delta systems, Tortonian, Southern Alps. Evidence of lateral deposits in delta slope channels: Memorie della Societa geologica italiana, n. 18, p. 93–124.

Mucsi, M., and Révész, I., 1975, Neogene evolution of the southeastern plain on the basis of sedimentological investigations: Acta Mineralogica Petrographica, Szeged, Hungary, v. 23, p. 29–49.

Nagymarosy, A., 1981, Chrono- and biostratigraphy of the Pannonian basin: A review based mainly on data from Hungary: Earth Evolution Sciences, n. 1, v. 3-4, p. 183–195.

Phillips, R. L., I. Bérczi, 1985, Processes and depositional environments of Neogene deltaic-lacustrine sediments, Pannonian basin, Southeast Hungary: Core Investigation: U.S. Geological Survey Open File Report 85-360, 66 pp.

Pogácsas, G., 1980, Neogén sullyedékeink fejlödéstörténeti viszonyai a felszini geofizikai mérések tükren [The evolutionary relations of Neogene subsidence as reflected by geophysical measurements]: Földtani Közlöny (Bulletin of the Hungarian Geological Society) n. 110, p. 485–497.

Royden, L., F. and B. C. Burchfiel, 1982, Transform faulting, extension, and subduction in the Carpathian Pannonian region: Geological Society of America Bulletin, v. 93, p. 717–725.

Royden, L., F. Horváth, Rumpler, J., 1983, Evolution of the Pannonian Basin system, I. Tectonics; American Geophysical Union, Tectonics, v. 2, n. 1, p. 63–90.

Rumpler, J., and F. Horváth, in press, Extensional tectonics on seismic sections and its significance in oil exploration within the Pannonian basin, Földtani Kutatás, A Központi Földtanti Hivatal Kiadványa, Budapest, Hungary.

Sangree, J. B., and J. M. Widmier, 1978, Seismic stratigraphy and global changes of sea level, Part 9: Seismic interpretation of clastic depositional facies: American Association of Petroleum Geologists Bulletin, v. 62, n. 5, p. 752–771.

Stefanovic, P., 1951, Pontische stufe im engeren Sinn [The Pontian stage in a strict sense]—Obere Congerienschich-ten Serbiens: Serbische Akademie der Wissenschaften und Khunste, Belgrade, Yugoslavia, p. I–187.

Vail, P. R., R. M. Mitchum, Jr., R. G. Todd, J. M. Widmier, S. Thompson, III, J. B. Sangree, J. N. Bubb, and W. G. Hatlelid, 1977, Seismic stratigraphy and global changes of sea level, *in* C. E. Payton, ed., Seismic stratigraphy—application to hydrocarbon exploration: American Association of Petroleum Geologists Memoir 26, p. 49–205.

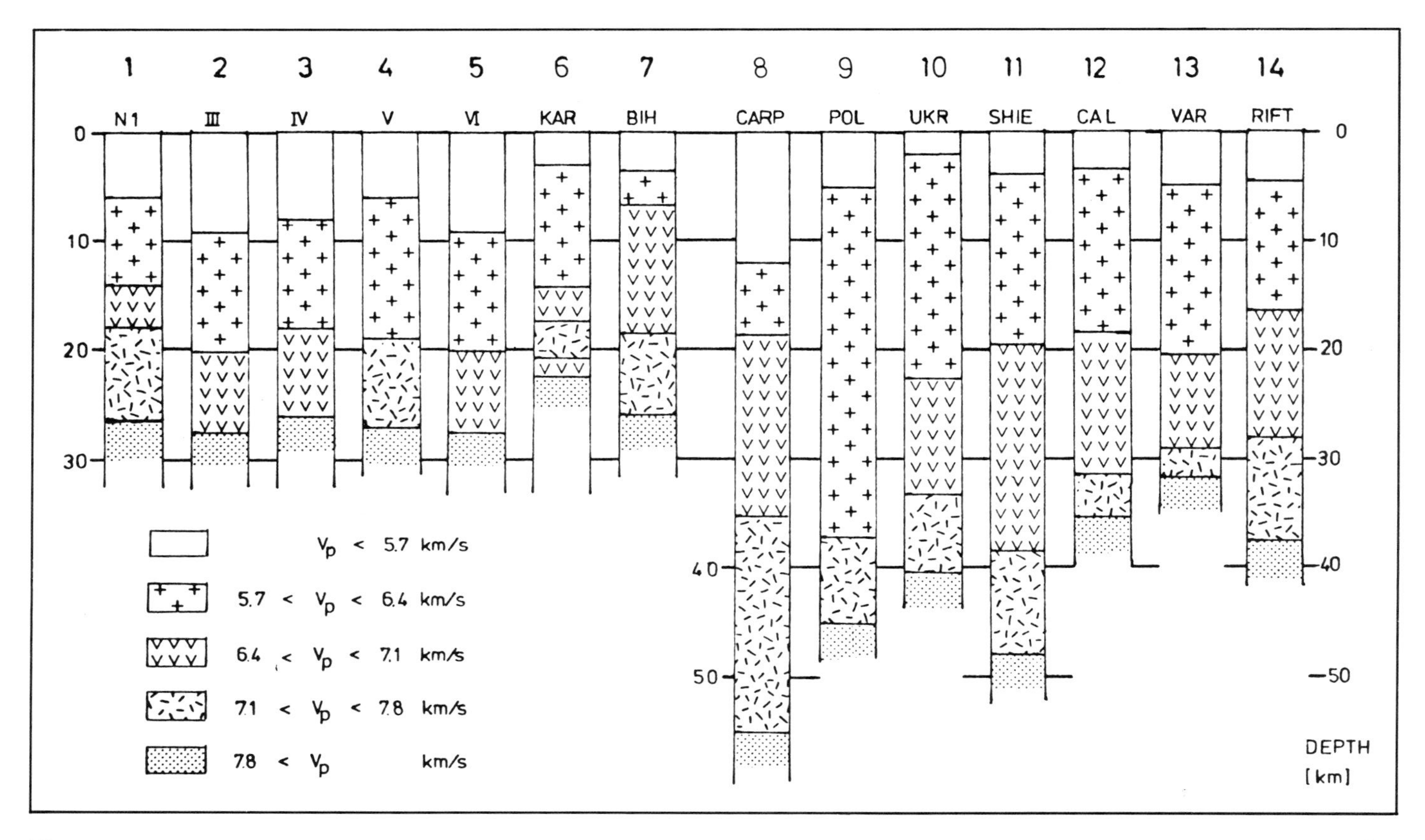

Figure 3. Standardized crustal velocity-depth profiles of the Pannonian basin, from the area of the national section N1 and the international sections III, IV, V, and VI. KAR, BIH (locations, see Figure 1): velocity determinations using wide angle reflection seismic techniques. For comparison: typical crustal velocity profiles for the Eastern Carpathians (CARP), for the Polish part of the Eastern European Platform (POL), and for the Ukrainian part of the East European Platform (UKR). Typical shield (SHIE) Caledonian (CAL), Variscan (VAR), and rift (RIFT) velocity structures, collected by Meissner (1986).

depths between 33 and 55 km, much deeper than the shallow Moho depth in the Pannonian basin. Third, the transition zone between gabbro and eclogite should cover a large depth interval of roughly 10 to 15 km and not show a rather sharp discontinuity such as that below the Pannonian basin, where no gradient zones are observed.

Concerning the second hypothesis (2), it is well known that many back arc and interarc basins have suffered stretching and crustal thinning in an extensional environment. In the Pannonian basin some stretching is indicated by the observations of fault patterns in the upper, rigid crust, but there are serious questions (Artyushkov and Baer, 1984) regarding the total amount of extensional tectonics in view of a very limited outward movement of the Carpathian arc. Moreover, the seismic observations show that the thickness of the upper crust did not change very much compared to that of the surrounding shield area (see Figure 3); the largest differences occur in the thickness of the lower crust. These observations suggest a mechanism by which the lower crust, generally supposed to be much more ductile than the upper crust, is removed by magmatic convective processes (Meissner and Strehlau, 1982). In other words, a differential stretching must have taken place with a maximum in the ductile lower crust and the lower lithosphere, transferring only a small amount of stretching to the rigid upper crust (Royden et al., 1983; Bott, 1976). Such a concept may well explain the reduced crustal and lithospheric thickness, with only minor thinning of the upper crust.

Certainly, stretching in an extensional environment (2) cannot be separated completely from diapirism (3). Both processes may be considered a consequence of, not the primary cause for, extensional processes at greater depth. Retreating subduction zones (predominantly at depth?), secondary pattern in connection with subduction, or simply a rising plume have been suggested as causes for stretching and diapirism (Toksöz et al., 1971; Li Yinting et al., 1983).

It is stressed again that both processes, stretching and diapirism, must be intimately connected with each other. Without diapiric movements (3) the high heat flow values in the Pannonian basin cannot be explained (Stegena et al., 1975, 1981; Schlater et al., 1980). For the modified stretching hypothesis, as suggested in the last paragraph, a very ductile lower crust, heated up by mantle diapirism, is a welcome contribution. The Tertiary volcanism is an additional indication of an extensive heating of the lithosphere and cannot be understood without diapiric processes. Even the reduced crustal thickness can be explained by a diapiric heating and melting of the lower crust, thereby forming granitic melts which enter the upper crust and mafic to ultramafic residues which now have become part of the upper mantle.

The primary extensional processes most probably are related

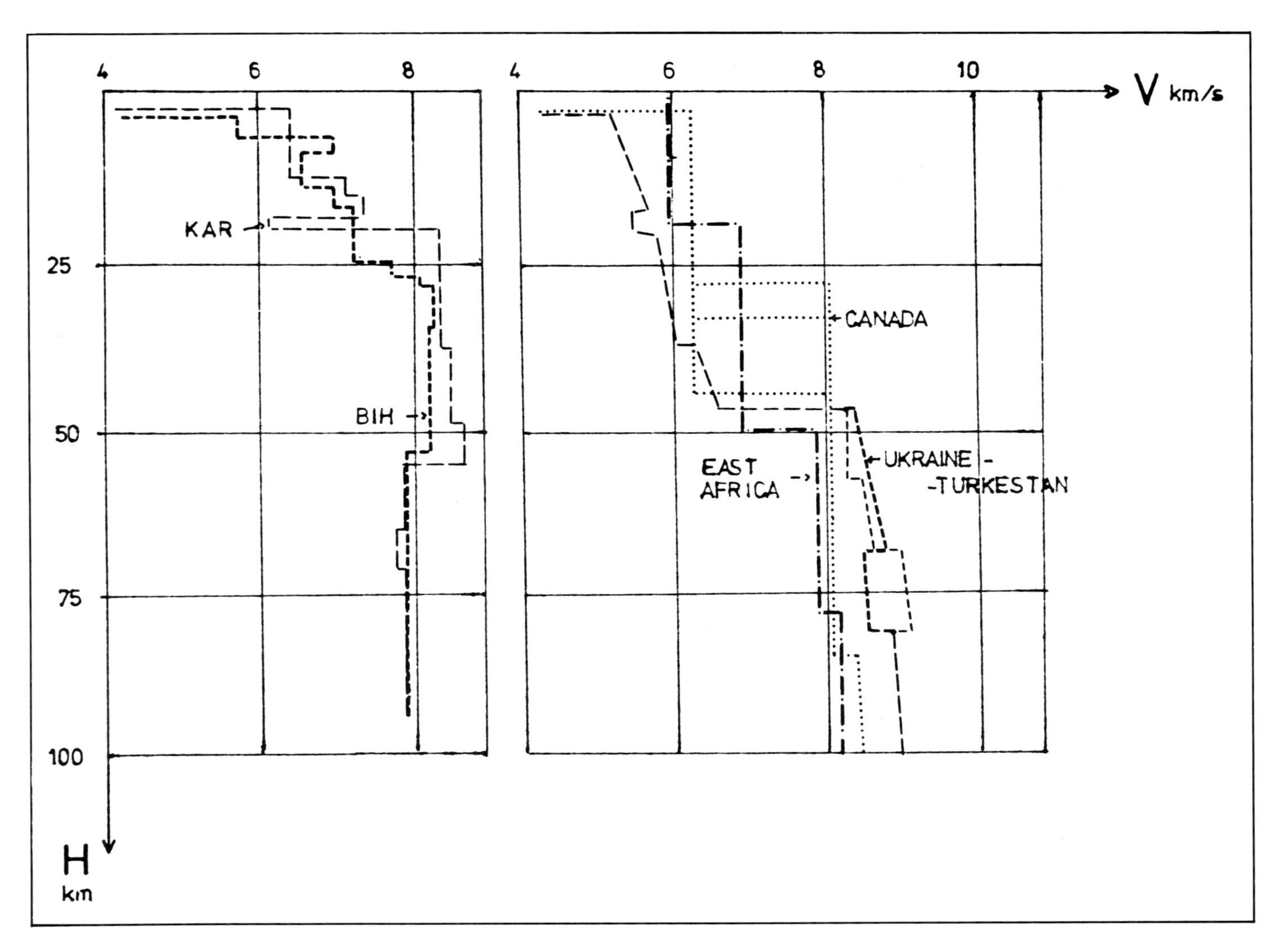

Figure 4. Actual velocity depth profiles in the Pannonian basin (KAR and BIH), modified after Posgay et al. (1981). For comparison, some curves from shield areas are shown.

to subduction processes in the Carpathian arc. An independent plume below the Pannonian basin can be ruled out. It would have caused an initial updoming for which there is no evidence. We know from the development of many basins that even the initial phase of formation may not be marked by updoming, but by subsidence (Meissner, 1986). Secondary flow patterns behind (retreating?) subduction zones might generate an extensional environment.

In conclusion, the Pannonian basin is considered to have been created by a combination of diapirism (3) and a differential stretching (2) in an extensional environment behind the Carpathian subduction zone.

REFERENCES

Artyushkov, E. V., and M. A. Baer, 1984, Mechanism of continental crust subsidence in the Alpine belt: Tectonophysics, v. 108, p. 193–228.

Bodri B., and L. Bodriné-Cvetkova, 1982, Geothermal relations of the thinning out of the Pannonian crust (in Hung.): Magyar Geofizika, XXIII, v. 3, p. 94–106.

Bott, M. H. P., 1976, Formation of sedimentary basins of grabentype by extension of the continental crust: Tectonophysics, v. 36, p. 77–86.

Burchfiel, B. C., 1980, Eastern Alpine system and Carpathian orocline as an example of collision tectonics: Tectonophysics, v. 63, p. 31–62.

Constantinescu P., E. Mituch, K. Posgay, and F. Radulescu, 1975, Deep seismic sounding in the eastern part of the Pannonian basin: Geophys. Transactions, Budapest, v. 23, p. 7–12.

Dövényi, P., F. Horváth, P. Liebe, J. Gálfi, and I. Erki, 1983, Geothermal conditions of Hungary: Geophys. Transactions, Budapest, XXIX, p. 1.

Gálfi, J., and L. Stegena, 1960, Deep reflections and the structure of the earth's crust in the Hungarian plain. Geophys. Transactions, Budapest, VIII, v. 4, p. 189–195.

Gálfi, J., and L. Stegena, 1963, A generalized method for the determination of crustal thickness by means of PP and PS waves: Geophys. Transactions, Budapest, XII, v. 1–2, p. 57–64.

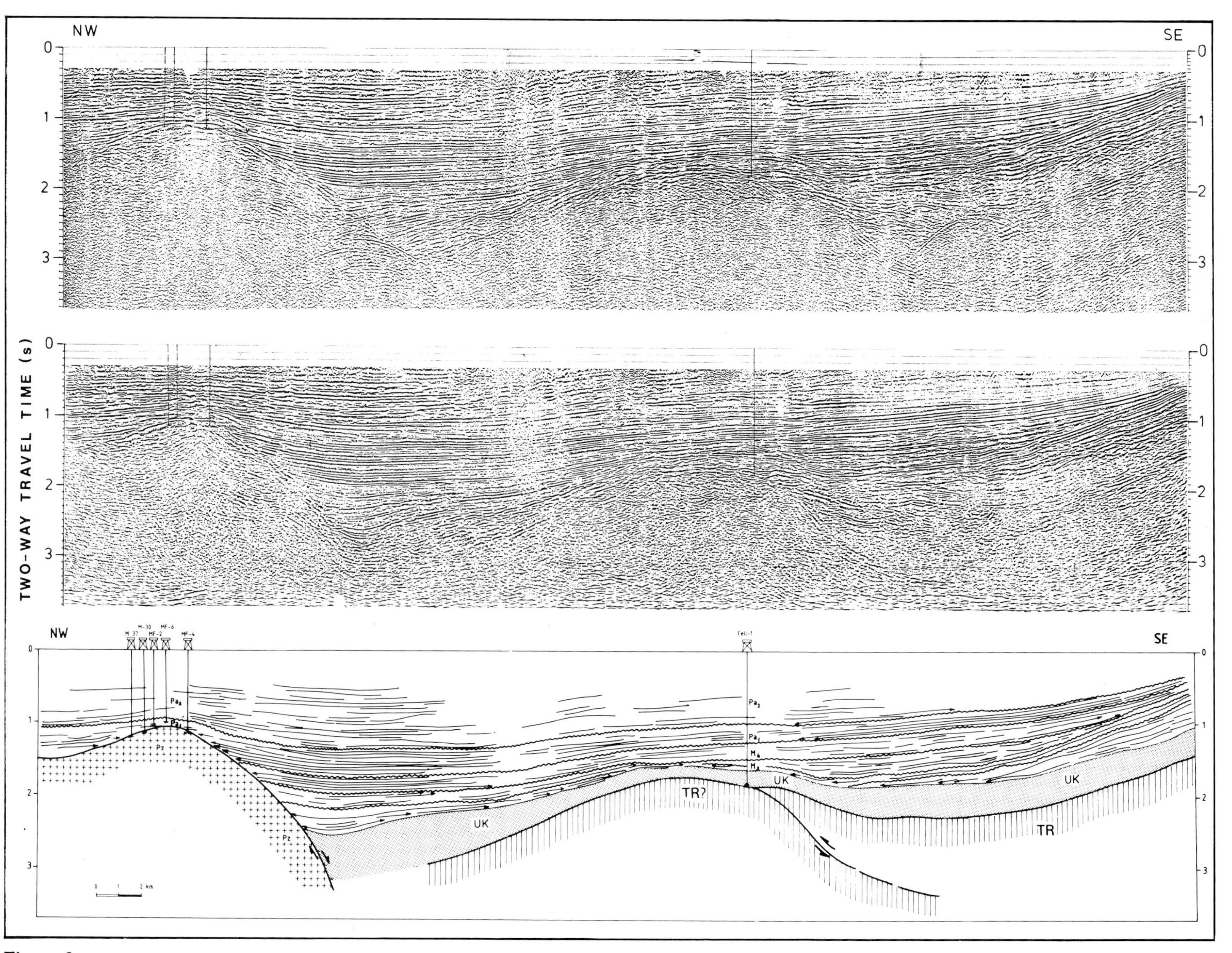

Figure 3. Seismic section VPá-38: 24-fold stacked section (upper), migrated section (middle), and line drawing interpretation (lower). Pa_2, upper Pannonian (*s. l.*); Pa_1, lower Pannonian (*s. l.*); M_4, Badenian; M_3, Karpatian; UK, Upper Cretaceous; TR, Triassic; Pz, Paleozoic.

(M_4, mostly volcanogenic) and Sarmatian (M_5, mostly marly) deposits above the Triassic rocks. Above the Paleozoic basement rocks no Badenian and only very thin Sarmatian strata are present.

Line VPá-38 is a longer reflection section located about 90 km northeast of line Zi-108, and it also crosses the Rába fault (Figure 3). The contact of the Paleozoic and Mesozoic basement rocks here is also a fault with obvious normal displacement. The tilted Upper Cretaceous and Triassic(?) strata in the hanging wall suggest rotation of this block, probably because of flattening of the fault plane at depth. We infer that in addition to the obvious dip-slip displacement, significant strike-slip displacement also occurred along this fault. Kázmér and Kovács (1985) correlated the facies and fauna in the Transdanubian Central Mountains with the Eastern and Southern Alpine Mesozoic rocks and concluded that about 450 km of sinistral motion had occurred along the Rába fault from middle Eocene until late Oligocene time. We suggest that the Rába fault was an oblique-slip fault during the middle Miocene, with a lateral component of some tens of kilometers.

An interesting structure is present in Figure 3 southeast of well Cell-1. A package of reflectors in the basement at about 2.5 seconds on the right edge of the section can be followed northwestward until they disappear at the bottom of well Cell-1. Horváth and Rumpler (1984) suggested that these reflectors represent a thrust fault. In the hanging wall are the same Mesozoic rocks that are exposed in the Transdanubian Central Mountains. In the footwall are the lower Austroalpine or perhaps Penninic complexes that may be present in the middle of line VPá-38. Thrusting occurred prior to the deposition of the Upper Cretaceous beds, which overlap the thrust fault. This part of the seismic section can also be interpreted in another way as is shown in the lower diagram of Figure 3. It is possible that the thrust fault was reactivated as a listric normal fault during the middle Miocene and resulted in subsidence and rotation of the hanging wall southeast of the fault.

Zala Basin

The Zala basin is represented by seismic lines Zi-I-K, D-3/d, and Kad-58. Line Zi-I-K crosses the Budafa anticline (Figure 4). This area has been thoroughly investigated because important oil fields have been found here in structural traps in Pannonian rocks (Dank, this volume). Line Zi-I-K has 48-fold coverage and shows well the structure of the Neogene sedimentary rocks. However, it does not provide good information about the structure of the basement. Interpretation of this line is constrained by data from four drill holes along the section, and a compressional structure is clearly present. The younger strata (M_4, Badenian; M_5, Sarmatian; Pa_1, lower Pannonian; and Pa_2, upper Pannonian) are folded, while the older Miocene strata (M_3, Karpatian) may contain thrust faults. The fold axes of the Budafa anticline and of several other similar structures in the neighboring parts of Hungary and Yugoslavia trend east–west. The involvement of very young rocks in the folding, together with the topography in the area, suggest that this north–south compression may still be active.

Line D-3/d (Figure 5) is another seismic line in the Zala basin and is located about 40 km east of line Zi-I-K. Triassic carbonates were penetrated in the Sáv-1 well in the northern end of

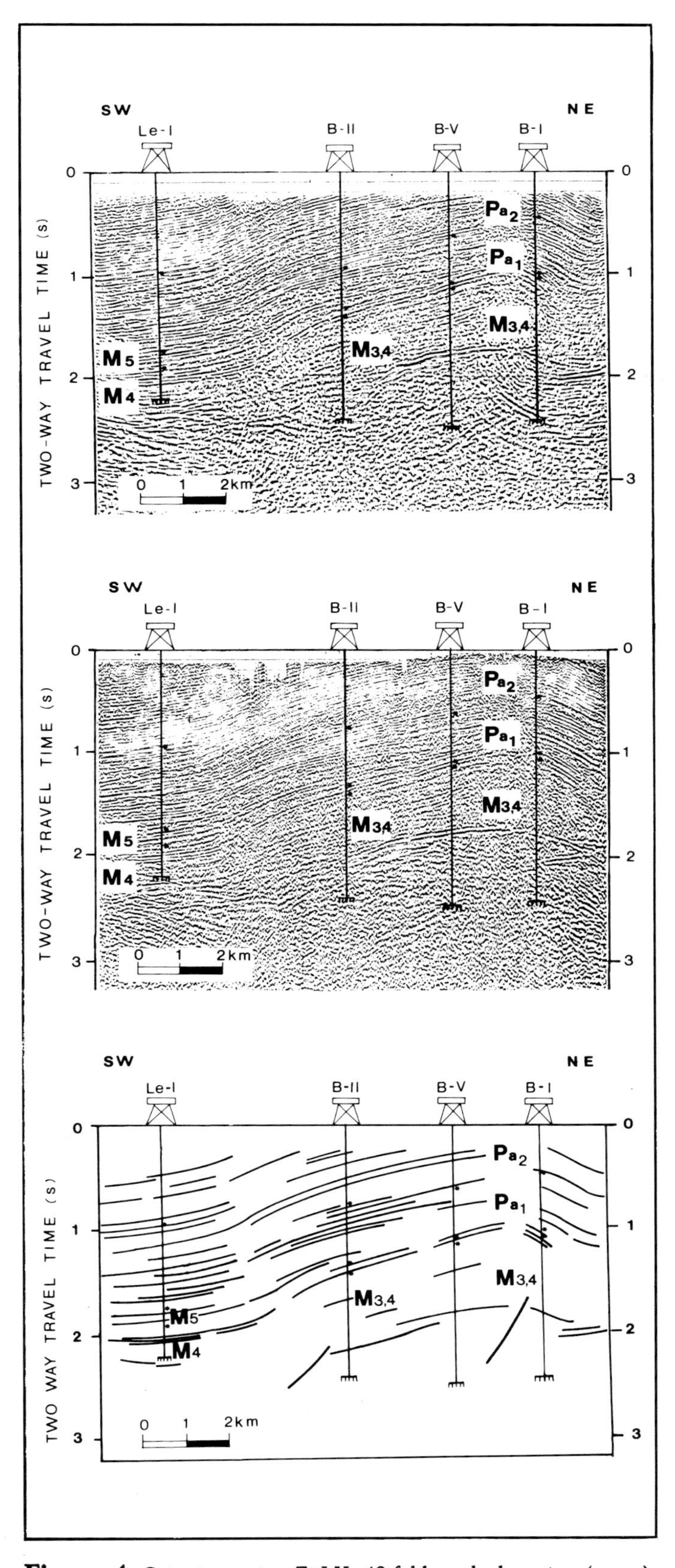

Figure 4. Seismic section Zi-I-K: 48-fold stacked section (upper), migrated section (middle), and line drawing interpretation (lower). Pa_2, upper Pannonian (*s. l.*); Pa_1, lower Pannonian (*s. l.*); M_5, Sarmatian; M_4, Badenian; $M_{3,4}$, Karpatian and Badenian.

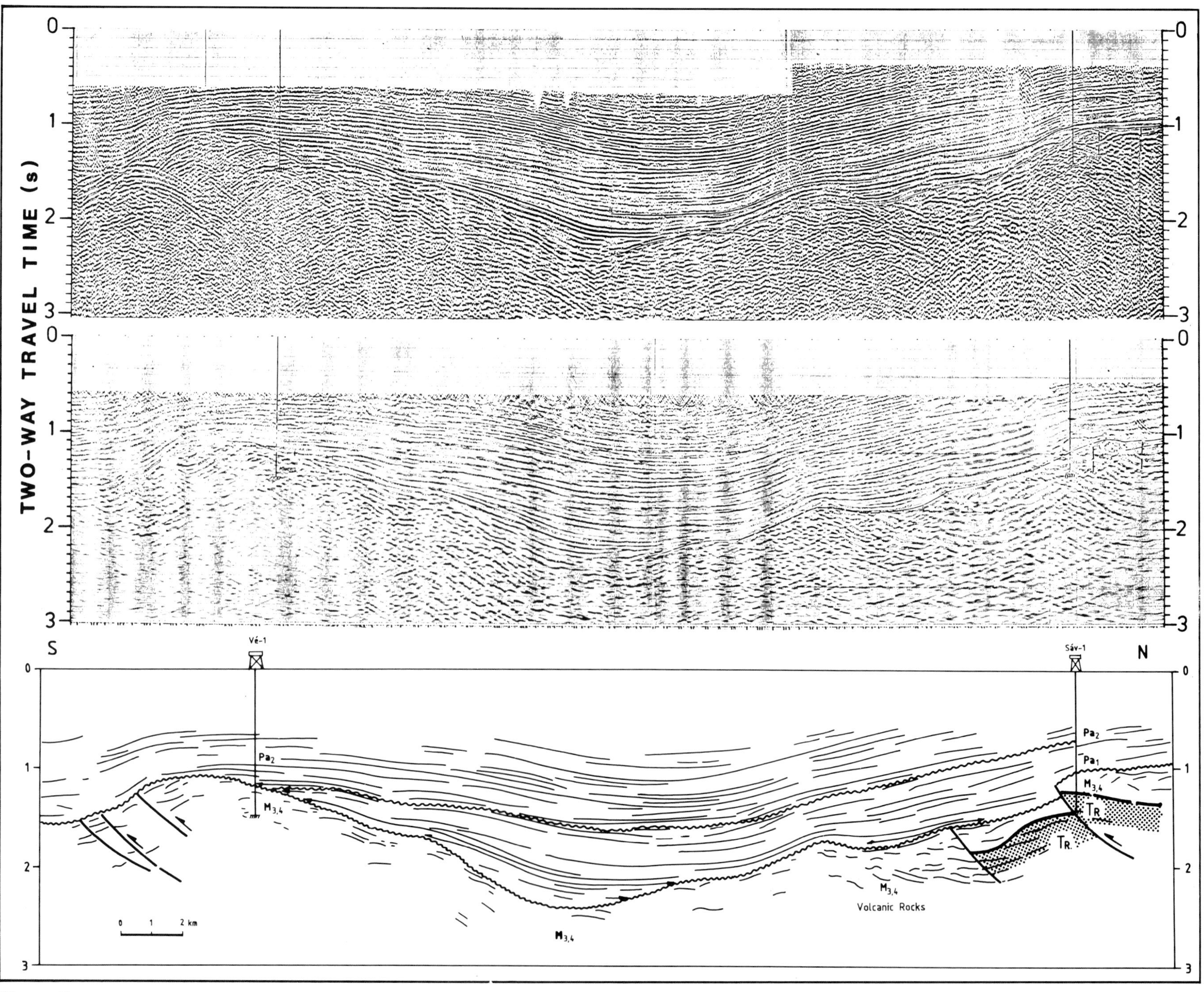

Figure 5. Seismic section D-3/d: 24-fold stacked section (upper), migrated section (middle), and line drawing interpretation (lower). Pa_2, upper Pannonian (*s. l.*); Pa_1, lower Pannonian (*s. l.*); $M_{3,4}$, Karpatian and Badenian; TR, Triassic.

the section, but the basement is not known elsewhere along the section. Karpatian and Badenian (M_3 and M_4) rocks are clearly folded, and their contact with the overlying lower Pannonian (Pa_1) or upper Pannonian (Pa_2) sedimentary complex is unconformable. Sarmatian rocks were not identified in either of the two wells along this section. Although upper Pannonian rocks south of Vé-1 well are slightly folded, it is clear that the main folding occurred during the middle Miocene (Badenian and Sarmatian).

Line Kad-58 (Figure 6) is located 20 km southeast of line D-3/d. There are no drill holes along the section, but the interpretation is constrained by a few boreholes that are located close to this section. The Paleozoic (and locally Mesozoic) basement is present at a shallow depth (1.5–2 km) on both sides of the section. In the middle part of the section, a set of listric normal faults bound characteristic half-grabens that formed during the deposition of middle Miocene ($M_{4,5}$, Badenian and Sarmatian) sediments. Study of other seismic sections shows that the deep grabens are localized within an area of about 12 by 20 km. Reconstruction of the prerift geometry suggests that the amount of extension in the graben is about 50% (Horváth and Rumpler, 1984). Note that in this region, not far from sections Zi-I-K and D-3/d, no compressional features were observed.

Danube–Tisza Interfluve Area

Line Ku-68 (Figure 7) crosses another localized sedimentary trough in the southern part of the Danube–Tisza interfluve region. This area has been studied intensively because some smaller hydrocarbon fields have been found here in middle Miocene traps (Dank, this volume). The basement is made up mainly of Mesozoic rocks. The seismic section can be interpreted as a deep half-graben with steeply (lower part) to gently (upper part) dipping strata of Karpatian (M_3) to Badenian (M_4) age. Sarmatian faunas were not found in the core samples. The unconformity at the top of the synrift sediments (M_3 and M_4) is overlain by the youngest subunit of the lower Pannonian (*s. l.*) (Pa_1^2). The lower part of the lower Pannonian (*s. l.*) (units Pa_1^{1a} and Pa_1^{1b}) is missing.

Figure 8 shows a seismic section (As-24) located about 25 km southeast of Ku-68. A set of rotated basement blocks, made up of Paleozoic and locally thin Mesozoic rocks, can be seen in this section. Drill hole data place good constraints on this interpretation. The basement blocks are separated by listric normal faults, and displacement apparently occurred during deposition of Badenian and Sarmatian ($M_{4,5}$) sediments. Lower and upper Pannonian rocks are not disrupted significantly by the normal faults. Undulation of the bedding planes is probably the consequences of differential compaction of these rocks.

Eastern Great Hungarian Plain

Seismic sections Ka-27 and De-5 (Figures 9 and 10) cross one another. A fairly broad fault zone is shown in Figure 9 consisting of a set of curved faults that in places can be shown to penetrate the basement. Faults shown in Figure 10 can be interpreted as part of a flower structure, typical for strike-slip fault zones. Drill hole data and interpretation of many other seismic sections suggest that this is a shear zone characterized by two principal left-lateral strike-slip faults. The faults are discontinuous and are related to a small pull-apart basin (Figure 1). The oldest sediments in the basin are of middle Miocene age. Note that some of the faults cut even the uppermost Pannonian strata, indicating that strike-slip motion, which started in the middle Miocene, has continued, probably with less activity, until recent times.

Figure 9 shows an interesting structure in the basement. Sáránd-I well penetrated more than 900 m of Paleozoic rocks and terminated in Triassic strata. This overthrust within the basement, and a similar one shown in Figure 3, indicate that the substratum of the Pannonian basin is part of the Alpine orogenic belt that foundered during Neogene extension.

TECTONIC MAP AND INFERRED STRESSES

A systematic reevaluation of all available reflection seismic data of good quality from Hungary has recently resulted in a new depth to basement map for the whole country. The map shows many details of basement relief that were not known before (Kilényi and Rumpler, 1984). During the same study, the most important faults that affect the Neogene basin fill were mapped. We used these maps as a primary source in constructing the tectonic scheme shown in Figure 11. In addition, we used the following data and interpretations:

1. Depth to basement map of the Pannonian basin (see Map 8).
2. Map of Miocene strike-slip faults and related thrust faults in the Bakony Mountains north of Lake Balaton (Mészáros, 1983; Kókay, 1976).
3. Neogene folds and thrusts in the Mecsek Mountains (to the north of Pécs), and contemporaneous left-lateral displacement of about 36 km along the Mecsekalja fault (Némedi Varga, 1983; Szederkényi, 1974).
4. Left-lateral displacement along the Darnó fault (to the west of Miskolc) which cut and displace by more than 25 km one-half of the Badenian volcanic caldera in the Mátra Mountains (Zelenka et al., 1983).
5. Neotectonic map of Hungary based on the interpretation of satellite images (Brezsnyánszky and Síkhegyi, 1985).
6. Pull-apart origin of the Vienna basin and the related left-lateral strike-slip faults (Horváth and Royden, 1981; Gutdeutsch and Aric, this volume).
7. Neogene to Holocene strike-slip faults in the northwestern Dinarides determined from geological mapping and interpretation of satellite images (Gospodaric, 1969/1970; Poljak, 1984).

Combining these data with interpretation of many seismic sections (some examples of which are shown in the previous section), we prepared a simplified tectonic map of the Pannonian basin and its surroundings (Figure 11). This map shows the most important faults that controlled the middle Miocene deformation. However, in some places deformation continues with greatly reduced intensity, and the present seismic activity in the Pannonian basin can be considered to be inherited from the middle Miocene.

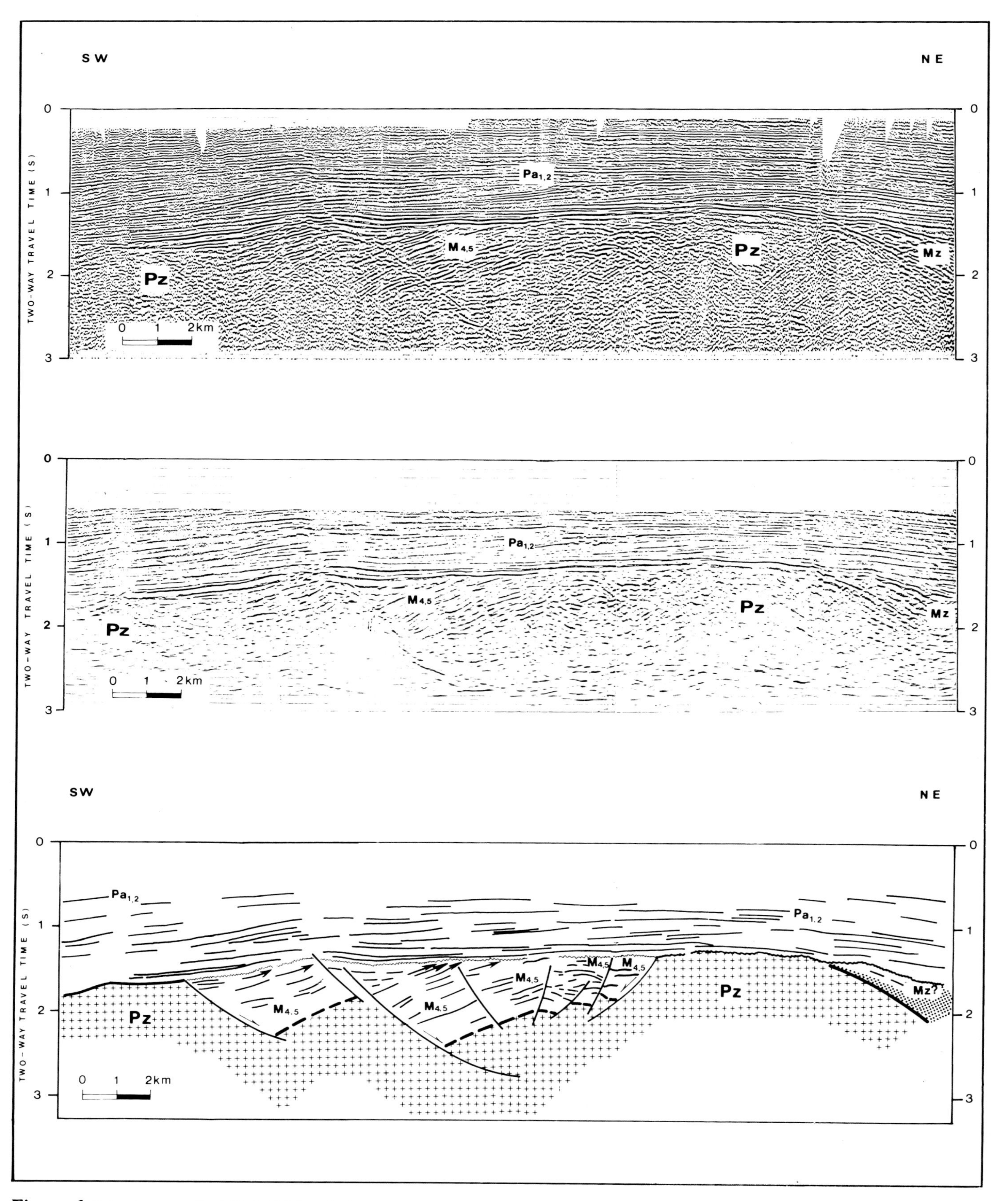

Figure 6. Seismic section Kad-58: 24-fold stacked section (upper), migrated section (middle), and line drawing interpretation (lower). $Pa_{1,2}$, lower and upper Pannonian (*s. l.*); M_4, Badenian and Sarmatian; Mz, Mesozoic; Pz, Paleozoic.

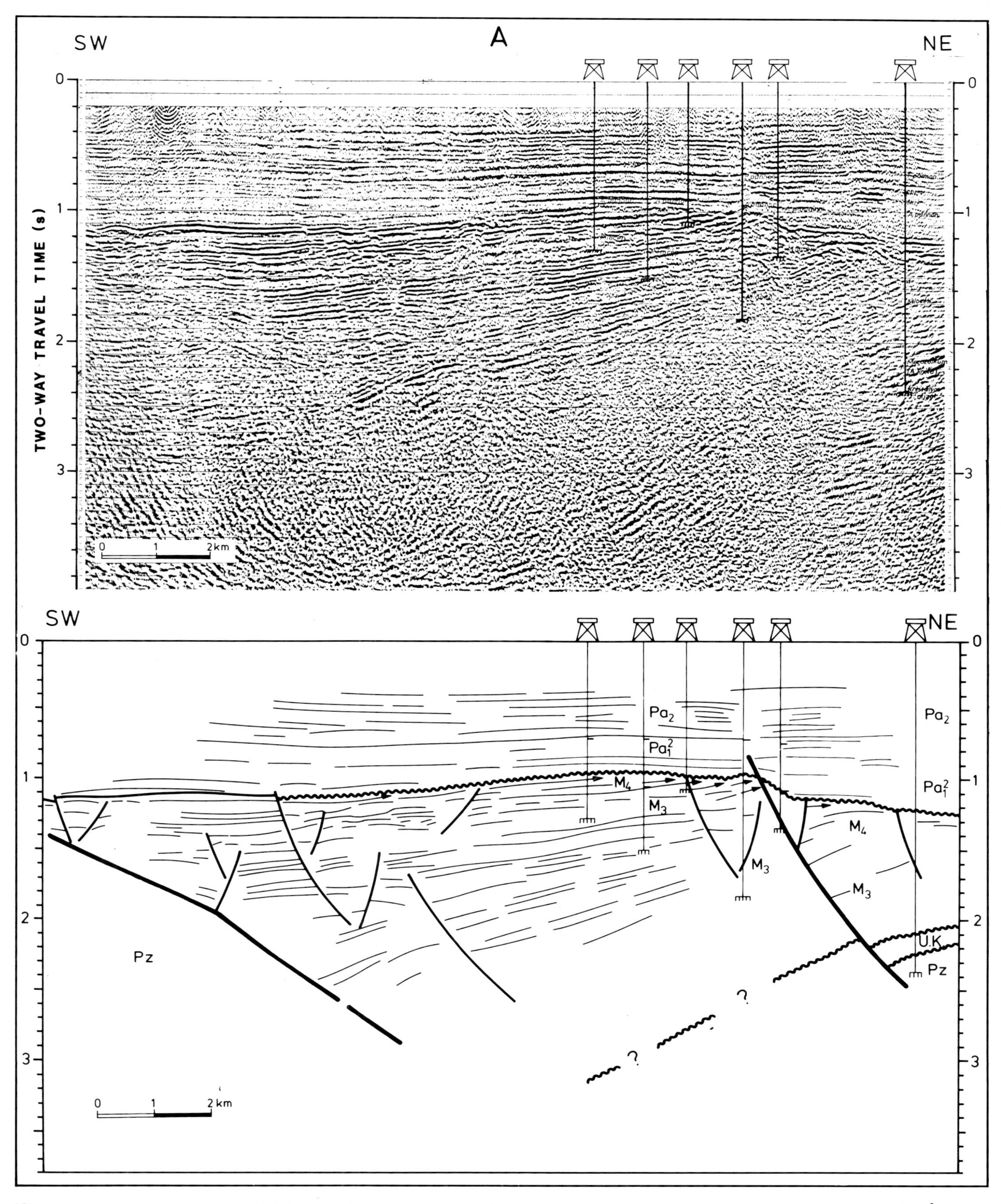

Figure 7. Seismic section Ku-68: 24-fold stacked section (upper), and line drawing interpretation (lower). Pa_2, upper Pannonian (*s. l.*); Pa_1^2, upper lower Pannonian (*s. l.*); M_4, Badenian; M_3, Karpatian; UK, Upper Cretaceous; Pz, Paleozoic.

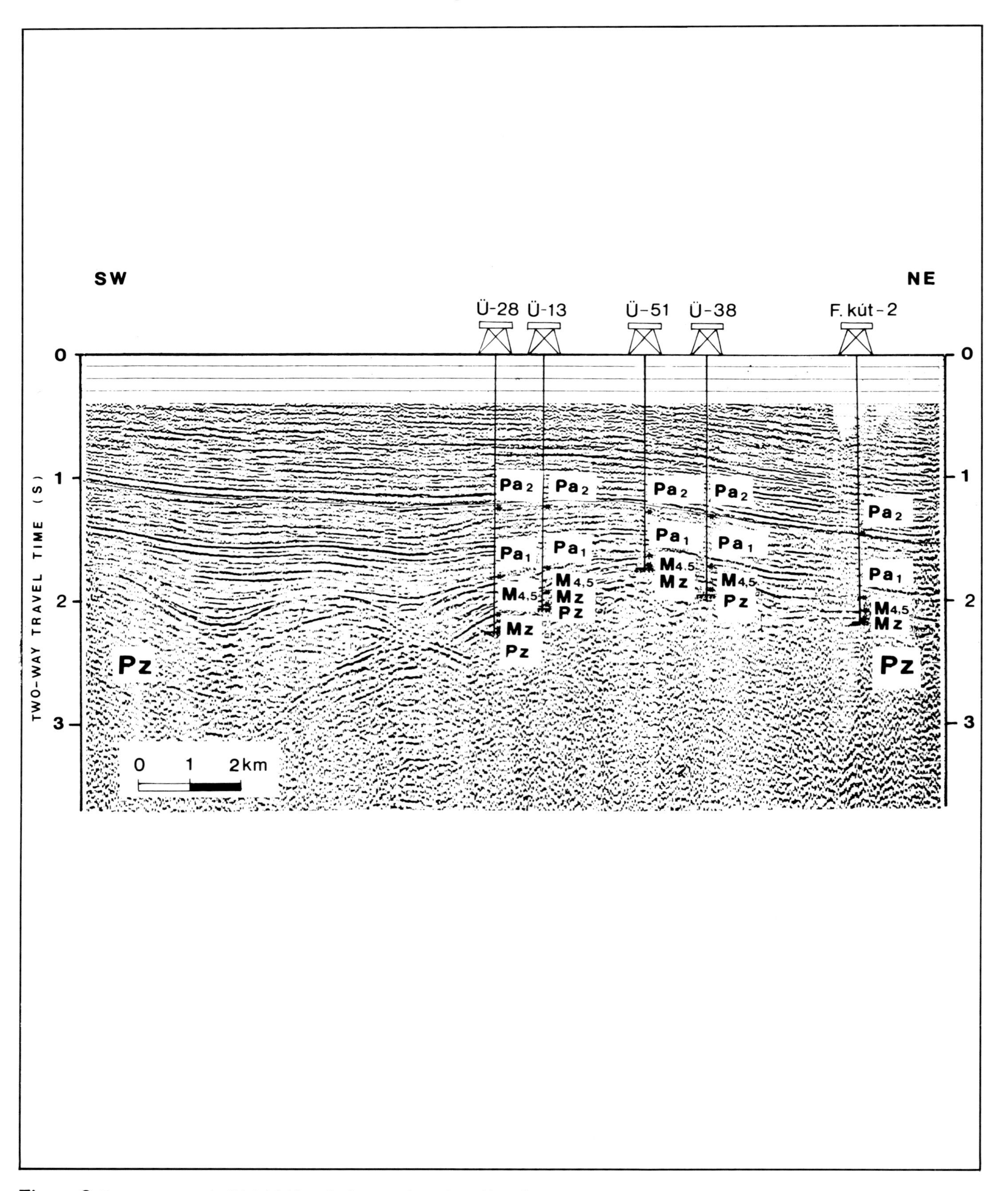

Figure 8. Seismic section As-24: 24-fold stacked section (upper), and line drawing interpretation (lower). Pa_2, upper Pannonian (*s. l.*); Pa_1, lower Pannonian (*s. l.*); $M_{4,5}$, Badenian and Sarmatian; Mz, Mesozoic; Pz, Paleozoic.

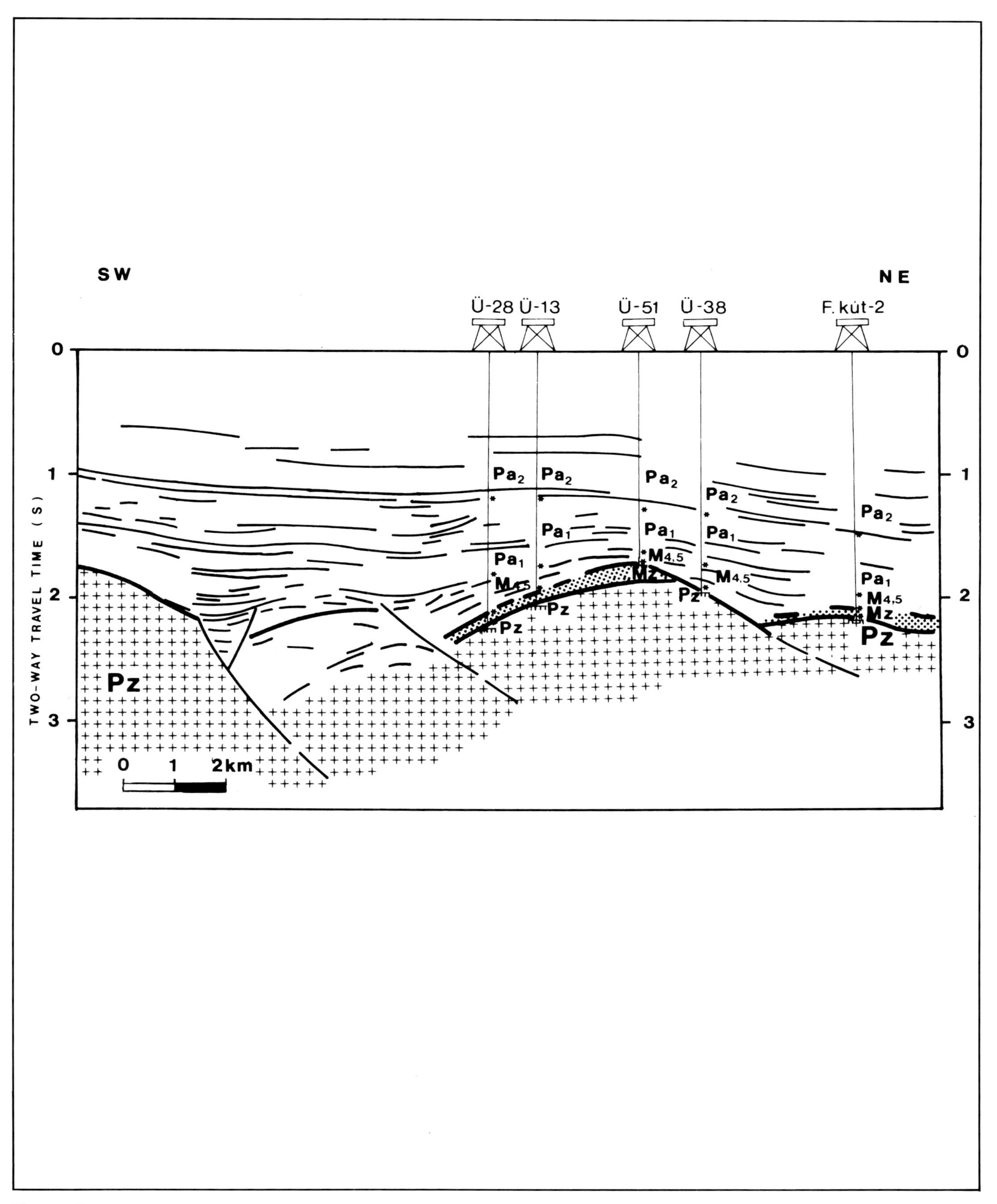

Figure 8. (continued)

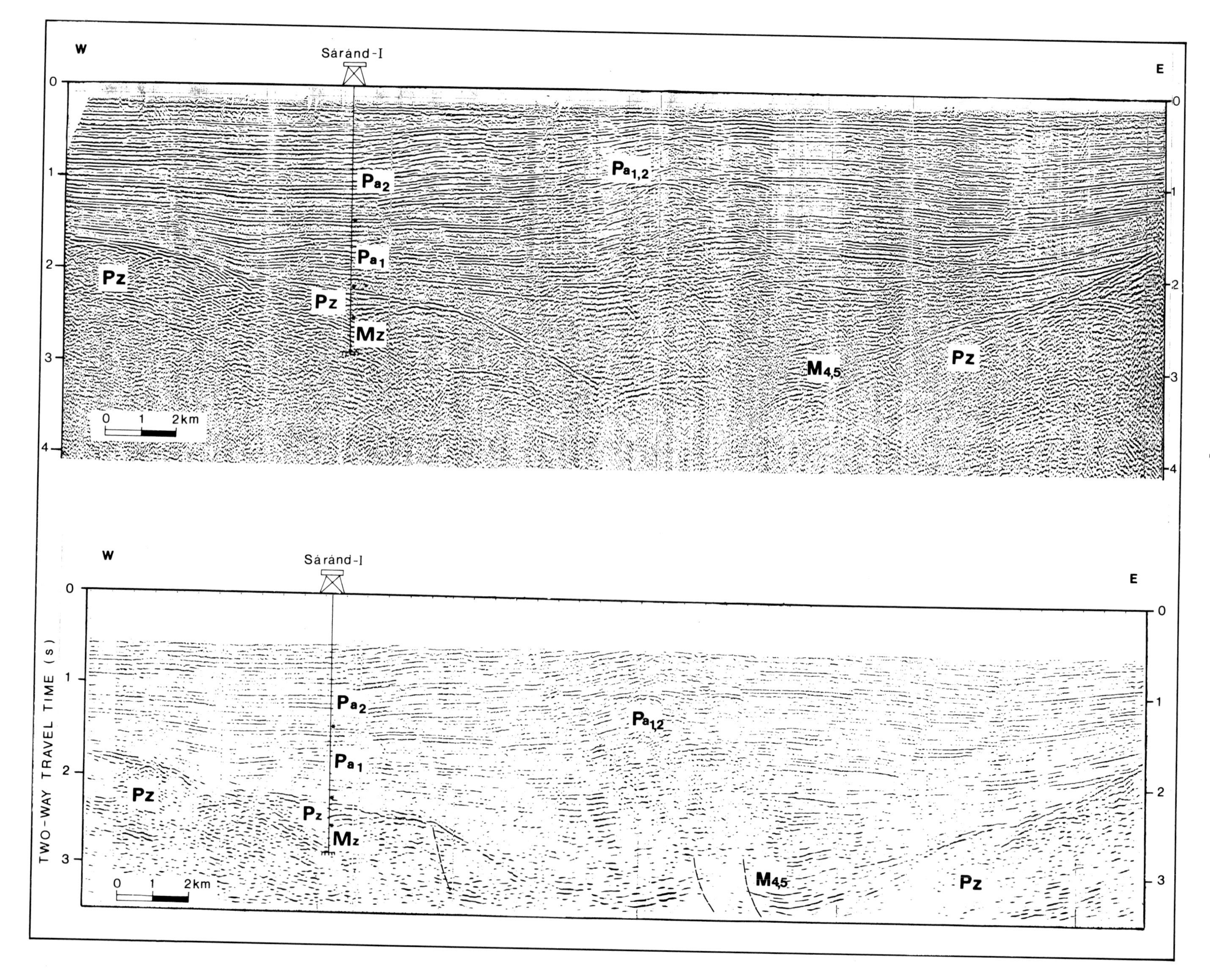

W
Sárand-I
E
Pa2
Pa1,2
Pa1
Pz
Mz
M4,5
0 1 2km
0
1
2
3
4
TWO-WAY TRAVEL TIME (s)

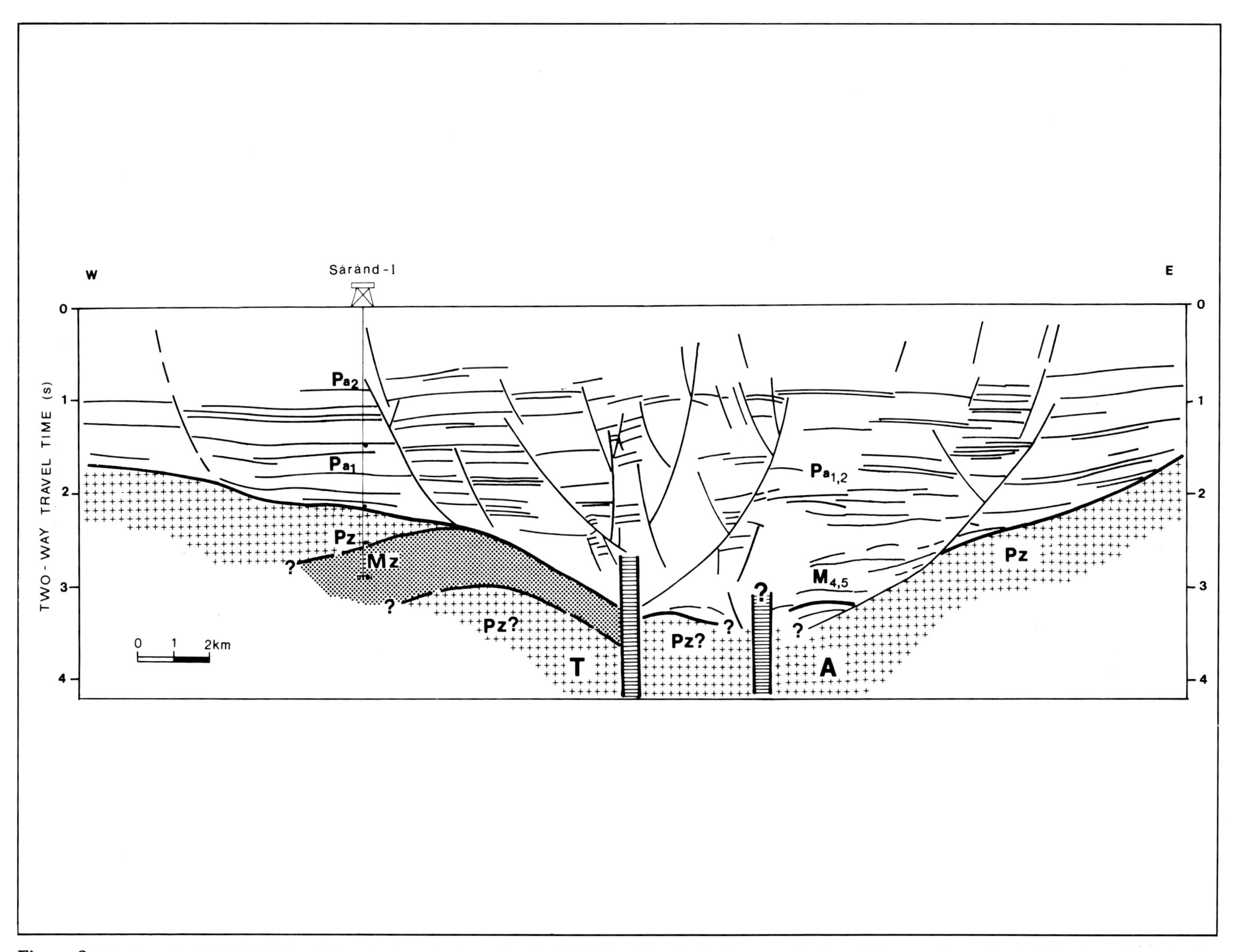

Figure 9. Seismic section Ka-27: 24-fold stacked section (upper), migrated section (middle), and line drawing interpretation (lower). Pa_2, upper Pannonian (*s. l.*); Pa_1, lower Pannonian (*s. l.*); $M_{4,5}$, Badenian and Sarmatian; Mz, Mesozoic; Pz, Paleozoic. T and A indicate strike-slip displacement toward and away from the viewer respectively.

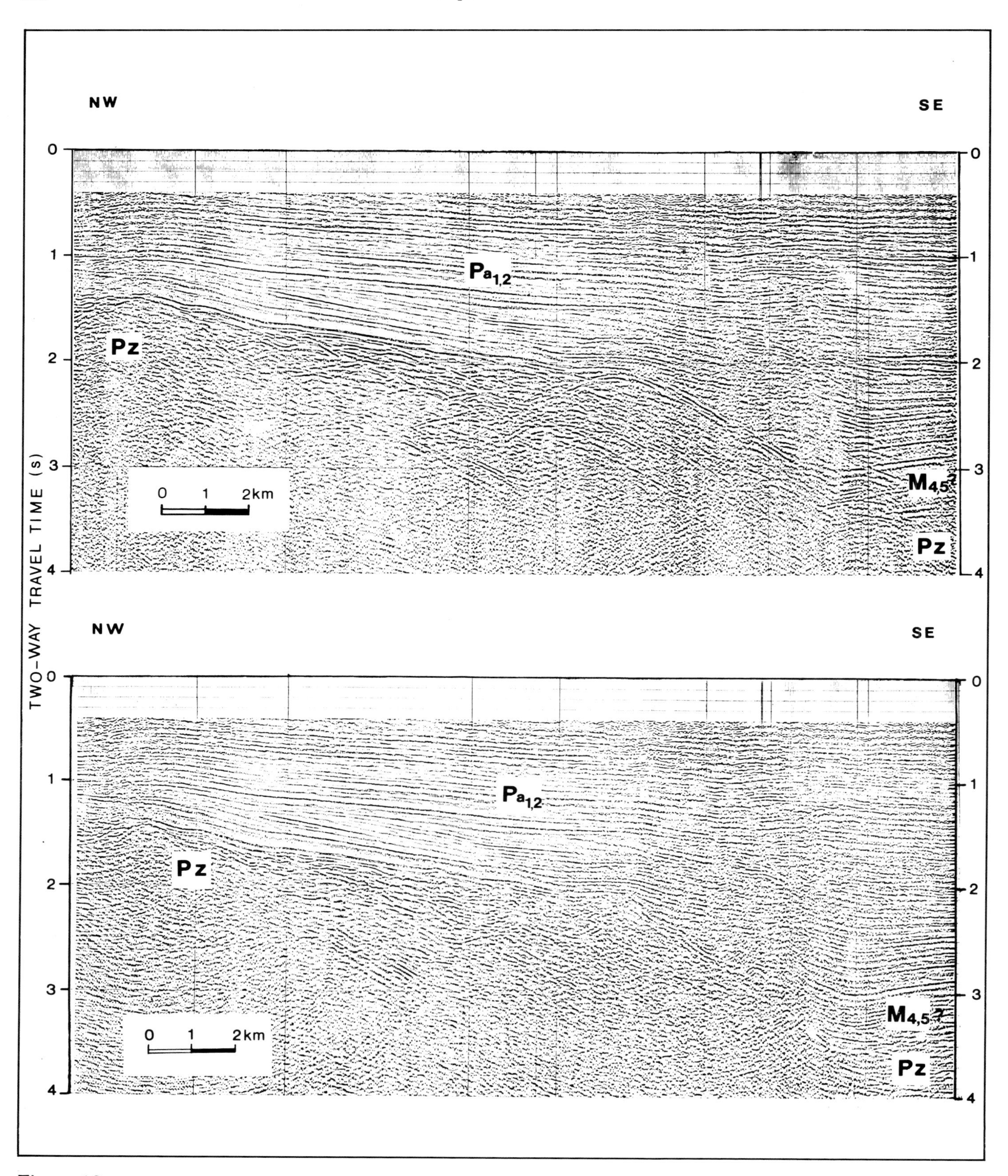

Figure 10. Seismic section De-5: 24-fold stacked section (upper), migrated section (middle), and line drawing interpretation (lower). $Pa_{1,2}$, lower and upper Pannonian (*s. l.*); M_4, Badenian and Sarmatian; Pz, Paleozoic. T and A indicate strike-slip displacement toward and away from the viewer respectively.

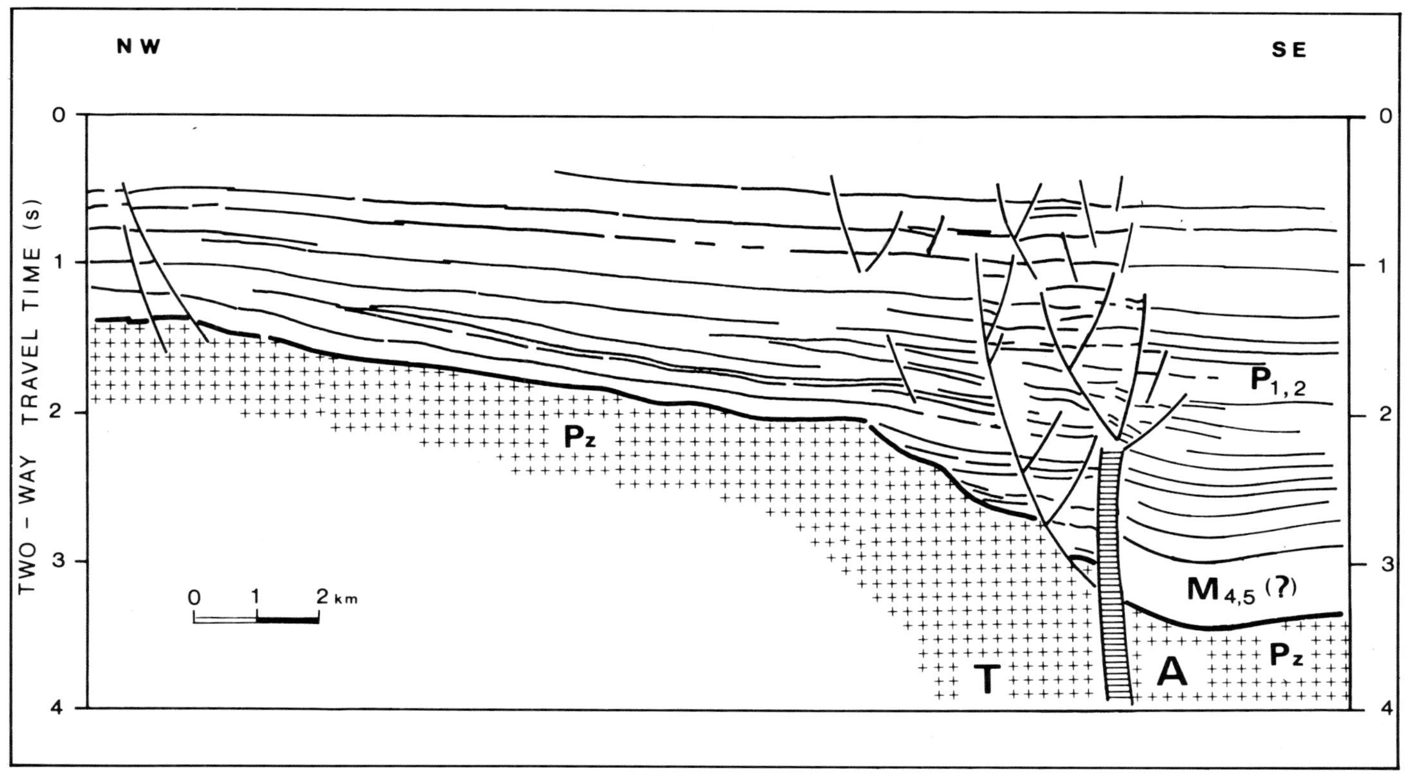

Figure 10. (continued)

The tectonic map shows two principal sets of conjugate strike-slip faults. All of the faults that strike east–northeast or northeast are left-lateral. All of the faults that strike northwest are right-lateral. Areas of extension and normal faulting are associated with discontinuous or divergent strike-slip faults and with fragmentation in zones bounded by two major strike-slip faults. Locally, thrust faults and folds (with east-west-trending fold axes) are present.

This kinematic pattern can be explained by a simple regional stress field. The Pannonian region has been characterized by a north–south maximum principal stress (σ_1), an east–west minimum principal stress (σ_3), and a vertical intermediate stress (σ_2), where:

$$\sigma_1 = \sigma_{N\text{-}S} \geq \sigma_2 = \sigma_{vertical} > \sigma_3 = \sigma_{E\text{-}W}$$

This stress field is compatible with two conjugate sets of strike-slip faults (if $\sigma_1 = \sigma_{N\text{-}S}$), and it provides the conditions for oblique slip along these faults (if $\sigma_1 \sim \sigma_2$) and for pure normal faulting elsewhere (if $\sigma_1 = \sigma_{vertical}$).

Microtectonic studies in the Transdanubian Central Mountains have led to similar conclusions (Bergerat et al., 1984). These authors, moreover, suggest that the period of strike-slip faulting and normal faulting are distinct and that extension occurred after strike-slip faulting. The seismic sections shown in this chapter, however, show that development of compressional and extensional structures occurred at about the same time (Zala basin), and that strike-slip motion spanned a long time interval from the middle Miocene to Holocene time (eastern Great Hungarian Plain). We think, therefore, that separate periods characterized by contrasting styles of deformation cannot be distinguished in the Pannonian basin for middle Miocene through Quaternary time. Some temporal variation of the stress field may have occurred, of course, but the spatial variation seems to have been a more important factor.

CONCLUSIONS

A systematic reevaluation of high-quality reflection seismic data in Hungary and their tectonic interpretation has led to the following main conclusions:

1. Strike-slip faults and listric normal faults are the most characteristic structural features of Neogene age in the Pannonian basin. Compressional features (mostly folds) can be also found in places.
2. The main period of extension is middle Miocene, but locally minor activity has continued until recent time. A distinctive unconformity separates the synrift sediments (Badenian and Sarmatian) from the postrift sediments (Pannonian through Quaternary) in most parts of the basin.
3. Two sets of conjugate strike-slip faults are present. The main areas of extension are associated with divergent or discontinuous strike-slip faults, and with fragmentation of units bounded by two major strike-slip faults. A normal component of displacement is commonly associated with the strike-slip faults.
4. The pattern of Miocene deformation in the Pannonian basin can be interpreted in terms of a fairly constant stress field. Distinct or episodic changes in this stress field over the whole basin are difficult to establish, and the spatial variation seems to have been more important.

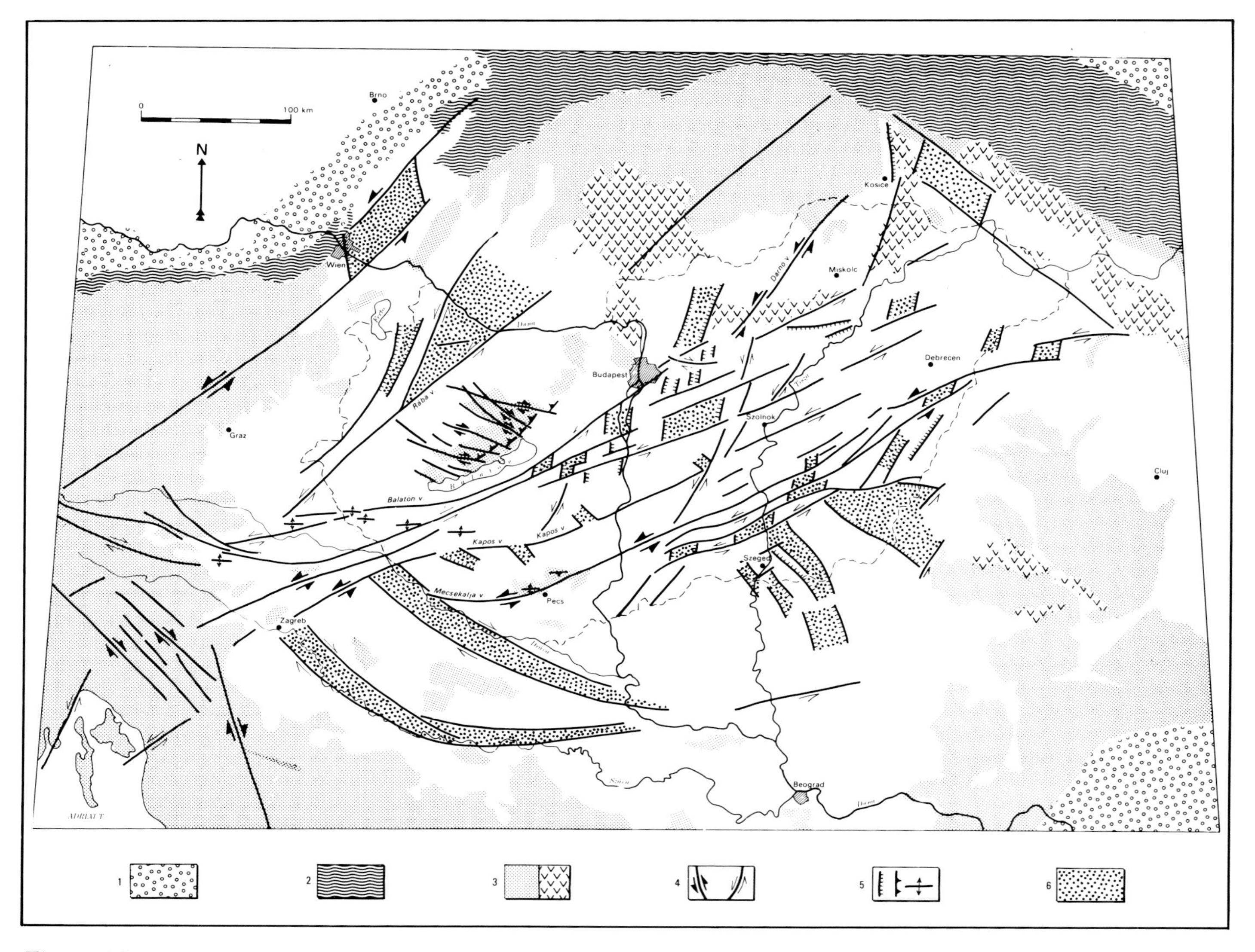

Figure 11. Tectonic map of the Pannonian basin and surrounding regions showing the main faults and folds of Neogene age. Tectonic activity culminated during the middle Miocene and has been greatly reduced since then. Legend: (1) Molasse foredeep; (2) Alpine–Carpathian flysch belt; (3a) Inner Alpine–Carpathian Mountain belt and the Dinarides; (3b) outcrops of Neogene calcalkaline volcanic rocks; (4) strike-slip faults; the sense (and usually the amount) of displacement is well constrained (thick arrows) or unconstrained (thin arrows); (5) normal fault, thrust fault, and fold; and (6) areas of major crustal extension and subsidence.

REFERENCES

Bergerat, F., J. Geyssant, and C. Lepvrier, 1984, Neotectonic outline of the intra-Carpathian basins in Hungary: Acta Geol. Hung., v. 27, n. 3-4, p. 237-249.

Brezsnyánszky, K., and F. Síkhegyi, 1985, Neotectonical evaluation of lineaments of Hungary, interpreted from satellite images: Paper presented at the Int. Symp. Comm. Recent Crustal Movements, Budapest, October 1–5, 1985.

Gospodaric, R., 1969/1970, Probleme der Bruchtektonik in der NW-Dinariden: Geol. Rundsch., v. 59, n. 1, p. 308–322.

Horváth, F., and L. Royden, 1981, Mechanism for the formation of the intra-Carpathian basins: a review: Earth Evol. Sci., v. 1, n. 3-4, p. 307–316.

Horváth F., and J. Rumpler, 1984, The Pannonian basement: extension and subsidence of an Alpine orogene: Acta Geol. Hung., v. 27, n. 3-4, p. 229–235.

Kázmér, M., and S. Kovács, 1985, Permian–Paleogene paleogeography along the eastern part of the Insubric-Periadriatic lineament system: evidence for continental escape of the Bakony-Drauzug unit: Acta Geol. Hung., v. 27, n. 1-2, p. 69–82.

Kilényi, É., and J. Rumpler, 1984, Pre-Tertiary basement relief map of Hungary: Geophys. Transact., Budapest, v. 30, n. 4, p. 425–428.

Kókay, J., 1976, Geomechanical investigation of the southeastern margin of the Bakony mountains and the age of the Litér fault line: Acta Geol. Hung., v. 20, n. 3-4, p. 245–257.

Körössy, L., 1965, Geologischer Bau der ungarischen Becken: Verh. Geol. B.-A., Sonderheft G., Wien, p. 36–51.

Mészáros, J., 1983, Structural and economical geological significance of strike-slip faults in the Bakony: Ann. Rep. Hung. Geol. Inst. 1981, p. 485–502.

Némedi Varga, Z., 1983, Structural development of the Mecsek Mts. during the Alpine orogenic cycle: Ann. Rep. Hung. Geol. Inst. 1981, p. 467–484.
Poljak, M., 1984, Neotectonic investigations in the Pannonian basin based on satellite images: Adv. Space Res., v. 4, n. 11, p. 139–146.
Szederkényi, T., 1974, Paleozoic magmatism and tectogenesis in south-east Transdanubia. Acta Geol. Hung., v. 18, n. 3-4, p. 305–313.
Zelenka, T., Cs. Baksa, Z. Balla, J. Földessy, and K. Földessy-Jarányi, 1983, The role of the Darnö line in the basement structure of north-eastern Hungary: Geol. Zborn. Geol. Carpath., v. 34, n. 1, p. 53–69.

CHAPTER 14

Interpretation of Seismic Reflection Profiles from the Vienna Basin, the Danube Basin, and the Transcarpathian Depression in Czechoslovakia

Cestmir Tomek
Geofyzika, n.p. Brno
P.O. Box 62, 61246 Brno
Czechoslovakia

Arnóst Thon
Moravské naftové doly
Úprkova 5, 69530 Hodónín
Czechoslovakia

Three seismic reflection profiles across three intra-Carpathian basins in Czechoslovakia, the Vienna basin, the Danube basin, and the Transcarpathian depression, suggest that each basin exhibits a different evolutional history. In the Vienna basin, seismic reflection lines 284 and 630 confirm that the Vienna basin is probably a pull-apart feature resulting from thin-skinned tectonics and extension within allochthonous thrust sheets of the Carpathians. Two fault systems in the northern Vienna basin, the Schrattenberg fault system and the Steinberg fault, were probably the sites of Badenian, and possibly also younger, left-slip. The whole Carpathian thrust complex probably has the same thickness (about 10–12 km) beneath the central part of the Vienna basin as beneath other parts of the Carpathian flysch belt along strike with the Vienna basin.

In the Danube basin, seismic reflection line 556 displays reflections not only from Neogene sedimentary layers but also from the buried Tatrides, which are exposed in the northern part of the inner West Carpathians. These northwest dipping events are considered to be the detachment surfaces of Badenian–Pannonian low-angle normal faults. A significant fault-bounded trough that trends northeast and is filled by Badenian (and lower Sarmatian) sediments has been discovered beneath the central part of the Danube basin. Pannonian age and younger sediments onlap unconformably onto these trough sediments and onto the Carpathian basement. Pannonian age and younger subsidence was not tectonically controlled in the basin except immediately adjacent to the Little Carpathians (west side of the basin), where probable Pontian and Pliocene low-angle normal and steep thrust faults (perhaps with left-slip displacement) have been observed.

The Transcarpathian depression also displays two sedimentary complexes corresponding to different stages of basin evolution. The older sedimentary complex formed during extremely fast late Badenian–middle Sarmatian subsidence, and sediments of this age lie within a northwest trending fault-bounded trough. The younger sedimentary complex formed during slower basement subsidence, which was not tectonically controlled.

The subsidence history of the Danube basin and the Transcarpathian depression agrees with the interpretation that evolution of the intra-Carpathian basins was dominated first by extensional faulting related to contemporaneous thrusting within the West and East Carpathians and, second, by cooling and thermal contraction of underlying crust and mantle. There are striking differences between these two phases in the Danube basin and Transcarpathian depression. This problem remains unexplained.

INTRODUCTION

The purpose of this paper is to discuss our interpretation of three regional seismic reflection lines that cross three intra-Carpathian basins, the Vienna basin, the Danube basin, and the Transcarpathian depression, and their implications for Neogene and Quaternary tectonic models for the northern parts of the Pannonian region. These important regional seismic lines were recorded under the auspices of the Czech and Slovak Geological Offices and Moravské naftové doly Hodonín (Moravian Oil Co.) in 1973, 1976, 1982, and 1983 using a dynamite source. The three profiles (284-630, 556, and 540) have provided valuable information about the structure and tectonic development of the Vienna basin and northern parts of the Pannonian region (Figures 1A, 2A, and 3A). The Vienna basin surveys consisted of two profiles, 284 and 630, that cross each other obliquely in the Mistelbach block. Both lines (24 km surface coverage, Figure 1A) began in the Zdanice–Waschberg nappe (outer flysch belt), crossed through the Schrattenberg and Steinberg fault systems, crossed into the central Moravian depression of the Vienna basin, and ended in the eastern part of this depression. The data were acquired using the field parameters given in Table 1.

Lines 630 (Vienna basin) and 556 (Danube basin, between km 60–80) were recorded to 12 sec; the other lines were recorded to 5 or 6 sec. Attenuation for the standard petroleum line 630 was very high and only weak, lower crustal or Moho reflections were observed (in many places no reflections from the Moho were seen). Part of line 556, between km 60–65, clearly shows Moho reflections at 9.2–9.3 sec (about 26 km depth). These results are not discussed in this paper and will be discussed in a future publication.

The processing procedures used are common to the petroleum industry and are discussed in the geophysics literature. Correction for crossed-line geometry was made during processing of lines 284, 630, and 556, so that common midpoint (CMP) locations of the processed data do not correspond to those of the original data.

In the Danube basin, line 556 (81 km surface coverage, Figure 2A) began north of Komarno near the important boreholes Kolarovo 1, 2, and 3. It then ran west-northwest through the deep part of the north Danube basin and up to the foothills of the Little Carpathians (a horst between the Vienna and Danube basins). Transverse line 553 (not interpreted here) connects the borehole Abraham-1 with line 556 and with the deepest part of the Danube basin in the south basement (at more than 3 sec two-way travel time).

Line 540 was recorded at the easternmost part of the Transcarpathian depression near the Soviet border and is 40 km long (Figure 3A). It crosses obliquely the whole northern flank of the Transcarpathian depression (see Rudinec et al., 1981), but does not cross the deepest part of the basin. This line was chosen as a model line because all other parallel seismic lines in the west have shown disturbed reflections due to buried Badenian and Sarmatian volcanic rocks. In the north the line approaches the northern boundary fault of the Transcarpathian depression.

In this chapter we will show that the sedimentation history and tectonic development of the Vienna basin, the Danube basin, and the Transcarpathian depression are different. The Vienna basin seems to be a result of thin-skinned tectonics (Royden et al., 1983a; Jiricek and Tomek, 1981), while extension of the Danube basin and the Transcarpathian depression involves rocks at greater depths (see also Royden et al., 1983a, b). Sedimentation in the Vienna basin has been mainly tectonically controlled throughout the last 20 m.y. Sedimentation within the Danube basin and the Transcarpathian depression was tectonically controlled during the Badenian and (lower) Sarmatian. Later sedimentation occurred within a tectonically quiet regime except in the area near the Little Carpathian horsts and the Vihorlat–Guten Mountains. Here, not only post-Sarmatian normal (listric) faults are observed, but also a very young Pliocene thrust fault with a probable strike-slip component is well documented (line 556).

Table 1. Parameters used in the acquisition of the reflection seismic data for the Vienna basin lines 284 and 630, the Danube basin line 556, and the Transcarpathian depression line 540

	284 and 630	556	540
Source (single shot dynamite)	10–30 kg	10 kg	10 kg
Shot depth	32–36 m	28–30 m	20 m
Shot interval	40 m	80 m	80 m
Geophone station interval	40 m	80 m	160 m
Geophones for group	24	24	24
Recording	96 channel split-spread	48 channel split-spread	48 channel split-spread
Nominal fold of stack	48	24	12

GEOLOGIC AND TECTONIC SETTING

In this chapter we interpret the history of three of the intra-Carpathian basins formed by extension. The geology of the Vienna basin has been described by many authors (for example, by Jiricek and Tomek, 1981), and its tectonic history has been interpreted more recently by Royden et al. (1983a). The basement of the Vienna basin is formed by Austroalpine (sometimes also inner West Carpathian) units in the south, and by the Pieniny Klippen belt and outer and inner flysch nappes in the north (see also Wessely, this volume). The thickness of the Neogene sedimentary fill of the Vienna basin reaches 6 km in its central part. The Vienna basin probably developed in three main stages. Eggenburgian to Karpatian (23–16.5 Ma) sedimentation and subsidence of the basement took place on the moving Carpathian thrust complex. The total thickness of the Eggenburgian to Karpatian sediments varies from site to site and reaches 2000–3000 m in the northwest part of the basin. Badenian (16.5–13.0 Ma) sedimentary rocks are developed mainly within the classic rhombohedral shape of the Vienna basin. Their thickness reaches 1200–2000 m in the main depressions. In partial agreement with Royden et al. (1983a), we believe that Badenian (but not Karpatian) extension is related to synchronous sinistral strike-slip motion along the Schrattenberg and Steinberg faults. The inner Magura and outer Silesian flysch nappes were displaced transcurrently north-northeast with respect to the foreland Brunovistulic Precambrian platform, and with respect to the outer Zdanice–Waschberg nappe, which had previously been thrust over the Precambrian platform. Roth (1980) was the first to propose this transcurrent motion, and estimated its magnitude to be at least 50 km.

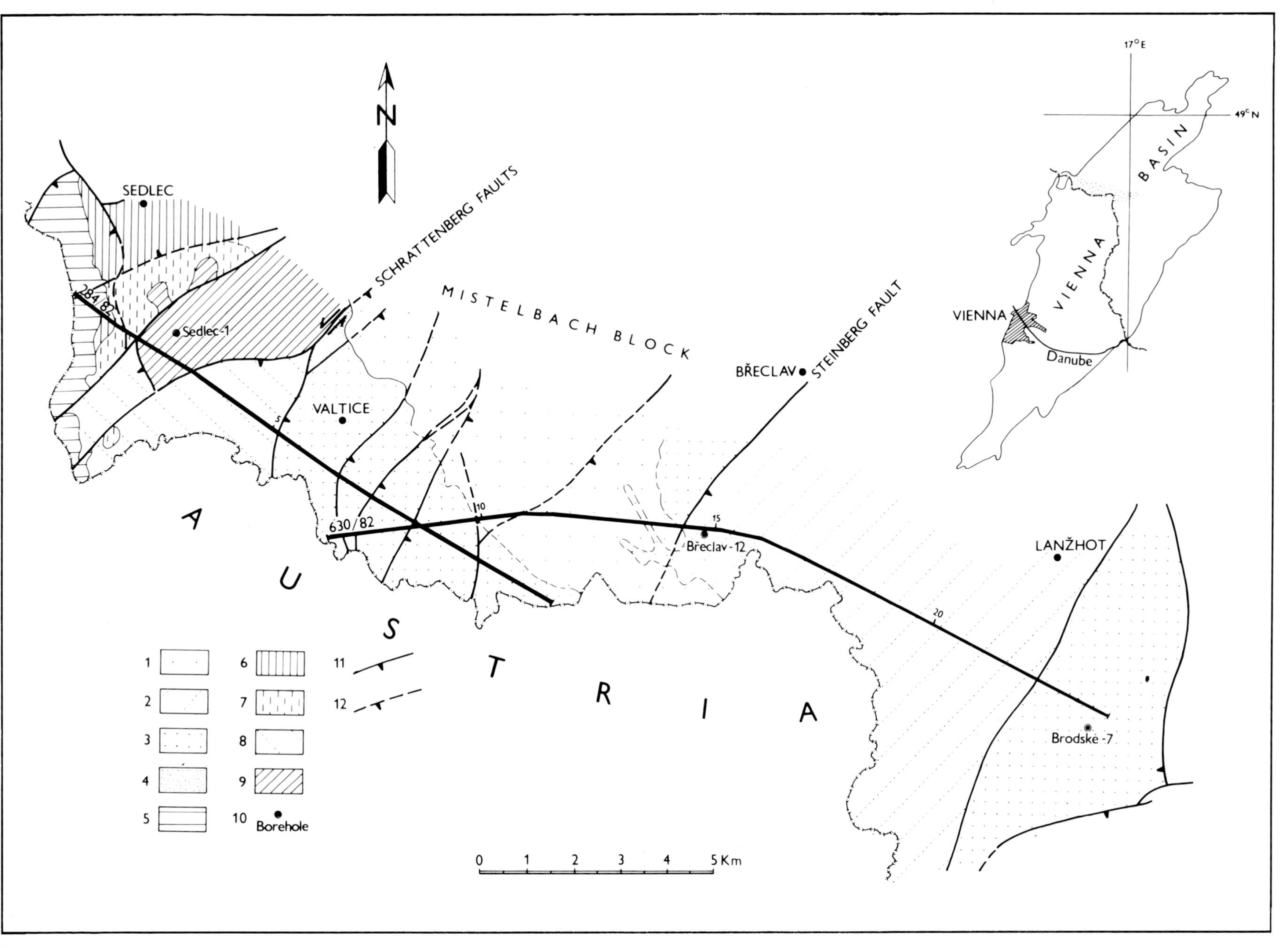

Figure 1A. Generalized geologic map showing rocks exposed at the surface of the western central part of the Vienna basin and positions of seismic reflection lines 284 and 630. (1) Levantian sedimentary rocks; (2) Pontian sedimentary rocks; (3) Pannonian age sedimentary rocks; (4) upper Sarmatian sedimentary rocks; (5) upper Badenian sedimentary rocks; (6) lower Badenian sedimentary rocks; (7) basal strata of the Badenian; (8) Eggenburgian sedimentary rocks; (9) Egerian sedimentary rocks; (10) borehole; (11) fault; and (12) inferred fault. (After unpublished report of Geofyzika Brno).

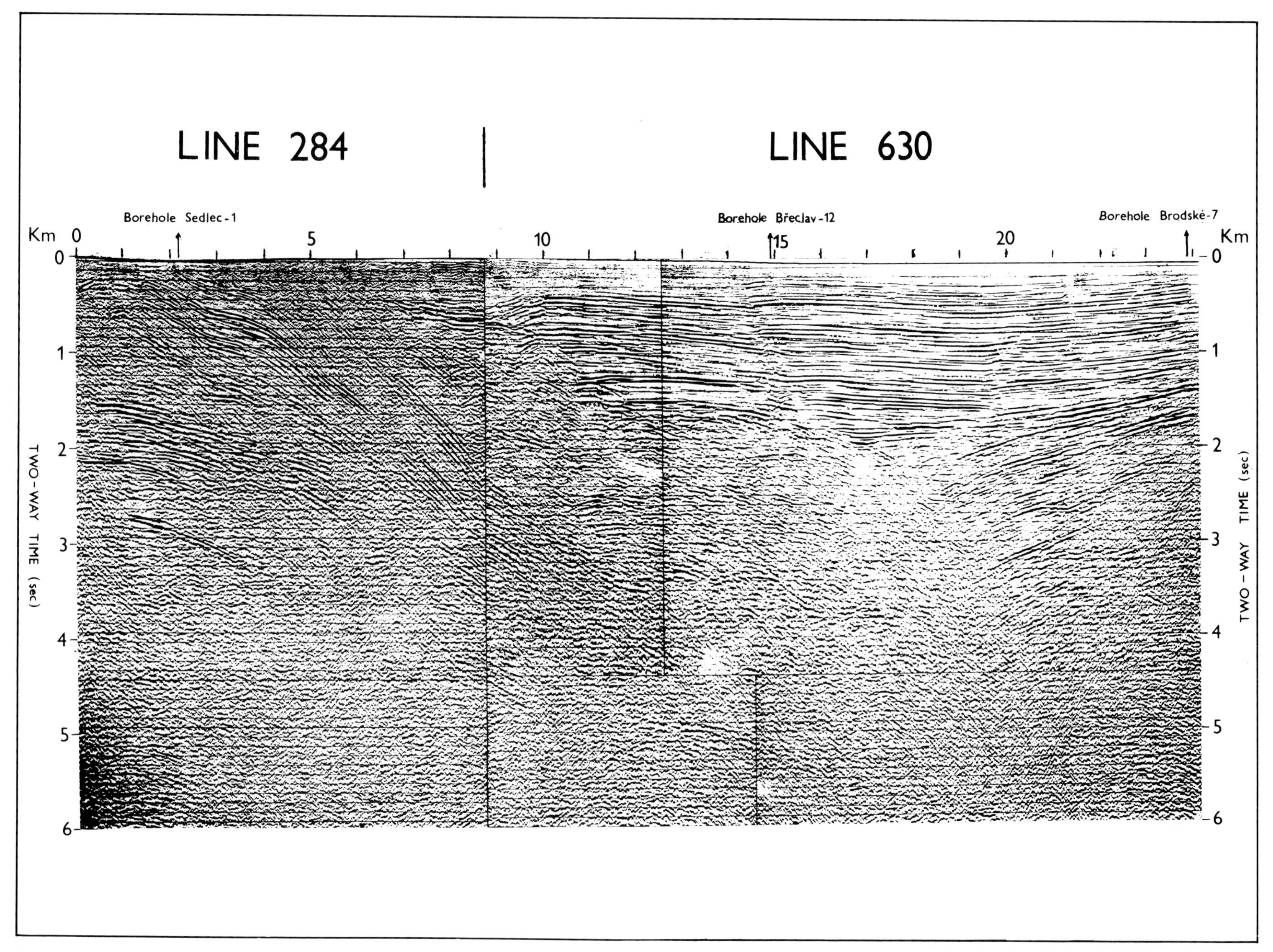

Figure 1B. Upper 6 sec of unmigrated seismic reflection profiles along lines 284 and 630. Vertical axis is two-way travel time. No attempt (such as by migration) has been made to place reflectors in their correct spatial location. Numbers along the top are kilometers. Tic marks are every 25 shot points. Section gives no vertical exaggeration for seismic velocity of 4.0 km/sec.

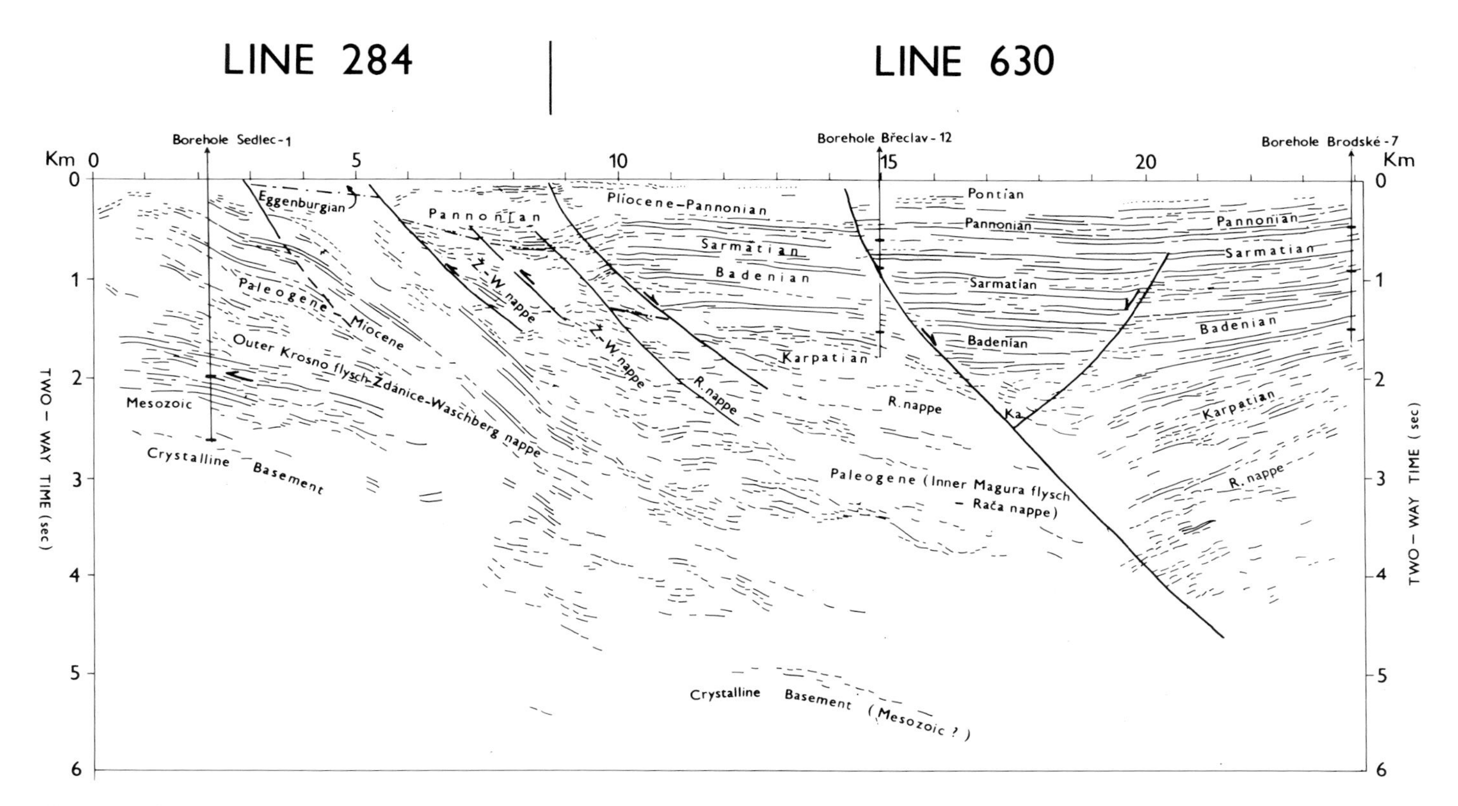

Figure 1C. Interpretation of principal reflectors on lines 284 and 630, prepared from unmigrated sections. Vertical axis is two-way travel time. Note the prominent east-dipping reflectors on the left side of the line. These may correspond to the outer (flysch) Carpathian thrust system. These reflectors are separated from other west-dipping reflectors by the Steinberg fault (km 14). The deepest reflectors may correspond to autochtonous crystalline basement. (The Neogene reflectors are modified slightly after Cahelova et al., unpublished report of Geofyzika Brno.)

Sarmatian to Recent (13.0–0.0 Ma) sedimentation was much slower than the preceding sedimentation. The total thickness of Sarmatian to Quaternary sediments reaches 1000–1500 m. Sedimentation has been and still is strongly tectonically controlled (for example, by the Steinberg fault, the Farsky fault, and so on).

The Danube basin lies southeast of the Little Carpathians and the East Alpine Leithagebirge and northwest of the Bakony and Gerecse Mountains (areas of exposed Mesozoic basement in western Hungary). The basement of the Danube basin is made up of the inner West Carpathian thrust complexes (Tatrides and Veporides) in the north (see, for example, Fusan et al., 1971) and by Austroalpine (and possibly also by Penninic) units in the south. The thickness of the Neogene sedimentary strata of the Czechoslovakian part of the Danube basin reaches 6 km near Gabcikovo (3.3 sec two-way time).

Extension within the Danube basin occurred primarily during Badenian and early Sarmatian time, when a large amount of andesitic volcanism occurred, probably along with strike-slip and dip-slip faulting. Badenian and early Sarmatian subsidence of the Danube basin is characterized by development of fault-bounded troughs. The thickness of Badenian–lower Sarmatian sediments in the central Danube basin trough reaches 2 km. Pannonian age and younger rocks within the Danube basin (except the westernmost part) are not restricted to fault-bounded troughs and onlap onto the pre-Neogene basement over a much wider area than that occupied by the Badenian and lower Sarmatian sedimentary rocks. Although the enormous thickness of Pannonian–Quaternary sedimentary rocks reaches 4 km, there has probably been only minor tectonic activity within the Danube basin in the last ten million years. It seems that this second stage of development of the Danube basin has been controlled by the cooling and thermal contraction of underlying crust and mantle after extension was mainly finished.

The Transcarpathian depression forms the most extensive promontory of the intra-Carpathian basin system into the northeastern part of the Carpathians. It acquired its present shape during upper Badenian–middle Sarmatian extension, which was accompanied by synsedimentary faulting, rapid subsidence and intense volcanic activity (Rudinec et al., 1981). The basement of the Transcarpathian depression belongs to the Paleo-Alpine (mid-Cretaceous) and Meso-Alpine (middle–late Paleogene) edifice of the Carpathians. Starting from the lowest structural units, the nappe pile is formed by the Zemplinicum, the Veporicum, and the Tatricum. The depth of the Transcarpathian depression in its deepest part is about 6 km.

Before the formation of the depression itself, relatively insignificant thicknesses of Egerian to middle Badenian sediments, which were related to acid volcanism, accumulated on the moving Carpathian thrust surface. In upper Badenian time, the main northwest trending fault-bounded trough formed. From upper Badenian (15–13.5 Ma) to the end of middle Sarmatian (11.5 Ma) time, a great thickness of sediments (3–4 km) accumulated in this trough. Formation of the trough was probably related to strike-slip movement along the Carpathian arc and

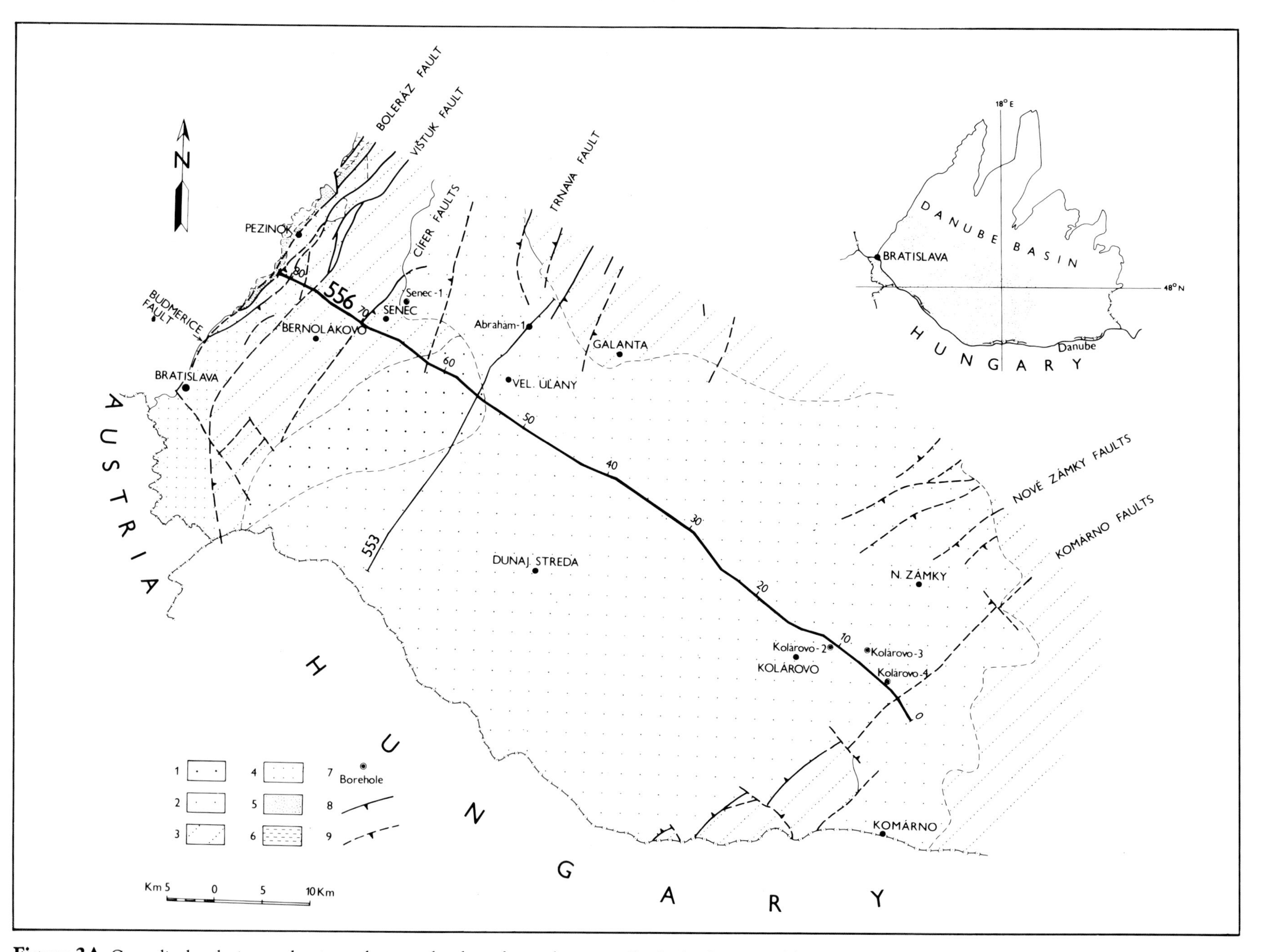

Figure 2A. Generalized geologic map showing rocks exposed at the surface in the western Czechoslovakian part of the Danube basin and the position of seismic reflection line 556. (1) upper Pliocene sedimentary rocks; (2) Levantian sedimentary rocks; (3) Pontian sedimentary rocks; (4) Pannonian age sedimentary rocks; (5) Sarmatian sedimentary rocks; (6) upper Badenian sedimentary rocks; (7) borehole; (8) fault; and (9) inferred fault. (After the Geological map of Czechoslovakia, 1962, scale 1:200,000.)

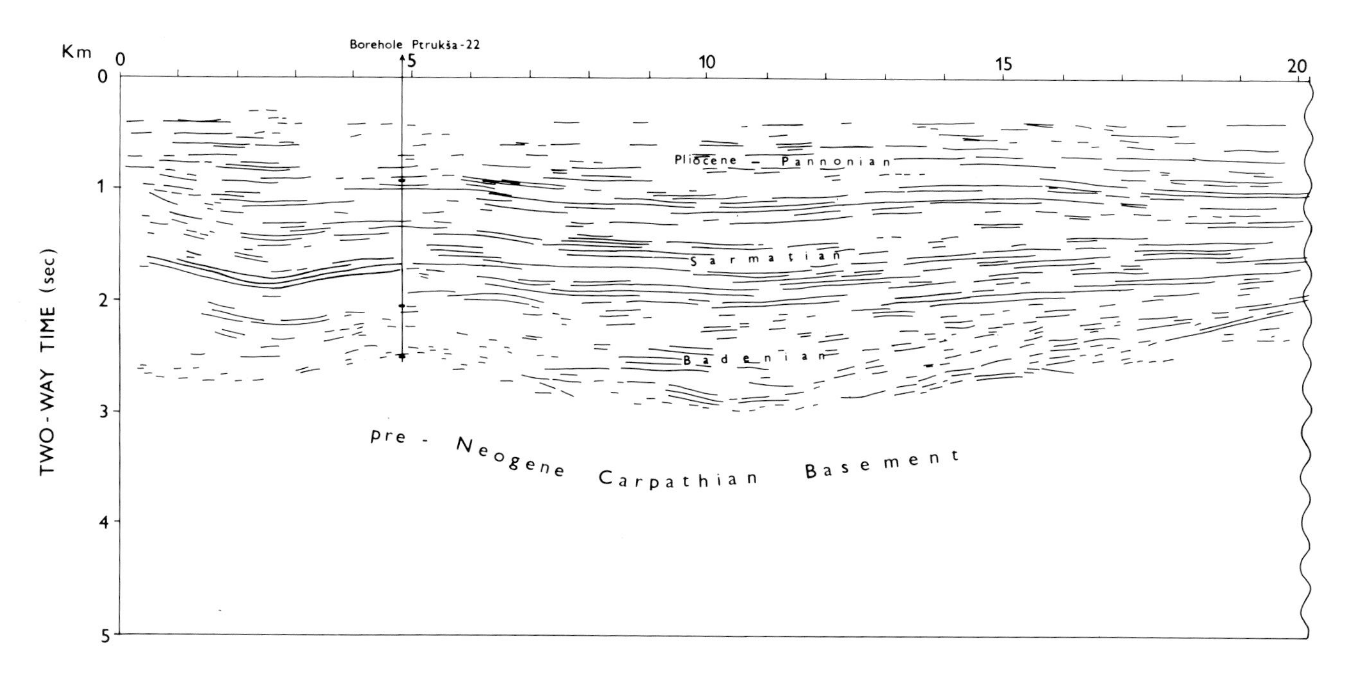

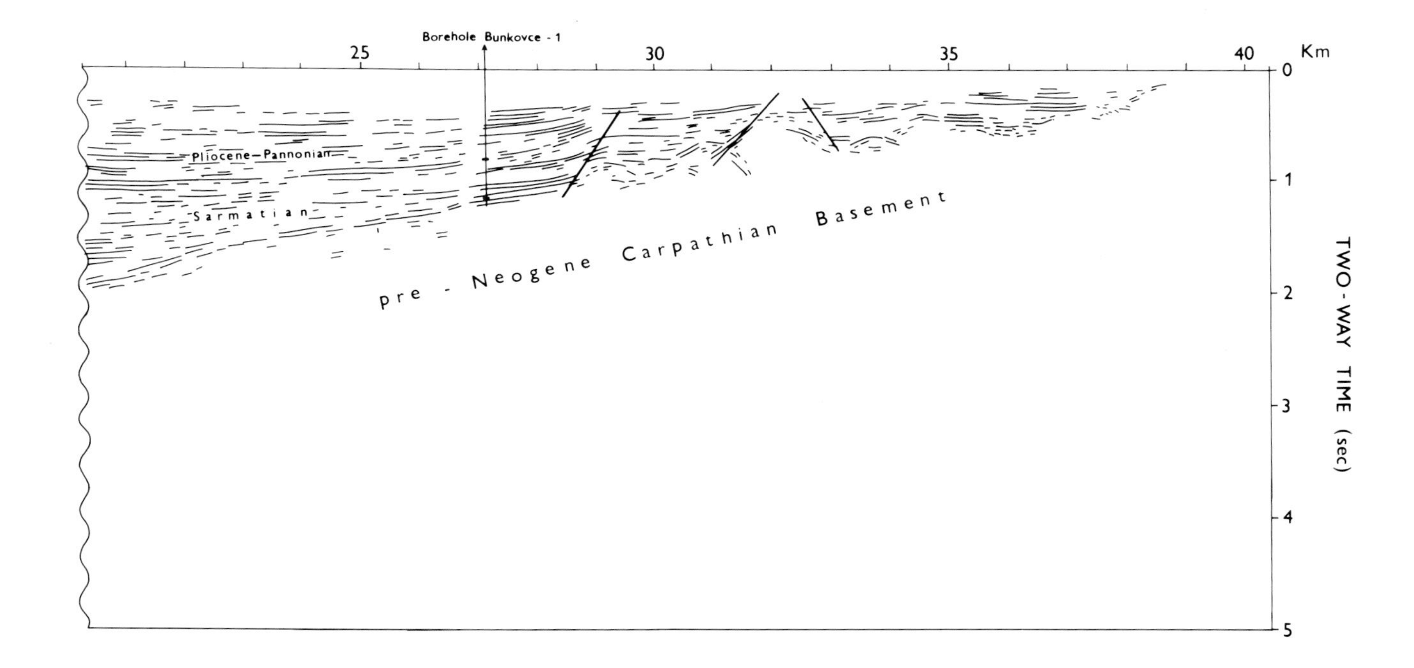

Figure 3C. Interpretation of principal reflectors on line 540, prepared from unmigrated section. Vertical axis is two-way travel time. Note only a few reflectors corresponding to Badenian and Sarmatian sedimentary rocks are warped or cut by minor faults. Reflectors corresponding to Pannonian age and younger sediments are subhorizontal and undeformed everywhere.

accompanied coeval thrusting in the East Carpathians (Royden et al., 1983a). Post-Sarmatian to recent sedimentation (10.5–0 Ma) was relatively quiet and only 1000 m of sediments were deposited. As in the Danube basin this late sedimentation was not restricted to the central trough area. The magnitude of post-Sarmatian subsidence was much smaller in the Transcarpathian depression than elsewhere, probably due to the proximity of the rapidly uplifting East Carpathians.

RESULTS

The Vienna Basin

An interpretative line drawing of the original unmigrated seismic sections from lines 284 and 630 (Figure 1B) is shown in Figure 1C. The steep dips of reflectors in the western part of the profile (beneath km 0–9) indicate that migration is badly needed and that the location of some steeply dipping reflectors has been shifted significantly. Nevertheless, even this unmigrated line should be adequate for the purposes of interpretation in this paper. The vertical and horizontal scales were chosen to minimize vertical exaggeration in the flysch parts of the seismic section (layers corresponding to average velocities of about 4.2 km/sec).

The relatively short seismic sections 284 and 630 are extremely complex at first glance and many strong reflectors can be identified on these lines. The section can be easily divided into two parts. The western (left) part between 0 and 8 km is dominated by the fairly steep east-dipping reflectors that flatten with depth. These reflectors are closely related to a series of thrust faults within the strongly imbricated outer Zdanice–Waschberg and inner Magura flysch zones. Flattening of the reflectors at 3–4 sec (km 7–10) demonstrates that the major, west verging thrust faults penetrated in boreholes and mapped at the surface in this part of the survey have listric geometries. Within the imbricate slices, reflectors are flatter and probably represent original layering of the flysch rocks. A major detachment surface separating allochthonous and autochthonous rocks was penetrated by borehole Sedlec-1 at 3489 m depth (Figure 1C). The autochton is formed of upper Jurassic platform deposits (limestones and marls). Beneath them, granitoids of the Precambrian Brunovistulicum (Dudek, 1980) were drilled at 4850 m depth. Both Mesozoic sediments and Precambrian granitoids remained undeformed during Neogene thrusting. Reflections from the master detachment are not very clear or distinct. Reflections from the boundary between Jurassic sediments and granodiorites are also weak. Continuous reflectors from both seismic boundaries are sometimes lacking and generally not very clear. We think, however, that the gently curved, east-dipping reflector beneath 12 and 16 km at 5.2–5.4 sec comes from the crystalline basement or, less probably, from Jurassic autochthonous sediments. Under other parts of the West Carpathians we found reflectors from Brunovistulic crystalline rocks at the same depth and at the same distance internal to the foreland. This is important because it implies that the depth to the base of the Carpathian thrust complex is not affected by the presence of the Neogene Vienna basin, and that autochthonous basement continues relatively undisturbed beneath the basin.

East of km 8.7 on line 630, the seismic character in the upper two seconds of the section changes drastically (Figures 1B and 1C). Unlike the seismic section within the frontal thrust zone to the west, strong continuous horizontal or only gently dipping reflectors are present. These reflectors appear east of the westernmost fault of the Schrattenberg fault system at km 5.2, but only at shallow depths. The main events in the central part of the Vienna basin begin beneath km 8.7 east of the easternmost faults of the Schrattenberg system.

Gently east-dipping reflectors, between the Schrattenberg faults and the Steinberg fault beneath km 8.7–15.2, represent Neogene fill of the Mistelbach block. Karpatian sedimentary rocks beneath this block are more tilted than younger sedimentary strata. This seismic section, combined with drillhole data shows that, from km 8.7 to the eastern end of line 630, Karpatian and Badenian subsidence was relatively rapid, Sarmatian subsidence was slow and Pannonian to Recent subsidence was very slow (Table 2).

The Steinberg fault is clearly identifiable on the seismic section as a listric normal fault (Figure 1B and 1C). The Steinberg fault truncates east-dipping reflectors in Raca nappe slices in the basement west of the fault. The same reflectors (extrapolated from surface mapping) dip west on the east side of the Steinberg fault. This suggests that not only dip-slip but also strike-slip movement probably occurred on the Steinberg fault. Significant strike-slip motion has been inferred by Roth (1980) to occur only along the Schrattenberg fault, but we think that left-slip occurred along both faults. In fact, we think that most of the Neogene left-slip within the Carpathian allochthon occurred along the Steinberg fault.

The Danube Basin

Profile 556 displays excellent reflections, which not only constrain the thickness and structure of Neogene sedimentary rocks, but also show low-angle normal faults of Neogene age, within the Tatric basement in the northwest part of the section (Figures 2B and 2C). Between km 60–66, the Moho occurs at 9.2–9.3 sec (not shown here), corresponding to depths of 26–27 km. Line 556 runs generally east southeast and reflectors are tied to boreholes Kolarovo 2, 3, and 4 in the eastern part of the section (Table 2). Except for about 200 m of Badenian sediments, only Pannonian age sediments and an extremely thick sequence of Pontian and younger sediments are present. The reflectors within the sedimentary sequence dip gently west. The thickest section of Pannonian age and younger sediments occurs beneath km 25 (at 3 sec or about 4 km depth). From this point, the Pannonian–Quaternary section thins westwards toward the Little Carpathian foothills. There are probably no Sarmatian rocks present east of km 35. There is no disruption of the Pannonian–Quaternary section by normal or reverse faults east of km 64.

An important trough 15 km wide is present between km 33 and km 48. We interpret this fault-bounded trough as a Badenian–lower Sarmatian structure filled by sediments and andesitic volcanic rocks, as indicated by seismically transparent zones between km 33 and km 43. A significant northeast-trending magnetic anomaly occurs above this trough, with amplitude more than 100 nT. Pannonian–Quaternary sedimentary rocks onlap the trough sediments unconformably.

Table 2. Sedimentation data for 10 selected wells in the Vienna basin, the Danube basin, and the Transcarpathian depression

	Depth to Base (m)									
	Breclav −12	Brodské −7	Grob −1	Senec −1	Abrahám −1	Kolárovo −2	Kolárovo −3	Kolárovo −4	Bunkovce −1	Ptruksa −22
Quaternary								50		
Pliocene				500	890			1060		
Pontian			240	784	1060	1395	1190	1170		
Pannonian	540	373	480	1238	1370	2950	2600	2500	800	880
Sarmatian	890	920	708	1875	1703	3048	2690	2640	1215	2707
Badenian	1865	1720	1283	2540	2239	—	—	—	—	3607
Karpatian	2204	1955	—	—	—	—	—	—	—	—
Basement	—	—	1336	2579	2251	3185	2834	2665	2100	3713

The Badenian–Pannonian section is offset by a listric normal fault beneath km 64 that extends into the basement (see following text). We infer a steeply dipping thrust fault of probable Pliocene age beneath km 70, with about 500 m of thrust separation. Our interpretation of both faults is controlled by boreholes Grob-1, Senec-1, and line 553, which is tied to borehole Abraham-1 (Table 2). To our knowledge, this thrust fault is the first well-documented Pliocene compressional event within the Pannonian area and may have a left-slip component of displacement as well. In addition to the faults within the sedimentary section described above, the westernmost part of section 556 contains three significant and continuous pre-Neogene reflectors. The first can be traced beneath km 55 to km 65 at 4–4.5 sec (approximately 9–10.5 km). The second continuous reflector dips west-northwest and is clearly identifiable between km 57 and km 80 between 2 and 4.4 sec (3–11 km). Note that these reflectors are probably cut by the Pliocene thrust fault inferred above. The third reflector can be traced beneath km 69–80 at 2.2–3.3 sec (approximately 3.3–8 km).

We think that the first of these reflectors may represent a thrust detachment within the Tatrides (in the sense of Andrusov, 1968). The second and third of these reflectors are probably the detachment surfaces of Badenian–Pannonian low-angle normal faults. These gently dipping faults were active during the formation of the horst and graben structure in the Danube basin, such as the Little Carpathians, Inovec and Tribec horsts. The entire system is structurally and seismically similar to that of the Basin and Range Province in the western U.S. (for example, Allmendinger et al., 1983 and Wernicke and Burchfiel, 1982).

The Transcarpathian Depression

Line 540 mainly exhibits gently dipping or horizontal reflectors within the Badenian to Pliocene fill of the central and northern part of the Transcarpathian depression (Figures 3B and 3C). Interpretation of these reflectors was facilitated by two boreholes, Ptruksa-22 and Bunkovce-1 (Table 2). The main part of the sedimentary fill is upper Badenian to middle Sarmatian in age. Beds are gently dipping and are cut by many minor normal faults, mainly in the upper Badenian (not interpreted here). Lower and middle Sarmatian sedimentation was quieter and was accompanied by eruption of a large volume of andesitic volcanic rocks in the Vihorlat Mountains and also in the main upper Badenian–middle Sarmatian trough. All upper Badenian–middle Sarmatian subsidence was confined to the main northwest-trending fault-bounded trough (see Rudinec et al., 1981).

Line 540 shows only the central and northern part of the trough. Its deepest part occurs beneath km 11 where about 3 km of upper Badenian–Sarmatian sedimentary rocks are present. The central trough of the Transcarpathian depression ends beneath km 29. The southern end of the trough cannot be seen on line 540.

The flat-lying Pannonian age and Pliocene rocks unconformably overlie older Miocene and basement rocks. These young rocks are not confined to fault-bounded troughs (the same thickness of post-Sarmatian was penetrated by wells Ptruksa-22 and Bunkovce-1). Pannonian and younger subsidence, however, was controlled by normal faults between km 29 and km 39, near the uplift of the Vihorlat–Gutin volcanic mountains.

CONCLUSIONS

Details of the interpretation of the seismic reflection profile described above have been presented in the context of detailed borehole data, surface geologic data, and other geophysical results (including other seismic lines). Interpretation of the Vienna basin, the Danube basin and the Transcarpathian depression provides the following conclusions:

1. The Vienna basin is probably a pull-apart feature that represents thin-skinned extension, as suggested by Royden et al. (1983a). The whole Carpathian thrust system is probably uniformly thick (about 10–12 km) beneath the central part of the Vienna basin and the central part of the flysch belt to the northeast. Thrust faults within the Carpathian flysch slices at about 6–7 km depth west of the Vienna basin have a listric geometry.
2. The Steinberg fault, and to a lesser extent the Schrattenberg fault system, probably accommodated Badenian left-slip combined with late-stage dip-slip. The subsidence of the Vienna basin was not very rapid during Sarmatian time and was relatively slow after the Sarmatian. All the subsidence in the Vienna basin has been tectonically controlled. We think that the later normal faulting, which is active even now on the Steinberg fault and some other faults, was to some extent accompanied by strike-slip movement even after Badenian time.
3. The northeastern Czechoslovakian part of the Danube basin contains reflectors not only from Neogene sedimentary

layers but also from the buried Tatrides (northern part of the inner West Carpathians). The reflectors in the buried Tatrides dip northwest and are considered to be detachment surfaces of Badenian–Pannonian low-angle normal faults. The resulting structures are similar to those of the Basin and Range Province in the western United States.

4. A significant northeast-trending, fault-bounded trough formed during Badenian (and lower Sarmatian) time beneath the central part of the Danube basin. The trough is about 2 km deep and filled with sediments and andesitic volcanic rocks that appear as seismically transparent zones.

5. The Pannonian–Quaternary sedimentary rocks in the Danube basin are 3 to 4 km thick and onlap unconformably onto the Badenian (and Sarmatian) sediments. Pannonian–Quaternary subsidence was not tectonically controlled except near the Little Carpathian and Inovec horsts where young, normal faults exist. One of the most surprising results of this survey was the existence of a Pliocene thrust fault in the northwest part of the basin.

6. Like the Danube basin, the Transcarpathian depression also exhibits two sedimentary complexes with different structures. The first consists of rapidly deposited, upper Badenian–middle Sarmatian sediments, and these sediments are confined to a northwest-trending, fault-bounded trough. The second consists of Pannonian and younger sedimentary rocks, which are not restricted to this trough. Except near its northern boundary, deposition of these rocks was not tectonically controlled.

7. Although the histories of subsidence and sedimentation in the Danube basin and the Transcarpathian depression are roughly similar (consistent with the interpretation of Royden et al., 1983a, b), significant differences exist between the evolution of these basins. The first phase of subsidence (in fault-bounded troughs), is dominant in the Transcarpathian depression, with about 3 to 4 km of sediments deposited in a large trough. This phase occurred sooner in the Danube basin than in the Transcarpathian depression (by about 2 m.y.), but the trough had a smaller areal extent and subsidence was slower. In contrast, the second phase of subsidence (controlled by cooling and thermal contraction of underlying crust and mantle) was much faster in the Danube basin than in the Transcarpathian depression. We cannot explain these differences (see Royden and Dövényi, this volume).

8. The seismic lines interpreted in this paper provide good evidence of extensional styles in three of the intra-Carpathian basins (the Vienna basin, the Danube basin, the Transcarpathian depression). While extension of the Vienna basin was probably restricted to the Alpine–Carpathian allochton that overlies the Precambrian Brunovistulicum and its Jurassic cover, extension of the other two basins probably involved rocks at deeper crustal levels and in the upper mantle.

ACKNOWLEDGMENTS

We wish to thank Petr Sedlák and Libuse Dvoraková for their help in the early stages of this work in 1983 and 1984. They contributed greatly to the interpretation of the seismic sections. We also thank Leigh Royden and Arnošt Dudek for their helpful comments.

REFERENCES

Allmendinger, R. W., J. W. Sharp, D. Von Tish, L. Serpa, L. Brown, S. Kaufman, J. Oliver and R. B. Smith, 1983, Cenozoic and Mesozoic structure of the eastern Basin and Range Province, Utah, from COCORP seismic reflection data: Geology, v. 11, p. 522–526.

Andrusov, D., 1968, Grundriss der tektonik der Nordlichen Karpaten: Slovak Academy of Sciences, Bratislava, 188 pages.

Dudek, A., 1980, The crystalline basement block of the outer Carpathians in Moravia: Brunovistulicum, Academia, Prague, 85 pages.

Fusan, O., J. Ibrmajer, J. Plancar, J. Slavik, and M. Smisek, 1971, Geological structure of the basement of the covered parts of southern part of inner West Carpathians: Zbornik geologickych vied, Zapadne Karpaty, v. 15, p. 5–173, Bratislava.

Geological Map of Czechoslovakia, 1962, Central Geological Prague, Prague.

Jiricek, R., and C. Tomek, 1981, Sedimentary and structural evolution of the Vienna basin: Earth Evol. Sci., v. 1(3), p. 195–204.

Roth, Z., 1980, West Carpathians–Tertiary structure of central Europe (in Czech): Library of the Central Geol. Inst. Prague, v. 55, 128 pages.

Royden, L, F. Horváth, and J. Rumpler, 1983a, Evolution of the Pannonian basin system: 1. Tectonics, v. 2, p. 63–90.

Royden, L., F. Horváth, A. Nagymarosy, and L. Stegena, 1983b, Evolution of the Pannonian basin system: 2. Subsidence and thermal history: Tectonics, v. 2, p. 91–137.

Rudinec, R., C. Tomek, and R. Jiricek, 1981, Sedimentary and structural evolution of the Transcarpathian depression: Earth Evol. Sci., v. 1(3), p. 205–211.

Wernicke, B. and B. C. Burchfiel, 1982, Modes of extensional tectonics: Jour. Structural Geol., v. 4, p. 105–114.

CHAPTER 15

Seismicity and Neotectonics of the East Alpine–Carpathian and Pannonian Area

R. Gutdeutsch and K. Aric
Institut for Meterology and Geophysics,
Univ. Vienna, A-1090 Vienna, Austria

A map of epicenter locations of the Eastern European Alpine system was compiled and, together with focal solutions and geologic data for recent deformation, was used to relate recent seismicity to a regional tectonic system. The seismicity data show a scattered distribution for most of the Carpathian–Pannonian region but, along the western edge of the Pannonian basin and in the Eastern Alps and Dinarides, several distinct seismic zones can be recognized. A linear zone of seismic activity in the Eastern Alps (Mur-Mürz-line) strikes northeast into the southern Vienna basin. Focal mechanisms indicate left-slip along this zone, and the southern Vienna basin may be extending east-west as a pull-apart basin. To the southwest, this transform-like fault zone and its associated seismicity end abruptly in two north-south trending grabens that exhibit east-west extension (Lavant and Metnitz Valleys). To the south, these grabens end in Yugoslavia in a dextral strike-slip fault system that trends southeast along the strike of the Dinaric Alps. A left-stepping offset between two dextral fault segments in this system results in roughly north-south crustal shortening and uplift near the ends of the two fault segments (Medvednica zone).

We show that a simple rigid block model can explain the most active seismic zones in the region, and most of this seismicity and associated deformation are consistent with regional north-south compression. From this we infer that roughly northward displacement of the Adriatic block relative to Europe may be the driving force behind recent deformation and seismicity within this part of the Alpine system.

INTRODUCTION

The East Alpine–Carpathian and Pannonian area comprises a portion of the collision zone between the Eurasian and African continent. At present, this area exhibits low seismic activity relative to other parts of the collision zone and earthquakes occur mainly in the upper crust. This, together with the complicated regional geologic history, makes it difficult to deduce the tectonic significance of the epicentral distribution. Seismological observations and geologic evidence suggest that the crust in this area has been disintegrated into crustal and lithospheric fragments of different sizes. The relative motions of these small fragments, which have been defined on the basis of seismological observations and geologic data alone, have not yet been satisfactorily explained by a workable tectonic model, As yet, only a few papers have attempted to use plate tectonic concepts to explain the distribution of seismic activity within various parts of the East Alpine and Pannonian area (for example, Beer and Szukin, 1978; Aric and Gutdeutsch, 1981). In this chapter, we attempt to construct a mechanical model to explain the seismicity of the East Alpine and Pannonian region that is consistent with tectonic theory and geologic observations.

PLATE TECTONICS AND SEISMICITY IN THE EAST ALPINE–CARPATHIAN AND PANNONIAN REGION

Plate tectonics was originally developed to explain the most active seismic and tectonic zones in the world, those associated

Date	Epicenter		Location	
	i	*q*		
1. 6 May 1976	13°18′	46°18′	Friuli, Northern Italy (Muller, 1977)	
2. 16 April 1972	16°09′	47°44′	Seebenstein, Austria (Gangl, 1975a,b)	
3. 29 Jan. 1967	14°19′	45°53′	Molln, Austria (Drimmel and Trapp, 1975)	
4. 12 Jan. 1956	19°04′	47°30′	Dunaharaszti, Hungary (Csomor in Gangl, 1975b)	
5. 22 June 1978	21°08′	46°47′	Békés, Hungary	D. Csomor and
6. 22 June 1978	21°02′	46°44′	Békés, Hungary	F. Horváth,
7. 26 Sept. 1978	18°51′	47°13′	Dunaharaszti, Hungary	pers. comm., 1984

with sea floor spreading, mountain building and volcanism. Within plate tectonic theory, deformation is concentrated along the boundaries of rigid plates of continental dimensions. There are three main reasons why the extension of plate tectonics to the East Alpine–Carpathian region should be performed with great caution. First, the earthquakes occur at shallow depths, so that seismic energy is only released in the crust. Second, the assumption of rigid plates, commonly used in plate tectonics, implies that the stress applied to a plate is completely converted into strain on preexisting plate boundaries. In this model any effect of internal deformation within the plates or crustal fragments is ignored. For example, a rigid block model probably does not hold for the Pannonian region which has undergone widespread deformation in recent geologic time (see also Burchfiel, 1980). Another consequence of the rigid plate concept is that plate boundaries and the associated seismic zones must form closed lines on the earth's surface. This should not be expected in the East Alpine–Carpathian area because distinct fragment boundaries may end in areas of flexure or complex and widespread faulting. Third, much more than 90 percent of global seismicity is associated with such highly active seismic zones of the earth as the circum-Pacific or the Himalayan–Mediterranean belts. Other areas such as the East Alpine–Pannonian region are seismically much less active and provide much less data. Very few events in this region are strong enough to give reliable estimates of important focal parameters and focal solutions. The scattered nature of the seismic activity makes it difficult to correlate epicenters uniquely with fault systems. One would need another thousand years of observation to gain a model of seismic activity with a statistical significance comparable to those established for the most active seismic zones of the world.

SEISMOTECTONIC DATA

Map 7 (in enclosure) shows the epicenters plotted onto the main fault systems which have been redrawn from the tectonic map of the Carpathian–Balkan Mountain system and adjacent areas (Mahel, 1973). The relationship between fault systems and earthquake zones is not one to one. Inactive faults are not now the site of earthquakes and may never be again (e.g., Bolt, 1978). Thus, the faults shown on a tectonic map can be used only as an aid to understanding the present earthquake pattern. Some of these faults may be still active, and the sense of displacement along some inactive faults may be helpful in explaining related active faulting and seismicity. However, many inactive faults may have nothing to do with the present tectonic system or seismic activity.

We will use the term "seismic lineament" in this paper in the sense of Shebalin et al. (1974). Seismic lineaments are elongated zones where earthquake epicenters are concentrated. Shebalin et al. (1974) also define the intersection of seismic lineaments as "knots," and propose that these knots are sites of enhanced earthquake risk. The term "tectonic lineament" will be used for large deep fault zones, flexures or systems of grabens, horsts and faults of regional dimensions (for example the Periadriatic lineament or the Elbe lineament). The term does not imply seismic activity. A tectonic lineament may or may not be seismically active.

It should be emphasized that the epicentral distribution data given in Map 7 are not homogeneous. The number of epicenters in Italy is much greater than indicated in Map 7 (Caputo and Postpischle, 1972). In Yugoslavia only earthquakes with magnitudes greater than 4.1 have been plotted between 45° and 46.5°N. The inhomogeneous nature of this data, however, does not seem to be a serious drawback, as the main features of the epicentral distribution are shown rather clearly. Therefore, we will use this map as the basis of our seismotectonic analysis.

INTERPRETATION OF SEISMIC PATTERNS

In Map 7 we have avoided interpretating the epicentral pattern in order to present an unprejudiced map of seismic activity. Figure 1 shows tectonic lineaments suggested by various authors (see following text), as well as our own interpretation of seismic lineaments. Our interpretation will be discussed later within a plate tectonic framework.

Seismicity in the Dinaric Area

Roughly northwards translation of the Adria Plate in recent time has been well documented (for example, McKenzie, 1972; Udias 1975; Ritsema 1974, 1975), These authors suggest that the present plate boundary between Adria and Europe runs roughly 40 km south of, and parallel to, the Periadriatic lineament (PE) in the Southern Alps. To the east its strike bends around to east-southeast at about 14°E longitude. Müller (1977) determined a thrust-type focal mechanism for the large Friuli earthquake of May 6, 1976 (magnitude 6.4). The rupture plane dips 18° northward below the southern Alps (Figure 2). As earthquakes of this magnitude occur extremely rarely in this region, this information is important in providing a tectonic framework for this area. Ritsema (1974) studied six focal solutions of earthquakes in the Dinaric area, north of latitude 45°. He interpreted these, together with many other first arrival

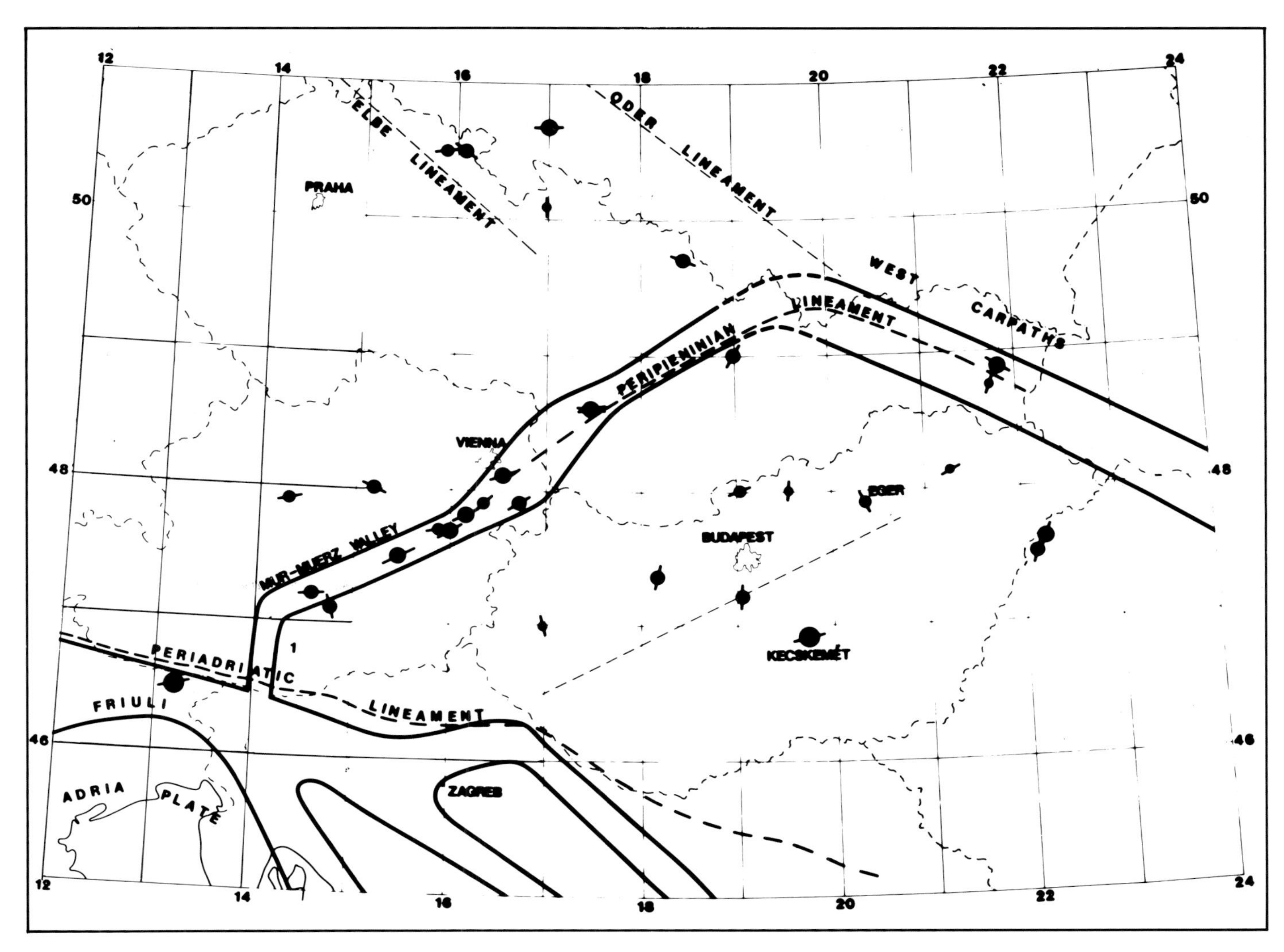

Figure 1. Main seismic lineaments inferred for Alpine–Carpathian–Dinaric region from Map 7 (bounded by dark lines). Dashed lines show a few tectonic lineaments. Solid circles with bars show epicenters and direction of the inner isoseismals derived from Prochaskova and Karnik (1978); Drimmel (1980); and Kowatsch (1911).

data, by constructing a composite focal solution for a larger area that includes the central part of Yugoslavia. From these data he concluded that the pressure axis trends predominantly northeast along the strike of the Dinarides. It turns slightly northward in the southern and eastern Alps. A northeast trending pressure axis in the Dinarides southeast of Zagreb was also found by Gangl (1975a) who studied focal solutions of two earthquakes at Banja Luka (Oct. 26–27, 1969). These solutions indicate a horizontal strike-slip displacement as opposed to a reverse or normal vertical rupture.

The seismic lineaments then form a wide zone that follows the coast of the Adriatic sea. Three more or less parallel seismic lineaments can be proposed within this zone (compare Map 7 and Figure 1). Near Zagreb, these lineaments are connected by a very active, east-west trending, seismic lineament (the Medvednica zone), that does not follow the southeast strike of the Alpine zone (see also Ribaric, 1982). To the east along the strike of this elongated zone, it passes into the central Pannonian plain where the seismicity is rather low. A detailed study of the active Medvednica zone was made by Skoko and Cvijanovic (1978). They argue that the present rapid uplift of the Medvednica mountains known from leveling measurements (Joo et al., 1981) reflects northwest directed compression. The rate of uplift increases from southeast to northwest and reaches a maximum roughly 100 km west of Zagreb.

Seismicity in the East Alpine–West Carpathian Area

North of latitude 46.5°, the seismicity drops drastically to much lower levels. The stress drop per year of the most active seismic zone north of 46.5° is less than 20% of the stress drop south of 46.5° in the Southern Alps (Aric, 1981). We interpret the seismicity in this northern region to reflect the collision of the Adria plate with the Eurasian continent, but only in an indirect way. The clearest seismic lineament runs from roughly 47°N, 14°E to 49°N, 19°E (Map 7 and Figure 1). It passes

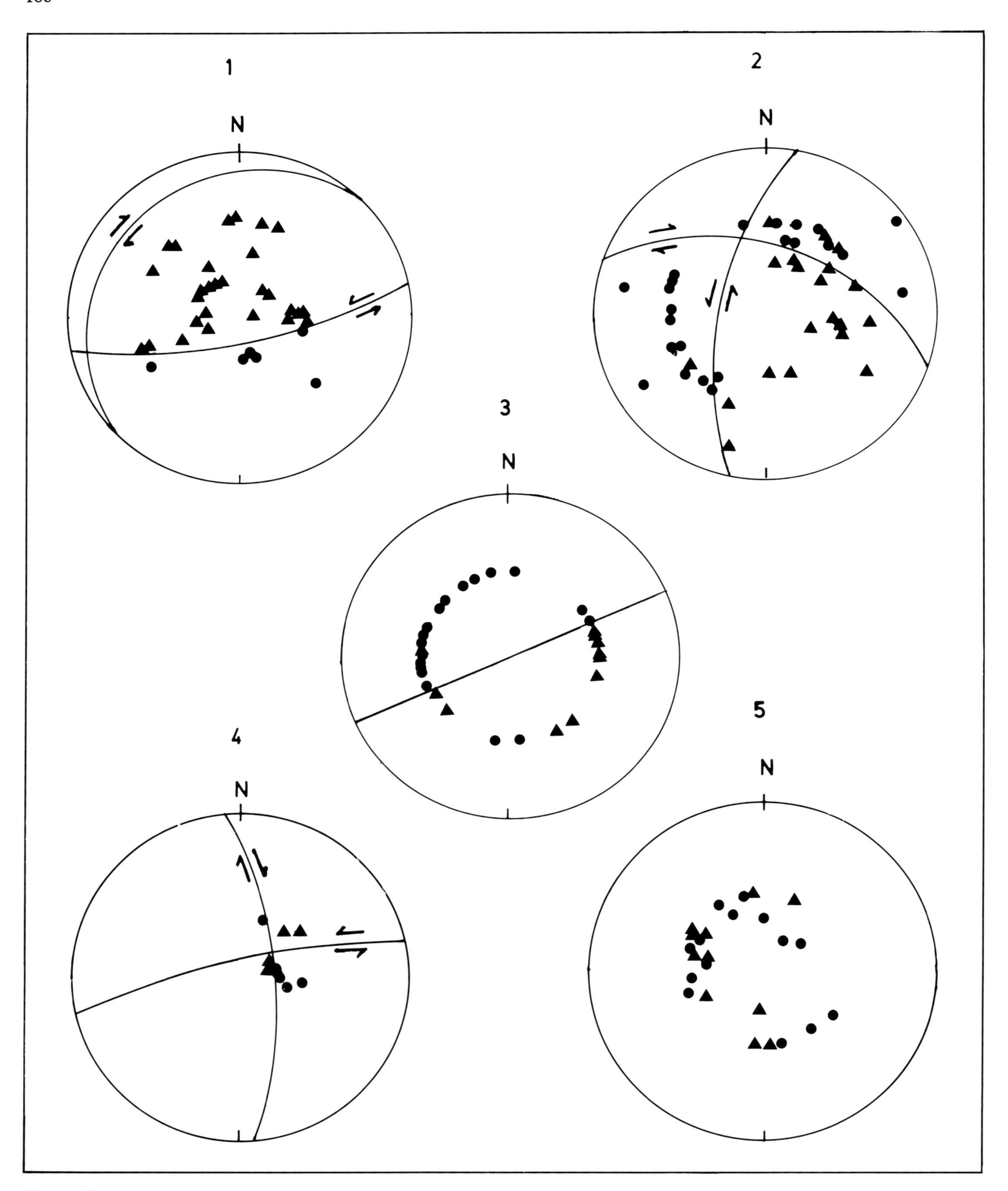

Figure 2. Focal solutions and first arrival data from earthquakes in the East Alpine and Pannonian area. Solid circles indicate dilatation, and triangles indication compression.

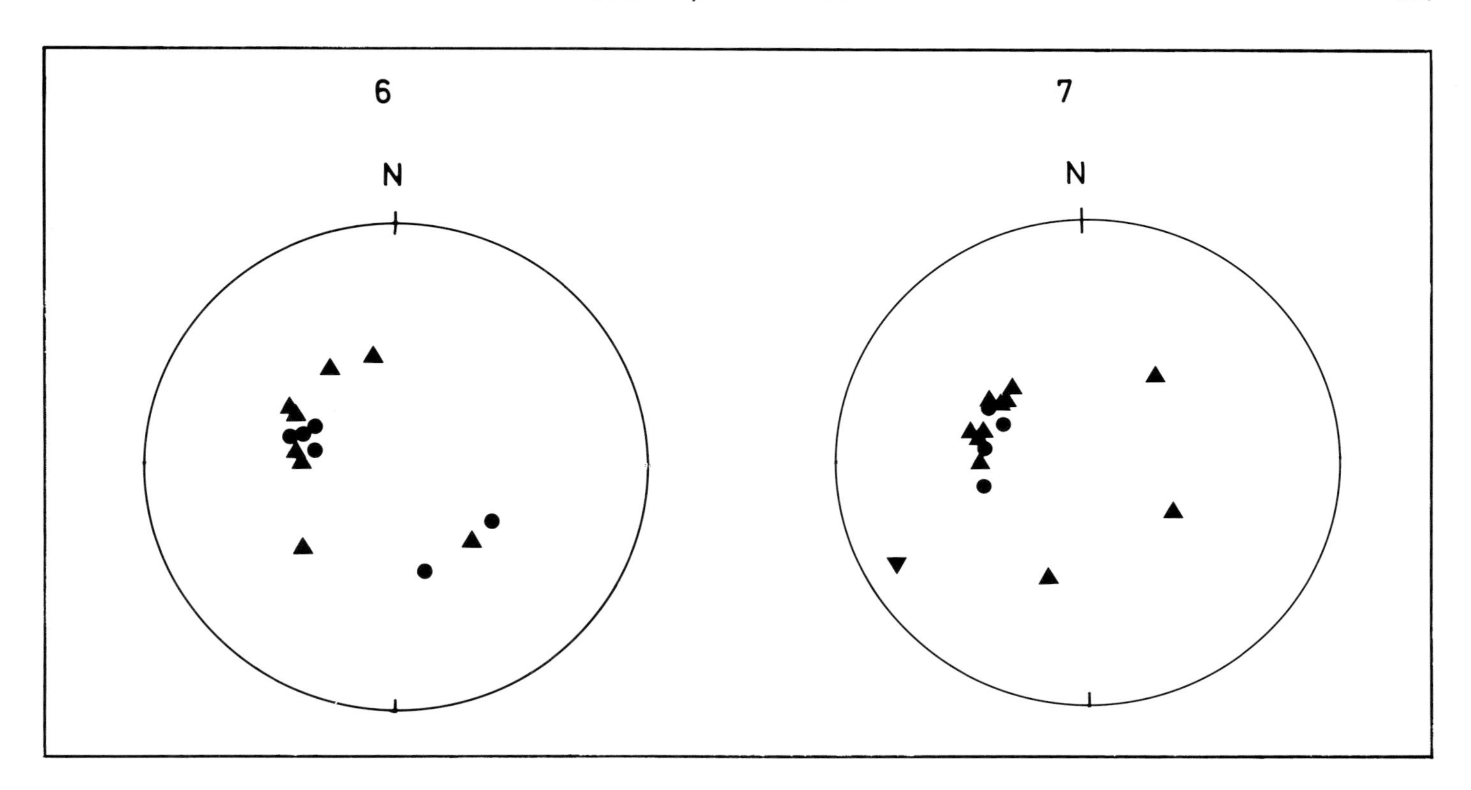

through the Mur–Mürz valley (MM), the Semmering mountains (S), the east Vienna basin (VB) the Little Carpathians and then runs roughly parallel to the strike of the Western Carpathians (see also Drimmel, 1980). Schenk et al. (1982) associate this seismic activity with the peri-pieninean lineament (Pieniny Klippen belt), The seismic pattern does not show clearly whether this seismic lineament has a continuation to the southwest in the North Italian earthquake zones (for example, Friuli). Schneider (1968) claims that this seismic lineament continues until Verona about 150 km to the southwest of Friuli. Aric (1981) finds strong arguments for a direct correlation between earthquakes north and south of the Periadriatic lineament from their energy build-up curves, which exhibit a similar structure.

Gangl (1975a,b) studied five focal solutions from the Vienna basin and found nearly vertical nodal planes. Three solutions have nodal planes that strike northwest and southeast with the pressure axis oriented north-south. Two solutions for earthquakes further to the southwest show north and east trending nodal planes with the pressure axes oriented northwest-southeast. Gangl (1975b) interpreted these two groups of earthquakes as unrelated phenomena, but in our view they are all related to the northeast trending zone of seismic activity south of the Vienna basin (Figure 2). The scatter in the orientation of the nodal planes probably results from the complex nature of the fault zone. That is, displacement may be taken up on many small faults that are not completely parallel, rather than on a single and continuous fault, The focal mechanisms of Gangl (1975b) indicate a more or less east-west direction of extension, and first arrival data of many smaller shocks in the same region are consistent with this interpretation (Aric and Gutdeutsch, 1981). This east-west extension is also consistent with the composite focal solutions of Ritsema (1974). The seismic lineament runs along the southeastern border of the Vienna basin. Leveling measurements (Senftl in Gangl, 1975b) and the occurrence of thermal springs in the Vienna basin indicate a subsiding neotectonic structure.

The southern end of this seismic lineament is at about 14 °E. A gap with low seismic activity occurs in the Metnitz and Lavant valleys, which strike roughly north-south, perpendicular to the strike of the seismic zones (M and L, Figure 3), Tollman (1970) interprets these valleys as graben structures, where vertical displacements of 5000 m have been observed. These are bounded by a system of en echelon north-south striking normal faults. The east-west directed extensional character of these graben areas is obvious.

Unfortunately, focal solutions of earthquakes along the eastern part of the West Carpathians are not available. The main direction of the inner contours of constant seismic intensity (isoseismals) roughly follow the direction of the ruptured fault (Shebalin et al,, 1976). However only two earthquakes have yielded reliable contour directions, as plotted in Figure 2. They do not show a clear orientation, in strong contrast to the inner isoseismals of earthquakes in the Mur–Mürz valley, which are clearly aligned in an east-northeast direction. It is tempting to associate this seismic activity in the West Carpathians with tectonic lineaments such as the Oder- or Elbe-lineament. (Schenk et al., 1982). The scatter and paucity of the epicentral distribution, however, is so great that this cannot be determined. The active zone along the eastern part of the West Carpathians seems to end at about 24 °E longitude.

Seismicity in the Pannonian Region

The seismicity in the Pannonian plain is weak. Bisztricsányi (1978) has shown the difficulty inherent in constructing any

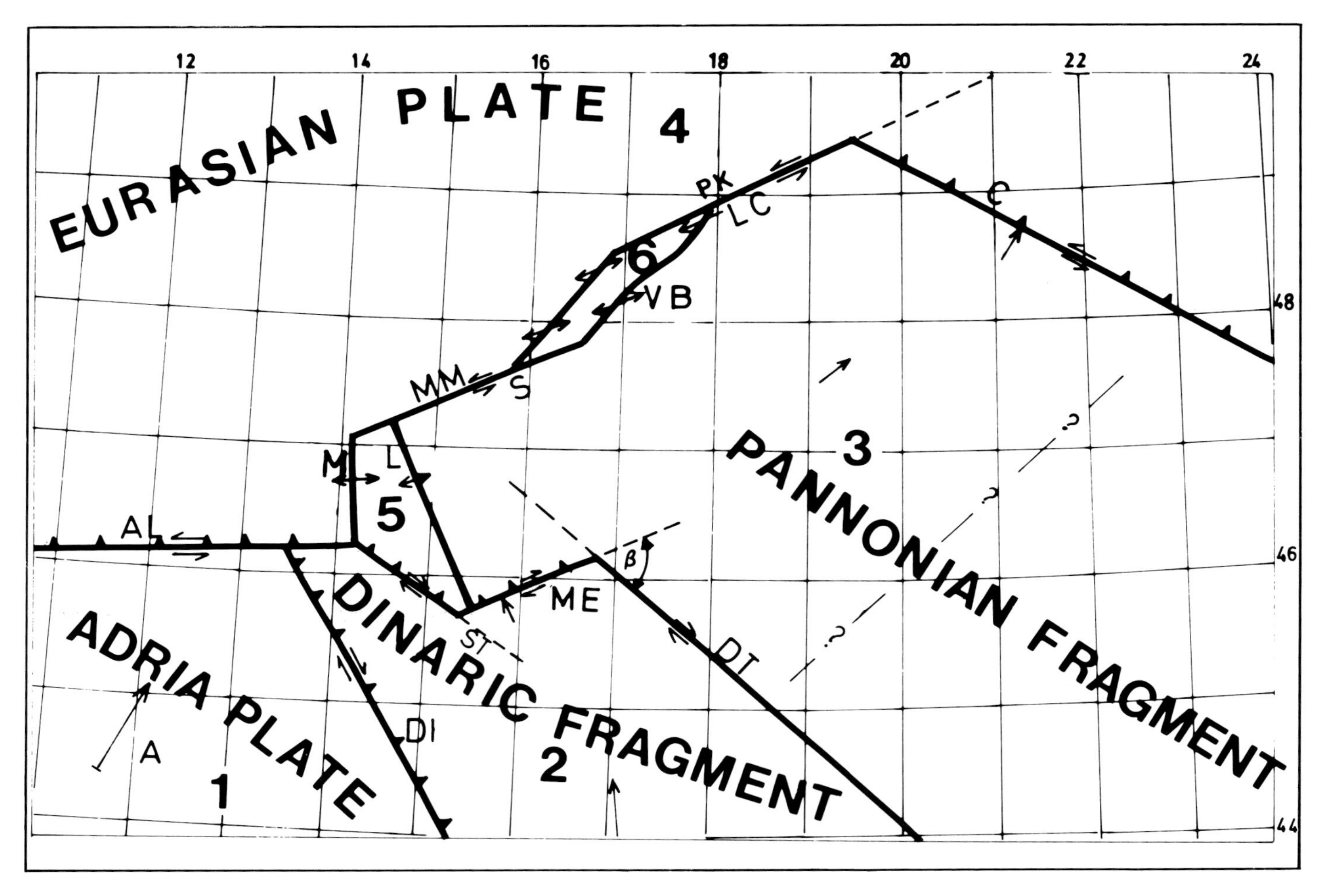

Figure 3. Schematic rigid block model of plates and crustal fragments within the Pannonian and adjacent areas. Large arrows show the direction of relative motions of various crustal blocks with respect to Europe (length of arrow does not indicate velocity). Barbs along fragment boundaries indicate overthrusting, outward pointing arrows indicate divergence, inward pointing arrows indicate convergence, and arrows parallel to the boundaries indicate the sense of strike-slip displacement. AL = Adriatic lineament, M = Metnitz valley, L = Lavant valley, DI = Dinarides, ST = Sava trough, DT = Drava trough, ME = Medvednica zone, MM = Mur–Mürz valley, S = Semmering area, VB = Vienna basin, PK = Pieniny Klippen belt (or peripienninean zone), and C = West Carpathians (eastern part).

meaningful geographical pattern of maximum seismic intensity when the statistical significance of the data is so low, It is particularly difficult to decide whether the epicenters in the Pannonian basin occur at isolated places or along elongated zones. The data available at present exhibit a more or less clustered distribution. Csomor and Kiss (1959) have noted that at several single places earthquakes occur repeatedly. For example, at 47.9 °N, 20.4 °E (near Eger) at least 16 earthquakes, with approximately 50 greater aftershocks, occurred over a time interval of 72 years. These strange phenomena, together with the high heat flow in the Pannonian basin (Bodri, 1981; Cermak, 1982), led Stegena (1980) to suggest that a thermal effect was responsible for the earthquakes. The great intensities of several shocks, however, seems to exclude pure thermal expansion (the earthquakes at Eger in 1908, Kecskemét in 1911, and Dunaharaszti have a maximum intensity of I_o = 8 °MKS and the great shock of Kecskemet in 1911 had I_o = 9 °MKS) (Csomor and Kiss, 1959).

Only a few earthquakes in the Pannonian plain are strong enough to yield reliable focal solutions. The focal solution of the Dunaharaszti earthquake in 1956 (Figure 1) indicates vertical nodal planes striking north-south and east-west (Csomor, in Gangl, 1975b). The inner isomals of several earthquakes plotted in Figure I show a rather strong scatter, so that no clear pattern can be seen.

It is possible that some of these earthquakes occur along older tectonic lines, such as those that bound the deep, extensional troughs in the Pannonian area. There is no obvious or simple correlation, however, between the seismic activity and the Neogene–Quaternary sediment thicknesses (Figure 4). Nevertheless, a few observations can be made. The seismic activity in the Medvednica zone occurs near the junction of the Periadriatic line with the Zala, Sava and Drava basins. We have the impression that there are two northwest-southeast-trending seismic lineaments that are subparallel to Sava and Drava troughs. A more scattered east-northeast-trending zone of seis-

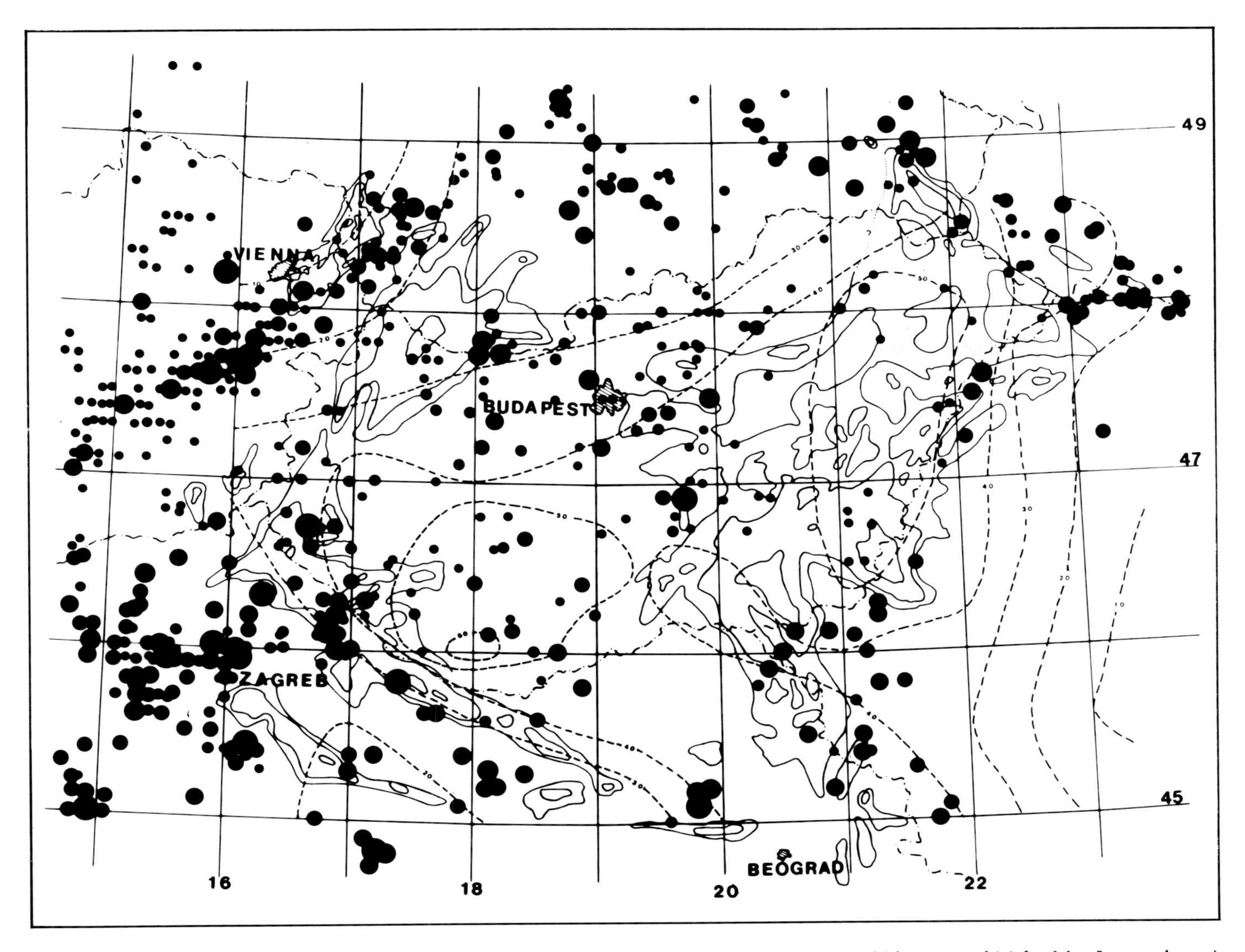

Figure 4. Isopach map of Neogene sediments in the intra-Carpathian region (solid lines show 1, 2, and 3 km intervals) (after Map 8, in enclosure) with epicentral data taken from Map 7. Dashed lines show the distribution of the estimated value of heat flow at the Mohorovicic discontinuity (after Cermak, 1982).

mic activity appears to connect them at their northern end. The Sava and Drava troughs were the sites of dextral strike-slip in Miocene time. The more scattered seismic activity in the Medvednica area is roughly superimposed on a region where east-northeast-trending fold axes indicate north-south compression in Miocene time (Sava folds). The present seismic pattern may indicate reactivation of this general tectonic system, but the displacement need not necessarily be taken up on exactly the same faults.

Seismicity East of the Pannonian Basin

East of the Pannonian basin, the Apuseni Mountains and the Transylvanian basin display very low seismic activity. They appear to be bounded to the west by a broad zone of seismic activity that follows the edge of the Pannonian basin. Throughout Neogene time this area (the Apuseni mountains and Transylvanian basin) has behaved more rigidly than the surrounding areas, which have been intensively deformed (Royden et al., 1983; Horváth and Royden, 1981). This pattern seems to be reflected in the present seismic activity.

East of about 26° longitude, a limited area (the Vrancea region) exhibits high seismic activity at depths down to 150 km (see Koch, 1982). These earthquakes have been interpreted as due to the presence of a subducted slab that dips steeply westward beneath the Carpathians at about 70° to vertical. Several authors have suggested that this seismic activity reflects the westward drift of the Black Sea subplate and its interaction with stable Europe and the Carpathian belt (Radu, 1974; McKenzie, 1972; Josif and Josif, 1975). Others have suggested that this seismic zone represents the final part of the mainly Miocene convergence between the Pannonian fragment and Europe (see Royden, this volume).

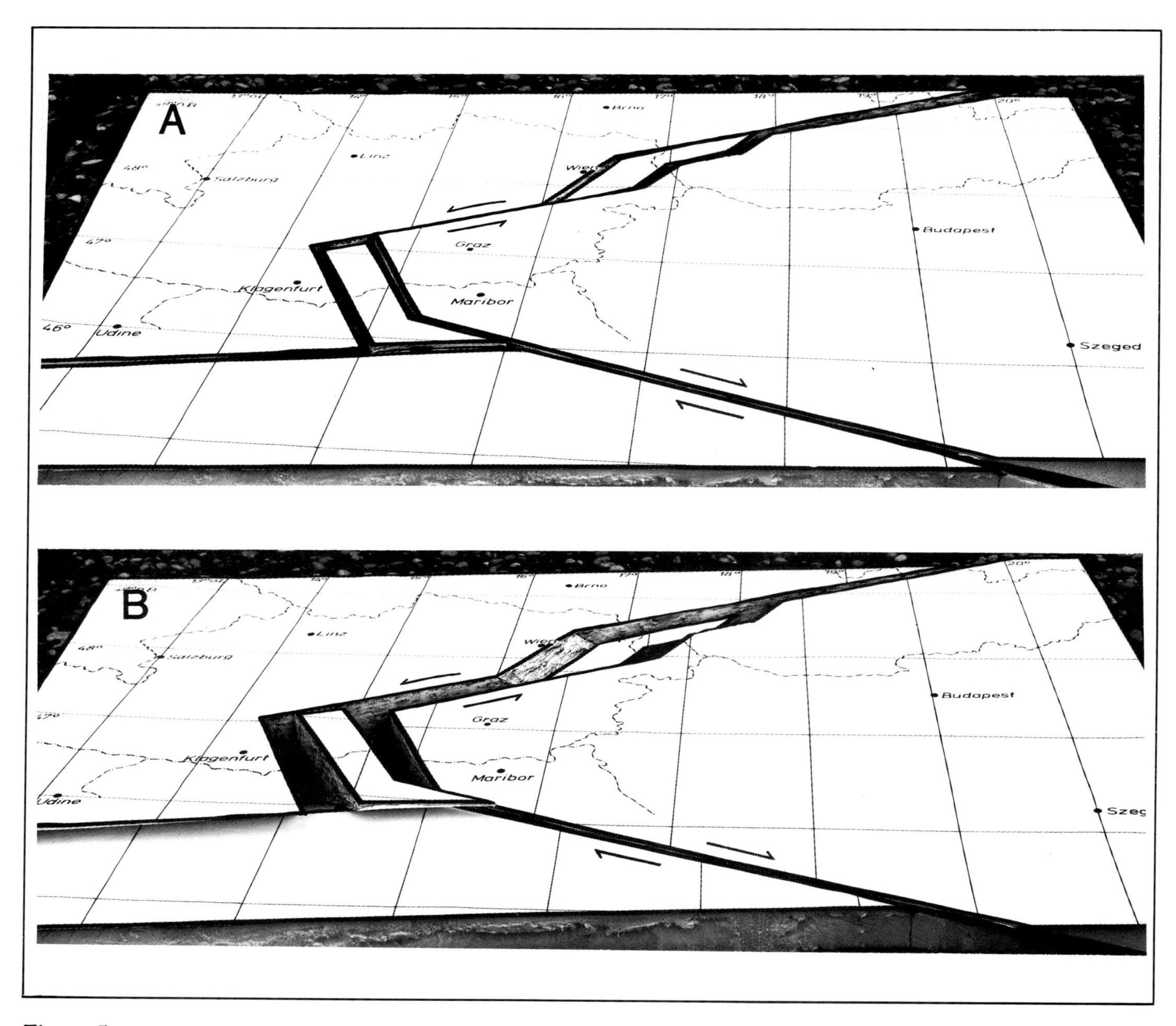

Figure 5. Three-dimensional block model for recent deformation in the western edge of the Pannonian basin, the West Carpathians and Eastern Alps, after Figure 3, (a) shows geometry before slip occurs on the simplified fault zones shown, and (b) shows geometry after slip has occurred. The Adria plate underthrusts the Eastern Alps. The Pannonian block moves eastward, and the Vienna basin and Lavant and Metnitz valley exhibit roughly east-west extension.

In any case, the intermediate depth earthquakes in this eastern part of the Carpathian loop seem to be spatially isolated from the seismicity within the Pannonian region and westward. We interpret the intermediate depth seismicity as a remnant of the Miocene subduction zone that was responsible for the present geometry of the Carpathian belt. We interpret the seismicity of the Alps and the Dinarides as a more direct consequence of the interaction between Adria and Europe. Thus, on a local scale these two regions of seismicity are mostly independent. On a very large scale, however, they are both related to the complicated pattern of plate interaction that has resulted from the convergence of Africa and Europe throughout the Mediterranean region.

NEOTECTONIC MODEL OF THE EAST ALPINE AND THE PANNONIAN-CARPATHIAN AREA

The neotectonic model presented in Figure 3 makes use of the assumed seismic lineaments denoted in Figure 1 and also of geologic data concerning late Miocene to recent deformation. We

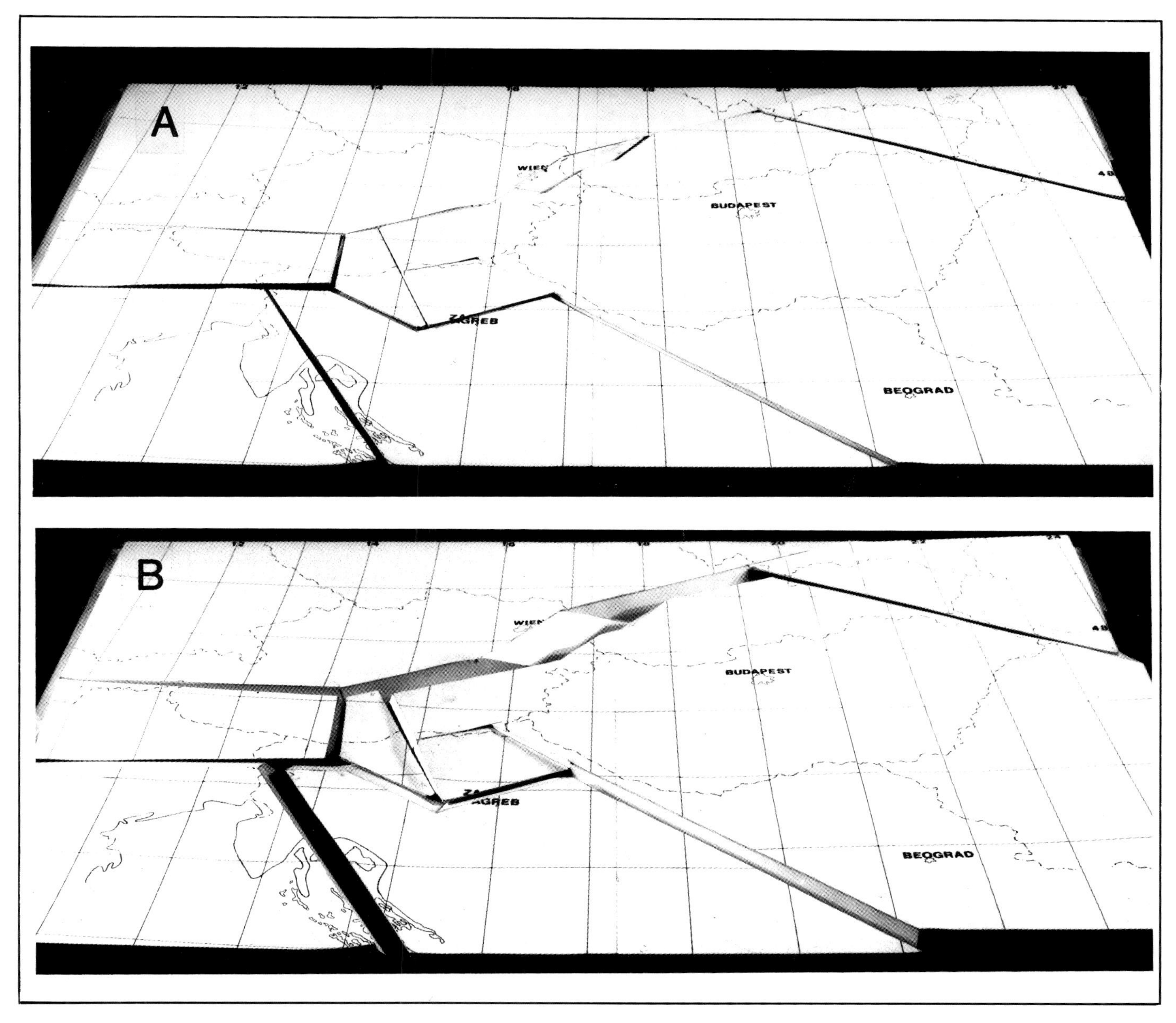

Figure 6. Three-dimensional block model for recent deformation for all tectonic lineaments shown in Figure 3. (a) shows geometry before slip occurs, and (b) shows geometry after slip has occurred (compare with Figure 5).

assume that the scattered seismicity in the Pannonian basin cannot be related to distinct seismic lineaments but reflects a penetrative fracturing of the entire region. We have neglected any influence of the deep seismicity in the East Carpathians as a broad, mainly aseismic region exists between the West Carpathians and the East Carpathians that is without significant seismicity. The character of the seismicity is so different in these two areas that we feel justified in this assumption. Despite these restrictions, a simple tectonic model with rigid blocks can account for most of the observed seismicity and recent tectonism in this area.

The simplified tectonic model presented here explains the seismic lineaments of Figure 1 as active fault systems or tectonic boundaries between these coherent crustal fragments, where any internal compression or stretching has been ignored. The model consists of six rigid blocks as indicated in Figures 3, 5 and 6: (1) Adria plate, (2) Dinaric fragment, (3) Pannonian fragment, (4) Eurasian plate, (5) area between the Lavant and Mettnitz valleys, and (6) Vienna basin. The direction of displacement of the Adria plate relative to Europe has been indicated by arrow A (Figure 3).

Most tectonic reconstructions of this region conclude that the roughly northward motion of the Adria plate relative to Europe has resulted in its collision with the European continent in early Tertiary time. Subsequently, the Adria plate has been broken into smaller crustal fragments. Horváth and Royden

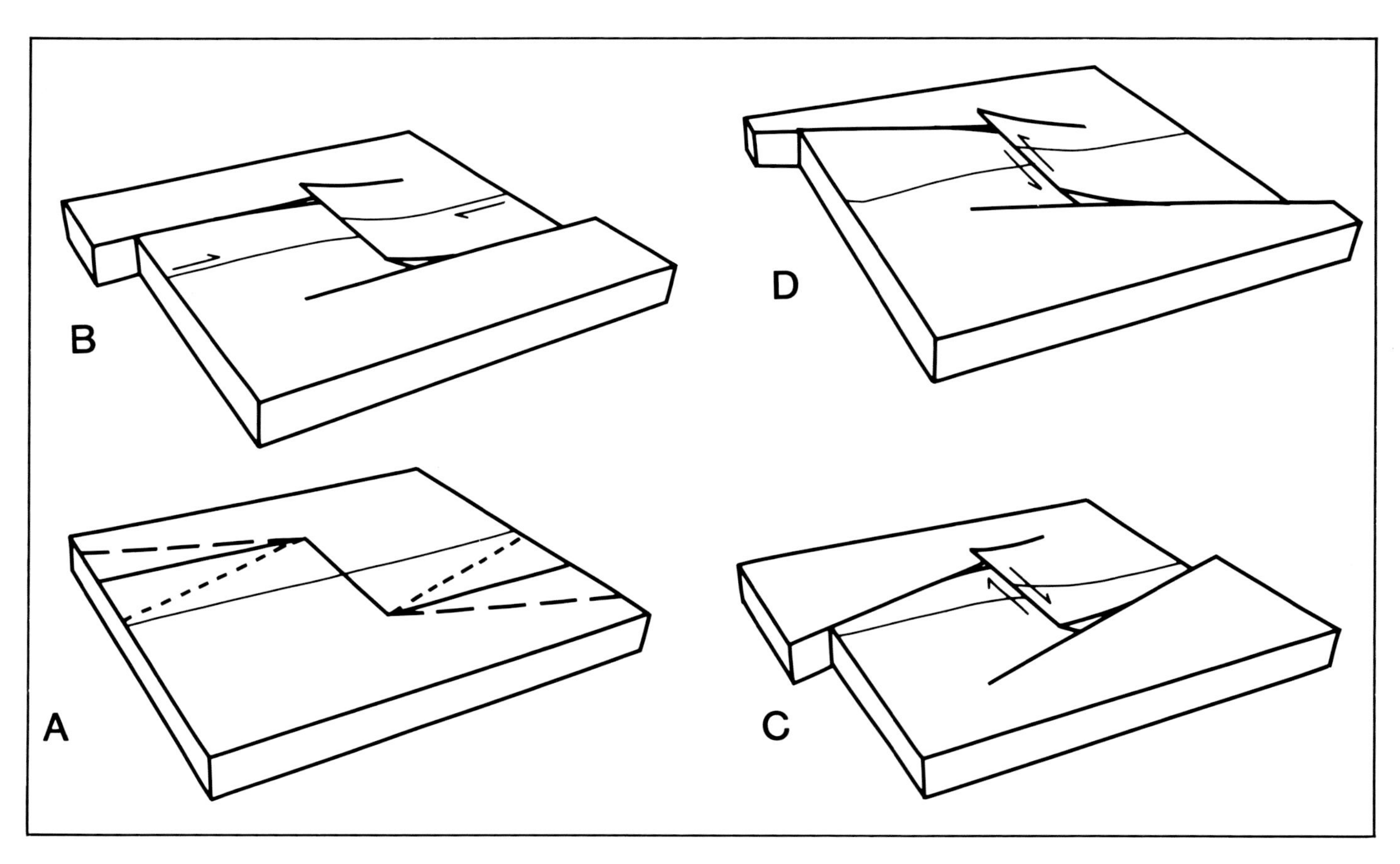

Figure 7. Cartoon of the Medvednica zone illustrating how the orientation of convergence and the relative angle of related strike-slip faults control the sense of lateral displacement: (A) indicates situation prior to thrusting, (B) shows pure overthrusting at ME, (C) depicts overthrusting and dextral shear at ME, and (D) indicates overthrusting and sinistral shear at ME.

(1981) and Royden and Báldi (this volume) infer that the relative motion of Adria with respect to Europe during and after collision was in large part responsible for the later Tertiary deformation of the Carpathian–Pannonian region. In the same way, we infer that the present motion of Adria with respect to Europe can account for most of the recent deformation and the observed seismicity in the region shown in Map 7 and Figure 3.

The assumed slip direction of Adria relative to Europe predicts northward or northeastward underthrusting of Adria beneath the Dinarides (along DI) and the Alps (along AL). One or both of these fault zones must have some strike-slip displacement as well, but the direction of the strike-slip displacement along AL is not well constrained by our data because focal solutions show only overthrust motion. The sense of lateral offset depends strongly on the angle of relative motion between Europe and Adria and on the orientation of the faults along DI. Focal solutions from the Dinaric Alps indicate overthrusting with a right-slip component along the strike of the belt. In this simple model we have assumed no rotation of Adria, which may not be realistic, For example, counterclockwise rotation of Adria about a nearby pole could generate right-slip along both AL and DI.

Within the same system, generally northward displacement of the Dinaric fragment results also in northward-directed underthrusting along its northern margin (lines ME and ST) and in right lateral slip along its northeastern boundary (DT). Note that the northern boundary faults may also exhibit strike-slip displacements, the direction of which depends upon the orientation of the faults relative to the direction of convergence (Figures 6 and 7). In Figure 3, right-slip occurs along the northwest-trending ST line, while left-slip occurs along the roughly northeast trending ME. We cannot accurately determine from this analysis, however, the sense of strike-slip offset occurring at present along AL, DI, ST, ME, and C. Small changes in convergence direction can reverse the sense of strike-slip displacement on these fault zones (Figure 7). The dashed lines in Figure 3 indicate the propagation of the rupture beyond the simplistic fragment boundaries shown, giving rise to seismic activity.

The fault zone MM through PK, and the zone AL through DT, form two major fragment boundaries and define a wedgelike fragment within the crust. These two fault systems form an angle of roughly 50 toward the east from their intersection near the extensional structures of the Metnitz and Lavant valleys. A wedge-shaped crustal fragment (the Pannonian fragment) is squeezed out to the east between the Adriatic (plus the Dinaric) plate and the stable Eurasian platform that begins north of the Peripienninean lineament. This crustal wedge is squeezed progressively eastward as convergence continues between Adria and Europe. This escape mechanism is similar to that when one squeezes a cherry pit out by pressing the cherry between two fingers. Left lateral slip occurs along the Peripienninean and Mur–

Mürz zone (PK and MM) and right lateral along the Periadriatic–Dinaric lineaments (AL through DT or DI).

In this overly simplified model, convergence between Adria and Europe is partly accommodated by underthrusting of the Adriatic plate beneath the Alpine area. Along the Periadriatic lineament, both underthrusting and horizontal slip occur. The Peripienninean lineament and the faults of the Mur–Mürz valley can be thought of as a transform fault system, with sinistral displacement. Associated extensional structures appear in the Lavant–Metnitz valleys and in the Vienna basin, The structure of the Lavant valley is complicated by north-south compression and uplift in the south while east-west extension prevails in the north.

The active seismic zone in the eastern part of the West Carpathians (C) may correspond to a zone of combined underthrusting and left-slip displacement, but no focal mechanisms are available. Another tectonic line trending roughly northeast along the eastern boundary of the Pannonian basin has been indicated by a dashed line on Figure 3. Rigid plate theory cannot easily explain this seismic zone and its significance is not clear.

CONCLUSIONS

The main pattern of seismic activity in the East Alpine, Pannonian and Carpathian region can be explained by a simple rigid block model (Figures 5, 6, and 7). The model proposed is consistent with the hypothesis that most of the active deformation results directly or indirectly from the convergence of the Adria plate and Europe. The region of deformation and seismicity extends a considerable distance from the Adria–Europe plate boundary, and involves several smaller crustal fragments, including the Pannonian fragment. The observed seismicity and deformation are mostly consistent with north-south compression, east-west extension, and related strike-slip displacements. Clearly the crustal fragments shown in Figure 3 are being deformed internally, so that a rigid block model can only describe their behavior in an approximate way. Nevertheless, such a model can provide insight into the overall pattern of deformation in this region, and into the relationship between the different seismic lineaments.

ACKNOWLEDGMENTS

This study is part of the Interdisciplinary Project S 15 (Geologischer Tiefbau der Ostalpen) and financially supported by the Fonds zur Förderung der wissenschaftlichen Forschung in Österreich and the Österreichischen Akademie der Wissenschaften (Geophysikalische Kommission).

REFERENCES

Aric, K., 1981, Deutung krustenseismischer und seismologischer Ergebnisse im Zusammenhang mit der Tektonik des Alpenostrandes: Osterr. Akad.d.Wiss. Abt. 1, Bd. 190, Wien, Springer-Verlag, p. 235–312.

Aric, K.,and R. Gutdeutsch, 1981, Seismotectonic and refraction seismic investigations in the border between the Eastern Alps and the Pannonian basin: Pageoph, v. 119, p. 1125–1133.

Beer, M. A., and U. K. Szukin, 1978, Analyse der Geodynamik und Seismizität des Karpato-Dynarischen Systems: Proceedings of the symposium on the analysis of seismicity and on seismic risk, Liblice 1977–Prague 1978, p. 37–46.

Bisztricsányi, E., 1978, Problems of seismic regionalization for Hungary: Proceedings of the Symposium on the analysis of seismicity and seismic risk, Liblice 1977–Prague 1978, p. 443–450.

Bodri, L., 1981, Three-dimensional modelling of deep temperature and heat flow anomalies with applications to geothermics of the Pannonian Basin: Tectonophysics, v. 79, p. 225–236.

Bolt, B. A., 1978, Earthquakes: A primer: W. H. Freeman and Company, San Francisco, 241 pages.

Burchfiel, B. C., 1980, Eastern European Alpine system and the Carpathian orocline as an example of collision tectonics: Tectonophysics, v. 63, p. 31–61.

Caputo, M., and D. Postpischl, 1972, (comp.) in structural model of Italy 1:1,000,000: Seismicity map, 1900–1970, C.N.R. 1 map.

Cermak, V., 1982, Crustal temperature and mantle heat flow in Europe: Tectonophysics, v. 83, p. 123–142.

Csomor, D. and C. Kiss, 1959, Die Seismizität von Ungarn: Studia Geophysica et Geodetica, p. 33–42.

Drimmel, J., 1980, Rezente Seismizität und Seismotektonik des Ostalpenraumes: Der Geologische Aufbau Österreichs, Springer-Verlag, Wien, New York, p. 507–527.

Drimmel, J., and E. Trapp, 1975, Das Starkbeben am 29. Januar 1967 in Molln. Oberösterreich: Mitt.Erdb.Komm.N.F. 76, Wien, p. 1–45.

Gangl, G., 1975a, Seismotectonik investigations of the western part of the inneralpine Pannonian basin (Eastern Alps and Dinarides): Paper on XIV Gen. Ass. of the ESC Trieste 74, Berlin (DDR), p. 409–410.

Gangl, G., 1975b, Seismotektonische Untersuchungen am Alpenostrand: Mitt. Geol. Ges. in Wien, 33–48, p. 66–67.

Horváth, F., and L. Royden, 1981, Mechanism for the formation of the intra-Carpathian basins: A review: Earth Evolution Sciences, v. 3–4, p. 307–316.

Joo, J., E. Casati, P. Jovanovic, M. Popescu, V. J. Somov, H. Thum, J. Thury, J. N. Totomanov, J. Vanko, and T. Wyrzyhowski, 1981, Recent vertical crustal measurements of the Carpatho-Balkan region: Tectonophysics, v. 71, p. 41–52.

Josif, T., and S. Josif, 1975, Some tectonic aspects of Vrancea region (Romania): XIV Gen. Ass. of the ESC, Trieste 1974, Berlin, 1975, p. 417–424.

Koch, M., 1982, Seismicity and structural investigations of the Romanian Vrancea region. Evidence for azimuthal variations of P-wave velocity and Poisson's ratio: Tectonophysics, v. 90, p. 91–115.

Kowatsch, A., 1911, Das Scheibbser Erdbeben vom 17. Juli 1976: Mitt. Erdb. Komm. N. F. 40, Wien, p. 1–54.

Mahel, M., 1973, Tectonic map of the Carpathian–Balkan mountain system and adjacent areas: Published by Stur's geological Institute in Bratislava/Unesco, Paris, 1 map.

McKenzie, D., 1972, Active tectonics in Mediterranean region: Geoph. J. R. Astron. Soc., v. 30, p. 109–185.

Müller, G., 1977, Fault-plane solution of the earthquake in Northern Italy, 6 May 1976, and implications for the tectonics of the Eastern Alps: J. Geophysics, v. 42, p. 343–349.

Prochaskova, D., and V. Karnik, 1978, Atlas of isoseismal maps, Central and astern Europe: KAPG, Prague, 135 pages.

Radu, C., 1974, Contribution á l'ètudie de la seismicité de la Roumanie et comparaison avec la seismicite du bassin mediterranéen, en particulier avec la seismicité de la France de SE: Thèse, Univ. Strasbourg, 404 pages.

Ribaric, V., 1982, Seizmicnost Slovenije, Katalog Potresov 792 n.e.: Catalogue of Earthquakes (792 A.D.–1981), Ljubljana, 649 pages.

Ritsema, A. R., 1974, The earthquake mechanism of the Balkan region: UNDP Rem 70/172, UNESCO, de Bilt, p. 1–35.
Ritsema, A. R., 1975, Seismic sequences in the European Mediterranean: XIV Gen. Ass. of the ESC, Triest 1974, Berlin 1975, p. 367-371.
Royden, L., F. Horváth, and J. Rumpler, 1983, Evolution of the Pannonian basin system: Tectonics, v. 2, no. 1, p. 63–90.
Schenk, V., V. Karnik, and Z. Schenkova, 1982, Seismotectonic scheme of Central and Eastern Europe: Studia Geophysica et Geodaetica, v. 26, p. 132–144.
Schneider, G., 1968, Erdbeben und Tektonik in SW-Deutschland: Tectonophysics, v. 5, p. 459–511.
Shebalin, N. V., V. Karnik, and D. Hadzievski, 1974, Catalogue of Earthquakes Part I, 1901–1970, Part II prior to 1901: UNDP/UNESCO Survey of the seismicity of the Balkan Region, Skopje, 56 pages.
Shebalin, N. V., G. I. Reisner, A.V . Drumea, J. Y. Aptekman, V. N. Shholpo, and N. Y. Stepanenko, 1976, Earthquake origin zones and distribution of maximum expected seismic intensity for the Balkan region: Proceedings of the seminar on seismic zoning maps, Skopje 27 October–4 November 1974, UNESCO, Skopje, p. 68–171.
Skoko, D., and D. Cvijanovic, 1978, Seismic activity based on the Neotectonics: Proceedings of the symposium on the analysis of seismicity and on seismic risk, Liblice 1977–Prague 1978, p. 82–91.
Stegena, L., 1980, Geothermics and seismicity in the Pannonian basin: Terrestrial and space techniques in earthquake prediction research, A. Vogel, ed., *in* Friedrich Vieweg & Sohn, Braunschweig/Wiesbaden, p. 467–471.
Tollmann, A., 1970, Ostalpen Tektonik I: Schweizerbart'sche Verlagsbuchhandlung (Nägele und Obermitter), Stuttgart, p. 1–90.
Udias, A., 1975, Seismicity and possible plate boundary relations in the Western Mediterranean: XIV Gen. Ass. of the ESC, Triest, 1974, Berlin 1975, p. 395–400.

CHAPTER 16

A Review of Temperature, Thermal Conductivity, and Heat Flow Data for the Pannonian Basin

P. Dövényi and F. Horváth
Geophysical Department, Eötvös University
H-1083, Kun B. tér 2, Budapest, Hungary

More than 2700 pieces of data have been selected from evaluation of all available temperature measurements made in boreholes in Hungary. They comprise steady-state temperatures, drill stem test data, and corrected outflowing water temperatures. These data have been used to construct average temperature versus depth profiles for different subunits of the central part of the Pannonian basin in Hungary.

The increase of thermal conductivity with depth in Neogene sedimentary rocks can be well described by simple relationships for different lithologies. Using such relationships we made more than 150 heat flow estimates for wells where good temperature data were available and the lithology was known. Four new heat flow determinations are reported for the Great Hungarian Plain. These were used for the construction of a new detailed heat flow map of the Pannonian basin and adjacent areas. Regional water circulation systems markedly disturb the conductive thermal field in the Transdanubian Central Range and in the mountainous area of northeastern Hungary, but such disturbances are absent or negligible in the basins. Accordingly, the significant heat flow contrast between some of the basins puts important constraints on the mechanism of their formation.

INTRODUCTION

Temperature and heat flow data are closely related to the structure and evolution of the earth's lithosphere. This is especially true for some areas of regional subsidence (e.g., oceanic ridges, passive continental margins, and extensional basins) because their subsidence is largely controlled by the temperature structure of the underlying lithosphere. Temperature conditions are also important in exploration for natural resources, particularly for hydrocarbon prospecting. Many temperature measurements have been made in boreholes in Hungary, but their spatial distribution (both lateral and vertical) is very inhomogeneous. Also, their reliability is strongly variable due to different measurement conditions. Therefore, the available data must be carefully evaluated for reliability when analyzing the regional features of the temperature field. All reliable data for Hungary are presented at the end of this paper in a data catalogue modified after Dövényi et al. (1983).

A total of 18 accurate heat flow determinations have been made in Hungary, 4 of which are published for the first time in this paper. However, heat flow can be estimated reasonably well in many other boreholes where temperature and average thermal conductivity values are known. We have derived a model for thermal conductivity variations in the Neogene sedimentary rocks of the Pannonian basin which has been used to determine about 150 estimates of heat flow in Hungary. Similar estimates and heat flow determinations outside Hungary were also used to construct a new detailed heat flow map of the Carpathian–Pannonian region.

The main purpose of this chapter is to present the basic geothermal data set, which represents a most important constraint on modeling the subsidence and maturation history of the Pannonian basin.

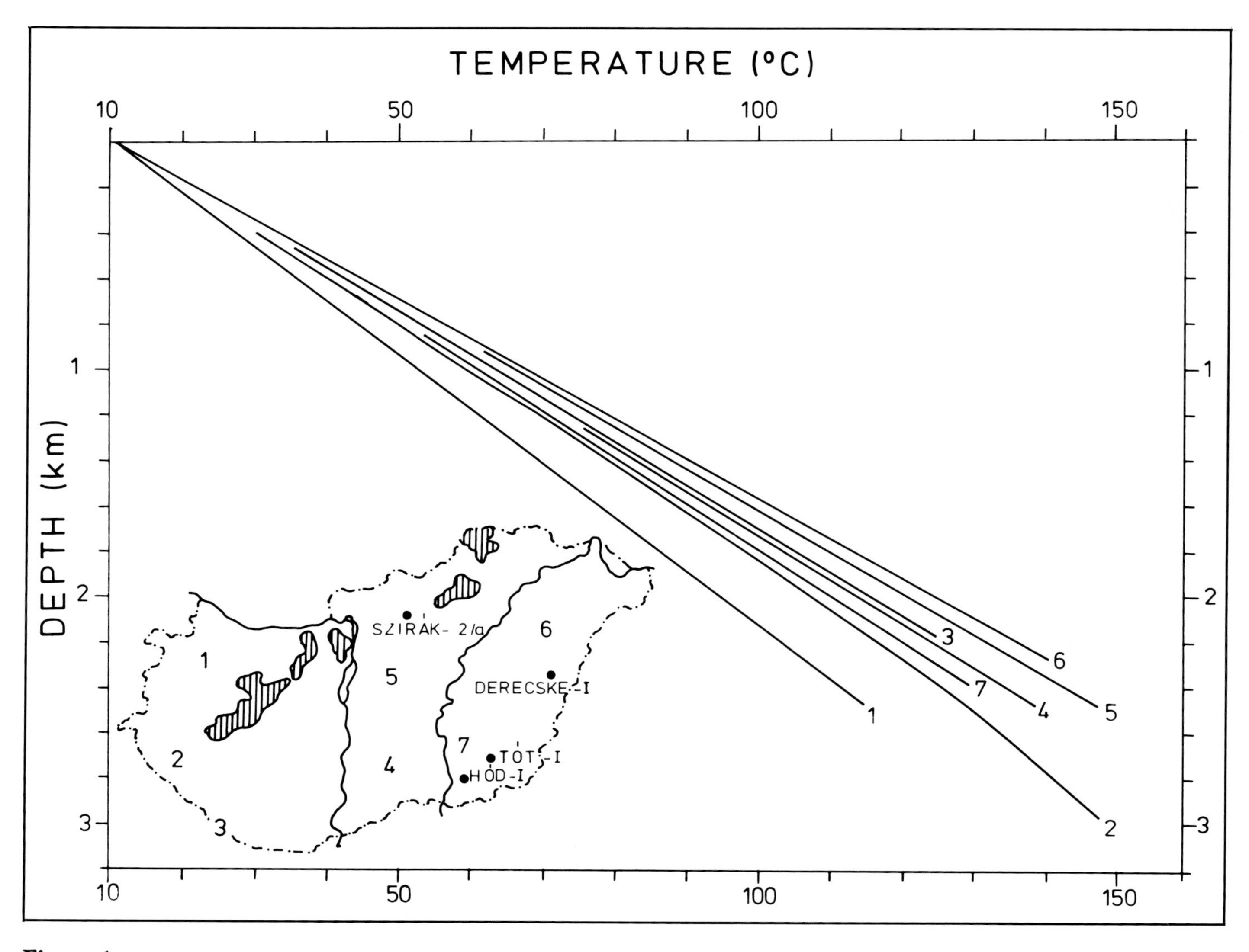

Figure 1. Average temperature versus depth profiles for major subunits of Hungary (central part of the Pannonian basin), and the location of new heat flow determinations. (1) Little plain, G = 41.6; (2) Zala basin, G = 47.8; (3) Dráva basin, G = 52.1; (4) Duna–Tisza interfluve region, south, G = 51.3; (5) Duna–Tisza region, north G = 54.6; (6) East of Tisza River region, north, G = 56.4; (7) Makó trough and Békés basin, G = 49.1. Average gradients (G) are given in °C/km and are for the 250 to 2250 m depth interval.

COMPILATION OF TEMPERATURE DATA CATALOGUE

We selected from about 10,000 pieces of data the reliable temperature values that reflect steady-state conditions and give good coverage for Hungary. The data catalogue in the appendix to this chapter contains 2734 data, which include:

1. Temperatures measured under steady-state conditions.
2. Temperatures measured during drill stem tests both in water and petroleum exploration wells, and
3. Temperatures calculated from outflowing water using an empirical formula derived from the data of 80 master wells in Hungary (Dövényi et al., 1983). These temperatures were corrected to a value $T(z_s)$ where:

$$T(z_s) = T_o + 5z_s Y^{-0.71}$$

where T_o is the outflowing water temperature (in °C), z_s is the depth of source layer (in km), and Y is the yield of the source layer (in m^3/min).

Shallow temperatures (depth <250 m) were rarely used. Corrected borehole temperatures were not used. The catalogue of temperature data contains the following columns.

Column 1. The position of the borehole (in km) in a rectangular grid system with an accuracy of ±1 km. The grid system is shown in the Appendix. In some cases (where the coordinates are given by two numbers) the accuracy is ±5 km. If "telep" occurs in the third column, the temperature data come from several wells in an oil field, and the coordinates give the center of the field. The distance of wells from the center of the field is always less than 5 km.

Column 2. Name of the locality, village, or town near the borehole.

Column 3. Number or symbol of the borehole.

Column 4. Depth of the temperature measurement. If the temperature of outflowing water was measured, the depth indicated is the mean depth of the layer yielding the water.

Column 5. Steady-state temperature. Measured values or corrected outflowing water temperature (in °C).

Column 6. Geothermal "gradient" (in °C/km) is equal to $(T_z - T_s)/z$, where T_s and T_z are the temperatures at the surface and at depth z, respectively. Surface temperatures were determined from a long series of average air and soil temperature data recorded at meteorological stations (Dövényi et al., 1983).

Column 7. Numbers 1, 2, and 3 categorize the reliability of temperature data and the letters **a, b, c,** and **x** refer to the data source and/or the method of measurement.

Category 1. Several measurements were made in one hole at about the same depth. These were measured by different methods, measured by the same method but at different times, or calculated from outflowing water temperature; they are consistent with one another. That is, the average geothermal gradients differ by less than 10%. At least one of the data points must be a true steady-state temperature or a drill stem test temperature. The catalogue contains only the best data.

Category 2. A single steady-state temperature or drill stem test data in one hole.

Category 3. Corrected temperature of outflowing water and any other data that may have errors of more than ±10%.

Categories **x** *and* **a**. Data measured in petroleum exploration wells. **a** indicates that several consistent data were available from a 50 m (or less) depth interval, and only the average value is listed.

Category **b**. Data obtained by the Hungarian Geological Institute.

Category **c**. Value belongs to a suite of temperature–depth measurements determined solely for special geothermal studies. The category of reliability depends on the time of thermal recovery (t_r):

category **c1** if $t_r > 45$ days
category **c2** if $20 < t_r < 45$ days
category **c3** if $5 < t_r < 20$ days

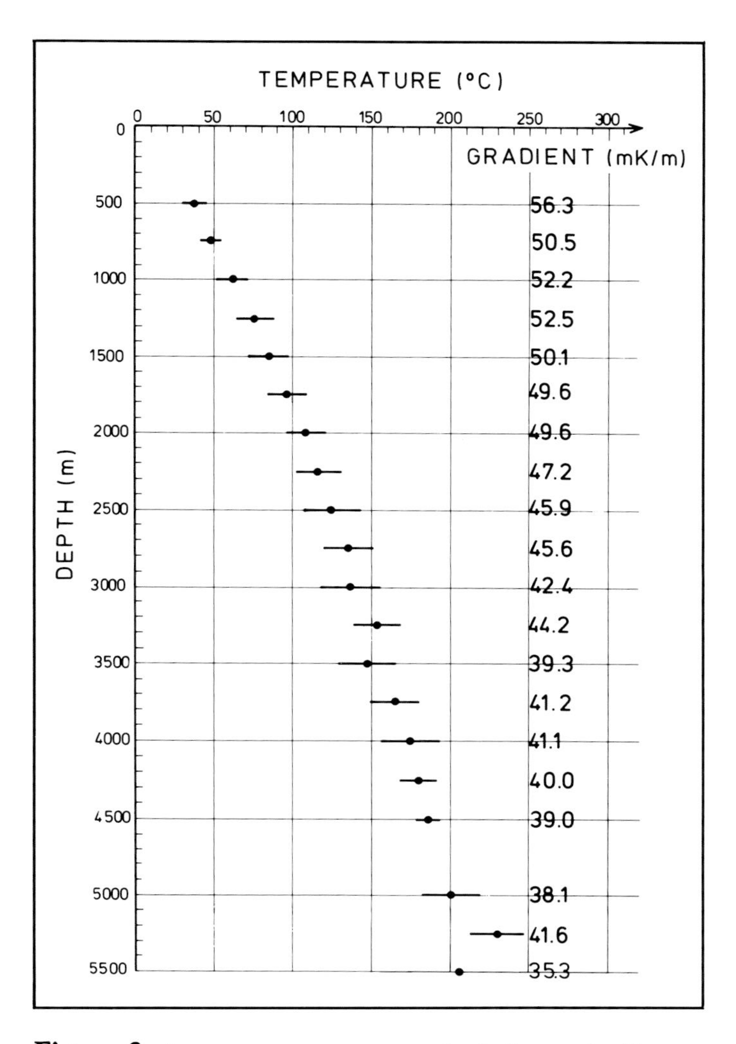

Figure 2. Average temperature versus depth diagram for Hungary (central part of the Pannonian basin). Average gradients represent the mean values of geothermal gradients for 250 m depth intervals.

The absence of a letter in column 7 indicates that the temperature measurement was made in a water exploration well.

Using all the data from this catalogue, we constructed a map of isotherms at 1 km depth (Dövényi et al., 1983) and plotted average temperature versus depth for major subunits of the Pannonian basin (Figure 1). Temperature is a fairly linear function of depth down to 2000–2500 m. Below this depth it is not linear, and usually the thermal gradient is lower because of the increased thermal conductivities at depth. The same features can be seen on the average temperature–depth plot for all of Hungary (Figure 2).

ANALYSIS OF THERMAL CONDUCTIVITIES

The first heat flow measurements in Hungary were made in the Mecsek Mountains, one of the few outcrops of pre-Neogene rocks in the central part of the Pannonian basin (Boldizsár, 1956, 1964). This site was chosen primarily because of the difficulty of measuring thermal conductivity of loose or semiconsolidated rock samples. Water saturated cylindrical samples of these semiconsolidated Neogene rocks are often destroyed by the pressure of the bars in the traditional divided bar instrument. In the last decade reliable measurements of thermal conductivity of Neogene sedimentary rocks in the Pannonian basin have been made using the differentiated line source method (Cull, 1974; Dövényi et al., 1983).

A fairly large number of such data have made it possible to establish a general relationship between depth (or porosity) and thermal conductivity for different lithologies. Excluding volcanic rocks and rare limestone beds, the sedimentary fill of the Pannonian basin can be divided into two broad lithological categories: psammites (sand, sandstone, and gravel) and pelites (clay, claystone, marl, and shale). Using 308 data points from 13 boreholes shows that the depth and porosity dependence of thermal conductivity can be described by:

$$k(z) = k_m^{1-\phi(z)} k_w^{\phi(z)}$$

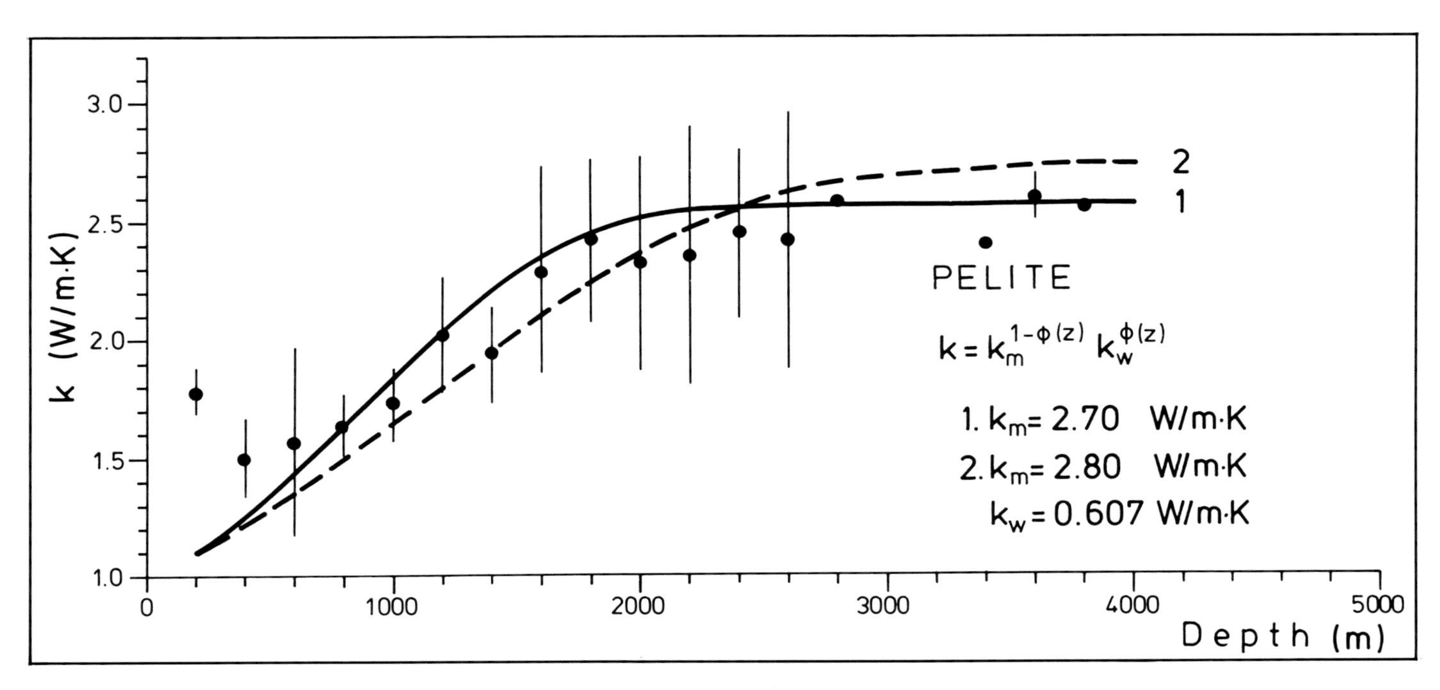

Figure 3. Conductivity averages determined from measured values on pelitic samples, and the theoretical conductivity versus depth functions calculated from the parameter k, as shown, and from the normal porosity trends shown in Figure 5.

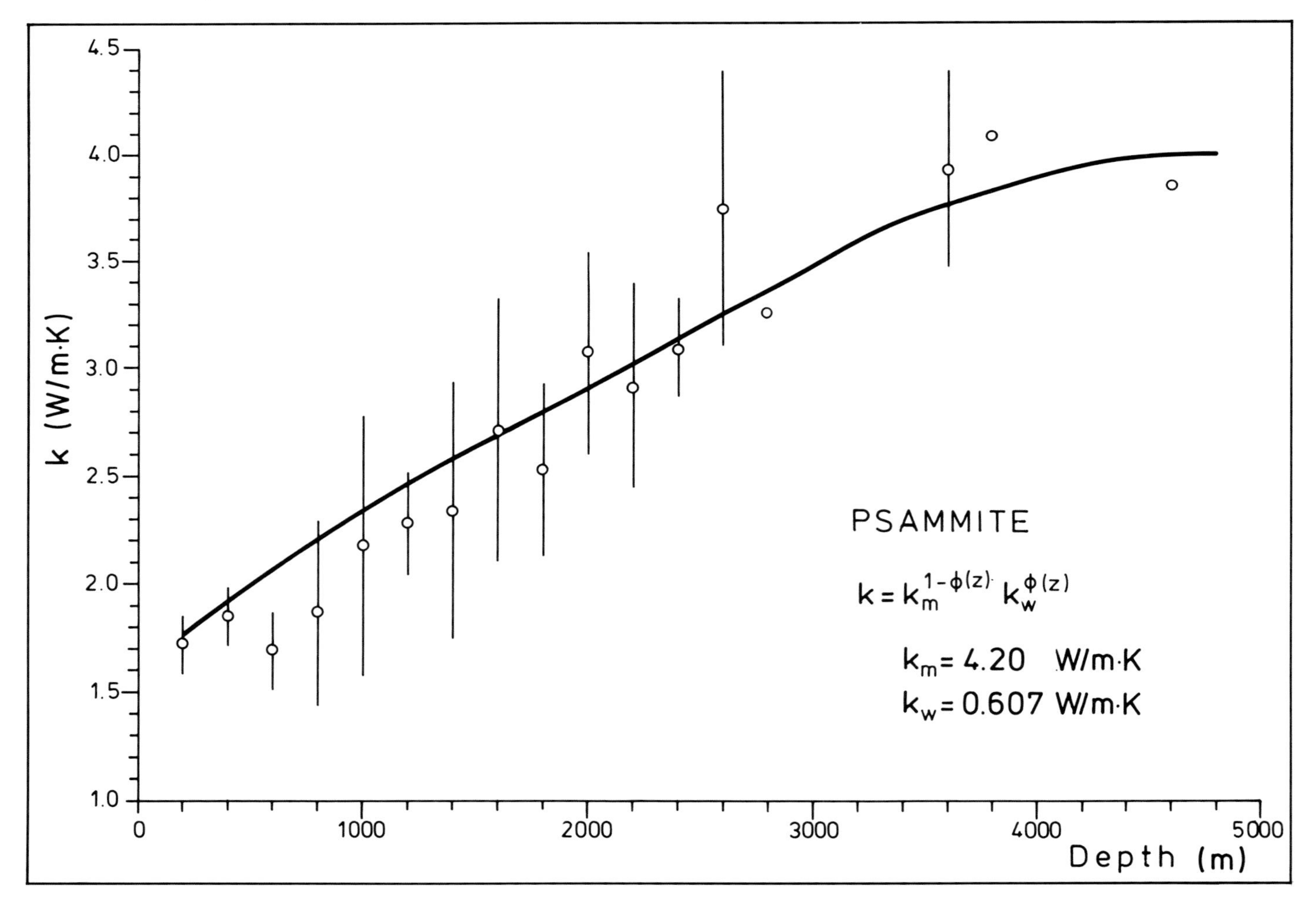

Figure 4. Conductivity averages determined from measured values on psammitic samples, and the theoretical conductivity versus depth function calculated from the parameter k, as shown, and from the normal porosity trend shown in Figure 6.

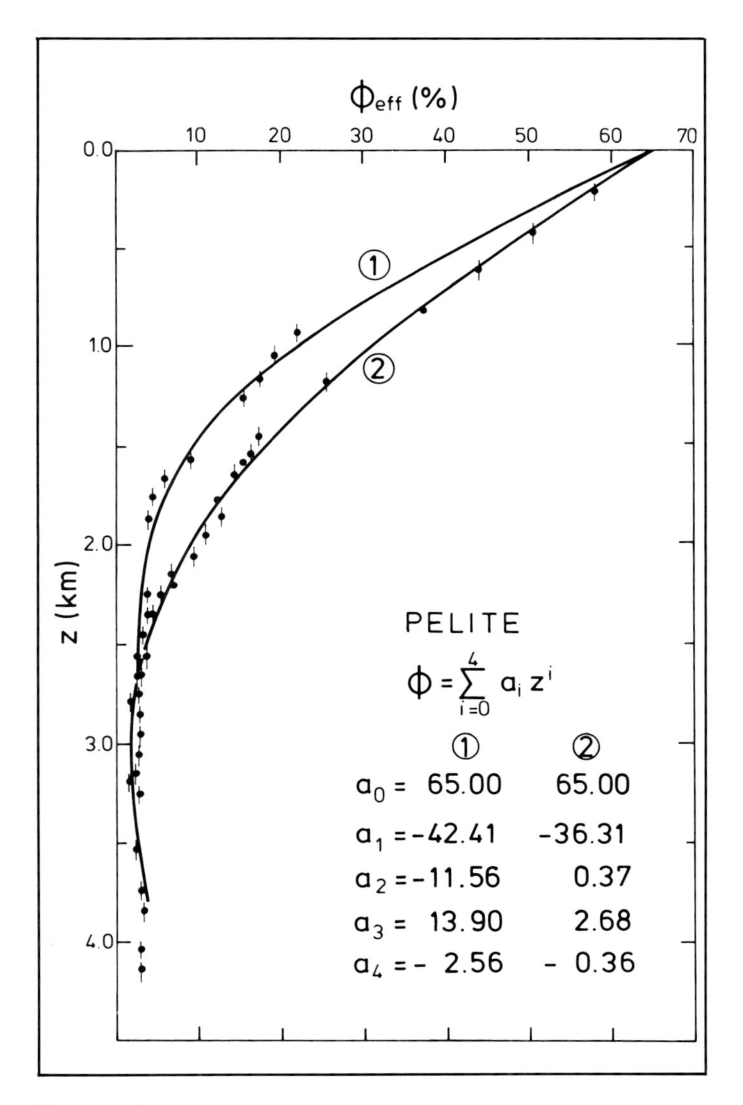

Figure 5. Normal porosity trend for pelite: (1) wells of medium depth and (2) deep wells. Solid circles with vertical bars indicate values averaged over 100 m depth intervals. (After Szalay, 1982.)

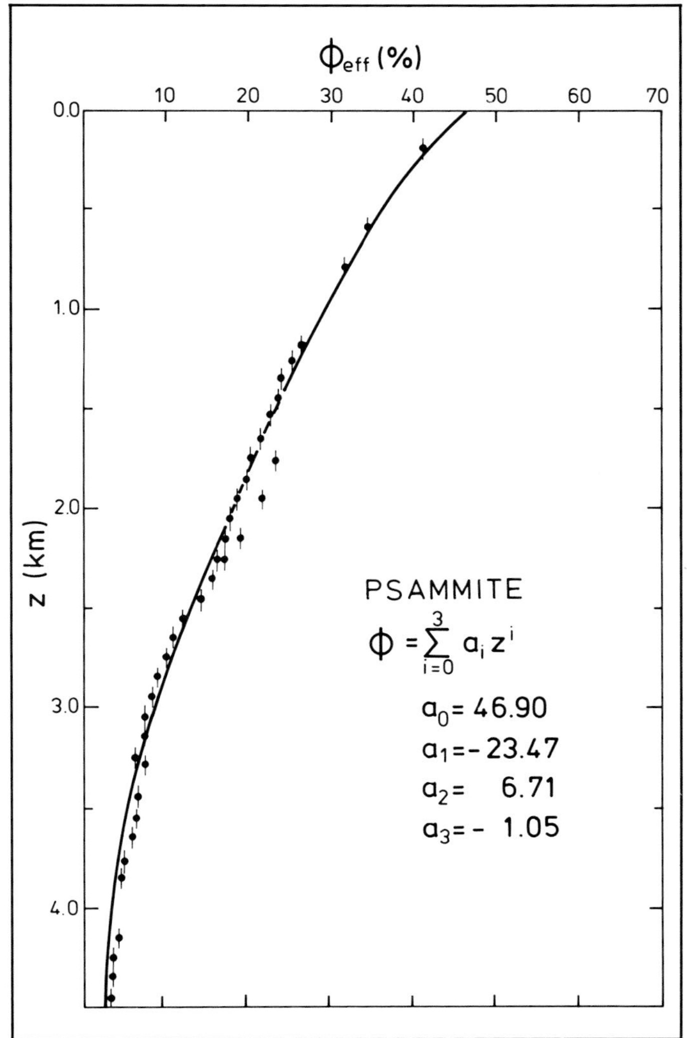

Figure 6. Normal porosity trend for psammites. Solid circles with vertical bars indicate values averaged over 100 m depth intervals. (After Szalay, 1982.)

for both pelitic and psammitic rocks (Figures 3 and Figure 4) where $k(z)$ is conductivity as a function of depth, k_w is the thermal conductivity of water, k_m is the thermal conductivity of pelitic or psammitic rock matrix, and $\phi(z)$ is the normal porosity trend for pelites or psammites as a function of depth (Figures 5 and 6).

Using this formula, heat flow can be estimated for wells where there are good temperature measurements available and the lithology (especially the pelite to psammite ratio) of Neogene strata is known from continuous well logs. In this way we estimated about 150 values of heat flow, which dramatically increased the areal coverage of the heat flow data for Hungary. Error analysis and comparison with direct heat flow determinations suggest that the error of these estimates never exceeds 25%, and in most cases it is better than 20%.

The thermal conductivity of rocks also depends on temperature. This effect is not negligible in the Pannonian basin where temperatures up to 200–250 °C are often measured in deeper wells. In our experience, Sekiguchi's (1984) formula:

$$k(T) = 365.75(k_{20} - 1.84)(1/T - 1/1473) + 1.84$$

adequately describes the change of thermal conductivity with temperature in this interval. In the formula T is the temperature in degrees Kelvin and k_{20} is the conductivity at 20 °C.

HEAT FLOW DETERMINATIONS IN THE PANNONIAN BASIN

All heat flow data measured in the Pannonian basin and adjacent regions are listed in Table 1. They come from five different countries: Czechoslovakia, Poland, Soviet Union, Romania, and Hungary. Most of the heat flow determinations are reviewed in the book *Terrestrial Heat Flow in Europe* (Cermák and Rybach, 1979), where the circumstances and accuracy of measurements are summarized. The new data (not included in Cermák and Rybach, 1979) are collected from recent publica-

Table 1. Heat Flow Determinations in the Pannonian Basin and Adjacent Regions

Site	Geographic Coordinates	G (°C/Km)	k (W/mK)	q (mW/m^2)	Q[a]	Ref.
Hungary						
Barszentmihalyfa-I	46°41 'N 16°36 'E	31.1	2.96	92	C	4
Nagylengyel-47,62	46°46 'N 16°45 'E	45.6	1.84	84	B	5
Bakonya MV-2142/a	46°07 'N 18°04 'E	37.9	2.72	103	B	6
Kovagotottos MP-19	46°06 'N 18°07 'E	35.0	3.09	108	A	7
Komlo-Zobak	46°10 'N 18°14 'E	49.4	2.80	138	B	8
Hosszuheteny	46°10 'N 18°22 'E	40.7	2.56	104	B	13
Val-3	47°22 'N 18°40 'E	38.7	2.79	108	A	4
Szentendre-2	47°41 'N 19°05 'E	39.2	2.14	84	B	9
Budapest (Nepliget)	47°27 'N 19°10 'E	13.0	3.99	52	C	10
Szirak-2/a	47°48 'N 19°31 'E	41.0	2.04	84	A	32
Recsk Rm-8,15	47°56 'N 20°07 'E	39.9	2.73	109	B	4
Sandorfalva-I	46°25 'N 20°01 'E	42.3	2.67	113	C	4
Hod-I	46°22 'N 20°26 'E	38.0	2.19	82	B	32
Totkomlos-I	46°27 'N 20°46 'E	41.0	2.55	106	B	32
Kaba EK-1	47°30 'N 21°25 'E	58.8	1.58	97	B	11
Hajduszoboszlo-6	47°32 'N 21°22 'E	58.8	1.83	108	B	11
Derecske-I	47°17 'N 21°41 'E	44.0	2.30	102	B	32
Edeleny E-475	48°18 'N 20°16 'E	57.5	2.28	131	B	13
Poland						
Lodygowice	49°43 'N 19°08 'E	20.2	2.06	42	A	1
Zakopane	49°58 'N 19°58 'E	18.8	2.93	56	—	1
Leszczyna	50°01 'N 20°23 'E	30.0	1.99	60	A	1
Siekieczyna	49°46 'N 20°52 'E	32.1	2.03	65	A	1
Zolcza	50°22 'N 20°59 'E	34.5	1.34	46	—	1
Sucha Rzeki	49°16 'N 22°30 'E	23.1	1.87	43	B	1
Cisowa	49°41 'N 22°33 'E	16.0	1.69	27	B	1
Czechoslovakia						
Dunajov-13	48°50 'N 16°34 'E	27.0	2.61	72	C	27
Dunajov-11	48°51 'N 16°36 'E	30.0	1.31	46	C	27
Mikulov	48°48 'N 16°36 'E	25.0	2.04	42	B	27
Musov	48°53 'N 16°37 'E	28.0	2.29	49	B	14
Nesvacilka	49°05 'N 16°44 'E	24.0	2.60	54	A	16
Vranovice	48°58 'N 16°41 'E	26.0	2.36	60	A	16
Ujezd	49°07 'N 16°45 'E	22.0	3.22	48	C	27
Nikolcice-6	49°00 'N 16°46 'E	26.0	2.24	51	A	14
Nikolcice-1	48°59 'N 16°46 'E	27.0	2.38	64	A	16
Kobyli	48°55 'N 16°54 'E	22.0	2.49	63	B	14
Zarosice-2	49°04 'N 16°57 'E	24.0	2.89	45	A	15
Zarosice-1	49°03 'N 16°58 'E	22.0	2.18	46	A	15
Hrusky-10	48°47 'N 16°58 'E	35.0	1.50	53	B	17
Lab-90	48°23 'N 16°59 'E	42.0	2.19	62	B	17
Slavkov	49°10 'N 16°52 'E	12.0	4.20	55	A	14
Kuty	48°39 'N 16°59 'E	29.0	1.99	47	B	17
Lab-93	48°23 'N 17°00 'E	37.0	1.50	56	B	17
Bratislava Rusovce	48°04 'N 17°01 'E	44.0	1.38	61	B	25
Malacky-20	48°27 'N 17°01 'E	31.0	2.33	44	B	17
Zdanice	49°04 'N 17°02 'E	21.0	2.28	51	C	27
Jezov-2	49°02 'N 17°09 'E	21.0	2.81	42	B	27
Zavod-57	48°33 'N 17°07 'E	25.0	2.70	41	B	17
Sastin-10	48°40 'N 17°09 'E	32.0	1.40	45	B	17
Rohoznik	48°29 'N 17°10 'E	32.0	1.50	49	B	17
Laksarska Nova Ves	48°34 'N 17°11 'E	39.0	1.40	54	B	17
Cilistov FGSC-1	48°01 'N 17°18 'E	34.0	2.00	68	B	26
Kralova pri Senci	48°11 'N 17°23 'E	42.0	1.97	83	B	25
Senec BS-1	48°13 'N 17°26 'E	39.0	2.19	84	B	26
Horna Poton FGHP-1	47°59 'N 17°29 'E	40.0	1.94	79	A	26

Table 1. (continued)

Site	Geographic Coordinates		G (°C/Km)	k (W/mK)	q (mW/m²)	Q[a]	Ref.
Rataje	49°16′N	17°20′E	30.0	2.25	51	B	14
Lubna-6	49°12′N	17°20′E	—	—	56	B	27
Lubna-8	49°11′N	17°22′E	35.0	1.80	76	B	14
Lubna-2	49°13′N	17°23′E	28.0	2.34	54	B	14
Gabcikovo FGG-1	47°54′N	17°36′E	38.0	1.96	75	B	26
Rusava	49°20′N	17°41′E	27.0	2.18	65	B	14
Galanta-FGG-2	48°10′N	17°42′E	42.0	1.70	79	B	26
Bohelov GPB-1	47°56′N	17°43′E	39.0	2.19	87	C	26
Galanta-FGG-1	48°13′N	17°43′E	39.0	1.61	63	C	26
Vizovice	49°12′N	17°52′E	26.0	3.01	71	B	27
Choryne	49°34′N	17°53′E	24.0	3.41	71	B	14
Topolniki	47°58′N	17°48′E	43.0	1.80	72	B	25
Vlcany-FGV-1	48°02′N	17°56′E	40.0	2.00	80	B	26
Kolarovo-2	47°56′N	18°02′E	40.0	2.27	75	B	17
Lidecko	49°13′N	18°03′E	21.0	3.53	75	B	18
Soblahov	48°51′N	18°05′E	26.0	2.95	80	B	27
Tvrdosovce FGT-1	48°06′N	18°05′E	46.0	1.70	80	B	26
Komarno FGK-1	47°45′N	18°05′E	38.0	1.76	67	B	26
Verovice	49°31′N	18°08′E	31.0	2.60	82	—	27
Roznov	49°31′N	18°10′E	23.0	2.73	63	—	27
Trojanovice	49°31′N	18°10′E	32.0	2.58	90	—	27
Trojanovice	49°32′N	18°11′E	30.0	2.39	71	—	27
Nove Zamky GNZ-1	48°00′N	18°10′E	42.0	1.82	77	C	26
Trojanovice	49°30′N	18°11′E	33.0	2.72	89	—	27
Trojanovice	49°31′N	18°12′E	30.0	2.76	82	—	27
Trojanovice	49°31′N	18°12′E	33.0	2.32	84	—	27
Trojanovice	49°31′N	18°12′E	23.0	1.96	45	C	27
Trojanovice	49°31′N	18°13′E	38.0	3.24	124	B	27
Trojanovice	49°31′N	18°14′E	18.0	3.48	61	B	27
Trojanovice	49°30′N	18°14′E	27.0	2.94	104	C	20
Kozlovice	49°36′N	18°15′E	30.0	3.12	86	B	20
Kuncice	49°32′N	18°15′E	27.0	2.98	104	C	20
Trojanovice	49°32′N	18°15′E	30.0	2.19	83	B	20
Trojanovice	49°31′N	18°15′E	30.0	2.86	86	B	27
Kozlovice	49°36′N	18°17′E	30.0	2.54	78	B	24
Kuncice	49°32′N	18°17′E	33.0	2.73	91	B	27
Frenstat	49°32′N	18°14′E	30.0	2.28	76	B	19
Ticha	49°33′N	18°14′E	32.0	2.42	75	A	19
Ticha	49°34′N	18°15′E	30.0	2.12	73	B	19
Kuncice	49°33′N	18°16′E	31.0	2.40	77	B	19
Kozlovice	49°34′N	18°17′E	30.0	3.25	112	C	20
Celadna	49°33′N	18°18′E	34.0	2.32	80	B	27
Kuncice	49°32′N	18°19′E	38.0	3.16	120	B	27
Pstruzi	49°34′N	18°20′E	33.0	2.64	87	B	27
Dvory n. Zitavou	47°59′N	18°15′E	33.0	2.12	70	B	26
Malenovice	49°35′N	18°25′E	29.0	2.94	85	B	27
Celadna	49°30′N	18°20°E	—	2.93	83	B	27
Ostravice	49°33′N	18°23′E	33.0	2.76	92	C	27
Skrecon	49°55′N	18°22′E	35.0	2.46	78	B	19
Stare Hamry	49°28′N	18°25′E	23.0	2.97	76	B	14
Stonava	49°48′N	18°32′E	31.0	2.99	91	A	24
Krasna	49°32′N	18°31′E	22.0	2.33	60	A	27
Louky	49°48′N	18°35′E	28.0	2.55	76	B	19
Bystrice	49°38′N	18°44′E	29.0	2.19	64	A	27
Gondovo	48°17′N	18°41′E	45.0	1.62	76	A	27
Rudno-9	48°26′N	18°41′E	31.0	2.70	85	B	27
Zlatno	48°25′N	18°47′E	—	—	101	A	27
Podhradie-86	48°40′N	18°41′E	—	—	74	B	27
Brehy-14	48°23′N	18°41′E	33.0	2.60	98	B	27
Banska Stiavnica	48°27′N	18°53′E	—	1.40	109	C	22
Podhorie	48°29′N	18°55′E	35.0	2.51	108	B	27
Kr. Bane	48°44′N	18°55′E	27.0	3.10	83	B	27

Table 1. (continued)

Site	Geographic Coordinates	G (°C/Km)	k (W/mK)	q (mW/m^2)	Q[a]	Ref.
Badin	48°42'N 19°02'E	31.0	2.50	80	B	27
Mikova-14	48°39'N 20°15'E	16.0	3.86	70	C	21
Ciz MJC-1	48°19'N 20°16'E	34.0	1.93	58	B	26
Liptovsky Mikulas	49°03'N 19°32'E	20.0	2.50	50	C	26
Ciz BCS-2	48°19'N 20°17'E	34.0	2.00	69	B	26
C. Tepl.-2	48°53'N 20°34'E	24.0	3.32	67	B	21
Zahora-2	48°53'N 20°54'E	20.0	2.31	51	B	21
Stretava-5	48°37'N 22°03'E	46.0	2.34	113	B	23
Stretava-7	48°36'N 22°03'E	49.0	1.94	113	B	23
Streda	48°23'N 21°44'E	56.0	1.70	108	B	27
Ptruksa	48°29'N 22°04'E	46.0	2.14	103	B	23
Remetske Hamre	48°53'N 22°11'E	34.0	2.12	73	B	21
Sobrance TMS-1	48°44'N 22°12'E	53.0	2.11	115	C	26
Russia						
Izhgorod-1	48°34'N 22°08'E	—	—	92	B	28
Vilikaya Dobron-1	48°33'N 22°18'E	—	—	84	A	28
Began-1164	48°15'N 22°32'E	54.1	—	96	—	28
Gorazdovka	48°25'N 22°29'E	—	—	88	B	28
Muzhievo	48°19'N 22°37'E	51.0	1.63	84	A	28
Zaluzhe-2	48°25'N 22°42'E	—	—	105	A	28
Borzhava-4	48°05'N 22°48'E	62.7	1.34	84	A	29
Lomna-2	49°12'N 22°50'E	21.8	2.62	57	B	29
Terebla-2	48°08'N 22°54'E	—	—	84	A	28
Svalava-2	48°40'N 22°58'E	26.0	2.92	75	A	28
Vola Blazhevskaya-4	49°24'N 23°04'E	20.0	3.45	69	A	29
Sadkovichi-3	49°38'N 23°08'E	26.6	1.84	54	B	29
Pinany	49°37'N 23°13'E	—	—	50	A	28
Pinany-1	49°36'N 23°15'E	23.3	2.08	45	A	29
Vola Blazhevskaya	49°23'N 23°15'E	21.0	3.87	80	C	28
Kohanovka	50°03'N 23°18'E	42.0	1.67	71	B	28
Zaluzhany-1	49°33'N 23°29'E	15.0	2.22	31	A	29
Skole-1	49°04'N 23°33'E	—	—	92	A	28
Orov-27	49°12'N 23°34'E	—	—	67	B	28
Orov-3	49°13'N 23°35'E	—	—	54	A	28
Rudki	49°33'N 23°38'E	37.0	1.76	63	B	28
Ulichnoe-16	49°13'N 23°40'E	—	—	33	A	28
Ulichnoe-4	49°13'N 23°40'E	17.6	1.80	38	A	28
Ulichnoe-15	49°12'N 23°41'E	—	—	38	B	28
Ulichnoe-14	49°11'N 23°43'E	—	—	42	B	28
S. Medynychi-4	49°24'N 23°45'E	—	—	46	B	28
S. Medynychi-10	49°24'N 23°49'E	—	—	54	B	28
S. Medynychi-16	49°25'N 23°50'E	32.4	1.76	59	B	28
S. Medynychi-8	49°18'N 23°55'E	—	—	42	A	28
Uzhnostryskaya	49°12'N 23°51'E	15.5	2.55	38	A	28
Severnaya Dolina-11	49°08'N 23°55'E	—	—	33	B	28
Severnaya Dolina-16	49°04'N 24°03'E	—	—	42	C	28
Bilche Volitsa-95	49°22'N 23°59'E	37.0	1.47	54	A	28
Dolina-60	49°04'N 23°59'E	—	—	50	B	28
Dolina-104	49°04'N 23°59'E	—	—	46	B	28
Derzhiv-3	49°23'N 24°00'E	—	—	50	B	28
Strutyn-12	48°57'N 24°04'E	—	—	42	A	28
Strutyn-23	48°57°N 24°03'E	—	—	50	B	28
Olhovka-15	48°52'N 24°11'E	—	—	54	B	28
Zagapol	48°32'N 25°18'E	20.8	1.66	35	B	29
Zagapol	48°32'N 25°18'E	19.7	1.66	33	B	29

Table 1. (continued)

Site	Geographic Coordinates	G (°C/Km)	k (W/mK)	q (mW/m²)	Q[a]	Ref.
Romania						
Siniob	46°12'N 21°20'E	45.0	—	85	—	3
Oradea	47°13'N 22°13'E	50.5	—	94	—	3
—	44°51'N 22°24'E	40.0	2.3	92	—	3
—	46°09'N 22°53'E	35.0	2.2	79	—	3
—	48°08'N 23°24'E	45.0	2.8	126	—	30
—	44°53'N 23°25'E	28.7	2.6	75	—	3
—	45°02'N 23°25'E	37.0	2.2	80	—	3
—	47°40'N 23°44'E	35.3	2.4	85	—	30
—	47°38'N 23°49'E	56.1	1.6	90	—	30
—	44°14'N 23°53'E	44.5	1.8	78	—	3
—	44°20'N 24°03'E	35.0	1.8	59	—	3
—	47°03'N 24°10'E	20.0	1.7	33	—	2
—	46°56'N 24°16'E	32.0	1.5	45	—	2
—	47°07'N 24°30'E	24.0	2.4	58	—	2
—	46°31'N 24°45'E	28.0	2.6	74	—	3
—	46°40'N 24°52'E	31.0	1.1	33	—	2
—	46°33'N 24°54'E	32.0	1.1	39	—	2
—	47°02'N 25°20'E	37.5	2.2	82	—	30
—	46°23'N 25°25'E	37.0	3.2	113	—	31
—	46°35'N 25°30'E	37.0	2.0	77	—	2
—	46°21'N 25°31'E	63.0	1.6	104	—	2
—	44°31'N 25°42'E	37.0	1.9	70	—	3
—	46°18'N 25°43'E	41.0	1.8	73	—	2
—	46°15'N 25°44'E	50.6	1.7	83	—	3
—	46°39'N 25°47'E	22.0	3.6	57	—	2
—	44°48'N 25°48'E	17.6	3.0	52	—	3
—	46°08'N 25°51'E	70.0	1.5	118	—	2
—	47°15'N 25°41'E	15.0	3.3	45	—	2
—	45°03'N 26°03'E	29.5	2.3	67	—	3
—	47°28'N 26°05'E	18.0	2.2	39	—	3
—	45°11'N 26°19'E	21.1	2.3	48	—	3
—	47°08'N 26°25'E	25.7	1.7	43	—	3
—	47°03'N 26°25'E	26.0	2.6	58	—	3
—	46°37'N 26°29'E	20.0	2.3	47	—	3
—	46°30'N 26°40'E	23.0	2.3	54	—	3
—	44°47'N 26°49'E	24.2	1.8	44	—	3

[a]Quality of measurements: A = error $<10\%$; B = $10\% \leq$ error $\leq 20\%$; C = error $>20\%$.

References: (1) Majorowicz and Plewa, 1979; (2) Veliciu and Demetrescu, 1979; (3) Veliciu et al, 1977; (4) Dovényi et al., 1983; (5) Boldizsár, 1959; (6) Boldizsár, 1967; (7) Salát, 1967; (8) Boldizsár, 1956; (9) Boldizsár, 1965; (10) Salát, 1968; (11) Boldizsár, 1966; (12) Boldizsár, 1964; (13) Greutter, 1977a; (14) Cermak, 1975; (15) Novák, 1975; (16) Cermák, 1968a; (17) Marusiak and Lizon, 1975; (18) Cermák, 1976a; (19) Cermák, 1968b; (20) Cermák, 1976b; (21) Cermák, 1977; (22) Boldizsár, 1965; (23) Cermák, 1968c; (24) Cermák, 1976c; (25) Greutter, 1977b; (26) Stegena et al., 1978–1984; (27) Cermák, pers. comm., 1982; (28) Lubimova et al., 1973; (29) Kutas et al., 1975; (30) Veliciu and Visarion, 1984; (31) Demetrescu, 1978; (32) this paper.

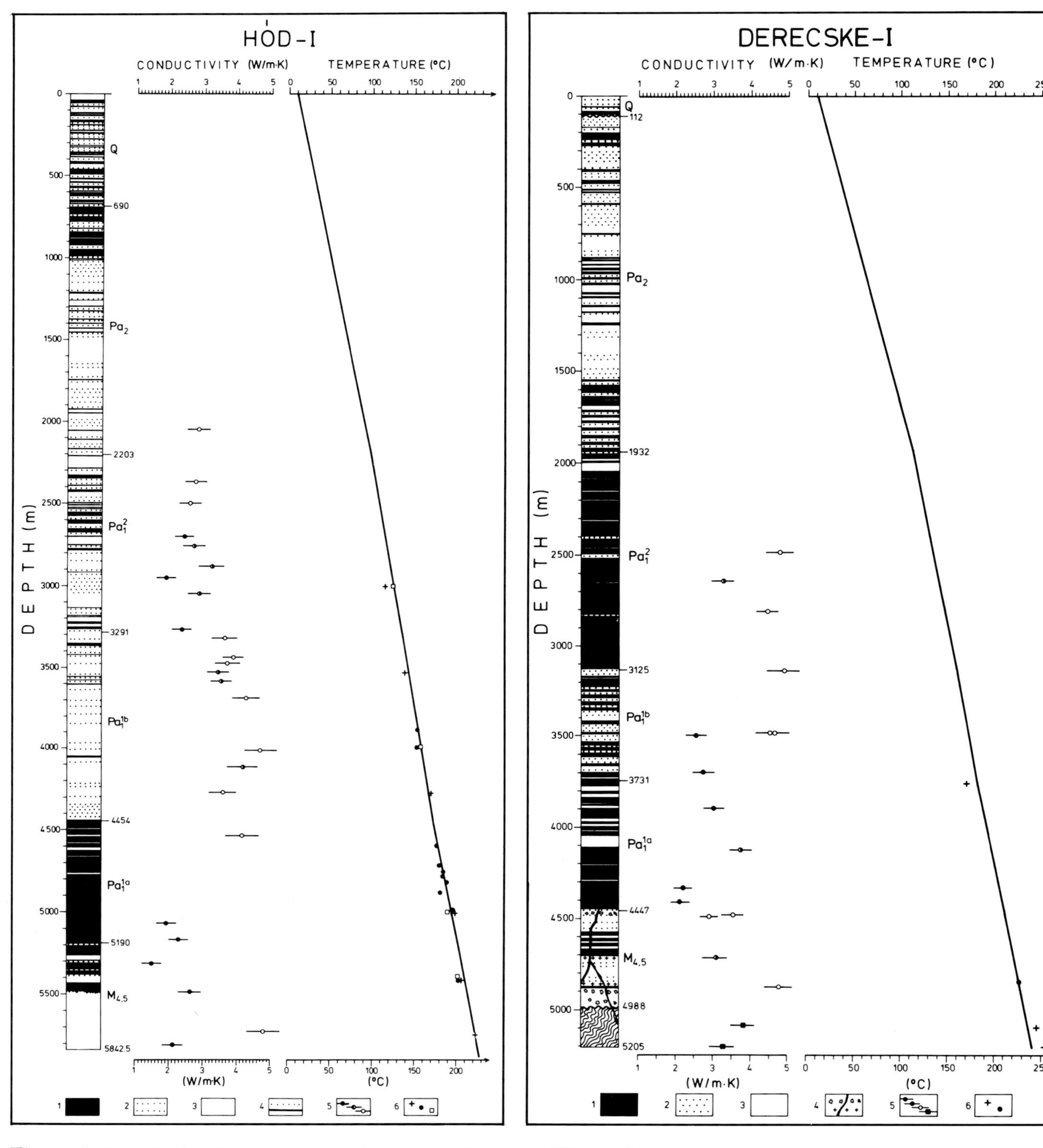

Figure 7. Simplified lithology, measured conductivities, and calculated temperature versus depth for the Hód-I well: (1) thick continuous shale bed; (2) thick continuous sandstone beds; (3) mixed development consisting of alternating shale and sandstone beds with thicknesses less than 5 m; (4) interbedded shale or sandstone layers, respectively; (5) thermal conductivities measured on shale, silt, and sandstone samples, respectively; (6) measured temperatures: corrected bottom hole temperature, drill stem test data, and temperatures measured by well logging, respectively.

Figure 8. Simplified lithology, measured conductivities, and calculated temperature versus depth for the Derecske-I well: (1) thick continuous shale beds; (2) thick continuous sandstone bed; (3) mixed development consisting of alternating shale and sandstone beds with thicknesses less than 5 m; (4) tectonic breccia including strongly deformed shales; (5) thermal conductivities measured on shale, siltstone, sandstone and Paleozoic schist samples, respectively; (6) measured temperatures: corrected bottom hole temperature and drill stem test data, respectively.

Table 2. Data Used to Determine the Heat Flow for the Hód-I Well

Lithogenetic or stratigraphic units[a]	Depth interval (m)	h_{pe} (m)[b]	h_{ps} (m)[b]	Heat conductivity values: Pelites N	k_{pe} (W/mK)	δ	$\bar{k}_{pe}$ (W/mK)[c]	$h_{pe}+h_{ps}$ Psammites $k=$ N	k_{ps} (W/mK)	δ	$\bar{k}_{ps}$ (W/mK)[c]	k = q = $(h_{pe}/k_{pe})+(h_{ps}/k_{ps})$	G = $\Sigma h_i / \Sigma h_i/k_i$	G (°C/km)[d]	G × k (mW/m²)	q/k (°C/km)[e]
Q + Pa_2	0–2203	1018	1185	—	1.72[f]	—	1.73	—	2.33[f]	—	2.27	1.98	2.10	38.4	80.6	40.6
Pa_1^2	2203–3291	496	592	3	2.25	0.30	2.13	6	2.95	0.33	2.62	2.37				33.9
Pa_1^{1b}	3291–4454	458	705	—	2.25[g]	—	2.09	9	3.90	0.43	3.09	2.60				31.9
Pa_1^{1a}	4454–5190	640	96	5	2.11	0.43	1.99	2	4.48	0.47	3.26	2.10	2.33	35.6	83.0	39.5
$M_{4,5}$	5190–5842.5	487.5	165	—	2.11[g]	—	1.98	—	4.48[g]	—	3.15	2.19				37.9
Total	0–5842.5												2.19		82.0	37.4

[a]Lithogenetic classification after Szalay and Szentgyörgyi (this volume).
[b]Summarized thickness of pelitic or psammitic layers (data from Szalay, this volume).
[c]Heat conductivity values corrected for temperature (after Sekiguchi, 1984).
[d]Average geothermal gradients calculated from the temperature values of 224°C, 126°C, and 10°C measured at 5750 m and 3000 m depth and on the surface respectively.
[e]Calculated "stationary" geothermal gradients.
[f]Estimated by equation shown on Figure 3 and Figure 4 respectively.
[g]Extrapolated from previous interval.

Table 3. Data Used to Determine the Heat Flow for the Derecske-I Well

Lithogenetic or stratigraphic units[a]	Depth interval (m)	h_{pe} (m)[b]	h_{ps} (m)[b]	Heat conductivity values: Pelites N	k_{pe} (W/mK)	δ	$\bar{k}_{pe}$ (W/mK)[c]	Psammites N	k_{ps} (W/mK)	δ	$\bar{k}_{ps}$ (W/mK)[c]	k = $\frac{h_{pe}+h_{ps}}{(h_{pe}/k_{pe})+(h_{ps}/k_{ps})}$	k = $\frac{\Sigma h_i}{\Sigma h_i/k_i}$	G (°C/km)[d]	q = G × k (mW/m²)	G = q/k (°C/km)[e]
Q + Pa_2	0–1932	912	1020	—	1.64[f]	—	1.67	—	2.26[f]	—	2.21	1.92				53.1
Pa_1^2	1932–3125	1056	137				2.49				3.73	2.59				39.3
Pa_1^{1b}	3125–3731	296	310	7	2.81	0.57	2.40	5	4.67	0.19	3.48	2.85	2.28	44.8	102.1	35.8
Pa_1^{1a}	3731–4447	627	89				2.34				3.32	2.43				41.9
$M_{4,5}$	4447–4988	235	306	1	3.12	—	2.45	3	3.74	0.95	2.74	2.61				39.0
Precambrian	4988–5205											2.61				39.0
Total	0–5205												2.30		102	44.3

[a]Lithogenetic classification after Szalay and Szentgyörgyi (this volume).
[b]Summarized thickness of pelitic or psammitic layers (data from Szalay, this volume).
[c]Heat conductivity values corrected for temperature (after Sekiguchi, 1984).
[d]Average geothermal gradients calculated from the temperature values of 10°C and 227°C measured on the surface and at 4848 m depth respectively.
[e]Calculated "stationary" geothermal gradients.
[f]Estimated by equation shown on Figure 3 and Figure 4, respectively.

tions. Four new heat flow measurements in Hungary are presented for the first time here and are reviewed below in some detail.

Hód-I

This borehole is situated in the ultradeep Makó trough (for location see Figure 1). The drill penetrated 5842.5 m of middle Miocene to Quaternary sedimentary rocks consisting of sand, sandstone, marl, shale, and siltstone, but it did not reach basement (Figure 7). The lithological trend diagram showing the division of strata in Hód-I into lithogenetic units and their average pelite to psammite ratios is given by Szalay (this volume). Conductivities were measured on 25 water saturated samples by a differentiated line source probe. Averages were calculated separately for pelites and psammites of each lithogenetic unit. Several temperature data were available (including bottom hole temperatures, drill stem test data, and temperature logs), and the two most reliable were chosen. The method used to determine heat flow and the results are shown in Table 2. The uncertainty is probably less than ±10% because heat flow values calculated independently for the upper and lower sections are in good agreement.

Derecske-I

This borehole is located on the flank of a narrow, deep trough. The drill hit the bottom of middle Miocene sediments at a depth of 4988 m (Figure 8). Calculations were made as described above and are shown in Table 3. The estimated uncertainty is greater than for Hód-I (±10–15%) because of less reliable temperature data.

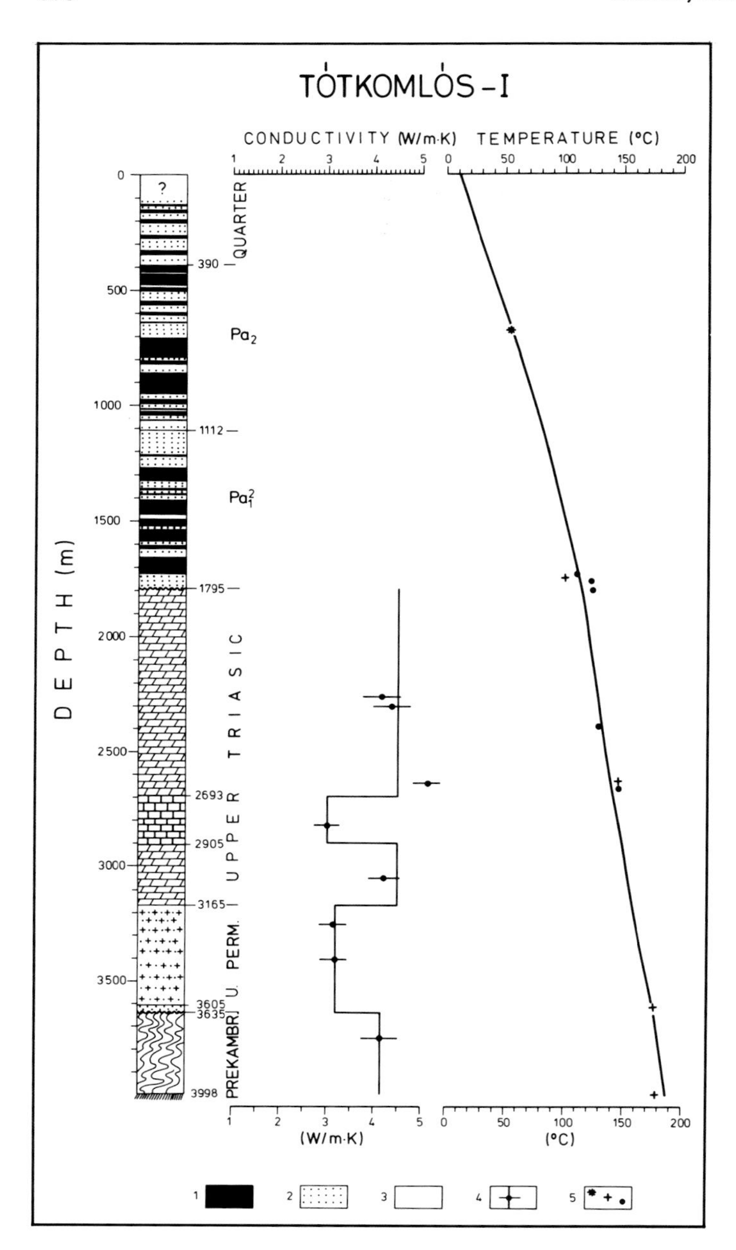

Figure 9. Simplified lithology, measured conductivities, and calculated temperature versus depth for the Tótkomlós-I well: (1) thick continuous shale beds; (2) thick continuous sandstone bed; (3) mixed development consisting of alternating shale and sandstone beds with thicknesses less than 5 m; (4) measured thermal conductivities and the conductivity trend; (5) measured temperatures: stationary values, corrected bottom hole temperature, and drill stem test data, respectively.

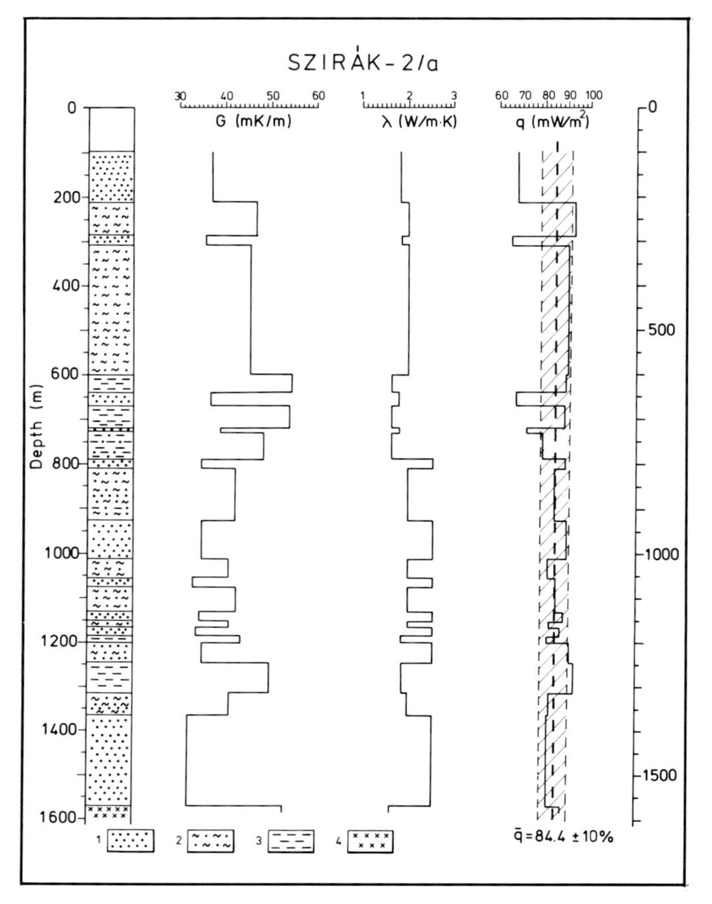

Figure 10. Simplified lithology, measured geothermal gradient profile (G), conductivity trend based on conductivity measurements (λ), and calculated heat flow (q) in the Szirák-2/a well: (1) sandstone bed; (2) siltstone bed; (3) shale bed; (4) volcanic rock.

Tótkomlós-I

This borehole is located on a basement high between the Makó trough and Békés depression. The drill hit the bottom of upper Miocene sediments at 1795 m and penetrated a Mesozoic, Paleozoic, and Precambrian(?) basement complex down to 3998 m (Figure 9). Because all of the drill cores came from the basement, the 1795–3998 m depth interval was used to determine the heat flow. In the Neogene sediments we could only estimate the heat flow. Calculations and results are shown in Table 4. The uncertainty of the heat flow value is about ±10–15%.

Szirák-2/a

This borehole is located southwest of the Mátra volcanic mountains. The drill bit penetrated Pliocene and upper Miocene strata and stopped in Karpatian schlier at the 2000 m depth. We made conductivity measurements on 39 drillcore samples and determined average conductivities for different rock types in different depth intervals, as shown in Table 5.

In this borehole, two continuous temperature logs were measured 1 year and 1.5 years after the drilling operation. Their

Table 4. Data Used to Determine the Heat Flow for the Tótkomlós-I Well

Lithogenetic or stratigraphic units[a]	Depth interval (m)	Lithology	N	k (W/mK)	δ	$\bar{k}$ (W/mK)[b]	$k = \frac{h_{pe}+h_{ps}}{h_{pe}/k_{pe}+h_{ps}/K_{ps}}$	$k = \frac{\Sigma h_i}{\Sigma h_i/k_i}$	G (°C/km)[c]	q = G×k	G = q/k (°C/km)[e]
Q + Pa_2	0–1112	60% pelite +	—	1.42[d]	—	1.46		1.82	59.4	108.1	64.6
		40% psammite		2.02[d]		2.00					
Pa_1^2	1112–1795	54% pelite	—	2.19[d]	—	2.10	2.22				47.8
		46% psammite		2.59[d]		2.39					
Upper	1795–2693	Dolomite	4	4.49	0.45	3.57	—				29.7
Triassic	2693–2905	Limestone	1	3.04	—	2.58	—				41.1
	2905–3165	Dolomite	—	4.49	—	3.44	—	3.80	27.4	104.1	30.8
Upper	3165–3601	Quartzic porphyry	2	3.19	0.04	2.62	—				
Permian	3605–3635	Quartzose sandstone	1	7.05	—	4.77	—				39.3
Precambrian	3635–3998	Granite	1	4.15	—	3.12	—				34.0
Total	0–3998							2.55		106	41.6

[a]Lithogenetic classification after Szalay and Szentgyörgyi (this volume).
[b]Heat conductivity values corrected for temperature (after Sekiguchi, 1984).
[c]Average geothermal gradients calculated from the temperature values of 111°C, 101°C, 123°C, 124°C, 177°C and 10°C at 1730 m, 1750 m, 1760 m, 1800 m, 3614 m depth and on the surface respectively.
[d]Estimated by equation shown on Figure 3 and Figure 4 respectively.
[e]Calculated "stationary" geothermal gradients.

Table 5. Data Used to Determine the Heat Flow for the Szirák-2/a Well

Lithology	Depth interval (m)	N	k (W/mK)	δ
clay, marl	0–1000	10	1.64	0.16
sand, sandstone	0–780	6	1.83	0.20
siltstone	0–1566	4	1.99	0.38
clay, marl	1000–1566	—	1.86[a]	—
sand, sandstone	780–1566	10	2.55	0.48
andezite	1566–1918	3	1.65	0.26

[a]Estimated by equation shown on Figures 4 and 5, respectively.

excellent agreement demonstrates that the well has already reached thermal equilibrium. Figure 10 shows that the temperature gradient changes greatly with depth because of the change in the average thermal conductivity. The mean value or heat flow determined for the different depth intervals is 84.4 mW/m^2.

The heat flow determination in Szirák-2/a is an exceptional example of an almost perfect determination of conductive heat flow with an error of ±5–10%. For the 0–1585 m depth interval the average gradient and mean conductivity are 41.5 °C/km and 2.04 W/mK, respectively.

HEAT FLOW MAP OF THE PANNONIAN BASIN AND ADJACENT REGIONS

The heat flow map (Map 5 in the enclosure) is based primarily on the data given in Table 1. These reliable values are shown on Map 5 in black. Less weight was given to the estimated data (shown in gray on Map 5). If more than one estimate could be derived for a small area (50–100 km^2), their average is shown. The Romanian data are from Negoita (1970) and Paraschiv and Cristian (1973). Other heat flow maps were also used: Cermák (1979), Cermák (1982, personal communication), Majorowicz and Plewa (1979), and Veliciu and Visarion (1984). If no data were available for a particular region, we took the patterns of temperature maps into consideration (Kolbah, 1978; Ronner, 1980; Dövényi et al., 1983). In such areas the heat flow isolines are broken to call attention to this uncertainty.

Map 5 is an uncorrected heat flow map, thus it reflects the heat flow disturbances due to rapid sedimentation, the effect of topography, and undulation of the basement.

DISCUSSION OF THE HEAT FLOW MAP

It has long been recognized that the thin crust, high heat flow, and formation of the Pannonian basin are genetically related (Stegena et al., 1975). This recognition was primarily based on the good correlation between the thin crust and elevated heat flow of the Pannonian basin. A comparison of our new map and a recent map of the Moho depth (Figure 11) lends further credence to this observation. In some cases, however, this correlation may be misleading. The low heat flow in the Transdanubian Central Range is not the consequence of a relative maximum in crustal thickness. Over much of this range Upper Triassic dolomite and limestone crop out. There is a great infiltration of meteoric water into these karstic carbonate rocks causing significant convective heat loss. The water infiltrates to a depth between 3 and 4 km and then comes back up to the surface in several hot springs at the margin of the Transdanubian Central Range (Deák et al., this volume). Simple heat balance calculations (Horváth et al., 1981) suggest that the conductive heat flow at depth should be as high as 80 to 100 mW/m^2. In other words, the deeper crust in this area is also characterized by high temperatures and heat flow. Similar convective processes also occur near Budapest in the Bükk Moun-

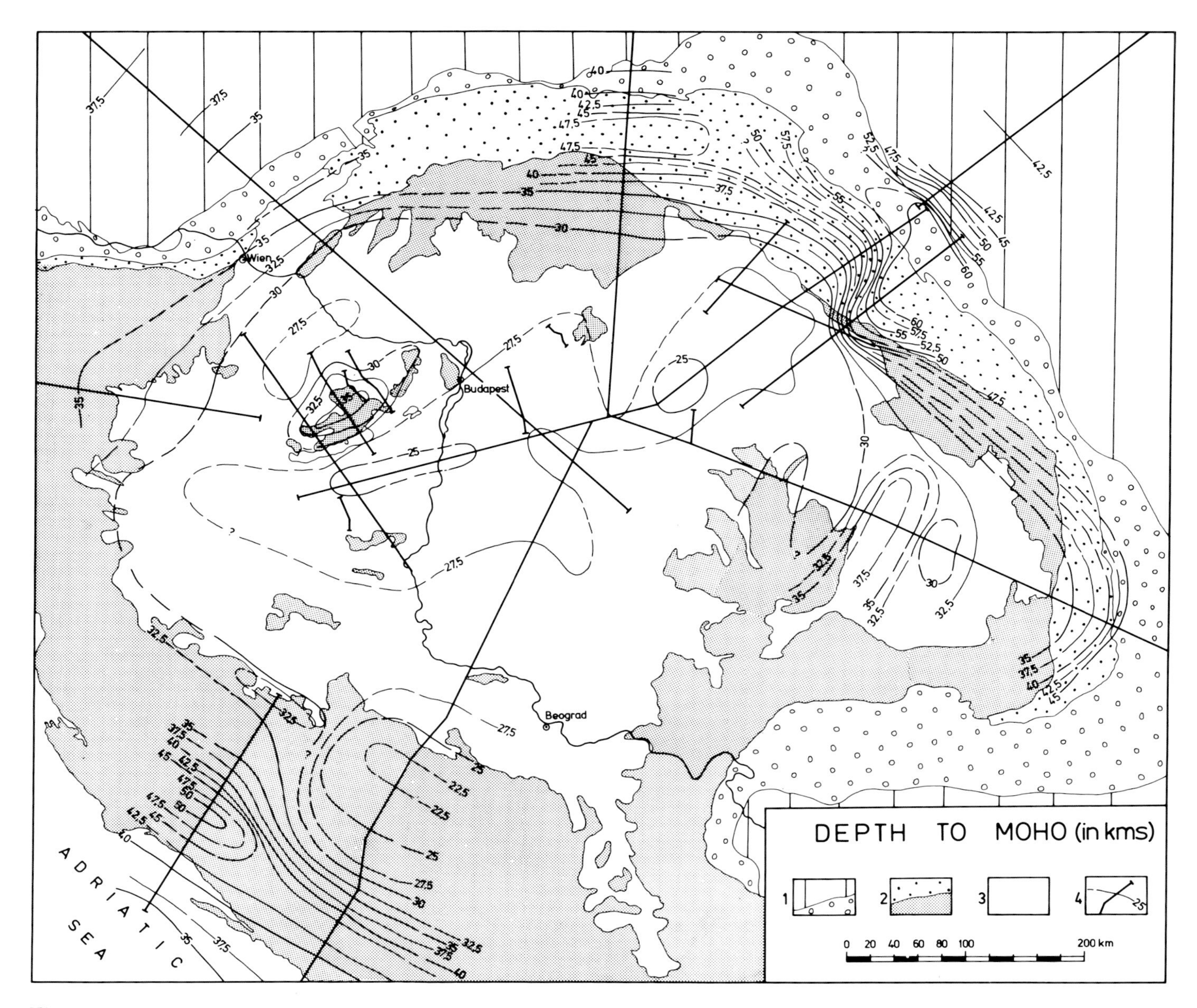

Figure 11. Depth to Moho in the Carpatho-Pannonian region: (1) European foreland and the foredeep molasse; (2) Alpine–Carpathian flysch belt and the inner units of the Alpine, Carpathian, and Dinaric mountains; (3) Tertiary basin areas; (4) Depth isolines to the Moho (in km) and the location of the deep seismic sounding lines. (From Horváth and Royden, 1981).

tains (to the west of Miskolc) and on the Aggtelek–Gemer karst plateau (southwest of Kosice).

Apart from these areas, which are associated with outcrops of Mesozoic carbonates in the Hungarian mountains, the conductive heat flow should be the dominant mechanism of heat transfer. This is because the hydraulic conditions of the Neogene sedimentary complex allow only local, shallow, or slow water movement (Horváth et al., 1981). From a geodynamic point of view, it is important to notice that the heat flow averages for the different subunits of the intra-Carpathian basins vary significantly (Figure 12).

One may expect a good correlation between basement depth and conductive heat flow because (1) the conductivity contrast between basement and sediments refracts the heat flow, and (2) thermal blanketing is less pronounced in areas with thinner young sediments. Such a correlation exists, but is not very significant (Dövényi et al., 1983). This indicates that this basement depth versus heat flow relationship is more complex and also depends on other factors, such as the heat generation of the basement, the magnitude of crustal extension, and the subsidence and sedimentation history. This problem is examined in detail by Horváth et al. (this volume) and Royden and Dövényi (this volume).

It is surprising that the Inner Carpathian calcalkaline volcanic regions are characterized by high heat flow anomalies. Simple model calculations (Horváth et al., 1986) show that the

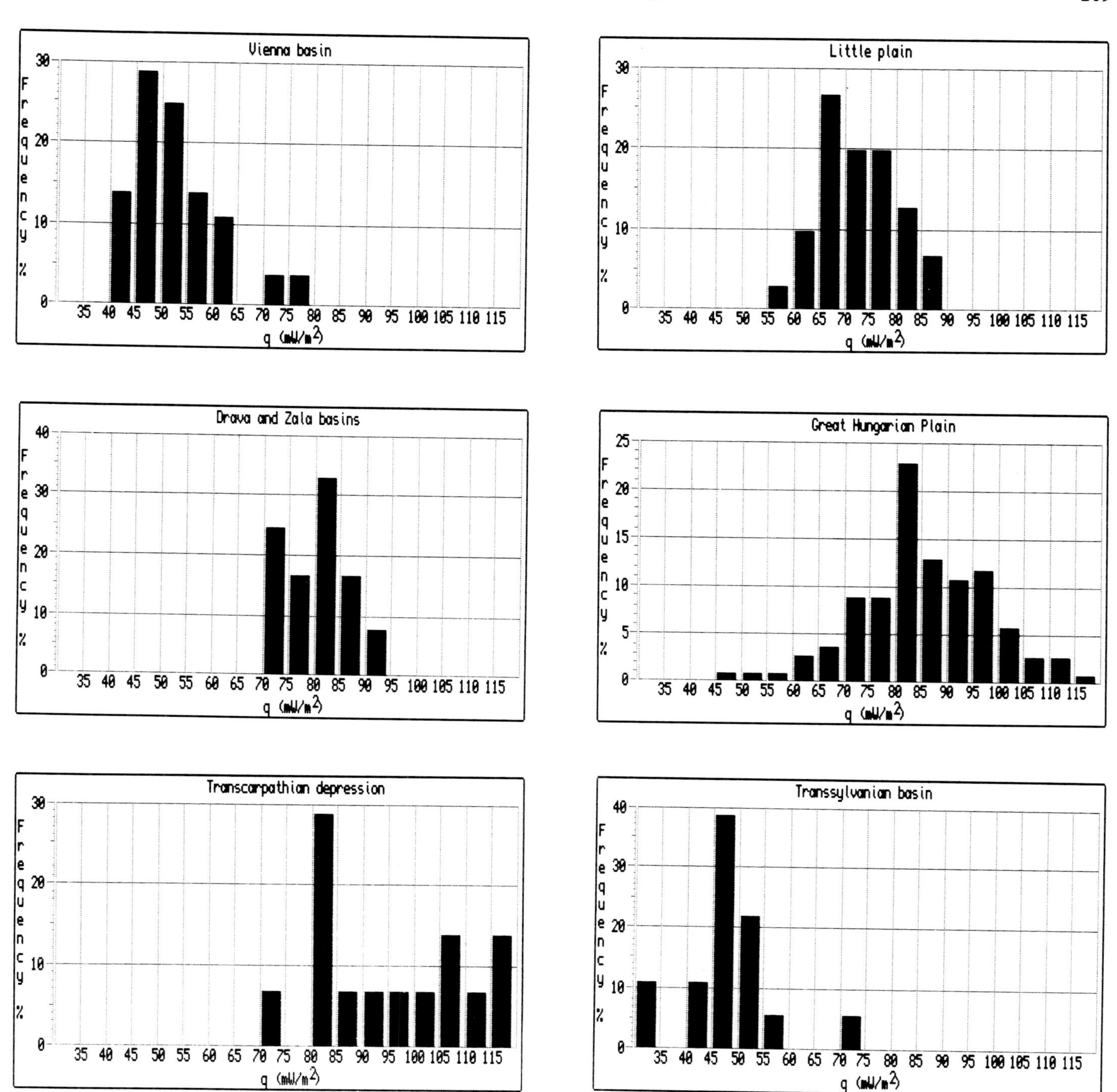

Figure 12. Frequency diagram of measured and estimated heat flow values for the intra-Carpathian basins based on heat flow data shown on Map 5 of the Enclosure.

extra heat cannot be derived from the cooling magmatic bodies intruded during Miocene and early Pliocene time. Such a local effect may be expected only in the East Carpathians where the final volcanic activity occurred as late as late Pliocene or early Pleistocene time. We conclude that the main reason for the heat flow highs in the neovolcanic areas, as well as in other parts of the Pannonian basin, must be related to elevated temperatures in the deeper crust. On the other hand, heat flow lows of the Vienna basin and Transylvanian depression are probably indicative of some important differences in the mechanism of basin formation.

REFERENCES

Boldizsár, T., 1956, Measurement of terrestrial heat flow in the coal mining district of Komló: Acta Techn. Acad. Sci. Hung., v. 15, p. 219–227.

Boldizsár, T., 1959, Terrestrial heat flow in the Nagylengyel oil field: Publ. Min. Fac. Sopron, v. XX, p. 27–34.

Boldizsár, T., 1964, Geothermal measurements of the twin shafts of Hosszuhetény: Acta Techn. Hung., v. 47/3-4, p. 467–476.

Boldizsár, T., 1965a, Heat flow in Oligocene sediments at Szentendre: Pure Appl. Geophys., v. 61, p. 127–138.

Boldizsár, T., 1965b, Terrestrial heat flow in Banska Stiavnica: Publ. Techn. Univ. Miskolc, v. 25, p. 105–108.

Boldizsár, T., 1966, Heat flow in the natural gas field of Hajduszoboszló: Pure Appl. Geophys., v. 64, p. 121–125.

Boldizsár, T., 1967, Terrestrial heat flow in Hungarian Permian strata at Bakonya: Pure Appl. Geophys., v. 67, p. 128–132.

Cermák, V., 1968a, Terrestrial heat flow in the Alpine-Carpathian foredeep in South Moravia: J. Geophys. Res., v. 73, p. 820–821.

Cermák, V., 1968b, Heat flow in the Upper Silesian coal basin: Pure Appl. Geophys., v. 69, p. 119–130.

Cermák, V., 1968c, Terrestrial heat flow in Eastern Slovakia: Trav. Inst. Géophys. Tchécosl. Acad. Sci., no. 275, Geofysikální Sborník 1967, NCSAV Praha, p. 303–319.

Cermák, V., 1975, Terrestrial heat flow in the Neogene foredeep and the flysch zone of the Czechoslovak Carpathians: Geothermics, v. 4, p. 8–13.

Cermák, V., 1976a, Zemský tepelný tok ve vrtu Likecko v magurském flysi ve onejsích Karpatech [Terrestrial heat flow in borehole Likeckoin the Magura flysch zone of the Czechoslovak Carpathians]: Cas. Min. Geol., v. 1, p. 193–198.

Cermák, V., 1976b, High heat flow measured in the Ostrava-Karvina coal basin: Stud. Geophys. Geol. v. 20, p. 64–71.

Cermák, V., 1976c, Terrestrial heat flow in two deep holes in the Ostrava-Karviná coal basin: Vestník Ustr. Us. Geol. v. 51, p. 75–84.

Cermák, V., 1977, Heat flow in five boreholes in Eastern and Central Slovakia: Earth Planet. Sci. Lett., v. 34, p. 67–70.

Cermák, V., 1979, Heat flow map of Europe, *in* V. Cermak and L. Ryback, eds., Terrestrial heat flow in Europe: Springer–Verlag, Berlin/New York, p. 3–40.

Cermák, V., and L. Rybach, 1979, Terrestrial heat flow in Europe: Springer–Verlag, Berlin/New York, 328 p.

Cull, J. P., 1974, Thermal conductivity probes for rapid measurements in rock: J. Phys E. Sci. Instrum., n. 7.

Demetrescu, C., 1978, On the geothermal regime of some tectonic units in Romania: Pure Appl. Geophys., v. 117, p. 124–134.

Dövényi, P., F. Horváth, P. Liebe, J. Gálfi, and I. Erki, 1983, Geothermal conditions of Hungary: Geophys. Transactions, v. 29/1, p. 3–114.

Greutter, A., 1977a, Földi hőáram Edelényben [Terrestrial heat flow in Edeleuy], Magyar Geofizika, v. 18/2, p. 15–25.

Greutter, A., 1977b, A kisalföld csehszlovák területéne k tervletenek geotermikus viszonyai [Geothermal conditions in the Czechoslovakian part of the Danube Lowland]: Geonómia és Bányászat, v. 10/3-4, p. 239–247.

Horváth, F., and L. Royden, 1981, Mechanism for the formation of the intra-Carpathian basins: a review: Earth Evolution Sci., v. 1/3-4, p. 307–316.

Horváth, F., P. Dövényi, and P. Liebe, 1981, Geothermics of the Pannonian basin: Earth Evolution Sci., v. 1/3-4, p. 285–291.

Horváth, F., P. Dövényi, and I. Laczó, 1986, Geothermal effect of magmatism and its contribution to the maturation of organic matter in sedimentary basins, *in* Lecture Notes in Earth Sciences, Vol. 5, G. Buntebarth and L. Stegena, eds., Paleogeothermics: Springer–Verlag, Berlin/Heidelberg, p. 173-183.

Kolbah, S., 1978, Znacaj mineralnih i termalnih voda ze prijenos i uskladistenje geotermicke energie [Importance of mineral and thermal waters in transfer and accumulation of geothermal energy]. Problemi hidrogeologije i inzenjerske geologije Jugoslavije, Zborn, radova 5: Jugoslovenskog simpozijuma o hidrogeologiji i inzenerskoj geologiji, Beograd, Knyiga 1, p. 125–140.

Kutas, R. I., Gordienko, V. V., Bevzyuk, M. I., and Zavgorodnaya, O. V., 1975, Novie apredelenia toplovaya potoka Karpatskom regione [New heat flow determinations of the Carpathian region]: Geofiz. Sbornik, "Naukova Dumka", v. 63, p. 68–71.

Lubimova, E. A., B. G. Polyak, Ya. B., Smirnov, R. I. Kutas, F. V. Firsov, S. I. Sergienko, and L. N. Liusova, 1973, Heat flow on the USSR territory, catalogue of data 1964–1972: Iss. of Soviet Geophysical Committee of the Academy of Sciences of the USSR, 64 p.

Majorowicz, J., and S. Plewa, 1979, Study of heat flow in Poland with special regard to tectonophysical problems, *in* V. Cermák and L. Rybach, ed., Terrestrial heat flow in Europe: Springer–Verlag, Berlin/New York, p. 240–252.

Marusiak, I., and I. Lizoń, 1975, Výsledky geotermického výsumku v Ceskoslovenskej casti viedenskej pánvy [The results of geothermic research in the Czechoslovak part of the Vienna Basin]: Geologicke práce, Správy 63 GUDS Bratislava, p. 191–204.

Negoita, V.,, 1970, Étude sur la distribution des températures en Roumania: Rev. Roum. Géol., Géophys. et Géogr., Série de Géophys., v. 14/1, p. 25–30.

Novák, V., 1971, Zemský tepelný tok v. hlubinných vrtech Zarosice 1 a 2 v oblasti Zdánického lesa [Terrestrial heat flow in deep borehole Zarosice 1 and 2 in the Zdánicke Forest region: Vesnik Ústr. Úst. Geol. V. 46, p. 277–284.

Paraschiv, D., and M. Cristian, 1973, Asupra particularitatilor regimului geotermic in nord-estul Depresiunii panonice [On the peculiarities of the geothermal regime in NE Pannonian basin]: Petrol si Gaze, v. 24/11, p. 655–660.

Ronner, V., 1980, Geothermische Energie, *in* R. Oberhauser, ed., Der Geologische Aufbau Österreichs: Springer–Verlag, New York, p. 574–579.

Salát, P., 1967, The measurements of terrestrial heat flow in the Mecsek Mts: Ph.D. thesis, University of Budapest, Hungarian.

Salát, P., 1968, The measurement of terrestrial heat flow at Budapest and Recsk: Unpublished paper.

Sekiguchi, K., 1984, A method for determining terrestrial heat flow in oil basinal areas: Tectonophysics, v. 103/1–4, p. 67–79.

Stegena, L., B. Géczy, and F. Horváth, 1975, Late Cenozoic evolution of the Pannonian basin: Tectonophysics, v. 26/1–2, p. 71–90.

Stegena, L., F. Horváth, I. Erki, and P. Dövényi, 1978–1984, Reports on heat flow measurements in boreholes Horna Poton FGHP-1, Tvrdosovce FGTv-1, Cilistov FGSC-1, Ciz MJC-1, Komarno FGK-1, Galánta FGG-1, Dvory nad Zytavou FGDZ-1, Galánta FGG-2, Nové Zamky GNZ-1, Bohelov GPB-1, Sobrance TMS-1, Liptovský Mikulás FGL-1, Vlcany FGB-2, Senec BS-1, Ciz BCS-2, Gabcikovo FGGa-1: Unpublished manuscripts.

Szalay, A., 1982, A rekonstrukciós szemléletü földtani kutatás lehetőségei a szénhidrogénperspektívák előrejelzésében [Possibilities of the reconstruction of basin evaluation in the prediction of hydrocarbon prospects]: PhD thesis, Hungarian Academy of Sciences, Budapest.

Veliciu, S., and C. Demetrescu, 1979, Heat flow in Romania and some relations to geological and geophysical features, *in* V. Cermák and L. Rybach, eds., Terrestrial heat flow in Europe: Springer–Verlag, Berlin/New York, p. 253–260.

Veliciu, S., and M. Visarion, 1984, Geothermal models for the East Carpathians: Tectonophysics, v. 103, p. 157–165.

Veliciu, S., M. Cristian, D. Paraschiv, and M. Visarion, 1977, Preliminary data of heat flow distribution in Romania: Geothermics v. 6, p. 95–98.

APPENDIX TO CHAPTER 16

Catalog and Locations of Temperature Measurements[a]

1	2	3	4	5	6	7
140-803	Nagylengyel	N1-259	2440	91	32.8	x2
			2484	92	32.6	x2
141-802	Nagylengyel	N1-353	2428	89	32.1	x2
141-801	Nagylengyel	N1-301	1800	66	30.6	x2
			2000	71	30.0	x2
			2250	109	43.6	x2
			2427	80	28.4	x2
			2427	81	28.8	x2
141-801	Nagylengyel	N1-244	2467	81.6	28.6	x2
141-804	Nagylengyel	N1-312	2300	87	33.0	x2
142-801	Nagylengyel	N1-427	2380	81	29.4	x2
143-805	Nagylengyel	N1-326	2260	97	38.1	x2
143-806	Nagylengyel	N1-317	2552	101.5	35.5	x2
			2580	102	35.3	x2
145-805	Nagylengyel	N1-307	2312	98	37.6	x2
145-806	Bak	Bak-5	500	28	34.0	c2
			1000	44	32.0	c2
			1500	63	34.7	c2
			2000	84	36.5	c2
			2500	106	38.0	c2
			2537	109	38.6	c2
145-807	Nagylengyel	N1-368	2340	98	37.2	x2
141-815	Söjtör	1	1690	90	46.7	x2
149-831	Magyarsztmiklós	S-4	2390	110	41.4	x2
149-841	Bajcsa	K-59	1380	90	57.3	3
146-844	Bajcsa	1	2087	106	45.5	a2
			3590	180	47.1	x2
147-843	Bajcsa	14-a	2240	114	46.0	x2
149-843	Bajcsa	Bj-17	2160	111	46.3	x2
144-854	Belezna	17	1905	108	50.9	x2
			2257	133	54.1	x2
145-854	Belezna	16	2290	128	51.1	x2
145-854	Belezna	19	2288	127	50.7	x2
146-852	Belezna	5	2448	133	49.8	x2
146-854	Belezna	13	2320	132.5	52.6	a2
146-853	Belezna	11	2303	122.5	48.6	x2
			2312	132	52.3	x2
146-853	Belezna	Be-2	2314	129	55.3	x2
146-854	Belezna	9	2300	124	49.1	x2
146-854	Belezna	4	2318	133	52.6	x2
147-853	Belezna	18	2457	135	50.5	x2
147-853	Belezna	20	1996	107	48.1	x2
			2005	117	52.9	x2
			2127	122	52.2	x2
			2152	114	47.9	x2
147-853	Belezna	21	2252	114	45.7	x2
			2256	112	44.8	a2
			2803	143	47.1	x2
147-853	Belezna	22	2192	120	49.7	a2
			2245	121	49.0	x2
147-853	Belezna	24	1710	99.5	52.1	x2
			1825	100	48.8	x2
			1970	110	50.3	a2
147-853	Belezna	25	1940	110	51.0	x2
147-853	Belezna	29	2264	119	47.7	x2
147-854	Belezna	15	2310	128	50.7	x2
148-853	Belezna	8	2293	117	46.2	x2
			2295	126	50.1	x2
148-854	Belezna	10	2465	123	45.4	x2
148-858	Zákány	2	2916	116	36.0	x2
157-712	Kapuvár	K-61	1780	81	39.2	1
155-731	Mihályi	Mf-1	1214	59.5	40.0	x2
156-730	Mihályi	M-26	1163	52	35.3	x2
			1393	63	37.3	x2
			1445	71.5	41.9	x2
152-736	Mihályi	M-37	61	11.5	8.2	c2
			114	11.3	2.6	c2
			175	11.7	4.0	c2
			237	14.0	12.7	c2
			305	16	16.4	c2
			365	20	24.7	c2
			425	22	25.9	c2
			485	26	30.9	c2
			545	29	33.0	c2
			605	34	38.0	c2
			665	41	45.1	c2
			726	44	45.5	c2
			787	47.5	46.4	c2
			845	50.3	46.5	c2
			907	53	46.3	c2
			967	56.5	47.1	c2
			1027	58.5	46.3	c2
			1087	61.5	46.5	c2
			1147	64.3	46.5	c2
			1207	67.5	46.8	c2
			1267	70.6	47.0	c2
			1327	74	47.5	c2
			1388	77	47.6	c2
			1448	80	47.7	c2
			1455	80	47.4	c2
150-751	Sárvár	17-43	1000	50	39.0	2
150-751	Sárvár	B-7	998	55	44.1	1
152-789	Vöckönd	Vö-1	620	42	51.6	b2
154-785	Vöckönd	Vö-2	544	39	53.3	b2
154-808	Pacsa	K-8	250	22	44.0	2
154-818	Kilimán		625	52	65.6	b2
156-831	Nagybakónak	Nab-1	2462	138	51.6	x2
			2387	123	46.9	x2
			2072	108	46.8	x2
158-839	Nagyrécse	3	3200	132	37.5	x2
159-841	Nagyrécse	4	2700	140	47.8	x2
150-842	Bajcsa	20	2112	108.5	46.2	x2
			2255	120	48.3	a2
			2304	127	50.3	x2
150-842	Bajcsa	23	2340	124	48.3	x2
150-842	Bajcsa	25	2295	123	48.8	x2
150-842	Bajcsa	28	2235	115	46.5	x2
150-842	Bajcsa	29	2118	110	46.7	x2
150-842	Bajcsa	31	2098	107	45.8	a2
150-842	Bajcsa	37	2065	110	47.9	x2

1	2	3	4	5	6	7	1	2	3	4	5	6	7
150–842	Bajcsa	38	2235	115	46.5	x2	163–848	Iharosberény	1	1278	78	52.4	a2
150–842	Bajcsa	19	2268	105	41.4	x2	163–849	Iharosberény	3	1364	82	52.1	x2
			2230	117	47.5	x2	168–849	Inke	22	528	39.5	54.9	x2
			2234	114	46.1	a2				742	50	52.6	a2
151–842	Bajcsa	8	2134	108	45.5	a2				822	54	52.3	a2
			2255	111	44.3	x2				887	55	49.6	x2
152–840	Nagykanizsa	B-62	1498	90	52.7	3				1174	68	48.6	a2
156–849	Liszó	Liszó-2	1749	96	48.6	x2	169–848	Inke	19-a	826	56	54.5	x2
			1826	99	48.2	x2				891	58	52.8	x2
			2519	139	50.8	x2				899	52	45.6	x2
154–852	Liszó	Liszó-1	1628	94	51.0	x2				946	60	51.8	a2
			1642	95	51.2	x2				1230	70	48.0	x2
			1720	98	50.6	x2				1145	65	47.2	x2
			2472	136	50.6	x2	167–849	Inke	?	1000	64	53.0	b2
159–861	Csurgó	B-7	401	30	44.9	1				1500	77	44	b2
168–694	Mosonszentjános	Mos-2	2106	100	42.3	x2	160–859	Csurgó	B-2	300	21	30	3
			2234	104.4	41.8	x2	163–865	Berzence	K-10	260	24	46.2	1
			2300	113.8	44.7	x2	169–880	Vizvár	Vi-7	1797	100	49.5	x2
			2370	115.5	44.1	x2				1782	99	49.4	x2
			2398	118.8	45.0	x2	169–880	Vizvár	9	1871	102	48.1	x2
161–719	Mihályi	29	1447	70	40.8	x2	169–881	Vizvár	22	1836	108	52.3	x2
			1458	67	38.4	a2	176–681	Mosonmagyaróvár	B-123	1850	83	38.9	2
			1605	73	38.6	x2				1956	87	38.9	1
162–719	Mihályi	M-21	1359	65	39.7	x2	171–693	Mosonszentjános	Mos-1	1446	91.1	55.4	x2
			1371	68	41.6	x2	173-702	Bosárkány	1	3890	173	41.4	x2
			1387	67	40.4	x2				3900	172	41.0	x2
			1403	69	41.3	x2				4063	185	42.6	x2
			1296	64	41.8	x2				4063	187	43.1	x2
			1249	62	40.8	x2				4474	225	47.6	x2
163–718	Mihályi	3	1068	58	44.0	a2				4485	226	47.7	x2
			1088	56	41.4	x2	173–709	Csorna	K-47	1640	76	39.6	2
165–716	Mihályi	2	1016	55	43.3	a2				1793	90	44.1	2
			1086	60	45.1	a2	172–711	Csorna	K-60	400	28	42.5	3
161–722	Mihályi	25	1607	76	40.4	x2	175–715	Pásztori	4	2040	88	37.7	x2
161–722	Mihályi	M-25	1415	70	41.7	x2				2807	123	39.9	x2
			1615	76	40.2	x2	172-710	Csorna	B-5	464	31	43.1	3
162-720	Mihályi	21	1120	58	42.0	a2	174–721	Pásztori	Pá-2	1374	62	37.1	x2
			1209	62	42.0	x2				1812	71	33.1	x2
162–720	Mihályi	26	1154	54	37.3	x2	177–720	Pásztori	Pá-1	1395	65	38.7	x2
			1402	65	38.5	a2				2003	86	37.4	x2
			1435	72	42.5	x2				2335	103	39.4	x2
166–750	Celldömölk	K-32	300	22.1	37.0	3				2562	115	40.6	x2
168–751	Celldömölk	1	2428	161	61.8	x2				2620	122	42.4	x2
			2505	168	62.7	x2	170-772	Ukk	K-4	471	33	48.8	3
161-753	Mesteri	17-14	1465	73	42.3	2	172-773	Ukk	1	812	37	32.0	x2
160–760	Borgáta	K-2	742	49	52.6	1				446	33	49.3	b2
167–786	Nagygörbo	Ng-1	750	31	28	b2	171–774	Ukk	2	463	31	43.2	b2
			1250	43	26.4	b2	174–779	Sümeg	B-1	280	34.3	85.7	3
			1500	49	26	b2	175–774	Csabrendek	K-2	397	43.3	83.1	3
161–828	Ujudvar	D-6	2307	120	47.2	b2	174–780	Sümeg		560	53	76.8	b2
162–828	Zalakaros	K-11	797	52	51.5	1	171–805	Keszthely	K-19	550	28	30.9	2
162–828	Zalakaros	19-18	2260	107	42.5	2	170–827	Sávoly	2	1960	116	53.6	x2
			2307	120	47.2	2	171–849	Vése	4	1190	71	50.4	a2
162–829	Zalakaros	K-8	2720	119	39.7	2	173–849	Vése	2	1265	72	48.2	a2
			2752	139	46.5	2	173–848	Vése	Vé-2	1270	74	49.6	x2
169–826	Sávoly	1	2340	100	38.0	x2	175–857	Somogyszob	B-8	264	27	56.8	1
			1844	106	51.5	x2	173–869	Tarany	12	2730	146	49.1	x2

1	2	3	4	5	6	7
178-868	Nagyatád	4	2500	136	49.6	x2
			2629	112	38.0	x2
176-869	Tarany	K-5	440	34	50.0	3
179-865	Nagyatád	B-53	490	41	59.2	2
			548	40	51.1	2
177-864	Nagyatád	K-45	349	31	54.4	1
178-864	Nagyatád	K-41	347	29	49.0	2
179-865	Nagyatád	B-17	542	36	44.3	3
177-878	Háromfa	B-6	300	24	40.0	3
171-881	Vizvár	10	1820	102.5	49.7	x2
			1950	114	52.3	a2
			2007	116	51.8	x2
171-881	Vizvár	26	1810	106	51.9	x2
171-881	Vizvár	7	1780	99	48.9	x2
171-881	Vizvár	29	1416	88	53.7	x2
			1842	106	51.0	a2
172-881	Vizvár	3	1832	105	50.8	a2
172-882	Vizvár	1	1872	105	49.7	a2
			1890	106	49.7	a2
176-883	Heresznye	12	2260	131	52.7	x2
176-883	Heresznye	13	1372	84	52.5	x2
			1472	89	52.3	x2
179-885	Babócsa	BF-5	1220	78	54.1	a2
			1310	84	55.0	x2
179-885	Babócsa	Bk-4	2099	108	45.7	x2
179-885	Babócsa	Bk-8	2125	109	45.6	x2
179-885	Babócsa	Bk-9	2314	114	44.1	x2
179-885	Babócsa	B-24	1760	98	48.9	x2
179-885	Babócsa	B-26	1890	109	51.3	x2
179-885	Babócsa	B-27	1765	101	50.4	x2
179-885	Babócsa	2	1605	96	52.3	x2
178-886	Babócsa	B-2	2383	119	44.9	3
189-683	Lipót	K-7	2201	111	45.4	2
185-698	Lébénymiklós	B-28	2200	103	41.8	3
188-736	Takácsi	1	1476	63	35.2	x2
189-737	Takácsi	2	1437	51	27.8	x2
189-741	Pápa	B-18	820	35	29.3	2
			825	40	35.2	2
185-791	Tapolca	18-27	540	39	53.7	3
185-791	Tapolca		480	34	50	b2
186-812	Balatonfenyves	B-18	473	33	46.5	2
			494	37	52.6	1
183-822	Marcali	K-24	716	53	58.7	1
183-822	Marcali	B-19	1270	87	59.8	2
185-825	Marcali	K-18	388	36	64.4	2
184-824	Marcali	B-25	270	27	59.3	3
184-824	Marcali	?	296	30	64.2	b2
183-835	Mesztegnyo	B-2	268	26	55.9	1
181-843	Böhönye	K-4	280	25	50.0	1
181-863	Nagyatád	K-54	400	33	52.5	2
			453	40	61.8	2
188-861	Nagykorpád	B-1	810	67	67.9	2
184-878	Rinyaujlak	K-2	398	25	32.7	1
185-880	Csokonyavisonta	K-2	1297	85	56.3	1
			1310	92	61.1	2
181-889	Komlósd	1	2964	136	41.8	x2
182-888	Görgeteg-Babócsa	GB-27	1773	101	50.2	x2
188-888	Görgeteg-Babócsa	GBK-2	2095	99	41.5	x2
189-889	Görgeteg-Babócsa	GBK-1	2124	109	45.7	x2
186-894	Barcs	B-2	700	42	42.9	3
190-684	Lipót	7-41	1798	73	34.5	2
197-716	Gyorszemere	K-7	1320	67	42.4	2
197-716	Gyorszemere	K-5	410	30	46.3	2
192-721	Tét	6	433	30	43.9	b2
199-727	Tét	2	2790	110	35.5	b2
192-721	Tét	B-12	433	30	43.9	2
191-726	Gyarmat	B-7	294	29	61.2	2
192-731	Vaszar	Vasz-3	1277	69	45.4	x2
191-741	Pápa	K-35	600	48	63.3	3
193-748	Tapolcafo	B-11	380	18	18.4	1
195-758	Magyarpolány	?	350	29	54.3	b
198-793	Kovágóörs	K-1	244	19.2	36.9	3
195-809	Fonyód	K-18	550	44	60.0	2
195-808	Fonyód	B-28	430	36	58.1	3
196-808	Fonyód	?	430	37	60.5	b2
196-814	Buzsák	K-2	540	49	70.4	3
195-825	Nikla	Ni-1	1748	100	49.9	b2
			1823	93	45.0	b2
192-821	Táska	K-5	670	80	103.0	b2
			1010	83	71.3	3
194-821	Táska	K-6	680	72	89.7	1
193-821	Táska	K-3	690	80	100.0	1
194-822	Táska	K-4	910	74	69.2	2
			930	70	63.4	1
198-825	Öreglak	1	2145	120	50.8	x2
198-838	Mezocsokonya	4	1722	100	51.7	x2
199-839	Mezocsokonya	6	1717	97	50.1	x2
			1731	92	46.6	a2
199-840	Mezocsokonya	15	1715	101	51.9	a2
			1764	97	48.2	x2
			1876	98	45.8	x2
198-844	Somogysárd		261	28	61.3	b2
198-845	Somogysárd	K-2	330	27	45.4	3
199-850	Kiskorpád	K-4	256	25	50.8	1
195-878	Homokszentgyörgy	B-7	277	24	43.3	3
199-884	Kálmáncsa	K-3	770	49	48.1	3
198-888	Darány	Dar-1	1999	108	48.0	x2
			1583	88	48.0	x2
			1566	86	47.3	x2
205-702	Gyor	?	540	34	42.6	b2
202-702	Gyor	B-109	512	34	44.9	1
204-703	Gyor	B-80	509	29	35.4	2
			582	35	41.2	2
201-703	Gyor	B-81	1973	87	38.5	1
202-702	Gyor	B-60	1998	94	41.5	2
205-707	Gyor	K-107	2000	86	37.5	3
200-704	Gyor	B-148	2028	82	35.0	2
207-714	Nyul	K-9	259	21	38.6	2
204-721	Gyorszemere	2	2280	94	36.4	x2
209-805	Kaposujlak	K-2	300	33	73.3	1
204-812	Szolosgyörök	2	267	27	59.9	b2
203-832	Osztopán	?	350	31	57.1	b2

1	2	3	4	5	6	7	1	2	3	4	5	6	7
201-839	Mezocsokonya	12	1588	93	51.6	x2	213-849	Kaposvár	B-40	432	39	62.5	3
			1616	100	55.1	x2	212-848	Kaposvár	144	400	38	65.0	b2
			1691	98.5	52.0	x2	215-851	Kaposvár	181	420	42	71.4	b2
			1734	94	47.9	a2	211-851	Kaposvár	B-55	268	30	67.2	3
201-838	Mezocsokonya	20	1713	98	50.8	a2	210-850	Kaposvár	K-160	252	30	71.4	2
202-839	Mezocsokonya	1	1615	99	54.5	a2				280	29	60.7	2
202-839	Mezocsokonya	5	1680	92	48.2	a2	214-851	Kaposvár	B-76	286	30	62.9	3
			1707	89	45.7	x2	210-850	Kaposvár	K-161	270	29	63.0	1
			1750	93	46.9	a2	215-851	Kaposvár	K-231	290	30	62.1	3
203-839	Mezocsokonya	7	1698	93	48.3	a2	219-850	Sántos	K-28	280	33	75.0	1
			1726	94	48.1	x2	218-850	Kaposvár	K-156	285	32	70.2	1
204-839	Mezocsokonya	9	1640	96	51.8	a2	213-852	Kaposvár	K-154	300	33	70.0	1
			1651	96.5	52.1	x2	218-851	Sántos	K-5	300	33	70.0	1
			1678	94	49.5	a2	219-851	Sántos	K-6	305	33	68.9	1
			1727	95	48.6	x2	219-850	Sántos	K-28	280	33	75.0	1
205-837	Mezocsokonya	8	2240	118	47.8	a2	215-851	Kaposvár	B-175	420	42	71.4	2
208-839	Mezocsokonya	10	1614	92	50.2	x2	213-851	Kaposvár	B-69	1000	70	58.0	1
202-840	Mezocsokonya	21	1685	97.5	51.0	x2	212-885	Szigetvár	K-19	298	32	67.1	3
			1734	99	50.2	a2	212-883	Szigetvár	K-60	992	64	52.6	1
			1769	105	52.6	x2	213-884	Szigetvár	B-45	305	37	82.0	2
204-840	Mezocsokonya	18	1718	96.5	49.5	x2				349	40	80.2	2
205-840	Mezocsokonya	14	1673	95	49.6	x2	213-884	Szigetvár	K-23	790	67	69.6	1
			1715	101	51.9	x2	214-885	Szigetvár	1-29	780	65	67.9	1
			1740	97	48.9	a2	216-904	Sellye	1	1952	81	35.3	x2
			1796	99	48.4	x2	216-906	Sellye	K-11	760	57	59.2	3
202-841	Mezocsokonya	K-1	291	28	55.0	1	225-700	Ács	1	1870	96	45.5	x2
209-846	Juta	B-5	330	32	60.6	1	226-745	Dudar	K-2a	250	13.3	12.0	3
209-850	Kaposnyilak	K-2	300	33	70.0	1	223-819	Andocs	B-2	260	28	65.4	1
209-850	Bárdudvarnok	K-2	309	32	64.7	1	224-830	Igal	B-6	360	54	119.4	3
208-886	Molvány	K-2	256	24	46.9	2	224-830	Igal	B-1	610	87	124.6	1
206-886	Merenye	K-2	258	28	62.0	2				620	83	116.1	2
			279	33	75.3	2	228-846	Mosdós	K-2	337	32	59.3	2
204-890	Kisdobsza	1	2700	124	41.5	x2	221-849	Taszár	K-3	464	35	49.6	3
204-890	Pettend	1-19	640	49	57.8	3	221-848	Taszár	13-150	506	40	55.3	2
205-906	Felsoszentmárton	1	3365	168	46.4	x2	220-853	Sántos	K-17	259	28	61.8	2
211-718	Pannonhalma	B-6	423	33	52.0	b2				266	31	71.4	2
210-733	Bakonyszentlászlo	Bszl-6	387	30	51.8	b2	220-851	Sántos	K-4	290	32	69.0	1
			533	33.2	42.6	b2	221-851	Sántos	K-7	290	31	65.5	1
215-798	Balatonöszöd	?	280	29.5	67.9	b2	220-850	Sántos	K-14	300	32	66.7	1
212-815	Somogytur	K-3	249	23	48.2	1	220-853	Sántos	K-20	347	32	57.6	1
215-839	Somogyaszaló	K-3	381	30	49.9	1	220-851	Sántos	K-10	210	25	61.9	1
210-844	Kaposvár	B-4/a	294	27	51.0	1	220-850	Sántos	K-29	309	27	48.5	1
212-848	Kaposvár	K-206	300	32	66.7	1	224-851	Kaposkeresztur	K-3	310	31	61.3	3
215-848	Kaposvár	K-218	250	31	76.0	1	224-855	Kaposkeresztur	K-10	280	29	60.7	3
215-848	Kaposvár	K-220	326	33	64.4	1	227-885	Szentlorinc	K-6	410	42	73.2	3
215-848	Kaposvár	K-15	312	32	64.1	1	228-886	Szentlorinc	K-7	410	38	63.4	2
215-849	Kaposvár	B-39	285	29	59.6	3				448	37	58.0	2
215-849	Kaposvár	K-204	280	31	67.9	1	238-698	Komárom	K-21	1263	62	40.4	1
211-848	Kaposvár	B-48/a	293	30	61.4	3	239-696	Komárom	B-62	1246	51	32.1	2
215-849	Kaposvár	K-208	310	34	71.0	1	237-712	Nagyigmánd	K-15	287	25	48.8	2
213-846	Kaposvár	K-216	330	33	63.6	1	231-748	Jásd	K-1	300	13	10.0	2
215-849	Kaposvár	K-196	347	29	49.0	2	233-824	Törökkoppány	?	248	30	76.6	b2
212-847	Kaposvár	K-210	360	36	66.7	1	235-838	Nak	K-3	270	35	88.9	2

1	2	3	4	5	6	7
235-838	Nak	K-3	272	29	66.2	2
233-837	Nak		348	28	48.9	b2
239-849	Dombóvár	B-46	400	38	65.0	1
239-849	Dombóvár	B-45	480	39	56.3	3
230-852	Nagyberki	K-2	265	28	56.4	1
233-852	Nagyberki	K-5	380	35	60.5	3
239-850	Kaposszekcso	K-9	323	29	52.6	2
			371	28	43.1	2
239-850	Kaposszekcso	K-7	408	36	58.8	3
239-851	Kaposszekcso	K-10	440	36	54.5	3
238-865	Oroszló	K-2	395	35	60.8	3
239-879	Kovágótöttös	MePe-19	100	16.0	50.0	c1
			200	21.0	50.0	c1
			300	24.5	45.0	c1
			400	28.0	42.5	c1
			500	32.0	42.0	c1
			600	37.0	43.3	c1
			700	41.5	43.6	c1
			800	45.0	42.5	c1
230-887	Szentlorinc	K-5	261	28	65.1	3
230-887	Szigetvár	K-26	250	31	80	2
231-886	Szentlorinc	K-9	330	26	45.5	1
240-697	Komárom	B-62	1090	51	36.7	1
249-698	Almásfüzito	K-65	328	25	42.7	2
247-733	Mór	K-82	457	25	32.8	2
249-803	Felsonyék	K-2	250	28	68	1
241-845	Dombóvár	?	270	34	81.5	b2
247-842	Döbrököz	K-12	250	31	76.0	2
244-844	Dombóvár	K-55	273	30	65.9	1
246-846	Mágócs	K-9	237	31	80.2	2
			298	28	53.7	2
243-846	Dombóvár	K-54	640	52	62.5	2
			788	61	62.2	1
241-846	Dombóvár	K-60	740	59	63.5	2
			1190	63	42.9	2
245-850	Mágócs	K-13	350	40	80.0	3
242-865	Liget	K-4	302	29	59.6	3
248-868	Komló	K-56	339	34	67.8	2
240-869	Magyarhertelend	K-1	480	43	66.7	3
244-869	Magyarszék	K-6	599	41	50.1	2
246-870	Mánfa	B-2	425	52	96.5	2
245-870	Magyarszék	K-4	400	36	62.5	3
248-872	Sikonda	K-6	599	42	51.8	2
240-879	Kovágótöttös	MePe-20	100	15	40.0	c1
			200	18	35.0	c1
			300	22	36.7	c1
			400	28	42.5	c1
			504	32	41.7	c1
			605	36	41.3	c1
			708	40	41.0	c1
			809	44.5	41.4	c1
			910	47	39.6	c1
			961	50	40.6	c1
241-879	Kövágótöttös	MePe-21	100	17	60.5	c2
			200	20	45.0	c2
			300	24	43.3	c2
			400	30	47.5	c2
			500	33	44.0	c2
			600	36	41.7	c2
			680	39	41.2	c2
241-879	Kovágótöttös	MePe-12	195	21.5	53.8	c3
			295	25.0	47.5	c3
			495	31.5	41.4	c3
			695	35.0	34.5	c3
			945	42.5	33.3	c3
241-880	Kovágótöttös	MePe-6	200	19.0	40.0	c3
			300	24.0	43.3	c3
			400	27.0	40.0	c3
			500	29.5	37.0	c3
			600	31.5	34.2	c3
			690	34.5	34.1	c3
242-880	Kovágótöttös	MePe-14	100	16.0	50.0	c3
			200	20.0	45.0	c3
			400	26.0	37.5	c3
			500	30.0	38.0	c3
			600	33.0	36.7	c3
			700	37.0	37.1	c3
			800	40.0	36.3	c3
			890	43.0	36.0	c3
242-880	Kovágótöttös	MePe-18	200	20.5	47.5	c3
			300	26.5	51.7	c3
			400	32.0	52.5	c3
			500	37.0	52.0	c3
			600	40.0	48.3	c3
			700	43.0	45.7	c3
			800	46.6	44.5	c3
			900	51.3	44.8	c3
			950	52.0	43.2	c3
243-880	Kovágótöttös	MePe-22	160	19.0	56.3	c3
			260	23.3	51.2	c3
			360	29.0	52.8	c3
			460	33.8	51.7	c3
			660	40.5	46.2	c3
			860	47.0	43.0	c3
			900	49.0	43.3	c3
244-880	Kovágótöttös	MePe-9	300	23.0	43.3	c3
			400	25.5	38.8	c3
			500	33.0	46.0	c3
			600	35.5	42.5	c3
			700	40.5	43.6	c3
			800	44.5	43.1	c3
244-880	Kovágótöttös	MePe-1	100	15.5	55.0	c3
			200	18	40.0	c3
			300	21	36.7	c3
			400	23	32.5	c3
			500	25	30.0	c3

1	2	3	4	5	6	7
244–880	Kovágótöttös	MePe-1	600	27.7	29.5	c3
			700	29.5	27.9	c3
			782	31	26.9	c3
244–881	Kovágótöttös	MePe-2	100	13	30.0	c3
			200	15	25.0	c3
			300	17	23.3	c3
			400	19.5	23.8	c3
			500	21.3	22.6	c3
			600	24.7	24.5	c3
			750	30.7	27.6	c3
246–881	Kovágótöttös	MePe-11	100	13.5	35.0	c3
			200	16.0	30.0	c2
			300	18.3	27.7	c2
			400	21.1	27.8	c2
			500	24.1	28.2	c2
			600	26.6	27.7	c2
			700	30.0	28.6	c2
246–881	Kovágótöttös	MePe-4	100	17.8	78.0	c3
			200	20.5	52.5	c3
			300	23.0	43.3	c3
			400	25.3	38.3	c3
			500	28.9	37.8	c3
			600	32.0	36.7	c3
			700	35.3	36.1	c3
246–883	Kovágótöttös	MePe-16	200	17.8	34.0	c3
			300	20.3	31.0	c3
			400	23.8	32.0	c3
			500	27.5	33.0	c3
246–883	Kovágótöttös	MePe-3	100	12	20.0	c3
			200	15	25.0	c3
			300	17.5	25.0	c3
			400	20	25.0	c3
			450	22	26.7	c3
246–884	Kovágótöttös	MePe-10	100	13	20.0	c3
			200	15	20.0	c3
			300	20	30.0	c3
			400	22	27.5	c3
			500	24	26.0	c3
246–884	Kovágótöttös	MePe-8	420	28.5	41.7	c3
			520	29.5	35.6	c3
			620	36.5	41.1	c3
			720	40.5	41.0	c3
			120	17.0	50.0	c3
			220	20.5	43.2	c3
			310	23.2	39.4	c3
			420	28.0	40.5	c3
			520	27.5	31.7	c3
			620	36.5	41.1	c3
247–884	Kovágótöttös	MePe-17	300	25.0	46.7	c3
			400	28.3	43.3	c3
			500	31.9	41.8	c3
			600	35.8	41.3	c3
			740	40.0	39.2	c3
244–880	Kovágótöttös	MePe-7	120	18.0	58.3	c3
244–880	Kovágótöttös	MePe-7	220	21.5	47.7	c3
			320	25.5	45.3	c3
			420	29.0	42.9	c3
			520	32.0	40.4	c3
			620	35.0	38.7	c3
			720	39.5	39.6	c3
			820	43.0	39.0	c3
			920	47.0	39.1	c3
244–880	Kovágótöttös	MePe-5	100	20.5	105.0	c3
			300	25.5	51.7	c3
			400	28.0	45.0	c2
			600	36.0	43.3	c3
			700	39.5	42.1	c3
			800	42.0	40.0	c3
244–880	Kovágótöttös	MePe-15	175	19.5	54.3	c3
			275	20.0	36.4	c3
			375	22.0	32.0	c3
			475	24.0	29.5	c3
			575	27.5	30.4	c3
			675	36.5	39.3	c3
247–884	Pécs	B-27	255	31	74.5	3
249–884	Pécs	K-80	249	28	64.3	1
241–891	Gyód	K-1	280	29	60.7	3
244–915	Drávaszabolcs	K-10	700	34	31.4	2
252–705	Tata	K-28	550	27	29.1	2
252–705	Tata	K-28a	1249	34	18.4	2
253–709	Tata	33	870	27	18.4	b2
255–706	Tata	30	546	34	42.1	b2
255–723	Oroszlány	K-6	739	20.9	14.9	3
251–747	Fehérvárcsurgó	K-5	270	26	59.3	3
251–799	Mezokomárom	8	250	27	64	b1
253–798	Mezokomárom	K-7	280	32	75	1
254–798	Lajoskomárom	B-11	426	41	70.4	1
259–805	Ozora	K-9	270	24	48.1	1
251–820	Tamási	B-41	415	37	62.6	1
252–823	Tamási	K-27	248	32	84.7	2
251–821	Tamási	K-35	1700	116	61.8	2
			1823	114	56.5	b2
255–838	Kurd	3	53	11.6	11.3	c2
			109	14.6	33.0	c2
			166	18.1	42.8	c2
			222	22.6	52.3	c2
			280	27.3	58.2	c2
			338	32	62.1	c2
			397	36	63.0	c2
			448	40	64.7	c2
256–848	Lengyel	K-6	500	36	48.0	2
253–868	Komló Zobák akna		347.7	25.5	41.7	c1
			361.7	25.6	40.4	c1
			368.4	26.9	43.2	c1
			402.8	28.1	42.5	c1
			412.8	28.5	42.4	c1
			418.4	28.9	42.8	c1
			426.6	29.1	42.4	c1

1	2	3	4	5	6	7
253-868	Komló Zobák akna		430.7	29.2	42.3	c1
			434.2	29.6	42.8	c1
			450.1	30.2	42.7	c1
			473.4	31.1	42.5	c1
			493.2	32.2	43.0	c1
			501.4	33.0	43.9	c1
			510.3	33.0	43.1	c1
			518.7	33.3	43.0	c1
			533.1	34.0	43.1	c1
			540.6	34.5	43.5	c1
			546.6	35.8	45.4	c1
			551.0	36.0	45.4	c1
			555.6	36.5	45.9	c1
			561.1	36.5	45.4	c1
			564.2	36.5	45.2	c1
			567.5	36.5	44.9	c1
			570.0	37.0	45.6	c1
			573.0	37.0	45.4	c1
			580.0	37.0	44.8	c1
258-872	Hosszuhetény felvonóakna		107.3	14.9	36.3	c2
			113.2	15.3	38.0	c2
			119.6	15.0	33.4	c2
			138.5	16.1	36.8	c2
			193.0	16.7	29.5	c2
			215.0	17.8	31.6	c2
			222.0	17.8	30.6	c2
			239.0	18.4	31.0	c2
			253.0	18.6	30.0	c2
			269.0	18.8	29.0	c2
			278.0	19.6	30.9	c2
			338.0	21.3	30.5	c2
			489.0	28.0	34.8	c2
			492.0	28.1	34.8	c2
			495.0	28.1	34.5	c2
			498.0	28.2	34.5	c2
			513.0	28.3	33.7	c2
			524.0	29.4	35.1	c2
251-873	Mánfa	B-42	337	26	44.5	3
259-887	Hasságy	K-1	298	34	73.8	2
257-883	Ellend-Romonya	K-1	1442	92	55.5	2
257-896	Ujpetre	K-2	450	44	71.1	1
255-889	Ujpetre	K-3	266	34	82.7	3
256-70	Tarján	T-9	468	32	44.9	b2
268-711	Tarján	B-5	552	18.2	12.7	3
268-712	Tarján	B-7	287	14.4	11.8	3
263-737	Csákvár	I	275	29	69.1	b2
266-735	Csákvár	K-17	415	25	33.7	2
260-758	Székesfehérvár	Csitaŕy-kut	1100	41	27.3	b2
261-792	Dég	K-11	250	25	56	1
261-803	Mezoszilas	K-9	270	24	48.2	1
268-805	Igar	K-6	275	36	90.9	2
263-815	Pincehely	K-7	250	31	80.0	1
263-843	Tevel	K-3	300	31	63.3	1
266-863	Hidas	B-3	304	28	55.9	3
264-906	Villány	B-12	599	22	16.7	1
278-701	Tokod	K-2	305	19.5	29.5	3
274-706	Nagysáp	?	422	39	66.4	b2
279-727	Bicske	18	426	33	51.6	b2
277-724	Bicske	B-2	391	15.5	14.1	3
276-721	Bicske	K-22	250	17	28.0	1
278-758	Agárd	K-149	907	64	58.0	2
274-768	Seregélyes	B-13	313	31	63.9	2
			360	28	47.2	2
275-767	Seregélyes	K-23	484	31	41.3	1
272-806	Simontornya	?	335	35	71.6	b2
273-807	Simontornya	B-21	282	29	63.8	3
271-807	Simontornya	B-29	270	35	88.9	3
272-806	Simontornya	16-12	415	40	69.9	2
279-810	Vajta	B-4	824	53	51	2
274-840	Felsonána	K-4	251	26	55.8	1
277-887	Székelyszabar	K-2	245	26	57.1	2
274-903	Majs	K-1	350	41	82.9	1
285-691	Esztergom	B-5	323	29.3	55.7	3
284-698	Dorog	B-1	374	20.2	24.1	3
284-699	Dorog	B-4	378	18.9	21.2	3
283-703	Csolnok	?	451	31	46.6	b2
286-725	Bia	K-4	355	32	61	1
287-724	Bia	K-3/a	800	43	41.3	3
284-728	Etyek	K-3	527	32	41.8	b2
286-726	Etyek	K-1	371	30.1	54.2	3
280-737	Vál	3	285	21	38.6	2
			783	38	35.8	2
			900	48.5	42.8	2
288-745	Martonvásár	47	260	28	65.4	b2
281-751	Kápolnásnyék	B-40	244	24	53.3	1
288-767	Pusztaszabolcs	K-20	270	23	44.4	1
287-811	Németkér	B-6	265	30	67.9	3
280-810	Vajta	B-4	824	53	49.8	2
284-854	Szekszárd	K-72	260	33	84.6	3
286-881	Dunaszekcso	K-10	267	25	48.7	3
287-886	Dunaszekcso	B-4	294	30	61.2	3
293-667	Perocsény	K-2	283	28	60.1	2
294-723	Páty	12-74	400	45	87.5	2
298-729	Törökbálint	19	470	33	48.9	b2
291-727	Biatorbágy	3	764	39	38	b2
299-730	Törökbálint	18	583	28	30.9	b2
290-740	Martonvásár	K-37	250	25	56.0	1
292-742	Martonvásár	K-47	260	26	57.7	1
298-741	Érd	31	250	29	72.0	b2
299-781	Dunaujváros	K-21	550	48	67.3	1
299-782	Dunaujváros	K-36	380	31	52.6	2
293-795	Eloszállás	K-6	260	26	53.9	1
299-801	Dunaföldvár	B-32	400	33	52.5	2
			599	32	33.4	2
294-822	Paks	B-54	345	33	60.9	3
298-886	Nagybaracska	2-100	400	42	75.0	3
298-890	Dávod	K-50	385	44	83.1	1
298-890	Dávod	K-47	400	41	72.5	2

1	2	3	4	5	6	7
298-890	Dávod	K-47	680	44	47.1	2
301-694	Visegrád	K-7	1301	74	49.2	1
308-714	Bp. Csillaghegy	B-5	500	24.1	28	3
309-719	Bp. Latorca u.	B-43	268	46	134.3	2
309-720	Budapest	B-43	260	43	126.9	2
			268	46	133.3	2
308-721	Budapest	B-20	310	74	206.4	3
308-727	Budapest	B-15	530	53	81.1	3
307-734	Bp. Budafok	B-68	2001	63	26.0	2
306-741	Szigetszentmiklós	K-30	290	19	27.6	2
305-740	Szigetszentmiklós	K-32	270	23	44.4	2
301-759	Ráckeve	K-59	940	47	38.3	2
			1059	71	56.7	2
300-782	Dunaujváros	K-21	750	48	48.0	2
305-805	Solt	K-166	270	32	81.5	3
300-800	Dunaföldvár	B-24	530	37	47.2	3
309-813	Harta	K-37	309	32	64.7	3
304-832	Kalocsa	K-107	318	34	69.2	3
310-699	Leányfalu	B-4	862	60.1	58.1	3
314-693	Vác	B-58	1098	42	29.1	2
			1177	42	27.2	b2
311-703	Szentendre	Sze-2	490	29.2	39.2	b3
311-703	Szentendre	K-35	1512	65	36.4	3
310-713	Bp. Békásmegyer	B-1	556	26.6	28.8	3
312-717	Bp. Tungsram		618	31	32.4	x2
313-726	Bp. Népliget	B-88	1883	66	29.2	2
314-725	Budapest	B-63	701	40	41.4	2
			524	35	45.8	2
317-726	Bp. X. ker.	B-104	415	36	60.2	2
311-722	Budapest	B-21	900	81	77.8	1
311-721	Budapest	B-13	1250	78	60.2	3
315-721	Budapest	B-24	1395	75	45.9	1
312-733	Budapest	B-10	1120	55.5	39.7	2
			1144	47	31.5	2
312-733	Budapest	B-19	1130	51	35.4	3
316-760	Bugyi	Bu-1	279	20	32.3	x3
325-716	Kerepes	B-8	410	25	34.1	2
325-718	Kistarcsa	K-7	296	25	47.3	2
325-716	Kerepes	B-9	310	30	61.3	2
325-719	Kistarcsa	B-9	520	35	46.2	1
324-735	Vecsés	B-35	298	22	36.9	2
327-812	Soltszentimre	Solti-3	1070	69	52.8	x2
327-820	Kiskorös	K-1081	920	64	56.5	2
			1018	81	67.8	2
329-828	Soltvadkert	/telep/	886	69	64.3	x3
			860	57	52.3	x3
			860	67	64.0	x3
327-830	Kecel	B-2	700	48	51.4	2
			941	65	56.3	1
328-830	Soltvadkert	9	976	65	54.3	x2
			996	70	58.2	x2
			953	70	60.9	x2
329-837	Kecel	/telep/	1458	91.1	54.3	x3
			1227	65.5	43.6	x3
			1128	67.7	49.4	x3
329-837	Kecel	/telep/	1059	58.3	43.9	x3
			1043	53.8	40.1	x3
			970	59	48.5	x3
338-705	Aszód	B-37	250	16	20.0	1
331-713	Gödöllo	B-74	1260	60	38.9	2
			1773	77	37.2	1
334-733	Gyömro	B-10	350	32	60.0	2
33-79	Izsák	/telep/	1005	72	59.7	x3
331-800	Izsák	K-1176	358	36	67.0	3
338-815	Kaskantyu	1	1153	77	56.4	2
337-849	Kiskunhalas	2	1069	71	55.2	2
335-867	Mélykut	B-21	671	43	46.2	1
341-660	Szécsény	1?	785	37	34.4	x2
348-711	Tura	K-50	1990	102.6	46.2	3
346-739	Bénye	B-4	470	40	61.7	2
			479	37	54.3	2
342-743	Monor	K-209	630	44	52.4	3
348-776	Lajosmizse	K-63	700	47	50.0	2
348-775	Lajosmizse	K-85	335	29	50.7	3
349-815	Orgovány	Org D-2	1381	91.6	57.6	x2
346-823	Bócsa	1	1520	96	55.3	x2
			1815	122	60.6	x2
345-837	Kiskunhalas	ÉK-64	2080	130	56.7	x2
346-837	Kiskunhalas	ÉK-66	2041	129	57.3	x2
349-835	Tázlár	1	1459	88	52.1	a2
			1871	116	55.6	x2
			1857	111	53.3	x2
349-835	Tázlár	I	2156	136	57.5	x2
347-835	Tázlár	5	2182	138	57.5	x2
347-835	Tázlár	6	114	14.6	22.8	c2
			222	16.6	20.7	c2
			323	20.4	26.0	c2
			419	24.2	29.1	c2
			521	28.2	31.1	c2
			622	32.6	33.1	c2
			720	36.4	33.9	c2
			818	39.6	33.7	c2
			912	44.6	35.7	c2
			1021	49.2	36.4	c2
			1120	53.0	36.6	c2
			1219	57.2	37.1	c2
			1318	64.4	39.9	c2
			1420	73.4	43.2	c2
			1520	79.4	44.3	c2
			1622	84.6	44.8	c2
			1722	92.4	46.7	c2
			1822	98.4	47.4	c2
			1922	104.6	48.2	c2
			2024	111.8	49.3	c2
			2074	117.2	50.7	c2
348-835	Tázlár	19	1838	118	57.7	x2
			1650	98	52.1	x2
			1930	116	53.9	x2
348-834	Tázlár	15	1465	87	51.2	x2
345-839	Kiskunhalas	EK-1	2015	118	52.6	x2

1	2	3	4	5	6	7	1	2	3	4	5	6	7
342–842	Kiskunhalas	B-20	973	65	54.5	2	357–830	Szank	12	1796	101	49.6	x2
342–842	Kiskunhalas	B-100	867	56	50.7	2	359–830	Szank	11	1780	102	50.6	x2
343–848	Kiskunhalas	7	1122	66	48.1	a2	359–831	Szank	17	1696	96	49.5	a2
342–848	Kiskunhalas	4	1128	66	47.9	x2				1685	104	54.6	x2
344–489	Kiskunhalas	3	1130	62	44.2	x2				1706	102	52.8	x2
342–849	Kiskunhalas	1	1075	62	46.5	x2	357–830	Szank	21	1839	109	52.7	x2
			1929	109	50.3	x2				1790	110	54.7	a2
			1660	91	47.6	x2	358–832	Szank	30	1883	114	54.2	x2
342–850	Kiskunhalas	5	1052	58	43.7	x2	359–821	Szank	29	1783	106	52.7	a2
357–657	Sóshartyán	3	1100	56	41.8	x2	356–831	Szank	45	1290	71	45.7	x2
354–657	Sóshartyán	2	1084	46	33.2	x2	350–835	Tázlár	4	1842	107	51.6	x2
			1121	53	38.4	x2				1874	113	53.9	a2
357–675	Pásztó	K-5	417	32	52.4	3	350–835	Tázlár	4	1842	114	55.4	x2
356–674	Pásztó	K-3	333	35	75.1	1				1444	87	51.9	x2
355–699	Lorinci	K-11	267	32	81.5	3				1853	111	53.4	a2
355–693	Lorinci	K-22	1198	78	56.8	2				1865	119	57.4	x2
354–706	Hatvan	B-145	1056	54	40.7	2				1870	116	55.6	x2
355–721	Tóalmás	K-19	820	53	51.2	3	351–835	Tázlár	9	1725	109	56.2	x2
356–721	Tóalmás	3	1851	101	48.6	x2				1845	116	56.4	x2
358–732	Nagykáta	1	2651	139	48.3	x2	351–834	Tázlár	11	1895	109	51.2	a2
			2667	105	35.3	x2	350–835	Tázlár	22	1795	114	56.8	a2
353–744	Pánd	1	1119	70	52.7	x2	354–847	Zsana	1	1849	85	39.5	x2
			1352	72	45.1	x2	355–843	Ereszto	8	1829	101	48.7	x2
352–751	Albertirsa	B-60	631	48	57.1	1				1875	110	52.3	x2
358–783	Nagykorös	NkU-1	1530	86	48.4	x2	355–844	Ereszto	6	1832	107	51.9	x2
			1625	84.4	44.6	x2				1905	115	54.1	x2
			1492	81.1	46.3	x2				1650	109	58.8	x2
358–783	Nagykorös	NkU-2	1197	69	47.6	x2	356–841	Ereszto	7	1828	113	55.3	x2
357–790	Kecskemét	B-783	900	55	47.8	1	356–852	Öttömös	3	1002	59	46.9	a2
358–790	Kecskemét	B-795	596	38	43.6	2				1424	77	45.7	x2
			699	38	37.2	2	356–853	Öttömös	7	952	52	42.0	x2
352–805	Jakabszállás	1	1640	83	43.3	x2	358–866	Kelebia	1	1018	73	59.9	x2
			950	54	44.2	x2	358–866	Kelebia	2	1040	73	58.7	x2
359–810	Bugac	1	1530	95	54.3	x2	354–870	Kelebia	21	806	54	52.1	a2
			1618	101	55.0	x2	354–870	Kelebia	22	809	56	54.4	a2
358–829	Szank	4	1789	109	54.2	x2	353–870	Kelebia	11	827	61	59.3	a2
356–829	Szank	3	1872	112	53.4	x2	354–870	Kelebia	12	764	56	57.6	x2
355–829	Szank	16	1890	115	54.5	x2	354–870	Kelebia	7	852	61	57.5	x2
353–818	Szank	14	1842	105	50.5	a2	362–659	Salgótarján	B-43	300	26	53.3	2
			1899	110	51.6	x2	362–656	Salgótarján	K-45	300	25	50.0	2
			2239	127	51.4	x2	362–655	Salgótarján	B-48	300	25	50.0	2
359–829	Szank	13	1885	116	55.2	x2	367–664	Kisterenye	K-14	373	27	45.6	2
357–829	Szank	9	1870	111	52.9	x2	365–663	Kisterenye	K-23	380	28	47.4	2
			1698	101	52.4	x2	364–702	Hort	B-13	470	35	55.3	2
			1868	108	51.4	x2	365–716	Pusztamonostor	K-33	368	30	51.6	3
357–829	Szank	B-25	460	32	43.5	1	365–740	Tápiószentmárton	K-34	860	60	55.8	3
359–828	Szank	52	1961	117	53.5	a2	366–740	Tápiószentmárton	1	2164	119	49.4	x2
358–828	Szank	25	1892	106	49.7	x2				2739	127	42.0	x2
359–829	Szank	24	1805	110	54.3	x2	366–758	Cegléd	B-209	360	32	55.6	1
358–829	Szank	23	1879	116	55.4	a2	368–753	Cegléd	K-282	380	31	50.0	2
358–830	Szank	8	1874	115	55.0	a2				412	37	60.7	2
356–830	Szank	7	1112	62	45.0	x2	366–756	Cegléd	B-280	418	33	50.2	1
357–830	Szank	6	1829	100	48.1	x2	367–759	Cegléd	B-288	553	43	56.1	1
357–830	Szank	6	1842	111	53.7	x2	368–754	Cegléd	K-272	705	50	53.9	1
			1850	107	51.4	x2	366–759	Cegléd	B-176	1166	71	50.6	1
357–830	Szank	12	1767	105	52.6	x2	368–754	Cegléd	K-259	1210	74	51.2	3

1	2	3	4	5	6	7	1	2	3	4	5	6	7
364–756	Cegléd	K-241	1350	78	48.9	1	360–830	Szank	77	1634	92	49.0	x2
369–757	Cegléd	K-320	289	29	58.8	1				1845	111	53.7	x2
365–769	Cegléd	2	1697	112	58.9	x2	360–830	Szank	73	1775	108	54.1	x2
365–761	Cegléd	B-302	493.7	40	56.7	1	360–830	Szank	72	1883	114	54.2	x2
369–773	Nagykorös	K-662	250	27	60	3	360–830	Szank	71	1890	114	54.0	x2
365–774	Nagykorös	B-648	900	55	47.8	1	360–830	Szank	93	1870	109	51.9	x2
365–779	Nagykorös	NkU-3	1105	57	40.7	x2	360–830	Szank	92	1860	109	52.2	x2
364–780	Nagykorös	NkU-4	1284	82	56.1	x2	360–830	Szank	90	1898	105	49.0	x2
			1423	82	49.2	x2	360–830	Szank	89	1844	108	52.1	a2
365–781	Nagykorös	NkU-5	1284	74	48.3	x2	360–830	Szank	109	1914	114	53.3	x2
			1327	81	52.0	x2	360–830	Szank	107	1900	111	52.1	x2
360–790	Kecskemét	K-860	994	46	34.4	2	362–833	Szank	106	1895	109	51.2	x2
361–797	Kecskemét	K-861	445	43	67.4	2	360–830	Szank	105	1880	111	52.7	x2
367–795	Nyárlorinc	B-20	298	23	36.9	1	360–830	Szank	104	1834	108	52.3	x2
367–795	Városföld	K-19	400	27	37.5	1	360–830	Szank	99	1870	101	47.6	x2
360–801	Kecskemét	KecsD-5	1543	76.6	41.9	x2	360–830	Szank	97	1885	107	50.4	x2
369–811	Kiskunfélegyháza	B-93	1380	63	37.0	3	360–830	Szank	96	1902	106	49.4	x2
360–829	Szank	19	1825	109	53.2	x2	365–853	Ruzsa	8	2300	135	53.5	x2
361–828	Szank	43	1875	111	52.8	a2				2700	139	47.0	x2
362–829	Szank	42	1859	112	53.8	x2	360–865	Kelebia	6	1090	79	61.5	x2
360–830		40	1876	115	54.9	a2	360–865	Kelebia	10	1074	80	63.3	x2
360–830	Szank	38	1770	108	54.2	x2	366–872	Ásotthalom	14	1072	81	64.4	x2
			1790	100	49.2	x2	365–873	Ásotthalom	10	1075	82	65.1	x2
			1848	111	53.6	x2	367–873	Ásotthalom	7	1076	83	66.0	x2
360–830	Szank	37	1750	110	56.0	x2	367–874	Ásotthalom	3	862	62	58.0	x2
			1833	109	52.9	x2				870	64	59.8	x2
360–830	Szank	34	1879	113	53.8	x2	365–873	Ásotthalom	2	1078	78	61.2	x2
			1784	109	54.4	x2	366–872	Ásotthalom	25	1073	78	61.5	x2
			1899	118	55.8	x2	365–872	Ásotthalom	24	1040	74	59.6	x2
360–830	Szank	32	1880	116	55.3	x2	367–872	Ásotthalom	22	1076	81	64.1	x2
360–830	Szank	31	1883	116	55.2	x2	366–873	Ásotthalom	21	1070	82	65.4	x2
360–830	Szank	20	1870	111	52.9	a2	366–873	Ásotthalom	19	1060	79	63.2	x2
361–831	Szank	27	1670	101	53.3	x2	366–873	Ásotthalom	18	1062	83	66.9	x2
360–832	Szank	28	1765	102	51.0	x2				1063	84	67.7	x2
361–833	Szank	44	1834	108	52.3	x2	366–873	Ásotthalom	17	1057	76	60.6	x2
360–830	Szank	49	1881	110	52.1	x2				1056	82	66.3	x2
362–831	Szank	46	1776	105	52.4	x2	365–872	Ásotthalom	16	1078	82	64.9	x2
360–830	Szank	68	1862	116	55.9	a2				1079	81	64.0	x2
360–830	Szank	67	1880	118	56.4	x2				1066	78	61.9	x2
360–830	Szank	65	1826	115	56.4	a2	366–873	Ásotthalom	15	1053	82	66.5	x2
360–830	Szank	64	1780	109	54.5	x2				1052	80	64.6	x2
360–830	Szank	59	1885	108	50.9	a2				1052	82	66.5	x2
360–830	Szank	58	1872	116	55.6	a2	373–663	Homokterenye	K-3	260	26	61.5	2
360–830	Szank	55	1843	114	55.3	x2	375–662	Mátranovák	B-9	390	28	46.2	2
360–830	Szank	41	1090	62	45.9	a2	374–662	Mátranovák	K-4	360	26	44.4	2
			1145	66	47.2	a2	374–696	Gyöngyöshalász	B-16	278	29	64.7	2
			1215	70	47.7	a2				280	33	78.6	2
360–830	Szank	85	1871	107	50.8	x2	373–696	Gyöngyöshalász	K-21	290	29	62.1	2
362–831	Szank	81	1915	111	51.7	a2				310	27	51.6	2
362–833	Szank	80	1856	113	54.4	x2	373–695	Gyöngyöshalász	K-17	345	30	55.1	1
360–830	Szank	79	1898	114	53.7	a2	372–698	Atkár	K-18	373	34	61.7	1
360–830	Szank	78	1805	113	56.0	x2	373–697	Gyöngyöshalász	B-24	382	34	60.2	1
			1876	107	50.6	x2	37–69	Gyöngyöshalász	K-13	810	59	59.3	2
			1876	114	54.4	x2	374–703	Vámosgyörk	B-17	370	34	62.1	3
			1892	103	48.1	x2	377–706	Jászárokszállás	K-59	320	34.1	72.2	2

1	2	3	4	5	6	7	1	2	3	4	5	6	7
377-706	Jászárokszállás	K-59	350	44	94.3	2	385-673	Recsk	K-3	998	48	38.1	1
378-701	Adács	B-20	390	34	59.0	1	385-675	Recsk	Rm-15	35	18.2	234.3	c2
371-718	Jászberény	K-556	270	25	51.8	3				100	20.9	108.5	c2
377-710	Jászárokszállás	B-61	319	30	59.6	3				200	24.3	71.3	c2
378-725	Jászberény	K-538	265	28	64.2	1				300	27.4	58.0	c2
371-725	Jászberény	K-553	385	32	54.5	3				400	30.7	51.8	c2
373-722	Jászberény	B-415	730	50	52.2	1				500	34.2	48.4	c2
			804	53	52.2	2				600	37.4	45.7	c2
374-728	Jászberény-1		110	16.8	52.7	c2				700	40.3	43.3	c2
			213	21.0	46.9	c2				800	43.2	41.5	c2
			312	25.2	45.5	c2				825	44.0	41.2	c2
			412	33.2	53.9	c2	386-673	Recsk	Rm-8	50	20.8	216.0	c2
			511	40.4	57.5	c2				100	22.6	126.0	c2
			610	45.4	56.4	c2				200	26.2	81.0	c2
			712	49.8	54.5	c2				300	29.7	65.7	c2
			811	55.4	54.7	c2				400	32.8	57.0	c2
			910	60.4	54.3	c2				500	35.2	50.4	c2
			1012	65.6	54.0	c2				600	38.9	48.2	c2
			1111	70.8	53.8	c2				700	42.0	45.7	c2
			1209	75.2	53.1	c2				800	45.0	43.8	c2
			1225	75.6	52.7	c2				876.5	47.0	42.2	c2
371-724	Jászberény	15-132	799	50	48.8	2	387-682	Kisnána	K-3	290	28	58.6	1
370-736	Farmos	1	1316	79	51.7	x2	388-686	Domoszló	B-6	260	33	84.6	2
379-749	Ujszilvás	K-17	295	38	88.1	2				514	41	58.4	2
377-740	Tápiógyörgye	K-7	332	30	54.2	3	389-688	Verpelét	VerpS-5	1252	68	45.5	x2
371-741	Tápiószele	B-15	340	33	64.7	3				940	51	42.6	x2
371-741	Tápiószele	B-12	339	30	53.1	3	387-695	Detk	B-17	320	29	56.3	1
374-749	Ujszilvás	B-18	335	30	50.7	1	388-702	Nagyfüged	B-8	330	33	66.7	1
376-741	Tápiógyörgye	B-9	440	37	56.8	3	381-708	Visznek	K-12	310	31	64.5	2
372-758	Cegléd	K-295	333	31	57.1	1				315	34	73.0	2
375-759	Cegléd	K-175	320	29	53.1	3	380-708	Jászárokszállás	B-62	696	60	70.4	1
377-766	Törtel	/telep/	1003	60	47.9	x3	389-706	Zaránk	K-6	260	28	65.4	1
			1520	86	48.7	x3	387-710	Erk	K-19	322	32	65.2	1
370-772	Nagykörös	K-658	300	31	63.3	1	386-710	Erk	B-18	330	29	54.5	2
373-793	Nyárlorinc	B-20	298	23	36.8	1	380-714	Jászdózsa	K-10	351	33	62.7	1
378-801	Tiszaalpár	K-19	306	30	58.8	3				376	36	66.5	2
373-829	Csengele	K-33	476	30	37.8	1	384-711	Tarnaörs	B-5	370	36	67.6	1
373-825	Pálmonostora	1	2221	114	45.9	x2	381-715	Jászdózsa	B-11	400	35	60.0	1
374-833	Kömpöc	1	2674	124	41.9	x2	380-724	Jásztelek	B-11	350	28	48.6	1
			2937	149	46.6	x2	383-725	Jásztelek	K-12	390	33	56.4	1
378-846	Forráskut	1	1778	85	41.1	x2	387-735	Jászalsósztgyörgy	K-19	540	44	61.1	1
			3500	147	38.6	x2	385-736	Jánoshida	B-12	548	44	60.2	1
370-853	Üllés-felso	ÜF-11	1111	72	54.0	x2	381-738	Jászboldogháza	B-31	557	44	59.3	1
370-853	Üllés-felso	ÜF-10	1117	74	55.5	x2	380-738	Jászboldogháza	K-32	580	45	58.6	3
370-852	Üllés-felso	ÜF-9	1144	72	52.5	x2	388-739	Jászalsósztgyörgy	K-21	604	41	49.7	3
370-854	Üllés	1	1960	102	45.9	x2							
372-852	Üllés	5	1808	118	58.6	x2	383-731	Alattyán	B-8	640	47	56.3	2
372-853	Üllés	7	2008	127	57.3	x2				686	54	62.7	2
			2002	134	60.9	x2	385-732	Alattyán	K-7/a	718	51	55.7	1
376-850	Forráskut	K-15	2586	165	59.1	2	389-737	Jászalsószt-györgy	B-20	756	57	60.9	1
378-854	Bordány	K-4	425	31	44.7	1							
377-850	Forráskut	K-11	1767	99	49.2	1	387-735	Jászalsószt-györgy	K-22	750	54	57.3	2
374-865	Mórhalom	B-13	620	43	50.0	3							
386-675	Susa	K-1	1530	80	45.8	2				845	74	74.6	2

1	2	3	4	5	6	7	1	2	3	4	5	6	7
386-747	Ujszász	B-20	290	31	65.5	1	399-704	Heves	K-43	620	45	54.8	3
387-746	Ujszász	B-21	640	48	56.3	2	391-719	Jászapáti	K-31	432	35	55.6	2
			666	64	78.1	2				460	40	63.0	2
387-743	Szászberek	K-10	745	50	51.0	1	393-713	Jászsztandrás	B-49	480	39	58.3	3
388-751	Zagyvarékas	K-7	620	47	56.5	3	399-719	Jászivány	K-43	590	41	50.8	2
389-751	Zagyvarékas	K-9	690	51	56.5	2				590	49	64.4	2
			720	59	65.3	2	393-713	Jászsztandrás	B-3	600	50	65.0	1
384-758	Abony	K-52	550	42	54.6	1	391-720	Jászapáti	B-16	460	40	63.0	2
383-758	Abony	B-51	629	52	63.6	1				560	40	51.8	2
383-757	Abony	B-44	630	48	57.1	1	390-723	Jászapáti	K-27	470	38	57.4	3
381-758	Abony	B-17	701	52	57.1	1	393-724	Jászapáti	K-30	530	41	56.6	3
381-760	Abony	K-56	277	27	54.1	2	393-723	Jászapáti	K-52	540	42	57.4	1
385-762	Abony	K-43	334	30	53.9	3	391-720	Jászapáti	B-53	747	52	54.9	1
382-761	Abony	1	2123	118	49.9	x2	391-722	Jászapáti	B-15	805	58	58.4	1
			2321	133	52.1	x2	394-726	Jászkisér	K-34	847	58	55.5	1
			2021	131	58.9	x2	395-727	Jászkisér	B-36	850	62	60.0	1
382-778	Jászkarajeno	/telep/	1414	80	48.1	x3	397-726	Jászkisér	K-38	1224	78	54.7	1
389-785	Tiszakécske	B-72	880	55	48.9	2	396-729	Jászkisér	15-131	795	60	61.6	2
			980	64	53.1	1	391-738	Jászladány	K-24	310	31	64.5	3
387-787	Tiszakécske	K-71	954	67	57.7	1	395-730	Jászkisér	K-23	610	47	59.0	2
386-789	Lakitelek	K-68	1160	66	46.6	2	395-730	Jászkisér	K-33	665	46	52.6	2
			1081	74	57.4	2	393-733	Jászladány	K-16	830	65	65.1	2
381-789	Lakitelek	K-26	2048	115	50.3	2	392-739	Jászladány	K-19	847	54	50.8	1
384-790	Lakitelek	K-29	880	61	55.7	2	396-737	Jászladány	K-21	864	56	52.1	1
			1025	63	49.8	2	394-738	Jászladány	B-23	867	58	54.2	1
381-790	Lakitelek	K-31	899	65	59.0	1	397-737	Jászladány	I	3645	196.1	50.8	x2
381-794	Lakitelek	K-28	1010	46	33.7	2		/Jászság/		2666	153.3	53.4	x2
			1200	58	38.3	2				2447	132.2	49.5	x2
385-802	Bokros	K-22	260	21	34.6	3	390-748	Zagyvarékas	B-8	680	51	57.4	2
384-834	Balástya	K-64	253	22	39.5	1				648	50	58.6	2
383-843	Sándorfalva	S-I	1650	75	38.2	x3	391-748	Szászberek	K-11	670	51	58.2	2
			2210	79	30.3	x3				730	58	63.0	2
			3710	160	39.9	x3	393-759	Szolnok	B-7	309	30	58.3	3
			4015	182	42.3	x3	398-751	Besenyszög	K-8	651	46	52.2	2
385-853	Szatymaz	K-43	360	26	38.9	1	395-759	Szolnok	B-5	870	59	54.0	3
389-855	Szeged	K-433	273	30	65.9	2	395-759	Szolnok	B-82	944	57	47.7	2
383-862	Domaszék	K-18	435	33	48.3	1	398-750	Besenyszög	K-21	990	61	49.5	3
389-864	Szeged	K-486	340	28	47.1	1	393-759	Szolnok	B-55	900	59	52.2	2
386-867	Röszke	K-46	1780	93	45.5	1				1001	69	56.9	1
389-864	Szeged	K-476	1849	101	48.1	2	392-757	Szolnok	K-77	1050	72	57.1	3
388-862	Kiskundorozsma	4	2993	161	49.8	x2	395-759	Szolnok	B-67	1094	74	56.7	2
386-863	Kiskundorozsma	1	2302	120	46.9	x2	396-759	Szandaszolos	K-9	1040	66	51.9	3
387-864	Kiskundoroszma	2	1628	93	49.8	x2	393-758	Szolnok	B-74	1089	77	59.7	2
392-667	Bükkszék	B-1	510	42	62.8	3	396-767	Szandaszolos	K-7	279	29	60.9	3
392-667	Bükkszék	B-8	530	42	60.4	1	392-763	Szolnok	K-75	385	34	57.1	1
393-663	Fedémes	19	410	25	36.6	x2	393-761	Szolnok	K-52	404	34	54.5	3
394-688	Verpelét	Verp-4	1095	62.2	46.8	x2	391-762	Szolnok	K-48	490	41	59.2	3
397-702	Tarnabod	B-3	314	29	57.3	1	397-760	Szolnok	K-79	640	51	60.9	2
397-701	Tarnabod	K-4	335	31	59.7	2	392-761	Szolnok	K-59	850	63	60.0	3
			340	37	76.5	1	396-760	Szandaszolos	K-8	990	65	53.5	3
392-703	Tarnazsadány	B-5	350	36	71.4	1	395-760	Szolnok	K-78	1000	65	52.0	3
397-706	Boconád	K-13	444	36	56.3	3	395-760	Szandaszolos	K-11	1010	66	53.5	3
393-707	Boconád	K-12	460	37	56.3	3	393-760	Szolnok	K-61	1030	65	51.5	3
392-705	Tarnaméra	B-11	600	48	61.7	1	393-760	Szolnok	K-63	1040	70	55.8	3

1	2	3	4	5	6	7	1	2	3	4	5	6	7
392-763	Szolnok	B-68	1073	61	45.7	2	397-854	Algyo	K-59/a	1270	72	47.2	1
393-760	Szolnok	B-58	1076	69	53.0	1	398-855	Algyo	K-49	1265	67	43.5	1
390-762	Szolnok	15-136	484	42	62.0	2	399-853	Algyo	K-58/a	1230	68	45.5	2
398-774	Vezseny	B-11	265	28	60.4	3				1278	65	41.5	2
399-770	Rákóczifalva	K-7	300	29	56.7	1	399-865	Algyo	K-60/a	1170	64	44.4	2
395-772	Rákóczifalva	6	1333	90	58.5	x2				1278	73	47.7	1
395-772	Rákóczifalva	7	1311	93	61.8	x2	396-855	Algyo	K-48	1298	65	40.8	1
			1366	93	59.3	x2	399-857	Algyo	K-31	1520	81	45.4	2
390-782	Tiszakécske	K-76	917	56	48.0	1				1603	81	43.0	1
391-785	Tiszakécske	K-70	1445	88	52.6	2	391-859	Algyo	K-452	1812	95	45.8	2
396-792	Cserkeszolo	B-1	2300	118	46.1	2	397-855	Algyo	1	2061	110	47.5	x2
			2311	143	56.7	2				1997	100	44.1	x2
397-793	Cserkeszolo	B-27	1147	73	53.3	1	398-856	Algyo	4	2467	116	42.2	x2
390-809	Csongrád	K-91	650	33	32.3	3				2628	127	43.8	x2
393-810	Csongrád	B-72	970	50	39.2	2	397-855	Algyo	8	2819	140	45.4	x2
			1000	53	41.0	2	398-854	Algyo	7	1924	97	44.2	x2
390-810	Csongrád	K-90	1040	49	35.6	2	397-856	Algyo	5	1943	98	44.3	x2
			1200	62	41.7	2				2085	116	49.9	x2
399-816	Szentes	K-558	1850	92	43.2	3	395-854	Algyo	18	1973	96	42.6	a2
397-816	Szentes	K-577	2498	117	42.0	1				2118	103	43.0	x2
399-820	Szentes	K-423	366	29	46.4	3	395-856	Algyo	21	1955	96	43.0	x2
399-825	Szegvár	B-99	480	27	31.3	1				2118	105	43.9	x2
390-829	Csanytelek	K-173	1997	68	28	2	399-856	Algyo	28	1950	98	44.1	x2
390-829	Baks	K-25	2139	86	34.6	2				1955	95	43.1	x2
398-830	Mindszent	K-89	1345	64	38.7	1				2050	103	44.4	x2
397-831	Mindszent	K-87	2524	108	38.0	1				2435	121	44.8	x2
399-849	Hódmezovásár-	K-877	720	44	44.4	1	398-867	Szoreg	1	1882	90	41.4	x2
	hely									1946	96	43.2	x2
390-851	Szeged	K-592	254	24	47.2	2	392-851	Szeged	2	2620	132	45.8	x2
394-859	Szeged	B-588	340	25	38.2	3				2647	140	48.4	a2
394-859	Szeged	B-589	400	28	40.0	3				2685	139	47.3	x2
394-859	Szeged	B-387	341	27	44.0	3				2688	144	49.1	x2
390-859	Szeged	K-493	365	31	52.1	1				2697	145	49.3	x2
399-859	Szeged	K-30	380	27	39.5	3				2735	144	48.3	x2
394-853	Szeged	B-412	387	29	43.9	3	392-862	Szeged	1	2240	116	46.4	x2
394-859	Szeged	K-392	440	34	50.0	1				2605	142	49.9	x2
392-859	Szeged	B-484	450	31	42.2	3				2560	139	49.6	a2
390-859	Szeged	B-402	470	35	48.9	1				2650	144	49.8	x2
398-852	Szeged	B-510	479	28	33.4	1				2590	144	51.0	x2
392-859	Szeged	9	2450	128	47.3	x2				2664	140	48.0	a2
393-859	Szeged	B-420	500	32	40.0	1				2667	149	51.4	x2
394-859	Szeged	K-386	505	34	43.6	3	393-861	Szeged-Felsováros	1	1970	99	44.2	x2
391-859	Szeged	K-457	512	32	39.1	1	394-864	Szeged	10	2770	139	45.8	x2
395-859	Szeged	B-434	525	35	43.8	2	392-864	Szeged	11	2775	148	49.0	x2
			530	38	49.1	2	395-862	Szeged	12	2847	144	46.4	x2
393-859	Szeged	B-419	531	36	45.2	1	391-861	Szeged	8	2672	142	48.7	a2
395-859	Szeged	K-445	530	37	47.2	1	391-862	Szeged	7	2808	150	49.1	x2
393-858	Szeged	K-414	531	36	45.2	3	393-863	Szeged	6	2651	137	47.2	a2
394-858	Szeged	B-449	542	33	38.8	1				2625	139	48.4	x2
391-859	Szeged	B-461	590	43	52.5	1	392-862	Szeged	5	2654	136	46.7	a2
391-859	Szeged	K-482	643	41	45.1	1	392-861	Szeged	3	2932	142	44.3	x2
395-856	Szeged	K-273	990	56	44.4	3				2750	139	46.2	x2
398-852	Algyo	K-61/a	1236	61	39.6	1				2440	121	44.7	x2
397-854	Algyo	K-59/a	1240	67	44.4	2	391-865	Szeged	K-590	270	23	40.7	3

1	2	3	4	5	6	7	1	2	3	4	5	6	7
392-865	Szeged	K-591	260	23	42.3	3	394-861	Szeged	B-415	1820	96	46.2	3
394-861	Szeged	B-467	335	27	44.8	3	398-867	Szoreg	K-48	1840	92	43.5	2
398-866	Szoreg	K-39	335	31	56.7	3				1860	97	45.7	2
392-862	Szeged	B-507	360	32	55.6	1	396-863	Szeged	B-474	1914	120	56.4	2
398-865	Szeged-Szoreg	K-52	360	30	50.0	1	396-869	Tiszasziget	K-24	1918	104	48.0	2
394-861	Szeged	B-141	360	29	47.2	3	396-870	Tiszasziget	K-22	390	28	41.0	2
392-863	Szeged	B-495	368	34	59.8	2				400	41	47.5	2
394-862	Szeged	B-473	370	30	48.6	1	397-870	Tiszasziget	K-25	1997	112	50.1	2
394-861	Szeged	B-399	380	32	52.6	1	405-647	Csokvaomány	K-4	400	28	45	2
391-860	Szeged	K-492	390	29	43.6	3	408-679	Eger	B-7	450	38	60.0	3
395-861	Szeged	B-409	410	32	48.8	1	404-682	Egerszalók	K-4	400	66	137.5	3
394-861	Szeged	B-466	415	28	38.6	1	406-683	Demjén	K-1	1520	72	40.1	x2
393-861	Szeged	B-129	428	32	46.7	3	407-683	Demjén	322	825	65	65.5	x2
394-862	Szeged	B-406	434	34	50.7	1	407-683	Demjén	357	1620	87	46.9	x2
395-861	Szeged	B-379	438	32	45.7	3	407-682	Eger	K-21	810	48	45.7	3
393-863	Szeged	B-508	440	32	45.5	2	401-699	Kál	B-13	307	30	61.9	2
			467	39	57.8	2	403-690	Kerecsend	B-3	257	24	50.6	1
391-860	Szeged	K-491	450	31	42.2	3	408-693	Füzesabony	K-15	345	33	63.8	2
393-860	Szeged	B-393	449	32	44.5	2				354	32	59.3	2
			474	36	50.6	2	402-709	Rákócziujfalu	K-12	315	29	58.1	1
394-860	Szeged	B-422	460	30	39.1	1	403-701	Erdotelek	B-25	374	33	58.8	2
392-862	Szeged	B-506	461	37	54.2	1	407-709	Átány	B-5	429	34	53.6	2
394-861	Szeged	B-400	464	34	47.4	1				450	33	48.9	1
396-862	Szeged	B-456	466	30	38.6	1	409-719	Tarnasztmiklós	B-9	430	35	55.8	1
394-862	Szeged	B-489	470	32	42.6	3	400-711	Heves	B-21	448	35	53.6	3
394-861	Szeged	B-138	475	33	44.2	3	407-715	Hevesvezekény	B-2	560	40	51.8	3
394-861	Szeged	B-133	485	33	43.3	3	401-710	Heves	B-44	653	46	53.6	1
392-863	Szeged	B-494	467	39	57.8	2	401-712	Heves	B-20	793	57	58.0	1
393-863	Szeged	B-504	490	35	46.9	2	405-723	Pély	K-5	626	46	55.9	2
			548	42	54.7	2	406-723	Pély	B-7	784	48	47.2	1
395-861	Szeged	K-372	492	33	42.7	3	406-721	Pély	K-9	810	60	60.5	2
392-862	Szeged	B-503	500	33	42.0	2	401-734	Jászkisér	K-35	926	60	52.9	2
			524	40	53.4	2	401-743	Besenyszög	K-17	1079	67	51.0	1
392-863	Szeged	B-367	501	35	45.9	1	403-759	Szajol	K-26	270	25	48.1	3
392-860	Szeged	B-483	510	31	37.3	3	406-755	Tiszapüspöki	1	1589	102	56.6	x2
399-86	Szeged	B-377	520	36	46.2	1	405-768	Kengyel	K-20	359	33	58.5	3
394-862	Szeged	B-398	530	36	45.3	1	403-761	Szajol	K-19	382	33	55.0	3
396-862	Szeged	B-455	535	31	35.5	1	402-761	Szajol	15-130	1290	68	43.4	2
392-863	Szeged	B-498	542	38	48.0	1	409-760	Törökszent-	K-89	250	24	48.0	3
392-862	Szeged	B-388	547	37	45.7	1		miklós					
391-860	Szeged	K-458	549	33	38.3	1	402-776	Martfü	B-1	310	31	61.3	3
394-861	Szeged	B-454	600	42	50.0	1	402-776	Martfü	B-12	256	27	58.6	3
394-861	Szeged	B-376	664	42	45.2	1	404-776	Martfü	K-14	291	30	61.9	3
395-861	Szeged	B-487	760	42	39.5	2	403-776	Martfü	B-13	302	29	56.3	2
			792	48	45.5	2				310	34	71.0	2
394-860	Szeged	B-88	940	54	44.7	3	404-779	Tiszaföldvár	K-58	300	31	63.3	1
397-861	Szeged	B-220	950	55	45.3	2	409-770	Kengyel	K-24	393	32	50.9	2
			1013	68	55.3	2				405	35	56.8	2
397-861	Szeged	B-453	1710	88	44.4	2	400-770	Rákócziujfalu	K-9	884	64	58.8	1
			1780	97	47.8	2	400-779	Tiszaföldvár	K-46	1046	78	63.1	1
391-860	Szeged	B-440	1790	97	47.5	3	400-783	Tiszaföldvár	K-61	280	30	64.2	3
395-861	Szeged	B-384	1800	96	46.7	2	401-781	Tiszaföldvár	K-52	334	30	53.9	3
			1900	106	49.5	2	404-797	Kuntszentmárton	K-33	244	30	73.8	3

1	2	3	4	5	6	7
403-796	Kunszentmárton	K-48	270	29	63.0	1
407-793	Öcsöd	K-35	273	28	58.6	1
407-793	Öcsöd	K-32	305	29	55.7	1
407-790	Öcsöd	K-33	325	32	61.5	1
402-795	Kunszentmárton	K-53	639	49	57.9	1
409-807	Nagytoke	K-17	363	31	52.3	1
408-808	Szentes	K-480	418	30	43.3	3
402-816	Szentes	B-631	542	33	38.7	1
407-819	Szentes	K-566	355	29	47.9	2
			460	28	34.8	2
402-816	Szentes	B-489	410	28	39.0	3
407-813	Szentes	K-591	425	29	40.0	1
402-816	Szentes	B-531	430	31	44.2	1
403-816	Szentes	K-509	440	32	45.5	1
403-819	Szentes	K-501	480	31	39.6	1
402-816	Szentes	K-559	1670	83	42.5	3
408-813	Szentes	K-562	1720	88	44.2	2
			1784	104	51.7	2
402-816	Szentes	B-17	1736	85	42.1	1
407-815	Szentes	K-564	1770	90	44.0	1
			1856	87	40.4	2
401-815	Szentes	K-533	1860	92	43.0	3
402-819	Szentes	K-557	1900	92	42.1	2
			2000	92	40.0	1
407-813	Szentes	K-498	1900	102	47.4	2
			1981	103	45.9	1
400-817	Szentes	K-505	1910	92	41.9	3
402-819	Fábiánsebestyén	K-59	1870	102	48.1	2
			1998	112	50.1	1
406-810	Szentes	K-645	1995	99	43.4	1
406-815	Szentes	K-630	473	32	42.3	2
407-810	Szentes	K-644	2249	110	43.3	1
407-811	Szentes	K-641	1983	101	44.9	1
407-811	Szentes	K-640	2235	109	43.4	1
407-814	Szentes	K-563	1987	91	39.8	2
407-811	Szentes	K-561	2013	100	43.7	1
407-815	Szentes	K-515	2200	109	44.1	2
401-811	Szentes	K-578	2170	103	41.9	2
			2401	102	37.5	2
406-813	Szentes	K-586	2140	102	42.1	1
			2180	110	45.0	2
407-825	Szegvár	K-90	358	28	44.7	3
409-826	Derekegyháza	K-14/a	373	29	45.6	3
409-824	Derekegyháza	K-44	380	29	44.7	1
401-824	Szegvár	B-87	915	53	44.8	2
402-824	Szegvár	K-94	2168	96	38.7	2
401-824	Szegvár	K-96	2490	109	39.0	2
400-830	Mindszent	K-90	1850	84	38.9	2
			1986	94	41.3	1
406-842	Hódmezovásárhely	B-41	343	29	49.6	3
407-843	Hódmezovh.	B-88	371	29	45.8	3
407-843	Hódmezovh.	B-881	522	31	36.4	3
406-844	Hódmezovh.	K-930	548	33	38.3	1
407-842	Hódmezovásárhely	B-951	556	34	39.6	1
405-845	Hödmezovh.	K-962	590	32	33.9	2
			628	36	38.2	2
405-853	Hódmezovh.	K-955	590	32	33.9	2
			616	37	38.4	2
405-844	Hódmezovh.	K-957	613	36	39.2	1
407-843	Hódmezovh.	B-107	1093	54	38.8	2
407-842	Hódmezovh.	B-913	1830	83	38.8	2
			1911	98	45.0	2
409-845	Hódmezovh.	K-1056	370	24	32.4	3
400-855	Algyo	K-57/a	1240	67	44.4	2
			1284	75	49.1	2
400-857	Algyo	112	2588	134	47.1	x2
			2621	136	47.3	x2
			2633	137	47.5	x2
			2565	133	47.2	x2
			1983	99	43.9	x2
			2001	100	44.0	x2
			2047	103	44.5	x2
			2033	102	44.3	x2
			2294	118	46.2	x2
			2052	102	43.9	x2
			2035	101	43.7	x2
			2052	102	43.9	x2
			2047	102	44.0	x2
			2064	103	44.1	x2
			2072	103	43.9	x2
			2125	108	45.2	x2
			2134	107	44.5	x2
			2115	106	44.4	x2
			2302	118	46.0	x2
			2348	121	46.4	x2
			2138	108	44.9	x2
			2397	123	46.3	x2
			2260	115	45.6	x2
			2125	108	45.2	x2
			2395	124	46.8	x2
			2364	122	46.5	x2
			2349	121	46.4	x2
			2247	115	45.8	x2
			2381	123	46.6	x2
			2300	118	46.1	x2
			2209	112	45.3	x2
			2451	127	46.9	x2
			2433	126	46.9	x2
			2485	128	46.7	x2
			2457	126	46.4	x2
			2463	128	47.1	x2
			2497	130	47.3	x2
			2467	128	47.0	x2
402-859	Algyo	11	1767	91	44.7	x2
			1860	96	45.2	x2
408-857	Maroslele	1	2330	83	30.5	x2

1	2	3	4	5	6	7	1	2	3	4	5	6	7
408-857	Maroslele	1	3127	112	32.6	x2	415-746	Nagykörü					
Nkö-10	1845	101	48.2	x2									
			3127	120	34.5	x2	414-755	Törökszentmiklós	K-90	380	32	52.6	3
400-862	Deszk	K-36	282	24	42.5	1	415-755	Törökszentmiklós	K-96	370	34	59.5	3
401-860	Algyo	3	1833	92	43.6	x2	415-785	Törökszentmiklós	K-87	240	26	58.3	3
403-861	Algyo	16	1827	95	45.4	x2	417-753	Surjány/Török-	1	1825	105	51.0	x2
404-861	Algyo	22	1931	103	47.1	x2		szentmiklós/					
			2488	138	50.6	x2				1985	124	56.4	x2
405-865	Deszk	1A	2540	136	48.8	x2				2223	130	53.1	x2
	Deszk		2574	134	47.4	a2	411-751	Nagykörü	3	1880	107	50.5	x2
			2336	116	44.5	x2				1925	110	50.9	x2
			2418	121	45.1	x2	411-753	Nagykörü	6	1815	110	54.0	x2
402-864	Deszk	B-33	390	30	46.2	1				1930	112	51.8	x2
401-865	Deszk	B-35	390	30	46.2	1				1832	112	54.6	x2
409-870	Ferencszállás	3	2199	123	50.5	x2	412-756	Törökszentmiklós	K-35	1146	76	55.9	2
			2278	126	50.0	a2	416-767	Kétpó	K-9	464	34	47.4	1
			2329	131	51.1	a2	416-766	Törökszentmiklós	K-71	400	32	50.0	3
			2380	134	51.3	x2	417-764	Törökszentmiklós	K-65	1190	80	57.1	2
417-677	Bogács	K-1	599	77	110.2	3				1254	79	53.4	1
419-676	Bogács	4-17	470	78	142.6	3	412-776	Mezohék	K-28	350	32	57.1	2
419-689	Mezokövesd	K-46	545	48	67.9	2				337	32	59.3	2
410-681	Andornaktálya	9-6	810	48	45.7	3	411-772	Mezohék	K-24	409	35	56.2	1
419-689	Mezokövesd	K-31	860	71	69.8	2	411-778	Mezohék	B-29	388	34	56.7	1
			850	71	70.6	1	410-778	Martfü	2	1770	107	53.7	x2
419-689	Mezokövesd	K-28	860	76	75.6	1	412-788	Öscöd	B-38	641	48	56.2	1
419-689	Mezokövesd	K-48	870	70	67.8	3	414-783	Mesterszállás	K-6	315	32	63.5	3
415-690	Szihalom	K-4	256	30	74.2	2	415-784	Mesterszállás	K-8	318	31	59.7	3
			345	33	63.8	2	414-782	Mesterszállás	B-12	380	35	60.5	3
416-699	Mezotárkány	K-7	338	28	50.3	3	415-799	Cserebökény	K-14	415	32	48.2	3
416-691	Szihalom	B-5	345	33	63.8	1	418-792	Békéssztandrás	B-11	652	49	56.8	1
411-698	Dormánd	B-5	371	32	56.6	2	411-803	Cserebökény	K-615	330	28	48.5	3
			395	27	40.5	2	412-804	Cserebökény	K-20	400	30	45.0	1
416-698	Mezotárkány	B-6	400	32	52.5	1	417-808	Cserebökény	K-16	510	37	49.0	3
412-700	Besenyotelek	B-5	368	32	57.1	2	416-807	Cserebökény	K-17	560	42	53.6	3
			404	31	49.5	2	412-804	Cserebökény	K-21/a	763	43	40.6	1
412-701	Besenyotelek	K-6	385	32	54.5	2	411-804	Cserebökény	K-18	2050	106	45.9	3
417-702	Mezotárkány	K-9	390	30	48.7	2	415-808	Fábiánsebestyén	1	1000	50	38.0	x2
			401	34	57.4	2				2000	92	40.0	x2
411-709	Kömlo	1	2900	165	53.1	x2				3000	140	43.7	x2
			3760	188	47.1	x2				3100	146	43.2	x2
413-710	Kömlo	K-19	600	44	55.0	3	419-819	Nagymágócs	K-76	307	30	58.6	3
412-711	Kömlo	K-20	715	62	71.3	2	412-813	Fábiánsebestyén	K-13	316	29	53.8	3
418-716	Tiszanána	K-22	930	59	51.6	2	416-813	Fábiánsebestyén	B-54	337	28	47.5	2
			950	68	60.0	2				379	27	39.6	2
417-722	Kisköre	B-5	412	31	48.5	3	419-812	Fábiánsebestyén	K-9	361	30	49.9	3
417-722	Kisköre	B-23	290	26	51.7	3	413-81	Fábiánsebestyén	K-56	348	28	46.0	2
416-733	Tiszaroff	15-109	1098	67	51.0	2				385	32	51.9	2
412-741	Kotelek	K-11	1250	75	50.4	1	416-813	Fábiánsebestyén	B-65	420	32	47.6	1
413-746	Kotelek-Nagy-	5	1833	100	48.0	x2	410-815	Szentes	K-504	433	33	48.5	1
	körü						414-813	Fábiánsebestyén	K-61	440	36	54.5	1
416-749	Nagykörü	7	1762	110	55.6	x2	416-814	Fábiánsebestyén	K-55	450	34	48.9	1
			1775	110	55.2	a2	410-815	Szentes	K-560	1650	93	49.1	3
			1805	108	53.2	x2	410-815	Szentes	K-503	1900	104	48.4	1
			1919	112	52.1	x2	412-813	Fábiánsebestyén	K-57	1964	109	49.4	2
415-746	Nagykörü	Nkö-10	1845	92	43.4	x2	413-811	Fábiánsebestyén	K-60	2086	117	50.3	2

1	2	3	4	5	6	7	1	2	3	4	5	6	7
411-816	Szentes	K-514	2199	122	50.0	2	429-746	Kisujszállás	19	1491	90	53.0	x2
416-827	Derekegyháza	K-45	332	28	48.2	1				1638	95	51.3	x2
419-823	Nagymágócs	K-184	359	31	52.9	3				1500	92	54.0	x2
418-825	Nagymágócs	K-191	387	29	43.9	3				1381	80	50.0	x2
417-824	Nagymágócs	K-194	430	31	44.2	1	428-747	Kisujszállás	18	1410	83	51.1	x2
418-835	Hódmezovásár-	K-887	269	28	59.5	3				1826	103	50.9	x2
	hely									1703	102	53.4	x2
418-832	Hódmezovh.	K-148	275	30	65.5	3	422-748	Fegyvernek	3	1670	95	50.3	a2
410-844	Hódmezovh.	K-1052	293	23	37.5	1				1597	97	54.0	x2
410-845	Hódmezovh.	K-1053	290	23	37.9	3				1379	78	48.6	x2
411-847	Hódmezovh.	K-883	310	28	51.6	3				1764	92	45.9	x2
410-842	Hódmezovh.	K-364	385	31	49.4	3				1764	102	51.6	x2
419-840	Hódmezovh.	K-934	430	33	48.8	1				1747	100	50.9	x2
411-844	Hódmezovh.	K-902	525	29	32.4	2				1712	108	56.7	x2
			515	33	40.8	2	421-748	Fegyvernek	1	1758	100	50.6	x2
411-845	Hódmezovh.	K-938	553	34	39.8	1				1663	97	51.7	a2
411-843	Hódmezovh.	K-915	595	35	38.7	1				1860	104	50.0	x2
411-844	Hódmezovh.	K-906	1890	103	48.1	2	424-747	Fegyvernek	K-1	1675	104	55.5	x2
411-842	Hódmezovh.	K-919	2347	105	39.6	2				1933	116	54.3	x2
416-849	Hódmezovh.	1	4886	181	34.6	x2				1948	96	43.6	x2
			5418	203	35.3	x2				2350	122	47.2	x2
410-859	Maroslele	4	1430	74	43.4	x2	429-748	Kisujszállás	20	1357	83	53.1	x2
418-853	Földeák	K-50	383	31	49.6	2				1380	79	49.3	x2
			452	30	39.8	2				1395	86	53.8	a2
419-855	Földeák	K-52	2187	98	39.3	1				1507	92	53.7	a2
419-866	Makó	K-194	374	30	48.1	2	427-756	Örményes	K-10	549	43	56.5	2
			420	29	40.5	2	427-757	Örményes	K-14	600	46	56.7	2
417-864	Makó	K-179	426	31	44.6	3	427-756	Örményes	15-138	527	43	58.8	2
417-864	Makó	K-205	470	31	40.4	1	429-778	Mezotur	B-87	402	33	52.2	3
417-864	Makó	K-201	500	30	36.0	1	428-778	Mezotur	B-111	430	30	41.9	2
419-864	Makó	B-189	2059	101	43.2	2				390	30	46.2	2
415-871	Kiszombor	1	2289	121	47.6	x2	428-775	Mezotur	B-4	482	39	56.0	3
			2290	123	48.5	x2	428-776	Mezotur	B-10	492	38	52.8	3
412-871	Ferencszállás	2	2448	136	50.7	x2	429-776	Mezotur	B-27	493	38	52.7	3
426-678	Tard	K-2	316	31	63.3	3	427-778	Mezotur	B-56	498	39	54.2	3
421-687	Mezokövesd	B-45	250	39	112	2	428-778	Mezotur	B-58	626	50	60.7	2
426-697	Egerlövo	B-6	327	32	64.2	2	428-778	Mezotur	B-110	645	52	62.0	1
			330	28	51.5	2	428-777	Mezotur	K-94	805	53	50.9	1
429-699	Borsodivánka	B-4	428	36	58.4	1	429-777	Mezotur	B-22	1100	78	60.0	2
425-708	Ujlorincfalva	B-5	372	32	56.5	2	428-779	Mezotur	K-120	644	43	47.9	2
			414	31	48.3	2	423-784	Mezotur	K-80	530	40	52.8	1
429-705	Poroszló	B-29	527	38	51.2	3	426-788	Szarvas	K-62	530	43	58.5	2
421-721	Kisköre	9-46	1190	66	46.2	2	423-792	Szarvas	B-75	580	41	50.0	3
428-736	Kunhegyes	B-41	328	28	51.8	1	429-794	Szarvas	K-69	304	32	65.8	3
429-747	Kisujszállás	13	1570	99	56.1	x2	429-793	Szarvas	K-68	440	31	43.2	3
			1480	89	52.0	a2	426-796	Szarvas	K-59	469	35	49.0	2
			1388	90	56.9	x2				460	40	60.9	2
			1806	111	55.4	x2	423-792	Szarvas	B-57	530	41	54.7	2
			1516	92	53.4	x2				520	42	57.7	1
429-747	Kisujszállás	15	1419	89	55.0	x2	425-793	Szarvas	B-20	530	44	60.4	3
			1514	90	52.2	x2	425-793	Szarvas	K-58	580	43	53.4	3
			1494	92	54.2	x2	422-793	Szarvas	B-66	618	40	45.3	1
			1354	82	52.4	x2	423-792	Szarvas	B-51	697	47	50.2	1
			1447	81	48.4	x2	421-793	Szarvas	K-35	800	53	51.3	2
			1548	95	54.3	x2	427-793	Szarvas	K-61	1740	97	48.9	2

1	2	3	4	5	6	7	1	2	3	4	5	6	7
427-793	Szarvas	K-61	2209	134	55.2	2	424-864	Makó	1	2000	110	49.0	x2
429-793	Szarvas	1	2600	115	39.6	x2				2500	123	44.4	x2
			3000	128	38.7	x2				3000	138	42.0	x2
			3020	154	47.0	x2				3500	150	39.4	x2
			3479	142	37.4	x2				4060	170	38.9	x2
425-809	Eperjeshedgyhát	3	1760	94	46.6	x2				3634	141	35.5	x2
			2000	105	46.5	x2				3709	150	37.2	x2
429-801	Csabacsüd	K-17	505	40	55.5	3	429-870	Apátfalva	K-31	2098	103	43.4	1
427-814	Gádoros	B-35	308	30	58.4	3	435-632	Edelény	E-475	626	47	59.1	b3
422-815	Eperjes	K-45	357	32	56.0	3	436-657	Miskolc	K-103	210	37	119.0	3
423-815	Eperjes	K-47	376	30	47.9	2	437-654	Miskolc	B-72	476	51	81.9	2
			360	32	55.6	2				480	49	77.1	2
428-816	Gádoros	B-34	390	35	59.0	2	438-654	Miskolc	B-10	550	49	67.3	2
			381	31	49.9	2				610	46	55.7	2
426-815	Gádoros	K-38	399	31	47.6	1	438-654	Miskolc	B-69	615	52	65.0	2
422-819	Árpádhalom	K-65	408	34	53.9	1				615	46	55.3	2
425-812	Eperjes	K-43	620	47	56.5	3	430-736	Kunhegyes	K-21	998	66	55.1	1
425-812	Eperjes	K-44	630	46	54.0	3	431-746	Kisujszállás	16	1398	82	50.8	x2
427-816	Nagyszénás	3	2887	138	43.6	x2				1751	112	57.7	x2
427-828	Székkutas	K-264	296	30	60.8	3	433-747	Kisujszállás	12	1355	91	59.0	x2
421-822	Nagymágócs	K-187	298	28	53.7	3				1488	90	54.9	x2
425-820	Árpádhalom	B-64	377	35	61.0	3				1388	89	56.2	x2
424-820	Árpádhalom	B-67	395	30	45.6	1	436-748	Kisujszállás	11	1237	81	56.6	a2
424-827	Székkutas	K-266	406	33	51.7	1	437-748	Kisujszállás	9	1232	84	59.3	x2
426-825	Orosháza	3	2766	165.5	55.5	x2	437-747	Kisujszállás	7	1260	85	58.7	x2
			2771	160	53.4	x2	432-746	Kisujszállás	21	1355	86	55.4	a2
			2772	155	51.6	x2	430-748	Kisujszállás	23	1399	82	50.8	x2
429-823	Orosháza	K-144	530	48	67.9	3				1656	96	51.3	a2
425-829	Székkutas	K-272	510	39	52.9	2	438-754	Kisujszállás	B-82	300	28	53.3	3
			460	41	63.0	2	439-753	Kisujszállás	B-18/a	470	43	66.0	3
427-828	Székkutas	K-271	1771	128	65.5	2	439-754	Kisujszállás	K-81	614	47	57.0	2
			1740	106	54.0	2				550	40	50.9	2
421,822	Nagymágócs	K-193	2001	118	53.0	2	437-752	Kisujszállás	B-63	1067	74	58.1	2
423-833	Székkutas	B-11	268	29	63.4	3	435-760	Turgony	2	1862	103	48.9	x2
423-831	Székkutas	K-270	280	28	57.1	1				2052	106	45.8	x2
423-833	Székkutas	B-269	369	32	54.2	1				2198	115	46.9	x2
425-842	Hódmezovásár-hely	K-731	363	33	57.9	3	439-778	Endrod	K-22	418	34	52.6	1
							437-781	Endrod	K-13	1520	96	55.3	3
420-845	Hódmezovh.	K-960	394	31	48.2	1	437-784	Endrod	2	2155	125	52.4	a2
421-853	Földeák	B-53	400	28	40.0	1				2382	141	54.2	x2
423-854	Makó	K-192	435	31	43.7	1				2178	133	55.6	x2
			372	29	45.7	2	430-794	Szarvas	3	2209	135	55.7	x2
423-864	Makó	K-191	405	31	46.9	3	430-792	Szarvas	4	1369	88	55.5	a2
420-866	Makó	B-210	437	31	43.5	1				1208	79	55.5	x2
423-862	Makó	K-187	450	30	40.0	1				2288	124	49.0	x2
421-864	Makó	B-66	539	42	55.7	1				2593	140	49.4	x2
			387	33	54.3	2	430-790	Szarvas	K-56	393	33	53.4	3
420-864	Makó	B-57	943	51	41.4	1	432-799	Csabacsüd	K-52	340	30	52.9	3
422-861	Makó	K-195	2081	102	43.2	1	431-797	Csabacsüd	B-50	450	38	57.8	2
424-864	Makó	1	3500	147	38.6	x2				435	32	46.0	2
			3260	138	38.7	x2	430-795	Szarvas	K-43	518	36	46.3	3
			4095	169	38.3	x2	436-800	Kardos	K-37	565	48	63.7	2
			3780	140	33.9	x2				490	35	46.9	2
			2765	124	40.5	x2	433-814	Nagyszénás	B-127	333	29	51.1	1

1	2	3	4	5	6	7	1	2	3	4	5	6	7
435-819	Orosháza	2	2600	164	58.5	x2	443-732	Tatárüllés	4	1130	71	53.1	x2
43-82	Orosháza	B-649	260	30	69.2	3		/Kunmadaras/					
431-822	Orosháza	1	2605	140	49.1	x2	444-734	Karcag	K-68	780	61	64.1	3
433-826	Orosháza	B-650	260	27	57.7	3	499-741	Karcag	B-127	1497	91	53.4	2
433-825	Orosháza	B-522	280	33	75.0	2				1250	80	55.2	2
			276	29	61.1	2	449-740	Karcag	K-165	550	42	56.4	3
435-825	Orosháza	K-474	450	41	64.4	3	446,764	Turkeve	/telep/	1170	73	52.1	x3
434-827	Orosháza	B-112	460	42	65.2	3				1160	71	50.9	x3
434-826	Orosháza	B-69	464	44	69.0	3				1147	71	51.4	x3
435-826	Orosháza	B-103	470	39	57.4	3				1130	71	52.2	x3
430-825	Orosháza	B-10	500	42	60.0	3	440-783	Gyoma	B-80	267	29	63.7	1
435-824	Orosháza	K-481	1580	117	66.5	2	443-781	Gyoma	K-82	350	30	51.4	2
			1540	100	57.1	2	444-789	Gyoma	1	2817	134	43.3	x2
430-824	Orosháza	K-524	1680	108	57.1	1	449-782	Gyoma	K-35	444	33	47.3	3
435-830	Orosháza	K-722	505	43	61.3	1	446-786	Gyoma	K-57	460	40	60.9	2
430-842	Békéssámson	B-26	296	32	67.6	3	446-783	Gyoma	K-67	669	46	50.8	1
439-842	Tótkomlós	B-139	317	33	66.2	3	440-783	Gyoma	B-4/a	1060	73	57.5	3
439-849	Ambrózfalva	B-6	415	38	62.7	1	448-780	Dévaványa	2	2710	140.5	47.4	x2
439-847	Nagyér	B-8	500	49	74.0	3	446-796	Hunya	K-21	250	25	52.0	2
438-843	Tótkomlós	B-19	540	47	64.8	3	447-799	Kondoros	1	3118	166	49.4	x2
438-842	Tótkomlós	K-97	1433	101	62.1	1				3176	167	48.8	x2
439-852	Pitvaros	B-12	361	31	52.6	3				3620	172	44.2	x2
430-859	Királyhegyes	K-10	500	33	42.0	3	449-805	Kétsoprony	K-102	321	32	62.3	3
436-852	Csanádalberti	K-4	526	38	49.4	3	445-804	Kondoros	K-104	350	29	48.6	3
439-851	Pitvaros	K-13	540	42	55.6	3	442-805	Kondoros	B-9	430	41	67.4	3
439-851	Pitvaros	K-15	1373	81	50.3	1	440-802	Kondoros	K-105	450	36	53.3	1
431-869	Magyarcsanád	B-20	350	28	45.7	2	440-802	Kondoros	K-93	550	39	49.1	1
			340	28	47.1	2	443-804	Kondoros	B-89	551	39	49.0	1
431-866	Apátfalva	K-27	400	27	37.5	2	445-808	Kondoros	K-90	753	46	45.2	2
			335	30	53.7	2	444-802	Kondoros	K-106	1500	92	53.3	2
439-867	Nagylak	B-18	700	48	51.4	1	443-828	Pusztaföldvär	6	111	23.6	104.5	c2
439-866	Nagylak	B-20	840	52	47.6	1				209	27.6	74.6	c2
432-870	Magyarcsanád	K-19	365	27	41.1	2				310	33.4	69.0	c2
			320	28	50.0	2				412	41.2	70.9	c2
447-683	Mezocsát	4-28	765	53	54.9	2				511	46.2	66.9	c2
			690	55	63.8	2				613	53.4	67.5	c2
445-709	Egyek	K-71	490	36	51.0	3				708	58.8	66.1	c2
442-724	Tiszaörs	K-10	1766	126	65.1	2				809	67.0	68.0	c2
441-729	Kunmadaras	4	1131	79	60.1	x2				911	77.2	71.6	c2
440-728	Kunmadaras	1	1122	79	60.6	x2				1010	83.6	70.3	c2
440-726	Kunmadaras	3	1110	76	58.6	x2				1109	90.8	71.1	c2
			1775	125	64.2	x2				1212	96.2	69.5	c2

1	2	3	4	5	6	7	1	2	3	4	5	6	7
443-828	Pusztaföldvár	6	1250	99.8	70.2	c2	450-808	Kétsoprony	B-103	360	32	55.6	3
445-829	Tótkomlós	23	1392	111	71.1	x2	457-803	Kamut	B-109	423	32	47.3	3
447-837	Kaszpaer-Dél	3	1615	100	54.5	x2	453-803	Kamut	K-111	491	29	34.6	2
448-838	Kaszaper-Dél	1	900	65	58.9	x2				420	30	42.9	2
			989	69	57.6	x2	459-803	Murony	K-67	750	56	58.7	2
			997	67	55.2	x2				650	43	47.7	2
			1028	68	54.5	x2	450-832	Csanádapáca	2	2154	144	61.3	x2
			1544	98	55.7	x2				2475	160	59.8	x2
			1544	105	60.2	x2				1034	83	68.7	x2
441-836	Pusztaszolos	2226	1070	77	60.8	x2	451-837	Magyar-	K-50	1193	79	56.2	1
	/Tótkomlós/							Nagybánhegyes					
446-839	Tótkomlós	19	1584	120	68.2	x2	457-831	Medgyesbodzás	1	2405	156	59.9	x2
449-842	Tótkomlós	K-1	1484	116	70.1	x2	454-843	Mezokovácsháza	K-64	970	72	61.9	3
			838	68	66.8	x2	454-841	Mezokovácsháza	K-63	1040	74	59.6	3
			1513	118	70.1	x2	458-855	Battonya	4	1030	73	59.2	x2
			900	70	64.4	x2	456-856	Battonya	21	1025	79	65.4	x2
			1591	125	71.0	x2	458-855	Battonya	53	1030	79	65.0	x2
443-841	Tótkomlós	29	1878	133	64.4	x2	457-854	Battonya	48	722	59	65.1	x2
443-841	Tótkomlós	22	1401	110	70.0	x2	453-851	Mezohegyes	12	1179	85	61.9	x2
443-841	Tótkomlós	24	1410	112	70.9	x2	454-850	Mezohegyes	17	1165	85	62.7	x2
440-841	Tótkomlós	K-145	1240	89	62.1	3	453-851	Mezohegyes	16	800	64	65.0	x2
440-854	Pitvaros	Pit-D-1	2513	114	40.6	x2	454-850	Mezohegyes	18	1178	86	62.8	x2
			2557	119	41.8	x2	452-850	Mezohegyes	15	803	63	63.5	x2
			2606	121	41.8	x2	453-852	Mezohegyes	K-165	577	48	62.4	1
444-853	Mezohegyes	B-167	405	37	61.7	1	467-652	Taktaszada	B-7	310	35	77.4	3
442-855	Mezohegyes	K-122	470	35	48.9	3	461-673	Leninváros	4-30	1174	65	46.0	2
445-854	Mezohegyes	K-170	485	43	63.9	1	463-678	Polgár	B-88	959	58	49.0	2
456-644	Megyaszó	B-8	280	27	60.7	3	466-712	Hortobágy	B-11	1094	82	64.9	2
453-666	Sajódhidvég	4-5 K-3	1877	108	51.7	2	468-728	Nádudvar	B-399	700	47	51.4	3
455-694	Tiszacsege	K-108	760	58	61.8	3	468-727	Nádudvar	Nu-Dk-4	1665	107	57.7	x2
452-698	Tiszacsege	B-119	1295	87	58.7	1	468-732	Nádudvar	Nu-Dk-3	2000	103	46.0	x2
454-749	Karcag-Bucsa	3	1446	97	59.5	x2	464-741	Püspökladány	B-31	650	52	63.1	1
			1446	108	67.1	x2	464-741	Püspökladány	B-179	1063	80	64.9	2
450-741	Karcag	K-163	260	27	61.5	3	461-741	Püspökladány	Ny-1	1800	123	62.2	x2
450-741	Karcag	K-162	300	29	60.0	3				1793	118	59.7	x2
458-754	Bucsa	K-14	446	38	58.3	2	467-750	Szerep	B-19	310	32	64.5	1
454-770	Dévaványa	K-41	375	34	58.7	2	464-750	Szerep	K-7	358	33	58.7	3
453-771	Dévaványa	K-39	402	32	49.8	3	462-757	Bucsa	K-13	380	33	55.3	1
457-776	Dévaványa	K-33	447	35	51.5	1	462-757	Bucsa	K-11	472	39	57.2	3
455-773	Dévaványa	B-48	530	41	54.7	3	463-761	Kertészsziget	K-15	390	30	46.2	1
453-773	Dévaványa	K-28	1130	80	60.2	2	466-769	Szeghalom	K-34	396	36	60.6	1
			1090	71	54.1	2	461-764	Szeghalom	K-44	500	38	52.0	1
453-773	Dévaványa	B-58	1791	99	48.3	1	467-763	Füzesgyarmat	2	111	15.4	30.6	c2
452-775	Dévaványa	1	2148	109	45.2	x2				212	21.4	44.3	c2
			2168	123	51.2	x2				308	25.2	42.9	c2
			2120	132	56.6	x2				410	33.8	53.2	c2
			2203	125	51.3	x2				509	39.8	54.6	c2
			2485	142	52.3	x2				612	45.6	54.9	c2
456-772	Dévaványa'	3	2200	127	52.3	x2				711	51.4	55.4	c2
455-783	Korösladány	K-46	415	33	50.6	3				812	56.2	54.4	c2
451-784	Gyoma	K-36	531	33	39.6	3				909	61.4	54.3	c2
459-790	Koröstarcsa	B-18	511	39	52.8	1				1010	68.2	55.6	c2
459-790	Koröstarcsa	B-23	650	42	46.2	2				1112	74.8	56.5	c2
			550	42	54.5	2				1211	81.4	57.3	c2
										1260	83.6	56.8	c2

1	2	3	4	5	6	7	1	2	3	4	5	6	7
468-769	Szeghalom	K-52	400	33	52.5	3	470-763	Füzesgyarmat	K-28	542	42	55.4	2
467-772	Szeghalom	K-54	560	45	58.9	2				500	44	64.0	2
467-772	Szeghalom	K-39	377	32	53.1	3	471-766	Füzesgyarmat	K-38	566	48	63.6	1
460-777	Körösladány	K-20	401	32	49.9	3	472-765	Füzesgyarmat	B-34	1198	80	56.8	1
464-772	Szeghalom	K-22	407	34	54.1	1				930	68	60.2	2
461-779	Körösladány	K-50	408	34	53.9	3	474-764	Füzesgyarmat	B-41	485	32	49.4	2
469-774	Szeghalom	B-41	550	40	50.9	3	479-773	Csökmo	B-9	394	31	48.2	3
464-784	Körösladány	K-47	404	33	52.8	1	471-776	Szeghalom	K-32	390	34	56.4	1
462-781	Körösladány	K-49	444	38	58.6	2	472-773	Szeghalom	K-35	430	38	60.5	2
			443	33	47.4	2				414	33	50.7	2
461-787	Körösladány	K-21	526	38	49.4	1	474-771	Szeghalom	K-21	500	42	60.0	3
466-782	Körösladány	K-55	255	23	43.1	2	474-775	Szeghalom	K-38	522	33	40.2	3
460-795	Mezoberény	B-76	520	40	53.8	3	470-786	Bélmegyer	K-24	367	31	51.8	3
460-795	Mezoberény	B-64	546	41	53.1	1	471-785	Vészto	K-45	360	37	69.4	3
461-797	Mezoberény	K-62	896	58	51.3	1	477-782	Vészto	K-40	479	35	48.0	3
460-795	Mezoberény	B-55	1030	65	51.5	2	476-784	Vészto	K-29/a	492	36	48.8	3
461-803	Murony	B-2	492	42	61.0	2	473-784	Vészto	K-39	498	37	50.2	3
469-801	Békés	B-112	62	50.0	3	477-							
						784	Vészto	B-42	550	35	41.8	3	
464-813	Békéscsaba	2	1212	74	51.2	x2	478-785	Vészto	K-38	483	43	64.2	2
			2380	134	51.3	x2				570	43	54.4	2
466-812	Békéscsaba	B-960	278	31	68.3	3	476-787	Vészto	K-37	582	43	53.3	1
464-813	Békéscsaba	B-880	433	37	56.4	3	471-790	Bélmegyer	B-2	510	44	62.8	3
466-812	Békéscsaba	B-282	1996	110	49.1	1	479-814	Gyula	B-454	427	33	49.2	1
464-812	Békéscsaba	B-953	2260	107	42.0	2	478-816	Gyula	K-413	1658	91	47.7	2
			2386	134	51.1	2	481-671	Tiszavasvári	K-78	1193	88	64.5	2
468-180	Békéscsaba	3-137	798	50	47.6	1				1050	72	58.1	2
460-822	Ujkigyós	B-70	310	29	54.8	1	486-683	Hajdunánás	K-114	1014	69	57.2	1
465-820	Szabadkigyós	B-80	355	32	56.3	1	480-707	Balmazujváros	B-38/a	420	42	73.8	3
460-821	Szabadkigyós	B-69	374	29	45.5	3	480-707	Balmazujváros	B-202	1100	72	55.5	2
460-821	Ujkigyós	B-56	520	36	46.2	3	480-719	Hajduszoboszló	Hsz-29	1300	87	58.5	x2
466-844	Kevermes	2	1710	121	63.7	x2	481-715	Hajduszoboszló	Hsz-30	1290	87	58.9	x2
461-856	Battonya	K-138	800	53	51.3	2	482-716	Hajduszoboszló	Hsz-13	1316	83	54.7	x2
465-858	Battonya	K-6	970	66	55.7	x2	483-716	Hajduszoboszló	Hsz-6	1270	85	58.3	x2
467-858	Battonya	K-20	973	66	55.5	x2	484-719	Hajduszoboszló	Hsz-5	1190	80	58.0	x2
467-859	Battonya	K-15	952	60	50.4	x2	485-714	Hajduszoboszló	Hsz-26	1080	75	59.3	x2
466-858	Battonya	K-9	978	60	49.1	x2	485-719	Hajduszoboszló	Hsz-14	1240	89	62.9	x2
471-656	Prügy	B-10	252	35	95.2	2	486-717	Hajduszoboszló	Hsz-33	1070	76	60.7	x2
473-720	Kaba	KabÉ-1	2120	134	58.0	x2	482-721	Hajduszoboszló	Hsz-4	1200	79	56.7	x2
477-736	Kaba	B-30	620	51	65.5	3	483-720	Hajduszoboszló	Hsz-8	1160	85	63.8	x2
478-732	Kaba	K-47	670	49	56.7	3	486-721	Hajduszoboszló	8-150	695	51	57.6	1
478-734	Kaba	Kab-2	1600	115	65.0	x2	485-721	Hajduszoboszló	Hsz-3	1215	87	62.6	x2
479-734	Kaba	Kab-1	1100	77	60.0	x2	485-723	Hajduszoboszló	Hsz-7	1175	84	62.1	x2
472-739	Püspökladány	3	1528	110	64.8	x2	485-725	Hajduszoboszló	K-295	250	29	72.0	3
479-737	Kaba	KabaD-3	2125	146	63.5	x2	486-726	Hajduszoboszló	HÁ-1	1250	87	60.8	x2
471-743	Püspökladány	Pü-14	1940	134	63.4	x2	486-726	Hajduszoboszló	HÁ-2	2032	127	57.1	x2
			1940	82	36.6	x2	486-725	Hajduszoboszló	B-319	400	40	72.5	1
474-752	Biharnagybajom	B-26	300	30	60.0	1	486-728	Hajduszoboszló	B-212	1010	75	63.4	3
474-754	Biharnagybajom	K-20	670	54	62.7	3	486-725	Hajduszoboszló	B-317	1030	75	62.1	3
470-768	Szeghalom	Sz-2	1837	97	46.3	x2	486-726	Hajduszoboszló	K-344	1096	77	60.3	1
			1959	121	55.6	x2	487-725	Hajduszoboszló	B-69	1050	74	60.0	3
474-763	Füzesgyarmat	K-21	328	29	51.8	3	487-726	Hajduszoboszló	B-318	1060	80	65.1	3
472-762	Füzesgyarmat	K-35	330	32	60.6	1	487-726	Hajduszoboszló	B-7	2000	127	58.0	3
474-764	Füzesgyarmat	B-30	360	33	58.3	1	480-739	Kaba	Kaba-D-22001		110	49.5	x2
473-763	Füzesgyarmat	B-3	500	41	58.0	3				2171	127	53.4	x2

1	2	3	4	5	6	7
482–742	Földes	K-25	1285	78	52.1	1
482–742	Földes	K-29	1325	83	54.3	1
483–741	Földes	K-38	1130	73	54.9	2
481–754	Nagyrábé	B-22	355	30	50.7	1
483–751	Bihartorda	B-7	375	33	56.0	2
			356	29	47.8	2
481–754	Nagyrábé	K-24	889	64	58.5	2
			810	51	48.1	2
480–768	Csökmo	K-14	377	30	47.7	3
489–761	Zsáka	B-10	378	33	55.6	3
489–760	Zsáka	B-11	378	32	52.9	1
488–765	Verkerd	K-6	470	33	44.7	1
480–772	Csökmo	B-4	406	28	39.4	2
			358	29	47.5	2
482–772	Csökmo	K-12	420	28	38.1	2
			403	32	49.6	2
483–776	Ujiráz	B-8	445	34	49.4	3
485–777	Ujiráz	B-9	540	40	51.9	1
484–787	Okány	B-1	424	32	47.2	3
485–788	Okány	K-19	475	35	48.4	3
481–783	Vészto	K-44	556	46	61.2	2
			540	43	57.4	2
485–793	Sarkadkeresztur	K-13/a	447	35	51.5	3
488–799	Korösujfalu	B-4	507	37	49.3	2
488–806	Sarkad	K-105	1110	62	45.1	1
489–806	Sarkad	K-117	280	25	46.4	1
480–815	Gyula	B-440	392	28	40.8	3
480–815	Gyula	B-395	950	52	42.1	1
480–815	Gyula	B-145	2004	112	49.9	2
481–815	Gyula	K-453	2498	135	49.2	2
496–627	Sárospatak	K-123	270	50	148.1	3
495–626	Sárospatak	K-104	287	50	139.3	1
496–627	Sárospatak	K-115	300	49	130.0	1
492–683	Hajdudorog	8-143	1073	73	57.8	2
492–699	Hajduböszörmény	K-243	745	51	53.7	2
			600	51	66.7	2
492–699	Hajduböszörmény	K-242	930	72	65.6	3
492–699	Hajduböszörmény	K-270	980	67	57.1	2
			880	55	50.0	2
495–700	Hajduböszörmény	B-271	250	32	84.0	3
496–707	Hajduböszörmény	K-259	298	32	70.5	3
493–712	Debrecen	K-2179	250	28	68.0	3
491–711	Debrecen	K-1889	341	32	61.6	3
491–726	Ebes	Eb-7	1355	96.5	63.1	x2
492–723	Ebes	Eb-5	1035	74	60.9	x2
492–724	Ebes	Eb-10	1347	94	61.6	x2
493–725	Ebes	Eb-6	1320	91	60.6	x2
492–725	Ebes	Eb-2	1330	95	63.2	x2
497–749	Berettyóujfalu	B-44	345	30	52.2	3
496–748	Berettyóujfalu	K-39	427	35	53.9	3
497–750	Berettyóujfalu	B-19	410	34	53.7	3
497–750	Berettyóujfalu	B-15	806	63	63.3	2
			750	46	45.3	2
497–750	Berettyóujfalu	B-54	1392	93	58.2	2
496–750	Berettyóujfalu	B-9	330	37	75.8	3
493–753	Berettyóujfalu	K-52	270	27	55.6	3
497–750	Berettyóujfalu	B-45	291	27	51.5	3
497–750	Berettyóujfalu	B-5	321	28	49.8	3
497–750	Berettyósztmárton	B-40	290	28	55.2	3
497–751	Berettyósztmárton	B-43	309	28	51.8	3
497–750	Berettyósztmárton	B-38	305	30	59.0	3
497–750	Berettyósztmárton	B-12	319	32	62.7	3
497–753	Berettyósztmárton	B-2	283	31	67.1	3
497–753	Berettyósztmárton	B-8	383	35	60.1	3
497–751	Berettyósztmárton	B-41	255	30	70.6	3
491–755	Bakonszeg	K-6	290	30	62.1	1
490–754	Bakonszeg	B-4	423	32	47.3	3
493–760	Furta	K-9	385	36	62.3	2
496–778	Komádi	K-40	993	61	49.4	2
494–774	Komádi	K-55	1111	65	47.7	2
495–796	Ujszalonta	K-3	395	32	50.6	1
491–791	Mezogyán	K-17	480	35	47.9	2
			475	35	48.4	2
490–794	Sarkadkeresztur	K-6	521	37	48.0	3
490–798	Sarkad	Sark-35	2720	138	46.3	x2
494–799	Sarkad	Sark-36	2680	141	48.1	x2
			2680	134	45.5	x2
			2940	157	49.3	x2
507–641	Tiszakarád	B-10	396	34	58.1	1
509–650	Nagyhalász	B-56	647	52	63.4	2
			560	42	55.4	2
506–669	Nyiregyháza	B-257	490	39	57.1	3
508–663	Nyiregyháza	B-392	602	47	59.8	1
507–667	Nyiregyháza	B-468	700	58	67.1	2
			670	50	58.2	2
508–663	Nyiregyháza	K-368	800	58	58.8	1
508–663	Nyiregyháza	K-365	820	57	56.1	3
506–669	Nyiregyháza	B-443	885	67	63.3	2
			780	53	53.8	2
508-669	Hajduböszörmény	B-508	870	64	60.9	2
			770	53	54.5	2
507–663	Nyiregyháza	K-496	900	55	48.9	1
			780	53	53.8	2
503–718	Debrecen	B-1771	640	49	59.4	3
506–714	Debrecen	B-208	970	69	59.8	1
506–712	Debrecen	K-1912	969	66	56.8	1
503–714	Debrecen	B-1998	1010	68	56.4	1
504–714	Debrecen	B-2188	1081	78	62.0	2
			960	66	57.3	2
501–718	Debrecen	B-2109	1095	72	55.7	1
503–748	Szentpéterszeg	B-8	398	34	55.3	3
509–747	Hencida	B-11	564	43	55.0	1
506–745	Derecske	I	3760	173	42.8	x2
			5090	246	46.0	x2
500–762	Mezosas	B-7	336	29	50.6	3
509–767	Berekböszörmény	B-37	377	30	47.7	1

1	2	3	4	5	6	7
503–778	Biharugra	K-20	1137	75	55.4	2
			1020	61	48.0	2
504–781	Biharugra	3	2255	115	45.8	x2
			2240	121	48.7	x2
			2300	125	49.1	x2
510–649	Nagyhalász	B-59	400	37	65.0	2
			353	31	56.7	2
513–657	Kemecse	B-16	480	37	54.2	3
518–677	Nagykálló	B-84	933	73	66.5	2
518–748	Kismarja	K-5	426	37	58.7	3
512–753	Bojt	B-7	461	34	47.7	3
513–761	Ártánd	2	109	17.6	51.2	c2
			211	20.4	39.8	c2
			313	26.4	46.0	c2
			408	31.2	47.1	c2
			510	35.4	45.9	c2
			612	39.8	45.4	c2
			713	47.6	49.9	c2
			811	50.2	47.1	c2
			909	52.2	44.2	c2
			1010	53.6	41.2	c2
524–627	Ricse	B-12	385	33	57.1	1
522–734	Álmosd	13	2643	138	47.7	x2
			2937	146.6	45.8	x2
526–732	Álmosd	Álm-11	2700	124	41.5	x2
520–742	Pocsaj	K-13	345	36	69.6	3
520–742	Kismarja	Kism-10	1070	74	57.9	x2
521–744	Kismarja	Kism-8	816	55	52.7	x2
			818	58	56.2	x2
			856	63	59.6	x2
			892	64	58.3	x2
			912	65	58.1	x2
			927	65.5	57.7	x2
			1150	70	50.4	x2
532–637	Kisvárda	K-119	598	49	63.6	1
532–637	Kisvárda	K-127	787	61	63.5	1
534–664	Baktalórántháza	B-26	862	53	48.7	1
542–647	Gemzse	K-7	1076	71	55.8	2
553–667	Mátészalka	B-98	1009	67	55.5	1
558–677	Nagyecsed	B-28	563	42	55.1	1
567–661	Fehérgyarmat	K-69	830	51	48.2	3

[a]Numbered column heads are explained in text.

CHAPTER 17

Variations in Extensional Styles at Depth Across the Pannonian Basin System

L. H. Royden
Department of Earth, Atmospheric and Planetary Sciences
Massachusetts Institute of Technology, Cambridge, Massachusetts
02139

P. Dövényi
Eötvös University, Geophysics Department
H-1083 Budapest, Kun Bela ter 2 Hungary

Extensional styles at depth beneath five basins within the Pannonian basin system (Vienna, Danube, Zala, and Transcarpathian basins and Great Hungarian Plain) were evaluated by applying a simple graphical technique to subsidence and heat flow data from wells in each basin. The average rate of thermal subsidence in each basin, excluding the Transcarpathian basin, indicates the same rate of cooling of the lithosphere as does the surface heat flow measured in the same basin. Thermal subsidence rates, heat flow, and effective mantle thinning or (heating) show a systematic increase with increasing distance from the Carpathian thrust front. Initial subsidence and crustal thinning show little to no correlation with distance from the thrust front. These results are consistent with thin-skinned extension (involving only crustal rocks) beneath the Vienna basin, which is located near the thrust front. At greater distances from the thrust front, extension involves both the crust and mantle–lithosphere. Beneath the basins located more than 200 km from the thrust front, the mantle lithosphere appears to be very thin or very hot. A transitional zone exists between the area of thin-skinned extension in the crust and the area of greatly thinned mantle lithosphere, and is roughly coincident with the Miocene calcalkaline volcanic belts.

INTRODUCTION

A common assumption in basin analysis is that basins belonging to a single basin system should exhibit similar geologic and geophysical characteristics, and should follow similar evolutionary paths. This is most frequently assumed when the basins within the system can be shown to be of the same age and general tectonic type. Such an assumption is not necessarily correct. The Pannonian basin system offers an excellent example of how genetically related and tectonically similar basins within a single basin system may exhibit important differences that control the evolution of these basins. This chapter focuses on the heterogeneous behavior of the lithosphere at depth during Miocene crustal extension and basin formation within the Pannonian basin system.

The Pannonian basin system lies adjacent to the Carpathian Mountain belt and formed as a Miocene back-arc basin relative to the south and west-dipping subduction zone beneath the West and East Carpathians (Figures 1 and 2). It consists of a

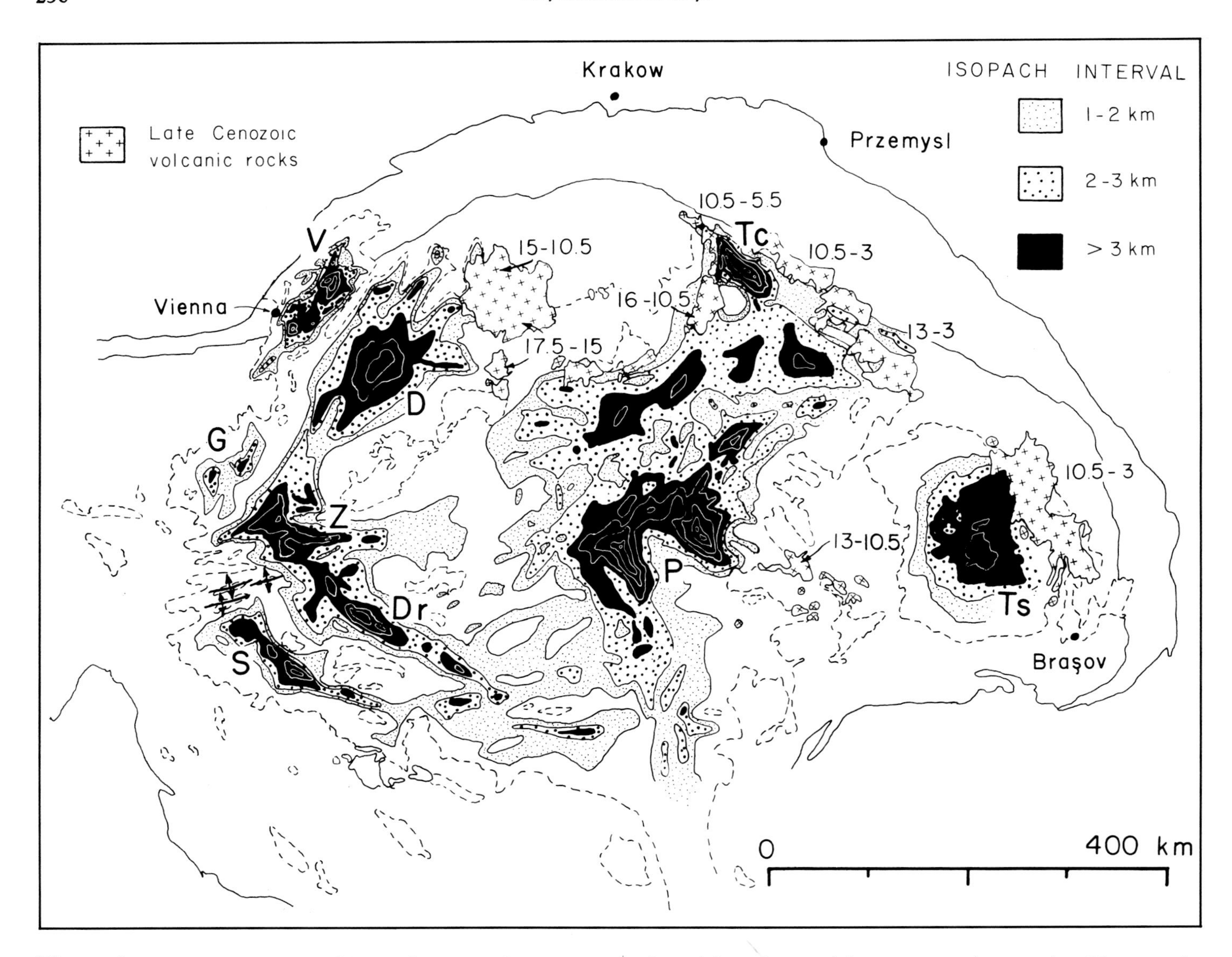

Figure 1. Isopach map showing thickness of Miocene–Quaternary rocks. Isopach lines for every kilometer interval greater than 3 km are indicated by white lines on black background (see also Map 8 in enclosure). Approximate ages of igneous rocks are shown. Basins: S, Sava; Dr, Drava; Z, Zala; G, Graz; D, Danube; V, Vienna; P, Great Hungarian Plain; Tc, Transcarpathian; Ts, Transylvanian. From Royden et al. (1983 a,b).

series of small, deep basins that formed by Miocene to recent subsidence and that are separated from one another by shallower areas (Figure 1). Except for the Transylvanian basin, these basins are areas of Miocene crustal extension that are connected to one another by a conjugate system of strike-slip faults (Horváth and Royden, 1981; Royden et al., 1982, 1983 a; Royden, this volume). Extension and initial subsidence of the basins was coeval with Miocene thrusting in the outer Carpathian thrust belt and basin extension appears to be intimately related to thrust-belt activity.

Geologic and geophysical data strongly indicate that the behavior of the lower crust and upper mantle during extension varied greatly across the Pannonian basin system, and there appears to be a close relationship between extension at depth beneath the basins and the position of each basin relative to the Carpathian thrust belt. Extension is thought to have involved the entire lithosphere of the Pannonian fragment, but not to have involved the subducted lithosphere of the European plate (Figure 2). Thus, extension beneath the most external basins located on top of the thrust belt would be thin skinned and involve only rocks at shallow depths. Extension beneath more internally situated basins and farther from the thrust belt would involve rocks to greater depth. Extension beneath the most internally situated basins would involve a complete Pannonian lithosphere underlain by asthenosphere. In this chapter, a simple graphical technique is used to examine subsidence and heat flow data from five basins within the Pannonian system: the Vienna, Danube, Transcarpathian, and Zala basins, and the Great Hungarian Plain. In this way, the transition from thin-skinned to whole lithosphere extension can be examined by defining the style of lithospheric extension present beneath each basin.

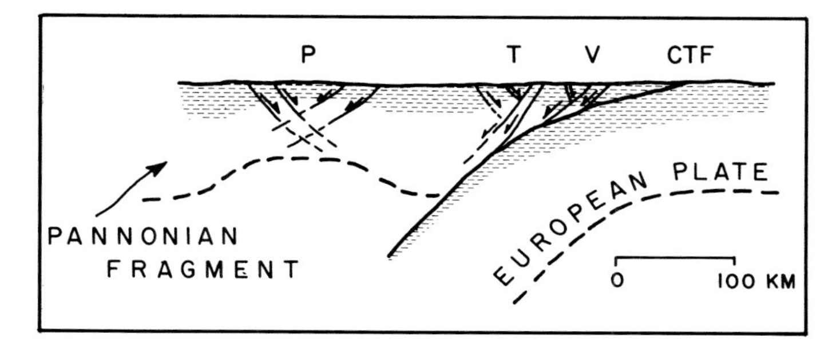

Figure 2. Schematic composite diagram illustrating how the thickness of the Pannonian fragment is related to distance from the thrust front, and showing the relative distances of the Vienna (V), and Transcarpathian (T) basins and the Great Hungarian Plain (P) from the Carpathian thrust front (CTF). Thus, if extension (and strike-slip faulting) has been confined mainly to the Pannonian fragment, the depth to which extension would have occurred should also increase with distance from the thrust front. Shaded area represents crustal rocks. Areas of extension indicated schematically by normal faults; the fault geometry shown here is not meant to be realistic. Modified after Royden et al. (1983 a).

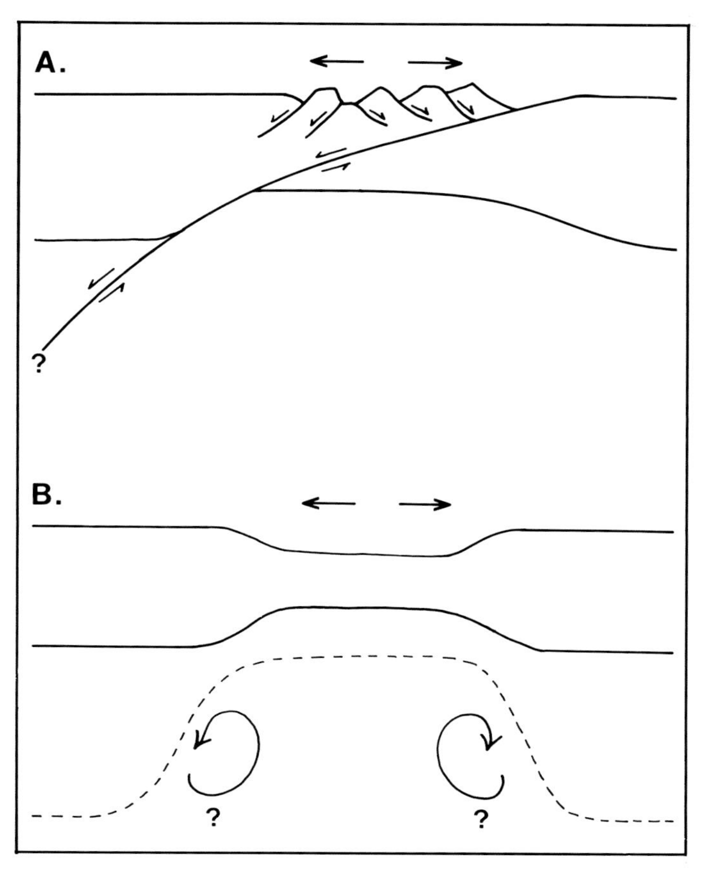

Figure 3. Two different ways in which modified (nonuniform) extension might occur. (A) Gently dipping detachment surface with extension occurring near the toe of the upper plate, but not in the lower plate immediately below. Upper plate extension must be balanced somewhere. Because material may also move perpendicular to the plane of the diagram, there are many ways to balance this geometry. (B) Poorly understood processes acting near the lithosphere–asthenosphere boundary effectively removes or thins lithospheric material and replaces it with hotter asthenosphere. Because this appears to occur rapidly, thermal conduction cannot be responsible. From Royden (1986). Reprinted with permission of Editions Technip, Paris, France.

HEATING AND SUBSIDENCE IN EXTENSIONAL BASINS

General Considerations

A graphical approach to extensional analysis can be derived from the modified extension model of Royden and Keen (1980) and Sclater et al. (1980). Modified extension (originally called nonuniform extension) differs mainly from simple uniform extension proposed by McKenzie (1978) in that it allows crustal thinning to occur somewhat independently of heating within the lithosphere. It is frequently assumed that modified extension implies different amounts of stretching within the crust and the mantle lithosphere, possibly leading to space problems in areas adjacent to the extended region. This is not strictly correct. Modified extension assumes only that crustal thinning and lithospheric heating do not necessarily have a one-to-one relationship within any particular column of lithosphere (Figure 3).

Subsidence and heat flow analysis of extensional basins is aimed at reconstructing the crustal thinning and the total lithospheric heating that have occurred during basin extension. Both crustal thinning and total lithospheric heating are recorded in the initial (synextensional) subsidence of a rifted basin, which occurs in isostatic response to the net density change within the lithosphere (Figure 4). The magnitude of lithospheric heating, and to a lesser extent the distribution of excess heat throughout the lithosphere, are recorded in the long-term thermal (postextensional) subsidence. Thus, by separating and comparing the initial and thermal subsidence of a rifted basin, one can compute both the crustal thinning and the lithospheric heating that occurred during basin formation. In young basins, the surface heat flow also reflects lithospheric heating, and is almost directly proportional to the thermal subsidence.

For the purposes of extensional analyses, lithospheric behavior need only be described by two variables that correspond to crustal thinning and to lithospheric heating. Note that these variables are only a convenient way to quantify total crustal thinning (or net density change of the crust) and mantle heating—they do not necessarily imply anything about physical processes.

In practice, crustal thinning can be described by a thinning factor δ (so that if the crustal thickness before stretching is h, the thickness after stretching is h/δ). This thinning factor does not necessarily imply anything about how thinning is distributed throughout the crust, but rather describes only the net crustal thinning (Figure 5).

The amount of heat distributed throughout the lithosphere is more difficult to describe by one parameter, because the resulting subsidence and heat flow will depend on exactly how temperature is distributed with depth. However, most geologically reasonable temperature structures can be approximated by two-legged geotherm such as that shown in Figure 5. Short wavelength differences from a two-legged geotherm will disappear

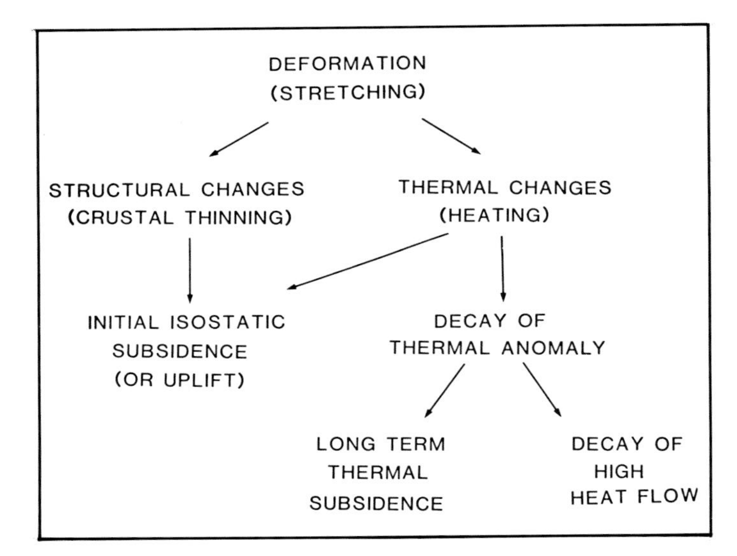

Figure 4. Simple flow chart diagramming the consequences of lithospheric stretching, which results in both thinned crust and a geotherm elevated above the background or equilibrium geotherm. Crustal thinning causes a density increase within a lithospheric column as lighter crustal rocks are replaced by denser mantle material. The raised geotherm causes a density decrease within the same column as colder, denser mantle lithosphere is replaced by hotter, less dense asthenosphere. If the net density of the lithospheric column decreases, initial uplift will occur; if the net density increases, initial subsidence will occur. As the elevated geotherm cools towards thermal equilibrium, the lithosphere cools, contracts, and subsides isostatically over about 200 to 300 m.y. Elevated surface heat flow present after stretching likewise decays. Note that surface heat flow and thermal subsidence both reflect how fast the lithosphere is cooling, and are thus linked. From Royden (1986). Reprinted with permission of Editions Technip, Paris, France.

quickly, and usually have little effect on subsidence and surface heat flow. Therefore, it is sufficient to describe heat distribution within the lithosphere with two parameters: δ to describe the assumed postrift geotherm in the crust and β to describe the assumed postrift geotherm in the mantle lithosphere (Figures 4 and 5). Note that δ also describes the net crustal thinning, so that crustal heating and crustal thinning are assumed to be linked. β (mantle heating) is independent of δ. This simple parameterization clearly does not describe all possible geotherms, but it comes close to describing most geologically reasonable geotherms. (Note that $\beta = \delta$ corresponds to uniform extension.) Analytical solution to the initial and thermal subsidence and the heat flow resulting from this parameterization are given by Royden and Keen (1980). These expressions can be simplified and calibrated to agree with oceanic data, as is described in the Appendix.

In the following sections, thermal subsidence data and heat flow data from five basins in the Pannonian system are first analyzed for consistency and then combined with initial subsidence data to investigate crustal thinning and net lithospheric heating during extension beneath each of the five basins.

Graphical Analysis

The simple, graphical approach to basin analysis used here is developed and described in greater detail by Royden (1986). In this chapter, we present only a graphical solution for analyzing extension in basins where 10 m.y. has elapsed since the end of extension. (A more complete analysis of wells in the Great Hungarian Plain is given by Horváth et al., this volume.) This is reasonable for basins in the Pannonian system, in which extension was slightly diachronous throughout the basin system and ended between about 15 and 10 Ma in all of the basins. The errors introduced by assuming that extension ended at 10 Ma are small using the approach, which follows.

Figure 6A shows the initial depth that is present immediately after instantaneous stretching of continental lithosphere with prerift elevation at sea level. Note that either initial uplift or initial subsidence can occur. Figure 6A also shows that even when prerift elevations and prerift thermal structure are speci-

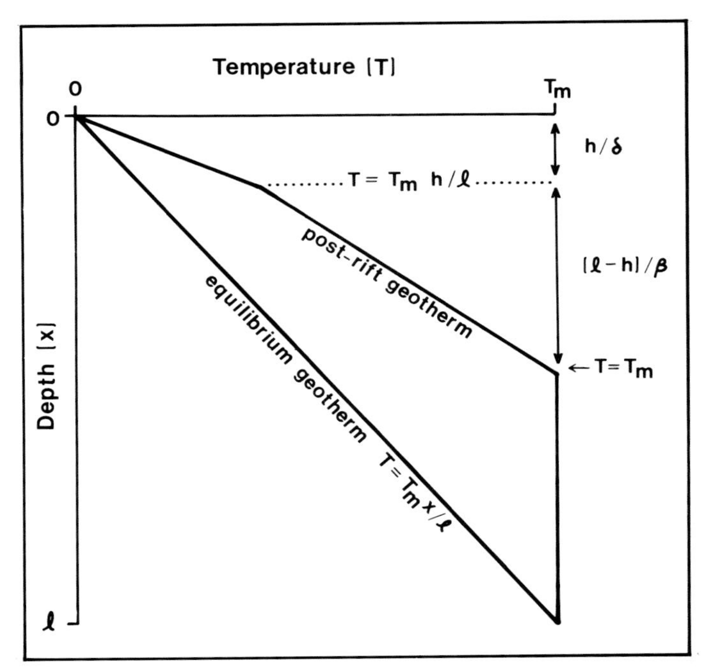

Figure 5. Schematic representation of the equilibrium (prestretching) geotherm, and the geotherm immediately after stretching. The value h is taken to be the crustal thickness before stretching, so that h/δ is the crustal thickness after stretching. Likewise, $(l - h)$ is the thickness of the mantle lithosphere before stretching, and $(l - h)/\beta$ is its thickness after. Symbols are listed in Table 1A. From Royden (1986). Reprinted with permission of Editions Technip, Paris, France.

Table 1 A. Symbols and Values Used

Symbols	Explanation
h	Crustal thickness before stretching (35 km)
d_o	Elevation before stretching (0 km)
$d(t)$	Depth after stretching
l	Lithospheric thickness (125 km)
$q(t)$	Surface heat flow
q_R	Radiogenic contribution to heat flow
s	Thermal subsidence
T	Temperature (°C)
T_m	Temperature at the base of the lithosphere
δ	Crustal stretching factor
β	Subcrustal stretching factor
x	Depth (m)
t	Time since stretching (m.y.)

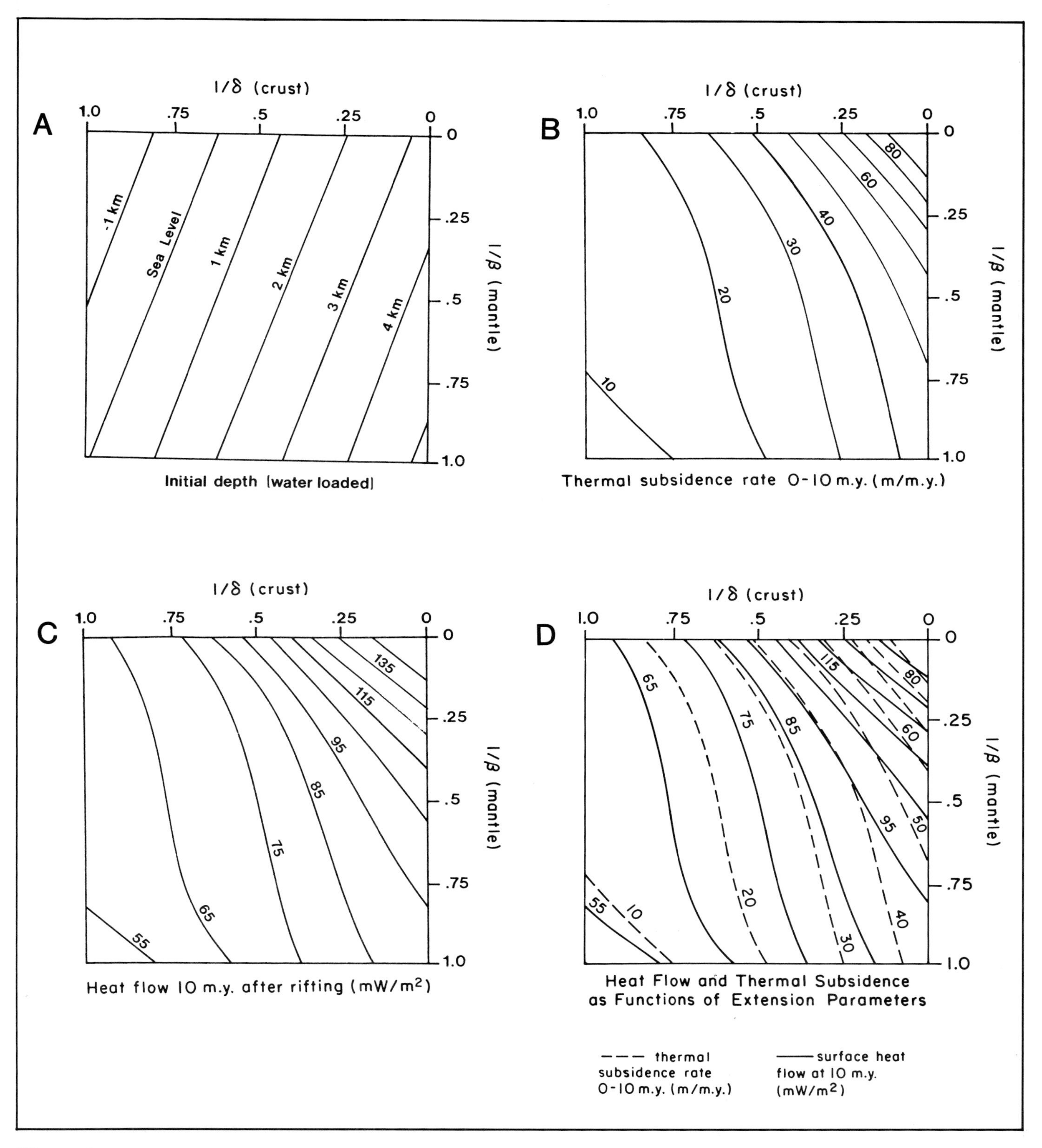

Figure 6. (A) Initial depth after stretching as a function of thinning parameters δ (crustal thinning) and β (subcrustal thinning). Elevation prior to stretching is assumed to be at sea level, and the prestretching geotherm is assumed to correspond to thermal equilibrium (see Figure 5). All elevations are water loaded, so that sediment loaded or unloaded uplift must be corrected. Extension is assumed to occur instantaneously. (B) Average thermal subsidence rate, as a function of β and δ, between $t = 0$ (immediately after stretching) and $t = 10$ m.y. (10 m.y. after stretching). Initial conditions are the same as (A). (C) Surface heat flow, as a function of β and δ, at $t = 10$ m.y. Initial conditions are the same as (A). Radiogenic heat production in the crust is assumed to contribute 12 mW/m^2 to the surface heat flow. This neglects possible effects of lateral heat conduction, thermal blanketing, and convection. (D) Surface heat flow at 10 m.y., superimposed on average thermal subsidence rate for 0–10 m.y. With small uncertainty, surface heat flow at 10 m.y. is a function of thermal subsidence rate and relatively independent of β and δ.

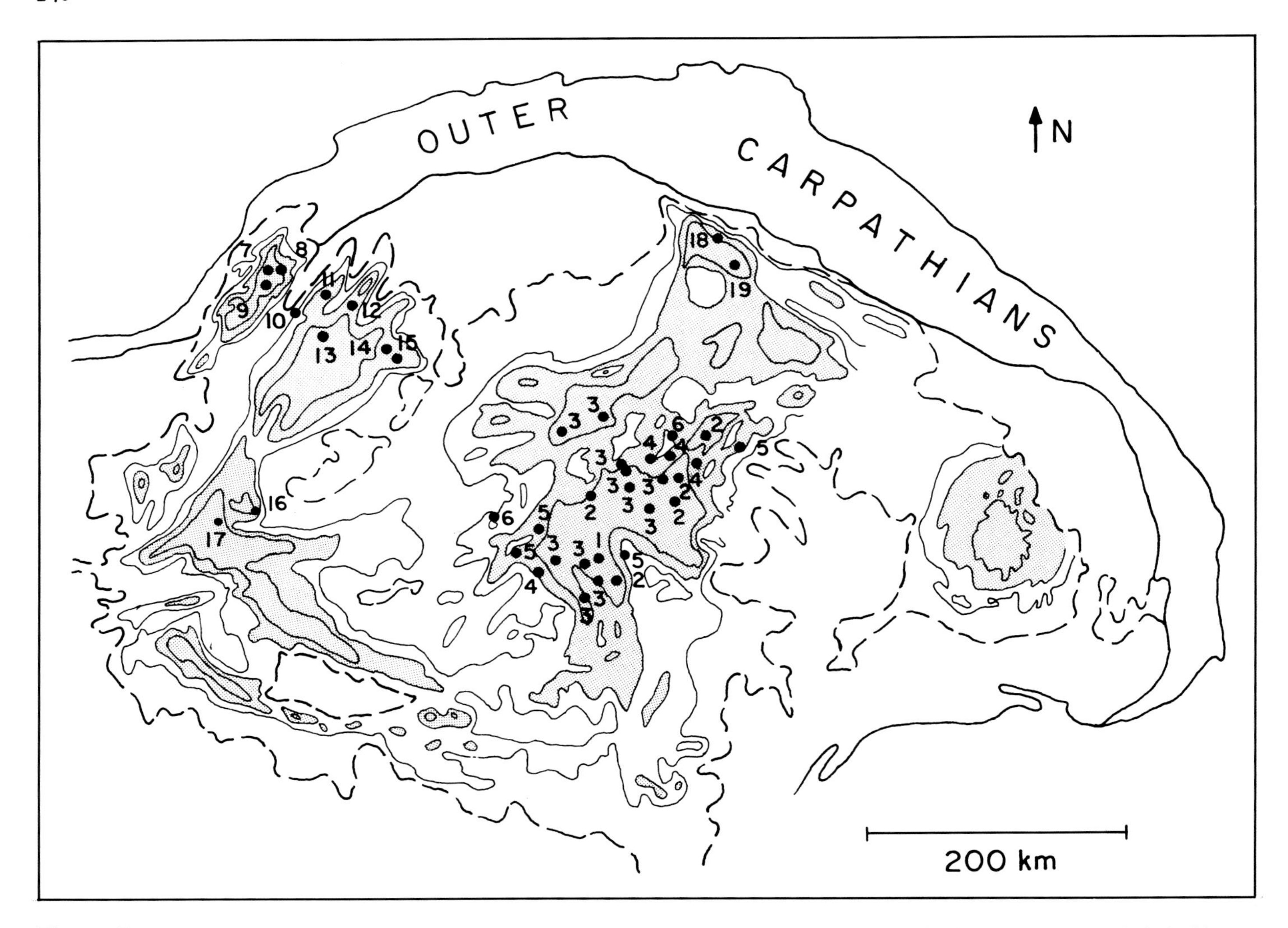

Figure 7. Location map for wells listed in Tables 1 and 2. Dark continuous line is boundary of outer (flysch) Carpathians. Dark dashed line is boundary of pre-Neogene outcrop. Light solid lines are isopachs for 1, 2 and 3 km of Neogene–Quaternary sedimentary rocks. Shading shows area where thickness of Neogene–Quaternary sedimentary rocks exceeds 2 km.

fied, there are a variety of different values of β and δ that give the same initial depth after stretching. Thus β and δ cannot be uniquely determined from the initial subsidence alone.

Figure 6B shows the average rate of thermal subsidence, as a function of β and δ, that occurs between 0 and 10 m.y. after instantaneous rifting. Figure 6B shows that even if thermal subsidence is reasonably well known, β and δ cannot be uniquely determined from the thermal subsidence alone.

Figure 6C shows surface heat flow, as a function of β and δ, at 10 m.y. after rifting. The background heat flow and radiogenic contributions were assumed to total 45 mW/m^2. Note that β and δ cannot be determined uniquely from heat flow data, even if heat flow could be determined precisely.

By superimposing 6A and 6B, or 6A and 6C, unique values of β and δ can be determined graphically from the subsidence history of a basin. For example, well group 1 in the Pannonian basin (Figure 7) has a water-loaded initial subsidence of about 1800 ± 300 m, depending on the decompaction parameters used, and a water-loaded thermal subsidence rate (for $0 \leq t \leq 10$ m.y.) of 44 ± 10 (m/m.y.) (see Tables 1, 2 and 3.) Assuming that prior to extension, the basement was near sea level, Figure 8A shows the initial subsidence (1800 m) plotted as a function of β and δ (taken from Figure 6A). Also plotted is the thermal subsidence rate (44 m/m.y.) for $0 \leq t \leq 10$ m.y. (taken from Figure 6B). The intersection of the initial subsidence curve and the thermal subsidence curve yield unique values of crustal thinning (δ = 2.8) and lithospheric heating (β = 7). When the uncertainties in the subsidence data are included, a field of permissible values of β and δ is defined, (δ = 2.4–3.6, β = 3–∞). One could also include uncertainties in prerift elevations by modifying Figure 6A using equation (3) (in Appendix), but we have neglected these uncertainties in this paper (see Royden et al., 1983b). (Note that different calibrations of the equations derived in the appendix can result in slightly different values of β and δ for the same well. See, for example, Horváth et al., this volume.)

Theoretical Consistency of Thermal Subsidence and Heat Flow

Thermal subsidence and heat flow can be evaluated for consistency regardless of initial subsidence or prerift conditions.

Table 1. Sedimentation Data for Selected Wells in the Pannonian Basin System

Basin	Well	Depth to Base (m)[a]								References
		Quaternary (2.5 Ma)	Pliocene or Dacian (5.5 Ma)	Pontian (8 Ma)	Pannonian (12 Ma)	Sarmatian (13 Ma)	Badenian (16.5 Ma)	Karpatian (17.5 Ma)	Miocene (24 Ma)	
Great Hungarian Plain[b]	(1) Group 1	690	1200	2200	5190	—	6500	—	—	1,2
	(2) Group 2	400[c]	1130[c]	2150[c]	4160[c]	—	4600[c]	—	—	1,2
	(3) Group 3	290	1000	1890	3390	—	3510	—	—	1,2
	(4) Group 4	480	840	1450	2380	—	2460	—	—	1,2
	(5) Group 5	280	640	1120	1790	—	1800	—	—	1,2
	(6) Group 6	270	600	840	1300	—	1310	—	—	1,2
Vienna	(7) Breclav-12	—	—	—	540	890	1865	2204	—	3
	(8) Brodské-7	—	—	—	373	920	1720	1955	—	3
	(9) deep basin	—	—	—	—	—	—	4500-5500	—	4
West Danube	(10) Grob-1	—	—	240	480	708	1283	—	1336	3
Central–Northern Danube	(11) Senec-1	—	500	784	1238	1875	2540	—	2579	1,3
	(12) Abrahám-1	—	890	1060	1370	1703	2239	—	2251	3
	(13) Ivanka pri Dunaji 1	—	—	—	1100	1480	1900	—	—	1
East Danube	(14) Kolárovo-3	—	—	1190	2600	2690	—	—	2834	3
	(15) Kolárovo-4	50	1060	1170	2500	2640	—	—	2665	3
Zala	(16) Nagylengyel 108	—	—	1200	2100	2200	2300	—	—	1
	(17) Cserstreg 1	—	—	1000	2350	2540	2800	—	—	1
Transcarpatian	(18) Bunkovce-1	—	—	—	800	1215	—	—	2100	3
	(19) Ptruksa-22	—	—	—	880	2707	3607	—	3713	3

[a]With approximate age at base.
[b]Averages for well groups.
[c]Excluding well Derecske 1 because Pliocene(?) beds are gently folded and eroded in this area.
[d]References: (1) Nagymarosy (1981); (2) Table 2 and Royden et al. (1983 b); (3) Thon and Thomek (this volume); (4) Based on stage isopach maps from Spicka (1972).

Superposition of Figures 6B and 6C show that, on a plot of β versus δ, contours of constant thermal subsidence rate and contours of constant heat flow are roughly coincident. Thus, with small uncertainties, heat flow can be determined directly from thermal subsidence and vice versa (Figure 6D). Ten million years after extension ends, this relationship is roughly:

$$\text{heat flow (mW/m}^2) = 45 \pm 10 + (\text{thermal subsidence rate, m/m.y.})/8$$

where the ± 10 mW/m^2 is estimated uncertainty due to variations in concentrations of heat producing elements within the crust. This relationship should hold reasonably well even where the thermal blanketing effect of sediments is significant, because both heat flow and thermal subsidence should be damped by comparable amounts. The correspondence between thermal subsidence rates and heat flow also holds for basins older than 10 m.y., but the equation given above changes with the age of the basin.

SUBSIDENCE AND HEAT FLOW DATA

Basement Subsidence from Sedimentation Data

Subsidence data from selected wells in five basins (Great Hungarian Plain, Zala, Danube, Transcarpathian and Vienna basins) are presented in Table 1 (see also Figure 7). The data shown from the Great Hungarian Plain are from six groups of wells; each well group contains averaged subsidence data for several wells with similar subsidence histories (Table 2). Data shown in Table 1 for the other basins are from individual wells.

The initial, or synextensional, subsidence of the basement was determined from the sum of (1) the thickness of sediment deposited during extension, and (2) the thickness of those postextensional sediments that record rapid filling of a deep basin created during extension. For example, in well group 1 in the Great Hungarian Plain, extension ended at about the beginning of the Pannonian stage (12 Ma), and a large thickness of

Table 2. Sedimentation and Heat Flow Data for Individual Wells in the Great Hungarian Plain

Group	Well Name	Depth to Base (m)[a]						Heat Flow[d] (mW/m^2)
		Badenian (16 Ma)	Sarmatian (13 Ma)	Pannonian (12 Ma)	Pontian (8 Ma)	Dacian[b] (5.5 Ma)	Quaternary[c] (2.5 Ma)	
1	Hódmezővásárhely I	6500	5350	5190	2200	1200	690	84 ± 7
2	Fábiansébtyen 3	5000	5000	3600	2170	1100	220	97 ± 11
	Derecske I	5000	5000	4450	1930	400	110	102 ± 8
	Makó 2	4900	4600	4600	2100	1100	550	—
	Doboz 1	4290	4290	4290	2450	1350	330	91 ± 10
	Vésztő	4200	?	?	1900	950	490	—
3	Hunya 1	3860	3860	3860	2020	1050	450	99 ± 10
	Kondoros 1	3700	3620	3620	1930	1000	350	100 ± 10
	Újszentiván 1	3670	3670	3670	2200	—	177	97 ± 11
	Jászladány 1	3640	3110	3110	1380	—	165	100 ± 10
	Algyő K-1	3560	3560	3560	2460	960	260	—
	Üllés DK-1	3480	3300	3300	1860	—	130	92 ± 9
	Gyoma 1	3450	3380	3380	1850	950	260	97 ± 10
	Békés 1	3450	3300	3230	1950	1000	350	—
	Köröstarcsa 1	3360	3230	3230	1800	1150	360	103 ± 11
	Abádszalók 1	3250	3130	3130	1395	—	374	87 ± 9
	Maroslele 1	3170	3170	3170	1950	880	250	95 ± 10
4	Körösladány 1	2790	2730	2730	1700	850	570	104 ± 10
	Dévaványa	2500	—	—	1550	880	420	—
	Üllés ÉNy-1	2320	2220	2220	1320	780	400	100 ± 11
	Furta 3	2240	2190	2190	1250	870	540	—
5	Szank 2	2000	1950	1950	1300	700	300	99 ± 10
	Tótkomlós 1	1790	1790	1790	1110	—	390	106 ± 9
	Harka 1	1770	1770	1770	1070	650	170	—
	Kismarja 2	1630	1630	1630	1000	580	280	—
6	Biharnagybajom 2	1440	1440	1440	900	680	290	—
	Kaskantyú 2	1180	1180	1160	780	520	240	—

[a]Mostly after Nagymarosy (1981) and Horváth et al. (this volume).
[b]Top of Nagyalföld variegated clay formation (Körössy, 1970).
[c]Regional unconformtiy dated isotopically at 2.5 Ma.
[d]Method used for heat flow determination described in Dövényi and Horváth (this volume).

Pannonian age rocks are postextensional. These postextensional sediments, however, are a deep water turbiditic to deltaic facies, and represent rapid filling of a deep basin (Pogácsás, 1984; Mattick et al., this volume; Berczi, this volume). By 8 Ma, the facies reflect shallow lacustrine conditions, indicating that the deep basin had been filled. Thus the total thickness of basin sediments deposited prior to 8 Ma (4300 m) represents the initial subsidence of the basement plus about 2 to 4 m.y. of thermal subsidence (uncorrected for sediment compaction, and so on). Because thermal subsidence rates are generally slow, and because the results of this paper are insensitive to small variations in initial subsidence, the 4300 m of sediment deposited prior to 8 Ma has been taken as the initial subsidence for well 1 in the Great Hungarian Plain. The age ranges and thicknesses of sediments used to calculate initial subsidence for each well are given in Table 3.

Thermal subsidence was calculated from the thickness of shallow water lacustrine (preferably swampy) to continental deposits that are clearly postextensional. Because thermal subsidence in most extensional basins occurs at a roughly constant rate for at least the first 15 m.y. after extension, and because extension occurred within the Pannonian system between about 15 and 10 Ma, it is sufficient to determine only the average rate of thermal subsidence for each well or well group. This can be determined by using that part of the thermal subsidence history which is best constrained, and over which the sedimentation history most accurately reflects true basement subsidence and not merely changes in the level of the deposition surface.

For example, within the Great Hungarian Plain, the postextensional sedimentation between 8 and 5.5 Ma represents gradual filling of a shallow lake, reaching a swampy to fluviatile environment by 5.5 Ma (about the end of the Messinian). Therefore, sedimentation between 8 and 5.5 Ma may have exceeded the rate of basement subsidence because of upbuilding of the deposition surface as the shallow lake filled. From 5.5 Ma until the present, the sedimentation rate should have been comparable to the rate of basement subsidence because the level of the deposition surface remained roughly constant. Therefore, in the Great Hungarian Plain, the sediment accumulation between 5.5 Ma and the present was used to constrain

Table 3. Initial and Thermal Subsidence and Heat Flow for Extensional Basins in the Pannonian System

		Initial Subsidence (m)		Thermal Subsidence			
Basin	Well	uncorrected[a]	water loaded[b]	uncorrected[c] (m)	water-loaded[b] subsidence rate (m/m.y)	Heat Flow (mW/m^2)	Heat Flow Reference[e]
Great Hungarian Plain	1	4300 (16–8 Ma)	1800 ± 300	1200 (0–5.5 Ma)	44 ± 10	84 ± 10	1
	2	2510 (16–8 Ma)	1200 ± 200	1050 (0–5.5 Ma)	39 ± 8	97 ± 15	1, 2
	3	1620 (16–8 Ma)	890 ± 150	100 (0–5.5 Ma)	45 ± 7	97 ± 10	2
	4	1040 (16–8 Ma)	640 ± 100	840 (0–5.5 Ma)	38 ± 4	102 ± 10	2
	5	670 (12–8 Ma)	450 ± 100	640 (0–5.5 Ma)	32 ± 2	104 ± 10	1, 2
	6	460 (12–8 Ma)	320 ± 70	600 (0–5.5 Ma)	36 ± 2	100 ± 15	3
Whole basin average			0–2000[d]		400 ± 100	95 ± 15[f]	
Vienna Basin	7	1700 (17–12 Ma)	600 ± 100	540 (0–12 Ma)	10 ± 2	50 ± 10	4
	8	1600 (17–12 Ma)	600 ± 100	370 (0–12 Ma)	6 ± 2	50 ± 10	
	9	4500-5500 (17–12 Ma)	1500–2000	no data	no data	50 ± 10	4
Whole basin average			0–2000[d]		8 ± 4	50 ± 10	
West Danube Basin	10	850 (24–12 Ma)	25 ± 5	480 (0–12 Ma) or 240 (0–8 Ma)	12 ± 4	60 ± 20	5
Central–Northern Danube Basin	11	1300 (16–12 Ma)	660 ± 100	1240 (0–12 Ma)	29 ± 5	84 ± 15	6
	12	900 (16–12 Ma)	550 ± 100	1370 (0–12 Ma)	33 ± 5	75 ± 15	6
	13	800 (16–12 Ma)	500 ± 100	1100 (0–12 Ma)	30 ± 5	75 ± 20	6
North–central basin average			0–1000[d]		31 ± 6	80 ± 20	
East Danube Basin	14	1700 (24–8 Ma)	900 ± 150	1190 (0–8 Ma)	35 ± 5	75 ± 15	7
	15	1500 (16–8 Ma)	830 ± 150	1170 (0–8 Ma)	35 ± 5	75 ± 15	7
East basin average			0–1000[d]		35 ± 5	75 ± 15	
Zala Basin	16	1100 (16–8 Ma)	600 ± 100	1200 (0–8 Ma)	42 ± 6	84 ± 15	8
	17	1800 (16–8 Ma)	1000 ± 150	1000 (0–8 Ma)	29 ± 4	92 ± 20	9
Whole basin average			0–1200[d]		35 ± 10	88 ± 15	
Transcarpathian Basin	18	1300 (24–13 Ma)	750 ± 150	1215 (0–13 Ma)	28 ⅓ ± 5	105 ± 15	3
	19	2900 (24–8 Ma)	1250 ± 200	880 (0–8 Ma)	24 ± 15	103 ± 15	10
Whole basin average			0–1500[d]		26 ± 6	105 ± 15	

[a]Uncorrected sediment thickness plus age at top and bottom of sediment sequence (in parentheses) used to determine water-loaded initial subsidence.

[b]Parameters from Table 4 used for decompaction and unloading of sediments.

[c]Uncorrected sediment thickness plus age at top and bottom of sediment sequence (in parentheses) used to determine water-loaded thermal subsidence rate.

[d]Range of initial (water-loaded) subsidence for whole basin estimated from individual well data and from seismic reflection profiles (for example, Horvath and Rumpler, this volume, and Thon and Tomek, this volume) and from stage isopach maps (Kőrössy, 1970, 1980; Spicka, 1972).

[e]References: (1) Dövényi and Horváth, (this volume) (measured heat flow); (2) Dövényi and Horváth, (this volume) (measured temperatures estimated conductivities); (3) Estimated from heat flow map in Dövényi and Horváth, (this volume); (4) Basin average (Cermak, 1975 and Dövényi and Horváth, this volume); (5) Greutter (1977) (heat flow measured as 70 mW/m^2 in top of well; 50 mW/m^2 in bottom of well); (6) Heat flow measurements in Senec BS-1 (Stegena et al., 1973) and adjacent Kralova pri Cenci (Greutter, 1977); (7) Heat flow measurement in Kolárovo-2 (Marusiak and Lizon, 1975); (8) Boldizsár (1959) (measured heat flow); (9) Heat flow measurement from adjacent well Bárszentmihalyta-1, Dövényi et al. (1983); (10) Heat flow measurement (Cermak, 1968).

[f]Average of values listed in Table 2.

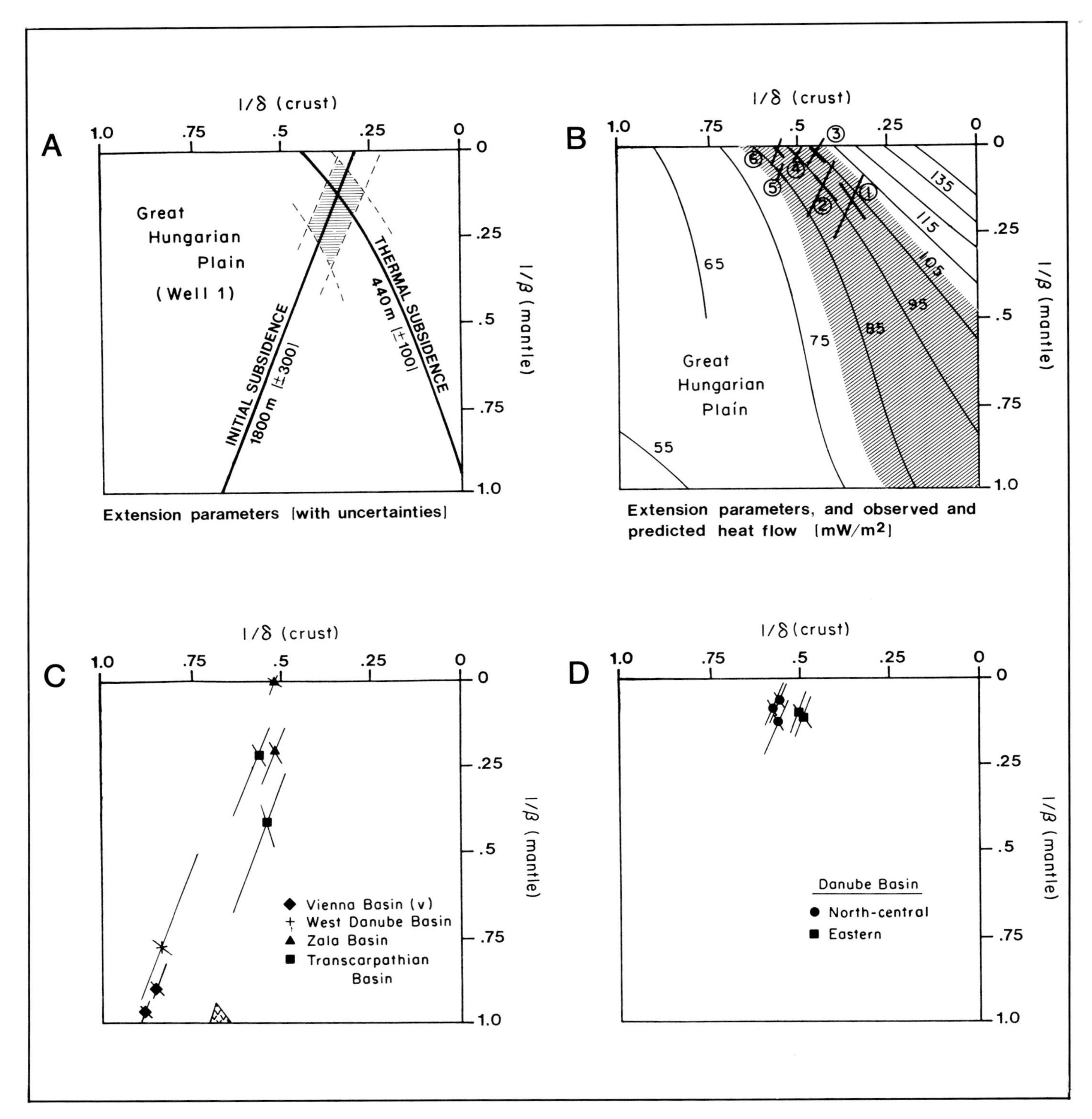

Figure 8. (A) Subsidence for well 1 in the Great Hungarian Plain as a function of δ (crustal thinning) and β (apparent subcrustal thinning). Dark lines show initial subsidence of 1800 m (from Figure 6A) and an average thermal subsidence rate 44 m/m.y. between 0 and 10 m.y. (from Figure 6B). Intersection of the dark lines gives thinning parameters $\delta = 2.8, \beta = 7$. Dashed lines show uncertainty of ± 300 m for initial subsidence and ± 10 m/m.y. for thermal subsidence rate. Shaded region thus indicates all values of β and δ that are consistent with the initial and thermal subsidence data, including uncertainties. (B) Great Hungarian Plain: Crosses show values of β and δ for wells 1–6, calculated by the method shown in (A). Also shown are the theoretical heat flow values taken from Figure 6C. The subsidence histories of wells in the Great Hungarian Plain are consistent with theoretical heat flow of about 80–110 mW/m^2. Shaded area between 80–110 mW/m^2 indicates the measured heat flow in the Pannonian Basin. Therefore the theoretically predicted heat flow and the observed heat flow are in good agreement (C–D). Other basins and subbasins in the Pannonian system: crosses show uncertainty range for β and δ as in (B). All values are plotted from the intersection of initial and thermal subsidence, except for the shaded box. The shaded box shows estimated values for the deep part of the (northern) Vienna Basin based on the intersection of the initial subsidence (1500–2000 m water loaded) and the surface heat flow (50 $\pm$ 10 mW/m^2).

the subsidence rate of the basement. The age ranges and thicknesses of sediments used to calculate thermal subsidence rates for each well are given in Table 3. Wells Kolárovo 3 and 4 in the eastern Danube basin and both wells in the Zala basin have little control on the sedimentation history after 8 Ma. These wells are located in the deeper, central parts of the basins and at some distance from the source of sediment supply in the outer Carpathians. They also show deltaic structures within the older (>8 Ma) postextensional sediments. Thus, these sedimentary rocks were deposited under conditions similar to those in the Great Hungarian Plain. Sedimentation rates observed here between 8 Ma and the present may have exceeded the rate of subsidence in the basement for the same reasons, as previously discussed for the Great Hungarian Plain. However, because no dates younger than 8 Ma were available, thermal subsidence was constrained by the thickness of sedimentary rocks of age 0–8 Ma.

The wells in the north-central Danube basin (wells 11, 12 and, from seismic data, probably 13) show a roughly linear rate of sedimentation from 12 Ma until the present, suggesting that over this time interval sedimentation kept pace with basement subsidence. These wells, plus well 10, are situated near the edge of the Danube basin and near a source of sediment supply from the outer Carpathians. Postextensional deposition in these areas occurred in a shallow water, coastal or fluviatile environment.

Within the Transcarpathian basin, the roughly 800 m of sediment deposited in wells Bunkovce 1 and Ptruksa 22 between 12 Ma and the present probably represents thermal subsidence of the basement because (1) the basin is located immediately adjacent to a sediment source area in the outer Carpathians, and (2) much of the deltaic sediments in the Great Hungarian Plain are derived from the northeast, and must have been first transported from the Carpathians across the area of the Transcarpathian basin. At worst, 800 m represents a rough upper bound on the true basement subsidence between 0 and 12 Ma.

Thermal subsidence rates in most parts of the Vienna basin are difficult to determine because strike-slip faulting (and some extension) extended into the Pliocene and is even active at the present. Therefore, estimates of thermal subsidence for this basin must come from areas where the tectonic activity ended early, leaving a sufficient record of posttectonic subsidence to provide reasonable accuracy in determining thermal subsidence rates. Wells Breclav 12 and Brodské 7 are both located in the northern and eastern part of the Vienna basin where little or no extension has occurred after Sarmatian time (~12 Ma). These two wells show only about 500 m of sedimentation since 12 Ma. This may overestimate the amount of basement subsidence that has occurred over the same time interval, because the surface elevation within the Vienna basin is now 200 m above sea level, while Badenian–Sarmatian (16.5–12 Ma) sediments were deposited near middle Miocene sea level. Thus some of the sediment deposited after 12 Ma may represent upbuilding of the deposition surface.

Decompaction of sediments and unloading of the basement was performed using the porosity–density relationships given in Table 4. The resulting initial subsidence and thermal subsidence rates include uncertainties that result from uncertainties in the various parameters used to decompact the sediments and unload the basement (Table 3). Significant differences between basins emerge after correction for sediment loading and compaction (Table 3). Thermal subsidence rates were very slow in the Vienna basin (8 ± 3 m/m.y.), intermediate in the Transcarpathian basin (26 ± 5 m/m.y.), and fast in the Great Hungarian Plain and Zala Basin (30 to 45 m/m.y.). The rate of thermal subsidence along the western edge of the Danube basin (Grob-1) is slow (14 ± 2 m/m.y.), but is rather fast in the north–central and eastern parts of the basin (30–35 m/m.y.). These values of thermal subsidence are consistent with the heat flow measured over the same area.

Table 4. Porosity and Density used for Decompaction and Unloading[a]

Parameter	Value
Porosity as a function of depth (x, in km)	$\phi(x) = 0.4e^{-0.65x}$ $\phi(x) = 0.07 + 0.43e^{-0.8x}$ $\phi(x) = 0.15 + 0.45e^{-1.0x}$
Surface porosity	0.4 – 0.6
Porosity at great depth ($x = \infty$)	0 – 0.15
Sediment matrix density (ρ_s at $\phi = 0$)	2.70 g/cm^3
Mantle density	3.20 ± 0.05 g/cm^3
Sediment density	$\rho_s + (1 - \rho_s)\phi$

[a]Airy isostasy assumed for decompaction. See Sclater and Christie (1980) for discussion of sediment decompaction method.

Heat Flow Data

One of the criteria used in selecting the well data presented here was the availability of reliable heat flow data or estimates from each well or from adjacent wells (Table 3). Average values of heat flow were used for each of the six well groups in the Great Hungarian Plain (Table 2). The uncertainties given to the heat flow values in Tables 2 and 3 are somewhat subjective: they partly reflect uncertainties in the measurements (see Dövényi and Horváth, this volume) and partly reflect scatter with each well group. The average heat flow for all the wells in the Great Hungarian Plain is about 95 ± 15 mW/m^2, and heat flow shows no increase with increasing depth to basement. All the well groups 1–6 had heat flow between 84 and 106 mW/m^2, and the scatter within each group was relatively small (Table 2).

Within the other basins, heat flow values given for individual wells were determined from heat flow measurements within each well or an adjacent well, or, in a few cases, estimated from average heat flow from the heat flow map of Dövényi and Horváth (this volume). One suspect value occurs in well 10, where heat flow measurements yielded 50 mW/m^2 in the lower part of the well and 70 mW/m^2 in the upper part of the well, indicating that either the measurements were of poor quality or that hydrothermal circulation was disturbing the conductive heat loss through the sediments. Heat flow in this well was thus estimated as 60 ± 20(?) mW/m^2.

Consistency of Thermal Subsidence and Heat Flow Data

Figures 9A and B show heat flow plotted against thermal subsidence rate for the data given in Table 3. For each basin, values of heat flow and thermal subsidence rate that are consistent

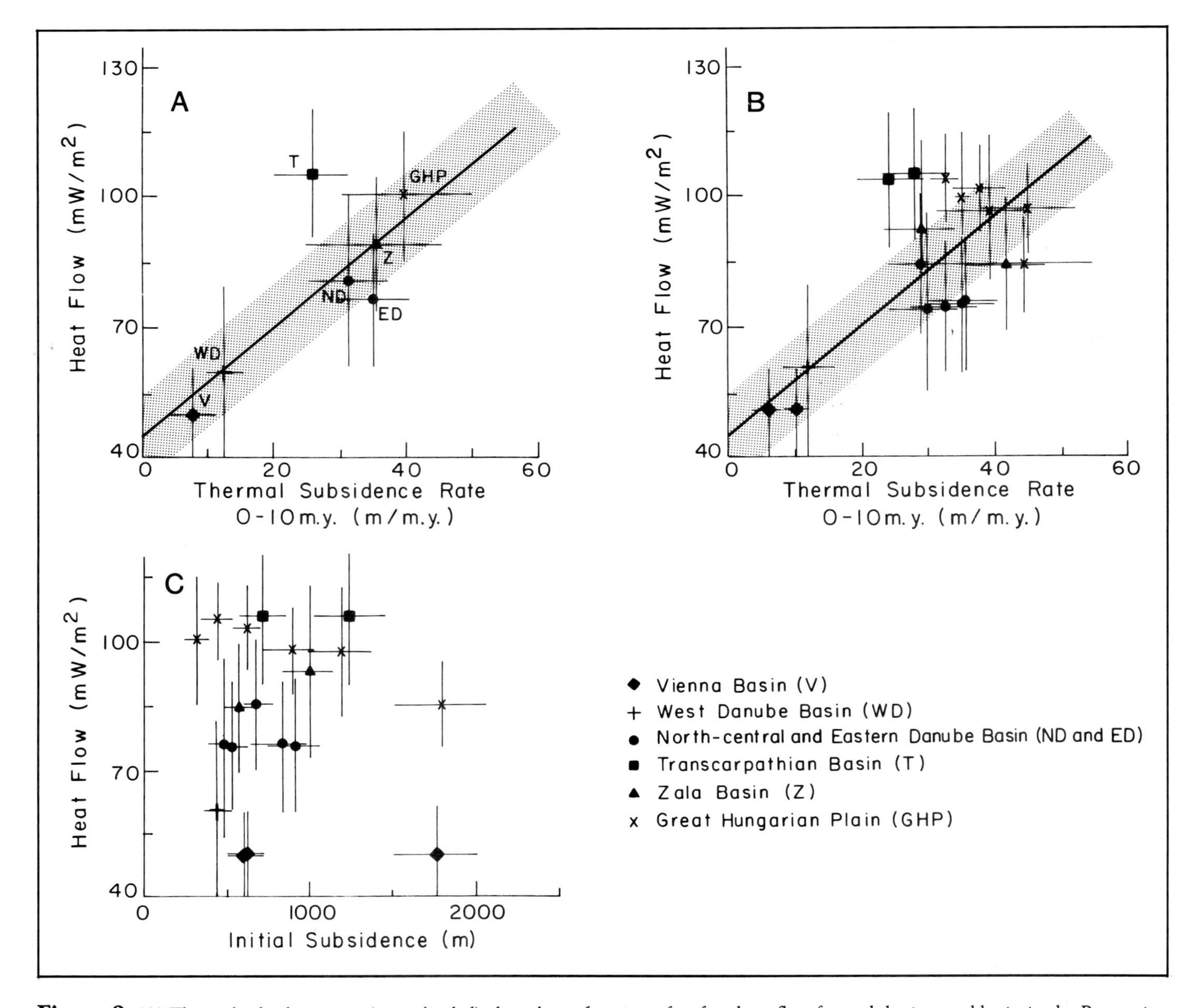

Figure 9. (A) Thermal subsidence rate (water loaded) plotted as a function of surface heat flow for each basin or subbasin in the Pannonian system. Dark line and shading show theoretical relationship between heat flow and thermal subsidence 10 m.y. after instantaneous stretching (from Figure 6D). (B) Same as for (A), except that heat flow and thermal subsidence are plotted for individual wells. (C) Initial subsidence (water loaded) plotted as a function of surface heat flow for individual wells.

with each other should fit the theoretical relationship described in the previous section and indicated by the diagonal line in Figures 9A and B. Consistent heat flow–subsidence pairs strongly suggest that both heat flow and subsidence are recording only conductive cooling of the lithosphere following extension. Apparently, inconsistent data indicate that processes other than conductive cooling are affecting either the surface heat flow or the thermal subsidence (such as convective heat transport or changes in elevation of the deposition surface) or that tectonic processes (like volcanism) have produced geotherms radically different from those shown in Figure 5.

Heat flow and thermal subsidence for the Vienna basin, western and north-central Danube basin, Zala basin and Great Hungarian Plain are self-consistent, although the uncertainties in heat flow estimates are significant. The Transcarpathian basin exhibits heat flow values much higher (105 ± 10 mW/m^2) than are consistent with the observed thermal subsidence. This discrepancy probably can be resolved by noting that the Miocene–Pliocene calcalkaline magmatic arc associated with Miocene subduction in the Carpathian occupies much of the area of the Transcarpathian basin (Figure 1), and some of the eruptions are as young as 3 Ma. Heat advected into the upper crust in this manner will greatly elevate the surface heat flow, and may, to a much lesser extent, also elevate the thermal subsidence rate. Indeed, the entire length of this magmatic arc ($\sim$1500 km) shows anomalously high heat flow relative to adjacent areas

(Dövényi and Horváth, this volume and Map 5). The thermal subsidence of this basin would be consistent with a heat flow of $\sim 75 \pm 10$ mW/m^2.

The eastern part of the Danube basin appears to be slightly colder (75 ± 10 mW/m^2) than would be expected from its thermal subsidence (35 ± 5 m/m.y.). Possible explanations are (1) the thermal subsidence may be overestimated because of building up of the depositional surface, or (2) the conductive heat flow values may be perturbed by hydrothermal circulation within the carbonate basement. Large hydrothermal systems are known from the Bakony–Buda hills when the karstic basement crops out just south and east of the Danube basin.

In contrast to the strong correlation between thermal subsidence and heat flow, heat flow plotted against initial subsidence for each well shows no correlation, as expected (Figure 9C).

CALCULATED EXTENSIONAL PARAMETERS

Figure 8B shows the values of β (mantle heating) and δ (crustal thinning) for all six well groups in the Great Hungarian Plain, constructed from the intersection of initial subsidence and thermal subsidence rates as illustrated in Figure 8A. The uncertainties in determining β and δ are much larger for the deep wells (for example, well group 1) than for the shallow wells (for example, well group 6). This occurs partly because uncertainties in decompaction parameters have a larger effect in deeper wells. All of the well groups plot in the vicinity of $\beta \geq \delta$, implying that the mantle lithosphere has been mostly replaced by hot asthenosphere or is very thin. This is particularly clear for the shallower well groups, 3 through 6.

These values of β and δ have been superimposed on the plot of surface heat flow at 10 m.y. after stretching (Figure 8B), to show heat flow values predicted directly from the subsidence history of each well group. All of the well groups yield predicted heat flow values between about 85 and 115 mW/m^2, with little systematic variation with total depth to basement. Thus, although the total water loaded depth to basement varies from 2200 m to 500 m, the predicted heat flow is nearly the same for all well groups.

The measured heat flow in the Great Hungarian Plain is about 95 ± 15 mW/m^2 (Table 2 and shaded region in Figure 8B), and shows a slight systematic decrease with increasing depth to basement. Thus the range of heat flow predicted from the subsidence history and the observed heat flow are nearly identical. This is expected because the thermal subsidence within the basin and the surface heat flow are consistent (see previous section).

β and δ can be determined for the other basins in the Pannonian system (Figures 8C and D) by combining initial subsidence with thermal subsidence rates as above. Subsidence data from the north-central and eastern Danube Basin and the Zala basin yield values of β (effective mantle thinning) greater than about 5. As for the Great Hungarian Plain, the data suggest that extension beneath these basins produced a very thin or very hot lithosphere relative to the magnitude of crustal extension. Subsidence data from the westernmost part of the Danube basin (well 10) indicate roughly uniform extension of the lithosphere.

Subsidence data from the Vienna basin are consistent with little or no heating of the lithosphere during crustal extension, and are thus consistent with a thin-skinned extensional origin for the basin (Royden, this volume). One determination of β and δ from the intersection of heat flow data with initial subsidence data is shown for the deep part of the northern Vienna basin (Figure 8C), where thermal subsidence cannot be constrained due to recent tectonic activity (of minor importance) within much of the Vienna basin (Gutdeutsch and Aric, this volume).

In each of these basins, the heat flow predicted from β and δ, which are in turn derived from the subsidence data, are in reasonable to good agreement with measured data, particularly for the Vienna and Zala basins (compare Figures 8c and D and Table 3).

Subsidence and heat flow data from the Transcarpathian basin are well constrained and are clearly inconsistent. The subsidence data from the Transcarpathian basin indicate moderate thinning or heating of the mantle lithosphere during extension (Figure 8C). The heat flow predicted from β and δ, as derived from the intersection of initial and thermal subsidence, is about 75 ± 10 mW/m^2 (Figure 8C). In contrast, observed heat flow values of 105 ± 15 mW/m^2 suggest very hot temperatures within the lithosphere. This discrepancy is not unexpected, because Figures 9A and B indicate that the surface heat flow through this basin is much higher than is consistent with the thermal subsidence. Because of the presence of a young (up to 3 Ma) magmatic belt beneath the basin, the advection of heat to the surface by volcanism will render invalid the assumptions used to construct Figures 6 A through D and to derive β and δ. This is likely, however, to have a much greater effect on surface heat flow than on thermal subsidence, because surface heat flow is sensitive to local and near-surface perturbations while thermal subsidence is mainly sensitive to temperature changes that are significant at the scale of the entire lithosphere. Therefore, the thermal subsidence should yield more reliable information about the net cooling of the lithosphere after extension. We infer that the Transcarpathian basin is probably not underlain by lithosphere as hot as that beneath the Great Hungarian Plain and Zala basin, but the data are ambiguous.

HEAT FLOW, SUBSIDENCE, AND EXTENSIONAL STYLE VERSUS DISTANCE FROM THE THRUST FRONT

The data presented in Table 3 and the graphically determined stretching parameters shown in Figure 8 can be used to interpret the general style of extension beneath the Pannonian basin system and to infer the extent of mantle-lithosphere involvement in the extension process. It must be kept in mind, however, that the stretching parameters β and δ are only a simplified, schematic representation of complex extensional events within the lithosphere. In this section, variations in subsidence and heat flow as a function of distance from the Carpathian thrust front are first evaluated, followed by interpretation of these data as variations in extensional style as a function of distance from the thrust front.

Figure 10 shows the variation of initial subsidence, thermal subsidence rate, and heat flow as a function of distance from the thrust front for each of the five basins. Distance from the thrust

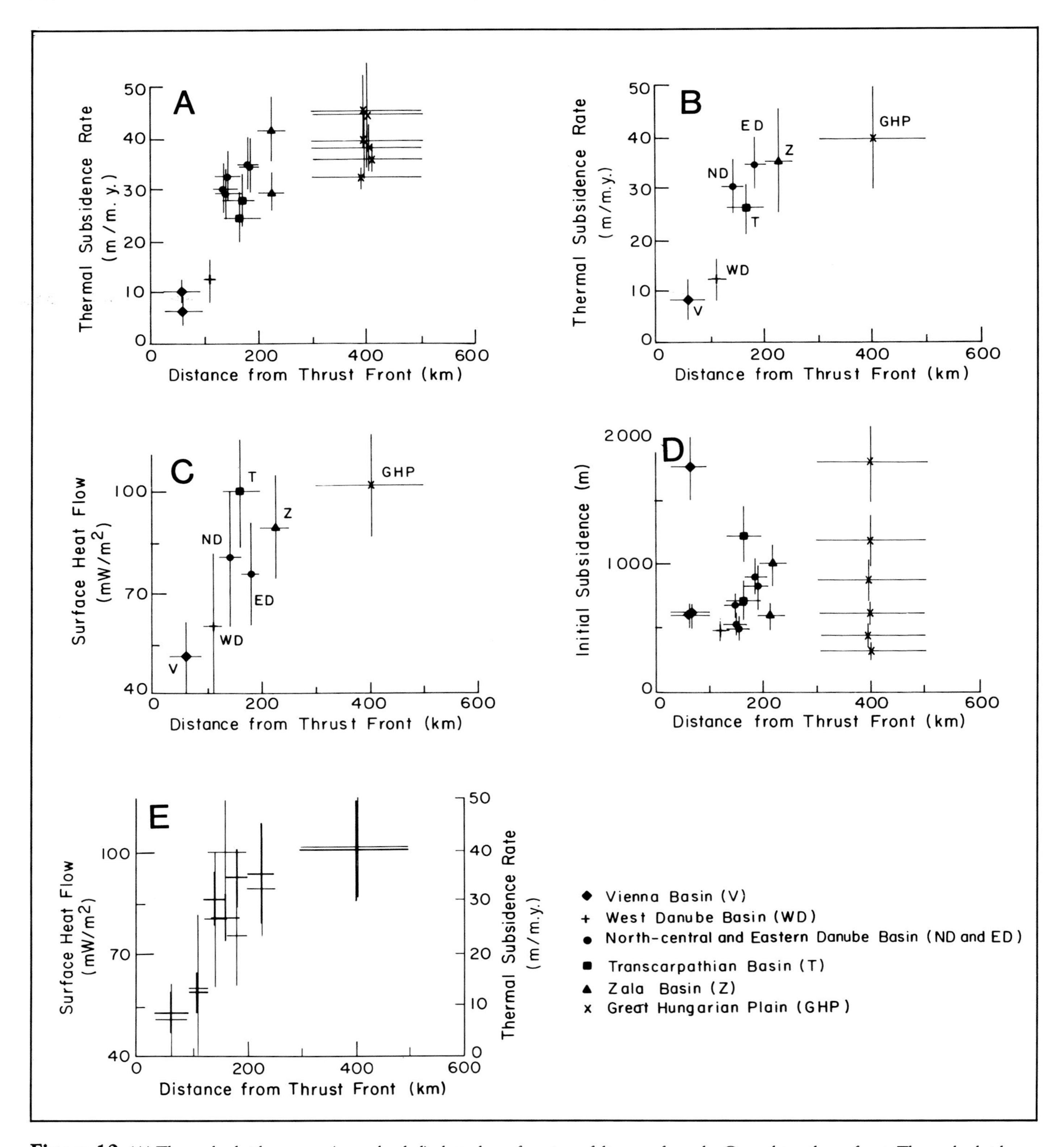

Figure 10. (A) Thermal subsidence rate (water loaded) plotted as a function of distance from the Carpathian thrust front. Thermal subsidence rates are for individual wells; solid symbols give best estimate and vertical error bars give uncertainty range (mainly resulting from systematic errors). Horizontal error bars show distance of entire basin (or subbasin) from nearest point on the Carpathian thrust front; solid symbol shows median distance of basin from thrust front. (B) Same as for (A), but thermal subsidence is whole basin (or subbasin) average as given in Table 3. (C) Surface heat flow plotted as function of distance from the Carpathian thrust front. Heat flow is whole basin (or subbasin) average. (D) Initial subsidence (water loaded) as a function of distance from the Carpathian thrust front for individual wells. (E) Heat flow and thermal subsidence rates from (B) and (C) plotted together. Consistent values of heat flow and subsidence in each basin should coincide.

front was determined by measuring the distance from each part of a basin to the closest point located along the external edge of the outer (flysch) Carpathian thrust belt. The range in distance shown for each basin reflects the closest and furthest point of each basin from the thrust belt. In constructing Figure 10, consideration was not given as to whether the segment of the thrust belt nearest to each basin was active or inactive at the time of basin formation. For example, the distance from the Vienna basin to the thrust belt was measured west-northwest of the basin, yet this segment of the belt was inactive by Badenian time (16.5–13 Ma) when much of the basin extension occurred.

The rate of thermal subsidence observed in each basin increases systematically with increasing distance from the Carpathian thrust front (Figures 10A and B). It is lowest in the Vienna basin, and highest in the north-central and eastern parts of the Danube basin, the Zala basin, and the Great Hungarian Plain. Subsidence rates in the western part of the Danube basin are low, but higher than those in the Vienna basin. The thermal subsidence rates from the Transcarpathian basin are intermediate; they are significantly higher than in the Vienna basin and lower than in the Great Hungarian Plain.

Similarly, surface heat flow within the basins increases systematically with increasing distance from the thrust front (Figure 10C). Excluding the Transcarpathian basin, the pattern observed in surface heat flow is similar to that observed in thermal subsidence rates, being lowest in the Vienna basin, slightly greater in the western part of the Danube basin, and greatest in the Pannonian basin (Figure 10E). The anomalously high heat flow through the Transcarpathian basin can be ascribed to the presence of a very young magmatic arc within the basin, and is probably not diagnostic of overall lithospheric heating during extension. If other parts of the basin area near the magmatic arc had been sampled, similar, anomalously high surface heat flow values would have been observed (Dövényi and Horváth, this volume).

In contrast, initial subsidence values show no correlation with distance from the thrust front (Figure 10D). Parts of the Great Hungarian Plain, including the uplifted areas between the Great Hungarian Plain and the Danube basin (Hungarian Mid-Mountains), appear to have undergone no initial subsidence and, in places, even initial uplift during extension (for example, Royden et al., 1983b). The initial uplift in these areas obviously is not recorded in the sedimentary record, and therefore initial elevation changes in these areas cannot be quantitatively evaluated.

These subsidence and heat flow data can be used to infer amounts of effective crustal thinning (δ) and effective mantle thinning or heating (β) as a function of distance from the thrust front. When the intersection of initial subsidence and thermal subsidence rate are used to determine mantle thinning (from Figure 8), there is a systematic increase in β (or decrease in $1/\beta$) with increasing distance from the thrust front (Figure 11A). The Vienna basin subsidence implies little or no involvement of the mantle lithosphere ($1/\beta = 1$–1.2). The north–central and eastern parts of the Danube basin, the Zala basin, and the Great Hungarian Plain show $\beta > 5$, indicating extreme thinning (or heating) of the mantle lithosphere during extension. The western part of the Danube basin and the Transcarpathian basin show intermediate values of β, with more effective thinning (or heating) beneath the Transcarpathian basin and less beneath the western Danube basin.

When β is determined from the intersection of the initial subsidence and the surface heat flow, a similar pattern is observed, except that β is more poorly constrained and the values determined for the Transcarpathian basin are probably not meaningful (Figure 11B). Nevertheless, within the resolution of the data, there is a systematic increase in β with increasing distance from the thrust front.

Effective crustal thinning (δ), calculated from the intersection of initial subsidence and thermal subsidence rate, shows a slight, apparent increase with increasing distance from the thrust front. Some of this apparent correlation may be real, but it is partly due to a sampling problem within the north–central to east Danube basin, Zala basin, and Great Hungarian Plain. Because no data was taken from wells with total (water-loaded) subsidence less than about 300 m (equivalent to about 1 km of sediment), areas with high heat flow, fast thermal subsidence and $\delta < 1.7$ were effectively excluded from analysis (Figure 6). However, the slight increase in maximum values of δ within each basin as a function of increasing distance from the thrust front is real.

It should be stressed that the values of crustal thinning given by δ in Figures 8 and 11 can be strongly affected by fault block geometries. In particular, the finite strength of the lithosphere is probably sufficient to support horst and graben structures, with horizontal dimensions up to at least a few tens of kilometers, without corresponding changes in the depth to Moho (Oliver, 1985). This implies that values of δ determined on top of shallow basement blocks will underestimate the amount of real crustal thinning, while those determined for the deep grabens will overestimate the amount of crustal thinning (Sawyer, 1986, unpublished work). Thus, the net crustal thinning is probably best determined from an areal average of δ over large areas ($\sim$100 km $\times$ 100 km). This approach yields about 100% extension ($\delta \approx 2$) or perhaps slightly less for the Great Hungarian Plain, and about 25% extension ($\delta \approx 1.25$) for the Vienna basin.

DISCUSSION

The magnitude of thinning (or heating) of the mantle lithosphere during extension within the Pannonian basin system shows a systematic increase with increasing distance from the Carpathian thrust front for distances between 0 and 200 km from the thrust front. These data can be used to subdivide the basin system into qualitatively different extensional domains (Figure 12). Four general domains are identified. (1) Thin-skinned extension with little involvement of mantle lithosphere occurred close to the Carpathian thrust front (Vienna and westernmost Danube basin). Seismic lines also show thin-skinned deformation (but not extension) within the outer East Carpathians east of the Transylvanian basin. (2) Considerable heating and/or thinning of the mantle lithosphere occurred within the central basin area far from the thrust belt (north-central and eastern Danube basin, Zala basin, and Great Hungarian Plain). (3) Between these two domains there appears to be a poorly defined transition zone that probably includes the Transcarpathian basin and passes through the Danube basin. (4) An area without middle–late Miocene extension exists beneath the Transylvanian basin.

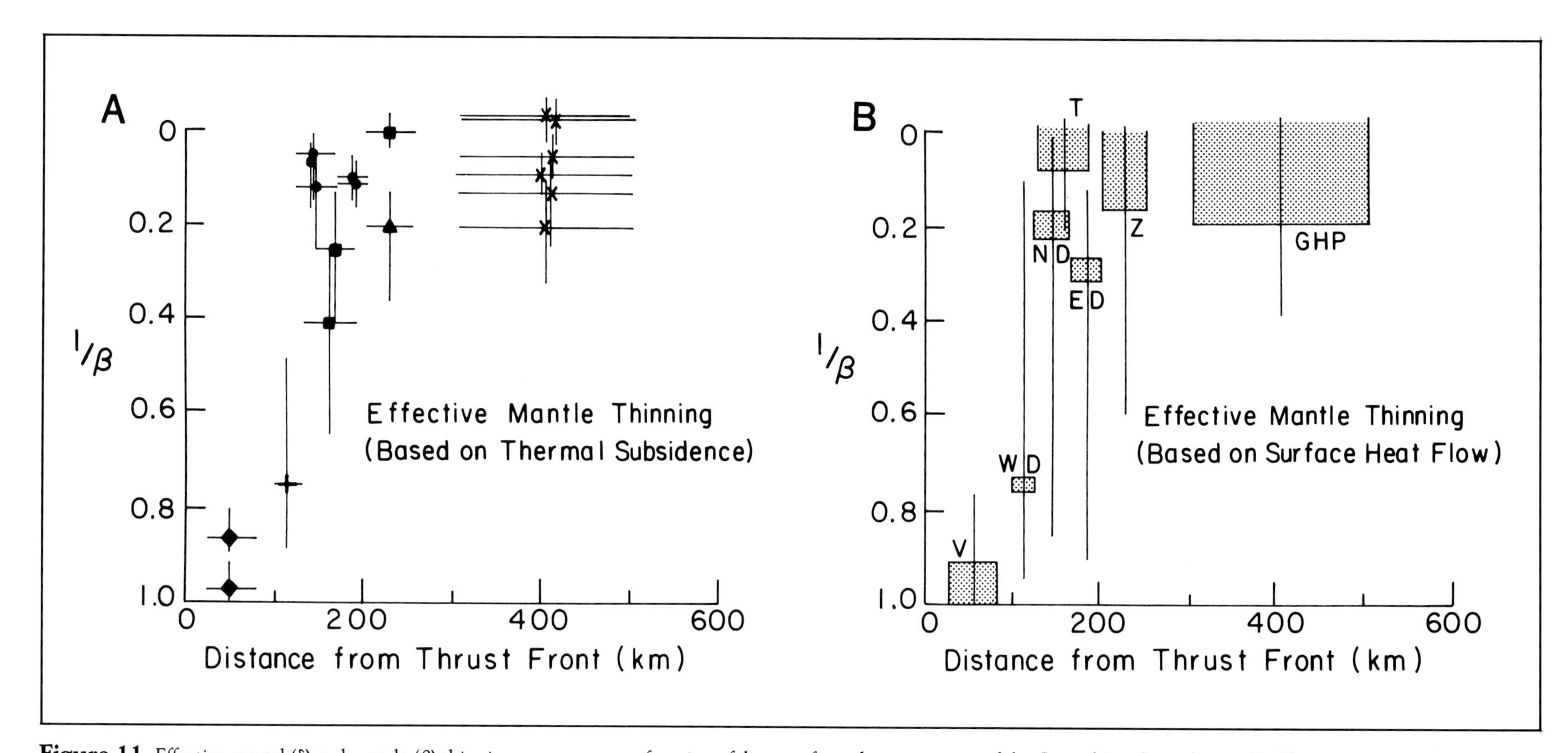

Figure 11. Effective crustal (δ) and mantle (β) thinning parameters as a function of distance from the nearest part of the Carpathian thrust front. (A) Effective thinning of the mantle lithosphere based on the intersection of initial and thermal subsidence values (Figure 8). In each basin or subbasin, solid symbols show mean distance of the basin from the thrust front, while horizontal error bars show the distance from thrust front to nearest and farthest part of each basin. Values of β are for individual wells in each basin. Solid symbols show best estimate of β, vertical error bars represent uncertainties. Note that effective mantle thinning increases systematically with increasing distance from the Carpathian thrust front. (B) Effective mantle thinning based on the intersection of initial subsidence and surface heat flow for each basin or subbasin. Vertical bars show uncertainty range in effective mantle thinning and correspond mainly to uncertainty in surface heat flow measurements. Shaded area shows estimated value of β for the mean estimate of heat flow in each basin or subbasin (Table 3). (C) Effective crustal thinning based on the intersection of initial and thermal subsidence values (Figure 8). Solid symbols and error bars are defined in the same way as in (A). Dashed box shows estimate for the deep part of the Vienna basin (~4500–5500 m) based on the intersection of initial subsidence (1500–2000 m water loaded) and surface heat flow (50 ± 10 mW/m^2). Effective crustal thinning also appears to increase with increasing distance from the thrust front, but this may be partly a sampling problem (see text for discussion).

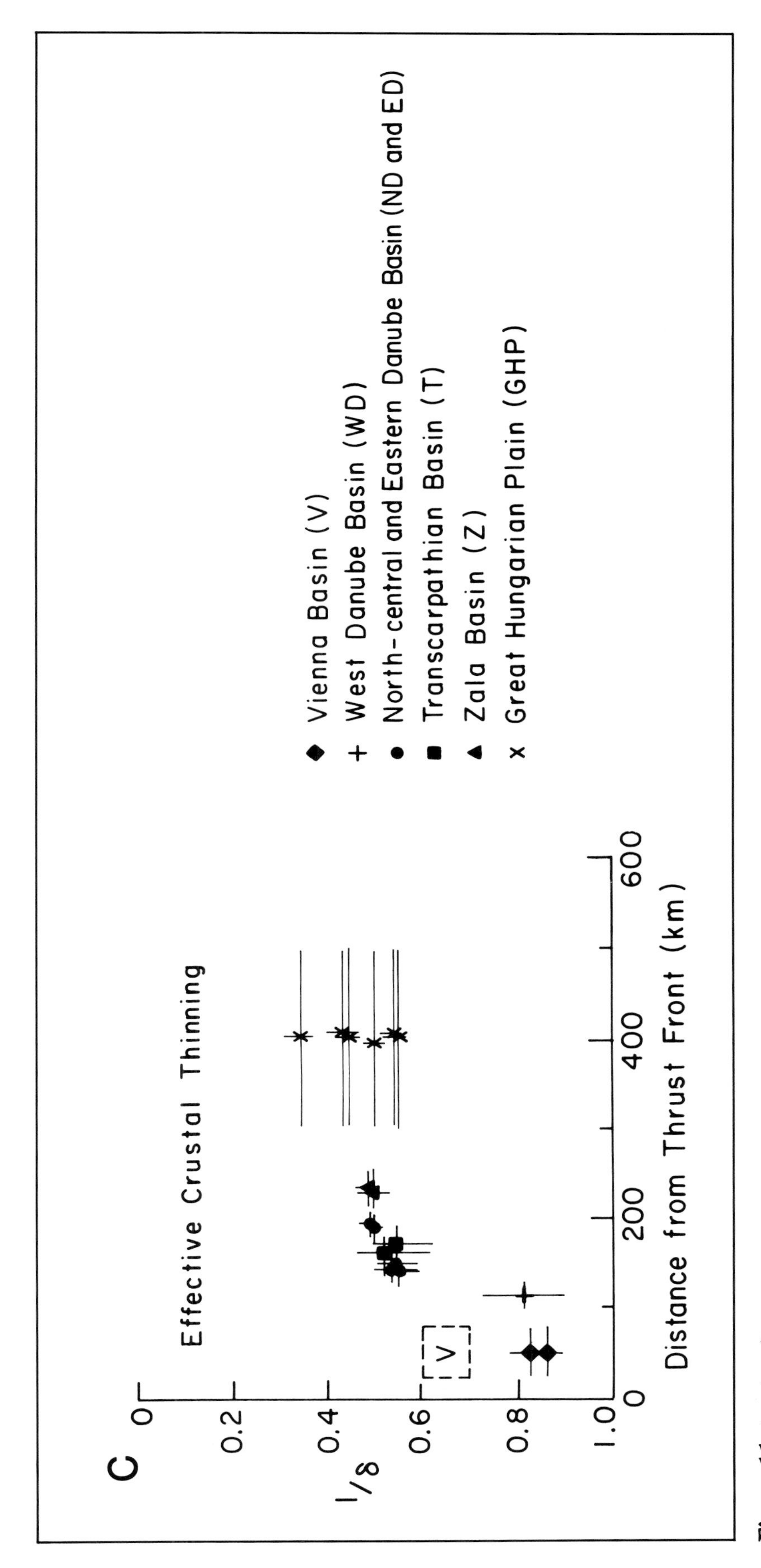

Figure 11. (continued)

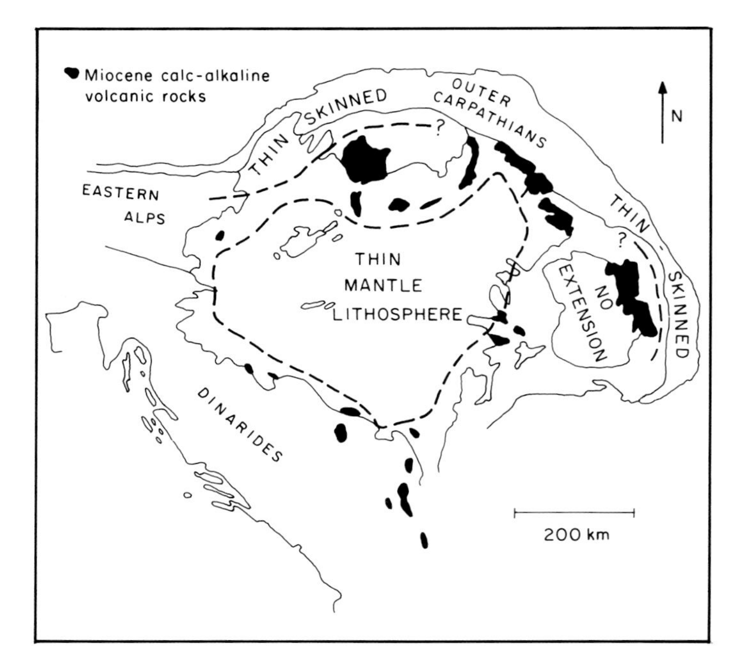

Figure 12. Tentative summary of the geographic distribution of extensional environments. Extension in the Vienna basin is considered to be thin skinned. The Great Hungarian Plain and the Zala and north-central to eastern Danube basins are considered to have very thin mantle lithosphere. The western part of the Danube Basin and the Transcarpathian Basin are considered to be transitional. The Transylvanian basin is considered to have little or no Miocene–Quaternary extension. Note that the boundary of the area of thin mantle lithosphere is roughly coincident with the exposures of Mio–Pliocene calcalkaline magmatic rocks (except in the southeast Carpathians).

The boundaries of the extensional domains shown in Figure 12 are only partly constrained by the data presented in this paper. Other considerations in constructing Figure 12 included the distribution of surface heat flow near the Apuseni Mountains, Dinarides and Eastern Alps (see Map 5 in enclosure), and seismic reflection lines south of the inner West Carpathians and northeast of the Dinaric Alps.

The data presented in this paper, and the extensional domains shown in Figure 12, are consistent with the hypothesis that extension involved the entire lithosphere of the overriding Pannonian fragment above the thrust-decollement and subduction zone beneath the Carpathians (Figure 2). It is difficult, however to document clearly the transition from thin-skinned extension at the thin leading edge of the Pannonian fragment to whole lithosphere extension and significant heating of the mantle lithosphere far from the thrust belt. This is due partly to a scarcity of available data over this transitional zone (Figures 7 and 12). There are however, several other inherent difficulties in this task. For example, there is only one major basin located very close to the thrust belt (Vienna basin). The presence of a Miocene magmatic arc within the transition zone renders heat flow data useless and subsidence data somewhat suspect near the magmatic arc (Transcarpathian basin). In other areas, the heat flow data have large uncertainties. Care must be taken in evaluating basement subsidence, because the basement subsidence rate and the sedimentation rate are not always equivalent. Other uncertainties are related to decompacting the sedimentary column and unloading of the basement (because of uncertainties in porosity, sediment density, mantle density, and so on). However, the latter are mainly systematic and should not affect any of the general trends across the basin system (Figures 10 and 11). One exception involves unloading of the basement, which was performed using Airy isostasy, although locally flexural effects may be important, particularly beneath the cold areas near the thrust belt (Vienna basin, westernmost Danube basin, and perhaps the Transcarpathian basin). Better areal coverage from well data will probably not resolve many of the uncertainties present in this study.

The locations of outcrops of Miocene calcalkaline volcanic rocks within and adjacent to the Pannonian basin system approximately coincide with the edge of the extended region (except in the East Carpathians, Figure 12). In particular, Figure 12 suggests that these magmatic rocks may delineate the edge of the area where roughly coeval Miocene extension resulted in a very thin or hot mantle lithosphere (Miocene igneous rocks are present beneath parts of the central basin area as well). Near the Carpathians, the magmatic arc is thought to be at least partly related to southward and westward subduction of Europe beneath the Carpathians . Along the southwestern edge of the Pannonian basin, Miocene andesitic volcanism appears to be related to right-slip faults of regional importance. It is not clear whether: (1) the thin, hot mantle lithosphere beneath the central basins is the result of the same processes that controlled the creation and location of the volcanic arc; (2) the thinning of the mantle lithosphere was genetically independent of the volcanism, but was restricted to zones where volcanism had previously heated and weakened the lithosphere; or (3) the locus of volcanism was controlled by and restricted to areas where the extensional process involved the deep mantle lithosphere.

CONCLUSIONS

The behavior of the Pannonian lithosphere at depth during Miocene crustal extension was heterogeneous, although it underlies genetically and tectonically similar basins within a single basin system. The extensional basins within this basin system exhibit increasing involvement of mantle lithosphere in the extension process with increasing distance from the Carpathian thrust front. This systematic variation occurs between about 0 and 200 km from the thrust front. Basins more than 200 km from the thrust front are underlain by very thin or hot mantle lithosphere.

Some of the results of this study can be applied to general questions in basin analysis, and are not specific to any particular basin system or basin type. This study provides one example of the type of complexities that can be present within one basin system and shows how heterogeneous processes may vary systematically across the basin system. It also illustrates how some features of basin evolution may be related to events outside of the basin system (for example, proximity to and geometry of the Carpathian thrust belt and subduction zone), and how variation of some features from one part of the basin system to another can shed light on evolution of the basin system as a whole.

ACKNOWLEDGMENTS

This work was supported by NSF grants #INT-7910275 and EAR-8115863, the Hungarian Academy of Sciences and Shell Research and Development, Inc. Additional support for LR was provided by PYI funds from NSF and from matching funds from Shell Companies Foundation, Inc.; Texaco, U.S.A.; Chevron Oil Field Research Co.; Kerr-McGee Corporation; and Intel Corp. Tina Freudenberger and Dorothy Frank typed the manuscript.

REFERENCES

Boldizsár, T., 1959, Terrestrial heat flow in the Nagylengyel oil field: Publ. Min. Fac. Sopron., v. 20, p. 27–34.

Cermak, V., 1968, Terrestrial heat flow in Eastern Slovakia: Trav. Inst. Géophys. Acad. Sci., no. 275, Geofysikálni Sbornik 1967, NCSAV Praha, p. 303–319.

Cermak, V., 1975, Terrestrial heat flow in the Neogene foredeep and the flysch zone of the Czechoslovak Carpathians: Geothermics, v. 4, p. 8–13.

Dövényi, P., F. Horváth, P. Liebe, J. Gálfi and I. Erki, 1983, Geothermal conditions of Hungary: Geophys: Transactions, v. 29, no. 1, p. 3–114.

Greutter, A., 1977, Geothermal conditions of the Slovakian part of the Little Hungarian Plain: Geonómia és Bányászat, v. 10, no. 3–4, p. 239–247.

Horváth, F. and L. Royden, 1981, Mechanism for the formation of the intra-Carpathian basins: A review: Earth Evol. Sci., v. 1, no. 3–4, p. 307–316.

Kőrössy, L., 1980, Neogén ősföldrajzi viszgálatok a Kárpát-medencében [Investigations into Neogene paleogeography in the Carpathian basin: Földt. Közl, v. 110, p. 473–484.

Kőrössy, L., 1970, Entwicklungsgeschichte der Neogene Becken in Ungarn: Acta Geol. Acad. Sci. Hung., v. 14, p. 42–49.

Marusiak, I. and I. Lizon, 1975, Výsledky geotermického výsumku v Ceskoslovenskej casti viedenskej panvy: Geologicke práce, v. 63, p. 191–204.

McKenzie, D., 1978, Some remarks on the development of sedimentary basins: Earth Planet. Sci. Lett., v. 40, p. 25–32.

Nagymarosy, A., 1981, Subsidence profiles of the deep Neogene basins in Hungary: Earth Evol. Sci., v. 1, no. 3–4, p. 218–222.

Oliver, J., 1985, Some observations of rifts by COCORP (abstract): Abstracts of the International Symposium on Deep Internal Processes and Continental Rifting (DIPCR), Sept. 9–13, 1985, Chenydu, China, China Academic Publishers, Beijing, p. 120.

Parsons, B. and J. G. Sclater, 1977, An analysis of the variation of ocean floor bathymetry with age: J. Geophys. Res., v. 82, p. 802–825.

Pogácsás, Gy., 1984, Seismic stratigraphy features of Neogene sediments in the Pannonian basin: Geophysical Transactions of Eötvös Loránd Geophys. Inst. of Hungary, v. 30, no. 4, p. 373–410.

Royden, L., 1986, A simple method for analyzing subsidence and heat flow in extensional basins: *in* J. Burrus (ed.), Thermal Modeling in Sedimentary Basins, Editions Technip, Paris, p. 49–73.

Royden, L. and C. E. Keen, 1980, Rifting process and thermal evolution of the continental margin of eastern Canada determined from subsidence curves: Earth Planet. Sci. Lett., v. 51, p. 343–361.

Royden, L., F. Horváth, and B. C. Burchfiel, 1982, Transform faulting, extension and subduction in the Carpathian–Pannonian region: Geol. Soc. Am. Bull., v. 73, p. 717–725.

Royden, L., F. Horváth, and J. Rumpler, 1983a, Evolution of the Pannonian basin system: 1. Tectonics: Tectonics, v. 2, p. 63–90.

Royden, L., F. Horváth, A. Nagymarosy, and L. Stegena, 1983b, Evolution of the Pannonian basin system: 2. Subsidence and thermal history: Tectonics, v. 2, p. 91–137.

Sawyer, D. S., 1986, Effects of basement topography on subsidence history analysis: Earth Planet. Sci. Lett., v. 78, p. 427–434.

Sclater, J. G. and P. A. F. Christie, 1980, Continental Stretching: An explanation of the post-mid-Cretaceous subsidence of the central North Sea basin: J. Geophys. Res., v. 85, no. B7, p. 3711–3739.

Sclater, J. G., L. Royden, F. Horváth, B. C. Burchfiel, S. Semken, and L. Stegena, 1980, The formation of the intra-Carpathian basins as determined from subsidence data: Earth Planet. Sci. Lett., v. 51, p. 139–162.

Spicka, V., 1972, Paleogeografie neogénu ceskoslovenských západních Karpat [Neogene Paleography of the Czechoslovakian West Carpathians]: Sborn. Geol. Ved., geol., v. 22, p. 65–108.

Stegena, L., F. Horváth, I. Erki, and P. Dövényi, 1978–1984, Reports on heat flow measurements in boreholes Horna Poton FGHP-1, Tvrdosovce FGTv-1, Cilistov FGSC-1, Ciz MJC-1, Komarno FGK-1, Galánta FGG-2, Nové Zámky GNZ-1, Bohelov GPB-1, Sobrance TMS-1, Liptovský Mikulás FGL-1, Vlcany FGV-2, Senec BS-1, Ciz BS-2, Gabcikovo GFFa-1: Manuscript.

APPENDIX: FORMULATION AND CALIBRATION OF SUBSIDENCE AND HEAT FLOW EQUATIONS

This appendix contains a simple derivation and calibration of initial and thermal subsidence and heat flow that result from modified extension as described in the text and Figure 5. In this formulation, the crust thins instantaneously to $1/\delta$ times its initial thickness, while the mantle lithosphere thins instantaneously to $1/\beta$ times its initial thickness. The temperature structure is similarly perturbed. Because the equations derived below are calibrated to oceanic subsidence and heat flow data, there are very few parameters that need to be defined. The expressions for initial and thermal subsidence given here are approximations accurate to within a few percent; exact solutions can be found in Royden and Keen (1980).

1. The final depth of a basin $d(t)$ at $t = \infty$, is a function of starting elevation and prerift elevation, and is linear in $1/\delta$ to within a few percent. Then:

$$d(\infty) = C_1 + C_2/\delta$$

C_1 can be evaluated by noting that in the oceans, where $\beta = \infty$, the final depth is about 7200 m. (This includes an 800 m correction for removing the oceanic crust isostatically). Uncertainty is estimated at a few hundred meters, so that 7200 ± 300 m should give a reasonable range of C_1 (Parsons and Sclater, 1977). If the prerift elevation was at a water loaded depth d_o below sea level, C_2 can be evaluated by noting that for $\delta = 1$ (no crustal stretching), the final depth must be the same as the prerift depth, d. Therefore:

$$d(\infty) = 7200\text{ m} + (d_o - 7200\text{ m})/\delta \qquad (1)$$

2. The total thermal subsidence (the thermal subsidence that occurs from immediately after rifting until an infinite time after rifting, denoted by s) must be proportional to the total amount of heat lost during cooling of the lithosphere. This is proportional to the difference in average temperature immediately after rifting and the average temperature a long time after rifting (the equilibrium geotherm). From analysis of Figure 5:

$$s = C_3[(1 - 1/\delta) + (1 - h/l)^2 (1/\delta - 1/\beta)]$$

C_3 can be evaluated by noting that, in the oceans, where $\beta = \delta = \infty$, the total thermal subsidence, s, is about 3900 m (Parsons and Sclater, 1977). Therefore:

$$s = 3900\text{ m}[(1 - 1/\delta) + (1 - h/l)^2(1/\delta - 1/\beta)] \qquad (2)$$

3. The initial depth immediately after rifting, $d(0)$, is equal to the final depth, $d(\infty)$ minus the total thermal subsidence, s or:

$$d(0) = d_o/\delta + 3300\text{ m}(1 - 1/\delta) - 3900\text{ m}(1 - h/l)^2 (1/\delta - 1/\beta) \qquad (3)$$

4. Time dependent depth, $d(t)$, as given by Royden and Keen (1980) can be written as:

$$d(t) = C_4 - C_5 \sum_{n=1\text{ n,odd}}^{\infty} [((\delta - \beta)/n\pi) \sin n\pi H + (\beta/n\pi) \sin n\pi G](1/n^2) \exp(-n^2t/C_6)$$

where t is time in millions of years and

$$H = h/l\delta$$
$$G = (h/l)(1/\delta - 1/\beta) + 1/\beta$$

C_4 can be evaluated by noting that at $t = \infty$, the expression after the summation sign is zero. Therefore, C_4 is equal to the final depth given by equation (1). C_5 can be evaluated by noting that for the oceanic case, where $\beta = \delta = \infty$, the expression inside the square brackets is equal to 1. Therefore, the total thermal subsidence, s, between time $t = 0$ and time $t = \infty$ is given by:

$$s = 3900\text{ m} = C_5 \sum_{n=1,\text{ n,odd}}^{\infty} 1/n^2 = C_5\pi^2/8$$

which gives $C_5 = 3200$ m. C_6, the thermal time constant, is assumed to be the same as that in the oceans, or approximately 60 m.y. Thus:

$$d(t) = d_o/\delta + 7200\text{ m}(1 - 1/\delta) - 3200\text{ m} \sum_{n=1\text{ n,odd}}^{\infty} [((\delta - \beta)/n\pi) \sin n\pi H + (\beta/n\pi) \sin n\pi G](1/n^2) \exp(-n^2t/60) \qquad (4)$$

5. Surface heat flow, $q(t)$, can likewise be written as:

$$q(t), = C_7[1 + 2\sum_{n=1}^{\infty}[((\delta - \beta)/n\pi) \sin n\pi H + (\beta/n\pi) \sin n\pi G] \exp(-n^2t/C_6)] + q_R$$

C_7 is the background heat flow and in the oceans is about 33 mW/m^2. The radiogenic contribution from crustal rocks, q_R, is variable, but in this paper we use a value of 12 mW/m^2 as a reasonable value for continental crust. Thus:

$$q(t), = 33\mathrm{mW/m^2}[1 + 2\sum_{n=1}^{\infty}[((\delta - \beta)/n\pi) \sin n\pi H + (\beta/n\pi) \sin n\pi G] \exp(-n^2t/60)] + 12 \mathrm{~mW/m^2} \quad (5)$$

Equations (1)–(5) thus give expressions for the water-loaded subsidence history and heat flow history of a sedimentary basin assuming that the basin was at thermal equilibrium prior to extension and knowing only the thinning factors β and δ, the prerift elevation d_o, and the prerift crustal thickness h. These expressions are convenient because, by calibrating all the parameters with the oceanic case ($\beta = \delta = \infty$), the subsidence and heat flow equations can be well calibrated even without knowing individual parameters, such as crustal density, lithospheric thickness, and so on. Equivalently, any combination of these individual physical parameters is acceptable and indistinguishable, provided that they give results consistent with observations in the oceans and thus yield Equations (1)–(5).

Equations (1)–(5) above can be easily presented in graphical form to allow simple and fast determinations of extension in a sedimentary basin. All of the plots presented in this paper assume $h/l = 0.3$ ($h \approx 35$ km, $l \approx 125$ km). Errors in $(1 - h/l)^2$ of about 20% are introduced when h is taken to be 25 or 45 km. This should not contribute significant errors to the results of Equations (2)–(5), and the error is zero when $\beta = \delta$. As in the previous section, all subsidence must be water loaded, so that corrections for sediment loading and compaction are already assumed.

CHAPTER 18

Neogene and Quaternary Volcanism of the Carpathian–Pannonian Region: Changes in Chemical Composition and Its Relationship to Basin Formation

T. Póka
Hungarian Academy of Sciences
Laboratory for Geochemical Research
H-1112 Budapest
Budaörsi út 45, Hungary

This chapter presents a statistical evaluation of the chemical variation in space and time of the Miocene–Quaternary volcanism of the Carpathian–Pannonian region. Three main genetic types of this volcanism can be distinguished as follows: (1) during Karpatian to late Pliocene time intermediate, mainly andesitic, stratovolcanic complexes formed; (2) acidic (mainly ignimbritic) volcanism developed in the inner part of the Pannonian basin from the Eggenburgian–Ottnangian boundary to late Sarmatian time, which partly overlaps the intermediate volcanism and; (3) alkali basaltic volcanism occurred in Pannonian to Quaternary time.

The data presented here show that the intermediate lavas became significantly more acidic from Karpatian to late Sarmatian time. At the Sarmatian–Pannonian boundary there was an abrupt change and in early Pannonian time the intermediate lavas which erupted were much more basic. The K_2O content of the intermediate lavas also increased from Karpatian to late Sarmatian time. This suggests that at the Sarmatian–Pannonian boundary a significant change may have taken place in the tectonics of the region.

The data presented in this chapter suggest that volcanism in the Pannonian basin can be largely explained by a mantle diapir model with related melting of crustal rocks. They further indicate that in the East Carpathians, volcanism may be largely related to the presence of a subducted slab. The geochemical data indicate that the main period of mantle diapirism lasted until the end of Sarmatian time, when very fast subsidence began in the Pannonian basin.

INTRODUCTION

Statistical evaluations of the evolution of the chemistry of volcanic rocks in specific tectonic settings can contribute toward our understanding of magma genesis on a global scale. Relationships between regional geochemical patterns and regional tectonic and structural evolution were constructed in the early days of plate tectonic theory, but now need revision. Scientists at the 1979 symposium of the International Association of Volcanology and Chemistry of the Earth's Interior (IAVCEI) on petrochemistry in different tectonic provinces concluded that these earlier views did not properly account for horizontal and vertical inhomogeneities in the mantle. This may be the primary reason why there does not appear to be a unique relationship between the depth of the Bennioff zone, the nature of the downgoing slab and the chemistry of the magmatic arc developed at subduction boundaries (Nicholls et al., 1980).

Regional differences in the chemistry of calcalkaline volcanites occur because of differences in the chemistry and structure of the underlying crust (Perfit et al., 1980). At the 1979 IAVCEI symposium it became clear that andesites need to be subdivided into a chemical series similar to those created for basaltic rocks. Then a relationship between mantle inhomogeneities, tectonic setting, and petrochemistry might be constructed by careful analysis of individual tectonic and magmatic systems. Yet, such a model developed for one tectonic system cannot be simply and directly applied to another tectonic system.

This conclusion motivated me to undertake a statistical evaluation of the variations in space and time of the chemistry of the Miocene–Quaternary magmatism of the Carpathian–Pannonian region. From this evaluation I hoped to relate the chemistry of volcanites to the tectonic and dynamic processes operating in this region of subduction.

SPACE-TIME DISTRIBUTION OF MIOCENE-QUATERNARY MAGMATISM: VOLCANIC AND PETROLOGIC CHARACTER

The Miocene–Quaternary magmatism of the Carpathian–Pannonian region can be divided into three main genetic types (Figure 1; see also Map 6 in enclosure).

1. From Karpatian to late Pliocene time, intermediate, mostly andesitic, stratovolcanic complexes formed within the basin area, usually during periods of basin extension (Balkay, 1960, 1962). This magmatism usually begins with a (shallow marine) volcano-sedimentary sequence, followed by development of terrestrial stratovolcanoes. These stratovolcanoes represent the most intense phase of volcanic activity. Following a period of quiescence, the calderas collapsed and less intense volcanism formed the volcanic cover of these structures.

This general sequence occurred in most of the volcanic mountains of Hungary, although not at exactly the same time. In the north part of the Pannonian basin, the intermediate volcanism migrated from south to north and from west to east. In the Cserhát–Mátra and Dunazug-Börzsöny mountains, intermediate volcanic activity began in Karpatian time. The most intense (stratovolcanic) period of volcanism occurred in Badenian time (Figure 1). In the Stavnica–Kremnica region and the Tokaj–Pressov mountains, intermediate volcanism began in Badenian time and ended in early Pannonian time. Intermediate volcanism in the Apuseni mountains (Transylvanian basin) occurred over the same time period, but was less intense. In the East Carpathians, intermediate volcanism migrated from north to south, beginning in late Badenian time and ending in Pliocene time. Peak volcanic activity in the East Carpathians is of Pannonian age.

2. Acidic volcanism (tuffs, rhyolites, rhyodacites, and often ignimbrites) developed in the interior of the Pannonian basin. This acid magmatism only partly overlaps the intermediate volcanism in space and time. The main manifestation of this acidic volcanism is the three large tuff horizons within the basin (Figure 1; see also Map 6 in enclosure). The "lower rhyolite tuff" horizon has been dated at about 21 Ma (about the Eggenburgian–Ottnangian boundary). The "middle rhyolite tuff" horizon is dated at about 17 Ma (about the Karpatian–Badenian boundary) and is actually composed of mostly dacitic tuffs. The "upper rhyolite tuff" horizon is dated at about 14 Ma (intra-Sarmatian) (Hámor et al., 1976). The rhyolite tuffs, together with the acid material in the stratovolcanic complexes and the 2 to 3 km of mostly Sarmatian ignimbrites in the mid-Tisza graben, have a volume five times greater than the intermediate volcanic material.

3. At about 10 Ma, alkali basaltic volcanism began in the central part of the Pannonian basin and continued until 3 Ma in the East Carpathians (Jámbor et al., 1980) (Figure 1). This sodic-type, alkali basaltic volcanism represents the final stage of magmatism in the Carpathian–Pannonian region. The total volume of these basalts is quite small. These low viscosity lavas reached the surface along faults associated with extensional tectonics, and formed flood basalts and basaltic cones.

METHOD

For statistical purposes, bulk analyses from 1760 samples were collected from the literature (references are given in Tables 1–10). Data were used only for those samples that were not altered or affected by hydrothermal or metasomatic processes. Moreover, only samples for which the age and locality were known precisely were used in this analysis. From this data set the following quantities for each volcanic unit (that is, for each eruptive phase) were determined for the arithmetic mean and standard deviation of the major elements: the $SiO_2/(Na_2O + K_2O)$; SiO_2/K_2O; Na_2O/K_2O; the (AFM) values of Wager and Deer (1956)[1]; the (FM) values of Simpson (1954)[2]; the Sugimura (1959)[3] indices; the Thornton and Tuttle (1960)[4] values, the serial indices of Rittmann (1957)[5] and the (SI) value of Kuno (1962)[6] (Tables 1–10). The values calculated for each volcanic

[1] AFM = $(Na_2O + K_2O) + (FeO + Fe_2O_3) + MgO = 100$.
[2] FM = $(Na_2O + K_2O)/(CaO + Na_2O + K_2O)$; M = $(FeO + Fe_2O_3)/(MgO + FeO + Fe_2O_3)$.
[3] $\theta = SiO_2 - 47\,(Na_2O + K_2O)/Al_2O_3)$.
[4] DI = Ne + Le + Ks + Pr + Q.
[5] $S = (Na_2O + K_2O)^2/(SiO_2 - 43)$.
[6] SI = (100 MgO)/(MgO + FeO + FeO + NaO + KO).

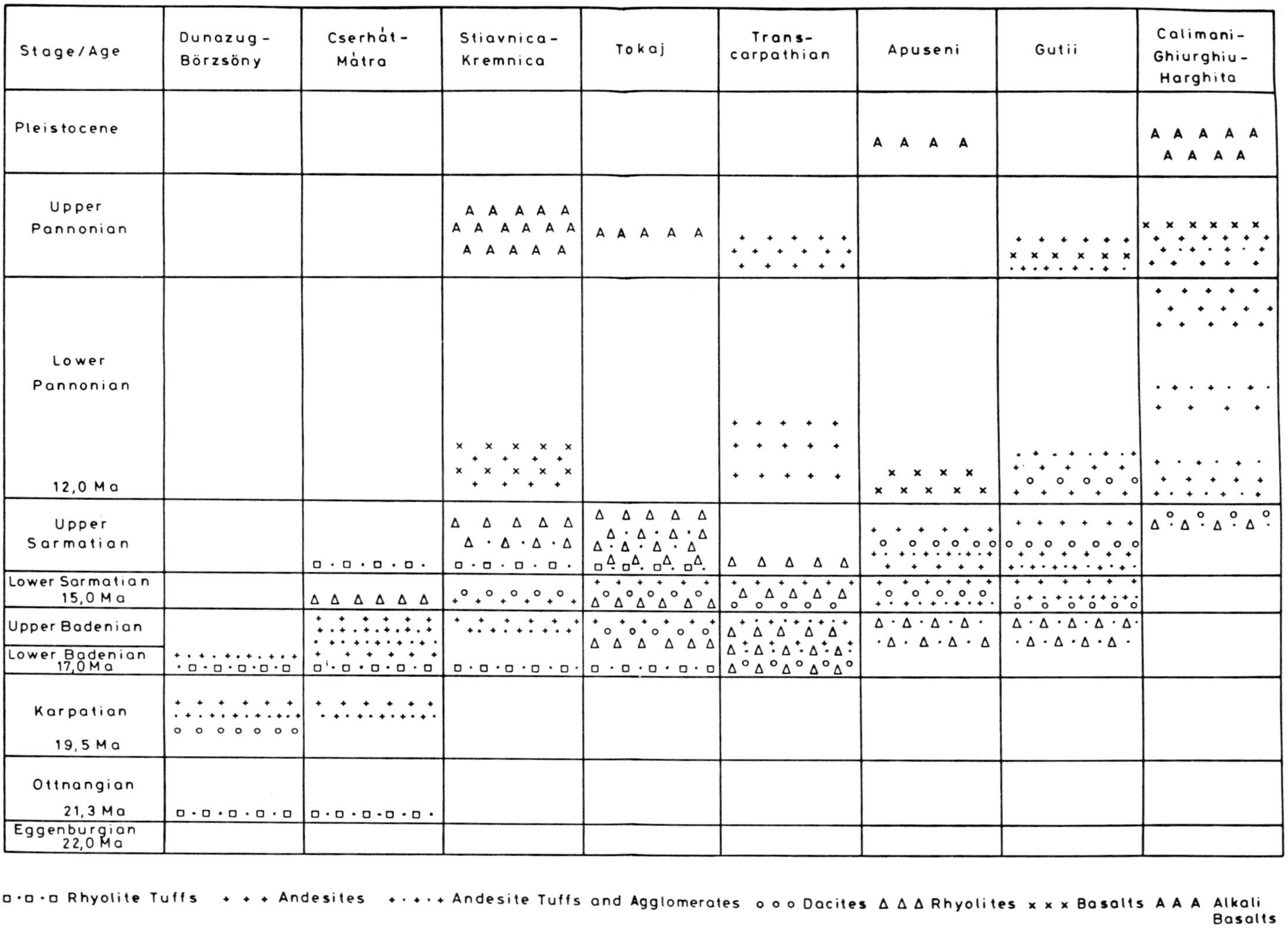

Stage/Age	Dunazug-Börzsöny	Cserhát-Mátra	Stiavnica-Kremnica	Tokaj	Trans-carpathian	Apuseni	Gutii	Calimani-Ghiurghiu-Harghita
Pleistocene						A A A A		A A A A A A A A A
Upper Pannonian			A A A A A A A A A A A A A A	A A A A A	+ + + + + +·+·+·+·+ + + + + +		+ + + + + + x x x x x x ·+·+·+·+·	x x x x x +·+·+·+·+ + + + + +
Lower Pannonian 12,0 Ma			x x x x x + + + + x x x x x + + + +		+ + + + + + + + + + + + + + +	x x x x x x x x x	·+·+·+·+ + + + + o o o o o + + + + +	+ + + + + + + + + + ·+·+·+· + + + + ·+·+·+· + + + + + +·+·+·+·
Upper Sarmatian		□·□·□·□·	Δ Δ Δ Δ Δ·Δ·Δ·Δ □·□·□·□·	Δ Δ Δ Δ Δ Δ·Δ·Δ·Δ Δ·Δ·Δ·Δ □Δ□Δ□Δ□Δ	Δ Δ Δ Δ Δ	+ + + + + + o o o o o o +·+·+·+·+ + + + + +	+ + + + + + o o o o o o o + + + + + +·+·+·+·+·	o o o o Δ·Δ·Δ·Δ·
Lower Sarmatian 15,0 Ma		Δ Δ Δ Δ Δ Δ	+o+o+o+o+	+ + + + + o o o o o o Δ Δ Δ Δ Δ Δ	+ + + + + + Δ Δ Δ Δ Δ o o o o o	+ + + + + + o o o o o +·+·+·+·+	+ + + + + +·+·+·+·+ o o o o o o	
Upper Badenian		+ + + + + + +·+·+·+·+ +·+·+·+·+ + + + + + □·□·□·□·□	+·+·+·+·+ □·□·□·□·□·	+ + + + + + o o o o o o Δ Δ Δ Δ Δ Δ □·□·□·□·□	+·+·+·+·+ Δ Δ Δ Δ Δ Δ·Δ·Δ·Δ·Δ ΔoΔoΔoΔoΔo	Δ·Δ·Δ·Δ· ·Δ·Δ·Δ·Δ	Δ·Δ·Δ·Δ·Δ· ·Δ·Δ·Δ·Δ·	
Lower Badenian 17,0 Ma	+·+·+·+·+·+·+ ·□·□·□·□·□							
Karpatian 19,5 Ma	+ + + + + + ·+·+·+·+·+· o o o o o o o	+ + + + + + ·+·+·+·+·+·						
Ottnangian 21,3 Ma	□·□·□·□·□	□·□·□·□·□·						
Eggenburgian 22,0 Ma								

□·□·□ Rhyolite Tuffs + + + Andesites +·+·+ Andesite Tuffs and Agglomerates o o o Dacites Δ Δ Δ Rhyolites x x x Basalts A A A Alkali Basalts

Figure 1. Ages of Neogene–Quaternary volcanic rocks in the Carpathian region.

cycle of the same geological age are mean values weighted according to the number of samples (Table 11). The andesite samples were also analyzed separately for each volcanic unit and for each eruptive phase. The depth to the subducted slab was determined using the method of Hatherton and Dickinson (1968)[7], and the thickness of the paleocrust (C) was determined using the Condie (1976) method[8] (Table 12).

MAGMA CHEMISTRY OF INTERMEDIATE VOLCANICS AND ITS SPATIAL VARIATION

The relationship between SiO_2 and total alkali content was plotted with the trend lines for tholeitic, the high Al, the calcalkaline and the alkaline series (Figure 2). The majority of the andesite samples fell into the category of basic andesites, while the dacitic samples are classified as least-acid dacites. More then half of the volcanics fall in the tholeitic field and the rest into the calcalkaline and high Al basalt field. I conclude from Figure 2 that the Miocene intermediate volcanism of the Carpathian–Pannonian region belongs to the basic part of the calcalkaline series, and exhibits a tholeitic affinity.

Other investigators have observed that the lavas of the Carpathians are alumina oversaturated (e.g., Karolus, 1975; Gyarmati, 1977). The average SiO_2 content of the 955 andesite samples varies between 54% and 62%, and 60% of the samples have an SiO_2 content between 56% and 58%. The K_2O content varies between 1% and 3.5%, with 58% of the samples having a K_2O content between 1.5% and 2.5%. According to Taylor (1969), these andesites are basic, potassium-rich andesites.

The degree of crustal contamination of these calcalkaline magmas can be partly inferred from the ratio of K to Si and Na. For the andesites alone, the ratio of SiO_2 to K_2O is the smallest (maximum 28.87) in the Dunazug–Börzsöny mountains and the largest (maximum 43.83) in the Gutin mountains. In general, the relative SiO_2 to K content of lavas in the volcanic mountains near the Transylvanian basin is smaller than that in the

[7] Sz (km) = 89.3 (K_2O) − 14.3.
[8] C (km) = 18.2 (K_2O) + 0.45.

Table 1. Average Chemical Composition and Some Petrochemical Indices of Volcanic Phases in the Dunazug–Börzöny Mountains

	Dunazug Mountains[a]		Börzsöny Mountains[b]		
	I. phase 9 samples	II. phase Lower Badenien 17 samples	I. phase Lower Badenien 18 samples	II. phase Lower Badenien 55 samples	III. phase Lower Badenien 21 samples
SiO_2	61.14 ± 5.94	55.49 ± 2.33	57.51 ± 5.84	56.06 ± 3.39	58.18 ± 2.19
TiO_2	0.23 ± 0.24	0.58 ± 0.31	0.69 ± 0.33	0.76 ± 0.24	0.52 ± 0.40
Al_2O_3	18.38 ± 2.57	18.76 ± 1.67	17.22 ± 3.22	18.19 ± 1.43	17.96 ± 1.44
Fe_2O_3	3.64 ± 2.45	4.38 ± 1.52	3.61 ± 1.81	3.06 ± 1.74	3.01 ± 2.05
FeO	1.20 ± 1.27	2.48 ± 1.30	2.67 ± 1.36	4.19 ± 1.70	3.74 ± 1.43
MnO	0.29 ± 0.29	0.22 ± 0.20	0.14 ± 0.21	0.16 ± 0.08	0.08 ± 0.07
MgO	1.23 ± 1.06	2.81 ± 0.87	2.01 ± 1.20	2.12 ± 0.95	1.80 ± 1.05
CaO	4.76 ± 2.70	6.57 ± 1.65	6.53 ± 3.27	6.82 ± 1.04	6.02 ± 0.95
Na_2O	3.04 ± 0.94	2.99 ± 0.47	3.15 ± 0.87	3.04 ± 0.94	3.33 ± 0.86
K_2O	2.86 ± 0.93	2.29 ± 0.62	2.32 ± 0.47	2.16 ± 0.67	2.26 ± 0.34
$SiO_2/Na_2O + K_2O$	10.36	10.51	10.51	10.78	10.41
SiO_2/K_2O	21.38	24.23	24.79	25.95	25.74
Na_2O/K_2O	1.06	1.31	1.36	1.41	1.47
Wager-Deer A	0.513 ± 0.413	0.353 ± 0.049	0.412 ± 0.112	0.360 ± 0.068	0.404 ± 0.088
F	0.393 ± 0.084	0.461 ± 0.052	0.449 ± 0.096	0.497 ± 0.060	0.468 ± 0.089
M	0.094 ± 0.070	0.186 ± 0.051	0.139 ± 0.073	0.144 ± 0.061	0.127 ± 0.004
Simpson F	55.36	44.56	49.88	43.26	48.15
M	79.73	73.01	75.75	77.37	78.94
Rittmann s	2.075 ± 0.653	2.443 ± 1.102	2.546 ± 1.737	2.232 ± 0.913	2.128
Kuno SI	10.27	18.80	14.29	15.83	12.72
Thornton-Tuttle DI	70.6	60.6	59.1	52.3	42.6
Sugimura θ	4.54	2.39	3.34	2.59	3.48
Sugimura SWS	14.85	12.31	11.86	11.48	11.86
Correlations					
SiO_2 − alk	0.593	−0.527	0.027	0.068	0.202
SiO_2 − K_2O	0.594	−0.353	0.545	0.007	0.362
Na_2O − K_2O	−0.600	0.221	−0.119	−0.305	−0.291
FeO − MgO	0.827	−0.057	0.421	0.175	0.362
TiO_2 − FeO/MgO	−0.352	−0.053	0.007	0.083	−0.182

[a]See Zelenka (1960) and Korpás et al. (1967).
[b]See Kubovics and Pantó (1970).

volcanic mountains near the Pannonian basin. The ratio of N_2O to K_2O is the smallest in the Tokaj mountains (≤ 1.0) and the largest in the East Carpathians ($>>1.5$), thereby supporting the inference that the magmatic evolution of the Pannonian and Transylvanian basins are significantly different. This may reflect the different evolution of the lithosphere beneath these two areas (Horváth and Royden, 1981; Royden et al., 1982 and Royden, this volume).

TEMPORAL VARIATION OF MAGMA CHEMISTRY

Earlier work on the magma chemistry of the Carpathian–Pannonian region (e.g., Vendel, 1944–47; Burri and Niggli, 1949; Karolus, 1965) suggested that the magmas became progressively more basic with time. The statistical compilation presented in this paper indicates that this widely accepted view should be revised. The data presented here, including the change in SiO_2 content as a function of time for all the intermediate rocks, show that the magma became significantly more acid from Karpatian to late Sarmatian time, even without including the rhyolite tuff horizons (Figure 3). At about the Sarmatian–Pannonian boundary there was an abrupt change, and in early Pannonian (s. l.) time the magma which erupted was much more basic. If only andesitic rocks are considered, the same trend may be observed, but it is less pronounced (Figure 4).

The ratio of Na_2O/K_2O was also plotted as a function of time (Figures 5 and 6). The Na_2O/K_2O increased from Karpatian to late Sarmatian time. This might imply that crustal contamination of the magma was more pronounced before the Sarmatian than it was later.

The ratio of Na_2O to K_2O is around 1 for magmas erupted from Karpatian to late Sarmatian time, and about 2 for lavas erupted after Sarmatian time. Thus, the relative K content decreased markedly after Sarmatian time. The same pattern is shown by the change of SiO_2 content, indicating that at the

Table 2. Average Chemical Composition and Some Petrochemical Indices of Volcanic Phases in the Cserhát and Mátra Mountains

	Cserhát[a]	Mátra Mountains[b]			
	Lower Badenien 50 samples	I. phase Karpathian 25 samples	II. phase Lower Badenien 109 samples	III. phase Upper Badenien 62 samples	IV. phase Upper Badenien Lower Sarmatian? 6 samples
SiO_2	54.87 ± 2.97	53.79 ± 5.93	56.38 ± 2.17	55.37 ± 2.15	70.56 ± 3.55
TiO_2	1.29 ± 0.64	0.88 ± 0.23	0.85 ± 0.21	0.85 ± 0.19	0.50 ± 0.24
Al_2O_3	17.67 ± 2.89	18.31 ± 2.88	18.18 ± 1.40	18.66 ± 1.21	14.52 ± 0.40
Fe_2O_3	2.98 ± 1.49	4.31 ± 2.55	3.07 ± 2.12	2.66 ± 1.03	1.36 ± 0.75
FeO	4.29 ± 1.77	2.62 ± 1.72	4.24 ± 1.58	4.46 ± 1.26	1.08 ± 1.69
MnO	0.12 ± 0.12	0.14 ± 0.09	0.20 ± 0.19	0.16 ± 0.06	0.04 ± 0.07
MgO	2.71 ± 1.20	1.70 ± 0.70	2.81 ± 1.15	2.79 ± 0.72	0.14 ± 0.03
CaO	7.85 ± 1.92	7.61 ± 4.38	7.38 ± 1.20	7.53 ± 1.46	1.93 ± 1.09
Na_2O	2.77 ± 0.86	2.36 ± 0.57	2.49 ± 0.89	2.28 ± 0.40	3.64 ± 0.32
K_2O	1.99 ± 0.68	1.96 ± 0.82	1.96 ± 0.46	2.02 ± 0.44	4.40 ± 0.45
$SiO_2/Na_2O + K_2O$	14.59	12.45	12.67	12.41	8.78
SiO_2/K_2O	27.57	27.44	28.77	26.42	16.04
Na_2O/K_2O	1.39	1.20	1.27	1.13	0.830
Wager-Deer A	0.317 ± 0.092	0.337 ± 0.090	0.309 ± 0.070	0.304 ± 0.044	0.073 ± 0.088
F	0.487 ± 0.123	0.533 ± 0.096	0.501 ± 0.059	0.499 ± 0.023	0.223 ± 0.089
M	0.197 ± 0.141	0.130 ± 0.048	0.190 ± 0.070	0.197 ± 0.045	0.014 ± 0.004
Simpson F	37.45	36.21	37.61	36.34	80.64
M	72.84	80.30	71.52	71.84	94.57
Rittmann s	2.103 ± 0.777	1.638 ± 0.852	1.551 ± 0.765	1.537 ± 0.455	2.376 ± 0.385
Kuno SI	18.38	13.12	19.28	19.63	1.32
Thornton-Tuttle DI	48.0	48.9	67.0	46.8	83.8
Sugimura θ	2.12	1.60	2.30	1.93	13.05
Sugimura SWS	11.28	9.48	10.82	10.53	14.61
Correlations					
SiO_2 − alk	−0.131	0.583	0.359	0.532	0.233
$SiO_2 - K_2O$	0.047	0.528	0.426	0.596	0.696
$Na_2O - K_2O$	0.441	0.361	−0.062	0.429	0.184
FeO − MgO	0.137	0.105	0.070	0.368	−0.299
TiO_2 − FeO/MgO	−0.058	−0.073	0.009	0.350	0.565

[a]See Póka (1968), Arkai (1974).
[b]See Kubovics and Pantó (1970), Varga et al. (1975), Póka (1968), and Póka and Simó (1966).

Sarmatian–Pannonian boundary a significant change occurred in the chemistry of the magmas being erupted.

The SI values of Kuno (1962), for the intermediate volcanics, are generally less than about 20, while they are around 35 for the Pliocene–Pleistocene alkali basalts. The latter value is characteristic for a primary, slightly fractionated magma. The change in the SI value through time indicates that crustal contamination might have been more important in Karpatian time (in the first stage of intermediate volcanism) than in Sarmatian time.

The average K_2O content of the andesites for each geologic stage was used to calculate the depth of magma generation (Sz) and the paleocrustal thickness (C) after the methods of Hatherton and Dickinson (1968) and Condie (1976) respectively (Table 11). Sz and C have been plotted as a function of time (Figure 7) and suggest that the depth of magma generation changed from 172 ±18 km in Karpatian and Badenian time to 140 ±30 km in the Quaternary, while the crustal thickness decreased from 38 ±4 km to 29 ±7 km over the same time interval. These values cannot, of course, be considered as absolute values because of crustal contamination, mixing of magmas, and so on, but they might indicate the general change fairly well. Thus, at the beginning of volcanism in the Pannonian basin the depth of magma generation and the paleocrustal thickness may have been greater than during the subsequent volcanism in the East Carpathians.

These data (Figure 7) suggest that crustal thinning beneath the Pannonian basin began about late Sarmatian time and was contemporaneous with the cessation of intense acidic volcanism. Since late Sarmatian time, there has been little partial melting within the crust. Instead, progressive thinning of the crust occurred beneath the Pannonian basin. At the same time, alkali basalts were generated at a depth of 60–70 km beneath the Pannonian basin as calcalkaline lavas erupted in the East Carpathians. The very young (3 Ma) basalts of the East Carpathians are also geochemically different from the alkali basalts in the Pannonian basin (Peltz et al., 1973; Embey-Istin et al., 1985).

Table 3. Average Chemical Composition and Some Petrochemical Indices of Volcanic Phases in the Banska Stiavnica and Kremnicka Mountains

	Banska Stiavnica and Kremnicka Mountains[a]			
	I. phase Upper Badenien 25 samples	II. phase Lower Sarmatian 11 samples	III. phase Upper Sarmatian 29 samples	IV. phase Pliocene 10 samples
SiO_2	57.33 ± 2.43	59.37 ± 3.31	71.75 ± 4.08	50.46 ± 3.20
TiO_2	0.70 ± 0.21	0.51 ± 0.22	0.23 ± 0.15	1.07 ± 0.60
Al_2O_3	15.99 ± 2.51	16.33 ± 1.33	12.86 ± 2.04	16.72 ± 1.66
Fe_2O_3	3.49 ± 1.70	3.95 ± 0.92	1.97 ± 1.30	4.03 ± 2.44
FeO	4.35 ± 1.26	2.56 ± 0.43	0.87 ± 0.53	4.31 ± 2.26
MnO	0.11 ± 0.08	0.13 ± 0.08	0.08 ± 0.18	0.14 ± 0.09
MgO	4.02 ± 1.28	2.56 ± 0.89	0.81 ± 0.59	6.43 ± 2.67
CaO	7.32 ± 1.06	6.14 ± 1.38	1.60 ± 0.71	8.85 ± 2.22
Na_2O	2.49 ± 0.66	2.52 ± 0.39	1.86 ± 0.89	2.80 ± 0.73
K_2O	1.94 ± 0.55	2.62 ± 0.48	4.27 ± 1.57	1.55 ± 0.36
$SiO_2/Na_2O + K_2O$	16.76	11.55	11.70	11.60
SiO_2/K_2O	29.55	22.66	16.80	32.55
Na_2O/K_2O	1.28	0.96	0.44	1.81
Wager-Deer A	0.281 ± 0.085	0.368 ± 0.077	0.634 ± 0.112	0.237 ± 0.068
F	0.473 ± 0.078	0.456 ± 0.054	0.285 ± 0.104	0.431 ± 0.103
M	0.246 ± 0.067	0.177 ± 0.050	0.082 ± 0.056	0.332 ± 0.091
Simpson F	37.10	45.56	79.30	32.95
M	66.10	71.77	77.80	56.46
Rittmann s	1.414 ± 0.600	1.638 ± 0.251	1.390 ± 0.729	3.218 ± 3.292
Kuno SI	24.67	18.01	8.28	33.62
Thornton-Tuttle DI	48.1	54.1	83.2	36.4
Sugimura θ	2.87	3.89	11.80	0.90
Sugimura SWS	11.99	11.81	11.84	14.26
Correlations				
SiO_2 − alk	0.539	0.773	0.008	0.455
$SiO_2 - K_2O$	0.796	0.805	0.030	0.546
$Na_2O - K_2O$	0.275	0.020	−0.575	0.179
FeO − MgO	0.170	0.551	0.123	0.278
TiO_2 − FeO/MgO	−0.318	−0.350	−0.369	0.104

[a]See Fiala (1962), Forgác and Kupco (1974, Konećny (1969), and Karolus (1965).

RB/SR MEASUREMENTS

Rb/Sr measurements on 6 samples from volcanites of different chemistry of the Calimani–Harghita Mts. (East Carpathians) yield an average Rb content of 62 ppm and an average Sr content of 333 ppm (Rb/Sr value of 0.19) (Peccerillo and Taylor, 1976). These data suggest a paleocrustal thickness of 25–30 km for the upper Pannonian (s. l.) rocks, in good agreement with the major element data.

Similar measurements for the central and eastern Slovakian Miocene (Badenian and Sarmatian) volcanic rocks yield an Sr content of 245 ppm and an Rb value of 87.6 ppm for andesites, and 410 and 94.6 ppm, respectively, for the rhyolites (Forgach and Kupco, 1974). An Rb/Sr ratio of 0.23 for these rhyolites suggests that they originated in the crust, while the 0.36 value for the andesites suggests a strong crustal contamination.

RARE EARTH ELEMENTS AND XENOLITES

Measurements of rare earth elements (REE) from Neogene calcalkaline volcanic rocks and Pliocene–Pleistocene basalts in Hungary (Pantó, 1981) indicate that the light rare earth elements (LREE) are strongly fractionated. The La/Yb ratio for andesites varies between 9 and 25, and is thus consistent with an active continental margin setting (Thorpe et al., 1976). The total REE content is higher in the andesites than is typical for island arcs and increases from southwest to northeast at the same time as the andesites also become younger from southwest to northeast. There is a marked Eu anomaly that excludes a direct mantle origin. The Eu anomaly of less than 0.7 indicates significant crustal contamination of the magma. Pantó (1981) states that the Eu anomaly of the acid volcanic complexes (rhyolite tuff horizons and acid phases of the Tokaj Mountains) is always less than 0.5, and often as low as 0.1 and that the LREE in these rocks are enriched ($La/Yb < 10$), indicating that the rhyodacites and ignimbrites are derived from the crust.

The REE of the Pliocene–Pleistocene alkali basalts are also characterized by relative enrichment of the light lanthanides. This feature, however, is much less pronounced than for the calcalkaline rocks. The total REE content of the basalts reaches 1,000 ppm. This is much higher than the total REE content of the Miocene lavas, which is typical of the volcanics with alkalic characteristics. Lherzolite and dunite inclusions within these alkali basalts also indicate a mantle origin (Embey-Isztin, 1976).

Table 4. Average Chemical Composition and Some Petrochemical Indices of Volcanic Phases in the Tokaj Mountains

	Tokaj Mts. (andesitic phases)[a]			Tokaj Mts. (acidic phases)[b]			
	I. phase Upper Badenien 57 samples	II. phase Lower Sarmatian 227 samples	III. phase Pliocene 7 samples	I. phase Upper Badenien 19 samples	II. phase Lower Sarmatian 23 samples	III. phase Upper Sarmatian 86 samples	IV. phase Upper Sarmatian 18 samples
SiO_2	59.87 ± 4.36	60.52 ± 3.28	47.92 ± 1.64	66.68 ± 2.49	70.00 ± 4.49	72.91 ± 3.14	71.69 ± 4.69
TiO_2	0.57 ± 0.22	0.62 ± 0.25	1.14 ± 0.27	0.11 ± 0.04	0.16 ± 0.18	0.08 ± 0.11	0.27 ± 0.24
Al_2O_3	16.38 ± 1.82	16.84 ± 1.43	17.12 ± 1.04	14.36 ± 1.36	13.77 ± 1.36	12.52 ± 1.26	13.57 ± 2.56
Fe_2O_3	2.35 ± 1.31	2.59 ± 1.40	3.22 ± 1.13	0.53 ± 0.33	1.12 ± 0.67	0.81 ± 0.49	1.44 ± 1.06
FeO	2.46 ± 1.26	2.89 ± 1.34	4.14 ± 1.13	1.46 ± 0.48	0.87 ±0.86	0.47 ± 0.31	0.79 0.55
MnO	0.11 ± 0.07	0.14 ± 0.10	0.17 ± 0.07	0.08 ± 0.05	0.07 ± 0.03	0.04 ± 0.05	0.08 ± 0.14
MgO	2.37 ± 1.02	2.81 ± 1.35	8.52 ± 1.67	1.33 ± 0.71	0.73 ± 0.43	0.42 ± 0.56	0.53 ± 0.53
CaO	5.19 ± 1.76	5.94 ± 1.69	9.38 ± 0.62	3.16 ± 1.21	2.33 ± 1.55	1.41 ± 0.72	1.73 ± 1.06
Na_2O	2.13 ± 0.92	2.57 0.79	2.57 ± 0.32	1.09 ± 0.92	2.04 ± 0.92	2.15 ± 1.03	2.37 ± 1.04
K_2O	2.76 ± 1.86	2.24 ± 0.80	1.44 ± 0.23	4.49 ± 1.32	3.87 ± 2.05	4.43 ± 1.34	3.86 ± 1.09
$SiO_2/Na_2O + K_2O$	12.24	12.58	11.95	11.95	11.84	11.08	11.51
SiO_2/K_2O	21.69	27.02	33.28	14.85	18.09	16.46	18.57
Na_2O/K_2O	0.77	1.15	1.78	0.24	0.52	0.49	0.61
Wager-Deer A	0.406 ± 0.115	0.374 ± 0.102	0.204 ± 0.036	0.626 ± 0.102	0.683 ± 0.132	0.793 ± 0.082	0.698 ± 0.101
F	0.401 ± 0.088	0.417 ± 0.069	0.371 ± 0.027	0.223 ± 0.052	0.227 ± 0.100	0.155 ± 0.051	0.249 ± 0.085
M	0.193 ± 0.074	0.210 ± 0.086	0.425 ± 0.055	0.150 ± 0.080	0.089 ± 0.059	0.052 ± 0.066	0.059 ± 0.050
Simpson F	48.51	44.33	29.94	63.84	71.72	82.35	78.26
Simpson M	67.17	66.58	46.52	59.69	73.43	75.29	80.79
Rittmann s	1.522 ± 0.842	1.380 ± 0.689	3.636	1.408 ± 0.654	1.390 ± 0.720	1.495 ± 0.525	1.491
Kuno SI	19.63	21.45	42.83	14.94	8.35	5.07	8.51
Thornton-Tuttle DI	63.4	55.7	30.5	73.9	77.1	83.7	84.10
Sugimura θ	3.82	3.86	0.022	7.65	9.87	13.63	11.01
Sugimura SWS	11.17	11.47	15.74	11.37	11.36	12.26	11.74
Correlations							
SiO_2 − alk	0.507	0.561	0.895	−0.109	0.425	0.419	−0.468
$SiO_2 - K_2O$	0.387	0.538	0.704	−0.143	0.503	0.240	0.161
$Na_2O - K_2O$	0.585	0.097	0.748	−0.392	−0.448	−0.389	−0.367
FeO − MgO	0.361	0.300	0.340	0.215	0.506	0.013	0.335
TiO_2 − FeO/MgO	0.240	0.020	0.611	0.283		−0.077	

[a]See Gyarmati (1977).
[b]See Ilkeyné–Perlaki (1973).

Table 5. Average Chemical Composition and Some Petrochemical Indices of Volcanic Phases in the Ukrainian Carpathians[a]

	Ukrainian Carpathians (acidic phases)					Ukrainian Carpathians (andesitic phases)			
	I. phase Paleogene 9 samples	II. phase Lower Badenian 7 samples	III. phase Upper Badenien 13 samples	IV. phase Sarmatian 12 samples	V. phase Pannonian 5 samples	I. phase Badenien 10 samples	II. phase Lower Sarmatian 10 samples	III. phase Lower Pannonian 36 samples	IV. phase Upper Pannonian 28 samples
SiO_2	70.62 ± 1.67	67.37 ± 2.71	67.45 ± 3.85	67.75 ± 9.43	64.72 ± 3.80	58.75 ± 4.50	69.59 ± 1.76	57.02 ± 3.37	61.52 ± 6.28
TiO_2	0.34 ± 0.17	0.32 ± 0.11	0.30 ± 0.17	0.28 ± 0.16	0.50 ± 0.33	0.71 ± 2.16	0.13 ± 0.04	0.70 ± 0.12	0.73 ± 1.22
Al_2O_3	11.73 ± 2.67	13.99 ± 1.37	12.20 ± 2.09	13.35 ± 2.58	16.24 ± 1.13	17.07 ± 1.33	15.91 ± 0.24	18.27 ± 1.14	17.25 ± 1.84
Fe_2O_3	2.65 ± 1.76	2.83 ± 1.24	2.87 ± 1.84	3.88 ± 3.11	3.72 ± 1.27	3.02 ± 0.81	1.39 ± 0.66	3.13 ± 1.07	2.61 ± 1.18
FeO	0.41 ± 0.41	0.97 ± 0.51	1.37 ± 1.21	0.53 ± 0.53	0.77 ± 0.63	3.85 ± 1.85	0.49 ± 0.23	3.91 ± 1.46	2.80 ± 1.88
MnO	0.04 ± 0.04	0.05 ± 0.05	0.08 ± 0.05	0.06 ± 0.13	0.04 ± 0.04	0.11 ± 0.06	0.03 ± 0.01	0.07 ± 0.05	0.08 ± 0.06
MgO	0.68 ± 0.43	2.79 ± 1.16	1.01 ± 0.67	0.80 ± 0.47	0.81 ± 0.43	2.71 ± 1.25	0.38 ± 0.06	3.17 ± 1.23	2.41 ± 1.73
CaO	1.39 ± 0.74	0.64 ± 0.31	2.26 ± 1.46	2.25 ± 0.90	3.15 ± 0.94	6.27 ± 1.68	2.72 ± 0.49	7.00 ± 1.39	5.74 ± 2.50
Na_2O	3.27 ± 1.18	2.88 ± 1.24	3.72 ± 1.36	2.32 ± 1.03	2.22 ± 1.38	2.92 ± 0.35	2.89 ± 0.33	2.69 ± 0.51	2.02 ± 0.55
K_2O	2.66 ± 1.82	2.63 ± 1.02	2.25 ± 1.32	2.24 ± 0.96	1.68 ± 0.87	1.86 ± 0.44	3.33 ± 0.53	1.91 ± 0.54	2.56 ± 0.91
$SiO_2/Na_2O + K_2O$	11.91	12.23	11.30	14.86	16.59	12.29	11.01	12.40	13.34
SiO_2/K_2O	26.55	25.62	29.98	30.25	38.52	31.59	20.90	29.85	23.88
Na_2O/K_2O	1.23	1.10	1.65	1.04	1.32	1.37	0.87	1.41	0.79
Wager-Deer A	0.626 ± 0.156	0.460 ± 0.117	0.548 ± 0.173	0.486 ± 0.149	0.410 ± 0.058	0.346 ± 0.103	0.739 ± 0.058	0.320 ± 0.096	0.434 ± 0.184
F	0.306 ± 0.130	0.317 ± 0.090	0.361 ± 0.127	0.422 ± 0.145	0.500 ± 0.054	0.473 ± 0.045	0.217 ± 0.057	0.472 ± 0.052	0.399 ± 0.105
M	0.069 ± 0.041	0.223 ± 0.068	0.091 ± 0.062	0.091 ± 0.057	0.090 ± 0.045	0.181 ± 0.064	0.045 ± 0.006	0.209 ± 0.066	0.167 ± 0.097
Simpson F	81.01	90.92	72.53	67.95	55.32	43.33	67.57	39.65	44.81
M	81.81	57.66	80.76	84.34	85.85	124.23	86.23	68.95	69.09
Rittmann s	1.377 ± 0.840	1.276 ± 0.468	1.561 ± 0.700		0.887 ± 0.689	1.477 ± 0.129	1.482 ± 0.453	1.536 ± 0.355	1.506
Kuno SI	16.70	23.10	9.00	8.19	8.80	18.87	4.48	21.40	19.43
Thornton-Tuttle DI	81.9	85.2	84.4	76.9	69.1	54.1	80.0	49.0	54.4
Sugimura θ	11.96	8.02	10.01	7.09	4.26	3.29	8.83	2.58	3.86
Sugimura SWS	11.35	12.78	11.76	9.01	7.83	11.31	11.76	11.45	10.65
Correlations									
SiO_2 – alk	0.034	0.805	0.479	−0.009	−0.664	0.944	−0.286	0.859	0.952
$SiO_2 - K_2O$	−0.400	0.163	0.619	−0.900	−0.167	0.938	−0.412	0.912	0.963
$Na_2O - K_2O$	−0.385	−0.375	−0.080	0.152	0.222	0.653	0.939	0.580	0.730
$FeO - MgO$	0.674	0.385	0.703	−0.184	0.515	0.937	0.911	0.680	0.926
$TiO_2 - FeO/MgO$		0.237	−0.007	0.210	−0.825	−0.803	0.873	0.029	−0.202

[a]See Tolsztoj et al. (1976) and Danilovics (1976).

Table 6. Average Chemical Composition and Some Petrochemical Indices of Volcanic Phases in the Gutii Mountains

	Gutii Mountains[a]			
	Inital phase Upper Badenien 6 samples	I. phase Lower Sarmatian 21 samples	II. phase Upper Sarmatian–Lower Pannonian (*s. l.*) 12 samples	III. phase Upper Pannonian (*s. l.*) 45 samples
SiO_2	70.88 ± 5.10	56.15	60.27 ± 4.86	54.54
TiO_2	0.09 ± 0.15	0.82	0.67 ± 0.43	0.73
Al_2O_3	14.62 ± 2.07	18.26	16.91 ± 1.49	18.73
Fe_2O_3	1.80 ± 2.29	3.44	3.36 ± 1.91	3.64
FeO	0.95 ± 0.56	4.83	2.32 ± 1.28	4.19
MnO	0.01 ± 0.02	0.19	0.10 ± 0.09	0.23
MgO	0.33 ± 0.24	2.84	1.87 ± 1.23	3.99
CaO	0.28 ± 0.20	7.12	5.28 ± 2.57	7.45
Na_2O	2.13 ± 1.64	2.30	3.03 ± 0.69	2.33
K_2O	6.97 ± 1.73	1.28	2.50 ± 2.46	1.42
$SiO_2/Na_2 + K_2O$	7.79	15.68	10.90	14.54
SiO_2/K_2O	10.17	43.87	24.11	38.41
Na_2O/K_2O	0.31	1.80	1.21	1.64
Wager-Deer A	0.757 ± 0.319	0.243	0.429 ± 0.155	0.225
F	0.215 ± 0.139	0.563	0.430 ± 0.087	0.476
M	0.028 ± 0.020	0.193	0.141 ± 0.081	0.239
Simpson F	97.01	43.94	51.15	29.53
M	90.16	74.44	75.23	61.08
Rittmann s	3.059	0.97	2.133	1.22
Kuno SI	2.70	19.37	14.30	25.63
Thornton-Tuttle DI	78.7	44.8	58.8	41.8
Sugimura θ	14.86	1.79	4.34	1.51
Sugimura SWS	16.68	9.28	11.82	10.74
Correlations				
SiO_2 − alk	0.432		0.262	
SiO_2 − K_2O	−0.087		0.099	
Na_2O − K_2O	−0.619		−0.310	
FeO − MgO	−0.066		0.681	
TiO_2 − FeO/MgO	−0.038	0.302		

[a]See Borcos et al. (1973) and Lang (1976).

Phase equilibria of minerals in these inclusions indicate that the basaltic magma originated at a depth of about 70–80 km, in good agreement with geophysical data that suggest a similar depth for the base of the lithosphere (Stegena and Horváth, 1982).

REE data from the Calimani–Harghita Mountains (East Carpathians) show a fractionated LREE and a less fractionated heavy REE distribution that is typical for basalts, basaltic andesites, and andesites (Peccerillo and Taylor, 1976). In contrast to lavas in the Pannonian basin, no marked Eu anomaly is present in the volcanic rocks of the East Carpathians. The average total REE content of these rocks is very similar to that of calcalkaline lavas in island arc complexes. Strong fractionation is characteristic for the light and heavy lanthanides in the rhyolites and dacites of the Calimani–Harghita mountains (Peccarillo and Taylor, 1976), but according to their data, there is no Eu anomaly in these rocks.

Peccarillo and Taylor (1976) suggest that the andesites and basalts of the Calimani–Harghita mountains formed by partial melting of garnet-pyroxenite mantle material, derived by mixing of the peridotic mantle with partial melt from the subducted lithospheric slab. These authors conclude that the acidic magmas formed as primary melt from the subducted material. This interpretation is consistent with the appearance of the East Carpathian acid volcanics mainly during the first volcanic phase (early Pannonian, *s. l.*) and with their negligible volume relative to the intermediate volcanic rocks.

DISCUSSION

The geochemical analysis of orogenic calcalkaline magmatic rocks by Ewart and Le Maitre (1980), based on 7000 samples from island arcs and active continental margins, indicates that the change of SiO_2 and K_2O in the volcanic rocks is characteristic of their tectonic setting and the nature of the lithosphere. They divided these rock samples into four suites: (1) low potassium content, (2) normal calcalkaline, (3) high potassium content, and (4) shoshonitic. According to Ewart and Le Maitre

Table 7. Average Chemical Composition and Some Petrochemical Indices of Volcanic Phases in the Apuseni Mountains

	Apuseni Mountains[a]			
	Inital phase Upper Badenien 22 samples	I. phase Lower Sarmatian 8 samples	II. phase Upper Sarmatian 107 samples	III. phase Pannonian (*s. l.*) 13 samples
SiO_2	66.06 ± 7.68	57.71 ± 0.96	57.91 ± 3.47	53.53 ± 6.30
TiO_2	0.38 ± 0.45	0.37 ± 0.54	0.64 ± 0.50	1.72 ± 1.30
Al_2O_3	15.29 ± 3.69	19.00 ± 2.72	18.51 ± 2.04	16.48 ± 2.83
Fe_2O_3	2.55 ± 1.71	3.75 ± 1.04	3.13 ± 1.78	4.04 ± 2.36
FeO	1.62 ± 1.47	3.25 ± 0.97	2.63 ± 1.56	4.24 ± 2.80
MnO	0.03 ± 0.06	0.03 ± 0.09	0.17 ± 0.18	0.39 ± 0.36
MgO	1.48 ± 1.21	2.38 ± 1.09	2.37 ± 1.21	5.00 ± 2.93
CaO	3.49 ± 2.39	6.04 ± 1.45	6.01 ± 2.05	7.34 ± 2.65
Na_2O	3.49 ± 1.40	4.16 ± 1.56	2.76 ± 0.93	2.87 ± 1.26
K_2O	2.30 ± 1.27	1.38 ± 0.59	1.95 ± 1.41	1.22 ± 0.92
$SiO_2/Na_2O + K_2O$	11.11	10.41	12.30	13.09
SiO_2/K_2O	28.72	41.82	29.70	43.88
Na_2O/K_2O	1.52	3.01	1.42	2.35
Wager-Deer A	0.540 ± 0.178	0.368 ± 0.069	0.371 ± 0.104	0.273 ± 0.173
F	0.339 ± 0.154	0.475 ± 0.074	0.450 ± 0.085	0.462 ± 0.132
M	0.121 ± 0.086	0.157 ± 0.056	0.179 ± 0.073	0.265 ± 0.118
Simpson F	74.78	47.48	43.93	33.44
M	72.95	74.62	70.84	62.34
Rittmann s	1.713 ± 1.024	2.211	1.704 ± 1.190	1.872 ± 1.893
Kuno SI		15.95	18.45	28.78
Thornton-Tuttle DI	68.8	56.7	55.1	42.9
Sugimura θ	7.22	2.56	2.78	1.62
Sugimura SWS	11.90	12.35	10.85	12.36
Correlations				
SiO_2 − alk	0.024	0.107	0.140	0.545
SiO_2 − K_2O	0.446	0.093	0.251	0.608
Na_2O − K_2O	−0.371	−0.419	−0.235	0.615
FeO − MgO	0.478	−0.175	0.353	0.640
TiO_2 − FeO/MgO	−0.54	−0.275	−0.147	−0.196

[a]See Radulescu (1960) and Borcos (1976).

(1980), island arc lavas belong to the low potassium and normal calcalkaline suites, while active continental margin magmatites belong to the high potassium suite. Figure 8 shows the distribution of Neogene calcalkaline lavas from the Pannonian and East Carpathian region in terms of these four divisions. Volcanic rocks from the northern margin of the Pannonian basin fall into the suite of high potassium content; that is, they show an affinity with active continental margin volcanism. In contrast, most of the data from the East Carpathian region plot in the normal calcalkaline field. Thus, they represent a transitional magmatic type between that of island arcs and active (convergent) continental margins.

The most distinctive difference between the volcanism of the Pannonian and East Carpathian regions is the eruption of acidic magmas in the Pannonian basin at the same time as the eruption of intermediate magmas.

A similar phenomena was observed in northwest Sardinia, (crustal thickness of 30 km), where calcalkaline (mostly andesitic) volcanics erupted in alternation with dacite and rhyolite ignimbrites between 29 and 13 Ma (Coulon et al., 1980). The acidic and calcalkaline magmatism in Sardinia was also followed by eruption of alkali basalts. On the basis of major and trace element geochemistry, and on the experimental work of Wyllie (1977), Coulon et al. (1980) have suggested that the acidic magmas are derived from intracrustal melting of tonalites and are of different origin from the intermediate magmas. Wyllie (1977) found that if the water content of tonalite is 2%, melting can occur at around 5 kb, or an equivalent depth of about 20 km, and 1050° C. More acidic crustal rocks, such as granite and granodiorite, melt at slightly lower temperatures. Coulon et al (1980) state that crystallization of the intermediate magmas in Sardinia occurred at low pressures (<5 kbars). They suggest that a secondary magma chamber containing the andesitic magma raised temperatures within the crust enough to cause partial or total melting and formation of an acidic magma.

The same scenario may be applicable to the Pannonian basin. Intense intermediate volcanism may have heated the crust beneath the Pannonian basin enough to create crustal melting. This acidic melt might interact with basalt parent magmas and produce contaminated hybrid lavas. This hypothesis is consistent with the increase in acidity and in potassium content in the intermediate magmas from the onset of volcanism until late Sarmatian time. After the cessation of acid volcanism, the magmas being erupted became abruptly more basic and lower in

Table 8. Average Chemical Composition and Some Petrochemical Indices of Volcanic Phases in the Calimani–Gurghiu–Harghita Mountains[a]

	Calimani		Gurghiu		Harghita	
	I. phase Upper Sarmatian–Lower Pannonian (*s.l.*) 41 samples	II. phase Upper Pannonian (*s.l.*)–Pleistocene 26 samples	I. phase Upper Sarmatian Lower Pannonian (*s.l.*) 6 samples	II. phase Upper Pannonian (*s.l.*)–Pleistocene 31 samples	I. phase Upper Sarmatian–Lower Pannonian (*s.l.*) 5 samples	II. phase Upper Pannonian (*s.l.*)–Pleistocene 46 samples
SiO_2	56.53 ± 5.37	57.00 ± 3.20	57.82 ± 4.10	56.75 ± 4.10	57.21 ± 2.15	59.11 ± 3.03
TiO_2	0.78 ± 0.36	0.87 ± 0.26	0.70 ± 0.32	0.82 ± 0.21	0.70 ± 0.15	0.68 ± 0.28
Al_2O_3	18.27 ± 1.90	18.66 ± 1.07	18.61 ± 1.41	18.66 ± 1.48	18.62 ± 0.64	18.39 ± 1.52
Fe_2O_3	3.90 ± 1.71	2.62 ± 1.33	4.76 ± 1.80	4.63 ± 1.48	1.61 ± 0.44	3.06 ± 1.93
FeO	2.98 ± 2.26	3.54 ± 1.14	2.13 ± 1.15	2.35 ± 0.97	3.95 ± 0.30	2.14 ± 1.41
MnO	0.15 ± 0.12	0.13 ± 0.03	0.11 ± 0.06	0.17 ± 0.16	0.10 ± 0.01	0.13 ± 0.11
MgO	3.61 ± 2.05	3.60 ± 1.26	2.86 ± 0.83	3.13 ± 1.09	4.43 ± 1.40	2.90 ± 1.16
CaO	7.37 ± 1.84	6.67 ± 1.19	6.97 ± 1.09	6.98 ± 1.06	7.52 ± 0.75	6.30 ± 1.38
Na_2O	3.19 ± 0.63	3.70 ± 0.63	3.38 ± 0.26	3.66 ± 0.49	3.14 ± 0.15	3.67 ± 0.64
K_2O	1.46 ± 0.73	2.07 ± 0.49	1.29 ± 0.30	1.33 ± 0.24	1.67 ± 0.41	2.07 ± 0.79
$SiO_2/Na_2O + K_2O$	12.16	9.88	12.38	11.37	11.89	10.30
SiO_2/K_2O	38.72	27.54	44.82	42.68	34.26	28.56
Na_2O/K_2O	2.18	1.79	2.62	2.75	1.88	1.77
Wager-Deer A	0.330 ± 0.131	0.379 ± 0.085	0.335 ± 0.083	0.336 ± 0.062	0.340 ± 0.059	0.419 ± 0.095
F	0.449 ± 0.076	0.392 ± 0.082	0.470 ± 0.061	0.460 ± 0.047	0.356 ± 0.018	0.376 ± 0.084
M	0.221 ± 0.095	0.229 ± 0.063	0.195 ±0.035	0.204 ± 0.059	0.304 ± 0.065	0.205 ± 0.068
Simpson F	38.64	46.60	34.49	41.37	39.01	47.67
M	65.58	63.11	30.66	69.66	53.62	64.19
Rittmann s	1.804	2.435 ± 0.567	1.527 ± 0.301	1.880 ± 0.557	1.650 ± 0.239	2.123 ± 0.786
Kuno SI	23.84	23.18	19.83	20.72	30.87	20.95
Thornton-Tuttle DI	49.5	64.6	53.5	69.8	50.4	57.3
Sugimura θ	2.40	3.10	2.72	2.61	2.64	3.78
Sugimura SWS	11.98	13.99	11.27	10.98	13.09	13.23
Correlations						
SiO_2 − alk	0.544	0.683	0.757	0.238	0.546	0.469
$SiO_2 - K_2O$	0.217	0.569	0.827	0.399	0.750	0.505
$Na_2 - K_2O$	0.067	0.280	0.779	0.215	−0.435	0.414
FeO − MgO	0.744	0.300	0.812	0.264	0.727	0.212
TiO_2 − FeO/MgO	−0.254	−0.255	−0.390	−0.133	−0.139	0.223

[a]See Radulescu (1960).

Table 9. Some Average Petrochemical Characteristics of the Neogene Acidic Volcanic Phases in the Carpathian Basin

						Tokaj Mts.				Ukrainian Carpathians				Gutii Mts.	Apuseni Mts.
	Eggenburgian–Ottnangian Boundary Lower Rhyolite Tuff Horizon 10 samples	Karpatian–Badenain Boundary Middle Rhyolite Tuff Horizon 12 samples	Lower and Upper Sarmatian Boundary Upper Rhyolite Tuff Horizon 10 samples	Mátra Mts. IV. phase Upper Sarmatian(?) 6 samples	Banska Stiavnica–Kremnicka Mts. III. phase Upper Sarmatian 29 samples	I. phase Upper Badenian 19 samples	II. phase Lower Sarmatian 23 samples	III. phase Upper Sarmatian 86 samples	IV. phase Upper Sarmatian 100 samples	I. phase Lower Sarmatian 7 samples	II. phase Upper Badenian 19 samples	III. phase Sarmatian 12 samples	IV. phase Pannonian 5 samples	Initial phase Upper Badenian 6 samples	Initial phase Upper Badenian 22 samples
SiO_2	71.23	67.01	72.09	70.56	71.75	66.68	70.00	72.91	71.69	67.37	67.45	67.75	64.72	70.88	66.06
TiO_2	0.22	0.37	0.08	0.50	0.23	0.11	0.16	0.08	0.27	0.32	0.30	0.28	0.50	0.09	0.38
Al_2O_3	13.99	14.87	12.92	14.52	12.86	14.36	13.77	12.52	13.57	13.99	12.20	13.35	16.24	14.62	15.29
Fe_2O_3	0.41	1.45	1.10	1.36	1.97	0.53	1.12	0.81	1.44	2.83	2.87	3.88	3.72	1.80	2.55
FeO	0.35	0.72	0.20	1.08	0.87	1.46	0.87	0.47	0.79	0.97	1.37	0.53	0.77	0.95	1.62
MnO	0.03	0.04	0.01	0.04	0.08	0.08	0.07	0.04	0.08	0.05	0.08	0.06	0.04	0.01	0.03
MgO	0.37	0.84	0.03	0.14	0.81	1.33	0.72	0.42	0.53	0.64	1.01	0.80	0.81	0.33	1.48
CaO	2.50	3.32	1.14	1.93	1.60	3.16	2.33	1.41	1.73	2.79	2.26	2.25	3.15	0.28	3.49
Na_2O	1.03	2.56	1.09	3.64	1.86	1.09	2.04	2.15	2.37	2.88	3.72	2.32	2.22	2.13	3.49
K_2O	5.47	3.56	4.97	4.40	4.27	4.49	3.87	4.43	3.86	2.63	2.25	2.24	1.68	6.97	2.30
$SiO_2/Na_2O + K_2O$	10.96	10.95	11.90	8.78	11.70	11.95	11.84	11.08	11.51	12.23	11.30	14.86	16.59	7.79	11.41
SiO_2/K_2O	13.02	18.82	14.51	16.04	16.80	14.85	18.09	16.46	18.57	25.62	29.98	30.25	38.52	10.17	28.72
Na_2O/K_2O	0.19	0.72	0.22	0.83	0.44	0.24	0.52	0.49	0.61	1.10	1.05	1.04	1.32	0.31	1.52
Wager-Deer A	0.852	0.670	0.820	0.763	0.634	0.626	0.683	0.793	0.698	0.554	0.548	0.486	0.410	0.757	0.540
F	0.099	0.238	0.176	0.223	0.285	0.223	0.227	0.155	0.249	0.382	0.361	0.422	0.500	0.215	0.339
M	0.049	0.092	0.004	0.014	0.082	0.150	0.089	0.052	0.059	0.064	0.091	0.091	0.090	0.028	0.121
Rittmann s	1.500	1.560	1.260	2.376	1.396	1.408	1.390	1.495	1.491	1.276	1.561	0.840	0.887	3.059	1.713
Simpson F	72.22	64.83	84.17	80.64	79.30	63.84	71.72	82.35	78.26	66.39	72.53	67.95	55.32	97.01	74.78
M	67.26	72.09	97.74	94.57	77.80	59.69	73.43	75.29	30.79	85.59	80.76	84.34	85.85	90.16	72.95
Kuno SI	4.78	9.20	0.41	1.32	8.28	14.94	8.35	5.07	8.51	6.43	9.00	8.19	8.80	2.70	68.8
															7.22
Thornton-Tuttle DI	86.2	75.90	82.5	83.8	83.2	73.90	77.1	83.7	84.10	85.2	84.40	76.9	69.10	78.70	11.90
Sugimura θ	13.15	9.88	13.64	13.05	11.80	7.65	9.87	13.63	11.01	8.02	10.01	7.09	4.26	14.86	13.05
Sugimura SWS	12.07	11.86	10.94	14.61	11.84	11.37	11.36	12.26	11.74	12.78	11.76	9.01	7.83	16.68	

Table 10. Average Chemical Composition and Some Petrochemical Indices of Different Regional Occurrences of the Pliocene–Pleistocene Alkali Basalts[a]

	Basalts of Transdanubia 183 samples	Basalts of Salgótarján District 36 samples	Basalts of Banska Stiavnica District 8 samples	Basalts of Central Slovakia 10 samples
SiO_2	46.91 ± 2.58	46.77 ± 1.87	47.93 ± 2.96	45.43 ± 2.00
TiO_2	2.04 ± 0.58	1.76 ± 0.51	1.60 ± 0.39	1.43 ± 0.37
Al_2O_3	18.86 ± 2.19	17.60 ± 1.44	15.80 ± 1.90	15.34 ± 2.03
Fe_2O_3	4.63 ± 2.52	3.61 ± 1.20	4.60 ± 1.16	4.46 ± 1.41
FeO	5.70 ± 1.94	5.66 ± 1.29	6.40 ± 1.72	6.07 ± 0.91
MnO	0.85 ± 0.08	0.15 ± 0.07	0.24 ± 0.16	0.25 ± 0.06
MgO	7.42 ± 1.99	5.87 ± 1.33	6.85 ± 3.70	7.14 ± 1.48
CaO	8.80 ± 1.49	9.40 ± 1.12	8.68 ± 1.02	10.51 ± 1.61
Na_2O	3.36 ± 0.96	4.17 ± 0.72	3.94 ± 0.93	3.38 ± 0.73
K_2O	1.96 ± 0.63	2.20 ± 0.47	1.79 ± 0.58	1.82 ± 0.54
$SiO_2/Na_2O + K_2O$	8.82	7.34	10.35	8.74
SiO_2/K_2O	23.93	21.26	26.78	24.96
Na_2O/K_2O	1.71	1.90	2.20	1.86
Wager-Deer A	0.239 ± 0.064	0.297 ± 0.043	0.253 ± 0.083	0.288 ± 0.049
F	0.433 ± 0.065	0.432 ± 0.036	0.467 ± 0.059	0.461 ± 0.064
M	0.327 ± 0.070	0.271 ± 0.049	0.280 ± 0.106	0.251 ± 0.059
Rittmann s	7.24	10.76	6.60	11.13
Simpson F	37.23	39.39	40.58	33.09
M	56.73	61.22	61.62	59.79
Kuno SI	32.16	48.79	29.10	31.21
Thornton-Tuttle DI	33.9	57.6	49.1	30.9
Sugimura θ	−0.02	−0.10	0.34	−0.54
Correlations				
SiO_2 − alk	0.194	0.006	0.716	0.376
$SiO_2 - K_2O$	0.137	−0.010	0.767	0.039
$Na_2O - K_2O$	0.302	0.540	0.625	0.579
FeO − MgO	0.00	0.246	0.275	−0.251
TiO_2 − FeO/MgO	0.110	−0.118	0.361	0.590

[a]See Jugovics (1978).

potassium. This change in character of the magma is roughly synchronous with the very fast initial subsidence within the Pannonian basin (Royden et al., 1982).

Geophysical data indicate that the deep structure of the Pannonian basin complex is very different from that of the European foreland (Stegena and Horváth, 1982). Within the basin area the deep structure is not uniform, and there are significant structural and tectonic differences between the Pannonian and Transylvanian basins (Horváth and Royden, 1981). For example, during the Miocene the Pannonian basin region underwent major extension and thinning of both the crust and upper mantle. In contrast, the Transylvanian basin region seems to have been largely unaffected by extensional tectonics. Both regions probably sat above the subducted slab of the European plate in Miocene times. These observations may provide some insight into the different character of the Miocene to Quaternary lavas erupted in these two regions.

The relationship between the structural and tectonic development of the Carpathian–Pannonian region has been discussed by many authors (Szádeczky-Kardoss, 1967; Szádeczky-Kardoss, 1976; Bleahu et al., 1973; Boccaletti et al., 1973; Radulescu and Sandulescu, 1973; Stegena et al., 1975; Channell and Horváth, 1976; Balla, 1980, and Konecny and Lexa, 1974). Their viewpoints may be divided into two categories: (1) that the Carpathian volcanism and tectonics are best explained by an island arc or a marginal basin model, or (2) that it is best explained by a mantle diapir beneath the basin region. However, the changes of magma chemistry in space and time do not make these models mutually exclusive. The data presented in this paper suggest that volcanism in the Pannonian basin can be largely explained by a mantle diapir model or at least by upwelling of hot asthenosphere with related melting of crustal rocks. They further indicate that in the East Carpathians, volcanism may be largely related to the presence of a subducted slab. These data are thus consistent with the Miocene tectonic events inferred for the Pannonian and East Carpathian (or Transylvanian) region.

The geochemical data indicate that the main period of mantle diapirism lasted until the end of Sarmatian time, at which point very fast subsidence began in the Pannonian basin. Such a picture of magma chemical evolution may help to explain the genetic differences between the acid, intermediate, and alkali basaltic volcanic rocks of this region and the space-time evolution of volcanism in the Carpathian basins.

Table 11. Average Chemical Characteristics of Different Volcanic Phases According to Their Geological Ages

Mts.	Phase	Number of Samples	SiO_2	K_2O	SiO_2/(Na_2O + K_2O)	SiO_2/K_2O	Na_2O/K_2O	Wager-Deer			Simpson		Kuno
								A	F	M	F	M	SI
Karpatian													
Dunazug	I.	9	61.14	0.93	10.36	21.38	1.06	0.513	0.393	0.094	55.36	79.73	10.27
Mátra	I.	25	53.79	1.96	12.45	27.44	1.20	0.337	0.533	0.130	36.21	80.30	13.12
average		34	55.74	1.69	11.90	25.84	1.16	0.380	0.500	0.120	41.28	80.15	12.37
Lower Badenian													
Dunazug	II.	17	55.49	2.29	10.51	24.23	1.31	0.353	0.461	0.186	44.56	73.01	18.80
Börzsöny	I.	18	57.51	2.32	10.51	24.79	1.36	0.412	0.449	0.139	49.88	75.75	14.29
Börzsöny	II.	55	56.06	2.16	10.78	25.55	1.41	0.260	0.497	0.144	43.26	77.37	15.83
Börzsöny	III.	21	58.18	2.26	10.41	25.74	1.47	0.404	0.468	0.127	48.15	78.74	12.72
Mátra	II.	109	56.38	1.96	12.67	28.77	1.27	0.309	0.501	0.191	37.61	71.52	19.29
Cserhát		50	54.87	1.99	14.59	27.57	1.39	0.317	0.487	0.197	37.45	72.84	18.38
Ukr. Carp. (Acid.)	II.	7	67.37	2.63	12.23	25.62	1.10	0.460	0.317	0.223	90.92	57.66	23.10
Ukr. Carp. (And.)	I.	10	58.75	1.86	12.29	31.59	1.57	0.346	0.346	0.181	43.33	124.23	18.87
average		287	56.56	2.12	12.19	27.23	1.35	0.312	0.500	0.188	42.08	74.77	17.72
Upper Badenian													
Mátra	III.	62	55.37	2.02	12.41	26.42	1.13	0.304	0.499	0.197	36.34	71.84	19.63
Tokaj (And.)	I.	57	59.87	2.76	12.24	21.69	0.77	0.406	0.401	0.193	48.51	67.17	19.63
Tokaj (Acid.)	I.	19	66.68	4.49	11.95	14.85	0.24	0.626	0.223	0.150	63.84	59.69	14.94
Bansk. Stiavn.	I.	25	57.33	1.94	16.76	29.55	1.28	0.281	0.473	0.246	37.10	66.10	24.67
Ukr. Carp. (Acid.)	III.	13	67.45	2.25	11.30	29.98	1.65	0.548	0.361	0.091	72.53	80.76	9.00
Apuseni	(Init.)	22	66.06	2.30	11.41	28.72	1.52	0.540	0.339	0.121	74.78	72.95	10.12
Gutii	(Init.)	6	70.88	6.97	7.79	10.17	0.31	0.757	0.212	0.028	97.01	90.16	2.70
average		204	60.30	2.64	12.54	24.40	1.02	0.410	0.390	0.200	50.63	69.93	17.61
Lower Sarmatian													
Mátra	IV.	6	70.56	4.40	8.78	16.04	0.83	0.763	0.223	0.014	80.64	94.57	1.32
Tokaj (And.)	II.	227	60.52	2.24	12.58	27.02	1.15	0.374	0.417	0.210	44.33	66.58	21.45
Tokaj (Acid.)	II.	23	70.00	3.87	11.84	18.09	0.52	0.683	0.227	0.089	71.72	73.43	8.35
Bansk. Stiavn.	II.	11	59.37	2.62	11.55	22.66	0.96	0.368	0.456	0.177	45.56	71.77	18.01
Ukr. Carp. (And.)	II.	3	69.59	3.33	11.01	20.90	0.87	0.739	0.217	0.044	67.57	86.23	4.48
Apuseni	I.	8	57.71	1.38	10.42	41.82	3.01	0.368	0.475	0.157	47.48	74.62	15.95
Gutii	I.	21	56.15	1.28	15.68	43.87	1.80	0.243	0.563	0.193	43.94	74.44	19.37
average		299	61.12	2.34	12.55	27.46	1.18	0.400	0.410	0.190	47.50	68.80	19.45
Upper Sarmatian													
Tokaj (Acid.)	III.	86	72.91	4.43	11.08	16.46	0.49	0.793	0.155	0.052	82.35	75.29	5.07
Tokaj (Acid.)	IV.	18	71.69	3.86	11.51	18.57	0.61	0.698	0.249	0.059	78.26	80.79	8.51
Bansk. Stiavn.	III.	29	71.75	4.27	11.70	16.80	0.44	0.634	0.285	0.082	79.30	77.80	8.28
Ukr. Carp. (Acid.)	IV.	12	67.75	2.24	14.86	30.25	1.04	0.486	0.422	0.091	67.95	84.34	8.19
Apuseni	II.	107	57.91	1.95	12.30	29.70	1.42	0.371	0.450	0.179	43.93	70.84	18.45
average		252	66.07	3.37	11.88	22.93	0.91	0.570	0.315	0.115	64.71	74.51	11.51
Upper Sarmatian–Lower Pannonian													
Gutii	II.	12	60.27	2.50	10.90	24.11	1.80	0.429	0.430	0.141	51.15	75.23	14.30
Calimani	I.	41	56.53	1.46	12.16	38.72	2.18	0.330	0.449	0.221	38.64	65.58	23.84
Gurghiu	I.	6	57.82	1.29	12.38	44.82	2.62	0.335	0.470	0.175	34.49	70.66	19.83
Harghita	I.	5	57.21	1.67	11.89	34.26	1.88	0.340	0.356	0.304	39.01	53.62	30.87
Apuseni	III.	36	57.02	1.91	12.40	29.85	1.41	0.320	0.472	0.209	39.65	68.95	21.40
average		100	57.26	1.76	12.09	33.92	1.87	0.309	0.501	0.190	40.27	67.65	21.42
Lower Pannonian–Pleistocene													
Selmec-Körmöc	IV.	16	50.46	1.55	11.60	32.55	1.81	0.237	0.431	0.332	32.95	56.46	33.62
Kárpátalja S	V.	5	64.72	1.68	16.59	38.52	1.32	0.410	0.500	0.090	55.32	85.85	8.80
Kárpátalja A	IV.	28	61.12	2.52	13.34	23.88	0.79	0.434	0.399	0.167	44.81	69.09	19.43
Gutin	III.	45	54.54	1.42	14.54	38.41	1.64	0.225	0.476	0.239	29.53	61.08	25.63
Apuseni	III.	13	53.53	1.22	13.09	43.88	2.35	0.273	0.462	0.265	33.44	62.34	28.78
Kelemen	II.	26	57.00	2.07	9.88	27.54	1.79	0.379	0.392	0.229	46.60	63.11	23.18
Görgény	II.	31	56.75	1.33	11.37	42.68	2.75	0.336	0.460	0.204	41.37	69.66	20.72
Harghita	II.	46	59.11	2.07	10.30	28.56	1.77	0.419	0.376	0.203	47.67	64.19	20.95
average		214	54.44	1.71	11.91	32.87	1.75	0.347	0.421	0.232	38.32	61.87	21.78
Pliocene–Pleistocene Alkali Basalts													
Transdanubia		183	46.91	1.96	8.82	23.93	1.71	0.239	0.433	0.327	37.23	56.73	32.16
Salgótarján		36	46.77	2.20	7.34	21.26	1.90	0.297	0.432	0.271	39.39	61.22	48.79
Bansk. Stiavn.		8	47.93	1.79	10.35	26.78	2.20	0.253	0.467	0.280	40.58	61.68	29.10
Central Slovakia		10	45.43	1.82	8.74	24.96	1.86	0.288	0.461	0.251	33.09	59.79	31.24
average		237	46.86	1.98	8.64	23.66	1.76	0.250	0.440	0.310	37.50	57.71	34.54

Table 12. Average Chemical Characteristics of Andesitic Samples of Different Volcanic Phases According to Their Geological Ages

Mts.	Phase	Number of Samples	SiO_2	K_2O	SiO_2/alk	SiO_2/K_2O	Na_2O/K_2O	Kuno SI	Simpson F	Simpson M	Sz	C
Carpathian												
Dunazug	I.	4	55.22	2.30	10.46	24.27	1.30	15.58	42.04	75.08	171.09	41.41
Mátra	I.	16	56.41	2.16	11.88	26.12	1.20	13.07	39.95	51.20	178.59	38.86
average		20	56.17	2.19	11.60	25.75	1.22	13.57	40.36	55.98	177.09	39.37
Lower Badenian												
Dunazug	II.	17	55.49	2.29	10.51	24.23	1.00	19.05	44.57	70.94	190.20	41.23
Börzsöny	I.	15	56.03	2.27	10.38	25.38	1.43	14.67	42.69	75.29	183.95	39.95
Börzsöny	II.	55	55.88	2.26	10.41	25.74	1.47	11.89	48.15	78.95	187.52	40.68
Mátra	II.	107	56.33	1.98	12.60	28.45	1.26	19.53	37.59	68.97	162.51	35.59
Cserhát		45	54.67	2.01	11.18	27.20	1.43	18.23	38.38	49.57	165.19	36.13
Ukr. Carp. (And.)	I.	8	56.91	1.71	12.54	33.28	1.65	19.55	39.51	46.05	138.40	30.67
average		248	57.05	2.07	11.53	27.19	1.34	17.39	40.50	69.91	170.30	37.16
Upper Badenian												
Mátra	III.	61	56.17	2.03	13.03	27.67	1.12	20.92	36.53	53.83	166.98	36.50
Tokaj (And.)	I.	38	58.26	2.42	12.50	24.07	0.93	21.12	44.55	41.56	201.81	43.59
Tokaj (Acid.)	I.	2	61.60	5.91	9.05	10.42	0.15	17.08	55.87	53.55	513.46	107.11
Bansk. Stiav.	I.	23	56.75	1.85	13.08	30.68	1.35	25.02	36.66	48.76	150.91	33.22
average		124	57.01	2.18	12.81	26.85	1.09	21.68	39.32	45.77	180.25	39.20
Lower Sarmatian												
Tokaj (And.)	II.	155	59.12	2.02	13.23	29.27	1.21	24.28	40.53	50.16	180.4	36.7
Bansk. Stiav.	II.	9	58.40	2.47	11.77	23.64	1.01	19.66	43.21	40.40	206.3	44.8
Apuseni	I.	8	57.71	1.38	10.42	41.82	3.01	15.95	47.84	74.63	123.2	25.6
Gutii	I.	21	56.15	1.28	15.68	43.87	1.80	19.21	33.46	37.64	100.0	22.9
average		193	58.70	1.93	13.64	30.08	1.29	23.17	40.21	52.86	170.5	35.1
Upper Sarmatian												
Apuseni	II.	92	57.61	1.83	12.55	31.48	1.51	18.18	42.54	71.74	149.1	33.3
Upper Sarmatian–Lower Pannonian (*s. l.*)												
Gutii	II.	8	57.38	2.45	11.14	23.42	1.10	16.95	40.10	73.53	204.49	44.14
Calimani	I.	24	56.88	1.33	12.50	42.77	2.42	22.15	37.70	66.02	104.47	23.76
Gurghiu	I.	5	56.65	1.22	12.48	46.43	2.72	20.94	43.49	69.54	94.65	21.75
Hargitha	I.	5	57.21	1.67	11.89	34.26	1.88	30.87	39.01	35.61	134.83	29.94
Ukr. Carp. (And.)	III.	32	57.13	1.92	12.18	29.76	1.41	21.20	43.91	47.35	157.16	34.49
average		74	57.05	1.72	12.16	34.73	1.82	21.68	41.12	56.97	139.4	30.89
Upper Pannonian–Pleistocene												
Bansk. Stiav.	IV.	2	54.34	3.06	8.80	17.76	1.00	23.60	39.13	63.46	258.96	55.24
Ukr. Carp. (And.)	IV.	17	57.41	2.08	12.93	27.60	1.13	22.62	38.28	46.39	171.44	37.41
Gutii	III.	45	54.54	1.42	14.54	38.41	1.64	25.63	38.93	66.24	112.51	25.39
Apuseni	III.	9	57.05	1.53	12.43	37.29	2.00	24.06	38.19	65.21	122.33	27.40
Calimani	II.	23	57.87	2.13	9.74	27.17	1.79	22.71	43.49	69.94	175.91	38.32
Gurghiu	II.	30	56.49	1.35	11.30	41.84	2.70	19.95	41.43	68.74	106.26	19.8
Harghita	II.	36	57.84	1.91	10.46	30.28	1.90	20.57	45.00	39.22	147.3	27.7
average		162	56.55	1.71	11.97	34.19	1.87	22.61	41.28	57.20	136.6	28.45

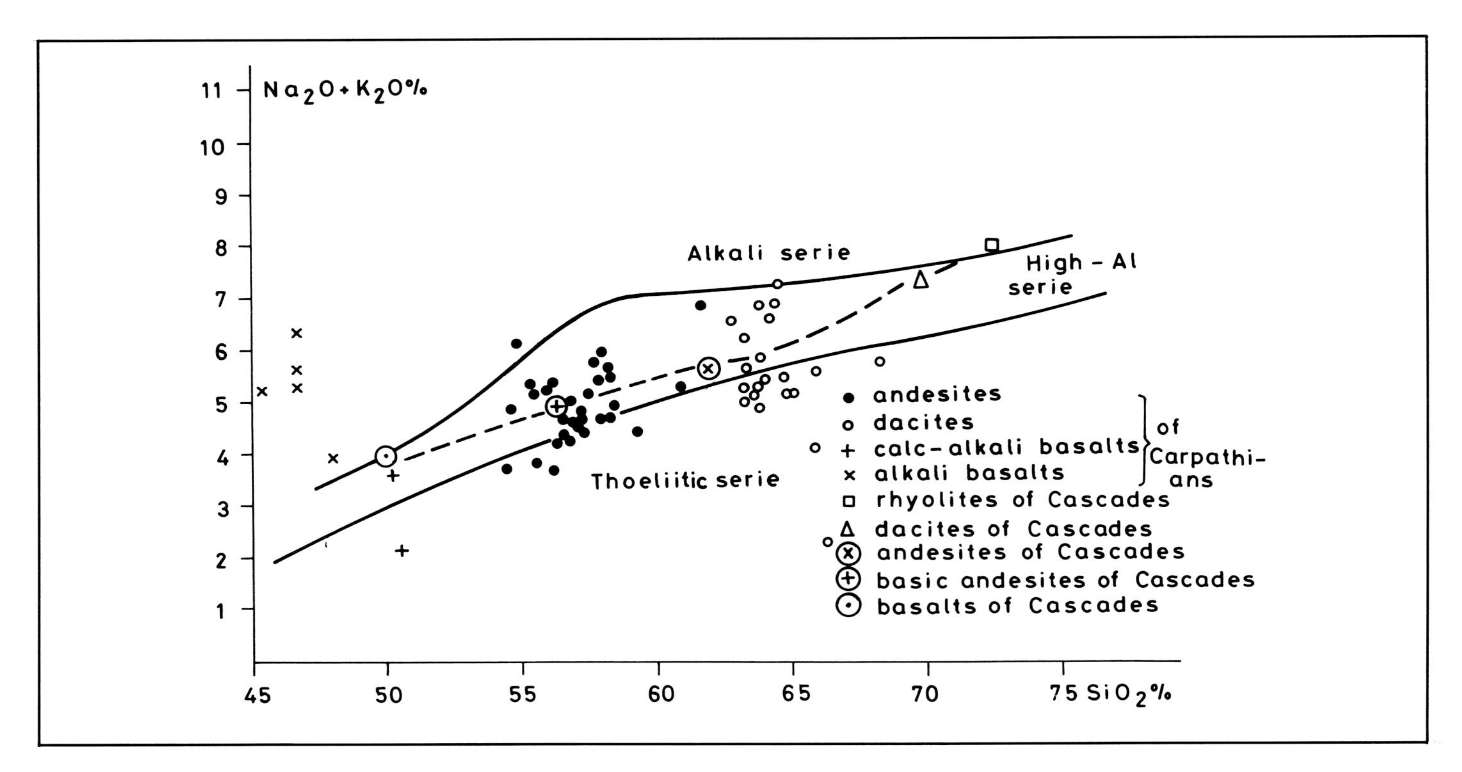

Figure 2. $Na_2O + K_2O$ versus SiO_2 for the volcanic series with the trend line of volcanics of the Cascades and the average content of the andesitic, dacitic, calcalkali and alkali basaltic volcanics of different volcanic phases of the Neogene and Quaternary volcanism in the Carpathian–Pannonian region.

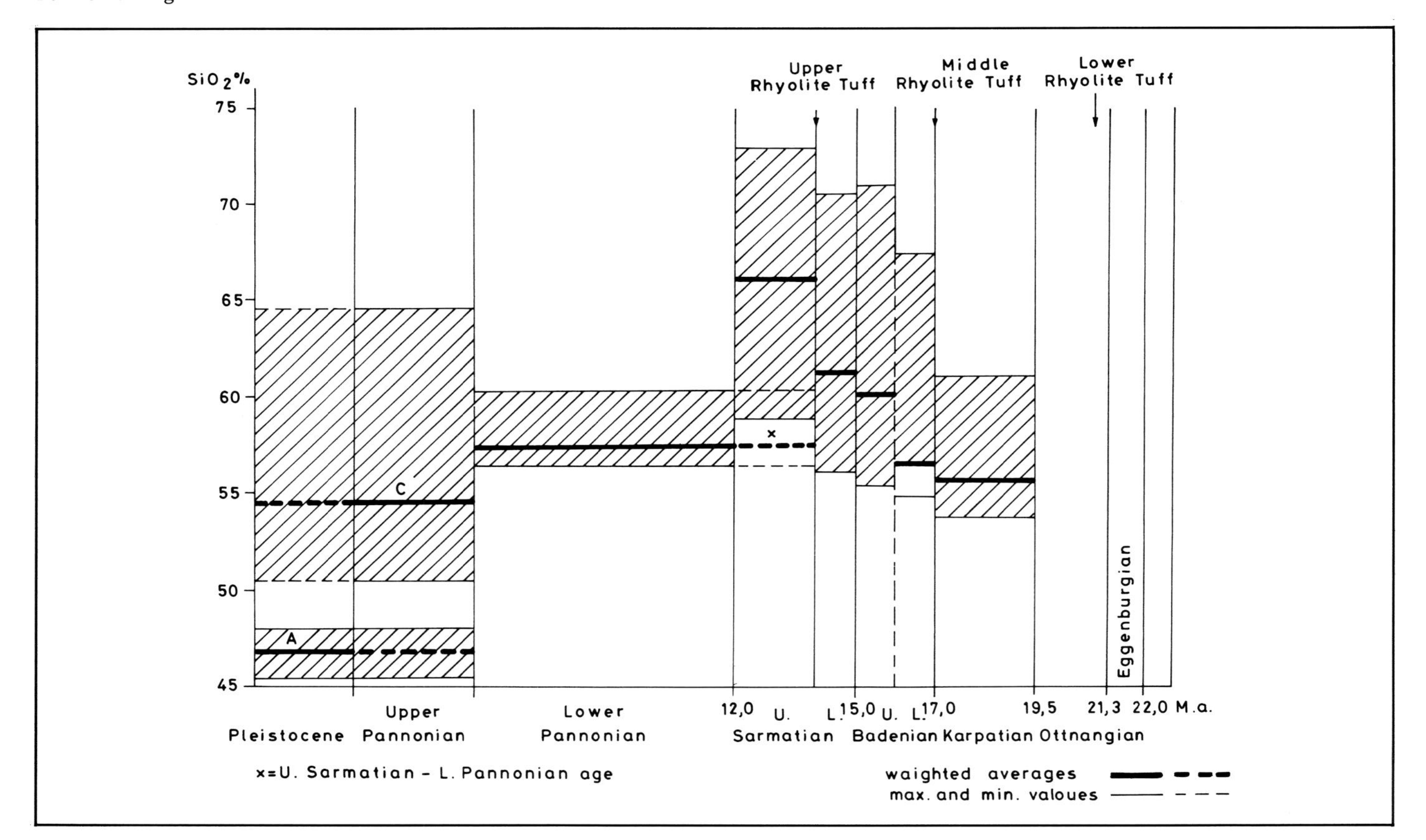

Figure 3. The average SiO_2 content as a function of time for the Neogene and Quaternary calcalkali and alkali volcanics of the Carpathian–Pannonian region. Hatched area shows ranges of values, and indicates maximum and minimum values. A is for andesites, C is for alkali basalts. "Pannonian" refers to Pannonian (*s. l.*).

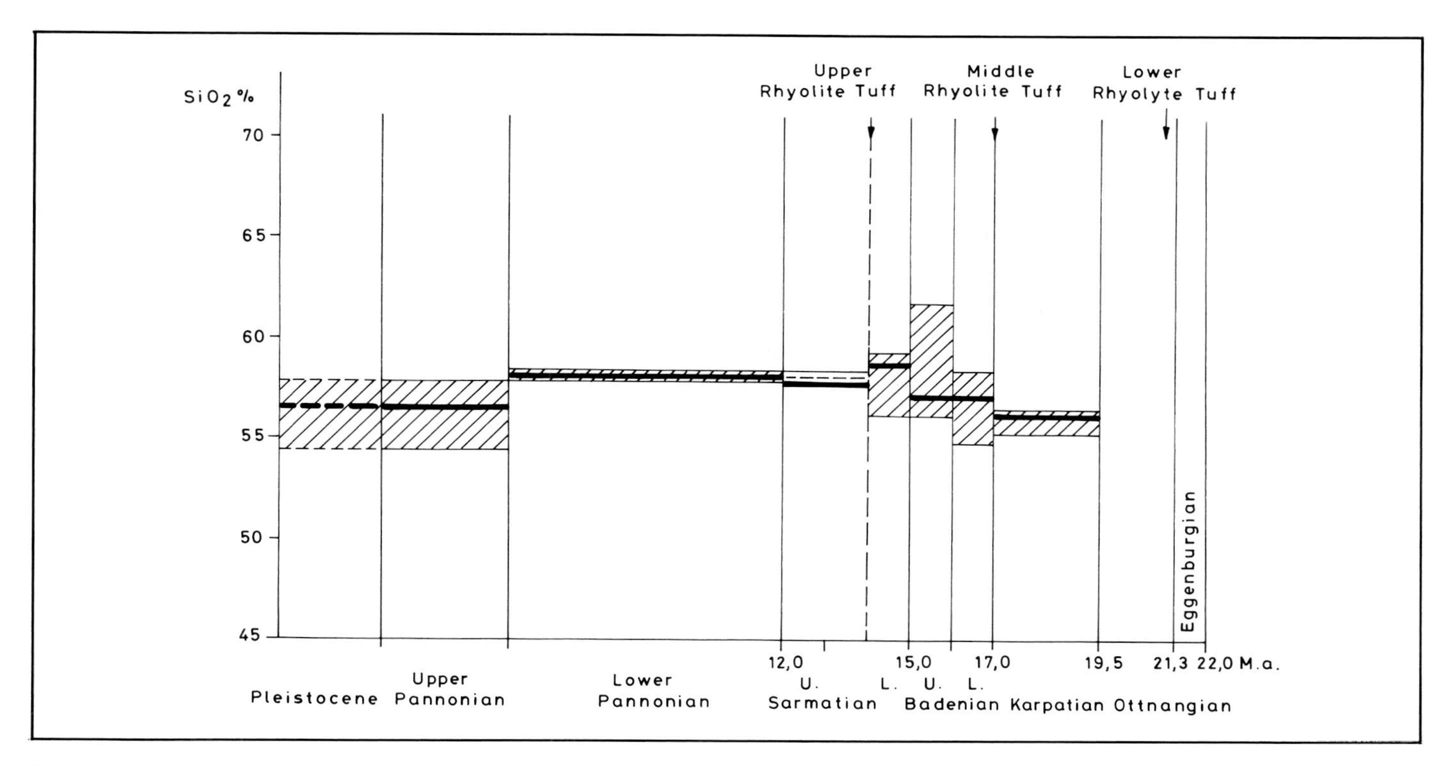

Figure 4. The average SiO_2 content as a function of time for the Neogene and Quaternary andesitic rocks of the Carpathian-Pannonian region. Hatched area shows range of values, and indicates maximum and minimum values. "Pannonian" refers to Pannonian (*s. l.*).

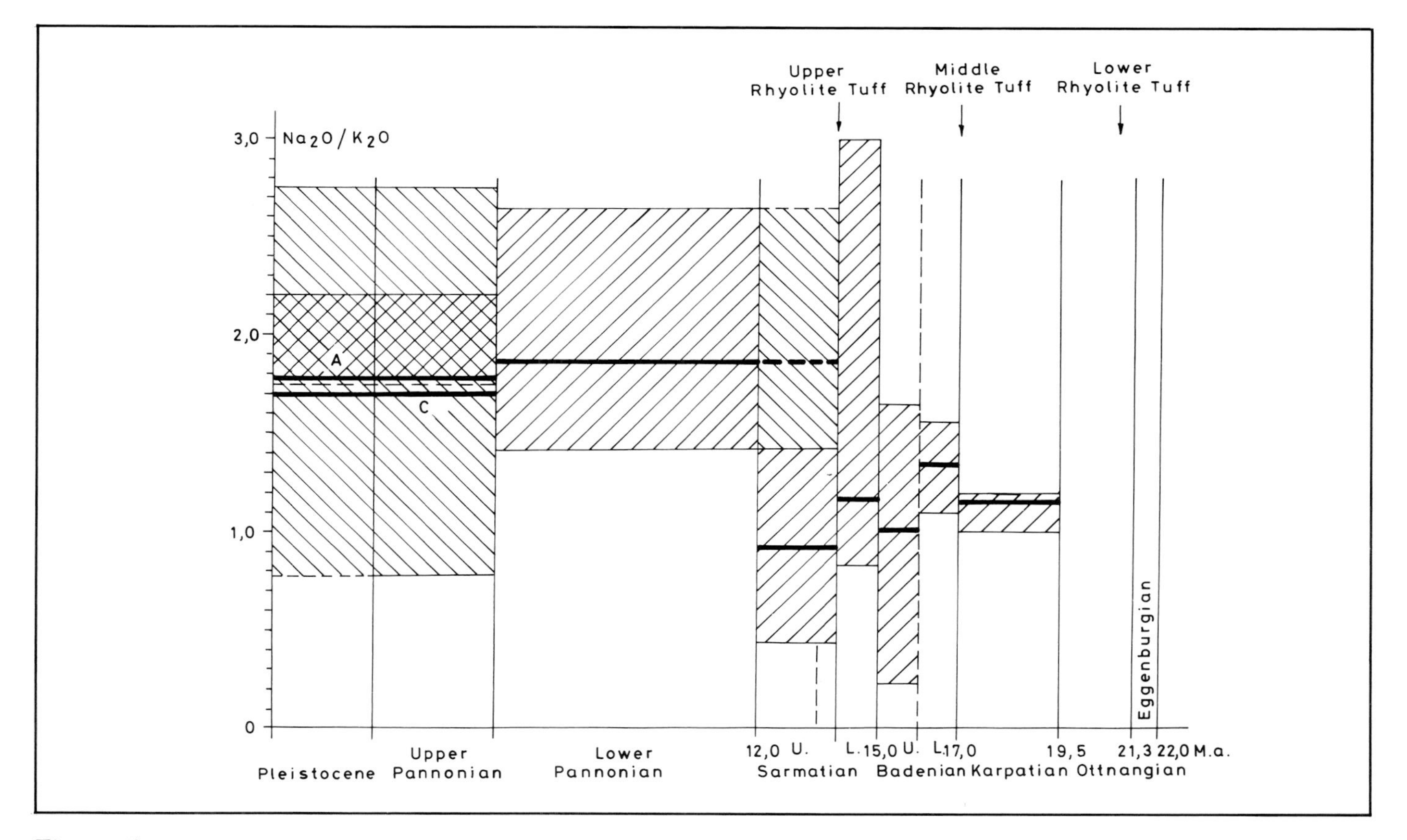

Figure 5. The average Na_2O/K_2O values of the calcalkali and alkali Neogene and Quaternary volcanic rocks in the Carpathian–Pannonian region as a function of time. A shows the average values for andesites, C shows the average value for alkali basalts. Hatched area shows range of values and indicates maximum and minimum values. "Pannonian" refers to Pannonian (*s. l.*).

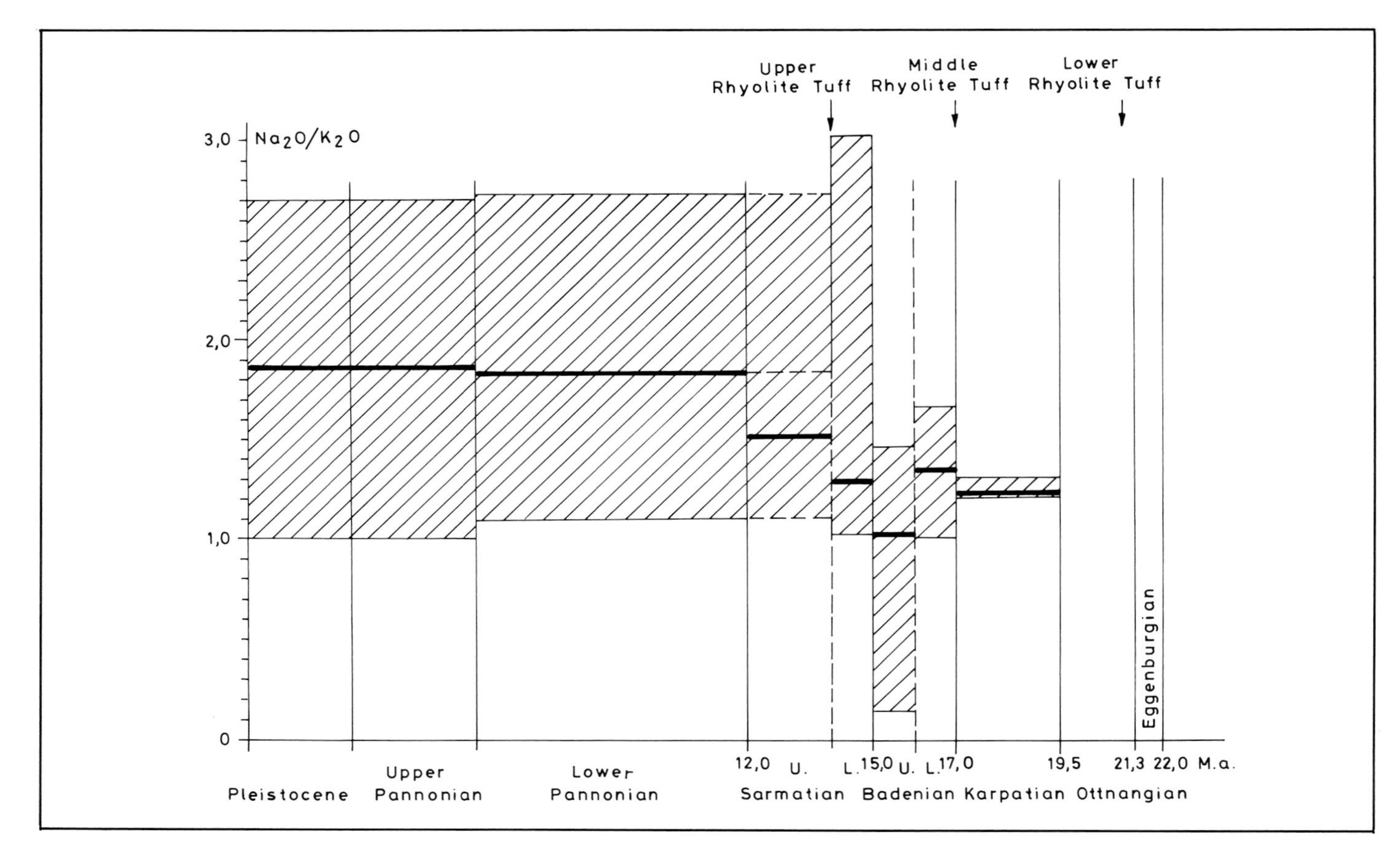

Figure 6. The average Na_2O/K_2O values of the Neogene and Quaternary andesitic rocks in the Carpathian–Pannonian region as a function of time. Hatched area shows range of values and indicates maximum and minimum values. "Pannonian" refers to Pannonian (*s. l.*).

REFERENCES

Árkai, P., 1974, Geochemical study on the early Tortonian Andesitic volcanism of the Central and Southwestern Cserhát Hills: Ann. Univ. Sci. Budap., Sec. Geol., v. 16, p. 19–31.

Balkay, B., 1960, Probleme der tektonischem Spannungverteilung im Karpatenraum: Geol. Rundschau., v. 50, p. 396–402.

Balkay, B., 1962, The tectonics of the Cenozoic volcanism in Hungary: Ann. Univ. Budap., Sec. Geol., v. III, p. 125–135.

Balla, Z., 1980, Neogene volcanites in the geodynamic reconstruction of the Carpathian Region: Geof. Közl., v. 26. 1–33.

Bleahu, M., M. Boccaletti, P. Manetti and S. Peltz, 1973, Neogene Carpathian Arc: A continental arc displaying the features of an island arc: Jour. Geophys. Res., v. 78, no. 23, p. 5025–5032.

Boccaletti, M., A. Peccarillo, and S. Peltz, 1973, Young volcanism in Calimani–Harghita Mts. (E. Carpathians): Tectonophysics, v. 19, no. 4, p. 299–315.

Borcos, M., B. Lang, S. Peltz, and N. Stan, 1973, Volcanism neogene des Montii Gutii: Rev. Roum. de Geol. Geophy. et Geogr., v. 17, no. 1, p. 81–95.

Borcos, M., 1976, Geological and metallogenic elements of the evolution of the Neogene volcanism (Apuseni Mts.): Rev. Roum. de Geol., Geophy. and Geogr., v. 20, no. 1, p. 85–103.

Burri, C. and P. Niggli, 1949, Die jungen Eruptivgesteine des Mediterranean orogens, I.,II.,III.: Schweizer Spiegel Verlag, Zurich, 680 p.

Channell, J. E. T. and F. Horváth, 1976, The African/Adriatic Promontory as a paleogeographical premise for alpine orogeny and plate movements in the Carpatho–Balkan region: Tectonophysics, v. 35, no. 1–3, p. 71–102.

Condie, K. C., 1976, Plate tectonics and crustal evolution: Pergamon Press, 288 p.

Coulon, C., J. Dostal, and C. Dupuy, 1980, Petrology and geochemistry of the ignimbrites and associated lava domes from N.W. Sardinia: Contr. Mineral. Petrol., v. 68, p. 89–98.

Danilovich, L. G., 1976, Kislij vulkanizm Karpat: Izd. Naukova, Dumka, Kiev, p. 1–182.

Embey-Isztin, A., 1976, Amphibolite-lherzolite composite xenolith from Szigliget, north of the Lake Balaton, Hungary: Earth Planet. Sci. Lett., v. 31, p. 297–304.

Embey-Isztin, A., S. Peltz, and T. Póka, 1985, Petrochemistry of the Neogene-Quaternary basaltic volcanism in the Carpathian basin: Fr. Miner. Paleont. Museum, Hist.-nat. Hung., Budapest, v. 12, p. 5–19.

Ewart, A. and R. W. Le Maitre, 1980, Some regional compositional differences within Tertiary-recent orogenic magmas: Chemical Geol., v. 3, no. 3, p. 257–285.

Fiala, A., 1962, Chemism of the Neogene volcanits of Kremnicke Hory: Geol Prace, v. 15, p. 26–32.

Forgach, J. K, and G. Kupco, 1974, Stopove prvky v neovulkanitoch Slovenska: Západné Karpaty, Ser. Miner., Petr., Geol., v. 8, p. 137–215.

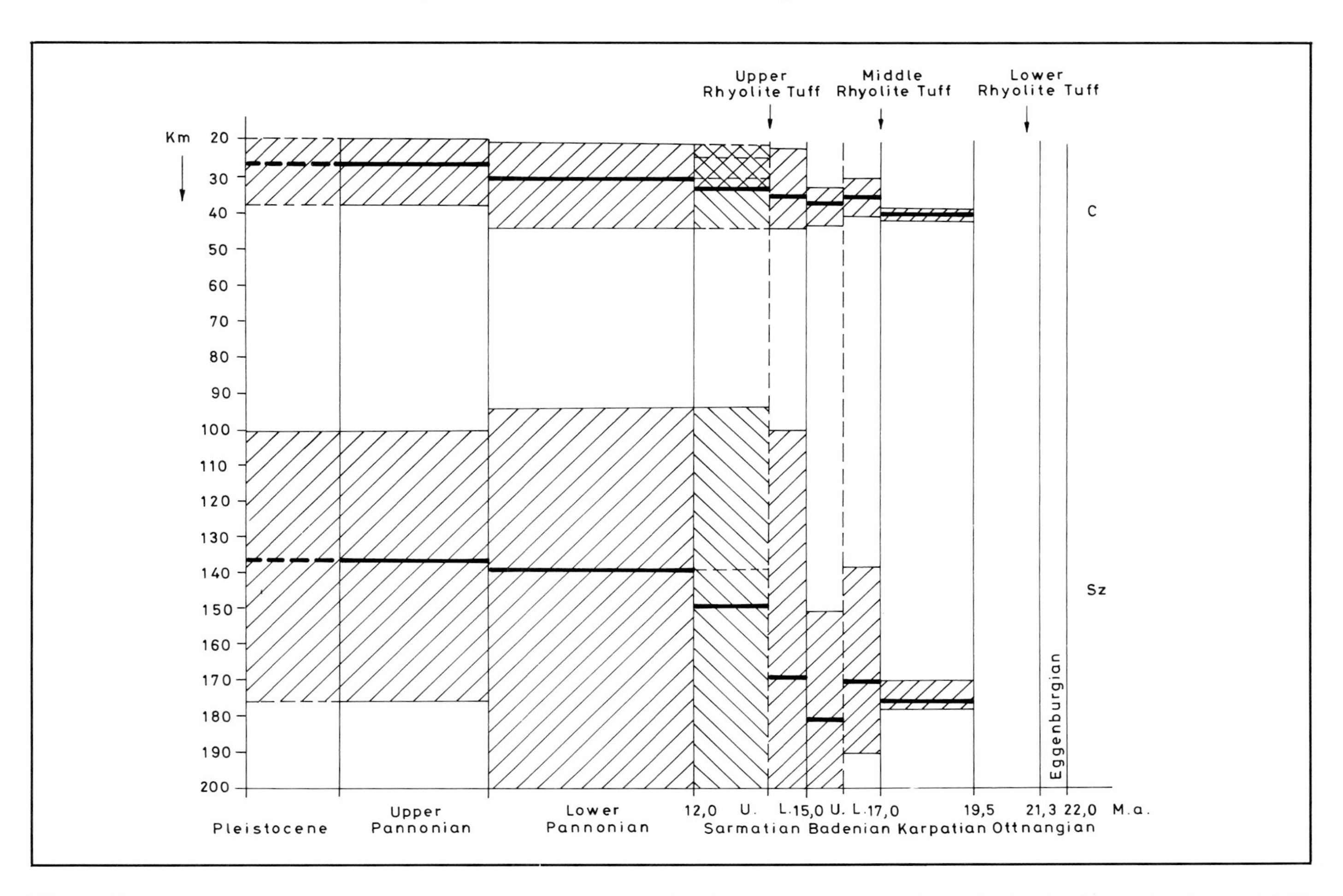

Figure 7. The average paleocrustal thickness (C) and paleosubduction depths (Sz) versus time calculated after Hatherthon and Dickinson (1968) and Condie (1976). Hatched areas show range of values and indicate maximum and minimum values. "Pannonian" refers to Pannonian (*s. l.*).

Gyarmati, P., 1977, The intermediary volcanism of the Tokaj Mts. (in Hungarian): MÁFI Évkönyv. LVIII, Müszaki Könyvkiadó, Budapest, 195 p.

Hámor, G., K. Balogh and L. Ravaszné Baranyai, 1976, Radioactive age of the tertiary formations in North Hungary: MÁFI Évi Jel., v. 15, p. 61–76.

Hatherton, T. and W. R. Dickinson, 1968, Andesitic volcanism and seismicity in New Zealand: J. of Geophys. Res., v. 73, n. 14, 206–308.

Horváth, F. and L. Royden, 1981, Mechanism of the formation of the Intra-Carpathian Basins: a Review: Earth Evol. Sci., v. 1, no. 3-4, p. 307–316.

Ilkeyné Perlaki, E., 1973, The Neogene acidic volcanism of the Tokaj Mts. (in Hungarian): Thesis for candidate degree, Budapest, 205 p.

Jámbor, A., Zs. Partényi, L. Ravaszné Baranyai, G. Solti and K. Balogh, 1980, K/Ar dating of basaltic rocks in Transdanubia, Hungary: ATOMKI Bulletin, v. 22, no. 3, p. 173–190.

Jugovics, L., 1978, Chemical features of the basalts in Hungary (in Hungarian): Annual Report of the Hungarian Geological Institute for 1976, p. 431–471.

Karolus, K., 1965, Contribution to the chemistry of Carpathian neovolcanic rocks: Geol. prace., Bratislava, v. 36, p. 18–62.

Karolus, K., 1975, An integral compositional characteristic of Carpathian neovolcanics and their comparison with other volcanic associations: Proc. of X. Cong. of KBGA, Bratislava, p. 38–40.

Konecny, V., 1969, Evolution of Neogene volcanism in Central Slovakia and its connection with absolute ages: Acta Geol. Acad. Sci. Hung., v. 2, p. 254–258.

Konecny, V. and J. Lexa, 1974, The Carpathian volcanic arc: a discussion: Acta Geol. Sci. Hung., vol. 18, no. 3-4, p. 279–293.

Korpás, L., Zs. Peregi and G. Szendrei, 1967, Petrological and geological investigations on the northern part of the Dunazug Mts.: Földtani Közlöny, v. 97, p. 211–233.

Kubovics, I. and Gy. Pantó, 1970, Volcanological investigations in the Mátra and Börzsöny Mts. (in Hungarian): Akad. Kiadó, Budapest, 302 p.

Kuno, H., 1962, Frequency distribution of rock types in oceanic, orogenic and cratonic volcanic association: Geophys. Monogr., no. 6, p. 133–239.

Lang, B., 1976, Mineralogy and geochemistry of the Neogene pyroxen andesites from the N-part of the Gutii Mts. (Romania): Anuarul Inst. de Geol. si Geoph., v. XXIX, p. 153–213.

Nicholls, I. A., D. J. Whitford, K. L. Harris, and S. R. Taylor, 1980, Variation in the geochemistry of mantle sources for tholeiitic and calcalkali mafic magmas, Western Sunda volcanic arc, Indonesia: Chemical Geol., v. 30, no. 3, p. 177–199.

Pantó, Gy., 1981, Rare earth element geochemical patterns of the Cenozoic volcanism in Hungary, Earth Evol. Sci., v. 1, no. 3-4, p. 249–256.

Peccarillo, A. and S. R. Taylor, 1976, Rare earth elements in East Carpathian volcanic rocks: Earth Planet. Sci. Lett., v. 32, p. 121–126.

Peltz, S., C. Vasiliu, C. Udrescu and A. Vasilescu, 1973, Geochemistry

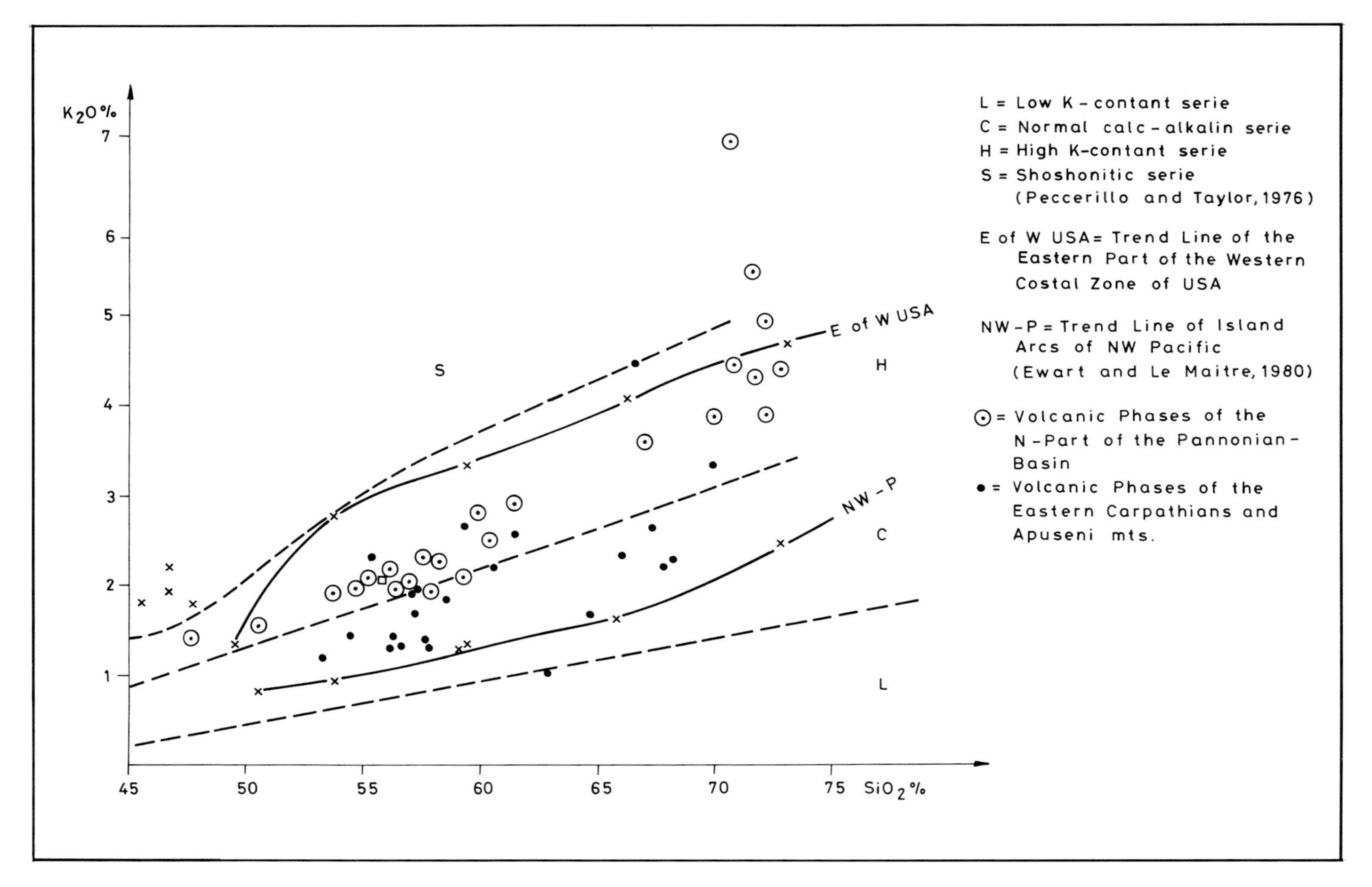

Figure 8. K_2O versus SiO_2 (according to Peccarillo and Taylor, 1976) for different volcanic phases of the Pannonian basin and the eastern Carpathians. Also plotted are the trend line of the volcanism associated with different tectonic environments (according to Ewart and Le Maitre, 1980).

of volcanic rocks from the Calimani, Gurghiu and Harghita Mts.: Ann. Inst. Geol. Bucuresti, v. 42, p. 339–393.

Perfit, M. P., D. A. Gust, A. E. Bence, R. J. Arculus, and S. R. Taylor, 1980, Chemical characteristic of island-arc basalts: Implications for mantle sources: Chem. Geol., v. 30, p. 227–256.

Póka, T. and B. Simo, 1966, De rolle des Nebangesteins in der entwicklung der subvulkanisches facies: Ann. Univ. Sci. Budap., v. 10, p. 52–81.

Póka, T., 1968, An undifferentiated stratovolcanic marginal facies of the intra-Carpathian volcanic girdle (Cserhát Hills): Ann. Univ. Sci. Budap., Sec. Geol., v. 12, p. 37–47.

Radulescu, D. P., 1960, K. izucsenyiju chimizma vulkanitseskich porod v predelach v nutrennij zoni Karpatskoj Dugi: Rev. de Geol., Bucuresti, v. 5, no. 2, p. 2–85.

Radulescu, D. P. and M. Sandulescu, 1973, The plate tectonics concept and the geological structure of the Carpathians: Tectonophysics, v. 16, p. 155–161.

Rittmann, A., 1957, On the serial character of igneous rocks: Egypt, J. Geol., Cairo, v. 1, p. 23–48.

Royden, L., F. Horváth and B. C. Burchfiel, 1982, Transform faulting, extension and subduction in the Carpathian–Pannonian region: GSA Bull., v. 93, p. 717–725.

Simpson, E. S. W., 1954, On the graphical representation of differentiation trends in igneous rocks: Geol. Mg., v. 91, p. 238–244.

Stegena, L., B. Géczy and F. Horváth, 1975, The late Cenozoic evolution of the Pannonian basin (in Hungarian): Földt. Közl., v. 105, no. 2, p. 101–123.

Stegena, L. and F. Horváth, 1982, Review of the Pannonian basin: Abstracts of Discussion Meeting: Evolution of Extensional Basins, Hungary, 1982, p. 19–25.

Sugimura, A., 1959, Geographische verteilung der charactertendenzen des magmas in Japan: Bull. Volcan. Sec. Jap. II, v. 4, p. 77–103.

Szádeczky-Kardoss, E., 1967, Considerations on the investigation of the deep structure and magmatectonics of the Carpathian Basin System, (in Hungarian): MTA X. Oszt. Közl., v. 1, no. 1–2, p. 41–65.

Szádeczky-Kardoss, 1976, Plattentektonik im Pannonisch-Karpatischen Raum: Geol. Rundschau, v. 65, no. 1, p. 143–161.

Taylor, S. R., 1969, Trace element chemistry of andesites and associated calcalkaline rocks: Proc. and Conf. Oreg. Dep. Geol. Mineral. Res. Bull., v. 65, p. 43–58.

Thornton, W. N. and O. Tuttle, 1960, Chemistry of igneous rocks: 1. Differentiation index: Am. J. Sci., v. 258, no. 9, p. 82–98.

Thorpe, R. S., P. J. Potts and P. W. Francis, 1976, Rare earth data and petrogenesis of andesite from Chilean Andes: Contr. Miner. Petr., v. 54, no. 1, p. 65–78.

Tolsztoj, M. I., Ju. L. Gazsanov, V. G. Molijavko, I. M. Osztafijchuk, G. T. Prodajvoda, A. Ju. Szerga and A. V. Suhorada, 1976, Geochimija, petrofizika i voprosi genezisa novejsich vulkanitov sovjetskich Karpatians [Geochemistry, petrophysics and problems of genesis of the Neogene volcanics in the Soviet Carpathians]: Id. Obj. "Visa skola", Kiev, p. 1–98.

Varga, Gy., E. Csillagné Teplánszky and Zs. Félegyházi, 1975, Geology of the Mátra Mts. (Monographie): MAFI Évkönyve, v. 57, 575 p.

Vendel, M., 1944–47, Studien aus den jungen Karpatischen Metallprovinz. Bány. es Koh. Közl., Sopron, v. 16, p. 194–280.

Wager, L. R. and W. A. Deer, 1956, A chemical definition of fractionation stages as a basis for comparison of Hawaiian, Hebridean and Alaska basic rocks: Geoch. Cosmoch. Acta, v. 7, p. 217–248.

Wyllie, P. J., 1977, Crustal anatexis: an experimental review: Tectonophysics, v. 43, p. 41–71.

Zelenka, T., 1960, Petrological and geological investigations on the SW-part of Dunazug Mts.: Földtani Közlöny, v. 90, p. 83–101.

CHAPTER 19

Origin of Late Cenozoic Volcanic Rocks of the Carpathian Arc, Hungary

Vincent J. M. Salters
S. R. Hart
Center for Geoalchemy, Earth, Atmospheric and Planetary Sciences
Massachusetts Institute of Technology
Cambridge, Massachusetts 02139

Gy. Pantó
Hungarian Academy of Sciences
Laboratory for Geochemical Research
H-1112 Budapest
Budaörsi út 45, Hungary

Medium to high K calcalkaline (CA) volcanics in the Carpathians are remnants of an Eocene and Miocene volcanic arc. This volcanism was followed by alkali basalts (AB) in the Quaternary. The CA rocks vary in composition from basaltic andesite to rhyolite, while the ABs are restricted in composition to 44–48% SiO_2.

CA rocks have $^{87}Sr/^{86}Sr$ = 0.70426–0.71125, $^{143}Nd/^{144}Nd$ = 0.51274–0.51221. The ABs have $^{87}Sr/^{86}Sr$ of 0.70320–0.70421 and $^{143}Nd/^{144}Nd$ of 0.51292–0.51270. The complete suite forms a narrow array on the Nd-Sr isotope correlation diagram, despite its large geographic and compositional variation. Pb isotopes exhibit a restricted range for $^{207}Pb/^{204}Pb$ = 15.61–15.72 and $^{208}Pb/^{204}Pb$ = 38.99–39.2. $^{206}Pb/^{204}Pb$ variation is larger, ranging from 18.69–19.42. On the Pb-Pb and Nd-Sr isotope diagrams the Carpathian volcanics overlap the array defined by oceanic island volcanics, and lie in the field of other arcs often associated with crustal contamination, such as the Banda Arc and the Lesser Antilles. However, on a Δ207/204 Pb-Δ208/204 Pb diagram all Hungarian volcanics except for one "pristine" alkali basalt fall outside the mantle array. The isotopic and trace element characteristics of the volcanics can be explained by two different three-component mixing models. In both cases, a pristine mantle ($^{143}Nd/^{144}Nd$ = 0.51292, $^{87}Sr/^{86}Sr$ = 0.7320, $^{206}Pb/^{204}Pb$ = 19.426) is one endmember. In one model this mantle melt mixes with the lower crust in order to produce the alkaline magmas and with crust with variable Pb content or Pb isotopic composition in order to generate the calcalkaline magmas.

In the second more preferable model the pristine mantle is modified by a component of the subducted slab. Melts from this modified mantle mix with a crustal component in order to generate both the calcalkaline and alkaline series.

INTRODUCTION

Two different types of magmatism, alkaline and calcalkaline, were active during the formation of the Carpathian orogenic arc. The tectonic setting of this Carpathian magmatism is well documented (Stegena et al., 1975; Royden et al., 1983; and Burchfiel, 1980). The calcalkaline volcanism predates the alkaline volcanism and is related to an extensional tectonic environment in which the Pannonian fragment extended while it overrode the European plate. The alkaline volcanism occurred in Pliocene to Quaternary time, and probably slightly postdates most of the extensional and convergent tectonics. Earlier studies of the Carpathian magmatism indicated heterogeneity in the source of the alkali basalts (Pantó, 1981; Vogl and Pantó, 1983) and suggested that the calcalkaline magmas contain a component from the subducted slab (Peccerillo and Taylor, 1976). The present study was undertaken to further study the nature of the source materials for these two magma types. Disagreement exists as to whether arc volcanism in general requires a component from the subducted slab. Stern (1982) and Morris and Hart (1983) argue that most island arc lavas are derived from an oceanic island-type mantle, without a significant slab-derived component. On the other hand, Gill (1981), Nicholls and Ringwood (1973), and Kay (1980) argue that a slab component is an important feature of most arc magmas. Study of calcalkaline and alkaline magmas in close spatial and age relationship, as in the Carpathians, may help elucidate this question.

This study is the first to report precise Sr, Nd and Pb isotope and trace elements analyses on the Carpathian volcanics. Samples were selected with the aim of covering as much of the observed chemical and geographical variations as possible. This chapter presents the results of this reconnaissance study, together with a first order interpretation of the results.

Analytical Techniques

Ba, Sr, K, Rb and Cs were separated by standard ion exchange column chromatography and the concentrations were determined by isotope dilution (Hart and Brooks, 1977). Nd-isotope ratios and Sm-Nd concentrations were determined following the technique of Zindler (1980), as adapted from Richard et al. (1976). Pb-isotopic compositions were determined using the method described by Manhes et al. (1978) and adapted at MIT by Pegram (1986).

Precision and normalization of the Sr, Nd, and Pb isotopic data is given in Table 1. Precision of the trace element data is better than 0.1%, except for Rb and Cs (0.5%). Blank levels are insignificant for any of the data reported here. Forty percent of the Nd and Sr isotope ratios were duplicated. All duplicates were within 2σ of the first analyses. The Pb isotopic ratios of critical samples were also duplicated, and were within the in-run precision of the first analyses.

Tectonic Setting

The Carpathians are an arcuate-shaped mountain belt forming the northern part of the eastern extension of the Alps (Figure 1). The mountain belt was a zone of convergence in close spatial relationship with a back-arc extensional basin: the Pannonian basin. Other Mediterranean examples of this type of paired system are the Appenine–Tyrrhenian system and the Alboran–Betic Cordillera–North African system. For a detailed tectonic history, see Royden (this volume).

Tectonic activity in the Carpathians began in Early Cretaceous time, with outwardly directed thrusting and folding. Probably southward subduction of the European plate under the northern margin of the Pannonian fragment during Paleocene and Eocene times is evidenced by a poorly developed Eocene calcalkaline arc located in the Eastern Alps and on the inner side of the West Carpathians (Burchfiel and Royden, 1982). There is no evidence for early Tertiary subduction beneath the eastern margin of the Pannonian fragment.

Subduction resumed during Miocene time and was, since middle Miocene, contemporaneous with extension of the Pannonian basin, which began roughly 16.5 m.y. ago (Royden et al., 1983). The subducted plate was probably partly oceanic, as indicated by deposition of sediments below the carbonate compensation depth, but some continental crust was probably also subducted (Burchfiel, 1980). During early Miocene time, the calcalkaline volcanism first developed in the western part of the Carpathians (Börzsöny, Mátra, and Cserhát Mountains, Figure 1), then migrated eastward and southward to the Tokaj Mountains and the East Carpathians, where volcanic activity ceased 10.5 m.y. ago (Póka, 1984; Royden et al., 1983). This eastward migration of magmatism was accompanied by eastward migration of thrusting within the outer Carpathian thrust belt.

Extension of the continental crust and formation of the Pannonian basin began in middle to late Miocene times and extension appears to have migrated in space and time in conjunction with the migration of the magmatism and thrusting. Active extension and related faulting and sedimentation ended in late Miocene time. One of the features accompanying the extensional tectonics was alkali basaltic magmatism. This new type of volcanic activity began in late Miocene (10.5 m.y.) and continued into the Quaternary, after extension had ceased (Póka, 1984; Pantó, 1981). The alkali basalts commonly contain lherzolite xenoliths, clearly indicating a mantle origin for this phase of the magmatism.

RESULTS

Samples for isotopic analysis were selected from a large collection covering all three periods of magmatic activity. Selection was based on freshness of the samples, and an attempt to cover a large range in chemical composition, geographic position and age. Sample localities are indicated on Figure 1, and the analytical data is given in Table 1.

The $^{87}Sr/^{86}Sr$ isotope ratios of the alkali basalts range from 0.703199–0.705950, while the $^{143}Nd/^{144}Nd$ isotope ratios vary from 0.512915–0.512546. On a $^{143}Nd/^{144}Nd$-$^{87}Sr/^{86}Sr$ isotope correlation diagram, all of the alkali basalts fall within the oceanic mantle field (Figure 2). The basalts are all strongly light REE enriched, with Sm/Nd ratios ranging from 0.184–0.211. Rb/Sr ratios are somewhat higher, and Sm/Nd ratios somewhat lower than typical oceanic alkali basalts (White et al., 1979; Clague and Frey, 1982; and Patchett and Tatsumoto, 1980).

Table 1. Trace Element and Isotopic Data, Carpathian Volcanics[a]

Sample	118	99	103	116	134	55	57	107	113	77
Location	Nógrád	Balaton	Balaton	Nógrád	Sárospatak	Börzsöny	Börzsöny	Börzsöny	Börzsöny	Velence
Age	"P-P"	"P-P"	"P-P"	"P-P"	unknown	Miocene	Miocene	Miocene	Miocene	Eocene
SiO_2	44.92	44.94	47.43	44.39	46.54	60.52	57.47	54.59	68.87	57.33
87/86 Sr	0.703199	0.704030	0.704204	0.703310	0.705950	0.707958	0.708103	0.708140	0.708915	0.707205
143/144 Nd	0.512915	0.512762	0.512700	0.512889	0.512546	0.512384	0.512400	0.512469	0.512338	0.512527
206/204 Pb	19.426	18.875	18.712	19.158	19.018	18.855	18.901	18.918	19.414	18.897
207/204 Pb	15.643	15.613	15.648	15.647	15.721	15.668	15.658	15.689	15.631	15.675
208/204 Pb	39.168	38.788	38.842	39.055	39.279	38.942	38.947	39.038	39.123	39.154
Δ207/204 Pb	4.6	7.6	12.8	7.9	16.8	13.3	11.8	14.7	3.6	13.6
Δ208/204 Pb	5.5	34.1	59.2	26.6	65.9	51.9	46.9	53.9	2.4	58.1
K (ppm)	19961	18809	16319	16491	29039	19476	19324	15001	25278	20253
Rb	67.9	64.1	50.5	58.2	133.5	87.2	88.9	71.9	116.4	77.4
Cs	0.74	1.25	1.10	1.00	1.04	4.89	4.07	4.06	6.11	1.55
Sr	841.2	937.4	621.4	622.9	375.4	630.2	527.3	344.1	326.5	317.8
Ba	788	836	531	562	547	1292	1031	380	776	533
Nd	42.48	44.64	29.73	32.27	18.83	26.30	22.91	17.55	18.58	26.76
Sm	7.81	8.20	6.11	6.47	3.97	4.96	4.60	3.98	2.93	5.35
K/Rb	294	293	323	283	223	223	217	209	217	262
K/Cs	26974	15047	14835	16491	27922	3982	4748	3695	4137	13066
Ba/Rb	11.61	13.04	10.51	9.66	4.10	14.81	11.60	5.28	6.67	6.89
Rb/Sr	0.0807	0.0684	0.0813	0.0934	0.3556	0.1384	0.1686	0.2090	0.3565	0.2435
Sm/Nd	0.1839	0.1837	0.2055	0.2005	0.2108	0.1886	0.2008	0.2268	0.1577	0.1999
87/86 Sr initial	—	—	—	—	—	0.70786	0.70798	0.70799	0.70865	0.70680

[a]P-P means Pliocene–Pleistocene Chemical analyses of the alkali basalts may be found in Pantó, 1981; SiO_2 contents of the calcalkaline rocks are from Pantó, unpublished data. Precision of the Sr and Nd isotopic ratios is better than $\sim$0.005% 2σ; values are relative to 87/86 Sr ∂ 0.70800 for the E&A standard, and 143/144 Nd ∂ 0.51262 for BCR-1; 143/144 Nd is normalized to 146/144 ∂ 0.72190. Pb isotope ratios have been normalized to the NBS 981 standard; the precision is better than 0.05% per amu. The trace elements were all determined by isotope dilution (see text). Sr isotope initial ratios were calculated assuming ages of 40 m.y. (#68, #77), 18 m.y. (Börzsöny), 16 m.y. (Mátra, Cserhát) and 15 m.y. (Tokaj) (Póka, 1984). Age corrections are insignificant for the Nd isotopic data.

Sample	68	61	92	87	91	128	132	126
Location	Mátra	Mátra	Mátra	Cserhát	Cserhát	Tokaj	Tokaj	Tokaj
Age	Eocene	Miocene	Miocene	Miocene	Miocene	Miocene	Miocene	Miocene
SiO_2	54.87	56.54	53.79	55.41	53.04	61.01	63.44	74.65
87/86 Sr	0.704464	0.709175	0.707936	0.706488	0.708016	0.709667	0.707364	0.713521
143/144 Nd	0.512741	0.512308	0.512380	0.512516	0.512417	0.512264	0.512420	0.512208
206/204 Pb	18.938	18.694	18.799	18.782	18.910	18.729	18.838	18.786
207/204 Pb	15.657	15.659	15.673	15.658	15.675	15.663	15.692	15.721
208/204 Pb	38.964	38.861	38.931	38.856	39.114	38.997	39.092	39.279
Δ207/204 Pb	11.3	14.1	14.4	13.1	17.2	14.2	15.9	18.9
Δ208/204 Pb	44.1	63.3	57.6	52.2	68.0	72.7	68.9	77.7
K, ppm	11712	18461	13005	17778	12626	19463	27226	35203
Rb	69.8	91.6	63.9	102.7	49.7	98.5	122.9	161.8
Cs	9.90	9.56	5.47	8.40	2.91	4.30	4.10	5.22
Sr	562.0	253.4	271.4	225.7	267.1	189.2	225.3	43.9
Ba	548	377	230	247	256	374	608	781
Nd	25.01	25.50	15.83	17.82	26.76	22.54	28.53	30.36
Sm	4.58	5.60	3.73	4.33	4.41	—	5.62	6.76
K/Rb	168	202	204	173	254	198	222	218
K/Cs	1183	1931	2378	2116	4339	4526	6640	6744
Ba/Rb	7.85	4.11	3.59	2.40	5.15	3.80	4.94	4.83
Rb/Sr	0.1242	0.3615	0.2354	0.4550	0.1861	0.5206	0.5455	3.69
Sm/Nd	0.1831	0.2196	0.2356	0.2430	0.2409	—	0.1970	0.2227
87/86 Sr initial	0.70426	0.70894	0.70778	0.70619	0.70789	0.70935	0.70703	0.71125

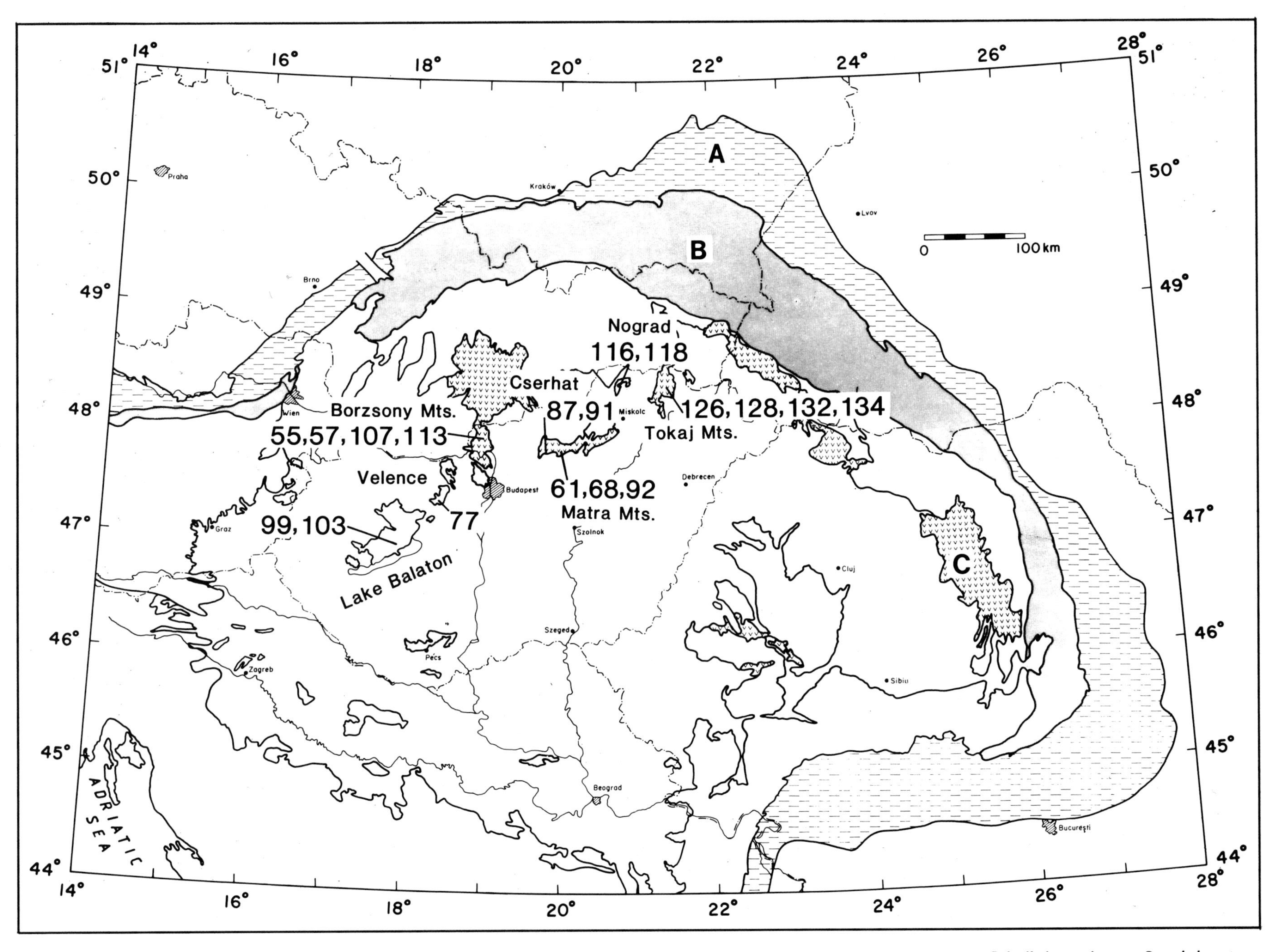

Figure 1. Simplified geological map of the Carpathians after Royden et al. (1983). A = Molasse Foredeep; B = Outer Carpathians; and C = Calcalkaline volcanics. Sample locations and numbers are indicated.

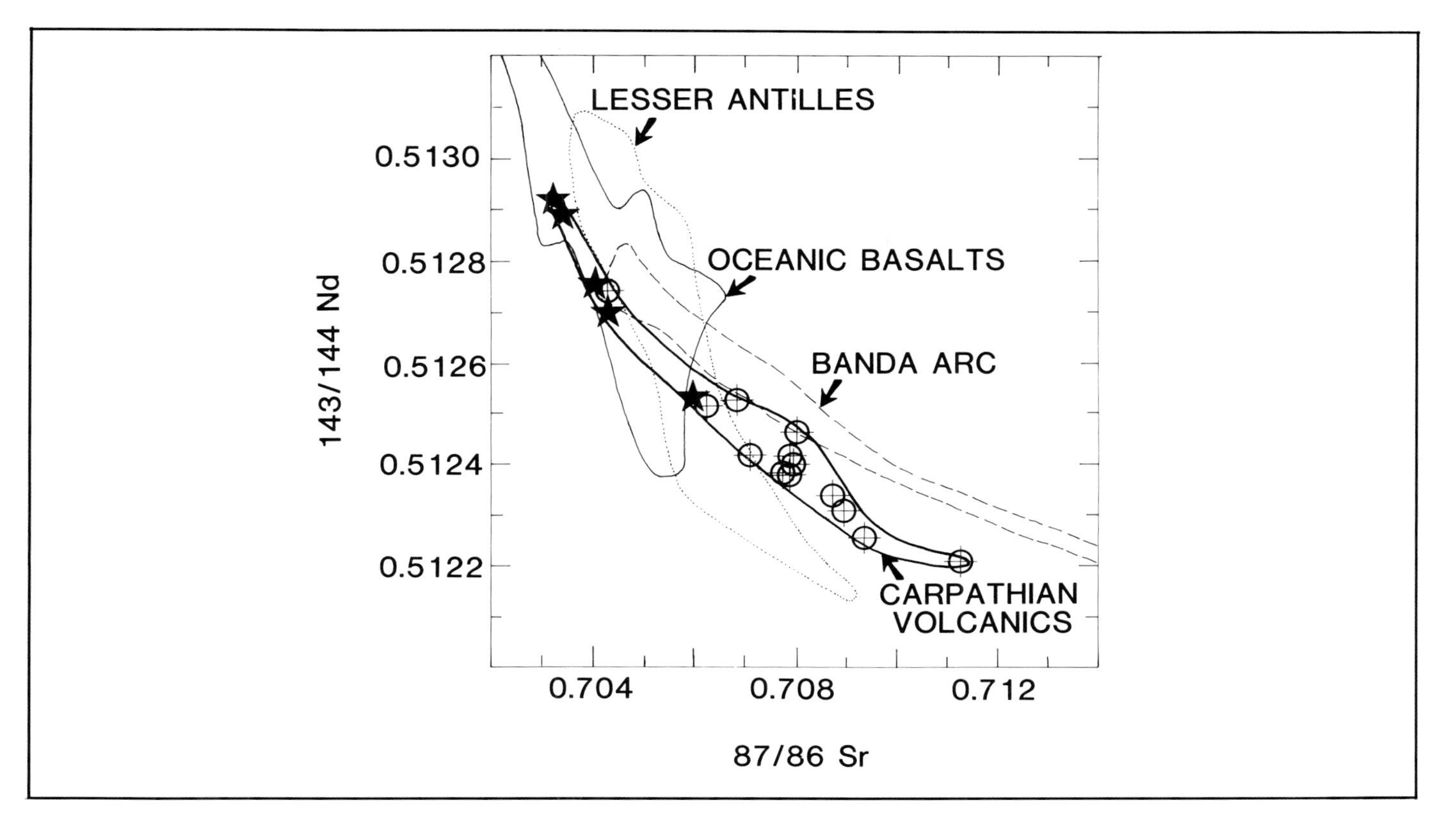

Figure 2. $^{143}Nd/^{144}Nd$ versus $^{87}Sr/^{86}Sr$ diagram for the Carpathian volcanics. The stars represent the alkali basalts and the open circles represent the calcalkaline volcanics. The initial $^{87}Sr/^{86}Sr$ ratio is used for the calcalkaline volcanics. This representation of the Carpathian volcanics will be maintained throughout this chapter. The array for the Carpathian volcanics is parallel to the Banda arc array. The alkali basalts form the high Nd and low Sr isotope ratio end of the array. Field for oceanic basalts from Staudigel et al., 1984. Field for the Banda arc from Morris, 1984 and Whitford and Jezek, 1979. Lesser Antilles array is compiled from Davidson, 1983; Hawkesworth and Powell, 1980; and Hawkesworth et al., 1979.

Pb-isotope ratios for the alkali basalts range from 18.712–19.426 for $^{206}Pb/^{204}Pb$; 15.613–15.721 for $^{207}Pb/^{204}Pb$; and 38.788–39.168 for $^{208}Pb/^{204}Pb$ (Figure 3). They fall on the high 207/204 Pb and high $^{208}Pb/^{204}Pb$ side of the array defined by oceanic volcanics from the northern hemisphere (Hart, 1984).

The analyzed calcalkaline rocks, both Eocene and Miocene, range in composition from basaltic andesite to rhyolite. SiO_2 contents range from 53 to 75 wt percent. In the classification of Gill (1981), the andesites lie close to the borderline between medium- and high-K andesites, with the majority lying within the high-K andesite field. The $^{143}Nd/^{144}Nd$ isotope ratios of the calcalkaline volcanics range from 0.512741–0.512208 and the initial $^{87}Sr/^{86}Sr$ ratios ranging from 0.70426–0.71125. On the Nd-Sr isotope correlation diagram (Figure 2), the calcalkaline volcanics define a narrow array, overlapping the high-Nd isotope ratio end of the oceanic basalt field, and extending to higher Sr and lower Nd isotope ratios. The narrow Nd-Sr isotope array defined by the Hungarian alkaline and calcalkaline volcanics resembles the Nd-Sr arrays found in the Lesser Antilles, the Banda arc (see Figure 2), and the Italian volcanic province. Sm and Nd concentrations in the calcalkaline lavas are in the same range as those reported by Peccerillo and Taylor (1976).

Although showing considerable scatter, the complete suite of Carpathian volcanics displays a positive correlation between $^{87}Sr/^{86}Sr$ and SiO_2 (Figure 4). Within the calcalkaline suite, SiO_2 contents show poor positive correlations with Rb and K and no correlation with Ba or Sr. Although the calcalkaline lavas, as a group, show a positive correlation between SiO_2 and Rb, K and $^{87}Sr/^{86}Sr$, the individual centers do not show any regular variations with SiO_2. The trace element ratios, Sm/Nd, Rb/Sr, K/Ba, Ba/Rb, and K/Rb also do not vary regularly with SiO_2, either within centers or as a whole.

With the exception of sample 113, the Pb isotope ratios of the calcalkaline volcanics show a more limited variation in $^{206}Pb/^{204}Pb$ (18.694–18.918) than the alkali basalts. $^{207}Pb/^{204}Pb$ (15.631–15.761) and $^{208}Pb/^{204}Pb$ (38.816–39:154) ratios for the calcalkaline volcanics are in general somewhat higher than those for the alkali basalts (Figure 3). The calcalkaline lavas straddle the high $^{207}Pb/^{204}Pb$ boundary of the oceanic basalt field and are clustered close to the field of pelagic sediment Pb.

A complete separation exists between the calcalkaline and alkaline volcanics in K/Cs and Rb/Cs ratios. Except for sample 134, the different magma types have nonoverlapping $^{87}Sr/^{86}Sr$ isotope ratios as well. Alkali basalt sample 134 is quite anomalous, with $^{87}Sr/^{86}Sr$; $^{143}Nd/^{144}Nd$; K/Rb ratios; and Nd, Sr, and Sm contents comparable to those of the calcalkaline volcanics, whereas the K/Cs ratio and major element contents are comparable to those of the other alkali basalts. The Rb content of sample 134 is the highest of all the lavas except for the rhyolite (#126). While the $^{206}Pb/^{204}Pb$ ratio of sample 134 is within the range of the other alkali basalts, the $^{207}Pb/^{204}Pb$ and $^{208}Pb/^{204}Pb$ ratios are the highest of all samples analyzed. Sample 134 was recovered from a drill hole near Sárospatak, where it was found

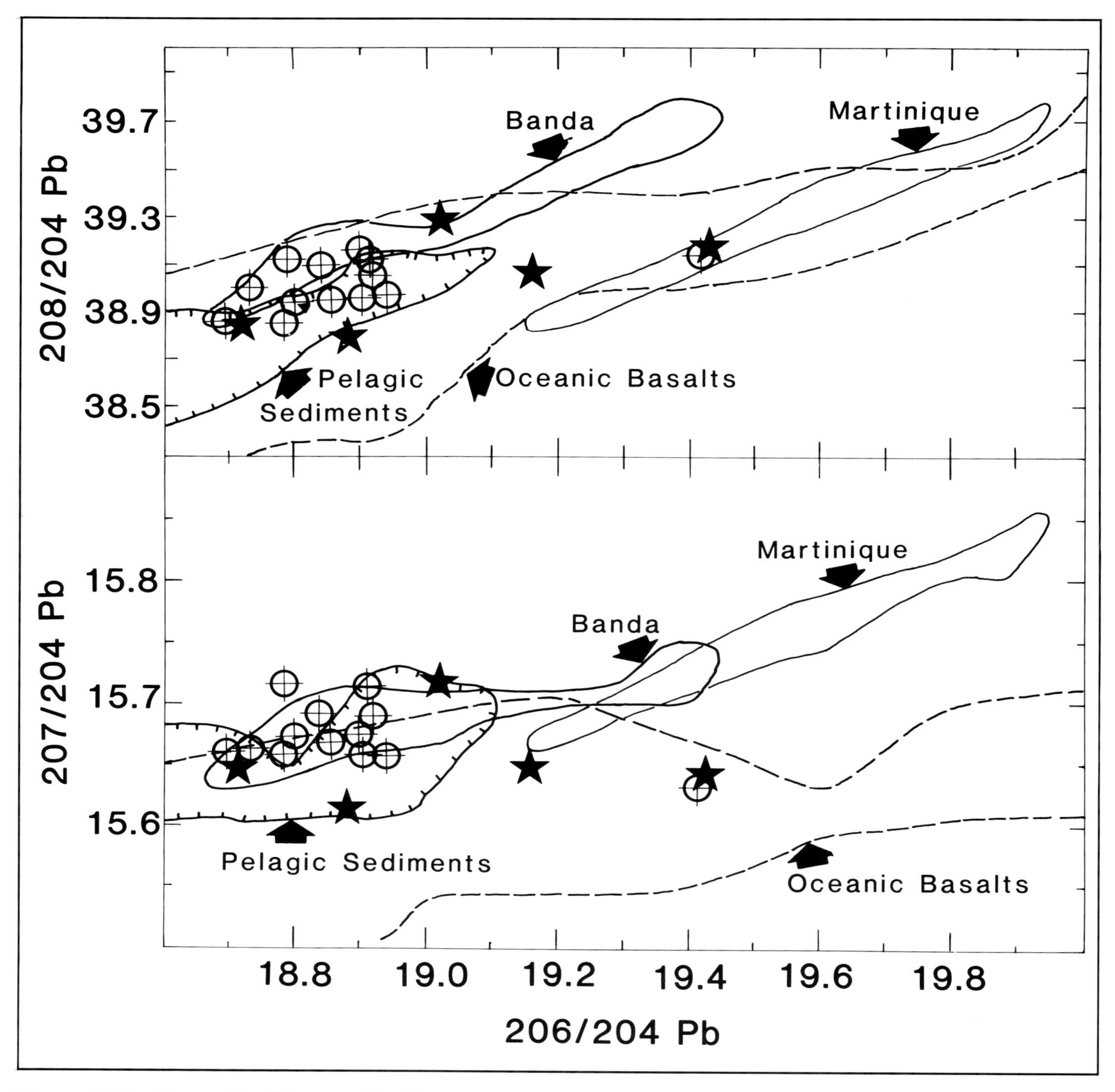

Figure 3. $^{207}Pb/^{208}Pb$ and $^{208}Pb/^{204}Pb$ versus $^{206}Pb/^{204}Pb$. Symbols as in Figure 2. The majority of the Carpathian volcanics fall inside the mantle array and overlap with pelagic sediments and the Banda arc. Field for the Banda arc from Morris, 1984. Field for Martinique, Lesser Antilles from Davidson, 1983. For the oceanic basalts reference used is Hart, 1984. Pelagic sediment data from Barreiro, 1983; Sun, 1980; Vidal and Clauer, 1981; Reynolds and Dasch, 1971; Church, 1976; and Hart and Sinha, unpublished data.

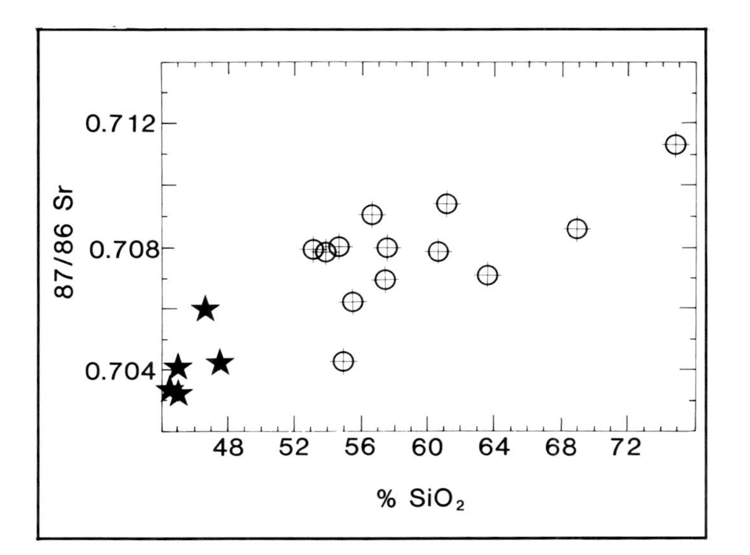

Figure 4. Initial $^{87}Sr/^{86}Sr$ versus SiO_2-content in weight percent. Symbols as in Figure 2. Note there is a weak positive correlation between Sr isotopic composition and silica content.

to overlie calcalkaline volcanics. While Pantó (1981) correlated this alkali basalt with the younger alkali basalt suite elsewhere in Hungary, the isotopic data suggest it is probably a member of the calcalkaline suite instead.

DISCUSSION

The fact that the alkaline and calcalkaline lavas occur in the same geographic province, and together show a very narrow array on the Nd-Sr isotope correlation plot (Figure 2), led us to attempt to explain the observed isotopic variations without making an a priori distinction between the calcalkaline and alkaline lavas. Considering that the analyzed samples represent an age spread of over 30 m.y. and a geographical extent of over 500 km makes this well-defined correlation between Nd and Sr isotopes even more remarkable and indicates a possible genetic relationship between the two types of volcanism. As quoted above, similar large and highly correlated ranges of Nd and Sr isotopic composition are only observed in the Lesser Antilles, in the Italian volcanic belt and in the Banda Arc, Indonesia. These three occurrences are all associated with some type of crustal contamination (Morris, 1984; Taylor et al., 1979; Davidson, 1983; and Hawkesworth and Vollmer, 1979). The high Sr and low Nd isotope ratios are typical for crustal material; this, together with the increase of $^{87}Sr/^{86}Sr$ with SiO_2 (Figure 4), leads to the conclusion that the observed isotopic variation in Nd and Sr was caused by crustal contamination, possible accompanied by crystal fractionation. The two endmembers of the Nd-Sr array are alkali basalt #118, a mantle-type component (with $^{87}Sr/^{86}Sr$ = 0.703199, $^{143}Nd/^{144}Nd$ = 0.512915, Sr = 841 ppm and Nd = 42.5 ppm) and rhyolite #126, a crustal-type component (with $^{87}Sr/^{86}Sr$ = 0.71125, $^{143}Nd/^{144}Nd$ = 0.512208, Sr = 44 ppm and Nd = 30 ppm). However, simple mixing of two magmas with these characteristics does not reproduce the observed array, but results in an array which is markedly more concave. The flattening of the observed array towards the right provides an accurate estimate of the Nd isotope ratio of the contaminant; however, the Sr isotope ratio of the contaminant is not well constrained by the shape of the array (see also Langmuir et al., 1978). Given the observed Sr and Nd concentrations of the endmembers, simple mixing will not reproduce the observed array, even if the $^{87}Sr/^{86}Sr$ of the contaminant is considered a free parameter. In a combined fractional crystallization and assimilation model like that described by Allegre and Minster (1978) and DePaolo (1982), the Nd and Sr isotope ratios of the contaminated end of the array can be generated when the mass of assimilated material is large. Consequently, the amount of material crystallized from the magma is large. This leads to unrealistically high Nd concentrations in the evolved magmas, unless unreasonably high crystal-liquid distribution coefficients for Nd are assumed.

We have derived a consistent (but not unique) set of model parameters for the Nd-Sr array as follows. We assume a hypothetical "parental" basalt for the calcalkaline suite having the isotopic characteristics of alkali basalt #118. The Nd isotopic ratio of the crustal endmember is chosen to be $^{143}Nd/^{144}Nd$ = 0.51204 (the average ratio of atmospheric dusts, continental sediments and river particulates; see Goldstein et al., 1984) and the $^{87}Sr/^{86}Sr$ of the crustal endmember is allowed to be a free parameter. A best-fitting curve to the calcalkaline data is obtained then with an "r-factor" of 2.39 and $^{87}Sr/^{86}Sr$ crust ~0.7155 [r-factor = (Sr/Nd) of parental basalt/(Sr/Nd) of crustal contaminant]. Using average crustal Sr and Nd of 350 ppm and 29 ppm (Allegre et al., 1983) then Sr/Nd ~12.1. This requires the parental basalt endmember to have Sr/Nd ~29. This is higher than the typical oceanic basalt value of ~20 (also observed in the Carpathian alkali basalts), but is similar to the range normally observed in island arc basalts (25–35; see DePaolo and Johnson, 1979; and White and Patchett, 1984). Although there is considerable scatter in plots of Sr and Nd versus SiO_2 for the differentiated calcalkaline rocks, the data are consistent with a parental basalt having 375 ppm Sr and 13 ppm Nd, which would provide the Sr/Nd of about 29 as required above by the curve fitting. The value of $^{87}Sr/^{86}Sr \approx 0.7155$ derived above for the crustal contaminant is consistent with the average crustal values of 0.715–0.716 determined by Allegre et al. (1983).

Clearly the two-component mixing model does not provide a unique solution, but only serves to show that the Carpathian Nd-Sr array is consistent with a simple model involving crustal contamination of relatively pristine mantle magmas (where the required Sr and Nd isotope ratios, and Sr/Nd concentration ratios, are quite typical of values for parental arc basalts, and average continental crust). Note that, using the model parameters established above, the most "contaminated" Carpathian volcanic (rhyolite #126) requires about 67% crustal contamination.

The Pb-Pb isotope correlation diagrams (Figure 3) again demonstrate some similarity of the Carpathian data with the Banda Arc, the Martinique, and the Italian volcanics (field not shown; see Vollmer, 1976; Vollmer, 1977; and Vollmer and Hawkesworth, 1980). Sample 126, which is the most contaminated in terms of Sr and Nd isotopes, also has the highest $^{207}Pb/^{204}Pb$ and $^{208}Pb/^{204}Pb$. The Sr-Nd-$^{206}Pb/^{204}Pb$ isotope correlation diagrams (Figure 5) show that the crustal component has high Sr, low Nd, and low $^{206}Pb/^{204}Pb$ isotope ratios. Whereas the Nd-

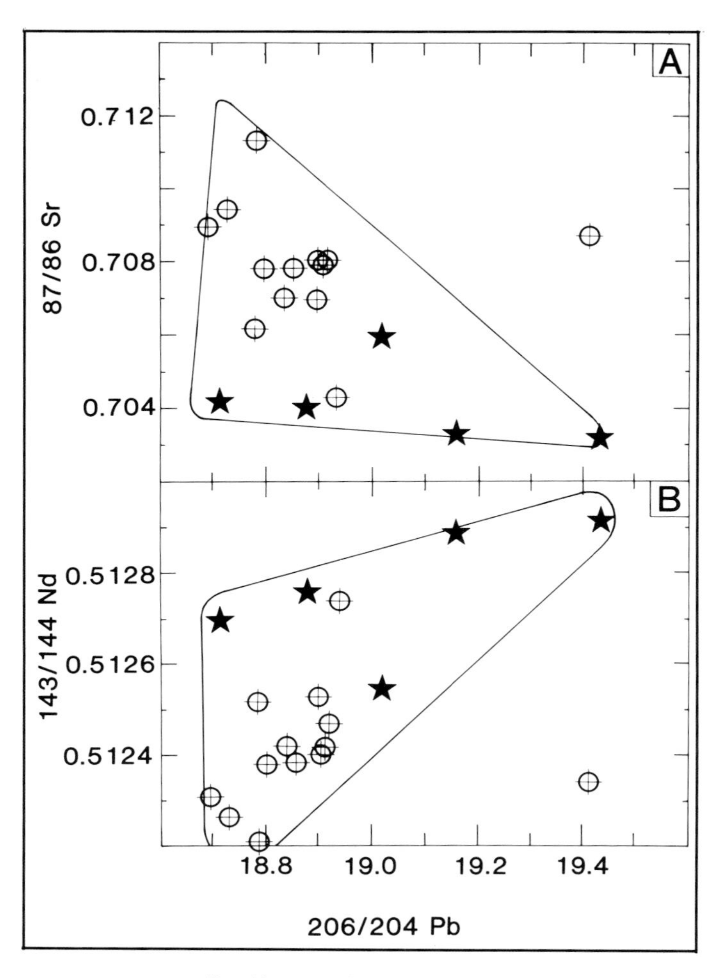

Figure 5. Initial $^{87}Sr/^{86}Sr$ and $^{143}Nd/^{144}Nd$ versus $^{206}Pb/^{204}Pb$. Alkali basalts (stars) and calcalkaline volcanics (circles) occupy a triangular-shaped field indicating three-component mixing. Sample 113 lies outside the three-component field. Excluding sample 134, the alkali basalt with the highest $^{87}Sr/^{86}Sr$ and lowest $^{143}Nd/^{144}Nd$ ratio, the alkali basalts form the low $^{87}Sr/^{86}Sr$ and high $^{143}Nd/^{144}Nd$ leg of the triangle.

Sr array resembles a two-component array, the combined Nd-Sr-Pb plots have the appearance of a three-component array, two of which are represented by the alkali basalts (samples 118 and 103). The "apparent" mantle endmember for the calcalkaline suite is intermediate between these two alkali basalts. We will consider first a model which produces the alkali basalt array (see Figures 2, 3, and 5) by crustal contamination of a pristine mantle endmember (note that alkali basalt #134 is not considered here as belonging to the Pliocene–Pleistocene alkali basalt suite). The mantle endmember is chosen to have the Sr, Nd, and Pb isotopic ratios and Sr, Nd concentrations of alkali basalt #118. The Pb content is estimated to be 5.6 ppm (based on the K content of #118, and assuming a K/Pb ratio ~3200, which is typical of oceanic alkali basalts). The crustal contaminant is assumed to have the Sr and Nd characteristics as derived above from the Nd-Sr array modeling. Using a $^{206}Pb/^{204}Pb$ ratio of 18.66 and a Pb concentration of 11 ppm, which are the values derived by Zartman and Doe (1981) for average continental crust, the resulting mixing curve (now shown) deviates markedly from the observed alkali basalt trend shown in Figure 5. A reasonable fit can only be achieved by invoking a very high Sr/Pb r-factor (> 100); this cannot be obtained without invoking unrealistically high (contaminant/basalt) Pb concentration ratios (e.g., 100 ppm in contaminant versus 2 ppm in basalt).

It is clear from the above that a crustal endmember with significantly lower $^{206}Pb/^{204}Pb$ is required as a contaminant. One extreme would be to assume that contamination is occurring in the lower crust, and utilize the fact that the lower crust is depleted in U (relative to Pb), leading to lower $^{206}Pb/^{204}Pb$ ratios. The plumbotectonics model of Zartman and Doe (1981) derives a $^{206}Pb/^{204}Pb$ ratio of 17.54 for the lower crust, with a Pb concentration of 6.4 ppm. As the lower crust presumably also is depleted in Pb relative to Sr, the $^{87}Sr/^{86}Sr$ probably is lower than the value of 0.7155 derived above. Based on the slightly steeper slope for the alkali basalt Nd-Sr array (Figure 2) than that generated by a crustal endmember with $^{87}Sr/^{86}Sr$ = 0.7155, a best fit to the alkali basalts above was derived using $^{87}Sr/^{86}Sr$ = 0.170 (and the observed Sr/Nd concentrations in alkali basalt #118). The alkali basalt trends in Figure 5 are now easily modeled using these lower crustal Pb and Sr isotopic parameters.

A summary of the fitting parameters is given in Table 2, and the model curves are shown in Figure 6. For our particular choice of parameters, this model requires a relatively large fraction of crustal contaminant (33%) in sample #103, and it is arguable as to whether the major element chemistry of sample # 103 will allow this amount of contaminant (permissible as long as the contaminant is no more siliceous than a diorite composition). The amount of contaminant can be decreased by a factor of two if a Pb concentration of 16 ppm is used for the lower crustal contaminant. To refit the data with the Pb value would require ad hoc but acceptable changes in the other fitting parameters.

Turning now to the Sr-Nd-Pb isotopic relationships of the calcalkaline rock suite (Figure 5), it is clear that the simple two-component mixing model which was adequate to describe the Nd-Sr array will not serve once the Pb data is added as a constraint. At first glance, Figure 5 suggests three-component mixing between a crustal contaminant and a variety of "parental" basalts lying along the alkali basalt trend. In light of our lower crustal contamination model for the alkali basalts, this would require a two-stage contamination process. First, variable contamination of pristine, mantle-derived parental basalts having the Sr-Nd-Pb isotopic character of sample #118 by lower crustal materials, followed by a second stage of contamination of these parental basalts by upper crustal material. The alkali basalts are obviously only an analogue for the first-stage process, since they post date and are certainly not the parental basalts to the calcalkaline suite. A variant model, which would provide a mantle source common to both the alkali and calcalkali suite, would envision a subarc mantle which was itself contaminated during the Cretaceous subduction event by subducted materials having a lower crustal Pb isotopic signature. This contaminated mantle source would then be tapped for both the Miocene calcalkaline suite and the later alkali basalt suite, with the former experiencing crustal contamination and the latter erupting through the crust with no interaction. Insofar as any of the contaminated alkali basalts can be shown to carry ultramafic xenoliths, this model would be preferable to one invoking lower crustal contamination during eruption of the alkali basalts. It

Table 2. Nd-Sr-Pb Mixing Parameters

	Alkali basalt array		Calc-alkaline array		
	Alkali basalt mantle melt	Lower crust contaminant	Calcalkaline parental mantle melt	Crustal contaminant low Pb	high Pb
Sr, ppm	841	350	375	350	350
Nd, ppm	42.5	29	13	29	29
Pb, ppm	6.2	8	4.5	12	80
87/86 Sr	0.7032	0.7100	0.7032	0.7155	0.7155
143/144 Nd	0.512915	0.51204	0.512915	0.51204	0.51204
206/204 Pb	19.426	17.54	19.426	18.660	18.660
Sr/Nd	19.8	12.1	28.8	12.1	12.1
Sr/Pb	135.6	43.8	83.3	29.2	4.4

seems clear from the nature of the data shown in Figure 5, and considering the large number of free parameters available in this model, that a satisfactory and qualitative, although ad hoc, description of the data could be achieved.

If we consider only the calcalkaline suite, a reasonable contamination model can also be constructed that involves only a single pristine parental basalt, and a crustal contaminant of slightly heterogeneous composition with respect to Pb. The same Sr and Nd parameters are used as were used for the Nd-Sr array model (see Table 2); the $^{206}Pb/^{204}Pb$ ratio of average crust (18.66) from Zartman and Doe (1981) is used. The parental basalt $^{206}Pb/^{204}Pb$ ratio is that of alkali basalt #118 (19.426). The Pb concentration of the parental basalt is fixed at 4.5 ppm (derived from the K-SiO_2 relationship of the calcalkaline suite, and a K/Pb ratio for typical arc basalts of 2050); the Pb concentration of the contaminant is allowed to be a free parameter.

As shown in Figure 6, the calcalkaline data suite can be bounded by two mixing curves, one using 12 ppm Pb and one using 80 ppm Pb for the crustal contaminant (mixing parameters are given in Table 2). The quantity and nature of the contaminant required for any given sample are consistent on all three isotope plots (Nd-Sr, Sr-Pb, and Nd-Pb). Rhyolite #126 requires 67% of the low-Pb contaminant, and andesites #68, # 87, #61 and #128 require 9%, 25%, 48% and 52% of the high-Pb contaminant respectively.

The Pb concentration used in the low-Pb case (12 ppm) is close to the average crustal Pb of 11.2 ppm from the plumbotectonics model (Zartman and Doe, 1981); the high-Pb case (80 ppm) is clearly extreme for any reasonable crustal materials, although partial melts of crustal materials with Pb concentrations in this range have been postulated (Vollmer and Hawkesworth, 1980). However, if the $^{206}Pb/^{204}Pb$ ratio of the contaminant is allowed to be a free parameter, the calcalkaline data can be bounded with a constant Pb concentration of 12 ppm and the $^{206}Pb/^{204}Pb$ ratio ranging from 18.1–18.66 for the contaminant. The amount of contaminant required in the limiting samples discussed above (e.g., #61 and #128) would not change. Since Pb concentration and $^{206}Pb/^{204}Pb$ ratio can be arbitrarily traded off against each other, a definitive choice of these parameters cannot be made from the present data set. The Pb concentrations in the resulting contaminated samples (such as #61) will differ radically as a function of this trade-off (from ~40 ppm Pb to ~8 ppm Pb). We are pursuing this question with Pb concentration measurements of selected samples.

Contamination of the type discussed above should be reflected in the major and trace element chemistry of the calcalkaline volcanics. However, conventional diagrams of major or trace elements and any ratios of these elements do not show any simple relationship with the isotopes, except for SiO_2 vs K, Rb and $^{87}Sr/^{86}Sr$. The $^{207}Pb/^{204}Pb$ and $^{208}Pb/^{204}Pb$ isotope ratios are not as well correlated with the Nd and Sr isotopes as $^{206}Pb/^{204}Pb$. Even within one eruption center a clear trend is not observed.

A different way to express the $^{207}Pb/^{204}Pb$ and $^{208}Pb/^{204}Pb$ isotope ratios is as vertical deviations from the so-called Northern Hemisphere mantle reference line, Δ207/204 Pb and Δ208/204 Pb (Hart, 1984).

Figure 7 shows the Δ208/204 Pb, Δ207/204 Pb variation of the Hungarian volcanics, in comparison to oceanic volcanics and selected other arcs. The positive correlation between the two ΔPbs simply indicates that samples which are high in $^{207}Pb/^{204}Pb$ are also high in $^{208}Pb/^{204}Pb$. This plot is more effective at distinguishing between oceanic and sediment leads than the conventional Pb-Pb plots (e.g., see Figure 3) as has also been noted by Hickey and Frey (1984) in plots of $^{207}Pb/^{204}Pb$ versus $^{208}Pb/^{204}Pb$. All of the calcalkaline samples, and four of the five alkali basalts, lie outside the oceanic field and within the field of pelagic sediments (crustal Pb). Only alkali basalt #118 therefore has the Sr-Nd-Pb isotope signature of pristine (oceanic) mantle; the other volcanics, including the four other alkali basalts, therefore, contain a clear crustal "contamination" signature (as already assumed above). Using the Δ207/204 Pb data as an index of contamination, correlations with other contamination-sensitive indices can be sought. Figure 8 suggests a weak, negative correlation between Ba/Rb ratio and SiO_2. On a plot of Ba/Rb versus Δ207/204 Pb, this correlation is considerably more convincing. An important aspect of the contamination can be shown on this combined diagram: the low Ba/Rb, low silica calcalkaline magmas have high Δ207/204 Pb signatures indicating large amounts of contamination even though the erupted products are still basaltic andesite.

On Δ207/204 Pb, $^{87}Sr/^{86}Sr$ and $^{143}Nd/^{144}Nd$ diagrams (Figure 9), the Carpathian volcanics can be bounded by triangular fields, where the apexes are the same as identified in Figure 5. The low Sr and high Nd isotope ratios for the "pristine" mantle endmember (118) is a reflection of time-integrated low Rb/Sr and high Sm/Nd ratio for the mantle source; similarly, the high Sr, low Nd isotopic ratios of the crustal endmember reflect time-integrated high Rb/Sr and low Sm/Nd ratios. These trace element ratios of the erupted lavas do not correlate with the

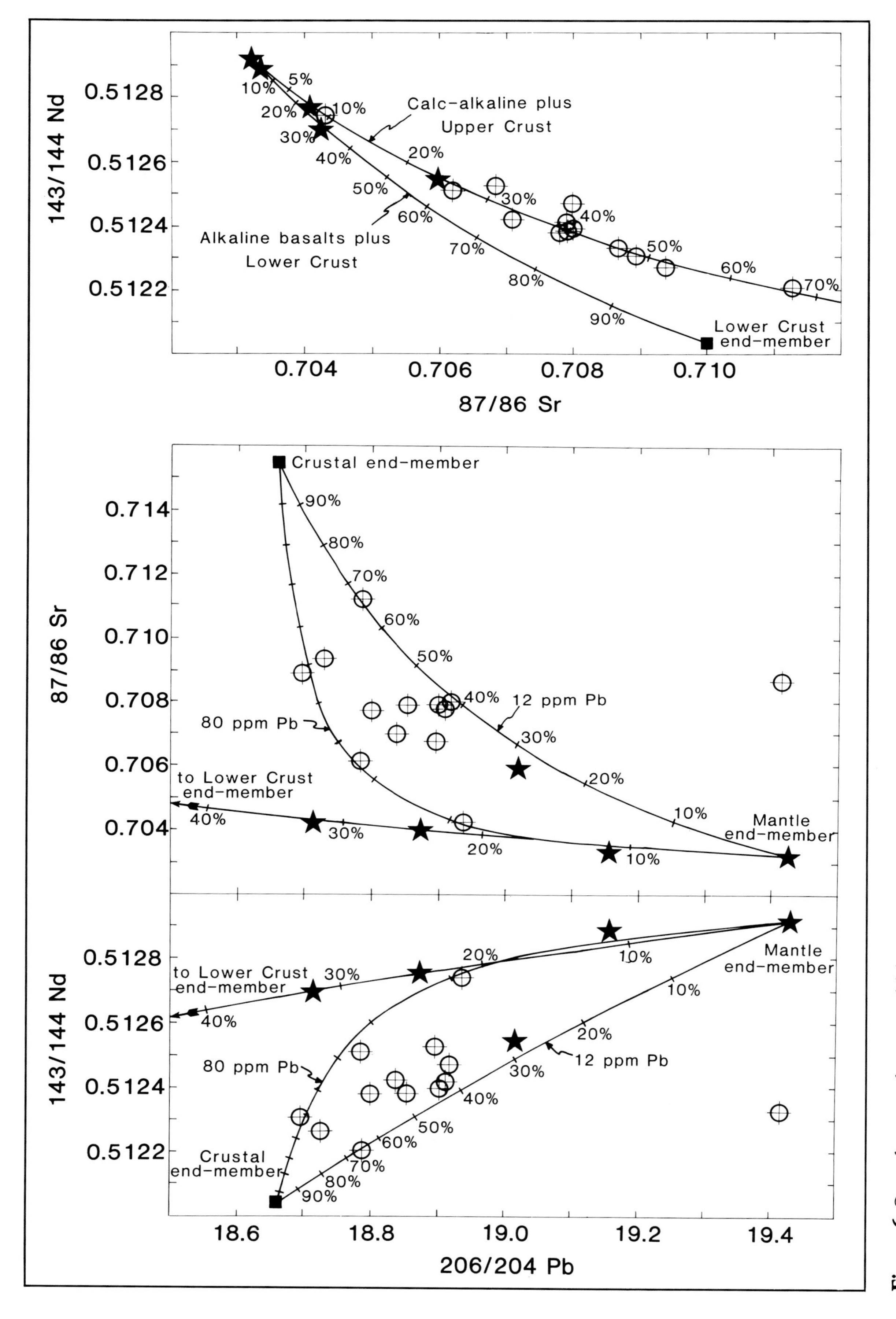

Figure 6. Curves for proposed mixing models bounding the Carpathian dataset for Sr, Nd, and Pb isotopic composition and content. Percentages on the mixing lines are percents of crustal endmember needed. Note that one mantle and two crustal endmembers can explain the data. The mixing curve labeled with 80 ppm Pb and 12 ppm Pb indicate the Pb content of the crustal endmember. The Pb content or the Pb isotopic composition of the crustal endmember has to vary in order to explain the Pb isotope variation in the calcalkaline volcanics.

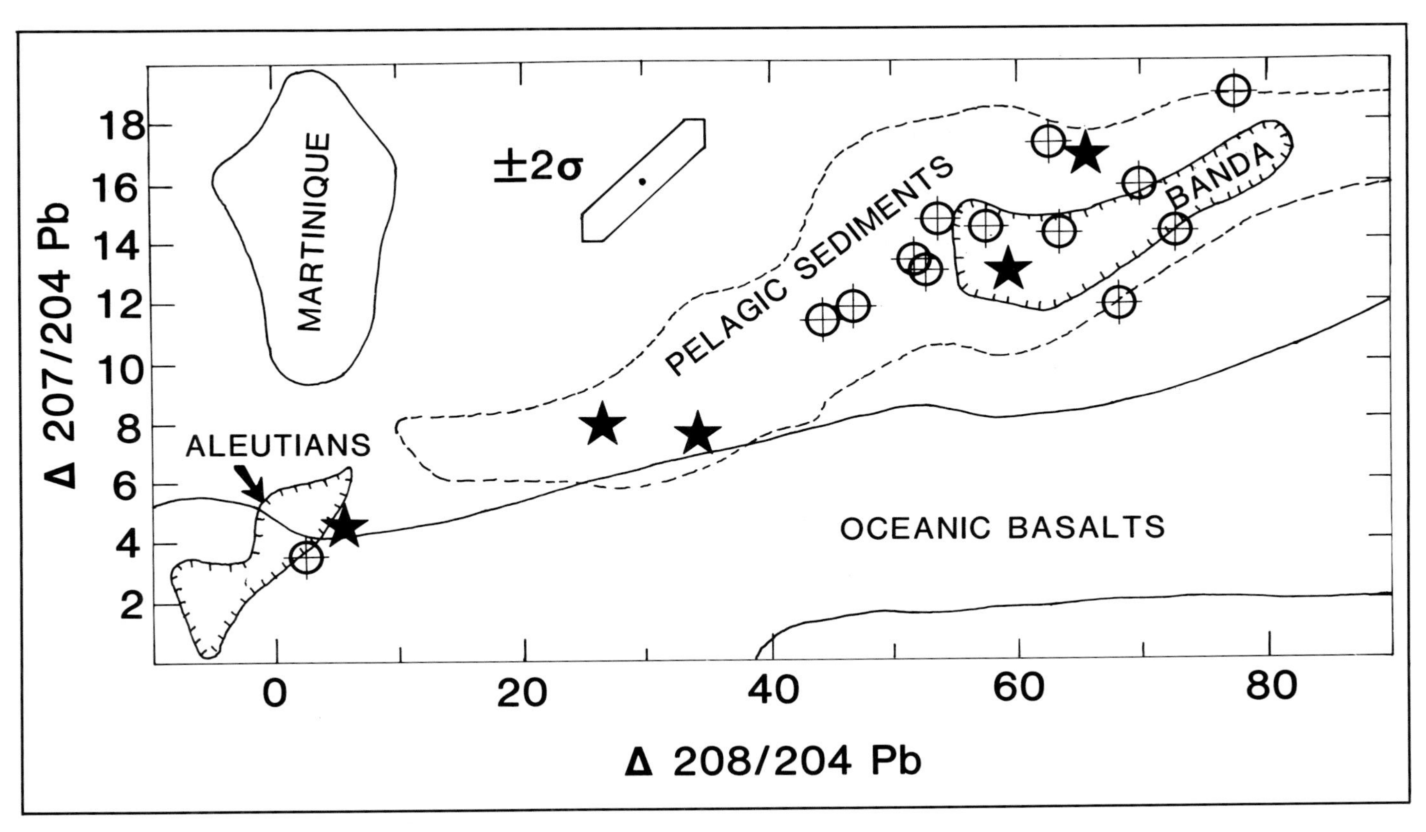

Figure 7. Δ207/204 Pb, Δ208/204 Pb diagram. ΔPbs are the deviation in $^{207}Pb/^{204}Pb$ and $^{208}Pb/^{204}Pb$ from the Northern Hemisphere Reference Line (NHRL). Equations for HNRL are:

$$(^{207}Pb/^{204}Pb)\,NHRL = 0.1084(^{206}Pb/^{204}Pb)m + 13.491$$
$$(^{208}Pb/^{204}Pb)NHRL = 1.209\,(^{206}Pb/^{204}Pb)m + 15.62$$

which are the $^{207}Pb/^{204}Pb$ and $^{208}Pb/^{204}Pb$ ratios of the NHRL for a measured $^{206}Pb/^{204}Pb$ ratio.

$$\Delta 207/204\ Pb = [(^{207}Pb/^{204}Pb)m - (^{207}Pb/^{204}Pb)NHRL] \times 100$$
$$\Delta 208/204\ Pb = [(^{208}Pb/^{204}Pb)m - (^{208}Pb/^{204}Pb)NHRL] \times 100$$

are the deviations from the NHRL, all according to Hart (1984). Compare this figure with Figure 3 and note that the pelagic sediments are separated from the oceanic basalts. The Carpathian volcanics lie almost entirely within the field for pelatic sediments. Data for oceanic basalts, Banda Arc, and Martinique as in Figure 3. Field for Aleutians from Kay et al., 1978 and Morris and Hart, 1983.

isotope ratios. There is, however, a positive correlation between Rb/Sr and Sm/Nd and **Δ207/204 Pb** (Figure 9). The positive correlation of Sm/Nd with **Δ207/204 Pb** is rather unusual, since it implies a higher Sm/Nd in the crustal component than in the mantle component. The results also in a positive correlation between Rb/Sr and Sm/Nd (not shown).

The crystalline phases in the calcalkaline magmas are plagioclase, clinopyroxene, orthopyroxene, amphibole, biotite and titanomagnetite. Amphibole and biotite contain inclusions of apatite. Published distribution coefficients for these mineral phases (Woerner et al., 1983; Nagasawa and Schnetzler, 1976; and Gill, 1981), lead to bulk distribution coefficients all higher for Sm than for Nd, with the consequence that fractional crystallization will always tend to decrease the Sm/Nd ratio, unless minerals such as apatite and monazite start to crystallize (> 1%) at a late stage. This implies that the basic calcalkaline lavas have Sm/Nd ratios which are minimum values for the calcalkaline parental magma. Therefore, the magmas parental to the calcalkaline suite have Sm/Nd ratios at least as high as the highest Sm/Nd ratio in the basaltic andesites, that is Sm/Nd = 0.250, and thus higher than the Sm/Nd ratios of the alkali basalts (< 0.21). The low Sm/Nd ratios and relatively high Nd isotope ratios of the alkali basalts indicate decoupling between the trace elements and the isotopes, as is true for most alkali basalts. The decoupling between the trace elements and the isotopes in the alkali basalts, and the shift to lower Sm/Nd ratios, can be explained by one of the following processes:

1. Lower degrees of partial melting of the same mantle source which generated the calcalkaline magmatism.
2. A recent LREE-enrichment event in the same mantle source that generated the calcalkaline magmas but after the calcalkaline magmas were produced.

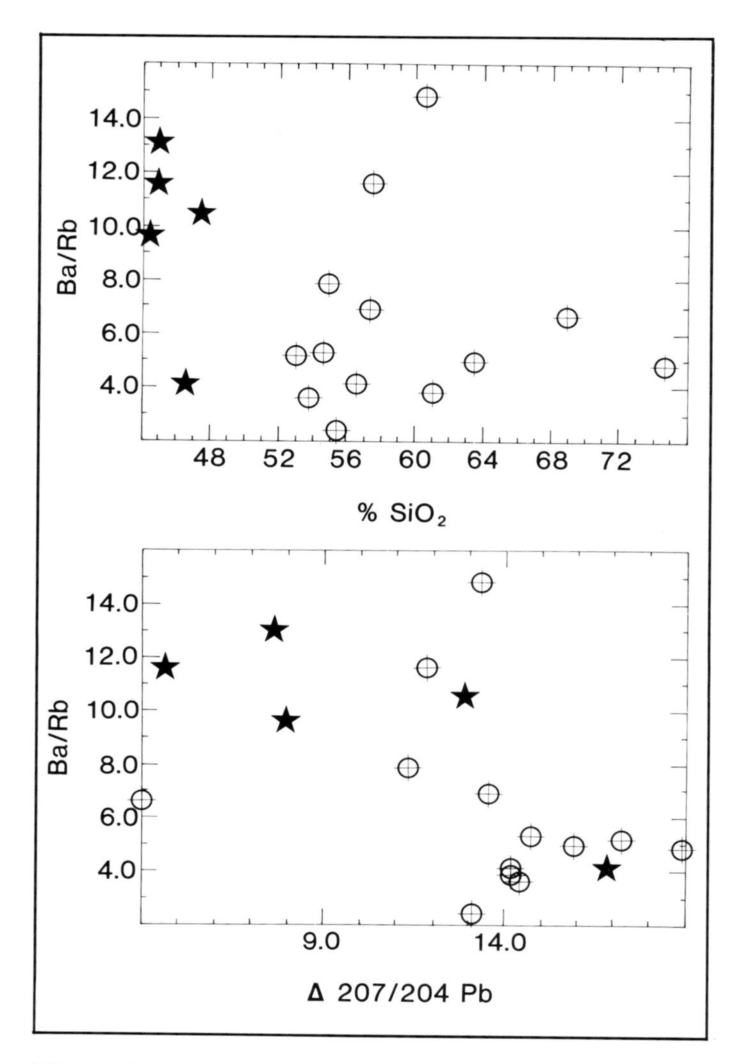

Figure 8. Ba/Rb versus Δ207/204 Pb and weight percent SiO_2. Symbols as in Figure 2. Note the lack of correlation on the SiO_2 diagram and the more coherent behavior of the Ba/Rb with Δ207/204 Pb.

3. Case 1 or 2, but in a mantle not related to the mantle source for the calcalkaline volcanism.
4. The stability of a phase rich in HREE somewhere during the history of the alkali basalts. The most obvious phase for this is garnet.

On the basis of REE data, Peccerillo and Taylor (1976) argue for this in order to explain the high LREE/HREE ratios of the calcalkaline magmas. Since the alkali basalts in general have higher HREE contents than the calcalkaline suite (Pantó, 1981; Peccerillo and Taylor, 1976), case 4 appears untenable. The close spatial relationship and the similar isotopic trends of the alkali basalts and the calcalkaline volcanics is an argument against case 3, although case 3 cannot be eliminated with certainty. It is even more difficult to distinguish between case 1 and 2 on the basis of the present data set. The occurrence of amphiobolite veins in the lherzolite nodules (Embey-Isztin, 1976) is evidence for a LREE enrichment event in the mantle. Case 2 is also compatible with the contamination and fractional crystallization models discussed earlier. On the basis of the limited evidence available we therefore tend to favor case 2.

CONCLUSIONS

In summarizing the above discussion, the following conclusions can be drawn:

1. The mantle underlying the Carpathians may consist of at least two components. Those two components are similar in Nd and Sr isotopic composition, but differ radically in $^{206}Pb/^{204}Pb$ isotope ratio. One of the mantle components has a lower crustal Pb signature.
2. The calcalkaline magmas are generated by contamination with crustal-type materials. Either one or both of the endmembers to this mixing process (mantle melts and crustal contaminant) are chemically and isotopically heterogeneous. The most silisic endmember of the calcalkaline suite (rhyolite) contains up to 67% of crustal material; the least contaminated member contains only 10% of crustal material.
3. The parental calcalkaline and alkaline magmas can be generated out of the same mantle reservoir.
4. Several lines of evidence indicate that the source for the alkaline magmas underwent a recent LREE enrichment event. This caused decoupling between the trace elements and the isotopes, especially Sm/Nd and $^{143}Nd/^{144}Nd$.

ACKNOWLEDGMENTS

L. Gulen is acknowledged for teaching V. S. the analytical intricacies and pitfalls which are part of the world of mass spectrometry. The graphical representation is the work of D. Walker. K. Burrhus somehow managed to keep the quirks of the mass spectrometers below irritation level. L. Gulen, N. Shimizu and B. Taras are thanked for their helpful discussions. This work was supported by NSF grant EAR 8219853.

REFERENCES

Allegre, C. J. and J.-F. Minster, 1978, Quantitative models of trace element behaviour in magmatic processes, Earth Planet. Sci. Lett., v. 38, p. 1–25.

Allegre, C. J., S. R. Hart, and J.-F. Minster, 1983, Chemical structure and evolution of the mantle and continents determined by inversion of Nd and Sr isotopic data, II, Numerical Results and Discussion: Earth Planet. Sci. Lett., v. 66, p. 191–213.

Barreiro, B., 1983, Lead isotopic compositions of South Sandwich island volcanic rocks and their bearing on magmagenesis in intraoceanic island arcs: Geochim. Cosmochim. Acta., v. 47, p. 817–822.

Burchfiel, B. C., 1980, Eastern European alpine system and the Carpathian orocline as an example of collision tectonics: Tectonophysics, v. 63, p. 31–61.

Burchfiel, B. C. and L. Royden, 1982, Carpathian foreland fold and thrust belt and its relation to Pannonian and other basins: AAPG Bull., v. 66, p. 1179–1195.

Church, C. E., 1976, The Cascade Mountains revisited: A reevaluation in light of new lead isotopic data: Earth Planet. Sci. Lett., v. 29, p. 175–188.

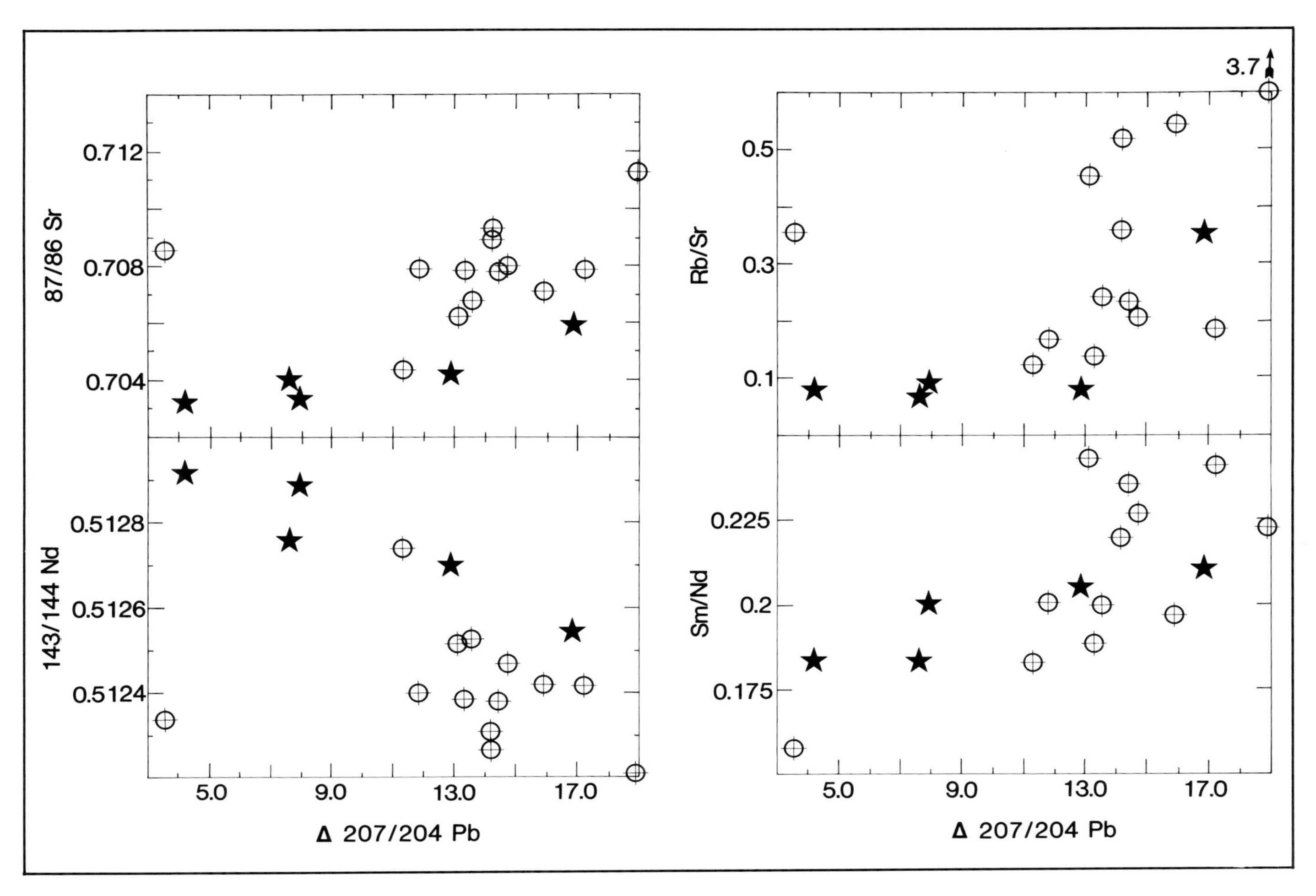

Figure 9. Initial $^{87}Sr/^{86}Sr$, $^{143}Nd/^{144}Nd$, Rb/Sr and Sm/Nd versus Δ207/204 Pb. Note the low Sm/Nd ratios in the alkali basalts (stars), which combined with the 143/144 Nd ratios indicate decoupling between the isotopes and the trace elements for the alkali basalts. Also the triangular shaped field as in Figure 5 is preserved. Crystal fractionation tends to decrease the Sm/Nd and increase the Rb/Sr ratio. The calcalkaline sample # 113 plots of trend on all the diagrams. The mantle endmember (alkali basalt #118) is the star plotting at the low Δ207/204 Pb side on the diagrams. A recent LREE enrichment event tends to lower the Sm/Nd and increase the Rb/Sr ratios. Present Rb/Sr ratios of the alkali basalts are somewhat higher than and Sm/Nd ratios are somewhat lower than typical oceanic alkali basalts.

Clague, D. A., and F. A. Frey, 1982, Petrology and trace element chemistry of the Honolulu volcanics, Oahu: Implication for the oceanic mantle below Hawaii: J. Petrol., v. 23, p. 447–504.

Davidson, J. P., 1983, Lesser Antilles isotopic evidence of the role of subducted sediment in island arc magma genesis: Nature, v. 306, p. 253–256.

DePaolo, D. J. and R. W. Johnson, 1979, Magma genesis in the New Britain island-arc: Constraints from Nd and Sr isotopes and trace element patterns: Contrib. Mineral. Petrol., v. 70, p. 367–379.

DePaolo, D. J., 1982, Trace element and isotopic effects of combined wallrock assimilation and fractional crystallization: Earth Planet. Sci. Lett., v. 53, p. 189–202.

Embey-Isztin, A., 1976, Amphibolite/Lherzolite composite xenolith from Szigliget, north of Lake Balaton, Hungary: Earth Planet. Sci. Lett., v. 31, p. 297–304.

Gill, J. B., 1981, Orogenic andesites and plate tectonics: Berlin, Springer-Verlag, 389 pages.

Goldstein, S. L., R. K. O'Nions, and P. J. Hamilton, 1984, A Sm-Nd isotopic study of atmospheric dusts and particulates from major river systems: Earth Planet. Sci. Lett., v. 70, p. 221–236.

Hart, S. R. and C. Brooks, 1977, The geochemistry and evolution of early Pre-cambrian Mantle: Contrib. Mineral. Petrol., v. 61, p. 109–128.

Hart, S. R., 1984, The DUPALL anomaly: A large scale isotopic mantle anomaly in the Southern Hemisphere: Nature, v. 309, p. 753–757.

Hawkesworth, C. J. and R. Vollmer, 1979, Crustal contamination versus enriched mantle: $^{143}Nd/^{144}Nd$ and $^{87}Sr/^{86}Sr$ evidence from the Italian volcanics: Contrib. Mineral. Petrol., v. 69, p. 151–165.

Hawkesworth, C. J., R. K. O'Nions, and R. J. Arculus, 1979, Nd and Sr isotope geochemistry of island arc volcanics, Grenada, Lesser Antilles: Earth Planet. Sci. Lett., v. 45, p. 237–248.

Hawkesworth, C. J. and M. Powell, 1980, Magma genesis in the Lesser Antilles island arc: Earth Planet. Sci. Lett., v. 51, p. 297–308.

Hickey, R. L. and F. A. Frey, 1984, Sources for arc volcanics: Evidence from central Chilean basalts, Proceedings of the ISEM Filed Conference on Open Magmatic Systems, edited by M. A. Dungan, T. L. Grove, and W. Hildreth, Institute for the Study of Earth and Man, Southern Methodist University, Dallas, Texas.

Kay, R. W., S. S. Sun, and C. N. Lee-Hu, 1978, Pb and Sr isotopes in volcanic rocks from the Aleutians and Pribilof Islands, Alaska: Geochim. Cosmochim. Acta., v. 42, p. 263–273.

Kay, R. W., 1980, Volcanic arc magmas: Implications of a melting-mixing model for element recycling in the crust upper mantle system: J. Geol., v. 88, p. 497–522.
Langmuir, C. H., R. D. Voche, Jr., G. N. Hanson, and S. R. Hart, 1978, A general mixing equation with applications to Icelandic basalts: Earth Planet. Sci. Lett., v. 37, p. 380–392.
Manhes, G., J.-F. Minster, and C. J. Allegre, 1978, Comparative uranium-thorium-lead and rubidium-strontium of St. Severin amphoterite: Consequences for early solar system chronology: Earth Planet. Sci. Lett., v. 39, p. 14–24.
Morris, J. D., and S. R. Hart, 1983, Isotopic and incompatible element constraints on the genesis of island arc volcanics: Cold Bay and Amak Islands, Aleutians: Geochim. Cosmochim. Acta., v. 47, p. 2015–2030.
Morris, J. D., 1984, Enriched geochemical signatures in Aleutian and Indonesian arc lavas: An isotopic and trace element investigation, Ph.D. Thesis, Massachusetts Institute of Technology, Cambridge, Massachusetts.
Nagasawa, H. and C. S. Schnetzler, Partitioning of rare earth, alkali and alkaline earth elements between phenocryst and acidic igneous magma: Geochim. Cosmochim. Acta., v. 35, p. 953–986.
Nicholls, I. A., and A. E. Ringwood, 1973, Effect of olivine stability in tholeiites and the production of silica undersaturated magmas in the island arc environment: J. Geol., v. 81, p. 285–300.
Pantó, Gy., 1981, Rare earth element geochemical pattern of the Cenozoic volcanism in Hungary: Earth Evolution Sciences, v. 3–4, p. 249–256.
Patchett, P. J. and M. Tatsumoto, 1980, Hafnium isotope variations in oceanic basalts: Geophys. Res. Lett., v. 7, p. 1077–1080.
Peccerillo, A., and S. R. Taylor, 1976, Rare earth elements in East Carpathian volcanic rocks: Earth Planet Sci. Lett., v. 32, p. 121–126.
Pegram, W. J., 1985, Geochemical processes in the sub-continental mantle and the nature of the crust-mantle interaction: Evidence from the Mesozoic Appalachian Tholeiite Province, Ph.D. Thesis, Massachusetts Institute of Technology, Cambridge, Massachusetts.
Reynolds, P. H. and E. J. Dasch, 1971, Lead isotopes in marine manganese nodules and the ore-lead growth curve: J. Geophys. Res., v. 76, p. 5124–5129.
Richard, P. N. Shimizu, and C. J. Allegre, 1976, $^{143}Nd/^{144}Nd$, A Natural Tracer: An application to oceanic basalts: Earth Planet. Sci. Lett., v. 31, p. 269–278.
Royden, L., F. Horváth, and J. Rumpler, 1983, Evolution of the Pannonian basin system, 1: Tectonics, 2, p. 63–90.
Staudigel, H., A. Zindler, S. R. Hart, T. Leslie, C.-Y. Chen, and D. Clague, 1984, The isotope systematics of a juvenile intraplate volcano: Pb, Nd, and Sr isotope ratios of basalts from Loihi Seamount, Hawaii: Earth Planet. Sci. Lett., v. 69, p. 13–29.
Stegena, L., B. Géczy, and F. Horváth, 1975, Late Cenozoic evolution of the Pannonian basin: Tectonophysics, v. 26, p. 71–90.
Stern, R. J., 1982, Strontium isotopes from circum-pacific intra-oceanic island arcs and marginal basins: Regional variations and implications for magmagenesis in island arcs: Geol. Soc. Am. Bull., v. 93, p. 477–486.
Sun, S. S., 1980, Lead isotopic study of young volcanic rocks from mid-ocean ridges, ocean islands, and island arcs: Phil. Trans. Roy. Soc. London, A297, p. 409–445.
Taylor, H. P., Jr., B. Giannetti, and B. Turi, 1979, Oxygen isotopes geochemistry of the potassic igneous rocks from the Roccamonfina volcano, Roman comagmatic region, Italy: Earth Plan. Sci. Lett., v. 46, p. 81–106.
Vidal, P. L. and N. Clauer, 1981, Pb and Sr systematics of some basalts and sulfides from the East Pacific rise at 21 °N (Project RITA): Earth Planet. Sci. Lett., v. 55, p. 237–246.
Vogl, M. and Gy. Pantó, 1983, Geochemistry of young alkaline basaltic volcanism in Hungary, *In*: the significance of trace elements in solving petrogenetic problems and controversies. p. 233–256. Theophrastus Publications S.A. Athens.
Vollmer, R., 1977, Isotopic evidence for genetic relations between acid and alkaline rocks in Italy: Contrib. Mineral. Petrol., v. 60, p. 109–118.
Vollmer, R., 1976, Rb-Sr and U-Th-Pb systematics of alkaline rocks from Italy: Geochim. Cosmochim. Acta., v. 40, p. 283–295.
Vollmer, R. and C. J. Hawkesworth, 1980, Lead isotopic composition of the potassic rocks from Roccamonfina (South Italy): Earth Planet. Sci. Lett., v. 47, p. 91–101.
White, W. M., M. D. M. Tapia, and J.-G. Schilling, 1979, The petrology and geochemistry of the Azores Islands: Contrib. Mineral. Petrol., v. 69, p. 201–213.
White, W. M., and P. J. Patchett, 1984, Hf-Nd-Sr isotopes and incompatible element abundances in island arcs: Implications for magma origins and crust-mantle evolution: Earth Planet. Sci. Lett., v. 67, p. 167–185.
Whitford, D. J. and P. A. Jezek, 1979, Origin of late-Cenozoic lavas from the Banda Arc, Indonesia: Trace element and Sr isotope variation: Contrib. Mineral. Petrol., v. 68, p. 141–150.
Woerner, G., J.-M. Beusen, N. Duchateau, R. Gijbels, and H.-U. Schmincke, 1983, Trace element abundances and mineral/melt distribution coefficients in phonolites from the Laacher Sea volcano (Germany): Contrib. Mineral. Petrol., v. 84, p. 152–173.
Zartman, R. E., and B. R. Doe, 1981, Plumbotectonics: The model: Tectonophysics, v. 75, p. 135–162.
Zindler, A., 1980, Geochemical processes in the earth's mantle and the nature of crust-mantle interactions: Evidence from studies of Nd and Sr isotope ratios in mantle-derived igneous rocks and lherzolite nodules, Ph.D. Thesis, Massachusetts Institute of Technology, Cambridge, Massachusetts.

CHAPTER 20

Helium Isotopes in Geothermal Waters from Northwest Hungary

J. Deák
Research Center for Water Resources Development, VITUKI
H-1095 Budapest, Kuassay J. út 1, Hungary

F. Horváth
Geophysical Department, Eötvös University
H-1083 Budapest, Kun Bela ter 2, Hungary

D. J. Martel
R. K. O'Nions
E. R. Oxburgh
Department of Earth Sciences,
University of Cambridge,
Cambridge, U.K.

L. Stegena
Department of Cartography, Eötvös University
H-1083 Budapest, Kun Bela ter 2, Hungary

$^3He/^4He$ ratios and helium concentrations have been measured for 15 thermal waters from northwest Hungary. All show a large enrichment in helium, relative to equilibrium solubility at recharge. The $^3He/^4He$ ratios ($0.22 > R/R_a < 1.07$) show an appreciable component of mantle-derived helium in all samples.

INTRODUCTION

This chapter presents the preliminary results of an investigation of the isotopic composition of helium dissolved in the groundwater systems of northwest Hungary. The area is one that is well known for its hot springs and the abundant availability of hot water from wells. It has been shown to be a region of high heat flow (Horváth et al., 1981).

For the interpretation of the tectonic history of the region, it is important to understand the origin of the thermal anomaly. It is to be expected that if the anomaly is produced by an enhanced mantle contribution to the surface heat flow, it will be associated with a locally enhanced release of mantle volatiles into the crust. Although a number of species contribute to the volatile flux, the only one that can be identified with certainty as mantle derived is helium with a $^3He/^4He$ (R) value of about 10^{-5}. This is in marked contrast to He generated in the crust ($R \approx 10^{-8}$). In many geothermal areas the helium is a mixture of gas from both sources (Figure 1). In a few, such as Yellowstone,

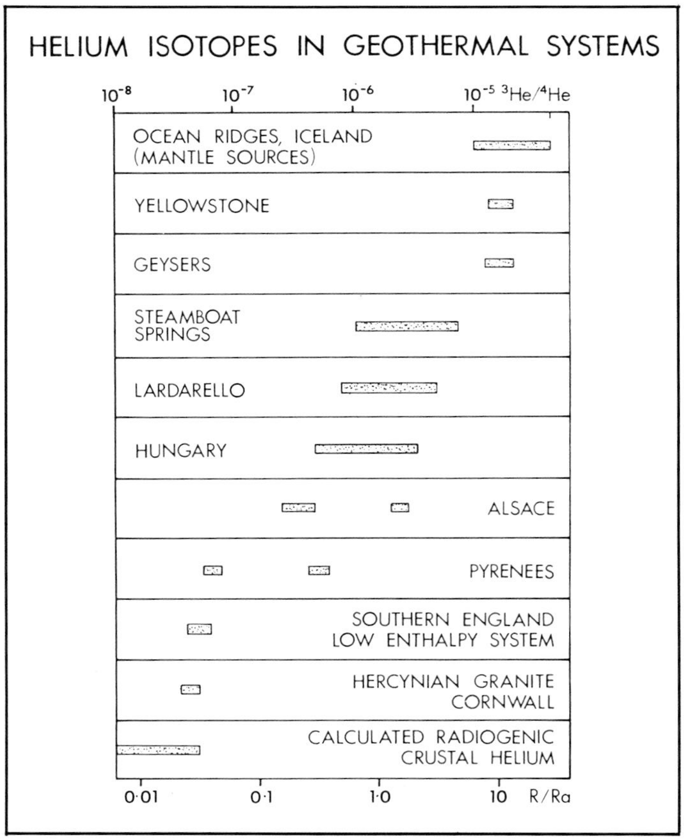

Figure 1. Comparison of $^3He/^4He$ ratios for a number of different geothermal groundwater systems. Ratios normalized to the atmospheric value ($R_a = 1.4 \times 10^{-6}$) are also shown.

the mantle component dominates and swamps the crustal component. In the majority of cases, however, there are significant amounts of both present.

GEOLOGICAL SETTING

The geology of northwest Hungary is characterized by a pre-Tertiary basement that comprises a Mesozoic sedimentary sequence overlying a metamorphic complex of variable grade. The basement is known from restricted areas of surface outcrop and a great many borehole observations. Within the basement the best aquifer is provided by Triassic carbonates with strong joint permeability that has been enhanced by karstification. The Mesozoic rocks are largely subhorizontal to gently inclined, but locally have undergone thrusting and strong folding. During the Tertiary there has been extensive differential subsidence of the basement and locally Tertiary clastics have accumulated up to a thickness of 4 km. At various times during the Tertiary there have been episodes of basaltic and andesitic igneous activity.

In the area of interest the main outcrop of the basement is provided by the Transdanubian Mountains that trend northeast-southwest from Budapest to Lake Balaton. These form a recharge area for a subsurface hydrological system that in the west flows southwestward toward the western end of Balaton, and in the east shows a more complex local pattern in the vicinity of Budapest. The map (Figure 2) shows the general characteristics of these flows.

It is to be expected that recently infiltrated rainwater will carry a largely atmospheric He signature, but during its circulation atmospheric component should be progressively diluted by He acquired from the rocks through which it is passing.

ANALYTICAL TECHNIQUES

Samples were collected in Weiss-type Cu-tube sample vessels, as described by Hooker et al. (1985). The extraction of helium and its purification and analysis are identical to those described by Hooker et al.

Helium concentrations were measured by peak height comparison with a standard volume of air helium, which was also used to normalize the $^3He/^4He$ ratios.

A correction was made for atmospheric contamination, by assuming that all the neon in the sample is atmospheric, and then correcting the ratio R/R_a with the following equation:

$$(R/R_a) = [(R/R_a) \cdot X - 1]/[X - 1]$$

where $X = (He/Ne)_{sample}/(He/Ne)_{10\,°C water}$.

RESULTS

Helium concentrations and isotopic analyses are reported in Table 1. Concentrations vary between 0.51 and 18.57×10^{-6} cm^3 STP g^{-1}, which is from 11 to 400 times the equilibrium solubility of helium in pure water at 10 °C. This excess helium has been acquired by the waters during their passage from the recharge area to the place of emergence.

$^3He/^4He$ ratios for each sample (R) are compared with the atmospheric $^3He/^4He$ ratio (R_a), and show a range of $0.22 < R/R_a < 1.07$, but with most values falling around 0.2 to 0.4 (Figures 1 and 2). This represents a significant input of mantle helium up to about 15%.

The samples from the Budapest area (H1–H9) are relatively uniform in isotope composition, whereas those from the area west of Balaton exhibit a wider range in both R/R_a value and He concentrations. In both areas there is a tendency for R/R_a to increase with decreasing contents of ^{14}C in the groundwater (Figure 3). The older waters, therefore, tend to show a greater contribution of mantle He. The ^{14}C concentration data (Deák, 1979) were measured some years earlier than the helium measurements. Ideally, both measurements should be made on the same samples at the same time.

DISCUSSION

The $^3He/^4He$ ratios measured in the samples indicate the acquisition of an He signature from the local aquifer, with the original atmospheric signal present from recharge quickly being swamped. This can be seen by the high values of helium concentration compared to the value of atmospheric equilibrium concentration in water at 10 °C, even in the youngest waters.

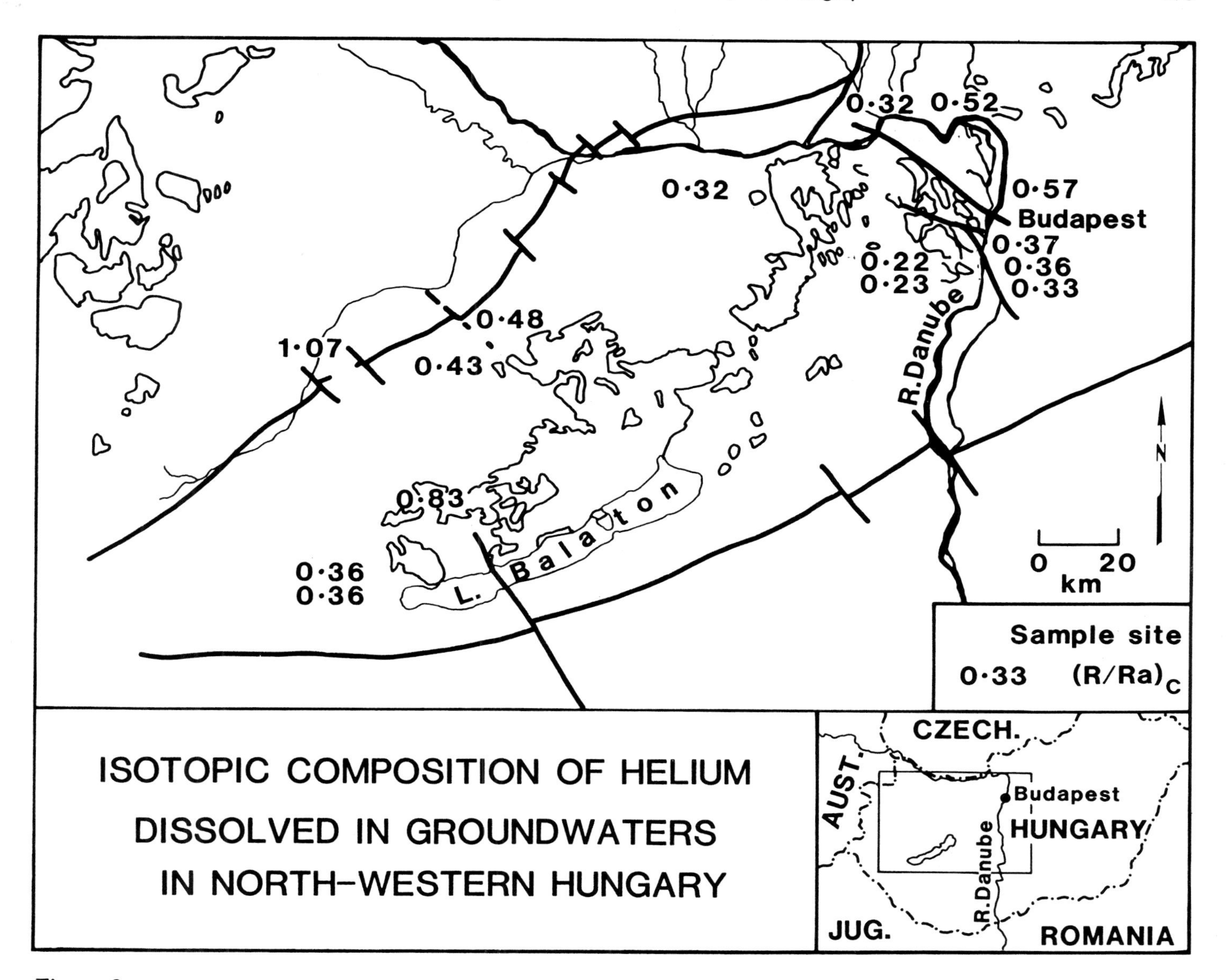

Figure 2. Map of northwest Hungary showing sample localities, with the associated R/R_a values. Stippled areas represent surface outcrop of basement; dashed areas are surface outcrops of Tertiary extrusives. Major faults and general characteristics of subsurface water flows are also marked.

The helium isotopic ratios clearly show an input of mantle volatiles, reaching a maximum of about 15%. The limited amount of data so far available seems to indicate a fairly general value for R/R_a of about 0.3 to 0.5, with superimposed areas of higher values. This can possibly be taken to represent an average contribution of from 3–7% mantle volatiles on a regional scale, with local areas of much more significant mantle contribution. These seem to occur in association with fairly major faults, although the paucity of data makes this correlation tentative. The idea of increased flux of volatiles associated with faults is an appealing one, as they should present a ready pathway for the transport of these volatiles, and evidence for such behavior has been observed in the Rhine Graben (authors' unpublished data).

The high R/R_a values seem to be a fairly local feature, and could be explained by the local injection of He rich in ^3He; as the flow continues, the strong signal decays towards the regional average by dilution with helium that has a lower ratio. This may take two forms: further helium flux from below with a larger crustal component, and/or mixing with waters that have a lower ratio.

CONCLUSIONS

The occurrence of $^3He/^4He$ ratios elevated above the value of radiogenic production in the crust is a clear expression of recent volcanism in northwest Hungary. The groundwater appears everywhere to carry traces of volatiles from the mantle.

As yet there is insufficient evidence concerning the process responsible for producing high values of R/R_a. This may be related either to volcanism, or to extensional faulting, or both. In the future, more detailed work should be able to resolve this problem.

Table 1. Isotopic Composition of Helium in Waters from Northwestern Hungary

Number	Location	(R/R_a)	X^a	$(R/R_a)_c$	[He] (μcc/g)	Mantle component (%)	Depth (m)	Temperature (°C)
H1	Gellért-táro I	0.335	563	0.334	10.4	4.5	spring	43
H2	Margit sziget II	0.391	185	0.388	3.5	5.2	310	70
H3	Városliget II	0.374	164	0.371	0.88	5.0	1257	73
H4	Herceghalom 1	0.248	33	0.224	1.53	3.0	850	39
H5	Herceghalom 2	0.250	31	0.225	0.99	3.0	350	28
H6	Tata, Fényesfürdö 1	0.324	109	0.318	9.2	4.3	600	22
H7	Esztergom Strand	0.325	192	0.322	4.3	4.3	223	28
H8	Visegrád Lepence Strand	0.527	50	0.517	2.6	7.1	1301	40
H9	Szentendre Papsziget	0.581	66.5	0.574	1.34	7.9	1630	65
H26	Héviz, Fontanlis	0.444	8.0	0.364	5.00	4.9	50	40
H27	Héviz, Rákóczu + György	0.461	6.2	0.356	6.70	4.8	<50	41
H64	Tapolcafö Vizmü 1	0.492	9.4	0.431	0.51	5.8	400	15
H65	Pápa-Uszoda	0.488	83.3	0.481	4.09	6.5	800	42
H67	Mesteri Strand 1	1.066	179	1.066	5.39	14.8	1650	64
H70	Tapolca-Strand	0.829	146	0.828	18.57	11.4	635	34

[a] $X = (He/Ne)_{sample}/(He/Ne)_{10°C water}$, where $(He/Ne)_{10°C water} = 0.230$.

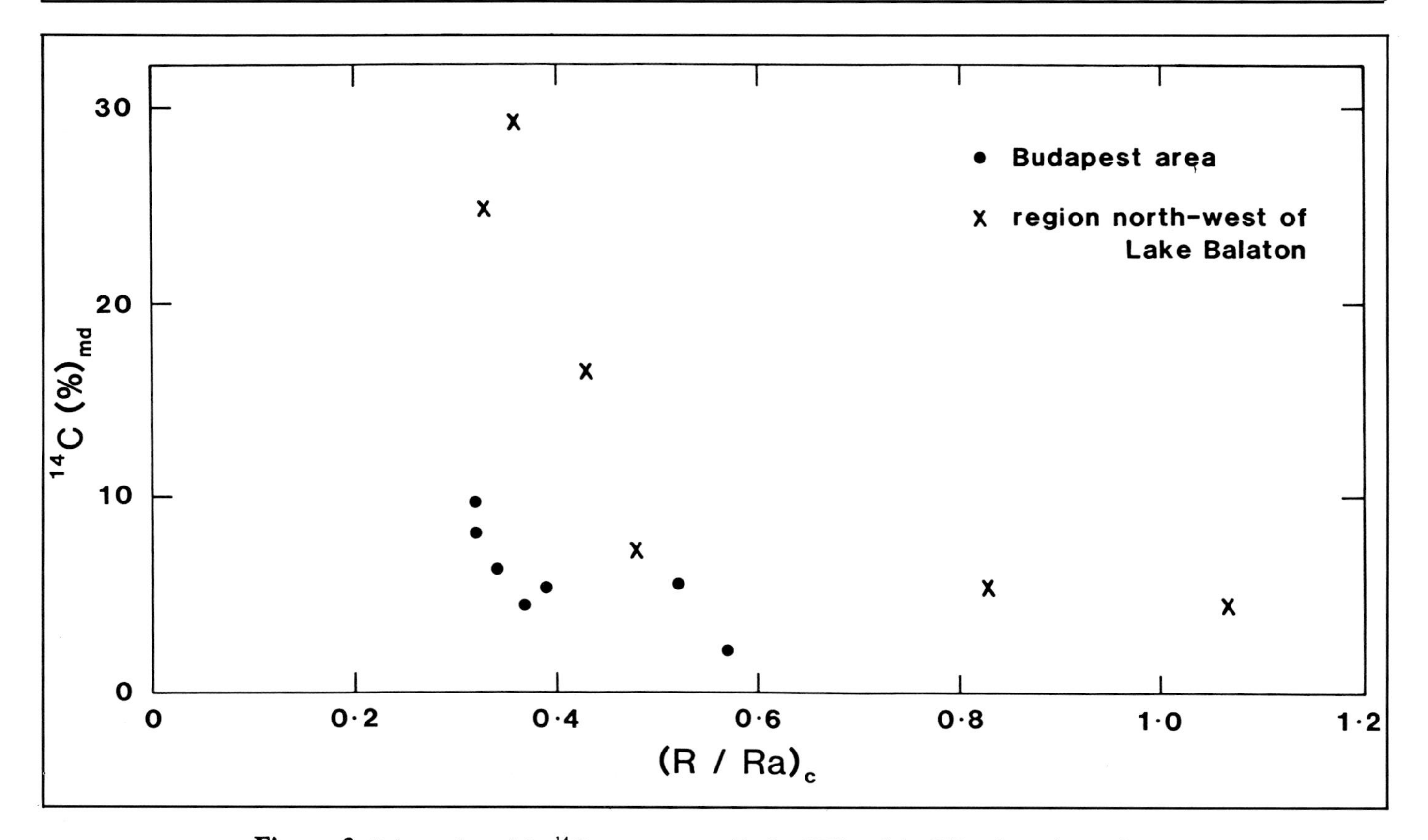

Figure 3. Relationship of the ^{14}C concentration (Deák, 1979) and the R/R_a values of groundwaters.

REFERENCES

Deák, J., 1979, Environmental isotopes and water chemical studies for groundwater research in Hungary: Isotope Hydrology 1978, IAEA Vienna, v. 1, p. 221–249.

Hooker, P. J., R. Bertrami, S. Lombardi, R. K. O'Nions, and E. R. Oxburgh, 1985, Helium-3 anomalies and crust-mantle interactions in Italy: Geochim. Cosmochim. Acta, v. 49, p. 2505–2513.

Horváth, F., P. Dövényi, and P. Liebe, 1981, Geothermics of the Pannonian basin: Earth Evol. Sci., v. 1, n. 3–4, p. 285–291.

CHAPTER 21

An Organic Maturation Study of the Hód-I Borehole (Pannonian Basin)

Cs. Sajgó
Z. A. Horváth
J. Lefler
Hungarian Academy of Sciences,
Laboratory for Geochemical Research,
H-1112 Budapest, Budaörsi út 45, Hungary

Several maturation parameters were determined for a 5800-m-thick Neogene sequence in the Pannonian basin of southeastern Hungary. The oil window was found to start at 3450 m corresponding to a vitrinite reflectance of about 0.7% R_o. This is typical of a mixture of Type II–Type III kerogen such as occurs in the U.S. Gulf Coast. The carbon preference index (CPI) of n-alkanes in the rock extracts also approaches 1.0 at this depth. Hydrocarbon generation appeared to continue to the bottom hole depth of 5800 m with an oil generation minimum or gap occurring in the 5000–5400 m interval that might be due to a change in the nature of the organic matter. Two isomerizations of biological marker compounds, the shift of steranes from 20(*R*) to 20(*S*) and of hopanes from 22(*R*) to 22(*S*) were found to reach equilibrium at 4000 and 3000 m, respectively, just prior to the threshold of intense oil generation. Also, the C_{29}-monoaromatic steranes were converted almost 100% to C_{28}-triaromatic steranes prior to this depth.

INTRODUCTION

Investigation of the maturity of organic matter in sedimentary rocks has received much attention in the last two decades and numerous methods have been developed to detect zones of oil and gas formation (see, for example, Philippi, 1965; Vassoyevich et al., 1970; Lopatin, 1971, 1976; Laplante, 1974; Hood et al., 1975; Dow, 1978; and Espitalié et al., 1977). Several thorough and comprehensive studies have been published in the last few years (Tissot and Welte, 1978; Hunt, 1979 and Héroux et al., 1979). In several instances, various maturation indices have been successfully applied to hydrocarbon prospecting, but a maturation index that is applicable to all geological situations has not been found. Vitrinite reflectance is the most frequently used maturation index, but is limited to sedimentary formations that contain vitrinite. Even when vitrinite-bearing rocks are present, difficulties in interpretation occur. Estimation of maturity from the color of spores is also not straightforward (e.g., see Raynaud and Robert, 1976). Analysis of changes in the composition of soluble organic matter is often complicated by the addition of migrated components.

Organic maturation does not occur by a single type of reaction: without an exact knowledge of the chemical changes that take place during maturation, maturity parameters can only be related to one another from field studies. For example, the model of oil generation outlined by Tissot and Welte (1978) is straightforward, but not universally applicable. The vitrinite reflectance and temperature ranges they assign to the oil generation zone disagree with some observations (Sajgó, 1980a,b; Price, 1982; Price et al., 1979, 1981; and Saxby, 1982).

In this chapter we discuss the variation in a number of maturation parameters in samples from the borehole Hódmezóvásárhely-I (Hód-I) in the Pannonian basin.

GEOLOGIC SETTING OF HÓD-I

Hód-I is located in southeast Hungary (about 25 km east of Szeged) in a Neogene sedimentary trough. Drilling by the National Oil and Gas Trust (OKGT) terminated at 5842.5 m depth in Badenian (middle Miocene) sedimentary rocks. Continuous Miocene to present sedimentation is assumed although the Sarmatian cannot be dated paleontologically (Mucsi, 1973; Mucsi and Révész, 1975; and Szentgyörgyi, 1975). The pre-Pannonian Miocene rocks consist of pelitic lime marl between 5100 and 5450 m; below 5450 m the sediments consist primarily of coarse-grained siltstones. The Pannonian (*s. l.*) strata were mainly deposited in nearshore delta and lacustrine environments and consist of alternating sandy, clayey, and marly layers 0.02 to 10.0 m thick and with a carbonate.

Several organic geochemical analyses have been reported on samples from this well. These are summarized in Sajgó (1980a,b) and Sajgó et al. (1987). In general they show the threshold of intense petroleum generation starting at 3450 m at a present sediment temperature of 142 °C.

THERMAL MATURITY OF DISPERSED ORGANIC MATERIAL USING VITRINITE DATA

Problems with Measurement

The vitrinite reflectance method adapted from coal petrology is based on the observation that the optical parameters of vitrinite change gradually as a function of progressive coalification (maturation). The reflectance of light by vitrinite grains in a polished section prepared from the sedimentary rock or from a separated kerogen concentrate, and covered with oil, can be used as a measure of the coal rank. For pure coals, the different macerals are found together and can be fairly well distinguished. About 80% of sedimentary rocks contain vitrinite, but organic rich rocks and carbonates may contain little or none. Identification of vitrinite in the latter rock types is sometimes difficult. Thus, the reflectance of other microcomponents (semifusinite, bitumen, pseudovitrinite) is occasionally measured by mistake instead of vitrinite.

Allochthonous or reworked vitrinite may also be present in the sedimentary rocks. This results in a bimodal population. Consequently, in addition to the average reflectance values, the reflectance histogram should also be presented so that anyone can reevaluate the vitrinite data. The measurements reported here were carried out with a precision of 0.01% and the histogram for each sample is constructed from at least 50 points. For coals the precision of the reflectance histogram is 0.05%; for dispersed organic matter, it is 0.10% because of the uncertainties mentioned above.

Forty core samples between 2050 and 5815 m depth were used for vitrinite measurements (Figure 1) (Horváth, 1980). Ten of the samples were also measured by J. R. Castaño at Shell Development Company. Graphite was present in all the samples studied indicating a metamorphic source area. In sample No. 1/3 of early Pannonian (*s. l.*) age, no allochthonous (reworked) vitrinite occurred in addition to graphite. The reflectance histogram shows small dispersion. Most of the upper Pannonian samples contained reworked vitrinite. Because of greater dispersion, the limits of the autochthonous vitrinites could only be determined with difficulty. Nevertheless, for samples No. 4/1 to 27/6, Castaño's measurements yielded systematically lower values than measurements we conducted on the same samples. For samples deeper than No. 27/6, Castaño's measurements produced higher average values. When comparing our reflectance histograms to those of Castaño, two phenomena are apparent. First, in measuring, many more grains of lower reflectance were identified: the dispersion is higher. Second, when calculating the average reflectance, Castaño excluded grains of higher and lower reflectance. According to Castaño, samples 35/1 and 39/9 contain considerable amounts of bitumen but no vitrinite. Shell assumes that in this maturity range ($R_o > 1.6$) the reflectance of bitumen is the same as that of vitrinite and bitumen was used to obtain vitrinite reflectance values. For sample 45/9, the agreement between the reflectance of bitumen and vitrinite is good. However, in our opinion, the reflectance of bitumen above 1.6 often exceeds that of vitrinite.

Because no depositional hiatus or erosion could be determined for the formations of the borehole Hód-I, it may be assumed that the present temperature of each sample site is its maximum temperature (see also Dövényi et al., 1983). Ammosov et al. (1977) collected all the available vitrinite reflectance data from basins in the Soviet Union where the sedimentation has been continuous. Based on these data, they constructed a scale that shows the minimum rock temperature required to reach a given reflectance value (Figure 2). Except for two samples, the average reflectance values follow the relationship determined by Ammosov et al. (1977) within the error range of ± 10% quoted previously.

Bitumen Analyses of Solvent Extract

The qualitative and quantitative changes of soluble organic matter during maturation have been studied by many authors (Brooks and Smith, 1967; Albrecht et al., 1976; Tissot et al., 1971, 1974, 1977; Allan and Douglas, 1977; and Radke et al., 1980). In the Hód-I borehole, Sajgó (1980a,b) proposed two oil generation zones from a plot of the chloroform soluble extract (in milligrams) divided by the total C_{org} (in grams) as a function of depth. Figure 3 shows more or less the same phenomenon for Σ CH mg/C_{org} g. Oil generation starts at about 3450 m depth. At about 5000 m depth, no significant oil generation could be detected, although there may be gas generation. Below 5450 m depth, a second oil generation zone may be observed that extends to the deepest sample. The first maximum is well-known to petroleum geologists as the main phase of oil generation. In the borehole Hód-I this appears to occur within a temperature range of 140–200 °C, which is higher than that observed elsewhere either because of the very short time that these sediments have been at high temperatures (due to rapid sedimentation) or because oil generation usually requires temperatures that are this high (Thompson, 1983). In this oil generation zone the vitrinite reflectance value varies between 0.69 and 1.5%.

We propose that a second oil generation zone starts at 218 °C and continues at least to the 233 °C measured at the well bottom. In this range the vitrinite reflectance value (Figure 2) var-

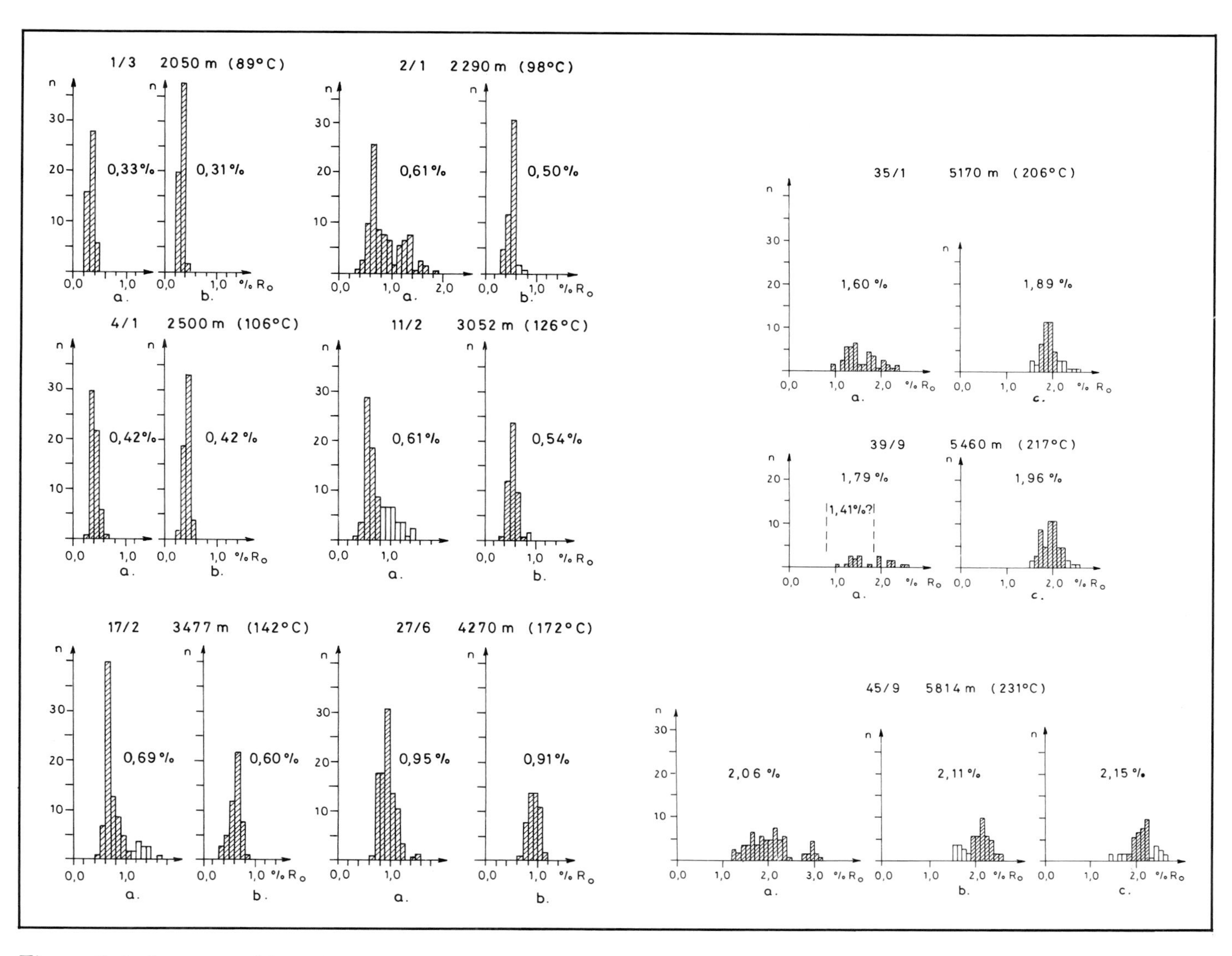

Figure 1. Reflectograms of the Hód-I samples. The reflectograms marked "A" were produced by Z. A. Horváth at LGR. Those marked "B" were produced by J. R. Castaño at Shell by measuring vitrite reflectance. Those marked "C" were produced by Castaño by measuring reflectance of bitumen.

ies between 1.6 and 2.15% (Sajgó, 1980b, later determined these values to be 1.41 to 1.57%).

Similar examples of a gap in generation in wells from other areas have been summarized by Sajgó (1980b). Other examples have been found by Yakovets et al. (1976) and Sajgó, unpublished work. We infer that our lower zone of oil generation is a similar phenomena to the deep zone of oil generation observed by Price et al. (1979) and Price (1982). Some of these apparent gaps may represent gas generation in preference to oil. In the Hód-I borehole, the H/C ratio of the kerogen in the gap drops to 0.76–0.78 (Figure 2), which is too low to sustain oil generation. The ratio increases to the 0.9 range in the deeper oil generating zone.

Figure 4 shows the distributions of *n*-alkanes and the relative quantities of three isoprenoid hydrocarbons (norpristane, pristane and phytane). The variety of alkane spectra may reflect the variety of organic matter types. In the deeper samples the relative quantity of lighter hydrocarbons (C-16 to C-20) increases at the expense of the heavier ones (> C-20), perhaps due to cracking of the heavier hydrocarbons at increased temperatures. The distribution may also be affected by migration, maturation and different types of organic matter.

The ratio pristane/phytane varies with the type of depositional environment (oxidizing versus reducing) and goes through a maximum with increasing maturity (Brooks et al., 1969; Leythaeuser, 1975; Flekken, 1978; and Radke et al., 1980). In Figure 5, this change with maturation is clearly shown, with a maximum around 3600 m.

All of the geochemical data shown in Figures 3 to 6 are sensitive to changes in the type of organic matter and to migration (both to migration out and migration in of oil). The lithology of the rock matrix should also be taken into account. The ratio Σ CH/NSO compounds also indicates the lack of oil generation between 5000 and 5400 m (Figure 6). However, this ratio is also dependent on the type of organic material and on the migration processes, as well as on the maturity.

The ratio of saturated hydrocarbons to saturated esters (E_{1470}/E_{1740}) (infrared extinction ratio) in the asphaltene fraction

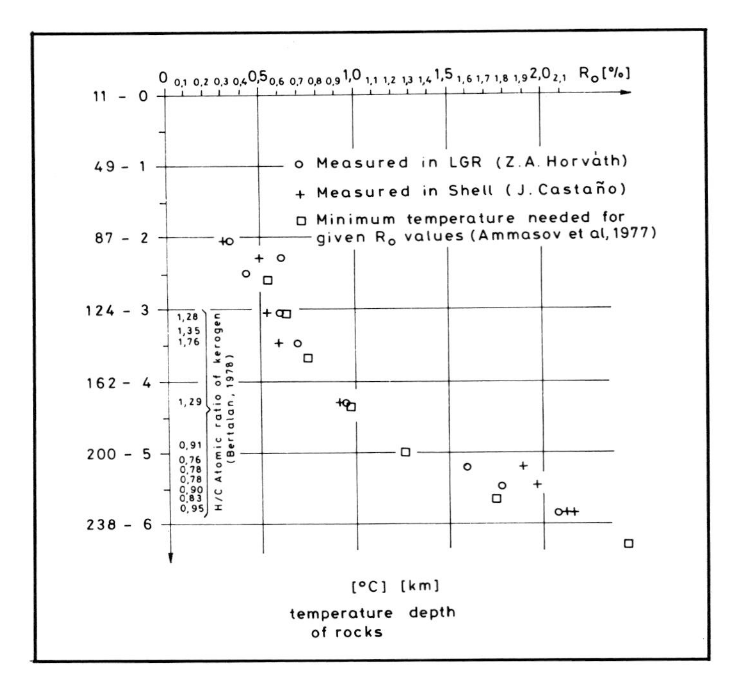

Figure 2. The average reflectance versus depth plot of the Hód-I borehole. The open squares correspond to the minimum temperature values needed to achieve a given vitrinite reflectance as determined for basins with no uplift in the Soviet Union (Ammosov et al., 1977).

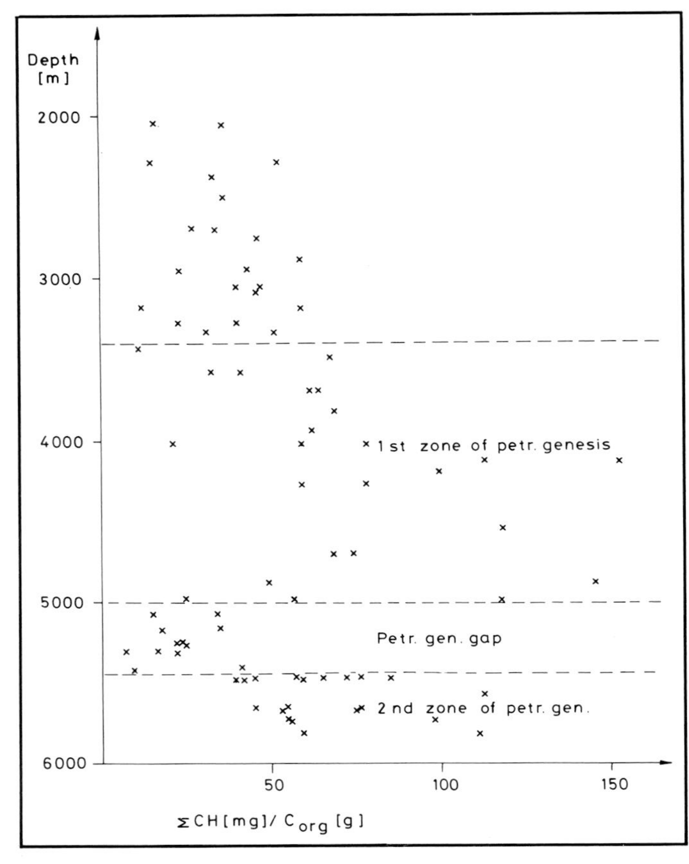

Figure 3. Variation of the amount of chloroform extract relative to total organic carbon as a function of depth.

shows a decrease in the hydrocarbon content of asphaltene below 4000 m due to asphaltene disproportionation (Figure 7). Below 4700 m the scatter in (E_{1470}/E_{1740}) is low, and the ratio Σ CH/NSO is usually below 4700 m as well (Figure 6). The low scatter in E_{1470}/E_{1740} thus implies that hydrocarbons in this depth range were not generated from asphaltane, and must therefore come from kerogen.

APPLICATIONS OF THE TRANSFORMATION OF BIOLOGICAL COMPOUNDS DURING HEATING

The coalification of plant material in sediments is a complicated process, so that one should not expect that a simple relationship can explain the coal rank of coals with different geological histories. For this reason, and because the sedimentation and thermal histories are often complicated, relationships between temperature, geologic age, and coal rank that work well in some basin cannot, in our opinion, be applied without restrictions to other regions (see, for example, Table 1). In our opinion, the ideal geochemical maturity parameter should be applicable to both oil and rock samples, independent of organic matter type and irreversible; or, if reversible, it eventually must attain an equilibrium state. The pressure dependence of most geochemical reactions is not well established; it is better, therefore, if no change in the volume occurs during the reaction. It is also important that the physical fractionation processes occurring during migration should not affect the maturity parameter. To the best of our knowledge the following three reactions more or less satisfy these requirements.

These reactions involve biological marker compounds (Speers and Whitehead, 1969). Biological marker compounds are organic compounds whose structure can be related to a biological precursor because of only minor alteration during sedimentation and diagenesis. In this chapter, the concentrations of four such compounds (steranes, hopanes, mono- and triaromatic steroid hydrocarbons) will be discussed as a function of depth. The concentrations of the starting material and of the products of the three reactions (Figure 8) were determined with a computerized gas chromatograph-mass spectrometer system (Mackenzie et al., 1980, 1981; and Sajgó and Lefler, 1986).

Reaction Kinetics

Steranoid and hopanoid structures experience several different types of chemical reactions during maturation. These reactions can be subdivided into: (1) isomerization reactions, (2) aromatization reactions, and (3) decomposition reactions. The third of these reactions accompanies generation of oil and gas, and the products of this reaction have not been identified. The first and second reaction types are intramolecular reactions whose products can be recovered and quantified. Thus the progress of the reaction can be followed by component ratios.

Reaction kinetics describe the progress of the chemical reactions with time. The rate at which the chemical reactions proceed is a function of the concentrations of the compounds involved in the reaction:

$$r = \frac{dc}{d\tau} = f(c_1, c_2, \ldots, c_i) \quad (1)$$

where r denotes the reaction rate, c_i the concentration of the ith compound, and τ the time.

The order of the chemical reaction is the same as the order of the differential equation. For first-order reactions, Equation (1) may be of the following form:

$$\frac{dc}{d\tau} = kc \quad (2)$$

We assume that for intramolecular transformations the reaction rate described by Equation (2). This expression, however, is not correct if the surrounding rock material also contributes to the intramolecular rearrangement (catalytic). The following equation:

$$\frac{dc}{d\tau} = kc_1, c_2 \quad (3)$$

would describe the reaction progress for second-order reactions. If the rock concentration (c_2) is constant in time, and is very high compared to the compounds studied, its effect can be included in the reaction rate constant k and Equation (2) applies. This is known as pseudo-first-order kinetics.

The reactions studied are of two kinds: (1) reversible reactions leading to equilibrium, for example, isomerization; and (2) irreversible reactions, for example, aromatization. The three reactions are shown in Figure 8. For a detailed explanation of them, see Mackenzie et al., 1980, 1981). The two isomerization reactions can be modeled as follows:

$$R \rightleftarrows S \quad (4)$$

When equilibrium is reached, the reaction rate equals zero, or:

$$\vec{k}\, c_R = \overleftarrow{k}\, c_S \quad (5)$$

The equilibrium constant is given by:

$$K = \frac{\overleftarrow{k}}{\vec{k}} = \frac{c_R^*}{c_S^*} \quad (6)$$

where c_R^* and c_S^* are the equilibrium concentrations of the R and S isomers. When the concentrations of R and S are far from equilibrium, the macroscopic reaction rate (the rate of transformation of c_S) can be written:

$$\frac{dc_S}{d\tau} = \vec{k}\, c_R - \overleftarrow{k}\, c_S \quad (7)$$

By means of the C-GC-MS system the relative concentrations of c_R and c_S can be measured, thus:

$$c_R + c_S = 1; \qquad \frac{c_S}{c_S + c_R} = \alpha \quad (8)$$

Applying the same transformation in the relationships (6), (7), and (8), it follows that:

$$\frac{dc_S}{d\tau} = \vec{k}(1 - \beta\alpha) \quad (9)$$

where $\beta = 1 + 1/K$ and K is the equilibrium constant. Rearranging Equation (9) and integrating with respect to c_S gives:

$$-\frac{1}{\beta} \ln(1 - \alpha\beta) = \int_0^\tau \vec{k}\, d\tau \quad (10)$$

There is no problem with the integration of the left side of Equation (10) (the concentration-dependent part). For the right side of Equation (1) (the time-dependent part):

$$\vec{k} = A \exp[-\frac{\Delta H^*}{RT}] \quad (11)$$

where A is the preexponential factor, ΔH^* is the activation energy of the reaction, R is the universal gas constant, and T is the absolute temperature.

The reaction rate coefficient $\vec{k}$, is not constant but depends on temperature. If the chemical reaction proceeds at constant temperature, $\vec{k}$ will be independent of time. If so, then in the right side of the Equation (10), $\vec{k}$ could be taken outside the integral. In our case this cannot be done because during basin evolution rocks are buried progressively deeper so that they pass through zones of different temperatures. This problem usually has been eliminated by introducing EHT (the effective heating time; Hood et al., 1975). When studying geological samples Hood et al. (1975) came to the conclusion that in the chemical reactions of vitrinite it is sufficient to take into account the time spent within 15 °C of the maximum temperature.

Another solution is to use the method of absolute times. Assuming that the functions describing the basin subsidence and the change of the geothermal gradient through time are known, we can write for each individual sedimentary layer:

$$T = g(\tau) \quad (12)$$

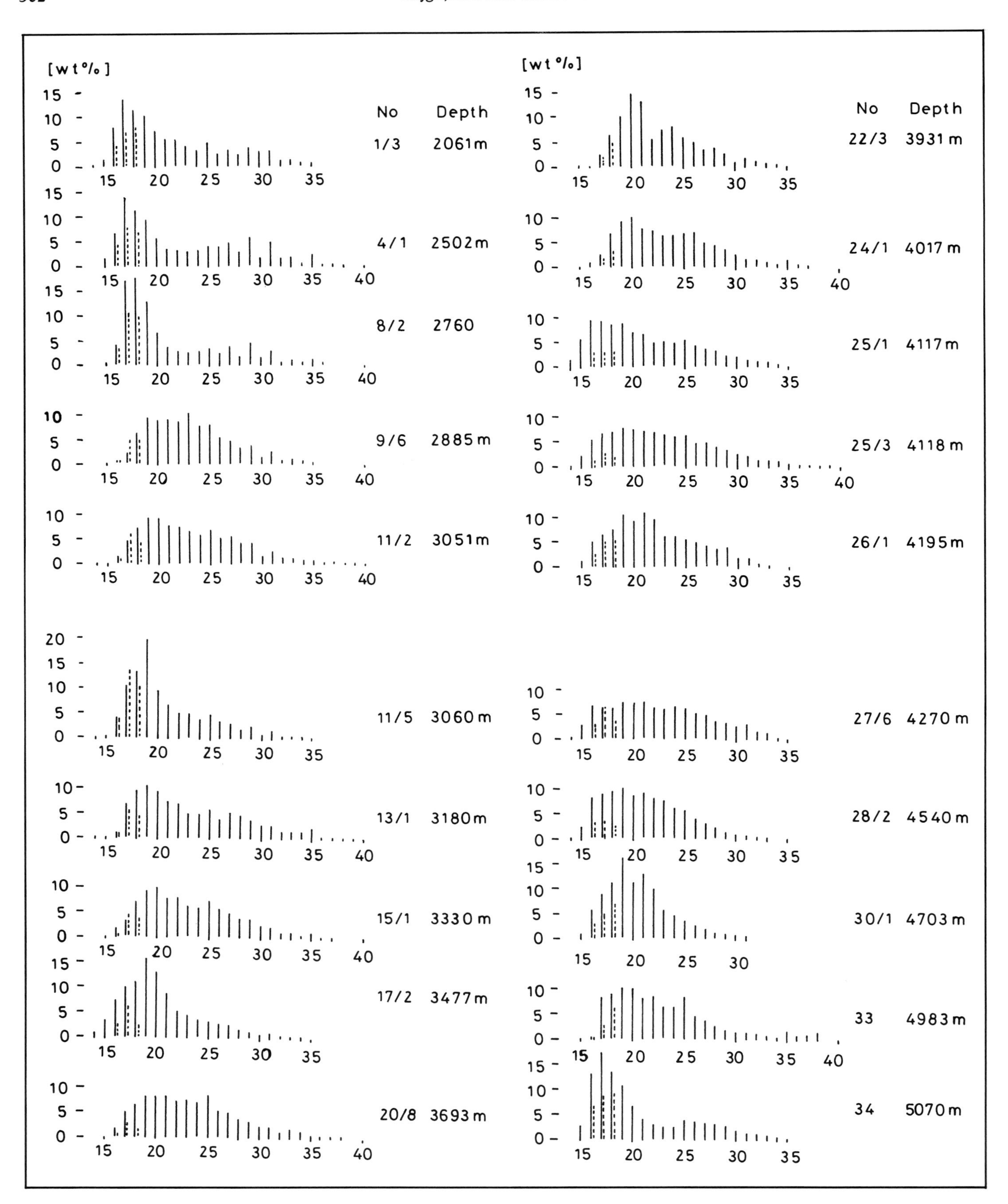

Figure 4. The relative distributions of *n*-alkanes (C_{15}–C_{40}) and isoprenoids (C_{13}–C_{20}) in chloroform extracts.

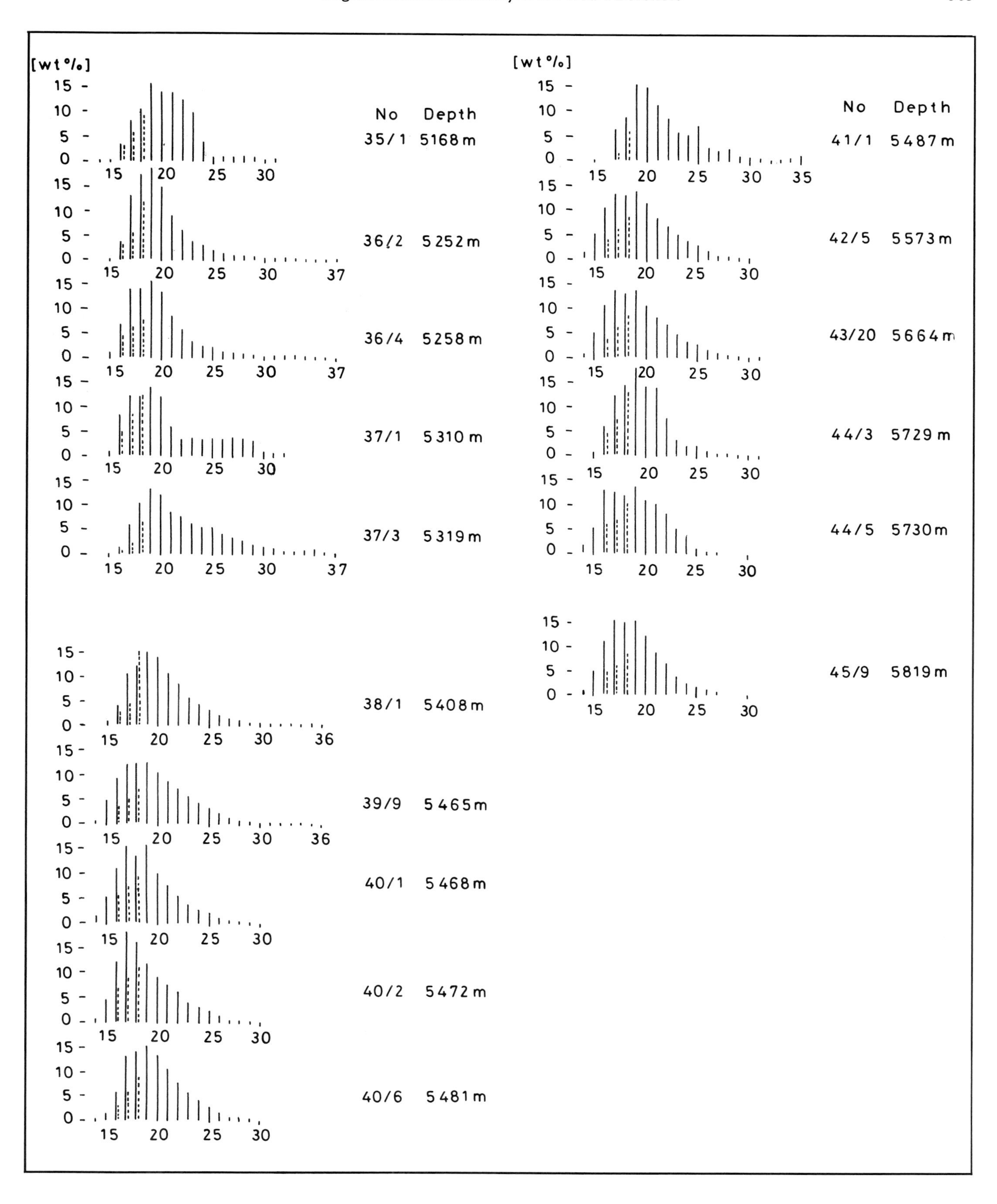
[wt °/o]
No Depth
35/1 5168 m
36/2 5252 m
36/4 5258 m
37/1 5310 m
37/3 5319 m
38/1 5408 m
39/9 5465 m
40/1 5468 m
40/2 5472 m
40/6 5481 m
41/1 5487 m
42/5 5573 m
43/20 5664 m
44/3 5729 m
44/5 5730 m
45/9 5819 m

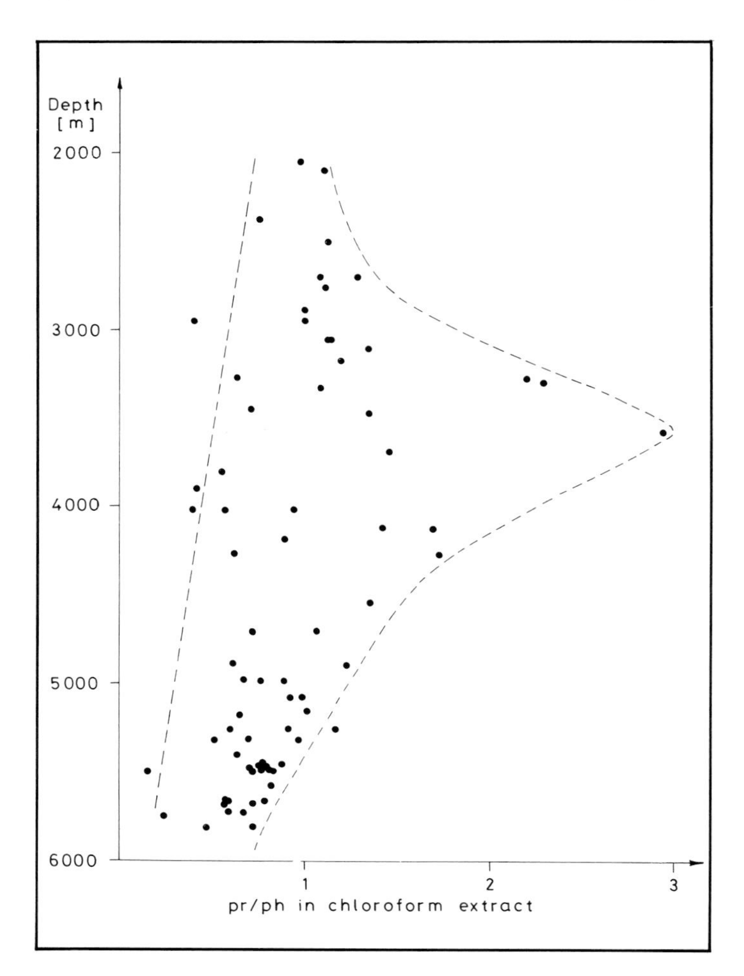

Figure 5. Variation of the pristane to phytane ratio (pr/ph) in chloroform extract as a function of depth.

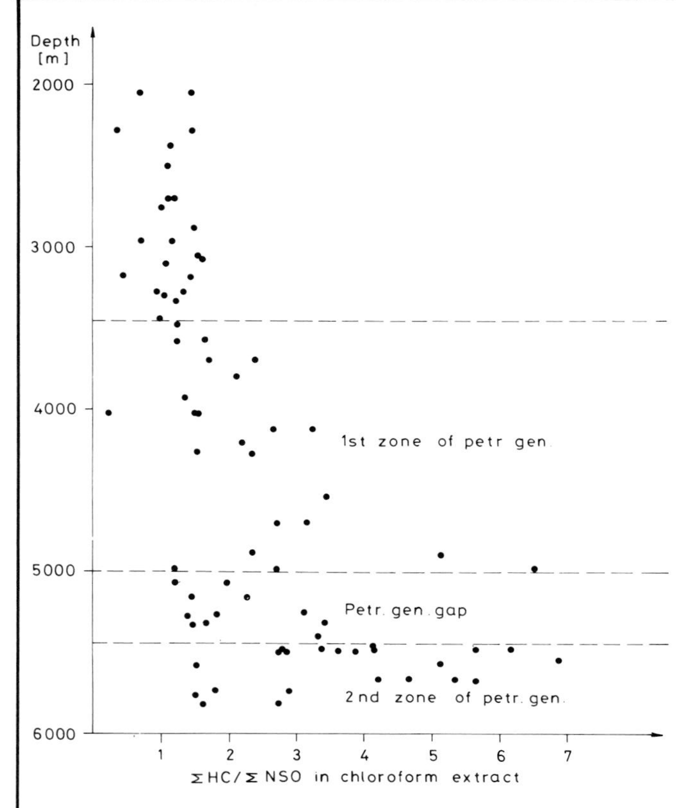

Figure 6. The ratio of hydrocarbons (HC) to nonhydrocarbons (NSO) in chloroform extracts as a function of depth.

Substituting the relationships (11) and (12) into Equation (10) we get:

$$-\frac{1}{\beta}\ln(1-\alpha\beta) = A\int_0^{\tau}\exp[-\frac{\Delta H}{Rg(\tau)}]\,d\tau \qquad (13)$$

The relationship given by Equation (13) forms the basis of the method of absolute times. Unfortunately, for most realistic temperature histories as described by Equation (12), the integral on the right of Equation (13) cannot be written in closed form and one of two numerical techniques must be used: (1) Assume a thermal history for Equation (12), substitute this relationship into Equation (13) and integrate. The expression on the left of Equation (13) can be calculated from measurements (c_S, c_R, c_S* and c_R*) and Equations (8) and (9). For aromatization reactions the mono- and triaromatic concentrations are used instead of c_S and c_R. This calculation is made step-by-step (between the adjacent measured points) so that the preexponential factor, A, can be determined as a function of depth or temperature. If this function is constant within the limits of measurement error and its fluctuation is random, the assumed thermal history is consistent with the maturity data. If the fluctuation of A is greater than expected or changes as a function of the number of samples, the assumed thermal history should be replaced by another model. In this way, the thermal history of a basin can be reconstructed. (2) A theoretically more satisfying method can be applied when the data for several different reactions (for example aromatization, hopane isomerization or sterane isomerization) are also available from the basin in question. In this case a system of equations can be established similar to Equation (13). The function g(τ) can be described as:

$$T = g(\tau) = h(\tau)g'(\tau) \qquad (14)$$

where $h(\tau)$ is the burial depth of the sample and $g'(\tau)$ is the geothermal gradient as a function of time.

By solving the system of equations, the functions $h(\tau)$ and $g'(\tau)$ can be determined. Although theoretically simple, this procedure is complicated by difficulties in solving the integral in Equation (13). For this reason, it is more convenient to apply a differential method as follows. Equations (10) and (11) can be transformed to derive the more general relationships:

$$-\frac{1}{\beta}\ln(1-\beta\alpha) = \int^{c} f(c)\,dc \qquad (15)$$

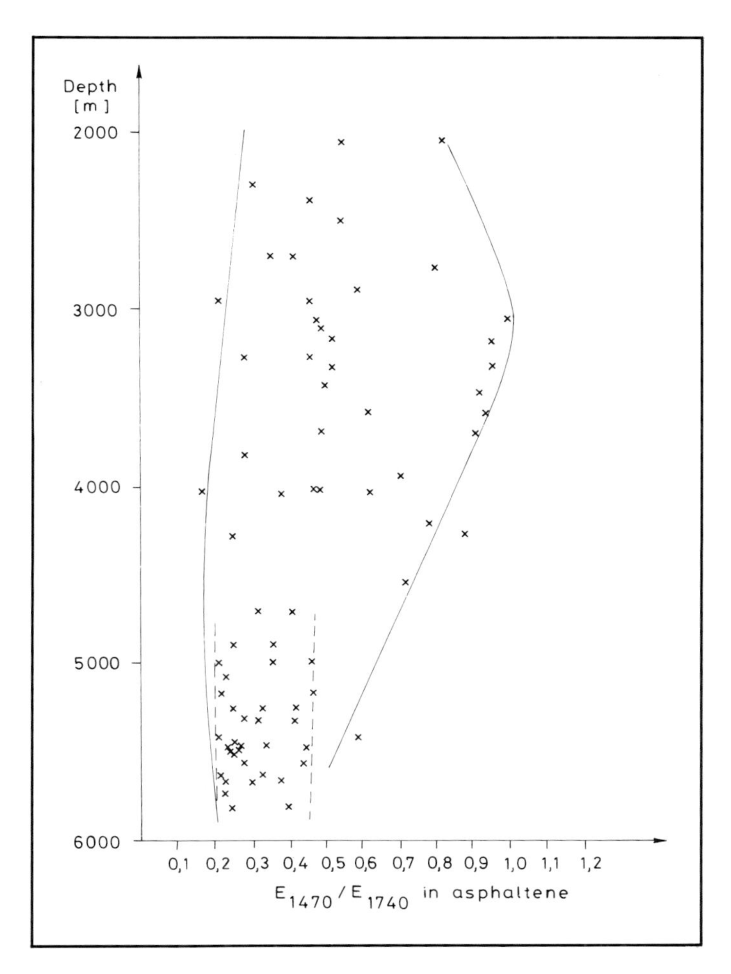

Figure 7. Variation of the E_{1470}/E_{1740} (infrared extinction ratio) in asphaltene as a function of depth.

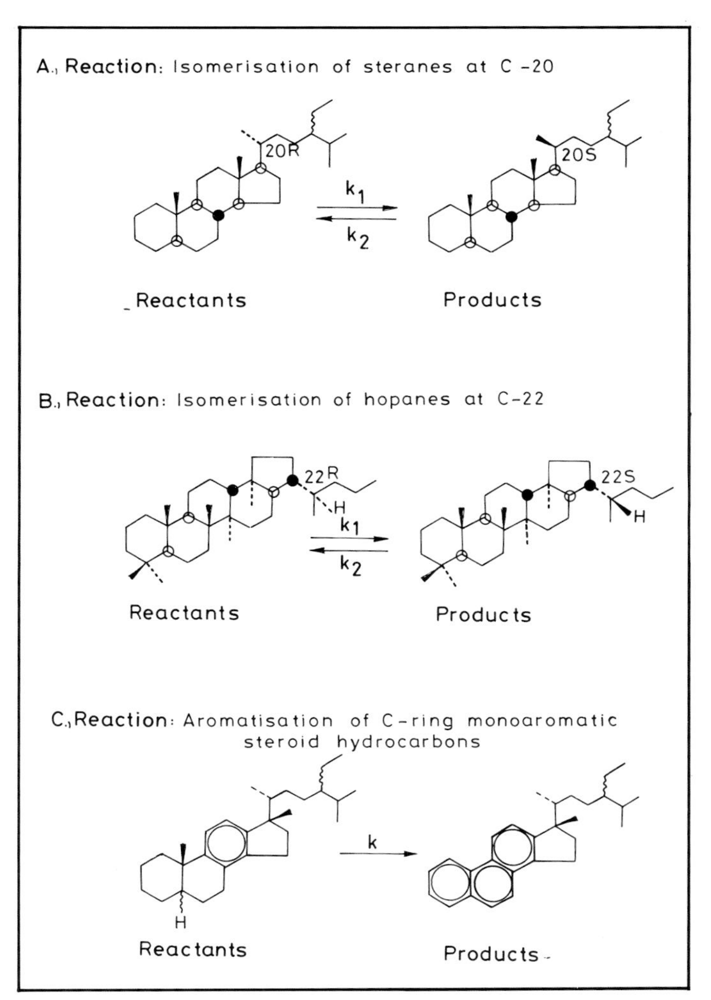

Figure 8. Reactions studied: (A) Configurational isomerization of the 5α(H), 14α(H), 17α(H), 20(R)–C_{29} steranes to 20(S) steranes. (B) Configurational isomerization of the 17α(H), 21β(H), 22(R)–C_{22} to 22(S) hopanes. (C) Aromatization of the 5α(H) and 5β(H) isomers of C_{29} C_{ring}-monoaromatic steroid hydrocarbons to C_{28}-triaromatic steroid hydrocarbons.

and

$$\vec{k}(\tau) = \frac{d}{d\tau}\int^{c} \ln f(c)\, dc \tag{16}$$

Taking the finite differences in Equation (16) and differentiating by h instead of r on the basis of Equation (14), and substituting Equations (12) and (14) into Equation (11), the thermal history can be obtained (a detailed discussion can be found in the papers of Sajgó and Lefler, 1986 and Lefler and Sajgó, 1986). The accuracy of this method is limited by the density of the available samples.

Either of these two methods (EHT and AT) are suitable for the reconstruction of thermal history using the six reactants and products of the three reactions discussed previously. If during a given time, the reaction (either an equilibrium or an irreversible transformation) reaches a point where changes in concentration cannot be measured, nothing further can be deduced about the thermal history from that reaction.

In Equations (9), (10), and (13), β depends on the temperature, and thus also on time. In case of isomerization reactions, the temperature-dependent part of β, K, is given by Equations (6) and (11). Based on both theoretical considerations and on measurement data, the temperature dependence of the two reaction rate constants for isomerization are equal, thus the temperature and time dependence of K can be neglected. In the aromatization reaction, the equilibrium is in practice displaced towards the triaromatic product, thus $\beta = 1$.

Calculations of Reactions of the Biological Marker Compounds

Using the results of the reactions and measurements shown in Figures 9–11, the reaction kinetic parameters of the sterane and hopane isomerization and of the aromatization reactions were calculated both by the EHT and by the AT methods. For the EHT method, the time which each sample spent within 15 °C of its maximum temperature was calculated using biostratigraphic data and temperature measurements within the bore-

Table 1. Estimated values of vitrinite reflectance (R_o)

Sample	Depth (m)	Age (Ma)	T (C)	Measured (T_o (%))	Estimated R_o (%)					
					1	2	3	4	5	6
17/2	3477	8.2	142	0.69	0.40	0.39	0.68	0.61	0.63	0.82
39/9	5460	15.0	217	(1.41?) 1.79	1.17	1.03	1.40	1.65	2.96	1.97

[a]References: (1) Lopatin (1971); (2) Bostick (1973); (3) Hood et al. (1975); (4) Bostick et al. (1979); (5) Waples (1980); and (6) Koncz (1983).

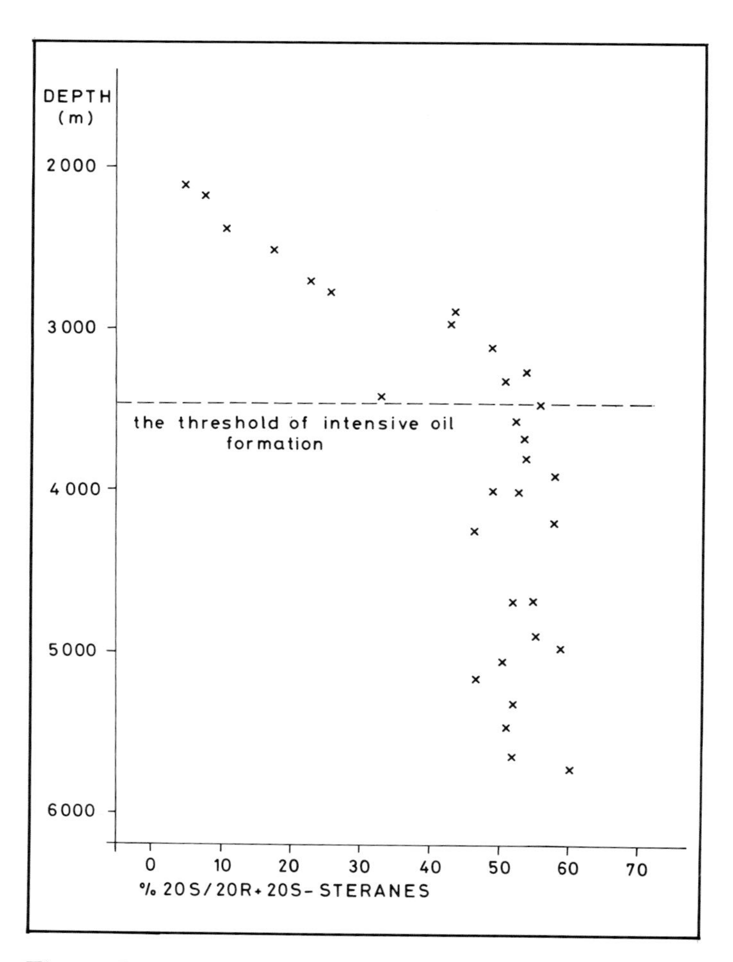

Figure 9. Increase in the extent of geochemical isomerization at C-20 of 5α(H), 14α(H), 17α(H), 20(R)–C_{29} steranes as a function of depth/temperature.

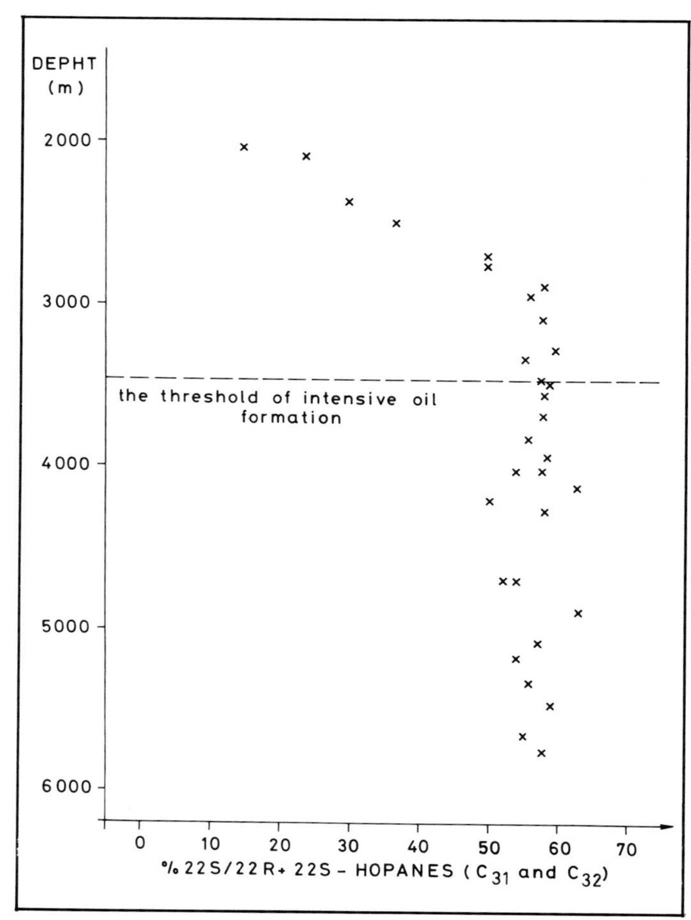

Figure 10. Increase in the extent of geochemical isomerization at C_{22} of 17α(H), 21β(H), 22(R)–C_{31}, and C_{32} hopanes (mean value) as a function of depth/temperature.

hole assuming a constant thermal gradient through time. For the AT method, the burial history was also determined from biostratigraphic data, and subsidence was considered to be constant between biostratigraphic markers. The activation energy of the reaction was calculated by the differential method as a first approximation, then refined by successive iterations.

Sterane Isomerization

The EHT Method. From measurements, $K = 1.38$ and $\beta = 1.724$. By means of the EHT method:

$$T_{eff} = 15\,^{\circ}C \rightarrow \tau_{EHT} = 1.16 \text{ m.y.}$$

Taking the logarithm of Equation (11):

$$\ln \vec{k} = \ln A = \Delta H^{*}/RT \qquad (17)$$

Substituting the measurement data into Equation (17):

$$\ln (k = -1.1768 \times 10^{4}/T + 29.32 \quad (\text{k.e.} = 0.9947)$$

Based on the correlation coefficient (k.e.), the measurements plotted onto a straight line. The activation energy was calcu-

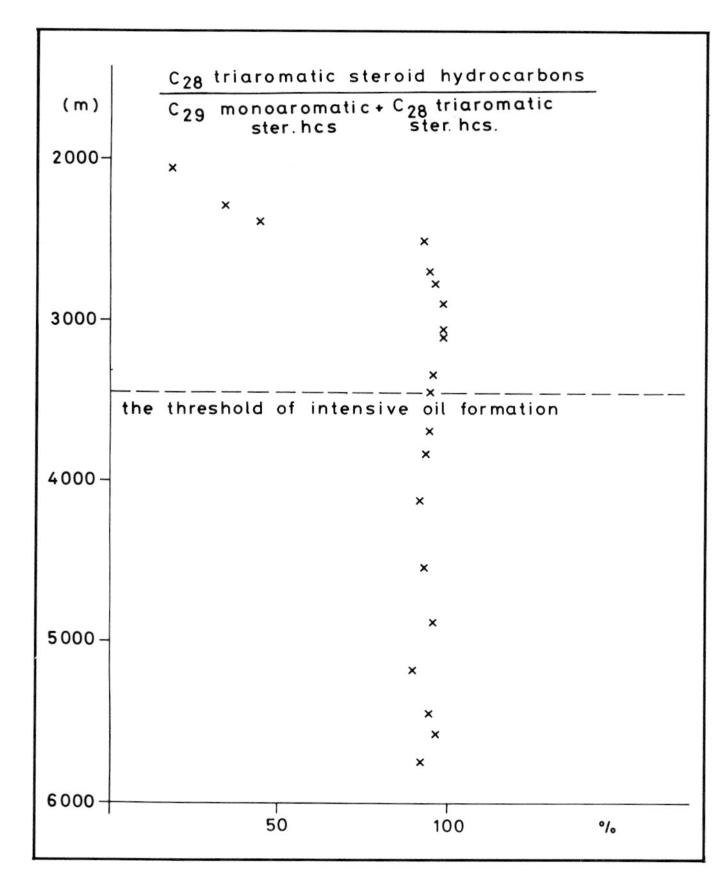

Figure 11. Increase in the extent of geochemical aromatization of C_{29} C-ring monoaromatic steroid hydrocarbons to C_{28}-triaromatic steroid hydrocarbons as a function of depth/temperature.

lated to be 97.84 kJ/mol, the preexponential factor, A, 5.43×10^{12} m.y.$^{-1}$.

The AT Method. In the relationship ($\ln \vec{k} - 1/T$), the values of k determined from the differential method produced a straight line corresponding to Equation (17):

$$\ln \vec{k} = -1.042 \times 10^4/T + 26.26 \quad (\text{k.e.} = -0.9851)$$

The corresponding activation energy, ΔH is 96.64 kJ/mol, preexponential factor, A is 2.548×10^{11} m.y.$^{-1}$. Using ΔH^* values as starting data obtained from the differential and EHT methods, successive iterations by the AT method gives

$$1.364 \pm 0.1605 \times 10^{12} \text{ m.y.}^{-1}$$

for the preexponential factor and 92.32 kJ/mol for the activation energy.

Hopane Isomerization

The EHT Method. From measurements, $K = 1.326 \rightarrow \beta = 1.754$. By the EHT method,

$$T_{eff} = 15\,°\text{C} \rightarrow \tau_{EHT} = 1.16 \text{ m.y.}$$

Substituting the measurement data into Equation (17) gives

$$\ln \vec{k} = 1.157 \times 10^4 \times 1/T + 30.30 \quad (\text{k.e.} = -0.996)$$

This fit is good. The value of the activation energy was calculated to be $\Delta H^* = 96.205$ kJ/mol, the preexponential factor was $A = 1.101 \times 10^{13}$ m.y.$^{-1}$.

The AT Method. The values determined by the differential method also plot along a straight line on a graph of $\vec{k}$ versus $1/T$. The relationship that corresponds to Equation (17) is

$$\ln \vec{k} = 1.162 \times 10^4 \times 1/T + 29.307 \quad (\text{k.e.} = 0.9873)$$

and the fit is good. The activation energy was calculated to be $\Delta H^* = 96.62$ kJ/mol, the preexponential factor:

$$A = 5.344 \times 10^{12} \text{ m.y.}^{-1}$$

Using these values as starting data obtained by the differential and the EHT methods, successive iterations by the AT method gives

$$A = (6.2406 \pm 0.5965) \times 10^{12} \text{m.y.}^{-1}$$

and

$$\Delta H^* = 92.89 \text{ kJ/mol}$$

Aromatization Reaction

The EHT Method. From measurements,

$$K = \infty, \quad \beta = 1$$

With the EHT method: $T_{eff} = 15\,° \rightarrow \tau_{EHT} = 1.16$ m.y. Substituting this data into Equation (17) gives

$$\ln \vec{k} = 1.4406 \times 10^4/T + 39.111 \quad (\text{k.e.} = 0.9589)$$

In this case the fit is poorer than for the isomerization reactions. The corresponding activation energy is 119.8 kJ/mol and the preexponential factor is 9.68×10^6 m.y.$^{-1}$.

The AT Method. For this reaction, the differential method produced scattered results because of the small number of measurements and their uneven distribution.

Successive iteration using the AT method gives

$$A = 1.430 \times 10^{17} \text{ m.y.}^{-1}$$

and

$$\Delta H^* = 122.4 \text{ kJ/mol.}$$

From the data in this paper, McKenzie et al. (1983) obtained the following: for sterane isomerization, 91 kJ/mol for activation energy and 1.89×10^{11} m.y.$^{-1}$ for the preexponential factor; for aromatization of the monoaromatic steroid hydrocarbon, 200 kJ/mol for activation energy and 5.68×10^{28} m.y.$^{-1}$ for the preexponential factor. The agreement is satisfactory for isomerization, but not for aromatization. Our activation energy for aromatization is one-half the value of McKenzie et al., and our frequency factor is at least ten orders of magnitude smaller.

Effective Heating Times

Using the values of activation energy and preexponential factor that were obtained above by the method of absolute times, we recalculated the effective heating time for each of the three reactions. We obtained effective heating times of 0.78 m.y. for sterane isomerization, 0.73 m.y. for hopane isomerization, and 0.567 m.y. for aromatization reaction. This indicates that these reactions reached equilibrium or completion within less than one million years in the well Hód-I.

ACKNOWLEDGMENTS

We thank the National Oil and Gas Trust of Hungary for providing samples and permission to publish. We would also like to thank Shell Development Company for their generosity in providing us with their vitrinite reflectance measurements for Hód-I. The computerized gas chromatograph-mass spectrometer measurements have been carried out at Bristol during the tenure of a fellowship of the Scientific Exchange Agreement by Cs.S. who is grateful to Professor G. Guiochon (Ecole Polytechnique Paris) and Professor G. Eglinton (University of Bristol) for his fellowship. Technical assistance from Mrs. A. Maròt and Mrs. M. Heltay in Budapest, and Mrs. A. Gowar in Bristol, is gratefully acknowledged.

REFERENCES

Albrecht, P., M. Vandenbrouke, and M. Mandengué, 1976, Geochemical studies on the organic matter from the Douala basin (Cameroon): I. Evolution of the extractable organic matter and the formation of petroleum: Geochim. et Cosmochim. Acta, v. 40, p. 791–799.

Allan, J. and A. G. Douglas, 1977, Variations in the content and distribution of n-alkanes in a series of carboniferous vitrinites and sporinites of bituminous rank: Geochim. et Cosmoch. Acta, v. 41, p. 1223–1230.

Ammosov, I. I., V. I. Gorshkov, N. P. Grechishnykov, and G. S. Kalmykov, 1977, Paleogeotermicheskiye kriteriyi rezmescheniyq neftyanykh zalezhey. (Paleogeothermic criteria of the location of petroleum deposits): Leningrad, Nedra Press, p. 158.

Bertalan, M., 1978, Qualification of kerogen of sedimentary rocks according to geochemical aspects (in Hungarian), PhD thesis, JATE, Szeged.

Bostick, N. H., 1973, Time as a factor in thermal metamorphism of phytoclasts (coaly particles): Congr. Int. de Stratigraphie et de Géologie du Carbonifere, C. R. 2, p. 183–193.

Bostick, N. H., S. M. Cashman, T. H. McCulloh and C. T. Waddell, 1979, Gradients of vitrinite reflectance and present temperature in the Los Angeles and Ventura Basins, California, *in* D. F. Oltz, ed., Low temperature metamorphism of kerogen and clay minerals: Los Angeles, SEPM, Pacific Section, p. 65–96.

Brooks, J. D. and S. W. Smith, 1967, The diagenesis of plant lipids during the formation of coal, petroleum and natural gas: I. Changes in the *n*-paraffin hydrocarbons: Geochim. et Cosmochim. Acta, v. 31, p. 2389–2397.

Brooks, J. D., K. Gould, and J. W. Smith, 1969, Isoprenoid hydrocarbons in coal and petroleum: Nature, v. 222, p. 257–259.

Dövényi, P., F. Horváth, P. Liebe, J. Gálfi and I. Erki, 1983, Geothermal conditions of Hungary: Lóránd Eötvös Geophysical Institute of Hungary, 114 p.

Dow, W. G., 1978, Kerogen studies and geological interpretations: Jour. Geochem. Exploration, v. 7, p. 79–99.

Espitalié, J., J. L. Laporte, M. Madec, F. Marquis, P. Leplat, J. Paulet, and A. Boutefeu, 1977, Methode rapide de caractérisation des roches méres de leur potential pétrolier et de leur degré d'évolution: Rev. Inst. Fr. Pétrole, v. 32, p. 23–42.

Flekken, P. M., 1978, Anwendung organisch geochemischer-kohlenpetrographischer-und isotopgeochemischer Untersuchungsmethoden in der Faziesanalyze und der Kohlenwasserstoffexploration am Beispiel des NE-Randes van Parisier Becken: Ph.D. thesis, RWTH, Aachen.

Héroux, Y., A. Chagnon, and R. Bertrand, 1979, Compilation and correlation of major thermal maturation indicators: AAPG Bull., v. 63, p. 2128–2144.

Hood, A., C. C. M. Gutjahr, and R. L. Heacock, 1975, Organic metamorphism and the generation of petroleum: AAPG Bull., v. 59, p. 986–996.

Horváth, Z. A., 1980, Optical studies on kerogen and application of results for hydrocarbon prospecting (in Hungarian): Ph.D. thesis, ELTE, Budapest.

Hunt, J. M., 1979, Petroleum geochemistry and geology: San Francisco, W. H. Freeman, p. 617.

Koncz, I., 1983, Comparison of the Lopatin methods and their critical evaluation: Acta Mineral. Petrog., Szeged, v. 26, p. 51–71.

Laplante, R. E., 1974, Hydrocarbon generation in Gulf Coast Tertiary sediments: AAPG Bull., v. 58, p. 1281–1289.

Lefler, J., and Cs. Sajgó, 1986, Limits of application of the reaction kinetic methods, *In:* Buntebarth, G. and L. Stegena (eds.), Paleogeothermics: Springer Verlag, New York, p. 153–173.

Leythaeuser, D., 1975, Erdölgenese in Anhängigkeit von der Art des organischen Materials im Muttergestein: Erdöl und Kohle, Compendium 74/75, p. 41–51.

Lopatin, N. V., 1971, Temperatura i geologicheskoe vremya kak faktori uglefikatsii. (Temperature and geologic time as a factor in coalification): Izv. An SSSR Ser. Geol., v. 3, p. 95–106.

Lopatin, N. V., 1976, Istoriko-geneticheskiy analiz nefteobrazovaniya s ispolzovaniem modeli ravnomernogo nepreryvnogo opuskanija neftematerinskogo plasta. (Historico-genetic analysis of petroleum generation: Application of a model of uniform continuous subsidence of the oil source bed): Izv. An SSSR Ser. Geol., no. 8, p. 93–101.

Mackenzie, A. S., R. L. Patience, J. R. Maxwell, M. Vandenbrouke, and B. Durand, 1980, Molecular parameters of maturation in the Toarcian shales, Paris basin: I. Changes in the configuration of acyclic isoprenoid alkanes, steranes and triterpanes: Geochim. et Cosmochim. Acta, v. 44, p. 1709–1721.

Mackenzie, A. S., C. F. Hoffmann, and J. R. Maxwell, 1981, Molecular parameters of maturation in the Toarcian shales, Paris basin, France: III. Changes in aromatic steroid hydrocarbon: Geochim. et Cosmochim. Acta, v. 45, p. 1345–1355.

McKenzie, D., A. S. Mackenzie, J. R. Maxwell, and Cs. Sajgó, 1983, Isomerization of hydrocarbons in stretched sedimentary basin: Nature, v. 301, p. 504–506.

Mucsi, M., 1973, Geological history of the southern Great Hungarian Plain during the late Tertiary (in Hungarian): Földtani Közlöny, v. 103, p. 311–318.

Mucsi, M., and J. Révész, 1975, Neogene evolution of the southeastern part of the Great Hungarian Plain on the basis of sedimentological investigations: Acta Miner. Petr. Szeged, v. 22, p. 29–49.

Philippi, G. J., 1965, On the depth, time and mechanism of petroleum generation: Geochim. et Cosmochim. Acta, v. 29, p. 1021–1049.

Price, L. C., 1981, Aqueous solubility of crude oil to 400 °C and 2000

bars pressure in the presence of gas: Journal of Petroleum Geology, v. 4, p. 194–223.
Price, L. C., 1982, Organic geochemistry of core samples from an ultra deep hot well (300 °C, 7 km): Chemical Geology, v. 37, p. 215–228.
Price, L. C., J. L. Clayton, and L. L. Rumen, 1979, Organic geochemistry of a 6.9 kilometer deep well, Hinds County, Mississippi: Gulf Coast Assoc. Geol. Soc. Trans., v. 29, p. 352–370.
Price, L. C., J. L. Clayton, and L. L. Rumen, 1981, Organic geochemistry of the 9.6 km Berta Rogers No. 1. well, Oklahoma: Organic Geochemistry, v. 3, p. 59–77.
Radke, M., R. G. Schaefer, D. Leythaeuser, and M. Teichmüller, 1980, Composition of soluble organic matter in coals: Relation to rank and liptinite fluorescence: Geochim. et Cosmochim. Acta, v. 44, p. 1787–1800.
Raynaud, J. F., and P. Robert, 1976, Les methodes d'étude optique de la matiére organique: Soc. Nat. Pétrole Aquitaine Cent. Rech. Pau Bull., v. 10, p. 108–127.
Sajgó, Cs., 1980a, Complex petroleum geochemical studies on core samples of the Makó-Hódmezövásárhely Trench (in Hungarian): D. T. Sc. thesis, BME.
Sajgó, Cs., 1980b, Hydrocarbon generation in a superthick Neogene sequence in south-east Hungary: A study of the extractable organic matter, *in* A. G. Douglas, and J. R. Maxwell, eds., Advances in geochemistry 1979: New York, Pergamon Press, p. 103–113.
Sajgó, Cs., and J. Lefler, 1986, A reaction kinetic approach to temperature-time history of sedimentary basins, *in* G. Buntebarth and L. Stegena, eds., Paleogeothermics: New York, Springer-Verlag, p. 119-151.
Sajgó, Cs., A. S. MacKenzie, and J. R. Maxwell, in press, Changes in the biological marker distributions in a thick Neogene sequence in Hungary: Org. Geochem.
Saxby, J. D., 1982, A reassessment of the range of kerogen maturities in which hydrocarbons are generated: Journal of Petroleum Geology, v. 5, p. 117–128.
Speers, G. C., and E. V. Whitehead, 1969, Crude petroleum, *in* G. Eglinton, and M. T. S. Murphy, ed., Organic geochemistry: Methods and results: Berlin, Springer-Verlag, p. 638.
Szentgyörgyi, K., 1975, Lithological petrophysical conditions of Neogene sediments encountered when drilling Hód-I well (in Hungarian): Kőolaj és Földgáz, v. 8, p. 172–175.
Thompson, K. F. M., 1983, Classification and thermal history of petroleum based on light hydrocarbons: Geochim. et Cosmochim. Acta, v. 47, p. 303–316.
Tissot, B., Y. Califet-Debyser, G. Deroo, and J. L. Oudin, 1971, Origin and evolution of hydrocarbons in early Toarcian shales, Paris basin, France: AAPG Bull., v. 55, p. 2177–2193.
Tissot, B., B. Durand, J. Espitalié, and A. Cobaz, 1974, Influence of nature and diagenesis of organic matter information of petroleum: AAPG Bull., v. 58, p. 499–506.
Tissot, B., R. Pelet, J. Rouchache, and A. Combaz, 1977, Utilisation des alkanes comme fossiles géochemiques indicateures des environments géologiques, *in* R. Campos, and J. Goñi, ed., Advances in organic geochemistry, 1975: Madrid, ENADIMSA, p. 117–154.
Tissot, B., and D. H. Welte, 1978, Petroleum formation and occurrence: Berlin, New York, Springer-Verlag, p. 538.
Vassoyevich, N. B., Yu. I. Korchagina, N. V. Lopatin, and V. V. Chernyishev, 1970, Principal phase of oil formations: Moscow Univ. Vestnik, 1969, v. 6, p. 3–27 (in Russian); Internat Geology Rev., v. 12, p. 1276–1296 (in English).
Waples, D. W., 1980, Time and temperature in petroleum formation: Application of Lopatin's method to petroleum exploration: AAPG Bull., v. 64, p. 916–926.
Yakovets, Yu. A., T. A. Safranov, and Ye. B. Yakovets, 1976, Organicheskoe veshchestvo v osadochnykh tolshchackh orogennoy oblasti yugo-vostoka sredney Azyi. (Organic matter in sedimentary sequences of orogen area of SE Middle Asia), *in* N. B. Vassoevich and P. P. Timofeyev, eds., Issledovaniya organicheskogo veshchestva sovremennik i iskopayemukh osadkov: Study of Organic Matter, p. 251–260.

CHAPTER 22

Secondary Heating of Vitrinite: Some Geological Implications

I. Laczó
Á. Jámbor
Hungarian Geological Institute, H-1143 Budapest
Népstadion út 14, Hungary

Vitrinite reflectance data in Hungary are mainly used for hydrocarbon exploration in the Neogene basins. However, enough data are available from secondary reworked vitrinite populations to help determine the source areas and maturation history of older reworked sediments within the Neogene basins. This reworked material is derived mainly from Paleogene to early Paleozoic deposits that were uplifted and eroded during Tertiary time. A simple model describing the maturation of reworked vitrinite gives results in good agreement with observations. This model explains how a discontinuity in vitrinite reflectance can occur across an erosional unconformity, and why this discontinuity may disappear with heating of the sediment pile over a sufficiently long time period. Vitrinite data are also sensitive to local heating caused by magmatic intrusions. This makes it possible to distinguish between intrusive and extrusive igneous rocks using vitrinite data, and to predict the existence and location of intrusive igneous rocks that have not been penetrated by drilling.

INTRODUCTION

For a long time it has been known that the quality of old coals is generally better than that of young coals. Since the nineteenth century, exceptions to this rule have been known because coalification was observed to occur much more rapidly in tectonically active orogenic belts than on stable platforms. This was originally thought to be the result of higher pressure in orogenic belts. Temperature was thought to play only a secondary role. More recently, Karweil (1956), Lopatin (1971) and others have demonstrated that coalification corresponds to a thermal maturation process. The amount of coalification or maturation is controlled by the cumulative temperature history of organic matter during its entire burial history.

Microscopic examination has shown that dispersed organic material with small grain sizes (1 to 1000 μm) is usually abundant in almost all sedimentary rocks. The coal rank of the organic matter can be adequately described by the reflectance of vitrinite. which is adequate for coalification ranging from immature peat to anthracite (Teichmüller and Teichmüller, 1950). The reflectance of primary vitrinite is initially about 0.2%, but may increase to several percent during burial. The rate of maturation increases rapidly with increasing temperature, so that the maturity observed in most rocks is largely a function of the peak temperatures experienced and the length of time that peak or near-peak temperatures are maintained. When particles of organic material experience large decreases in temperature, for example during erosion, maturation of these particles proceeds at a negligible rate compared to the earlier maturation rate at higher temperatures. Thus, the level of maturity attained while the particles are near their peak temperatures is effectively frozen into the rock until such time as temperatures become nearly as great as those experienced previously.

This paper deals mainly with the phenomena of secondary heating in vitrinite, when vitrinite particles that have been through a peak metamorphism, followed by significant cooling, are subsequently reheated. Two important settings for such secondary heating are: (1) reworking of vitrinite during burial, erosion and subsequent resedimentation; and (2) vitrinite located below a significant erosional unconformity and later covered by younger sediments. Changes in vitrinite reflectance, in material that is subject to temperature increases during magmatic or tectonic activity, are also discussed. In this paper we shall present some examples of each of these phenomena, using vitrinite reflectance data from Hungary.

Table 1. Minimum and Maximum Values of Vitrinite Reflectance (1, in percent) and the Depth Interval of the Measured Samples (2, in meters) in Hungary

	Sopron–Kőszeg		Zala Basin Central Range		Transdanubian Mountains		Mecsek Range		North Hungarian Plain		Great Hungarian Plain	
	1	2	1	2	1	2	1	2	1	2	1	2
Early Paleozoic	4.77 4.90	460 1955	—	—	4.39 4.82	11 990	— 4.65	— 915	3.52 4.94	40 1662	4.66 4.82	3740 3800
Carboniferous	—	—	2.91 4.55	3700 4303	0.70 3.61	133 399	3.60 4.34	91 1230	3.87 4.46	38 622	—	—
Permian	—	—	1.83 1.95	3369 3525	1.85	1249 1509	2.23 2.94	87 1448	3.17 4.55	surface 426	—	—
Triassic	—	—	0.69 1.36	2760 3864	0.50 1.36	30 1416	1.03 1.54	61 1835	2.80 4.50	surface 440	1.22 1.40	2255 4430
Jurassic	4.62 4.90	surface	0.85	3320	—	—	0.81 1.27	56 1262	2.31 3.08	surface	—	—
Cretaceous	—	—	0.47 1.99	2206 4675	0.41 0.65	150 849	—	—	—	—	106	4260
Eocene	—	—	0.59 0.60	2833 3078	0.40 0.53	148 497	0.46	1089	0.63 0.69	220 382	—	—
Oligocene	—	—	—	—	0.48	50 120	—	—	0.26 0.60	33 631	—	—
Pre-Pannonian Miocene	—	—	0.51 1.80	2006 4590	0.55	640	0.25 0.42	318 744	0.21 0.47	50 2000	0.37 2.97	1632 4485
Pannonian (*s. l.*)	—	—	—	—	0.30 0.54	24 600	0.21 0.52	13 307	0.23 0.61	51 2860	0.29 2.79	320 4424

In Hungary, the first vitrinite reflectance measurements were made in 1977 at the Geological Institute. About 3000 measurements of vitrinite reflectance have been made on Paleozoic to late Cenozoic sedimentary rocks (Laczó, 1982, 1984; see also Table 1). Most of these data are generated for hydrocarbon exploration in the Neogene basins and for coal studies in the Mecsek Mountains. In this chapter we summarize some of the applications of this large data set to the problem of secondary and tectonic heating of vitrinite.

VITRINITE DATA IN HUNGARY

Table 1 and Figure 1 give a summary of vitrinite reflectance data from Hungary. The data are classed according to age and area. Table 1 shows the maximum and minimum values of vitrinite reflectance in each class, as well as the depth range of samples for which vitrinite reflectance was measured. In general, the largest values of vitrinite correspond to the greatest depths, and the smallest values to the least depths, but there are a few exceptions.

Early Paleozoic to Carboniferous rocks are everywhere highly mature, and have vitrinite reflectance values between about 3.5% and 5%. In most places, Triassic and Jurassic rocks are much less mature, displaying values of vitrinite reflectance between 0.5% and 1.6% in the Zala basin, Transdanubian Central Mountains and the Mecsek Mountains. Cretaceous–Tertiary rocks are relatively immature in the mountainous areas of Hungary (Transdanubian Mountains) where the strata are usually thin. The same rocks display very high values of vitrinite reflectance (up to 3.0%) in and below the deep Neogene troughs of the Zala basin and the Great Hungarian Plain.

In two parts of Hungary, Triassic–Jurassic rocks display very high vitrinite reflectance values, and in both cases these high values can be attributed to heating of the rocks after they were overthrust in the Mesozoic by nappes of the Alpine belt. In the Kőszeg Mountains, Jurassic rocks now exposed at the surface in a Penninic-type tectonic window (the Rechnitz window) have vitrinite reflectance values of 4.6% to 4.9%. A rough estimate indicates that this level of maturation was attained at depths of 5 to 6 km. This gives a minimum estimate for the thickness of the overriding nappe stack, which has now been mostly destroyed by erosion. Anomalously high values of vitrinite reflectance (2.3% to 4.5%) also occur at or near the surface in Triassic and Jurassic rocks in North Hungary (in the Bükk, Uppony, Szendrő and Rudabánya Mountains). The structural setting of these rocks is under debate, but their vitrinite reflectance values argue that they were once overlain by a great thickness (5 km?) of rock, and thus probably formed the lower plate below one of the Alpine thrust faults.

HYDROCARBON GENERATION AND VITRINITE REFLECTANCE

Hydrocarbons are generated in suitable source rocks at depth when kerogen in dispersed organic material undergoes thermal alteration and maturation. Different phases of hydrocarbon

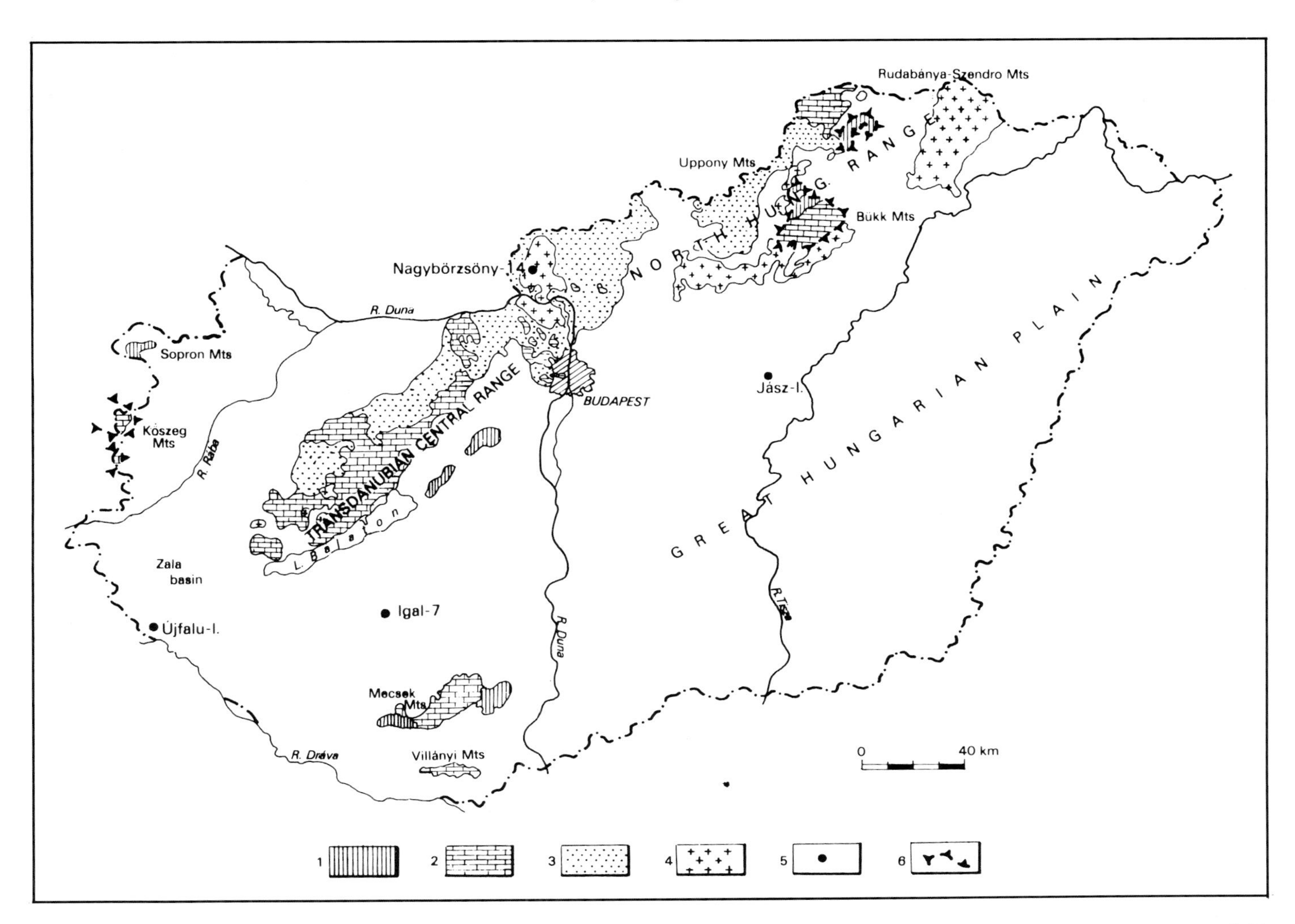

Figure 1. Geological sketch map with the locations of vitrinite reflectance measurements referred to in the text. Map shows outcrops of: (1) Paleozoic rocks, (2) Mesozoic rocks, (3) Paleogene rocks, (4) Miocene volcanic rocks, (5) location of vitrinite reflectance measurements in boreholes, and (6) tectonic windows.

generation can be directly related to different threshold values of maturity and vitrinite reflectance in the source rocks (Tissot and Welte, 1978; Hunt, 1979; Durand, 1980). The oil generation window is generally placed between vitrinite reflectance values of 0.6% and 1.2%, and samples from boreholes make it possible to determine if potential source rocks have passed into or through the oil generation window. More sophisticated analyses of maturation history can be made if the temperature history of the potential source rocks can be reconstructed (Waples, 1980; Lopatin, 1971). New techniques for reconstructing temperature and maturation histories of sediment packages from the subsidence and sedimentation histories of basins were first applied in Hungary by Stegena et al. (1981) and Horváth and Dövényi (1983). The most recent results of such analyses are given by Horváth et al. (this volume). In the following sections of this paper we will use some of the methods for calculating maturity from thermal histories to address the problem of secondary heating of vitrinite reflectance in several different settings.

VITRINITE REFLECTANCE ACROSS AN EROSIONAL UNCONFORMITY

When significant erosional unconformities occur within sedimentary sequences, a discontinuity in measured vitrinite reflectance often occurs across the unconformity (Dow, 1977). In other cases, there is no discontinuity in vitrinite reflectance across the unconformity, but in plots of vitrinite reflectance a break in slope occurs across the unconformity. In many cases, both the absolute value of vitrinite reflectance and the slope of vitrinite reflectance as a function of depth may change across the unconformity. For example, in the borehole Igal-7, Neogene basin sediments were found to overlie Triassic carbonates across an erosional unconformity (Figure 2). By comparing the section found in this drillhole with rocks exposed in the Transdanubian Central Mountains, one may infer that the Triassic strata in Igal-7 were originally overlain by an uppermost Triassic

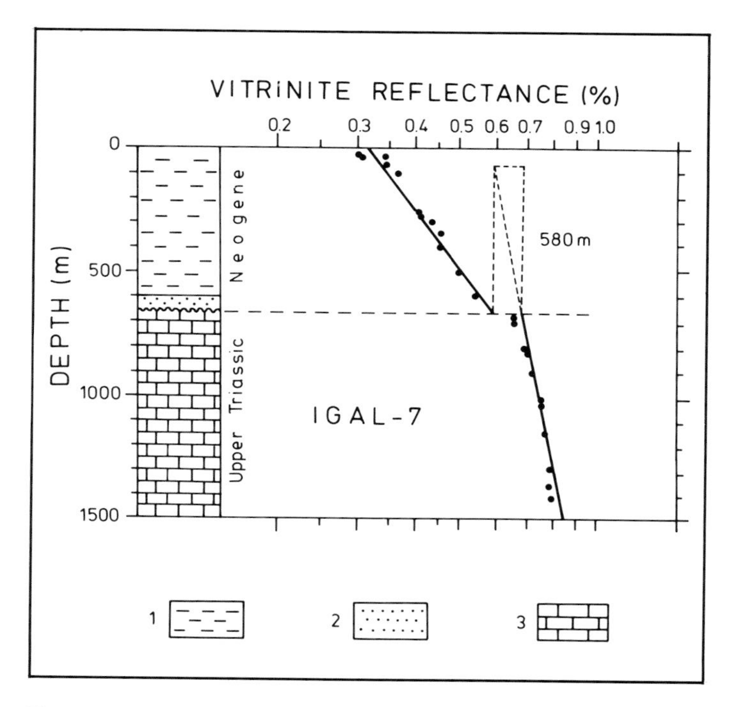

Figure 2. Vitrinite reflectance versus depth in the Igal-7 borehole. An erosional unconformity exists at about 650 m depth. A rough estimate of the thickness of the missing strata yields about 580 m, according to the method of Dow (1977).

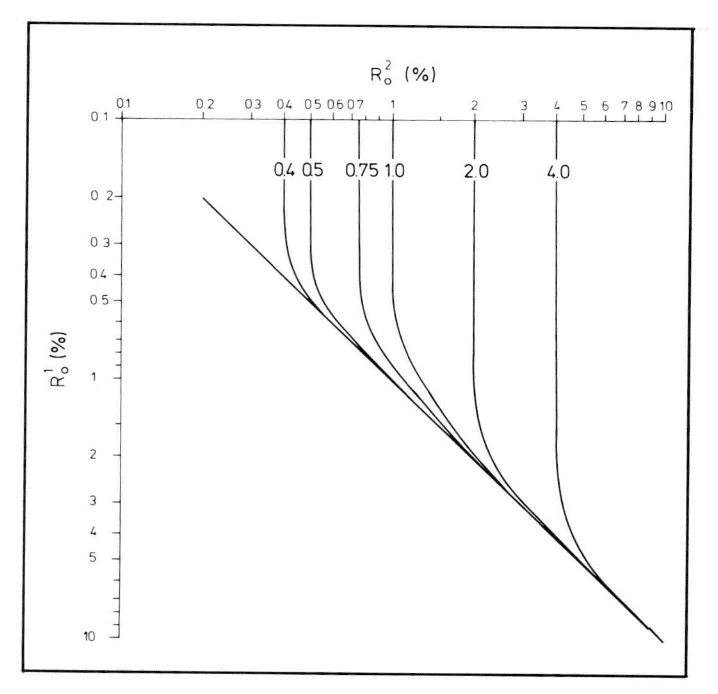

Figure 3. The reflectance history of a vitrinite particle with a low initial maturity (R_o^1) relative to the reflectance history of vitrinite particle with higher initial maturities (R_o^2). The particles experience the same temperature conditions and the paths indicate the increase in maturity with time. For example, consider one particle (1) with original vitrinite reflectance $R_o^1 = 0.2\%$, and a second particle (2) with original vitrinite reflectance $R_o^2 = 0.5\%$. During heating, the vitrinite reflectance of particle (2) remains nearly constant while the vitrinite reflectance of particle (1) increases from 0.2% to 0.4%. Then the vitrinite reflectance of particle (2) increases until particle (1) and particle (2) have about the same reflectance (at about $R_o^1 \approx R_o^2 \approx 0.8\%$). Above $R_o \approx 0.8\%$, the vitrinite reflectance of the two particles are effectively indistinguishable. The vitrinite reflectance paths shown here were calculated using the Lopatin-Waples method (Waples, 1980). The maturity paths shown are independent of the temperature history assumed (modified after Horváth and Dövényi, 1983.)

to Upper Cretaceous sequence, which was probably eroded during Late Cretaceous (Senonian) to Paleogene uplift.

As a first approximation, the thickness of the strata eroded off of the Triassic rocks in Igal-7 can be estimated using the method of Dow (1977). Dow (1977) extrapolated the vitrinite reflectance gradient, measured below the unconformity, upwards into the overlying sequence. The thickness of the eroded strata are then considered to be equal to the depth interval between the erosional unconformity and the depth where this extrapolated gradient gives vitrinite reflectance values equal to those measured just above the erosional unconformity. Applying this method to rocks in the drillhole Igal-7 gives about 580 m for the thickness of the eroded section (Figure 2).

This method for estimating the thickness of the eroded strata is probably not very reliable, because it is dependent on the size of the break or discontinuity in vitrinite reflectance values across the unconformity, as well as on the gradient of vitrinite reflectance in the rocks below the unconformity. In the following analysis we show that the size of the discontinuity in vitrinite reflectance across the unconformity decreases during progressive burial after erosion, and may disappear altogether if the unconformity is covered by a sufficiently thick sedimentary sequence. In order to address the maturation of the more mature vitrinite below the unconformity, relative to the maturation of the less mature vitrinite above the unconformity, we consider two neighboring vitrinite particles, 1 and 2, with vitrinite reflectance R_o^1 and R_o^2, respectively. Let particle 1 be initially immature vitrinite and particle 2 be initially mature vitrinite. Particle 1 would thus correspond to a vitrinite particle immediately above an erosional unconformity, while particle 2 would correspond to one immediately below the unconformity. The increase in maturity of both particles during heating (beginning with low temperatures) can be easily calculated using the Lopatin-Waples method (Waples, 1980) (Figure 3). The calibration between cumulative temperature history and vitrinite reflectance was taken to be that described in Horváth et al. (this volume).

Figure 3 shows that, during the progressive maturation of particle 1, the maturity of particle 2 remains practically unchanged until the degree of maturation of particle 1 approaches that of particle 2. Subsequently, the maturation paths of particles 1 and 2 merge, and become indistinguishable during the rest of their maturation history. The relationship shown in Figure 3 is independent of the details of the temperature history assumed. The degree of maturation in particle 2 can be calculated directly from R_o^1 and the initial (posterosional) maturity of particle 2. Such an analysis explains why a discontinuity in the vitrinite-depth relationship is sometimes not present across large erosional unconformities. It also demonstrates that the size of the break across an erosional unconformity can decrease and disappear during burial of the unconformity and heating of the adjacent rocks. Thus, estimates of the original thickness of strata eroded off of the unconformity, as determined by the method of Dow (1977), will change as the level of maturity of particles near the unconformity increases, and should not be regarded as particularly reliable.

RECYCLED VITRINITE

Erosion and resedimentation of old sedimentary rocks usually supplies relatively mature vitrinite and other clastic particles to a younger basin. These recycled vitrinite particles may be abundant in sedimentary rocks and, depending on the source area, more than one population of recycled vitrinite may be present. It is often difficult to distinguish primary and recycled vitrinites, and both are considered when calculating the mean reflectivity. This biases the mean value of the reflectance measurements and increases the standard deviation. Partly for this reason, different people or laboratories may determine different vitrinite reflectances for the same rock (Issler, 1984). Such need not be the case, however, if one can understand the behavior of recycled vitrinites during the new cycle of maturation.

The problem of maturation in recycled vitrinite can be addressed in exactly the same way as that described above for vitrinite below an unconformity (Figure 3). In this case, particle 1 is the primary organic material and particle 2 is the recycled material. Thus, the vitrinite reflectance of the recycled vitrinite remains essentially unchanged until the primary vitrinite is almost as mature as the recycled material. Subsequently, the maturities of R_o^1 and R_o^2 converge and become virtually indistinguishable.

Vitrinite reflectance measurements from borehole Jász-I give a good example of this process (Figure 4). Cores and drilling chips from 35 different depths were analyzed carefully and three populations of recycled vitrinite were distinguished. The first population was found only in the upper 1000 m of the borehole and has a vitrinite reflectance between about 0.5% and 0.6%. The second population is present throughout the borehole and has a vitrinite reflectance between 0.7% and 0.8%. The third population occurs in the deep part of the borehole, and has vitrinite reflectance between 1% and 1.5%. Vitrinite reflectance as a function of depth is, for all populations, in good agreement with the maturation diagram shown in Figure 3. In each case, vitrinite reflectance of the reworked populations remains unchanged with depth wherever the maturity of the primary vitrinite is significantly lower than that of the reworked vitrinite. For population 2 (originally R_o = 0.7% to 0.8%), the vitrinite reflectance of the reworked material and of the primary vitrinite are distinct above about 2.5 km depth. Below 3 km, the reworked vitrinite of population 2 becomes indistinguishable from the primary vitrinite. The vitrinite reflectance of population 3 (originally R_o = 1.0% to 1.5%) appears to merge with that of the primary vitrinite near the bottom of the borehole. Population 1 (originally R_o = 0.5% to 0.6%) was probably not deposited below about 1 km depth in this borehole.

Observations such as these help to identify the source areas for recycled vitrinite and other clastic particles. For example, vitrinite with reflectance R_o = 1.0% to 1.6% commonly

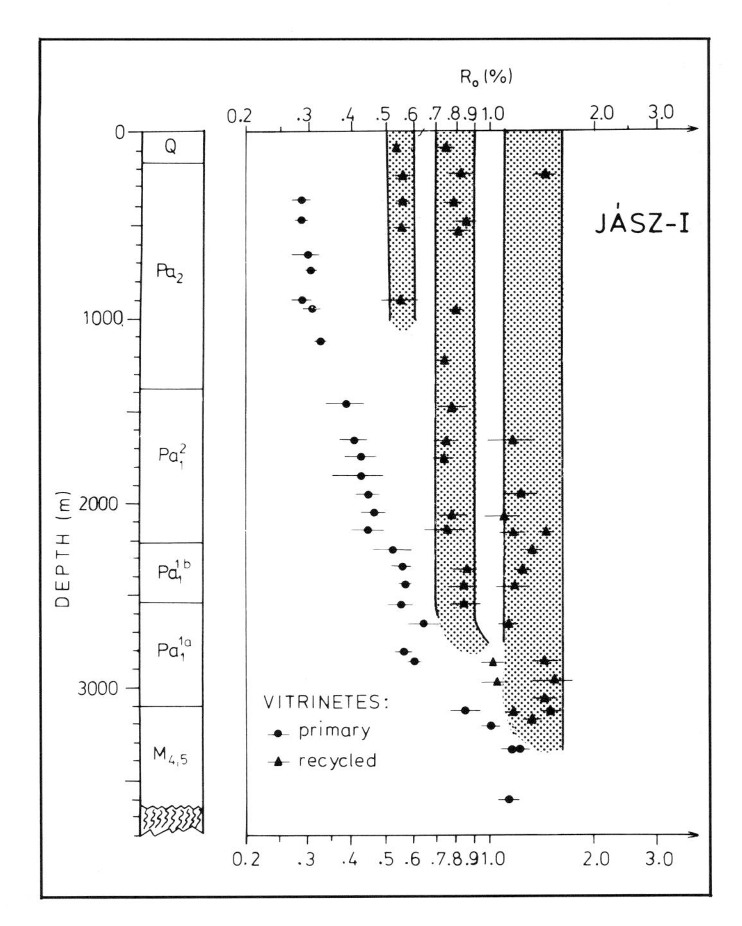

Figure 4. Vitrinite reflectance data from the borehole Jász-I. Maturity of the primary organic matter increases with depth. Three populations of recycled vitrinite were observed and their maturation paths agree with those predicted in Figure 3 and are discussed in the text.

occurs in the outer Carpathian Flysch belt of early Tertiary age. It is also found as reworked vitrinite in most parts of the Neogene basins of the Great Hungarian Plain (Figure 1). This strongly supports the view of Telegdi (1929), who proposed that the Carpathian Flysch belt was one of the primary distant sources of the Miocene basin fill. The Carpathian Flysch belt consists of Late Cretaceous to Oligocene sedimentary rocks that were deformed (by folding and thrusting) during Oligocene–early Miocene time and subsequently eroded. This provided a source for the Neogene basins until recent time. Clastic material was probably transported by rivers and wind.

MATURITY ADJACENT TO MAGMATIC BODIES

Intrusion of hot magmatic bodies greatly increases the temperature of the intruded country rock, at least for a short time, and creates an aureole of increased maturation around the magmatic body (Horváth et al., 1986). The maturity of sediments is increased near both top and bottom of the intrusion. This differs from the case of stratovolcanoes and other extrusive lava flows, which were buried by sediments only after they had cooled. Observations of maturity of organic material above and below igneous bodies may thus help to distinguish between intrusive and extrusive rocks, if it is not otherwise clear.

Data from a borehole in the Börzsöny Mountains (Nagybörzsöny-14) shows very high vitrinite reflectance values next to intrusive dikes (R_o = 4%) (Figure 5). Vitrinite reflectance values increase towards the contact of these sediments with the igneous bodies, indicating that the lower dacite intruded and heated the sedimentary rocks.

Other measurements near sedimentary-igneous contacts show that one group of vitrinites exhibited very high reflectance values near the contact, with decreasing reflectance values away from the contact (Figure 6). Another group of vitrinites were apparently not influenced by the magmatic heating because they show low and constant values of vitrinite reflectance. We speculate that this may be related to the presence of recycled vitrinite which was not sensitive to the rapid temperature increases caused by the magmatic body. Another example is given in Figure 7, where the Újfalu-I borehole penetrated a thick Neogene sequence and penetrated more than 1 km into the underlying basement. In this borehole, there is an abnormally rapid increase of maturity with depth, with an average gradient of about 0.27%/100 m. We think that it is caused by a large plutonic body located below the bottom of the borehole. If so, a similarly high vitrinite gradient should be observed in nearby boreholes.

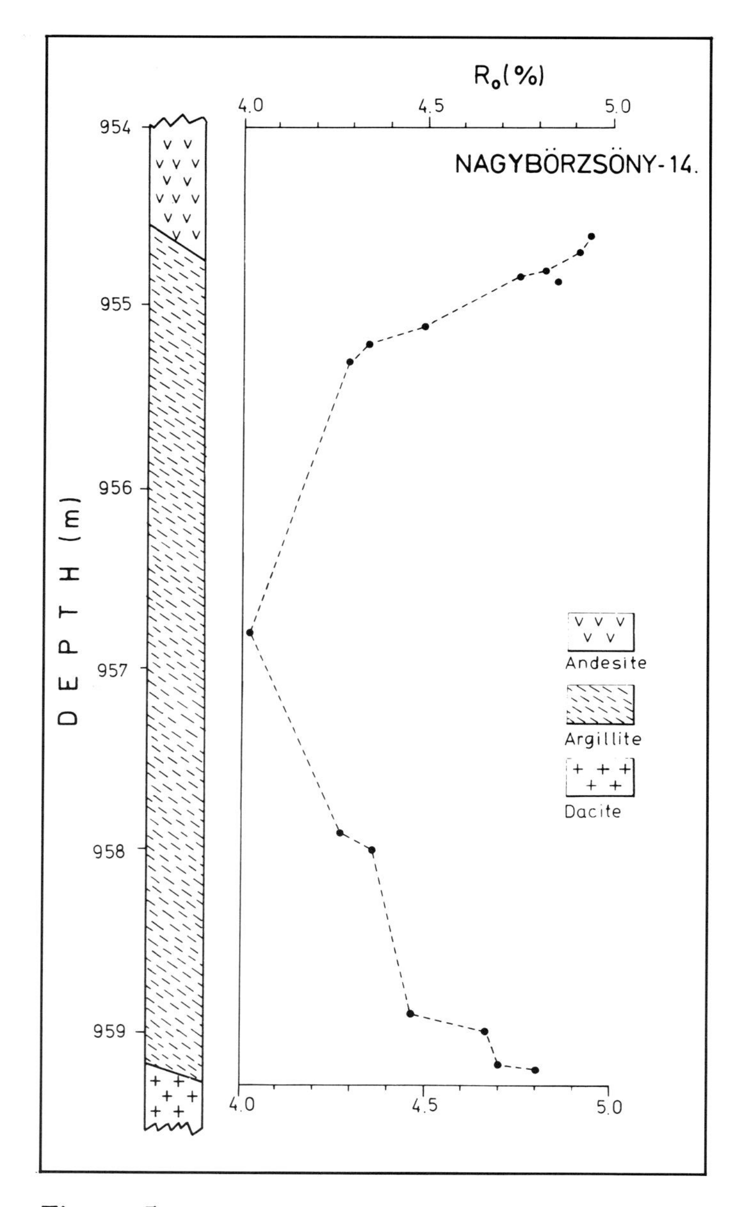

Figure 5. Some vitrinite reflectance data from the borehole Nagybörzsöny-14. An increase of measured vitrinite reflectance near the dacite body suggests that it was intruded as a sill.

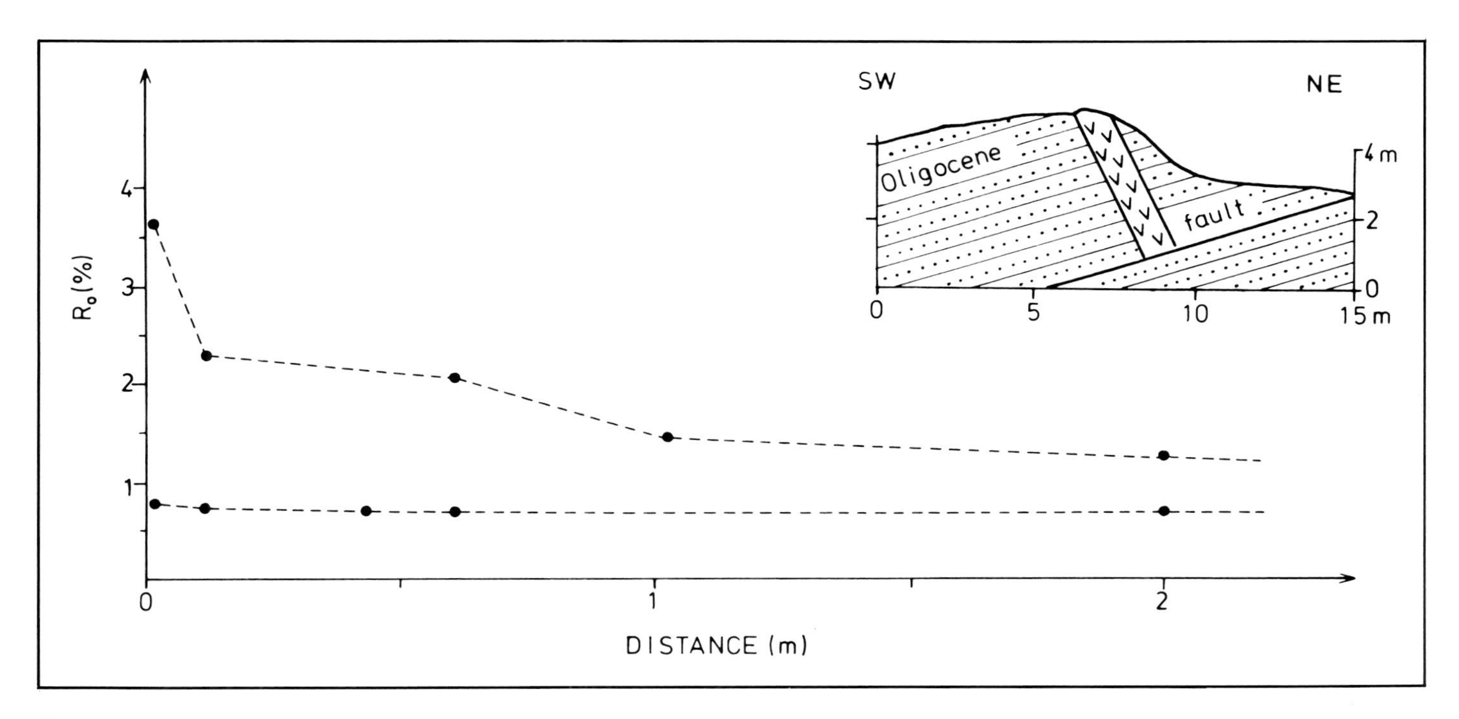

Figure 6. Vitrinite reflectance plotted as a function of distance from the contact of sediments with a basaltic dike at Salgótarján. Two populations of vitrinite were distinguished. One shows increased reflectance near the contact with the basalt. The second shows low and constant values of vitrinite reflectance.

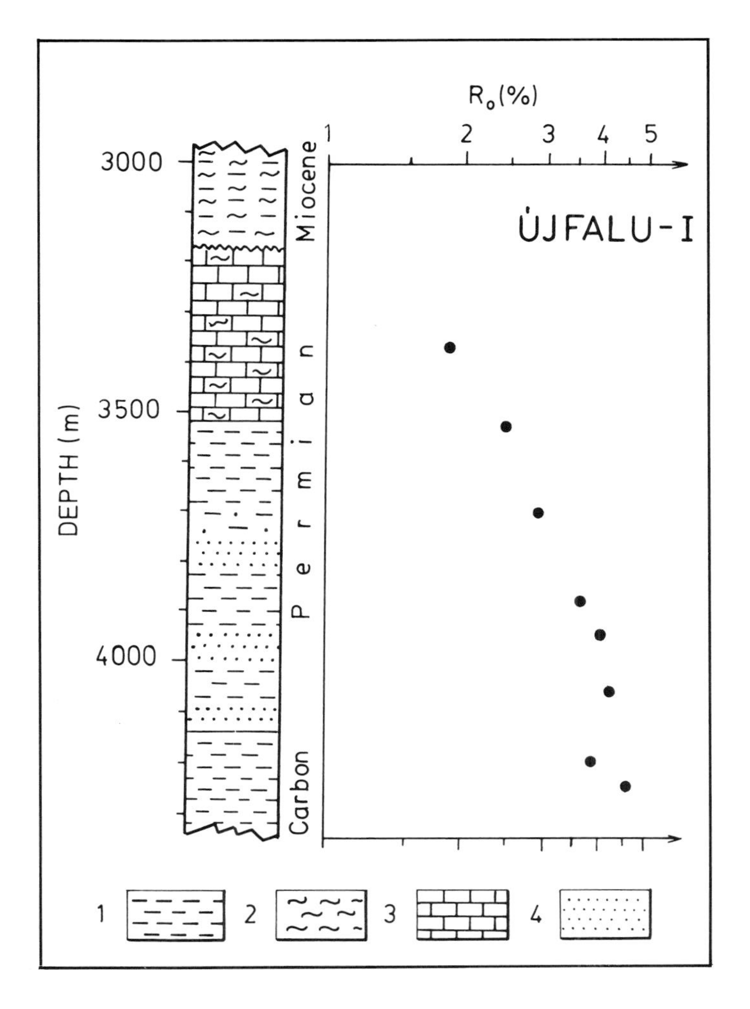

Figure 7. An unusually rapid increase of vitrinite reflectance with depth in basement rocks penetrated by the borehole Újfalu-I. This is thought to be related to heating from a large plutonic body below the bottom of the borehole.

REFERENCES

Dow, W. G., 1977, Kerogen studies and geological interpretations: J. Geochem. Explor., v. 7, p. 79–99.
Durand, B., 1980, Kerogen. Edit. Technip, Paris, 519 pages.
Horváth, F. and P. Dövényi, 1983, Evaluation of the maturity and thermal conditions of the Szirák-2 borehole: Internal. Rep. to the Hung. Geol. Inst. (in Hungarian), 37 pages.
Horváth, F., P. Dövényi and I. Laczó, 1986, Geothermal effect of magmatism and its contribution to the maturation of organic matter in sedimentary basins, *in* G. Buntebarth and L. Stegena, eds., Paleogeothermics: Springer-Verlag, p. 173–183.
Hunt, J. M., 1979, Petroleum geochemistry and geology. Freeman, San Francisco, 617 pages.
Issler, D. R., 1984, Calculation of organic maturation levels of offshore eastern Canada: Implications for general applications of Lopatin's method: Canad. Journ. Earth Sci., v. 21, n. 4, p. 477–488.
Karweil, J., 1956, Die Metamorphose der Kohlen von Standpunkt der physikalischen: Chemie. Z. deutsch. geol. Ges., v. 107, p. 132–139.
Laczó, I., 1982, Magyarországi vitrinitreflexió adatok földtani értékelése (Geological interpretation of vitrinite reflection data in Hungary): Ann. Rep. Hung. Geol. Inst. of 1980, p. 417–437.
Laczó, I., 1982, A magyarországi triász képzödmények vitrinitreflexió (R_o) értékei és földtani jelentőségük [R_o values of the Triassic formations of Hungary and their geological implications]: Ann. Rep. Hung. Geol. Inst. of 1982, p. 403–416.
Lopatin, N. V., 1971, Temperature and geologic time as factors in coalification (in Russian): Akad. Nauk. SSSR, ser. geol. Izv., v. 3, p. 95–106.
Stegena, L., F. Horváth, J. G. Sclater, and L. Royden, 1981, Determination of paleotemperature by vitrinite reflectance data: Earth Evol. Sci., v. 1, n. 3–4, p. 292–300.
Teichmüller, M. and R. Teichmüller, 1950, Das Inkohlungsbild des Niedersachsischen Wealden-Beckens: Z. deutsch. geol. Ges., v. 100, p. 498–517.
Telegdi Roth, K., 1929, Geology of Hungary: Tud. Gyüjt., Pécs, 250 pages.
Tissot, B. P., and D. H. Welte, 1978, Petroleum formation and occurrence: Berlin, Springer-Verlag, 539 pages.
Waples, D. W., 1980, Time and temperature in petroleum formation: Application of Lopatin's method to petroleum exploration: Amer. Assoc. Pet. Geol. Bull., v. 64, p. 916–926.

CHAPTER 23

Petroleum Geology of the Pannonian Basin, Hungary: An Overview

V. Dank
Central Geological Office of Hungary
H-1011 Budapest, Fö utca 44-50, Hungary

Source rocks in the Pannonian basin are dominantly lower Miocene to lower Pannonian shales that have matured during rapid Neogene subsidence and burial. Ages of reservoir rocks vary from Paleozoic to Pliocene and the reservoirs are often of mixed types. Structural and unconformity traps related to basement highs are common and have been the subject of much exploration in the last few decades. Future prospects are related to fractured and fissured basement rocks at greater depths and to stratigraphic and unconformity traps within the Neogene sediments in deep troughs.

INTRODUCTION

One of the first deliberate and successful searches for structural traps was conducted in the Vienna basin using gravity data at the beginning of the century. Since then about 160 oil and gas fields have been discovered in Hungary. Although the traps are small, the oil and gas produced from them play an important role in Hungary's economy.

Hungary occupies the central part of the Pannonian basin, and over about 70% of the country's territory (approximately 93,000 km^2) there is a medium to thick Neogene sedimentary cover. Except where local deposits of Paleogene rocks occur (Royden and Báldi, this volume), these early Miocene to Quaternary sedimentary rocks lie unconformably on Mesozoic or Paleozoic basement. The subsidence rate of the basement varied significantly in space and time and resulted in a characteristic horst and graben morphology of the basement (Rumpler and Horváth, this volume). The thermal history of the Neogene sediments has been appropriate for generation of hydrocarbons (Horváth et al., this volume), which accumulated in different kinds of traps. However, the small lateral extent of the deep basins, the relatively low concentration of organic matter in the source rocks, and complicated structural and stratigraphic conditions hampered the formation of large oil fields. Hence, the search for hydrocarbon reservoirs in the Pannonian basin has always been a difficult task. The total length of wildcat and production wells in Hungary drilled since 1935 is more than 11,000 km, while that of the multifold (12 to 48) stacked seismic profiles amounts to 30,000 km.

In this chapter, I review the history of petroleum prospecting in Hungary, present some characteristic examples, and draw some general conclusions about the strategy of current exploration.

HISTORY OF PROSPECTING FOR HYDROCARBONS IN HUNGARY

True hydrocarbon prospecting started in Hungary in 1909, when H. Böckh and coworkers applied the torsion balance, invented by R. Eötvös, to structural exploration for the first time. This led to discovery of an oil field in the Vienna basin (Gbely, 614 on Map 4 in enclosure). In 1923 the Anglo-Persian Oil Company carried out drilling in Transdanubia, but because it proved to be barren, exploration was not continued. In subsequent years, large-scale prospecting came to a standstill, although several wildcat wells were drilled in the Great Hungarian Plain. These yielded encouraging results at Hajdúszoboszló

(482 on Map 4) where, along with hot water, considerable quantities of methane gas were obtained.

In 1935 a concession in Transdanubia was given to the European Gas and Electric Company, which belonged to Standard Oil's sphere of interest. They first found pools with 94 to 96% carbon dioxide content in the Little Hungarian Plain (Mihályi, 413 on Map 4). Then, in 1937 considerable natural gas and oil reservoirs were discovered in lower Pannonian structural anticlines in the Zala basin (Budafa, 427 on Map 4). (In this chapter, the term Pannonian means Pannonian (*s. l.*) and includes all post-Sarmatian, pre-Quaternary sedimentary rocks.) In the same year a less important oil field was found in Oligocene strata at Bükkszék (439 on Map 4 in enclosure). Discovery of new oil and gas fields continued in the Zala basin with the discovery of the Hahót and Lovászi fields (421 and 426, respectively, on Map 4). In the period between 1937 and 1945, Hungarian oil and gas production was almost entirely from the Zala basin fields. Production amounted to a maximum of 800,000 tons/year in 1943.

As a result of stepped-up production during World War II, the condition of the oil and gas fields deteriorated and production decreased to a considerable extent. It reached only 483,000 tons/year in 1948. In this year, the Hungarian oil industry was nationalized and underwent a complete reorganization. The discovery of the Triassic–Cretaceous carbonate reservoirs at Nagylengyel (417 on Map 4) in 1951 marked a milestone in the history of Hungarian hydrocarbon prospecting. These reservoirs supplied 3000 tons of oil in 1951, 90,000 tons in 1952, 380,000 tons in 1953 and a maximum of 1.2 million tons in 1955.

The National Oil and Gas Trust of Hungary was born in 1957, and the first comprehensive hydrocarbon prospecting plan for the whole country was worked out. As a result of extensive geophysical (mainly seismic) surveys, exploration in the Great Hungarian Plain came into the limelight. The best discoveries were made in the southeastern part of the Great Hungarian plain, e.g., Üllés in 1962, Szank in 1964, Algyő in 1965, Ásotthalom in 1967, and Tázlár in 1968 (524, 513, 526, 530 and 515, respectively, on Map 4). Among these, the Algyö field represents the most important hydrocarbon occurrence in the country.

The 1970s saw major technological developments in seismic data acquisition and processing that gave rise to deeper penetration and higher resolution. This, combined with deeper boreholes and more reliable well-logging surveys, led to a better understanding of the structural and sedimentary features (e.g., Kiskunhalas, 582 on Map 4). Furthermore, it turned out that fractured and fissured Mesozoic and Paleozoic basement rocks have trapped hydrocarbons that have migrated upwards and sideways from the deeper Miocene source rocks. Recent discoveries in such traps include the Sarkadkeresztúr, Szeghalom and Kismarja fields (589, 590 and 591, respectively, on Map 4).

Since the beginning of this decade, reconstruction of subsidence, burial, and temperature history during the Neogene has become a key feature of hydrocarbon prospecting in Hungary. A glimpse of the applied methods and some of the first results are presented in this volume.

Table 1 shows the oil and gas production of Hungary in 1980 from different reservoirs. The present figures for total production are similar.

Table 1. Oil and Gas Production of Hungary in 1980.

Age of Reservoir Rocks	Oil (million tons)	Gas ($10^9 m^3$)	Co_2 ($10^9 m^3$)
Upper Pannonian (*s. l.*)	0.944	2.5	—
Lower Pannonian (*s. l.*)	0.236	1.8	0.053
Late and Middle Miocene	0.105	0.325	0.206
Oligocene	0.022	0.002	—
Eocene and Cretaceous	—	0.339	—
Jurassic and Triassic	0.490	0.106	—
Paleozoic	0.267	1.3	—
Total	2.064	6.372	0.259

EXAMPLES OF OIL AND GAS FIELDS IN HUNGARY

Map 4 is based on a data set compiled mainly within the framework of petroleum geological cooperation between the European socialist countries and is designed to show the distribution of oil and gas fields in the Carpathian–Pannonian system. The bulk of the fields are related either to the Carpathian foredeep or the intra-montane basins. In this paper, examples will be given only from the Hungarian part of the Pannonian basin (Figure 1). Interested readers can find examples from the Vienna basin in Wessely (this volume), and from the Transylvanian basin and Romanian Carpathians in Paraschiv (1979).

The Budafa Field

The Budafa field was the first significant discovery during early hydrocarbon prospecting in Hungary. Structures favorable for trapping hydrocarbons were discovered as early as 1919 by geological surface mapping. Upper Pliocene layers dip 3° to 10° and form a closed anticline. The first successful drilling was performed here in 1937. Subsequent surveys revealed an oil and gas field at a depth of 900 to 1300 m over an area of 18 km^2. The reservoir rocks are lower Pannonian sandstone within the upper part of an east-west striking folded anticline. The seismic section and geological profile in Figure 2 are perpendicular to the fold axis. Source rocks are almost certainly the mature shale members of the thick underlying Miocene series. Because of rapid production, particularly during the second World War, the field has been almost completely exhausted. However, production has been renewed recently by special techniques, and the Budafa field still yields about 1% of the total hydrocarbons produced in Hungary.

The Nagylengyel Field

Discovery of the Nagylengyel field in 1951 was a remarkable event in hydrocarbon exploration in Hungary because it covers a large area (75 km^2) and it was, and still is, the largest field discovered within Mesozoic rocks. Most of the reservoirs are in

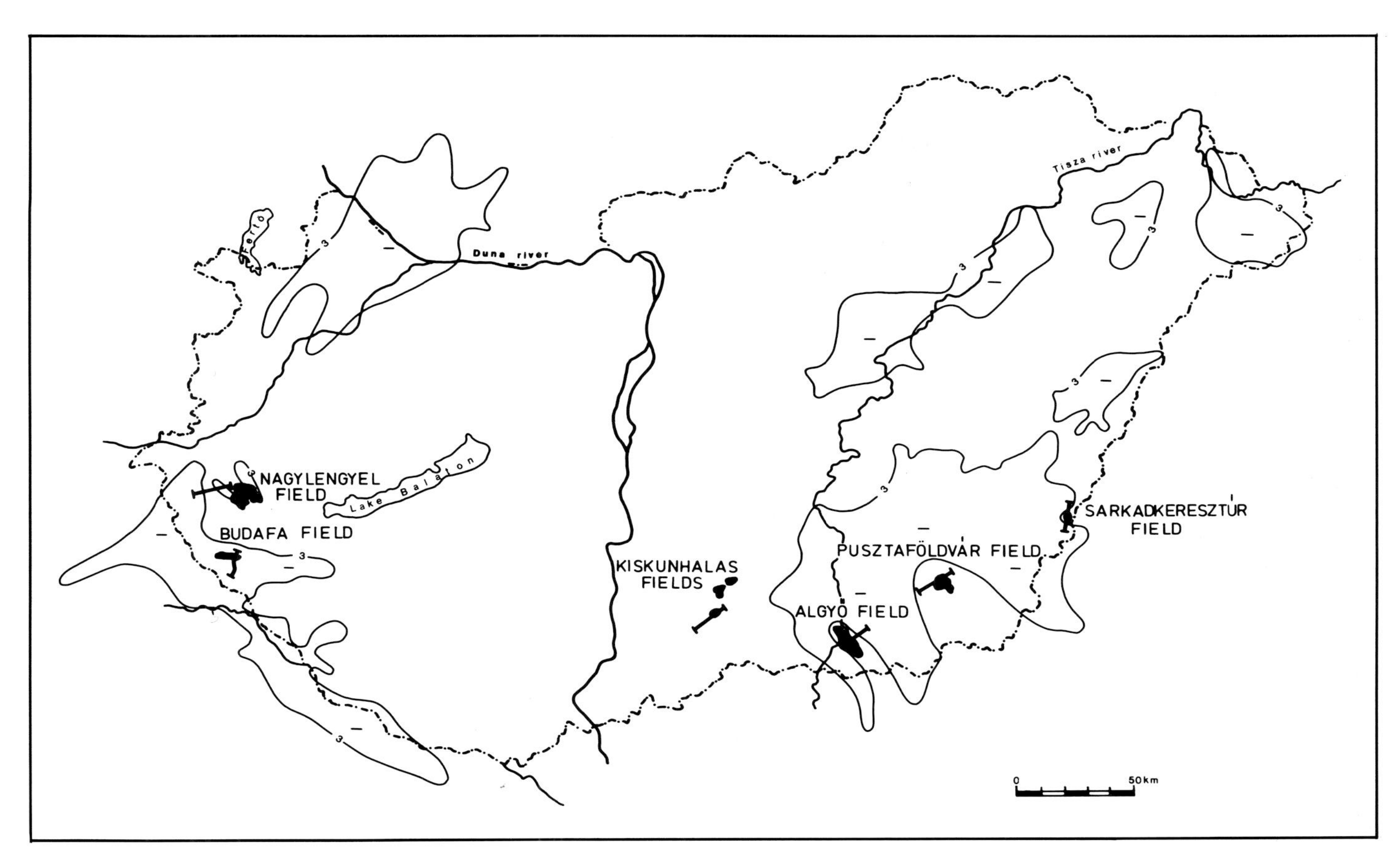

Figure 1. Location map for profiles in Figures 2–7, showing the national boundaries of Hungary, rivers, and the 3-km isopach lines for Neogene-Quaternary sedimentary rocks.

Upper Cretaceous limestone, and individual pools are 7 m to 90 m thick. The rest of the reservoirs occur in Upper Triassic (Norian) dolomite, and the thickness of pools varies from 20 m to 160 m. One horizon in early Miocene (Karpatian) sandstone also contains some oil. The seismic section and geological profile across the field (Figure 3) show that pools are associated with tilted blocks separated by normal faults of early Miocene age. The structural setting and geochemistry of the oil suggest that they were derived from Mesozoic source rocks (Koncz, 1987), which reached the oil generation window during Neogene burial and heating. This field is also practically exhausted. Its present contribution to the total hydrocarbon production of Hungary is 1.3%.

The Pusztaföldvár Field

The Pusztaföldvár field was discovered in 1958 and was the first significant result of geophysical and geological exploration in the Great Hungarian Plain. Oil and gas are trapped in a Pannonian compaction anticline above a basement high between the Makó trough to the west, and the Békés depression to the east. Source rocks are thought to be Miocene shale in these troughs, which have passed through the oil generation window (Horváth et al., this volume). The areal extent of the field is 57 km^2 and the thickness of the reservoir beds may reach 100 m. The deepest pool occurs in the lower Pannonian basal conglomerate that overlies Paleozoic basement (Figure 4). A series of reservoirs exist in the upper part of the lower Pannonian (Pa_1^2) and lower part of the upper Pannonian (Pa_2). They represent slightly deformed sand bodies associated with prograding delta lobes and delta plains. The present contribution of this field to Hungarian production is about 2.5%.

The Algyő Field

The Algyő field is the biggest oil and gas field of Hungary and was discovered in 1965. Its areal extent is 80 km^2 and the thickness of reservoir beds varies from 5 m to 70 m. A seismic section and geological profile through the field is shown in Figure 5. Structural, stratigraphic and genetic conditions are similar to those of the Pusztaföldvár field. The only apparent difference is that, in the Algyő field, the basement core of the anticline is fractured locally and contains some oil. This field yields 52% of the total hydrocarbons produced in Hungary.

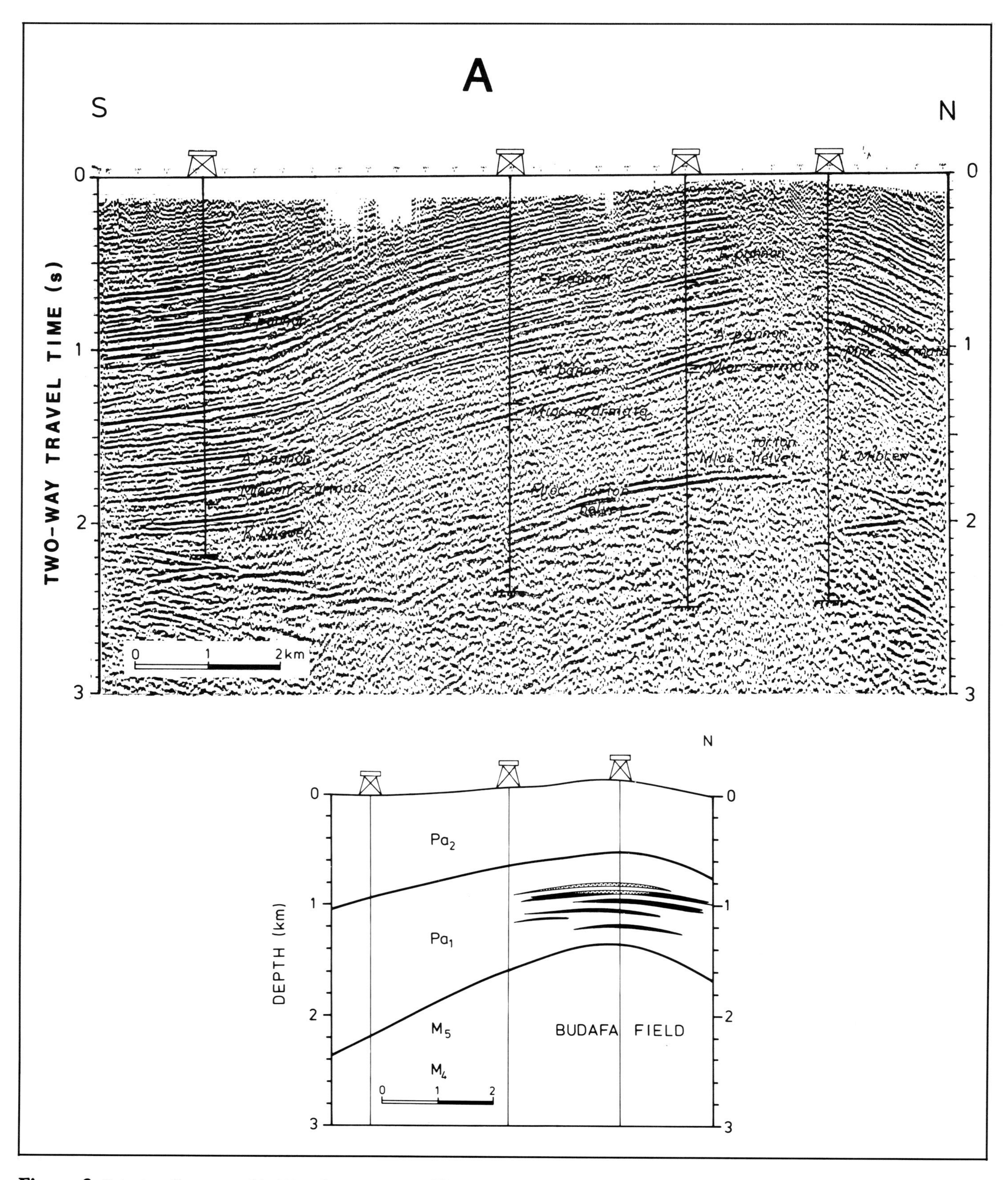

Figure 2. Seismic reflection profile (A) and interpretation (B) across an east-west striking, folded anticline of the Budafa field (see Figure 1 for location). Oil (black) and gas (dots) pools are also shown in B. M_4 = Badenian, Sarmatian; Pa_1 = lower Pannonian (*s. l.*); and Pa_2 = upper Pannonian (*s. l.*).

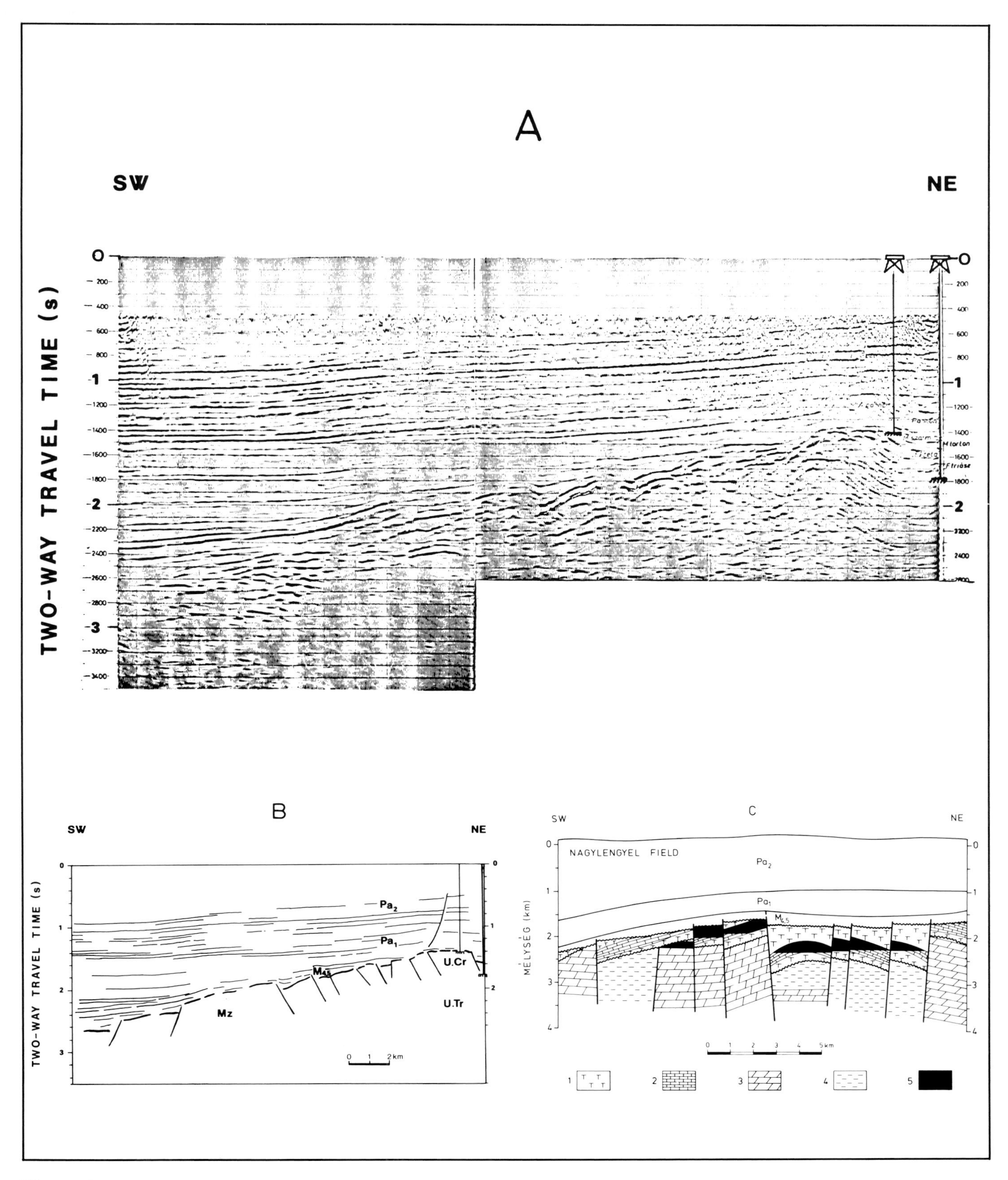

Figure 3. Seismic reflection profile (A), line drawing (B) and interpretation (C) across the Nagylengyel field located within Mesozoic reservoir rocks (see Figure 1 for location). 1 = Upper Cretaceous (U.Cr.) marls, 2 = Upper Cretaceous limestone, 3 = Upper Triassic Hauptdolomite, 4 = Upper Triassic shale, 5 = Oil reservoirs, $M_{4,5}$ = Badenian–Sarmatian, Pa_1 = lower Pannonian (*s. l.*), and Pa_2 = upper Pannonian (*s. l.*).

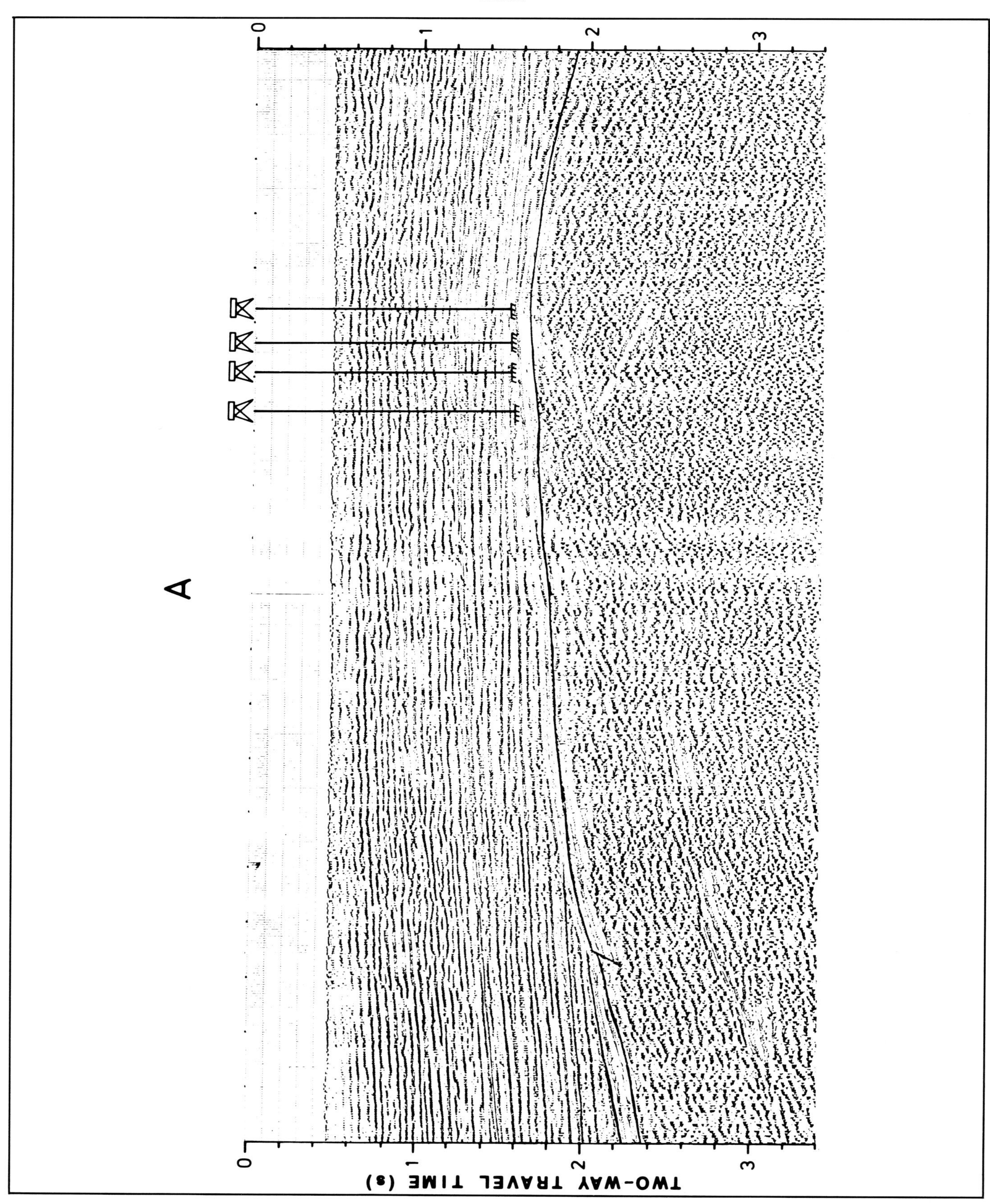

Figure 4. Seismic reflection profile (A), line drawing (B), and interpretation (C) across a Pannonian (*s. l.*) compaction anticline of the Pusztaföldvar field (see Figure 1 for location). Pa_1^2 = upper lower Pannonian (*s. l.*), and Pa_2 = upper Pannonian (*s. l.*). Oil (black) and gas (dots) pools are shown in C.

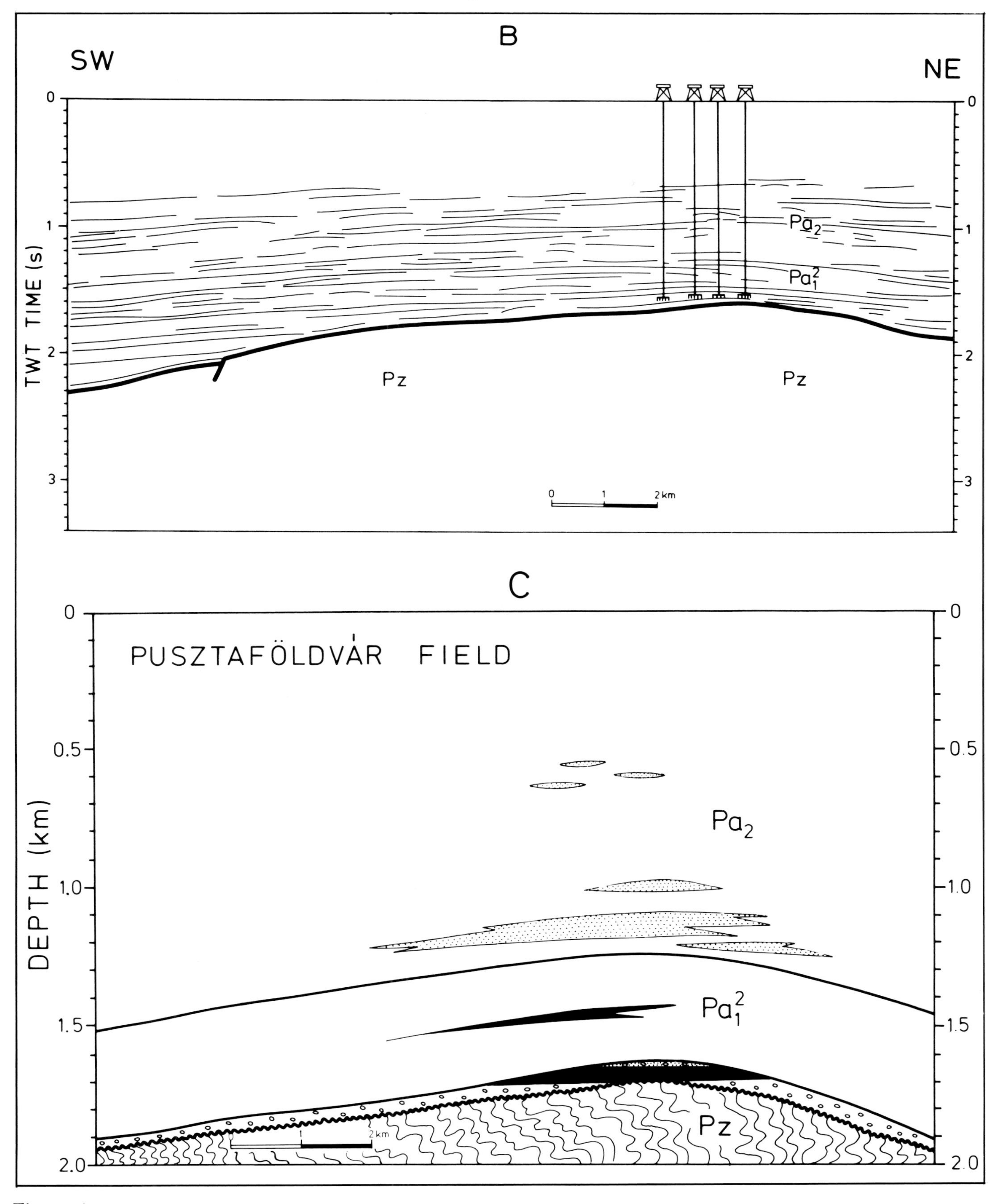

Figure 4. (continued)

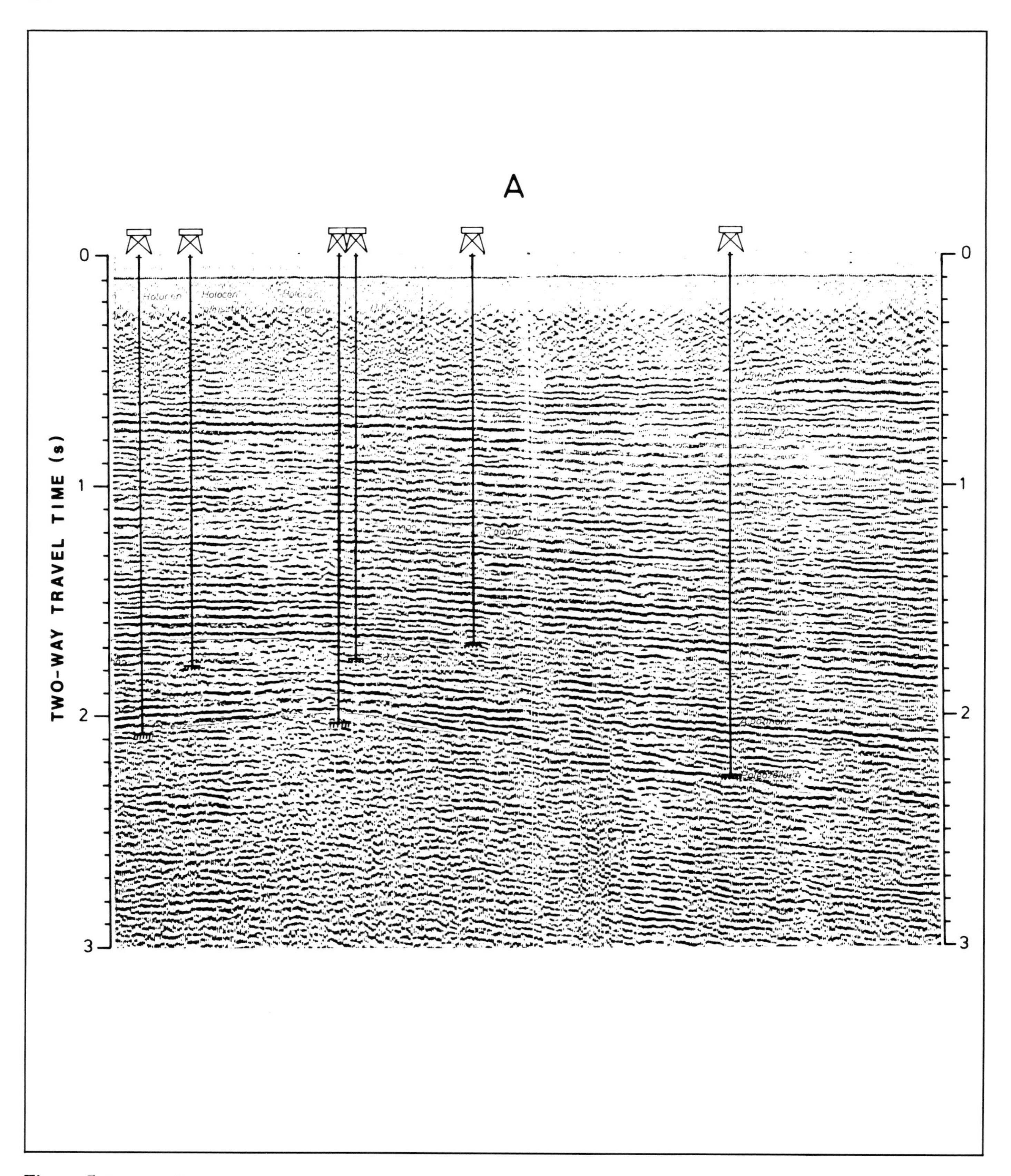

Figure 5. Seismic reflection profile (A), line drawing (B) and interpretation (C) across a Pannonian (*s. l.*) compaction anticline of the Algyő field (see Figure 1 for location). $M_{4,5}$ = Badenian–Sarmatian, Pa_1^2 = upper lower Pannonian (*s. l.*), and Pa_2 = upper Pannonian (*s. l.*). Oil (black) and gas (dots) pools are shown in C.

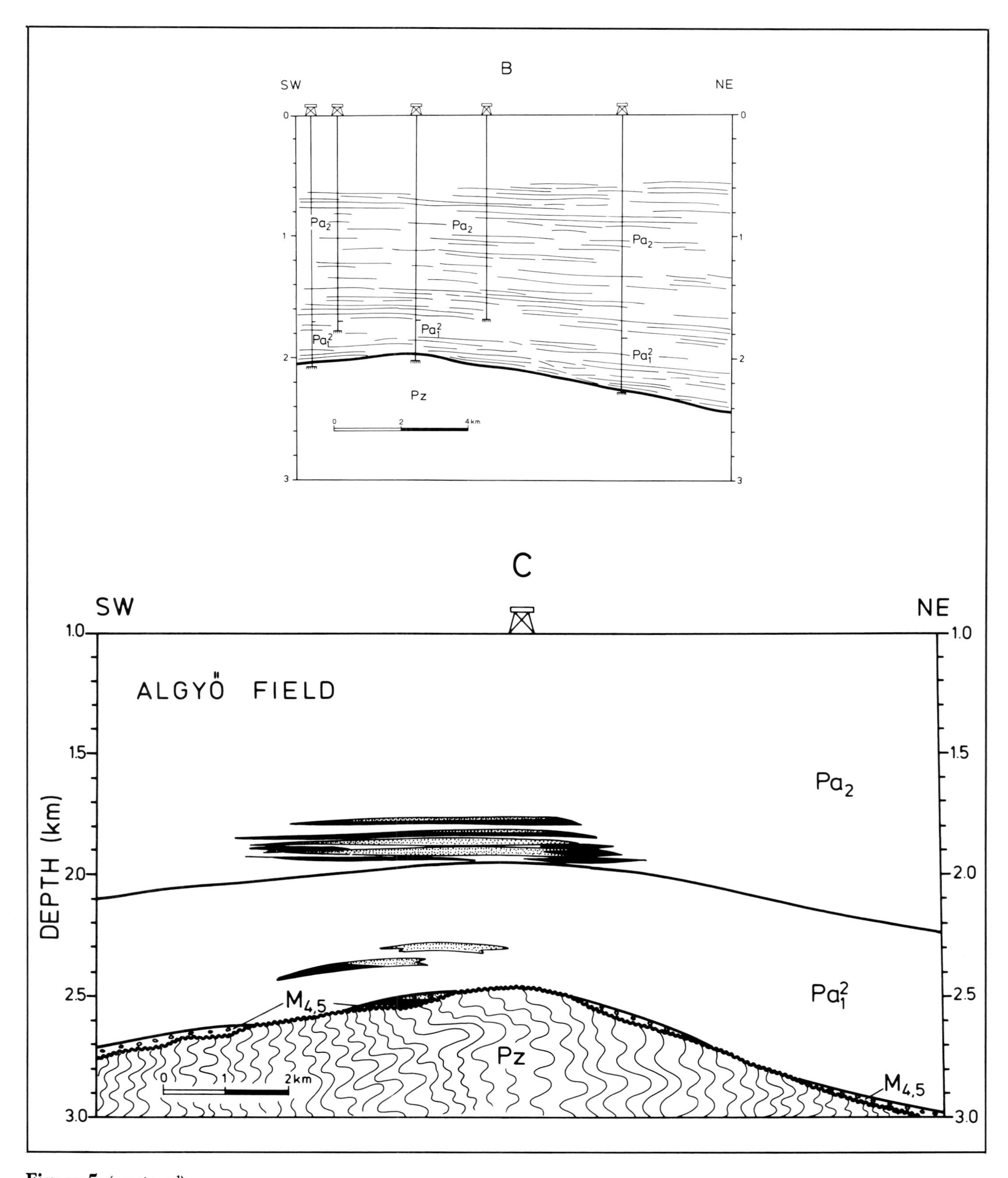

Figure 5. (continued)

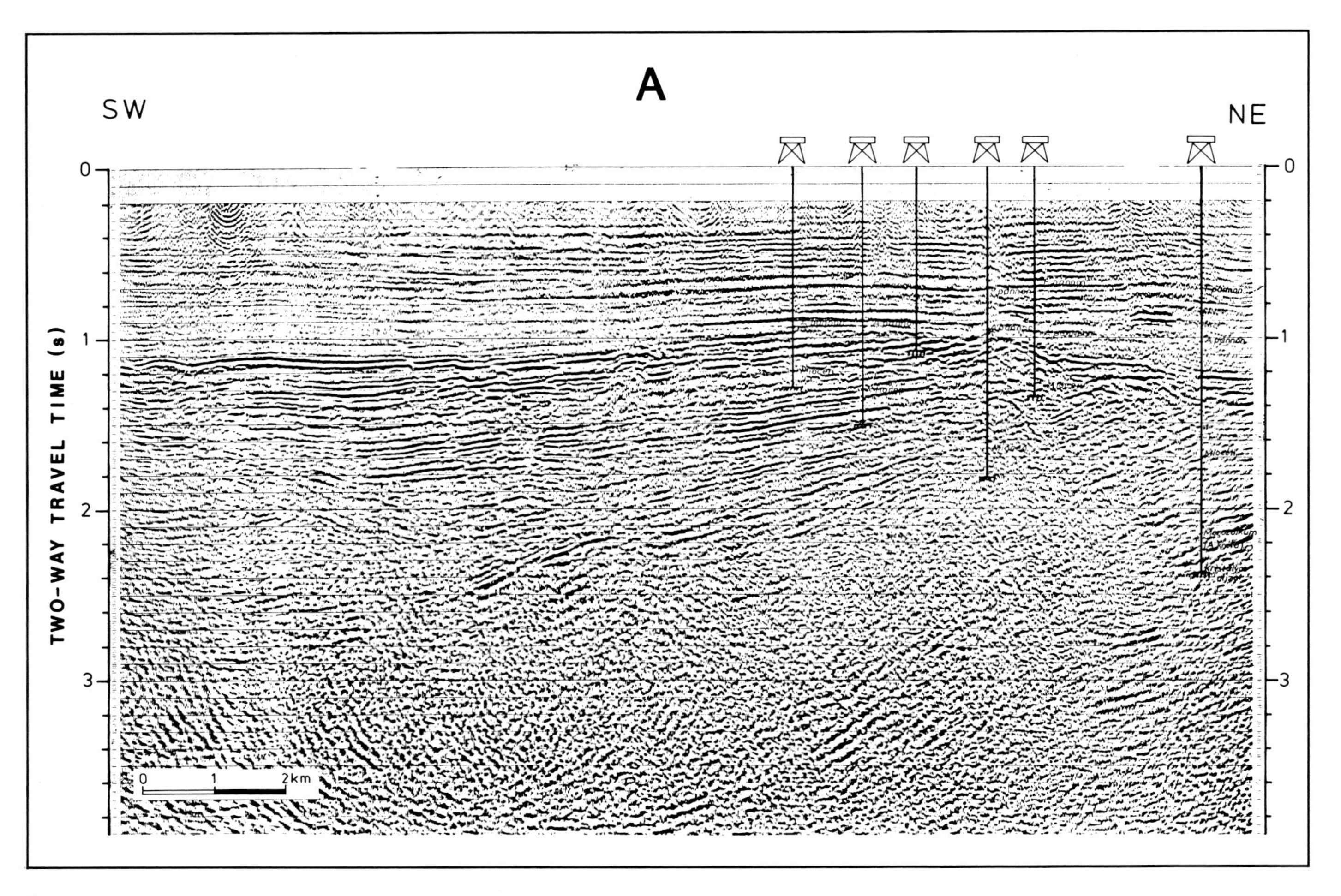

Figure 6. Seismic reflection profile (A), line drawing (B), and interpretation (C) across the Kiskunhalas-southwest field (see Figure 1 for location). M_3 = Karpatian, M_4 = Badenian, M_5 = Sarmatian, Pa_1^2 = upper lower Pannonian (*s. l.*), and Pa_2 = upper Pannonian (*s. l.*). Gas pools (dots) are shown in C.

The Kiskunhalas Field

The first successful drilling in the Kiskunhalas area was performed in 1974. Subsequent drilling between 1974 and 1981 discovered four larger oil and gas fields in a structurally complex region. The seismic section and geological profile in Figure 6 cross the Kiskunhalas-southwest gas field, and show some characteristic structures. Listric normal faults cut the Mesozoic basement and bound deep troughs filled with middle Miocene (Badenian and Sarmatian) sediments. The troughs are covered by posttectonic lower Pannonian rocks deposited above a major unconformity. Gas was found within the uppermost part of the middle Miocene troughs in tilted sand beds. In other parts of the Kiskunhalas field, oil and gas were found in the Mesozoic basement as well. Additional seismic surveys of this area have begun recently. The present contribution of the Kiskunhalas fields to the national production of Hungary is about 2%.

The Sarkadkeresztúr Field

The Sarkadkeresztúr oil and gas field was discovered in 1976. It is related to a very narrow horst of Precambrian(?) metamorphic basement, overlain by thin middle Miocene conglomerates and Pannonian strata (Figure 7). Ninety-nine percent of the hydrocarbons are trapped in the fractured-fissured basement and the overlying conglomerates. The thickness of the hydrocarbon pool reaches 370 m. This discovery is important conceptually because it shows that the metamorphic basement is locally permeable and can be a good reservoir for hydrocarbons. In addition, measurements and model calculations have shown that, in the deep troughs, overpressuring in middle Miocene and lower Pannonian shales has favored downward and/or lateral fluid migration since the late Miocene (Szalay, this volume). This enhances the exploration prospects of the deep basement. The Sarkadkeresztúr field contributes about 5% of Hungary's total production.

CONCLUSIONS

After more than 70 years of extensive hydrocarbon exploration in Hungary, a few basic working principles have been established and can be summarized as follows:

1. The main source of hydrocarbons in Hungary is lower Miocene to lower Pannonian pelitic rocks, which may be

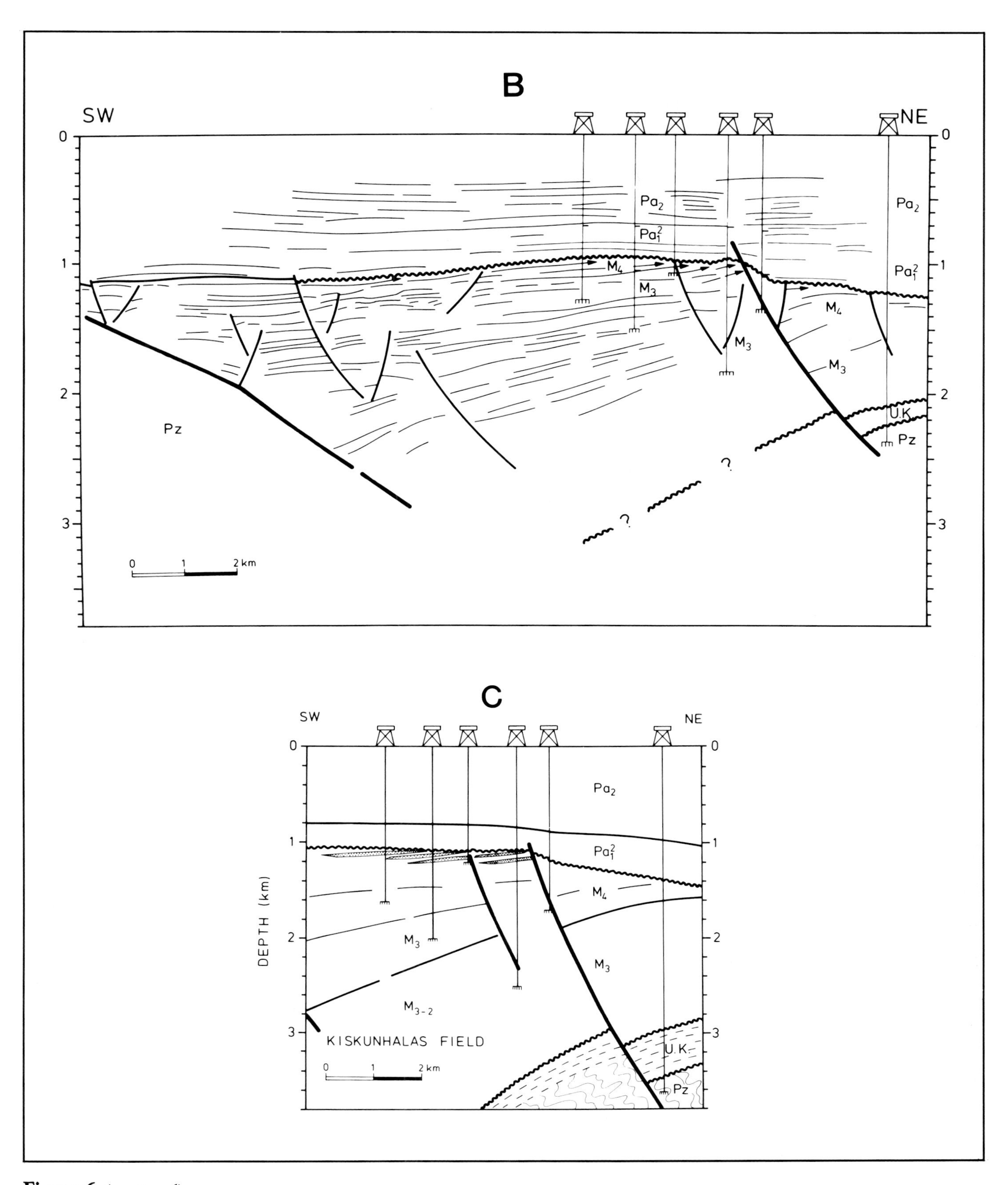

Figure 6. (continued)

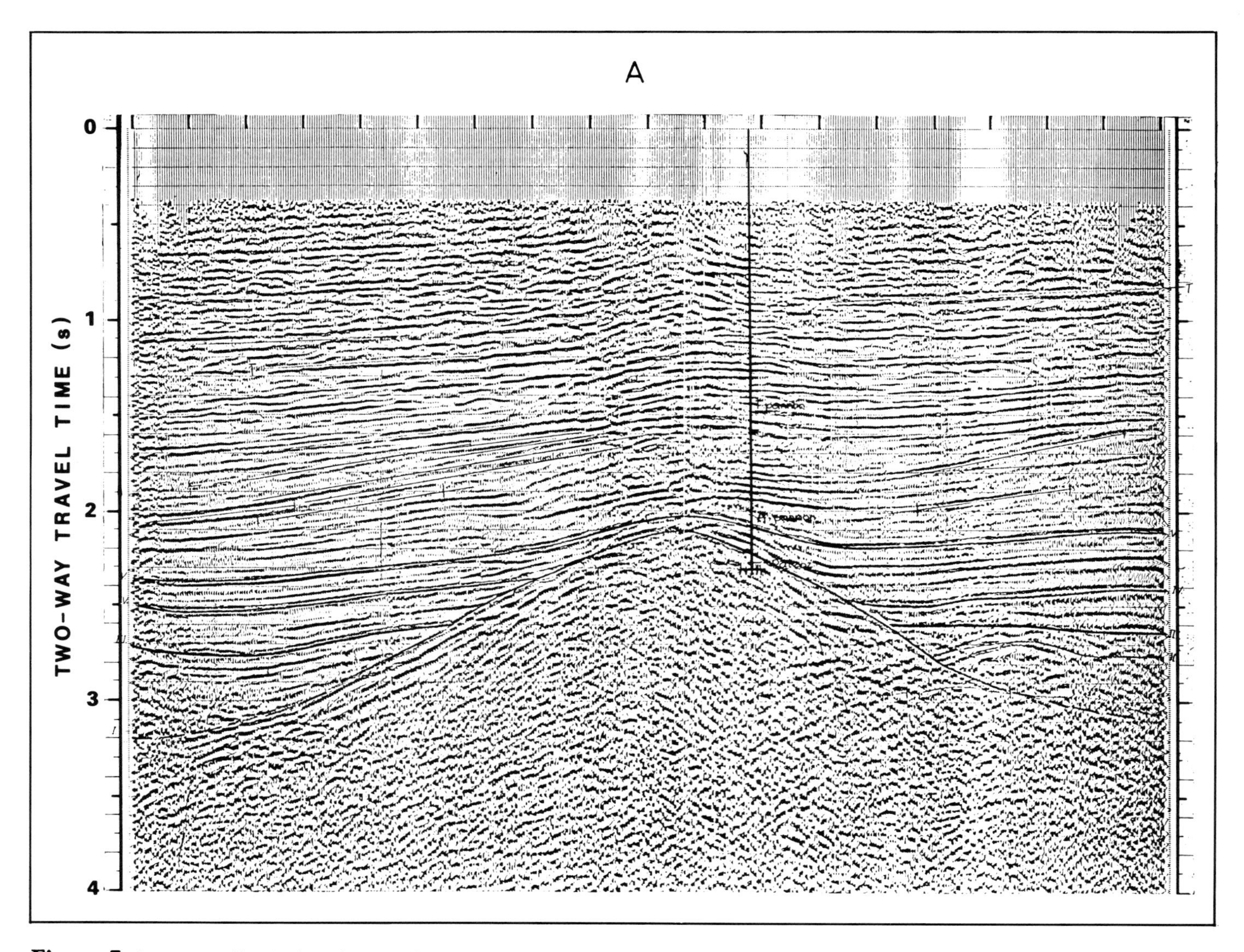

Figure 7. Seismic profile (A), line drawing (B), and interpretation (C) across the Sarkadkeresztur field located within fractured Paleozoic rocks and overlying conglomerates (see Figure 1 for location). M_{45} = Badenian–Sarmatian, Pa_1^{1a} and Pa_1^{1b} = lower lower Pannonian (*s. l.*), Pa_1^2 = upper lower Pannonian (*s. l.*), and Pa_2 = upper Pannonian (*s. l.*). Oil (black) and gas (dots) pools are shown in C.

several hundred to a few thousand meters thick in the deepest troughs of the Pannonian basin.

2. In the northeastern part of the country, thick Paleogene sedimentary rocks (up to 2 km thick) that occur below thinner Neogene cover have reached the oil generation window. They may have produced a small amount of exploitable fluid hydrocarbons.
3. Previously immature Mesozoic rocks might have become source rocks during fast Miocene subsidence and burial.
4. Fissured and fractured domains of the Mesozoic and Paleozoic basement may be traps for hydrocarbons that migrated out of Miocene source rocks in deep troughs. Such traps have not been sought extensively, and are promising prospects.
5. Neogene sand beds in folded and compaction anticlines over basement highs are the most obvious hydrocarbon traps, but the best of these have probably been found.
6. The best prospects for future exploration within the Neogene sediments are the stratigraphic and unconformity traps in deeper parts of the basin and along the flanks of the deep troughs.

REFERENCES

Koncz, I., 1987, Nagylengyel és környéke kőolajelőfordulásainak eredete (The origin of the oil at the Nagylengyel and nearby fields): To be published in Hungarian in Földt. Közl.

Paraschiv, D., 1979, Romanian oil and gas fields: Inst. Geol. Geophys. Bucharest, Geol. Prospect. Expl., series A, n. 13, p. 1–382.

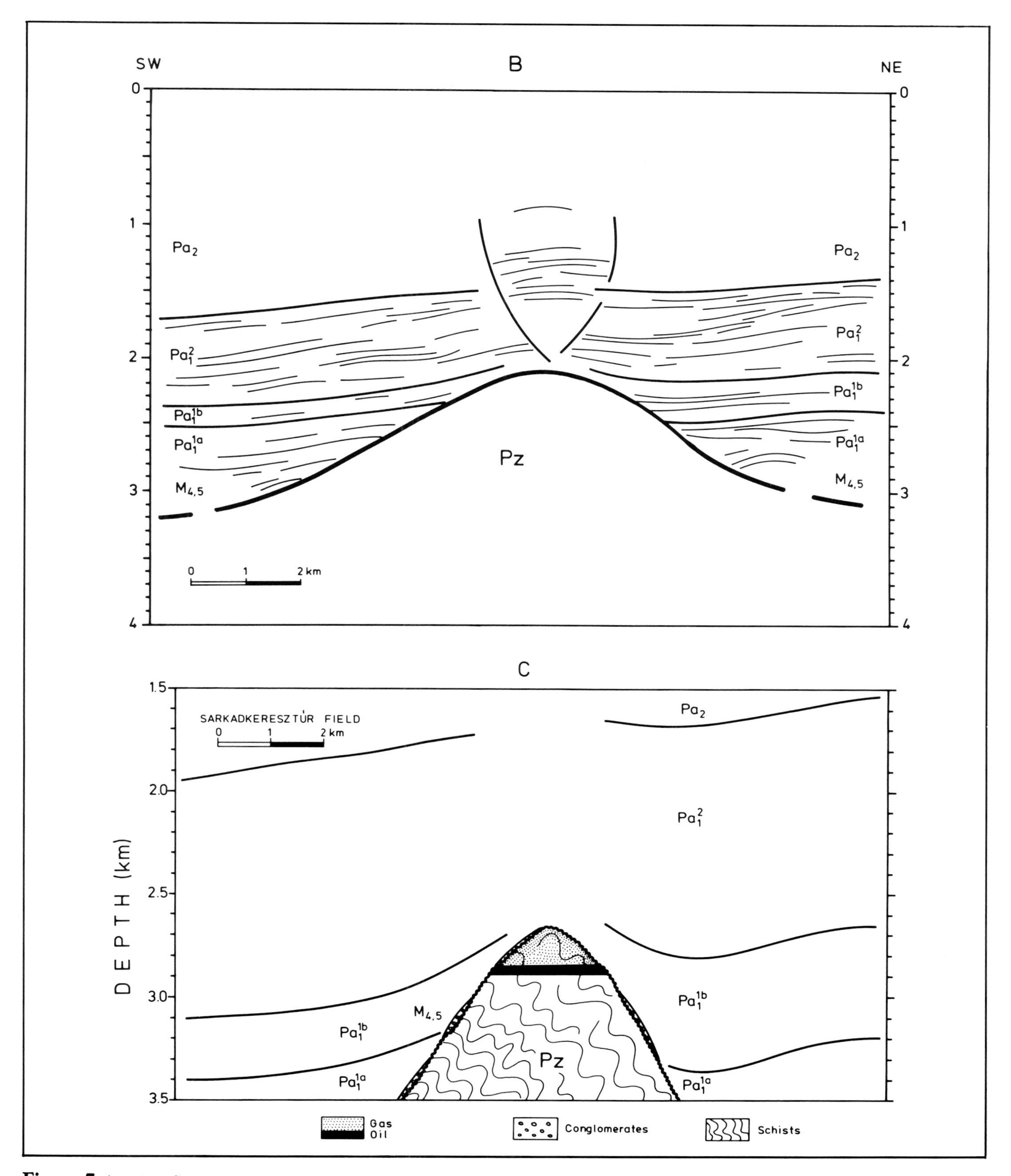

Figure 7. (continued)

CHAPTER 24

Structure and Development of the Vienna Basin in Austria

Godfrid Wessely
ÖMV-AG, ZG.,
Gerasdorfer Strasse 151,
A-1210 Vienna, Austria

The Vienna basin is mainly a Miocene feature superimposed on allochthonous nappes of the Alpine–Carpathian thrust belt. It is a classic area of exploration for oil and gas. The most recent target of exploration is the autochthonous sub-Alpine complex underneath the basin and thrust sheets. The autochthonous basement consists of crystalline and Paleozoic rocks overlain by a Jurassic–Upper Cretaceous cover. The Alpine–Carpathian units of the Waschberg zone, Flysch zone, and upper and lower Austro-Alpine–Tatride zone strike from the Alps underneath the Vienna basin and into the Carpathians, and form a bend in the vicinity of the Vienna basin. A structural synopsis of the basin shows a distinct arrangement of elevated and subsided blocks. Shallow, marginal horsts are bordered by deep depressions. A series of horsts forms a slightly sigmoidal median ridge.

The structural and sedimentary evolution of the Vienna basin started in the northern part of the basin in Eggenburgian (early Miocene) time. In Karpatian (middle Miocene) time the basin extended to the south. Badenian (middle Miocene) sediments transgressed over partly eroded and tilted older layers, and unconformably overlie them. An older generation of faults and tectonic structures must have been transported on the back of the Alpine–Carpathian thrust complex during thrusting until Karpatian time. The largest areal extent of the basin occurred in Badenian time, but was nearly as large in Sarmatian, Pannonian, and Pontian time (late Miocene to Pliocene). During and after Badenian time, growth faults, with displacements ranging from hundreds of meters up to 6000 m, were active. En echelon fault patterns, a sigmoidal shape of the entire system of faults and related structures, and distinctive kinds of sedimentary accumulation suggest that a pull-apart mechanism created the basin. It is inferred that this pull-apart and related horizontal deformation was caused by tectonic movements involving the sub-Alpine autochthonous basement below the thrust sheets as well as the Alpine–Carpathian thrust complex.

INTRODUCTION

The Vienna basin is one of the most important oil and gas provinces of Europe. The basin, located in Austria and Czechoslovakia, is 200 km long and 60 km wide, and strikes roughly northeast-southwest (Figure 1). It is an area of Miocene subsidence, and is situated northeast of a bend in the Alpine chain, where the east-west strike of East Alpine units changes into the northeast strike of Carpathian units. The basin is superimposed on the Alpine–Carpathian nappes (Figure 2). These nappes were imbricated and stacked beginning in Early Cretaceous time, and were thrust over the autochthonous basement of the foreland as late as early Miocene time.

In Austria alone, production from the basin sediments and from its Alpine–Carpathian basement has exceeded 79×10^5 tons of oil and 43×10^9 m^3 of gas since 1934, when the first commercial production started. About 3500 drill holes within Austria and numerous seismic data have made it possible to obtain extensive knowledge about the structure of the basin (Figures 3 and 4). Most of the early exploration activity was restricted to shallow and geologically simple plays within the

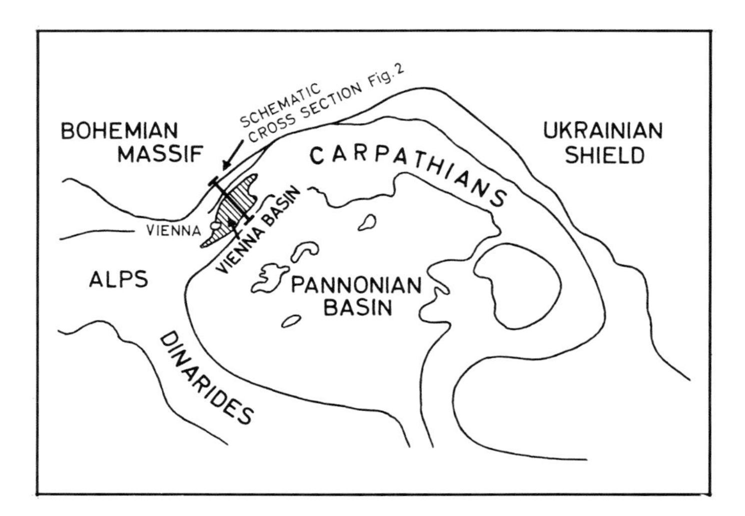

Figure 1. Location of the Vienna basin.

Neogene sedimentary rocks of the basin (referred to as the "first floor"). The greatest production of hydrocarbons is from sediments of Badenian and Sarmatian age and, to a smaller degree, from sediments of Pannonian and pre-Badenian age (for time scale see Figure 5). The most important fields are situated around Matzen, Aderklaa, Zwerndorf, and along the Steinberg fault near Zistersdorf (Figure 4). Exploration later was extended to the "second floor," the Alpine–Carpathian thrust complex underneath the basin fill. Hydrocarbon exploration focused on the flysch sediments of the thrust complex and finally on limestone Alpine rocks. Hydrocarbons were discovered in the Hauptdolomit on top of the basin floor in buried hill positions and traps were found within the limestone units of the Alpine thrust complex itself. The latter are also formed by Hauptdolomit, discordantly covered by Upper Cretaceous–Paleocene sediments of the Giesshübel furrow (between 4800–6100 m depth in the area of Schönkirchen).

At present, exploration of the "third floor" is just beginning in the autochthonous sedimentary cover of the European basement underneath the nappes. Such exploration requires very deep drilling. The well, Zistersdorf ÜT2A, reached a final depth of 8553 m (Figures 4 and 6). Two other wells, Maustrenk ÜT1 and Aderklaa UT1, terminated at depths of 6563 m and 6630 m, respectively. All of these wells penetrated Malm platform carbonate and a basin facies of Malmian marl up to 1000 m thick with good source conditions (Kratochvil and Ladwein, 1984). Very high pore concentrations of gas were found at depths below 6000 m, as well as a considerable amount of methane within sheared carbonate lenses above the autochthonous Malm in well Maustrenk ÜT1. This suggests economically important hydrocarbon occurrences may be found if good reservoirs can be located.

From a geological point of view, a substantial increase of knowledge about the deep structure and development of the basin is to be expected from such exploration. Together with results from the Czechoslovakian part of the basin, and its Carpathian and sub-Carpathian basement (Jiricek and Tomek, 1981), this data will lead to a better understanding of the relationship between the Vienna basin and the tectonics of the Alpine–Carpathian orogenic belts and their substratum, as well as to other intra-Carpathian basins, especially the Pannonian basin (Royden et al., 1983a, and this volume).

AUTOCHTHONOUS SUBALPINE–SUB-CARPATHIAN BASEMENT

Most interpretations of the sub-Alpine–Sub-Carpathian autochthonous basement under the Vienna basin are extrapolations of well data from wells that penetrated the basement underneath the Molasse zone (Figure 6). Some information about the basement is provided by lenses of basement within the Waschbergzone that were sheared off of the basement areas now situated under the Vienna basin. Only the wells, Zistersdorf ÜT2A, Maustrenk UT1 and Aderklaa ÜT1 (Figures 4 and 6), were drilled directly into the autochthonous sedimentary cover beneath the Vienna basin.

Blocks of crystalline rocks found together with blocks of littoral Eocene sediments in the Waschbergzone indicate that crys-

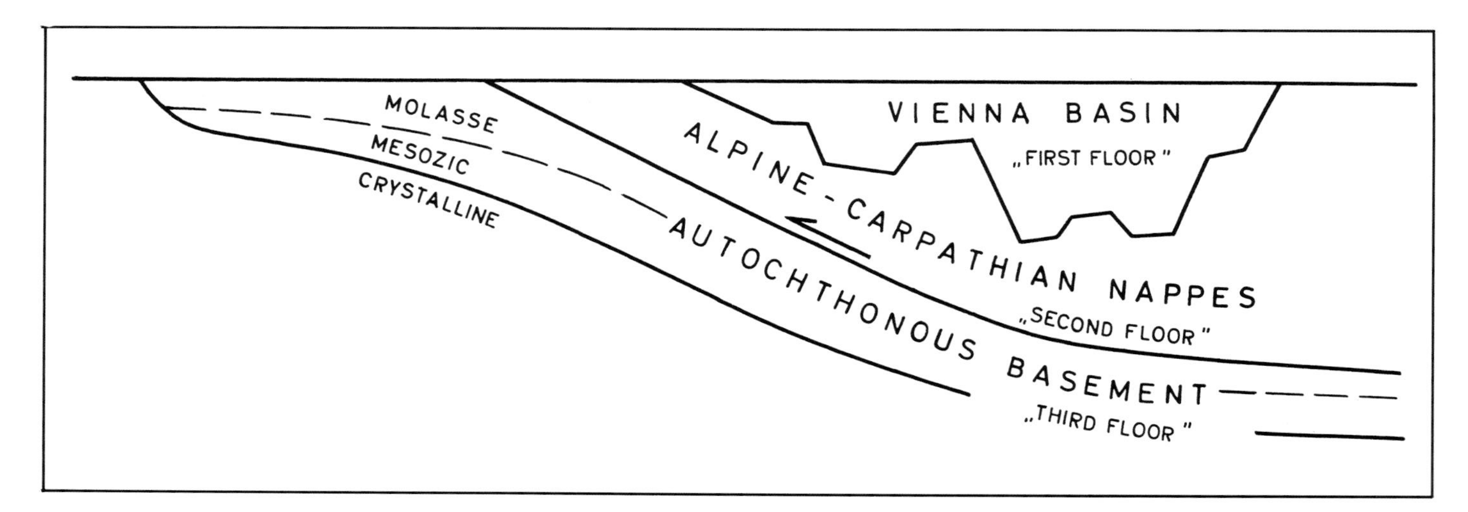

Figure 2. Diagrammatic cross section through the Vienna basin, illustrating that the basin is superimposed on allochthonous nappes of the Alpine–Carpathian thrust belt. The thrust belt, in turn, overlies the autochthonous basement of the European platform.

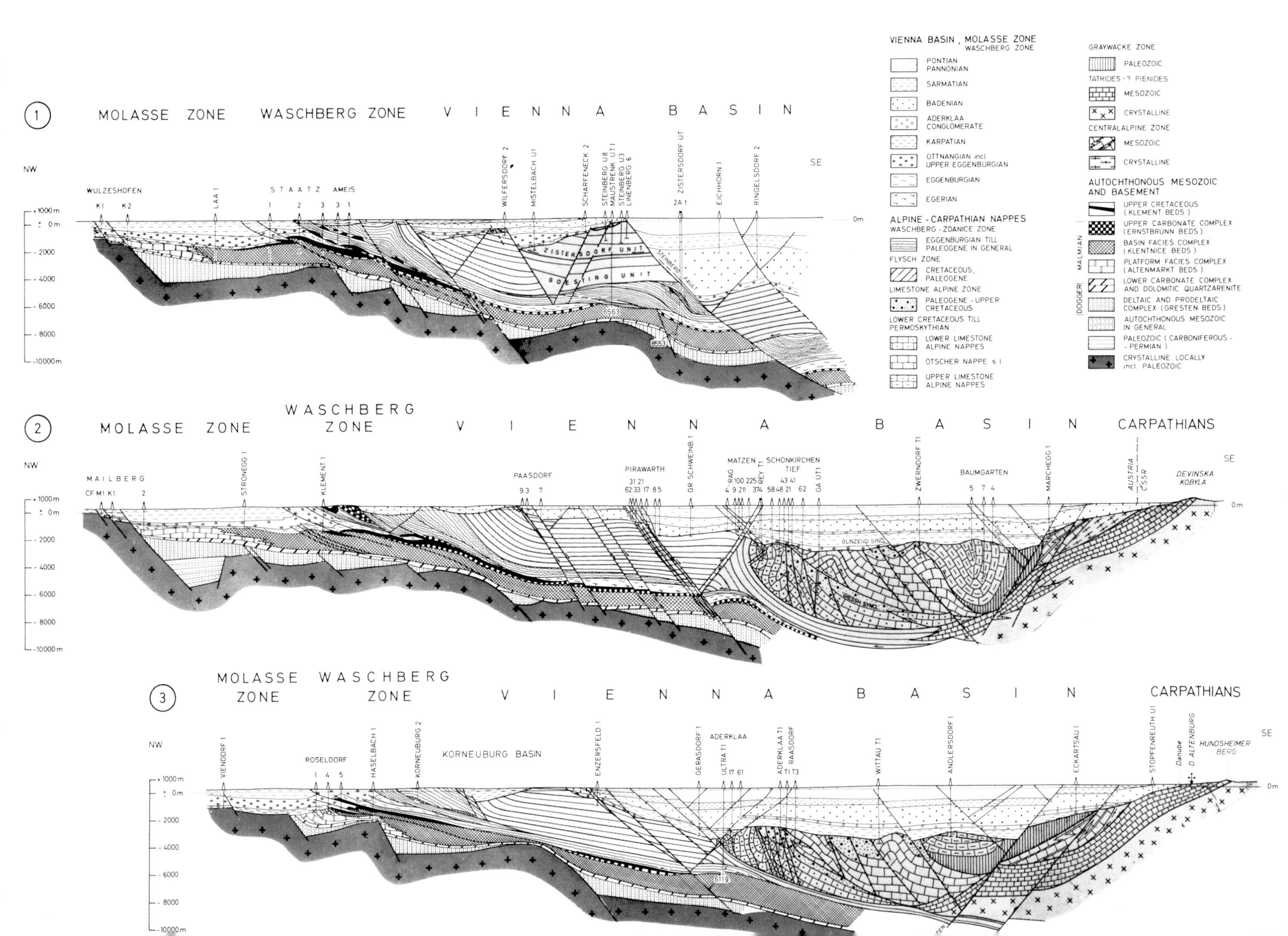

Figure 6. Geological sections through the Vienna basin and its allochthonous and autochthonous basement. Locations are shown
No vertical exaggeration. (According to R. Heller, A. Kröll, H. Unterwelz, G. Wachtel and G. Wessely. Drafted by D. Dvorak and R.

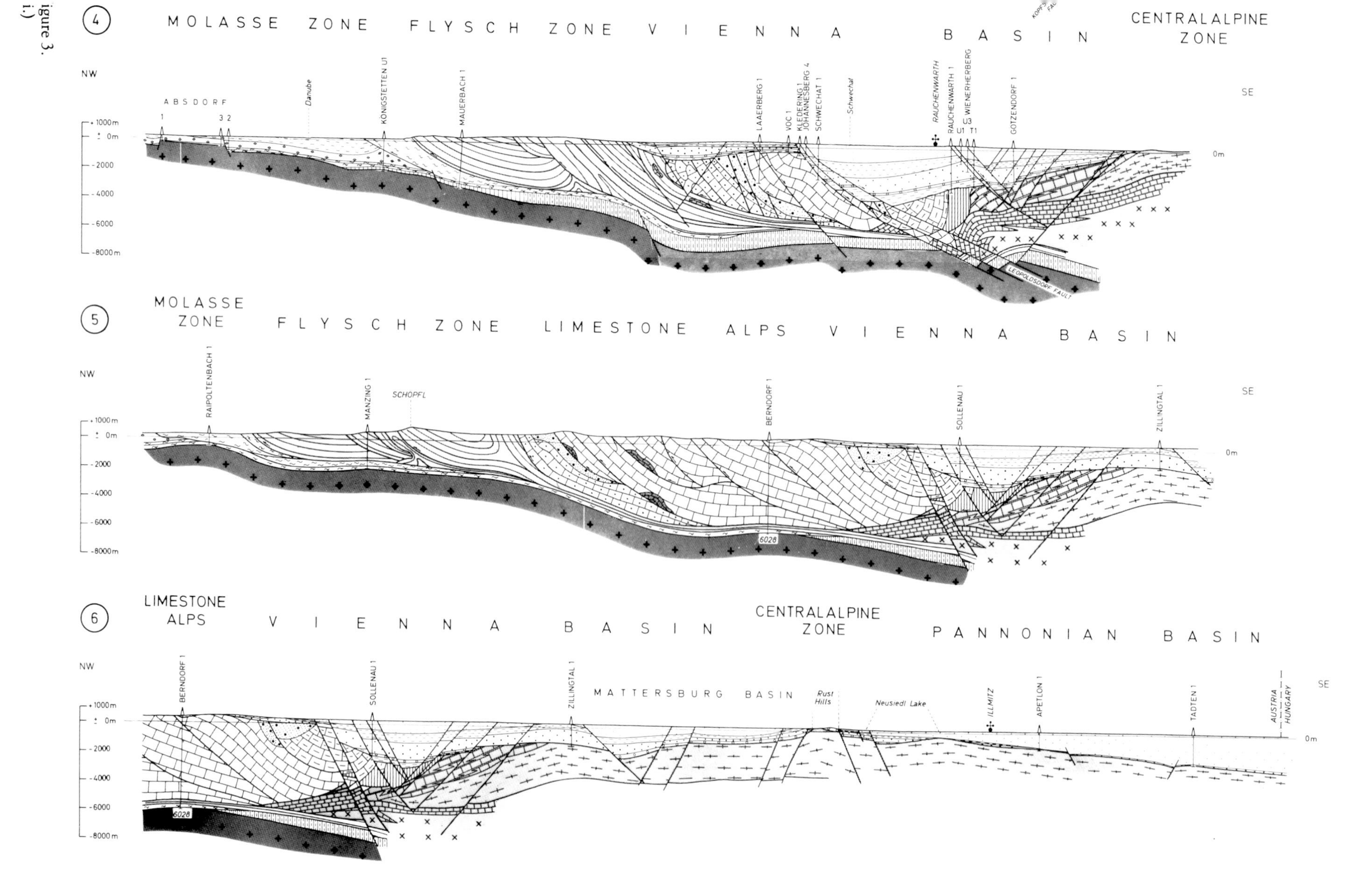
4
MOLASSE ZONE
FLYSCH ZONE
VIENNA BASIN
CENTRALALPINE ZONE
NW
SE
ABSDORF
1
3 2
Danube
KONIGSTETTEN U1
MAUERBACH 1
LAAERBERG 1
VOC 1
KLEDERING 1
JOHANNESBERG 4
SCHWECHAT 1
Schwechat
RAUCHENWARTH
RAUCHENWARTH 1
U1
U3
T1
WIENERHERBERG
GOTZENDORF 1
LEOPOLDSDORF FAULT
+1000m
0m
-2000
-4000
-6000
-8000m
5
MOLASSE ZONE
FLYSCH ZONE
LIMESTONE ALPS
VIENNA BASIN
NW
SE
RAIPOLTENBACH 1
MANZING 1
SCHOPFL
BERNDORF 1
6028
SOLLENAU 1
ZILLINGTAL 1
0m
6
LIMESTONE ALPS
VIENNA BASIN
CENTRALALPINE ZONE
PANNONIAN BASIN
NW
SE
BERNDORF 1
SOLLENAU 1
ZILLINGTAL 1
MATTERSBURG BASIN
Rust Hills
Neusiedl Lake
ILLMITZ
APETLON 1
TADTEN 1
AUSTRIA
HUNGARY
6028
0m

igure 3.
i.)

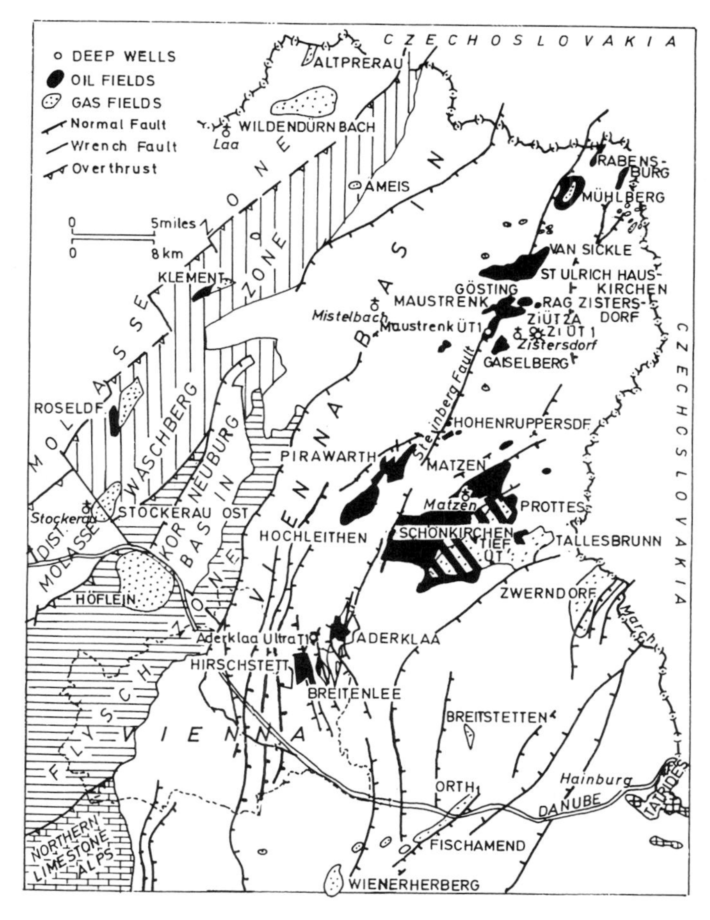

Figure 4. Oil and gas fields in the Austrian part of the Vienna basin and surrounding regions.

talline basement occurs far beneath the Alpine–Carpathian thrust sheets (Wieseneder et al., 1976; Dudek, 1980). Recently, well Aderklaa ÜT1 penetrated crystalline rocks beneath the Vienna basin. Autochthonous Paleozoic sediments, such as those found in Moravia (CSSR), may also exist in the Vienna basin.

The autochthonous Mesozoic cover of the European plate starts with Dogger deltaic sediments, consisting of coarse- to fine-grained sandstone with an intercalation of dark clay from a marine prodelta environment within the sequence and another on top of the sequence (Brix et al., 1977; Elias, 1979; Adamek, 1986) (Figure 7). In the uppermost Dogger, the Dolomitic Quartzarenite beds (called the Nikolcice beds in Czechoslovakia) underlie a widespread lower carbonate horizon of Malm reef limestone (called the Vranovice limestone in Czechoslovakia). Above this horizon, the carbonate rocks may be divided into a platform facies along the crystalline border zone to the west, called Altenmarkt beds (Ladwein, 1976), and into a basin facies complex to the east, called Klentnice beds, that has marly limestone and dark marl (called the Mikulov marl in Czechoslovakia) and regressive dark carbonate arenites (called the Kurdejov limestone in Czechoslovakia). The Klentnice beds become very thick toward the Vienna basin (eastward) where they exceed 1000 m thickness. Above them, light platform carbonates, locally containing reef complexes, indicate a regressive environment (upper carbonate horizon or Ernstbrunn beds). Lower Cretaceous rocks are missing and Upper Cretaceous rocks rest upon deposits of Malm, beginning with glauconitic sandstone and marl of the Klement beds. In the Vienna basin, Malm Ernstbrunn beds with a sequence of 130 m thick and a basin facies complex up to 1000 m in thickness were encountered by well Zisterdorf ÜT2A, which terminated at a depth of 8553 m. Very recently, well Maustrenk ÜT1 penetrated dark marl of Malm age in a basin facies at 6400 m depth before terminating at 6563 m. Well Aderklaa UT1 penetrated marl of Malm age between 6050 and 6223 m, Malm limestone (Altenmarkt beds) between 6223 and 6245 m, and crystalline schist of the Bohemian massif from 6245 to 6630 m, where drilling was terminated. The lower part of the Malm and the Dogger are missing in these wells. (Note that these very new results have not been incorporated into sections 1 and 3 in Figure 6.)

Tectonic activity during the Jurassic, possibly caused by rifting, is characterized by northeast trending synsedimentary faults formed during Dogger time (Figure 6). Westward tilting of the fault blocks resulted in greater thicknesses of Jurassic rocks on the western side of these asymmetric fault blocks and lesser thicknesses on the eastern side (sections 1–3, Figure 6; see also Grün, 1984). Diabase intrusions, such as those present in well Porrau 2 (Figure 7), may have accompanied this tectonic event and caused steepening and overturning of Dogger sediments in this well. Obviously, zones of weakness were generated along these faults. Later, these faults appear to have influenced the strike of alpine thrust units and the strike of Miocene normal faults that were active during the development of the Vienna basin. Seismic and drilling results show that these tectonic units are discordantly covered by a Callovian transgressional sequence.

Wells Zistersdorf ÜT1 and 2A encountered a molasse of Eocene–Neogene age that is transgressive on the Malm carbonates. The molasse consists of breccia, sandstone, and marls, and the breccia contain components of Malm carbonate. Olistoliths of Upper Cretaceous glauconitic sandstone also occur near the base of the molasse.

ALPINE–CARPATHIAN UNITS

The Alpine–Carpathian thrust sheets, from structurally lower to structurally higher units, are made up of the Flysch zone, the Tatride-Central Alpine zone, and the Limestone Alpine zone with its Paleozoic basement. Neogene cover overlies these Alpine–Carpathian thrust sheets, and in places also overlies the deepest and most external thrust sheet, the Waschberg zone.

The Waschberg zone consists of strongly folded and thrusted complexes of flysch-like sediments and molasse of Paleogene to Karpatian age. During thrusting, blocks of Jurassic, Upper Cretaceous, and littoral Paleogene rocks were sheared off of the autochthonous basement and transported westward to the surface. One lense of the Klentnice beds was encountered in Zistersdorf ÜT1 and 2A (Figure 6, section 1).

The northern part of the Vienna basin floor is occupied by the Flysch zone, which has several subunits and ranges from uppermost Early Cretaceous to Eocene in age (Figure 8).

The thrust plane below the Limestone Alpine zone is planar and subhorizontal except at its frontal part, where it is steep or

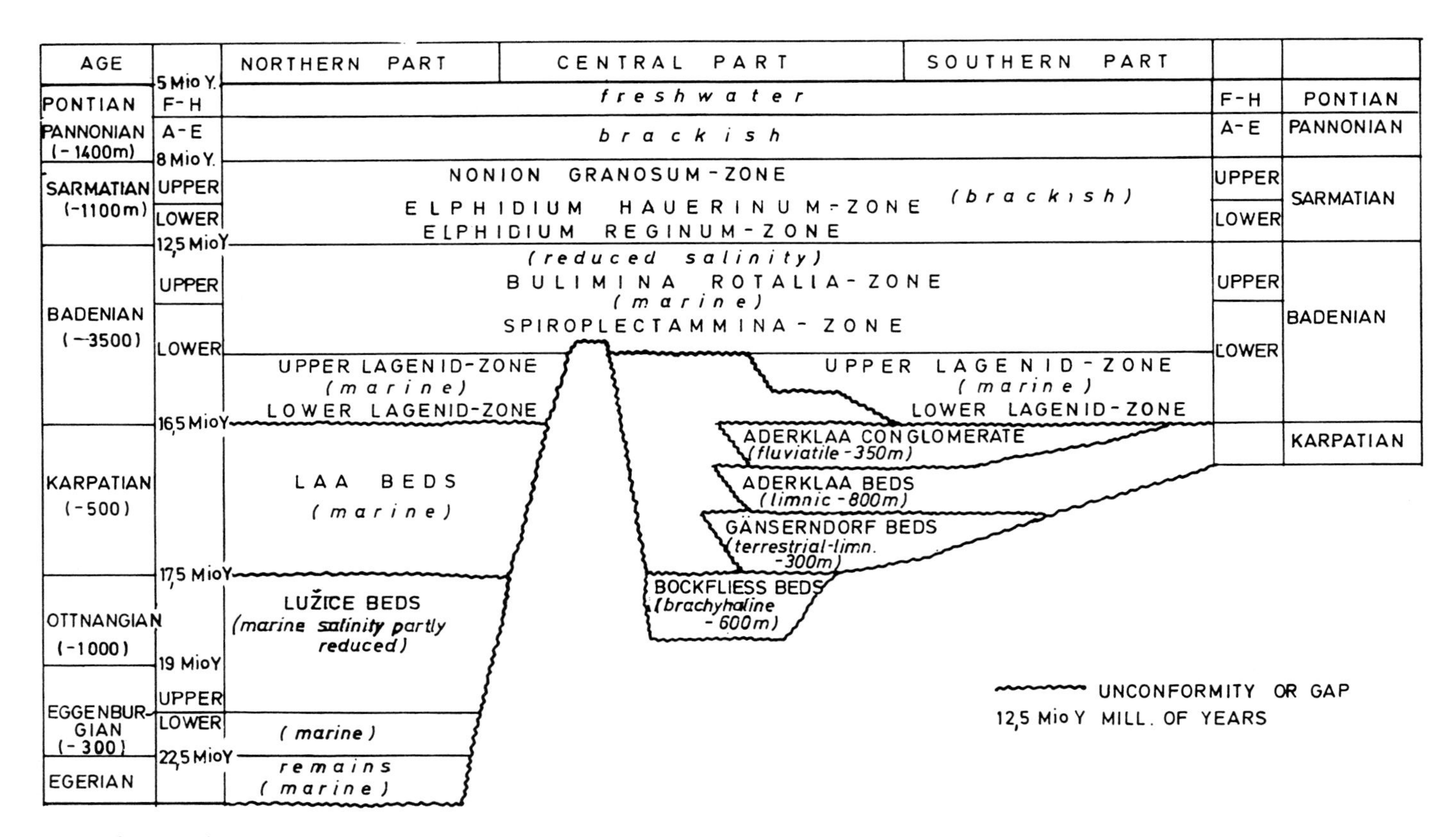

Figure 5. Schedule of the Neogene stratigraphy of the Vienna basin in Austria. (According to W. Krobot, R. Fuchs, and A. Kröll.)

overturned. The different units of the Limestone Alps can be followed along strike from the surface northeastward underneath the Neogene basin fill (Figure 9). There exist three main tectonic units of the Limestone Alpine zone (Figure 6): (1) the strongly folded and sliced Frankenfels–Lunz Nappe, which is overthrust by (2) the Ötscher nappe system, and which is in turn overlain by (3) the upper Limestone Alpine nappes (Kröll and Wessely, 1973; Wessely, 1983, 1984). The nappes are separated by Gosau beds that discordantly overlap folded and sliced Eo-Alpine structures of the underlying nappes. Upper Cretaceous–Paleogene sediments rest on the Frankenfels–Lunz nappe and form an asymmetric syncline, the Giesshübl furrow. Only the northern flank of the Giesshübl furrow is preserved; its southern flank is disturbed by syn- and postsedimentary (with respect of Paleocene sedimentation) overthrusting of the Ötscher nappe. Remnants of Cretaceous and Paleocene sediments also cover the front of the Ötscher nappe. Upper Cretaceous rocks also form the Glinzendorf syncline, which is situated in front of the structurally highest Limestone Alpine thrust units, and is in many respects similar to the Giesshübl furrow. Overthrusting of upper Limestone Alpine units occurred partly before sedimentation of Upper Cretaceous sediments, and partly afterwards.

The stratigraphic extension of the Limestone Alpine sedimentary complex and its facies beneath the Vienna basin corresponds with that exposed at the surface (Figure 8). It includes rocks of Permoscythian to Paleocene age. The oldest beds of the Limestone Alpine zone are of Permo–Scythian age and include the evaporitic-dolomitic Haselgebirge complex and the pelitic-clastic Werfen complex. These are overlain by Middle Triassic strata consisting partly of a basin facies of Gutenstein and Reifling limestone and partly of a platform facies of Wetterstein limestone and dolomite. Shale and sandstone of the Lunz beds form a distinct marker horizon between the carbonate of Middle and Late Triassic age. Facies of the Late Triassic change continuously from northwest to the southeast. In Late Carnian time, an evaporitic-dolomitic sequence developed in the northeast, which grades into a calcareous-marly facies in the southeast (Opponitz beds). The Norian Hauptdolomit, deposited in a lagoonal environment, shows a continental influence in the north, which decreases to the south (Ötscher nappe). The thickness of Hauptdolomite increases from north to south. Together with the Middle Triassic platform carbonates, it forms a structurally rigid mass. The Hauptdolomit is conformably overlain by Norian-Rhaetian Dachstein limestones. They are composed of Lofer cyclothemes and show transitions to a reef facies toward the south. Especially in the north, the Rhaetian is represented by the dark limestone and marl of the Kössen strata.

In Liassic time, a basin facies (represented by Allgäu beds and Kieselkalk) changes to a shallower water facies (Hierlatz limestone). Limestone of Dogger and Malm age consists of thin-bedded, condensed, red and brownish colored layers that are well stratified by microfacial elements. It changes from north to south into a thicker basin facies with red radiolarites and gray, sometimes cherty Oberalm beds. Neocomian beds with spotted limy marl are restricted to the Frankenfels–Lunz system. During Albian–Early Cenomanian time, clastic sediments with sandstone, conglomerate and marl of Losenstein beds were deposited.

Sediments of the Giesshübl syncline (see Figures 8 and 9 and Figure 6, section 2) include Cenomanian marl and Turonian to Maastrichtian conglomerate, breccia, sandstone, and marl. The Campanian is characterized by variegated pelagic marly limestone. Turbidites of upper Maastrichtian–Paleocene Giesshübl beds up to one km thick form the main sedimentary fill in the syncline. The Glinzendorf syncline (Figure 9 and Figure 6, section 2) consists of Upper Cretaceous limnic marl, sandstone, and conglomerate with some coal (Gosau beds). The highest Limestone Alpine nappes are stratigraphically underlain by dark schists and terrigenous Paleozoic sediments of the Graywacke zone.

The Tatride-Central-Alpine zone (cross sections in Figure 6) is represented by the lower Austro-Alpine unit. On the surface this unit covers tectonically the deepest alpine tectonic complex, the Penninicum. The lower Austro-Alpine unit below the Vienna basin consists of crystalline rocks and a cover of Permian to Lower Triassic quartzite, Middle Triassic platform carbonate, a series of Upper Triassic continental sediments (Keuper) built up of variegated shale, quartzite, and some dolomite, and finally Rhaetian dark limestone and Liassic sandy limestone (Figure 8). This cover series is called Semmeringmesozoikum. The Tatricum in the Carpathians has a tectonic position similar to the lower Austro-Alpine unit in the Alps. Its Mesozoic cover also contains a similar facies. The relationship between the Tatricum and the lower Austro-Alpine unit has not been clarified by drillings. A kind of digitation is drawn in the sections in Figure 6, but this is only hypothetical. The question of the presence of the Penninicum in the Carpathians remains open (Mahel, 1982). No evidence has been presented to show that the tectonic elements situated north of the Tatricum in the Carpathians occur in the Austrian part of the Vienna basin, for example, the Manin unit, Klape unit, and the Pieniny Klippen belt, in the strict sense (see Mahel, 1982).

The age of the orogenic movements within the Alpine–Carpathian thrust belt generally becomes younger from higher to lower units. In the Limestone Alpine zone beneath the Vienna basin, the oldest movement can be documented in its northernmost part, where Albian to lower Cenomanian synorogenic clastic sediments discordantly filled troughs of obvious tectonic origin. Unconformities occur within the Cenomanian, before the Coniacian, which locally overlaps folded Jurassic and Triassic rocks, and especially within the Maastrichtian before turbiditic sedimentation began (Giesshübl beds). In Paleocene time the Ötscher nappe overthrust the Giesshübl beds. Thrusting of higher nappes onto the Ötscher nappe occurred before Coniacian time, and Coniacian sedimentary rocks seal the thrust plane between these two units, but thrusting was reactivated in the Late Cretaceous or afterward. Above the Ötscher nappe and these uppermost Limestone Alpine nappes, Upper Cretaceous layers older than the Coniacian-transgressive sequence are missing.

The earliest thrusting of the Limestone Alpine zone over the Flysch zone occurred in late Eocene time. During thrusting, the Limestone Alpine zone was separated from its crystalline basement, and advanced far to the north, pushing the main part of the Flysch zone ahead of it. Attenuation of the flysch beneath the overlying thrust complex is demonstrated in well Berndorf 1 (Figure 6, section 5) (Wachtel and Wessely, 1981).

Well Berndorf 1, which is 35 km from the northern edge of the overthrust zone, penetrated Oligocene–Miocene molasse and crystalline rocks beneath the Flysch-Limestone Alpine nappes as proven by Miogypsina in conglomerates of the molasse at a depth of 5941 m. Oligocene–Miocene thrusting can be clearly seen at the northern edge of the Alpine overthrust zone (Waschbergzone). Successively younger molassic beds of Egerian to Karpatian age were involved in the deformation process. Older, more internal thrust planes were successively covered and sealed by synchronous sedimentation (Figure 6, section 3). Sediments of Eggenburgian to Karpatian age also rest on top of the Alpine nappe complex in the area of Vienna basin. Although thrusting was going on at that time, these sediments show no compressive strain. At the beginning of Badenian time all thrusting was finished and the youngest thrust planes were sealed by posttectonic sedimentary rocks.

In the eastern part of the Alps the east-west strike of the Alpine units changes eastward to a northeast strike (Figure 9). In all tectonic units this change in strike occurs near the border of the Vienna basin along the southern prolongation of the subcrop of the western boundary of the autochthonous Mesozoic underneath the Molasse (see Figure 9). This change in strike may be explained by the more stable character of the crystalline platform to the west, while an area of Jurassic and post-Jurassic subsidence time existed to the east (Kröll et al., 1981). This is also indicated by more steeply dipping and overturned structures and thrust planes of nappes near and within the Vienna basin. This zone of subsidence is probably related to the northeast-striking Jurassic normal faults in the molasse basement. Thus, the relatively stable foreland region in the west was not easily overridden by advancing thrust sheets in the Eastern Alps, while the more easily subducted basins to the east allowed the thrust sheets of the Carpathians to advance further to north. Beneath the Vienna basin the change in strike of Alpine thrust units occurs without transform faulting (Figure 9). The Alpine system, especially the Limestone Alpine and Flysch units, continues underneath the Vienna basin into the Carpathian Mountains. Investigations by Wessely (1975), Jiricek (1981), and Nemec and Kocak (1976, 1978), show that there is no interruption in these units beneath the Vienna basin and into the Carpathians except for the erosion and disappearance of higher units, including the Graywacke zone, towards the northeast.

STRUCTURE AND SEDIMENTARY FILL OF THE VIENNA BASIN

Structural Synopsis

The Vienna basin contains a system of horsts and troughs in a distinctive arrangement (Figures 3 and 9). Uplifted blocks along the eastern and western border of the basin are separated from deep depressions by normal faults, some with large displacements. The approximate axis of the basin is marked by a series of elevated blocks arranged in a slightly sigmoidal fashion and bordered by curved faults. This ridge disappears in the southernmost part of the Vienna basin, where only one depression exists.

The largest elevated marginal blocks are the Mistelbach and Mödling blocks (Figure 3). The Mistelbach block is separated

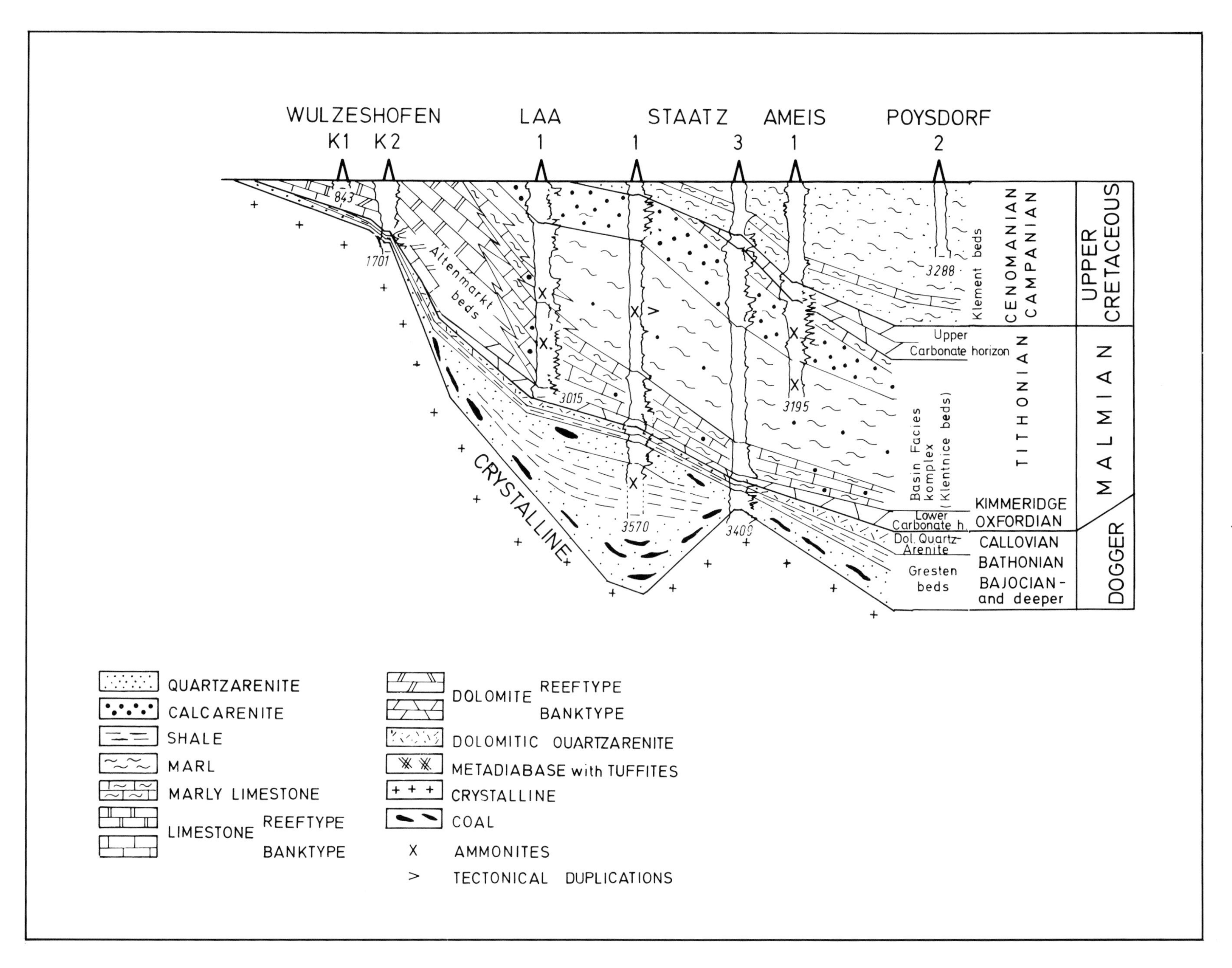
WULZESHOFEN K1 K2
LAA 1
STAATZ 1 3
AMEIS 1
POYSDORF 2
843
1701
Altenmarkt beds
3015
3570
3409
3195
3288
CRYSTALLINE
Klement beds
Upper Carbonate horizon
Basin Facies komplex (Klentnice beds)
Lower Carbonate h.
Dol. Quartz-Arenite
Gresten beds
CENOMANIAN CAMPANIAN
TITHONIAN
KIMMERIDGE
OXFORDIAN
CALLOVIAN
BATHONIAN
BAJOCIAN- and deeper
UPPER CRETACEOUS
MALMIAN
DOGGER
QUARTZARENITE
CALCARENITE
SHALE
MARL
MARLY LIMESTONE
LIMESTONE REEFTYPE
BANKTYPE
DOLOMITE REEFTYPE
BANKTYPE
DOLOMITIC QUARTZARENITE
METADIABASE with TUFFITES
CRYSTALLINE
COAL
X AMMONITES
> TECTONICAL DUPLICATIONS

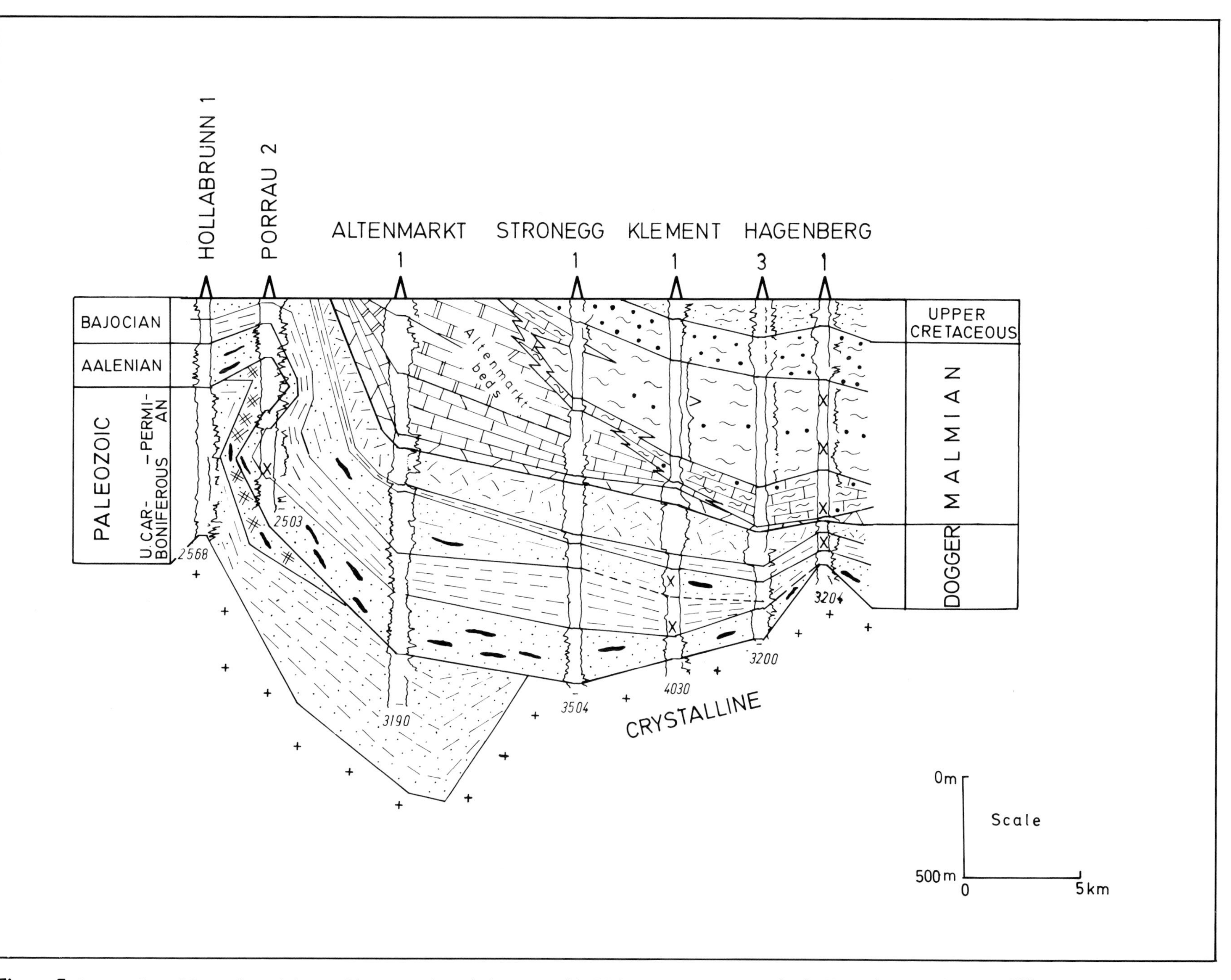

Figure 7. Stratigraphy and facies of autochthonous Mesozoic rocks in the basement of the Molasse zone continuing under the Vienna basin. (G. Wessely, 1977).

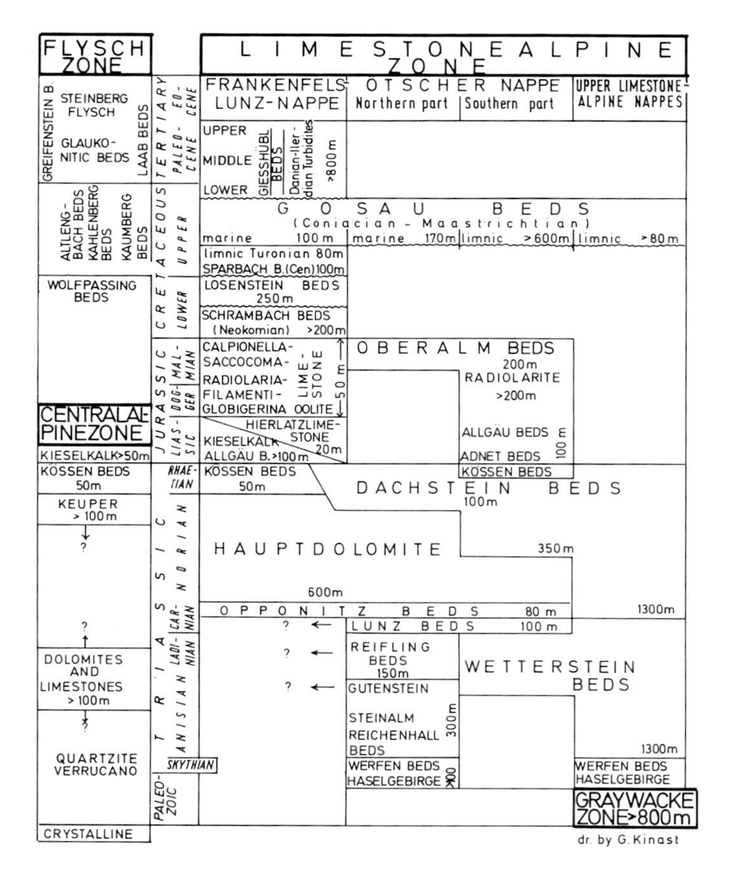

Figure 8. Stratigraphy of the Alpine–Carpathian basement of the Vienna Basin in Austria (simplified). (G. Wessely, 1983.)

from the Zistersdorf depression by the Steinberg fault; the Mödling block is separated from the Schwechat depression by the Leopoldsdorf fault. These deep oval depocenters are obviously related to the elevated blocks on the upthrown side of the faults (Steinberg and Oberlaa blocks). The Zistersdorf depression continues into the Moravian central depression. Between the Zistersdorf and Schwechat depressions is the Gross Engersdorf depression. In the east, the elevated blocks are bordered by the Marchfeld depression and the Drösing–Kuty depression. A series of young grabens, such as the Wiener Neustadt depression in the southernmost Vienna basin, the Mitterndorf–Lassee graben, and the Zohor Plavecky graben, seem to intersect or to widen this system. The median ridge is represented by the Hodonin–Gbely horst, Eichhorn ridge, Matzen–Aderklaa elevations, and probably the Wiener Herberg-Enzersdorf elevations.

Sedimentation

The Vienna basin is a classic region of Neogene stratigraphy and its sedimentary units are well known and documented in many monographs and special papers. Fundamental contributions were given by R. Janoschek, 1951, 1964; Grill, 1941, 1943, 1968; Friedl, 1937; Papp, 1963; and Kapounek et al., 1965. Recent comprehensive studies are published by Papp et al., 1974; Fuchs, 1980; Steininger et al., 1975, Janoschek and Matura, 1980; Buday and Cicha, 1968; Jiricek and Tomek, 1981; Kreutzer, 1971, 1984, 1986; and Kröll, 1980. The lower Neogene exists mainly in the northern part of the Vienna basin (Figures 3 and 5), suggesting another phase of extension of the basin during early Neogene time (Rögl, 1980; Jiricek, 1978). The oldest sediments that are generally referred to as "basin sediments" (Figure 5), are of Egerian age (Fuchs et al., 1980). It is possible that these Egerian sediments, however, belong to the Carpathian thrust sheets. A better-preserved series consists of Eggenburgian and Ottnangian molasse-like Schlier, subdivided in beds with Cyclammina–Bathysiphon, Cibicides–Elphidium, and fish fossils. Where the Schlier transgresses directly upon flysch, rubble beds occur at the base of the Schlier.

The southward continuation of these older Neogene beds is uncertain because of later erosion, especially prior to Badenian time (Figure 6, section 2). Erosion was particularly pronounced in the region of the Matzen Spannberg ridge (Figures 3 and 5), a morphological structure, which strongly influenced sedimentation and erosion. Only remnants of Ottnangian sediments were encountered in drill holes south of this ridge in a brachyhaline facies (Bockfliess beds) containing an Ammonia, Elphidium, Cytheridea microfauna. These older Neogene sediments, as well as the Karpatian sediments, are considered to be molasse overlapping the Alpine–Carpathian nappes (Jiricek, 1978; Janoschek and Matura, 1980). They were deposited on top of moving thrust sheets.

In the northern Vienna basin, the Karpatian is developed in marine facies (Laa beds, containing *Uvigerina graciliformis*). To the south, this facies is replaced by the freshwater development of gray, sandy, and marly Aderklaa beds. The Aderklaa beds are underlain by Gänserndorf beds containing terrestrial gray and variegated shale, sandstone, some anhydrite, and a basal conglomerate (Figure 5). In some places limnic fauna (large Ostracods) are mixed with small fauna of Globigerinids, Bolivinas, and some Uvigerinas. The Karpatian rocks generally overlie an unconformity and transgress in many places upon pre-Neogene rocks. The areal extent of Karpatian sedimentary rocks was nearly as great as that of the later basin sediments. The freshwater facies of the Aderklaa beds overlies the fluviatile Aderklaa conglomerate, with eroded detritus from the Alpine nappes. This conglomerate may be of Badenian age. Unconformities exist underneath and on top of the Aderklaa conglomerate. On the top there is an interfingering of marine and limnic sediments.

The boundary between the Karpatian and the Badenian is marked by an important paleogeographic and tectonic event. Older sedimentary rocks and structures were discordantly covered leading to a widespread marine development after spatially differentiated phases of erosion, emergence, and fluviatile sedimentation. Locally, as along the Matzen ridge, early Badenian sediments were deposited onto a pre-Neogene surface with considerable topographic relief, so that only by middle Badenian

Figure 9. The Austrian part of the Vienna basin floor showing the subcrop of Alpine tectonic units and contour lines of the Neogene basin floor in meters. (Based on maps of A. Kröll, S. Köves, H. Klob, H. Unterwelz, K. Schimunek, and G. Wessely. Compiled by G. Wessely and G. Gohs, January 1984.) (opposite, right)

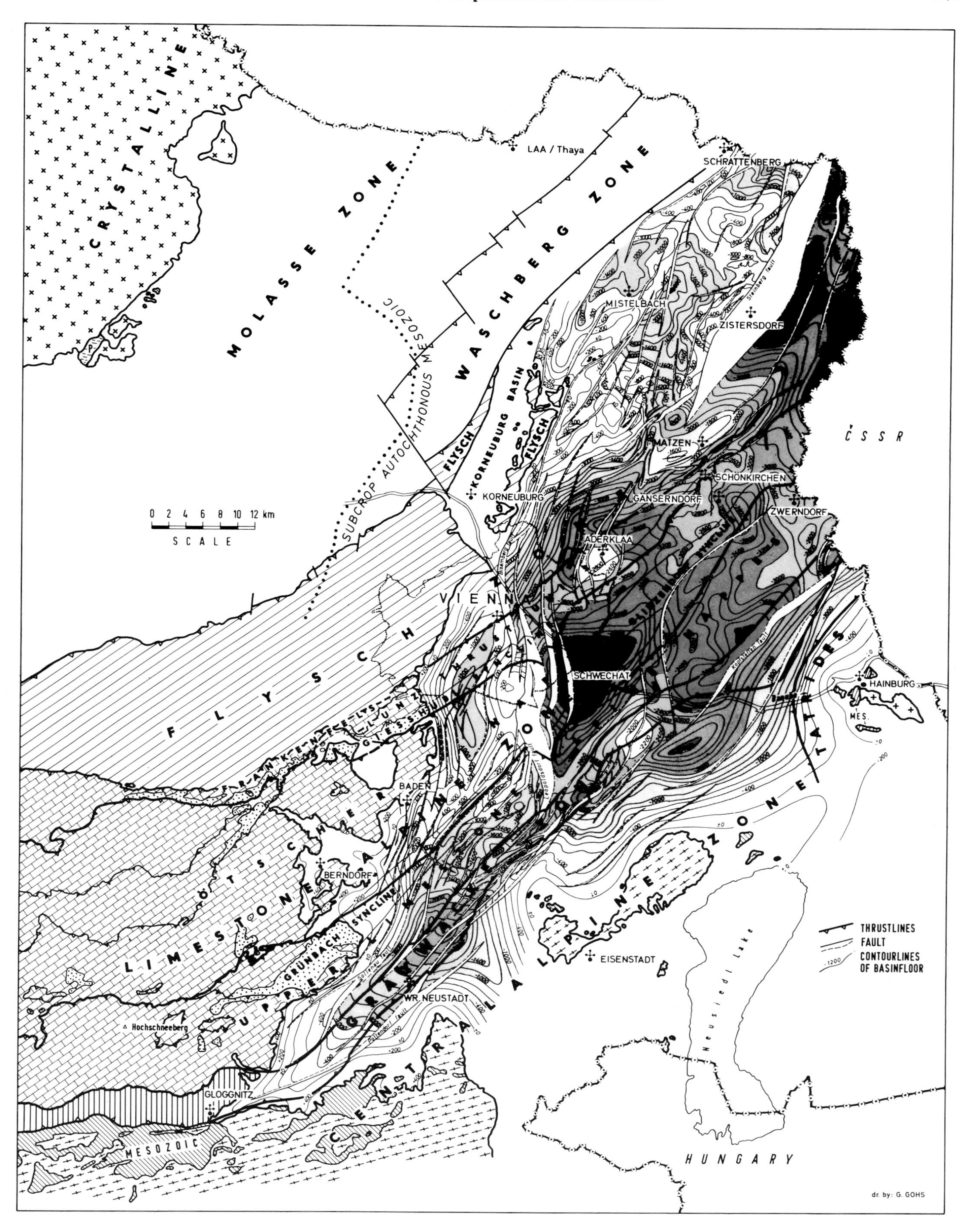
CRYSTALLINE
MOLASSE ZONE
WASCHBERG ZONE
SUBCROP AUTOCHTHONOUS MESOZOIC
KORNEUBURG BASIN
FLYSCH
FLYSCH
LAA / Thaya
SCHRATTENBERG
MISTELBACH
ZISTERSDORF
MATZEN
SCHÖNKIRCHEN
GÄNSERNDORF
ZWERNDORF
KORNEUBURG
ADERKLAA
VIENNA
SCHWECHAT
HAINBURG
MES.
BADEN
BERNDORF
EISENSTADT
WR. NEUSTADT
GLOGGNITZ
MESOZOIC
Hochschneeberg
ČSSR
HUNGARY
0 2 4 6 8 10 12 km
SCALE
F L Y S C H
FRANKENFELS
LUNZ
GIESSHÜBL
ÖTSCHER
LIMESTONE
UPPER
GRÜNBACH SYNCLINE
GRAUWACKEN ZONE
CENTRAL ALPINE ZONE
TATRIDES
Neusiedl Lake
Steinberg fault
THRUSTLINES
FAULT
CONTOURLINES OF BASINFLOOR
dr. by: G. GOHS

time did sedimentary layers extend over the total area of the basin. The Badenian sedimentary sequence varies in thickness from hundreds to thousands of meters depending on its position on downthrown or upthrown blocks (in the Zistersdorf depression, the thickness reaches up to 3500 m). The rates of sedimentation for sandstone and marl were high, so that even in areas of rapid subsidence the water depth remained relatively shallow. Lithothamnian biostromata grew along the basin margins and on the elevated blocks, or their detritus was deposited as on the Steinberg, Matzen, and Laxenburg elevations (the latter is the southern continuation of the Oberlaa high). The Badenian can be subdivided by microfaunal zonation (Grill, 1941, 1943), established by distinct assemblages as well as by a phylogenetic raws of foraminiferas (Figure 5). Near the top of the sequence, the Badenian rocks are brackish water deposits but brackish influences also occur in deeper parts of the Badenian sequence, particularly where the deposition rates of sandy sediments were very high.

In Sarmatian time the sandy development of the Upper Badenian continues with brackish water faunas containing Ammonia and Cibicides. In the higher part of the lower Sarmatian, sandstone and marl are interlayered and individual layers can be correlated over large distances. The Sarmatian may be divided into three microfaunal zones by distinct species of Elphidium and Nonion. Along the basin margin and on the elevated blocks, lumachelles, oolites, limestone built by Algal stromatolites, Bryozoas, and Nubecularias are present.

In Pannonian time, interlayered sequences of sandstone and marl continue to be deposited, but in a less saline environment as shown by Ostracods. Freshwater conditions prevailed by Pontian time, when deposition of sandstone becomes predominant. As in Badenian and Sarmatian time, the thickness of Pannonian and Pontian sediments depends on their position with respect to downthrown or upthrown basement blocks. The thickness of the Pannonian and Pontian sediments may be only tens of meters above the elevated blocks and up to 1500 m for each unit in the deep depressions. During Pannonian time, the main area of subsidence shifted eastward into the Pannonian basin. In particular, the thickness of Pontian sediments increases rapidly to the east.

In Pliocene and Quaternary time, sediments were mostly deposited on fluviatile fans derived from alpine source areas and from the Danube. They are of importance only in areas of young subsidence, such as in Mitterndorf and Lassee graben, where they reach thicknesses of up to 200 m (Figure 3) (Küpper, 1953; Boroviczeny and Brix, 1981; and Brix, 1981).

No volcanic activity took place in the Vienna basin but there are distinct horizons containing volcanic tuffs and their altered products (bentonite). They are used as stratigraphic markers in the deeper parts of the Neogene (Kapounek and Papp, 1969; Kreutzer, 1986). The contour map of the base of Neogene sediments gives a rough picture of the thickness of Neogene sediments and its variation above different fault blocks (Figure 9).

Tectonics

Structural Development

Information about sedimentation and structural deformation during early Miocene time is restricted to areas of intense exploration in the Mistelbach block, in the Matzen area, and in isolated wells north of the Matzen area (Figure 3). In addition, deposits of this age are often preserved only as remnants after several phases of erosion and tilting prior to the Badenian transgression. Particularly in the region of the Matzen–Spannberg ridge, the lower Miocene cover, which certainly existed there, has been removed by erosion. East-west trending tectonic features seem to have played a role as well as the northeast-southwest trending features common in the Vienna basin. East of drill hole Wilfersdorf 2, a growth fault displaces Eggenburgian, Ottangian and Karpatian sediments, which are thin on the upthrown block and thick on the downthrown block (Figure 6, section 1). Because thrusting of the Alpine–Carpathian nappes occurred until the end of the Karpatian, these synsedimentary structures and faults must have also moved relative to the sub-Carpathian autochthonous basement (overthrust Egerian molasse in the well Berndorf, gives a minimum overthrusting of 35 km from Egerian to Karpatian time). Thus alternation of shortening and extensional events must be assumed.

During Karpatian time, the present structure of the Vienna basin began to evolve. Southward tilting of the basin floor, beginning south of the Matzen–Spannberg ridge, resulted in subsidence of a large area in the southern part of the Vienna basin. Synsedimentary faulting occurred in Karpatian time as shown by the large differences in thickness of Karpatian sediments on the upthrown and downthrown fault blocks (for example, along the Leopoldsdorf fault; see Figure 3 and Figure 6, section 4).

Because of pre-Badenian erosion, structural development of the basin is better known for later Miocene and Pliocene time. In Badenian time the main phase of basin subsidence began and the lateral extent of the basin increased. Normal faults acquired large displacements, and the deep depressions were filled by thick accumulations of sediment with rapid depositional rates. This process continued during Sarmatian and Pannonian time, and in some deep depressions subsidence lasted until Pliocene and Quaternary time. Because of locally different subsidence rates and synsedimentary faulting, older structures are more accentuated than younger (see, for example, Figure 10). Beyond the east side of the Vienna basin, towards the Pannonian basin, Badenian and Sarmatian sediments are thin and were deposited in a littoral environment. They dip to the east and are overlain by Pannonian and Pontian sediments that thicken eastwards. At the same time as subsidence took place in the Vienna basin, uplift occurred along the margins of the basin, particularly along the eastern margin (Wessely, 1961).

Characteristics of Faults

Several phases of faulting occurred in the Vienna basin. The oldest event consists of lower Miocene synsedimentary faulting that has been transported within the moving thrust complex. More details are known about the younger Miocene (Badenian and younger) faults which started after thrusting had ended in lower Austria. They are commonly growth faults, in some cases with large displacements. The fault planes commonly dip at a constant angle of 40–50° and displacement can change rapidly along strike. In many cases an en echelon arrangement of faults can be observed, especially along curved fault zones. The large fault zones often contain horse lenses.

The faults along the margin of the basin do not have large displacement or persistent strike and only in a few distinct places

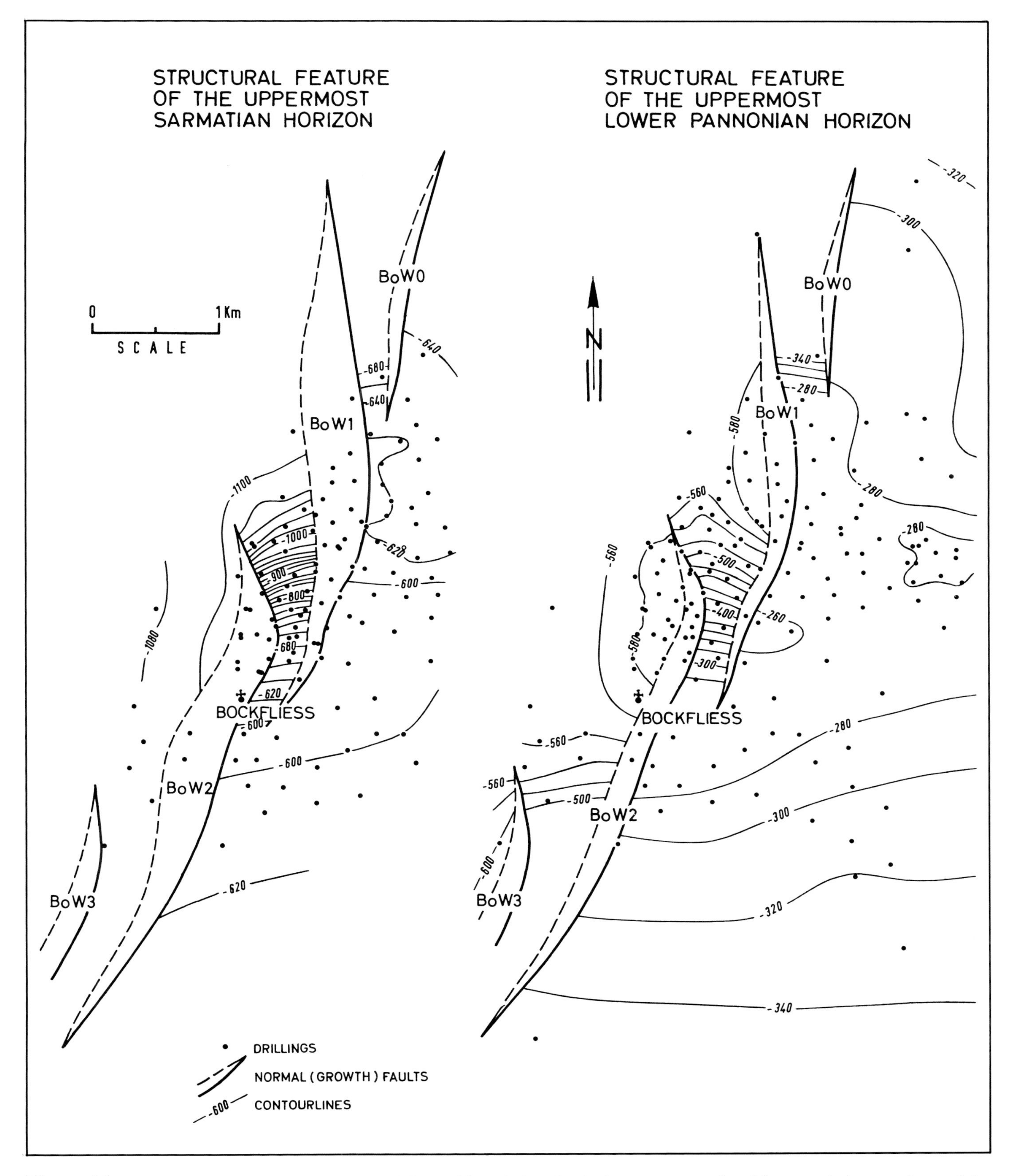

Figure 10. Vienna basin: en echelon arrangement of the Bockfliess faults. The displacement is transferred from one fault to another, and the upthrown and downthrown blocks are connected by steeply dipping strips between the faults. The synsedimentary faulting is evident by an increase of structural accentuation with depth. (G. Wessely, 1984.)

determine the morphology of the basin border. The most important faults within the Vienna basin bound the elevated marginal basement blocks toward the basin interior. They include the Steinberg fault, the Leopoldsdorf fault system, the Sollenau fault, the Pottendorf fault, and the Kopfstetten fault (Figures 3, 6, and 9).

The lateral extent of the Steinberg fault is about 55 km. Its maximum vertical throw is nearly 6000 m near Zistersdorf. Towards both ends of the fault the vertical throw decreases rapidly. In Czechoslovakia it converges with the Schrattenberg fault, which probably takes up most of the displacement. Fourteen km southwest of Zisterdorf, the Steinberg fault is replaced by the en echelon faults of the Hohenruppersdorf–Pirawarth–Hochleiten area, but without the large vertical displacement of the Steinberg fault. In some places along the Steinberg fault, smaller faults strike parallel or oblique to the main fault trace. Drag and rollover structures are described by Friedl (1937) and Stowasser (1958, 1966). Strike-slip displacement has not yet been proven.

The Leopoldsdorf fault system strikes north-south (Figures 3 and 9) and consists of several faults that partly replace each other. One of them acts as a marginal fault of the Vienna basin in its continuation north of the Danube. In the area of Schwechat, the total throw on the Leopoldsdorf fault system (Figure 6, section 4) reaches a maximum of 4000 m. Along both the Steinberg and Leopoldsdorf faults, the highest part of the upthrown block of the foot wall is situated opposite the deepest part of the depression on the downthrown block of the hanging wall. These depressions are oval shaped. Other faults that border the eastern marginal elevated blocks, such as the Kopfstetten and Engelhartstetten faults, show displacements of up to several hundred meters. A series of faults that dip in opposite directions bound a series of young grabens like the Mitterndorf and Lassee graben and the Wiener Neustadt basin, along the southeastern edge of the Vienna basin (Figures 3 and 9). The latter is bounded by the Sollenau and Pottendorf faults.

The elevated horsts along the axis of the Vienna basin, such as the Matzen and Aderklaa blocks, are generally separated from adjacent depressions by faults. The Bockfliess–Aderklaa fault system separates both of these elevated blocks from the Gross-Ebersdorf depression to the west (Figure 3). This fault system, which has a vertical throw of up to 500 m, consists of en echelon fault segments (Figure 10) that strike north-northeast and bend around the western part of the Aderklaa block in its southern continuation. The transfer of displacement from one fault to another is well known from data from many drill holes. Upthrown and downthrown blocks are directly connected by steeply dipping narrow strips between two faults. A large difference in the thickness of Neogene sediments on the upthrown and downthrown blocks and an increase in thickness within the narrow strips accentuate the synsedimentary character of the Bockfliess–Aderklaa fault system.

The Matzen fault system in the northwestern part of the Matzen elevated block is different from those mentioned above. It is post-Pannonian in age and the faults show only small displacements (10–80 m). These faults may be explained by tension above an updoming of the deeper basement (Figure 6, section 2) which causes the Matzen Spannberg elevation (Figure 3). This structural element seems to continue to the northeast as the Eichhorn ridge and Hodonin spur in Czechoslovakia. The median zone of elevated blocks is bounded against the Marchfeld depression to the southeast and east by the Markgrafneusiedl fault (Figure 3). The Enzersdorf and Wienerherberg blocks may represent an isolated southernmost element of the axial ridge system. The Zwerndorf high (Figures 3 and section 2 of Figure 6) belongs to another system of elevated blocks and depressions developed in Czechoslovakia.

The Enzersdorf, Wienerherberg, and Zwerndorf highs are bordered to the southeast by faults dipping toward a young series of grabens—the Wiener Neustadt basin, the Mitterndorf graben, and in Czechoslovakia, the Zahor Plavecky graben (Buday and Cicha, 1968). These narrow grabens are bordered by faults that were active until Pliocene and Quaternary time.

Remarks on the Tectonic Origin of the Vienna Basin

Two differing points of view exist about the cause and mechanism of the formation of the Vienna basin. At present these seem to be incompatible, but after some modification they may be reconciled with one another. According to the first point of view, the formation and evolution of the Vienna basin is related to the preexisting tectonic behavior of the autochthonous European basement beneath the Alpine thrust sheets, which had already influenced the strike of the Alpine–Carpathian thrust complex. According to the second point of view, basin evolution is the result of thin-skinned extension within the allochthonous thrust sheets, and can be related to contemporaneous thrusting of the sheets. The relationship between basin development and deep-seated tectonic features has been already discussed by Stowasser, 1958; Tollmann, 1978; Kröll et al., 1981; and Jiricek, 1981.

Studies of the tectonic origin of the Vienna basin have been stimulated by more regional studies of basin evolution in the Carpathian–Pannonian region (Horváth and Royden, 1981; Burchfiel and Royden, 1982; Royden et al., 1982, 1983a, b, and this volume). These regional studies are based on both tectonic and thermal considerations within the basins and on the relationship of basin evolution to the Carpathian thrust belt. The authors conclude that most of the basins in the Carpathian–Pannonian region, including the Vienna basin, resulted from or are related to strike-slip faulting within the crust, and that the basins are pull-apart basins (see also Mann et al., 1983).

Royden et al. (1982) pointed out that the en echelon arrangement of fault segments within the Vienna basin suggests a strike-slip component to the basin extension, and suggested that many of the faults within the basin may have both a strike-slip and a normal component of displacement. In addition, they postulate a pair of major left-lateral, strike-slip fault zones that bound the northwest and southeast margins of the basin. The authors suggest that the northern fault zone may correspond to a sinistral strike-slip fault within the flysch zone of Moravia as reported by Krs and Roth (1979) and Roth (1980). They suggest that the southern fault zone may correspond to a seismically active fault that strikes southwest out of the Vienna basin into the Central Alpine zone (Gutdeutsch and Aric, 1976, and this volume).

According to Royden et al. (1983a), the extension of the Vienna basin can be related to continued overthrusting and shortening in the Carpathian belt northeast of the Vienna basin, particularly in Badenian time, while the Alpine thrust sheets west of the Vienna basin had already become fixed with respect to stable Europe. According to these authors, extension within the Vienna basin was mainly restricted to the Alpine–Carpathian thrust complex. Active and inactive parts of the thrust belt thus move relative to one another along large strike-slip faults, which flatten at depth into the base of the thrust sheets.

The sedimentation, subsidence, and tectonic history of the Vienna basin as reported in this chapter support the interpretation of the Vienna basin as a pull-apart feature, with extension occurring first in the northern part of the basin, and later in Karpatian and especially in Badenian time in the whole basin. Along the southwestern continuation of the Steinberg fault and along the Leopoldsdorf and Aderklaa–Bockfliess fault systems, a striking number of classic en echelon features can be observed (Figures 3, 9, and 10). The faults and structures form a sigmoidal bend, which, according to Mann et al. (1983) suggests extension by a pull apart mechanism between two sinistral faults. The maximum curvature occurs in the area east of Vienna (Figures 3 and 9). Some deep oval depocenters with great thicknesses of sedimentary fill may also be rhombohedral pull-apart features of the basin (Figure 9).

I would like to suggest a modification to the interpretation of Royden et al. (1983a) as regards the tectonic level to which extension occurs. The data presented here suggest that extension is not restricted to the allochthonous thrust sheets, but also extends into the authochthonous basement. As mentioned above, the authochthonous basement is characterized by its tendency to subsidence during the Jurassic and most probably also during alpine thrusting and later. The Jurassic synsedimentary faults have a strike that is roughly subparallel to the large faults known in the Vienna basin, such as the Steinberg fault. In addition, the area of maximum thickness of autochthonous Malm in basin facies corresponds with the location of the Vienna basin (compare isopach map of Jurassic shown by Spicka, 1976, that may be completed by new drilling data in Austria and Czechoslovakia). The behavior of the alpine, especially the limestone alpine, thrust units within the Vienna basin suggests a thrusting during subsidence of the basement by a more steep to overturned arrangement of structural elements and by their bending near the border of the Vienna basin. Many faults of the Vienna basin do end above or within the plane of detachment of the Alpine–Carpathian thrust sheets but there are some important faults, such as the Steinberg fault and the Leopoldsdorf fault, which most probably extend down into the autochthonous basement and displace it (Figure 6, sections 1 and 4). Although the Steinberg fault has been penetrated by wells Zistendorf ÜT1 and 2A at about 4900 m depth, no significant flattening of the fault was observed.

In summary, I conclude the following:

1. The Vienna basin is a pull-apart feature.
2. There is a causal connection between the extension of the basin, the tectonic behavior of the allochthonous stack of nappes and the subsidence of autochthonous basement.
3. Within the nappes, strike-slip faults that would cause pull-apart motion may be replaced by bending of units. The mechanism that causes the subsidence of the allochthonous basement is not yet known. Possible strike-slip faults within the autochthonous basin may be covered by the allochthonous mass.
4. There are signs that the Vienna basin formed during several extensional events. For example, the series of grabens along the eastern side of the basin remained tectonically active much longer than other structures within basement. Eastward, toward the Pannonian basin, a quieter style of deformation can be observed, and is represented only by young eastward tilting without significant faulting.

ACKNOWLEDGMENTS

I would like to thank Professor dr. A. Kröll for appointing me to work on deep exploration in the Vienna basin and for cooperation in many studies. In addition, I express many thanks to all the ÖMV geologists, paleontologists, and geophysicists who contributed a lot of material to this recent picture of the Vienna basin, and to the coworkers who were engaged in drawing and typing this chapter. I am obliged to Dr. L. Royden and Dr. F. Horváth for their invitation to contribute to their scientific documentation in this volume and especially to Dr. L. Royden for the linguistic correction and to her advice for completing this presentation. Finally, I wish to acknowledge the ÖMV Aktiengesellschaft for permitting publication of this chapter.

REFERENCES

Adamek, J., 1986, Geologicke poznatky o stavbé Mezozoika v úseku jih jihovýchodnich svahů Ceskeho masivu. (Die geol. Erkenntnisse über den Bau des Mesozoikums im Süden der südöstl. Abhänge der Böhmischen Masse). Zemny plyn a Nafta, Hodonin. v. 31, n. 4, p. 453–484.

Boroviczeny, F., and F. Brix, 1981, Die Hydrologie auf Blatt Wiener Neustadt: Arbeitstagung der Geologischen Bundesanstalt, Blatt 76, Wiener Neustadt in Lindabrunn. Geologische Bundasanstalt Wien.

Brix, F., 1981, Der tertiäre und quartäre Anteil auf Blatt 76, Wiener Neustadt: Arbeitstagung der Geologischen Bundasanstalt, Blatt 76, Wiener Neustadt in Lindabrunn, Geologische Bundasanstalt Wien.

Brix, F., A. Kröll, and G. Wessely, 1977, Die Molassezone und deren Untergrund in Niederösterreich: Erdöl-Erdgas-Zeitschrift, 93. Jg., ÖGEW Sonderausgabe 1977, p. 12–35, Hamburg-Wien.

Buday, T., and J. Cicha, 1968, The Vienna basin, in Regional Geology of Czechoslovakia, Part II, The West Carpathians: Geological survey of Czechoslovakia in Akademia, Praha.

Burchfiel, B. C. and L. Royden, 1982, Carpathian foreland fold and thrust belt and its relation to Pannonian and other basins: The American Association of Petroleum Geologists Bulletin, v. 66, n. 9.

Dudek, A., 1980, The crystalline block of the outer Carpathians in Moravia, Brno-Vistulicum: Rozpravy Cechoslovenske Akademie ved, Rada mat. a prirodnich ved, Praha.

Elias, M., 1979, Facies and paleogeography of the Jurassic of the Bohemian Massif: Sbornik geologickych ved Geologie, 35, Praha.

Friedl, K., 1937, Der Steinberg-Dom bei Zistersdorf und sein Ölfeld: Mitteilungen der Geologischen Gesellschaft Wien, 29, p. 21–290.

Fuchs, W., 1980, Das Inneralpine Tertiär in: Der Geologische Aufbau Österreichs: Geologische Bundesanstalt, Springer Verlag Wien-New York.

Fuchs, R., W. Grün, A. Papp, O. Schreiber und H. Stradner, 1980, Vorkommen von Egerien in Niederösterreich: Verhandlungen der Geologischen Bundesanstalt, Jg. 1979, H. 3, Wien.

Grill, R., 1941, Stratigraphische Untersuchungen mit Hilfe von Mikrofaunen im Wiener Becken und den benachbarten Molasseanteilen: Öl und Kohle, 37, nr. 31, p. 595–602, Berlin.

Grill, R., 1943, Über mikropaläontologische Gliederungsmöglichkeiten im Miozän des Wiener Beckens: Mitt. Reichsstelle f. Bodenf. 7, Wien.

Grill, R., 1968, Erläuterungen zur geologischen Karte des nordöstlichen Weinviertels und zu Blatt Gänserndorf: Geologische Bundesanstalt Wien.

Grün, W., 1984, Die Erschliessung von Lagerstätten im Untergrund der alpin-karpatischen Stirnzone Niederösterreichs, Erdöl-Erdgas-Zeitschrift, Jg. 100, H9, Wien–Hamburg.

Gutdeutsch, R., and K. Aric, 1976, Eine Diskussion geophysikalischer Modelle des Grenzbereiches zwischen Ostalpen und Pannonischen Becken: Acta Geol. Acad. Sci. Hungary, 21, p. 287–309, Budapest.

Horváth, F., and L. Royden, 1981, Mechanism for the formation of the intra-Carpathian basin: A review: Earth evolution sciences, v. 1, n. 3–4, p. 307–317.

Janoschek, R., 1951, Das Inneralpine Wiener Becken in: Geologie von Österreich (Schaffer, F. X., Hrsg.) 2. Auflage, Wien Deuticke.

Janoschek, R., 1964, Das Tertiär in Österreich: Mitt. Geol. Ges., Wien 65 (1963).

Janoschek, W. R., and A. Matura, 1980, Outline of the Geology of Austria: Abh. Geol. BA, 26e C.G.I., 34, p. 7–98, Wien.

Jiricek, R., 1978, Paleogeografie spodniho miocenu v zapadnich Karpatech (Paleogeography of the lower Miocene in the West-carpathians): Zemny plyn a nafta, 23, cislo 1, p. 31–38, Hodonin.

Jiricek, R., 1981, Vyvoj a Stavba podlozi Videnske Panve (Evolution and structure of the base of the Vienna basin): Zemny plyn a nafta, 26, cislo 3, Hodonin.

Jiricek, R., and C. Tomek, 1981, Sedimentary and structural evolution of the Vienna basin: Earth evolution sciences, v. 1, n. 3–4, p. 195–204.

Kapounek, J., A. Kröll, A. Papp, and K. Turnovsky, 1965, Die Verbreitung von Oligozän, Unter- und Mittelmiozän in Niederösterreich: Erdöl-Erdgas-Zeitschrift, 81, p. 109–116, Wien-Hamburg.

Kapounek, J., and A. Papp, 1969, Der Vulkanismus in der Bohrung Orth und die Verbreitung von Grobschüttungen zwischen dem Spannberger Rücken und der Donau: Verh. Geol. BA, H 2, p. 114–124, Wien.

Kratochvil, H. und W. Ladwein, 1984, Muttergesteine der Kohlenwasserstofflagerstätten im Wiener Becken und ihre Bedeutung für die zukünftige Exploration: Erdöl-Erdgas-Zeitschrift, Jg. 100, H 3, Wien–Hamburg.

Kreutzer, N., 1971, Mächtigkeitsuntersuchungen im Neogen des Ölfeldes Matzen, NÖ: Erdöl-Erdgas-Zeitschrift 87, H 2, Wien–Hamburg.

Kreutzer, N., 1984, Die Produktionsgeologie in der ÖMV AG: Erdöl-Erdgas-Zeitschrift, Jg. 100, H 10, Wien–Hamburg.

Kreutzer, N., 1986, Die Ablageringssequenzen der miozänen Badener Serie im Feld Matzen und im zentralen Wiener Becken: Erdöl, Erdgas, Kohle, v. 102, n. 11, p. 492–504.

Kröll, A., 1980, Das Wiener Becken in: Erdöl und Erdgas in Österreich. Herausgeber: F. Bachmayer, Verlag: Naturhistorisches Museum Wien und F. Berger, Horn.

Kröll, A., and G. Wessely, 1973, Neue Ergebnisse beim Tiefenaufschluss im Wiener Becken: Erdöl-Erdgas-Zeitschrift, Bd. 89, H 11, p. 400–413, Hamburg–Wien.

Kröll, A., K. Schimunek, and G. Wessely, 1981, Ergebnisse und Erfahrungen bei der Exploration in der Kalkalpenzone in Ostösterreich: Erdöl-Erdgas-Zeitschrift, 97. Jg., H 4, Hamburg–Wien.

Krs, M., and Z. Roth, 1979, The insubric Carpathian Tertiary bloc system: Its origin and disintegration: Geol. Zborn. Geologica Carpathica, 30, v. 1, p. 3–17, Bratislava.

Küpper, H., 1953, Uroberfläche und jüngste Tektonik im südlichen Wiener Becken: Kober Festschrift 1953, Skizzen zum Antlitz der Erde, Wien.

Ladwein, W., 1976, Sedimentologische Untersuchungen an Karbonatgesteinen des autochthonen Malm in NÖ (Raum Altenmarkt-Staatz): Diss. Phil. Fak. Universität Innsbruck.

Mahel, M., 1982, Alpine structural elements: Carpathians–Balkan–Caucasus–Pamir Orogenese Zone: VEDA, Publishing house of the Slovak Academy of Science, Bratislava.

Mann, P., M. R. Hempton, D. C., Bradley, and K. Burke, 1983, Development of pull-apart basins: Journal of Geology, v. 91, p. 529–554.

Nemec, F. and A. Kocák, 1976, Predneogení podlozí slovenské casti vídenské pánve (The pre-Neogene basement of the Slovakian part of the Vienna basin): Mineralia Slovaca, v. 8, n. 6, p. 481–560.

Nemec, F. and A. Kocák, 1978, Nove vysledky pruzkumu vnitrokarpatsheho podlozí slovenské cásti vídenske pánve (New results of investigation of the inner-carpathian base in the Slovakian part of the Vienna basin): Zemny plyn a nafta, 23, Cislo 1, p. 39–71.

Papp, A., 1963, Die biostratigraphische Gliederung des Neogen im Wiener Becken: Mitt. d. Geol. Ges. Wien, 56, Bd., 1963, H 1, p. 225–317.

Papp, A., W. Krobot and K. Hladecek, 1974, Zur Gliederung des Neogen im Zentralen Wiener Becken: Mitt. Ges. Geol. Bergbaustud. 22 (1973).

Rögl, F., 1980, Biostratigraphie und Paläogeographie in: Erdöl und Erdgas in Österreich. Herausgeber: F. Bachmayer, Verlag: Naturhistorisches Museum Wien und F. Berger, Horn, Wien.

Roth, Z., 1980, Západni Karpaty-Tercíerní struktura Stredni Evropy [West Carpathians–Tertiary structure of central Europe]: Ustred, Ustav Geol., 55, 128 pages, Praha.

Royden, L., F. Horváth, and B. C. Burchfiel, 1982, Transform faulting, extension and subduction in the Carpathian Pannonian region: Geol. Soc. Am. Bull. 73, p. 717–725.

Royden, L., F. Horváth, and J. Rumpler, 1983a, Evolution of the Pannonian Basin System; 1. Tectonics: Tectonics, v. 2, n. 1, p. 63–90.

Royden, L., F. Horváth, A. Nagymarosy, and L. Stegena, 1983b, Evolution of the Pannonian basin system, 2. Subsidence and thermal history: Tectonics, v. 2, n. 1, p. 91–137.

Spicka, V., 1976, Hlubinná geologická stavba autochtonu na jizní Morave a jeho perspektivnost pro ropu a plyn (Geologischer Tiefbau des Autochthon in Südmähren und seine Erdöl- und Erdgashöffigkeit): Sbor. geol. Ved., Geologie, 28, p. 7–113, Praha.

Steininger, F., A. Papp, J. Cicha, J. Senes, and D. Vass, 1975, Excursion A, marine Neogene in Austria and Czechoslovakia, VI th Congress of the Regional Committee on Mediterranean Neogene stratigraphy Bratislava, Czechoslovakia, 96 pages.

Stowasser, H., 1958, Einige Bausteine Zur Tektogenese des Wiener Beckens: Erdöl-Erdgas-Zeitschrift, p. 395–400, H 12, Jg. 74, Wien-Hamburg.

Stowasser, H., 1966, Strukturbildung am Steinbergbruch im Wiener Becken: Erdöl-Erdgas-Zeitschrift, p. 188–191, Jg. 82, H 5, Wien-Hamburg.

Tollmann, A., 1978, Plattentektonische Fragen in den Ostalpen und der plattentektonische Mechanismus des mediterranen Orogens: Mitt. Österr. Geol. Ges. 69/1976 c, p. 291–351.

Wachtel, G., and G. Wessely, 1981, Die Tiefbohrung Berndorf 1 in den östlichen Kalkalpen und ihr geologischer Rahmen: Mitt. Österr. Geol. Ges. Wien, 1974/75, p. 137–165.

Wessely, G., 1961, Zur Geologie der Hainburger Berge: Jb. Geol. B. A. 104, Wien.

Wessely, G., 1975, Rand und Untergrund des Wiener Beckens-Verbindungen und Vergleiche: Mitt. Geol. Ges. Wien, 66–67 (1973-1974), p. 265–287.

Wessely, G., 1983, Zur Geologie und Hydrodynamik im südlichen Wiener Becken und seiner Randzone: Mitt. Österr. Geol. Ges., Wien.

Wessely, G., 1984, Der Aufschluss auf kalkalpine und subalpine Tiefenstrukturen im Untergrund des Wiener Beckens: Erdöl-Erdgas-Zeitschrift, Jg. 100, H 9, Hamburg-Wien.

Wieseneder, H., G. Freilinger, G. Vetters, und G. Tsambourakis, 1976, Der kristalline Untergrund der Nordalpen in Österreich: Geol. Rundschau 65, p. 512–525, Stuttgart.

CHAPTER 25

Maturation and Migration of Hydrocarbons in the Southeastern Pannonian Basin

Á. Szalay
Petroleum Exploration Company,
H-5001 Szolnok, Munkásör út 43, Hungary

Two potential source rock zones occur within the Neogene sedimentary fill of the Pannonian basin. These units may be up to one kilometer thick, but the average organic carbon content is relatively low (0.5–1.5%). The upper zone (in the upper part of the lower Pannonian, *s. l.*) has just entered the oil generation window, but the lower zone (in middle Miocene to lower Pannonian rocks) passed into the oil generation window at around 8 Ma, and in the deepest parts of the basin it is already overmature. This second zone is significantly overpressured, and is estimated to have become overpressured initially at about 8 Ma. Combining maturation histories with coeval estimated pore pressures in the Makó–Hódmezővásárhely trough indicate that within this deeper zone of potential source rocks: (1) migration out of the upper part of the source rock zone was upwards and laterally, while (2) migration out of the middle and lower parts of this zone was probably downwards out of the source rocks and then laterally within more permeable rocks below.

INTRODUCTION

Determination of the subsidence history of the Pannonian basin and the compaction of the sedimentary fill makes it possible to calculate paleo-pore pressures. By combining hydrodynamic conditions that prevailed in the past with the location and maturation history of the source rocks, one may reconstruct the probable timing and direction of migration of hydrocarbons.

In this chapter we wish to answer the following three questions about generation of hydrocarbons within the Pannonian basin:

1. Where are the potential source rocks within the southeastern Great Hungarian Plain?
2. What were the pressure conditions when the potential source rocks passed into and through the oil generation window?
3. Which areas are favorable for hydrocarbon prospecting?

To answer these questions, we examine four subbasin areas within the southeastern Great Hungarian Plain: (1) Makó–Hódmezővásárhely, (2) Nagykunság, (3) Békés, and (4) Derecske (Figure 1).

ORGANIC CONTENT AND MATURITY OF NEOGENE SEDIMENTS

The Neogene sedimentary fill of the Pannonian basin consists mainly of pelites and psammites, the distribution of which is highly variable throughout the basin. The trend of the lithological development, however, can be used to define lithogenetic units which approximate time units fairly well (Horváth and Pogácsás, this volume). These lithogenetic units consist of Badenian and Sarmatian strata ($M_{4,5}$), lower Pannonian strata (Pa_1^{1a}, Pa_1^{1b}, Pa_1^2, from bottom to top) and upper Pannonian strata (Pa_2). For a description and discussion of lithogenetic units see Szalay and Szentgyörgyi (1979 and this volume).

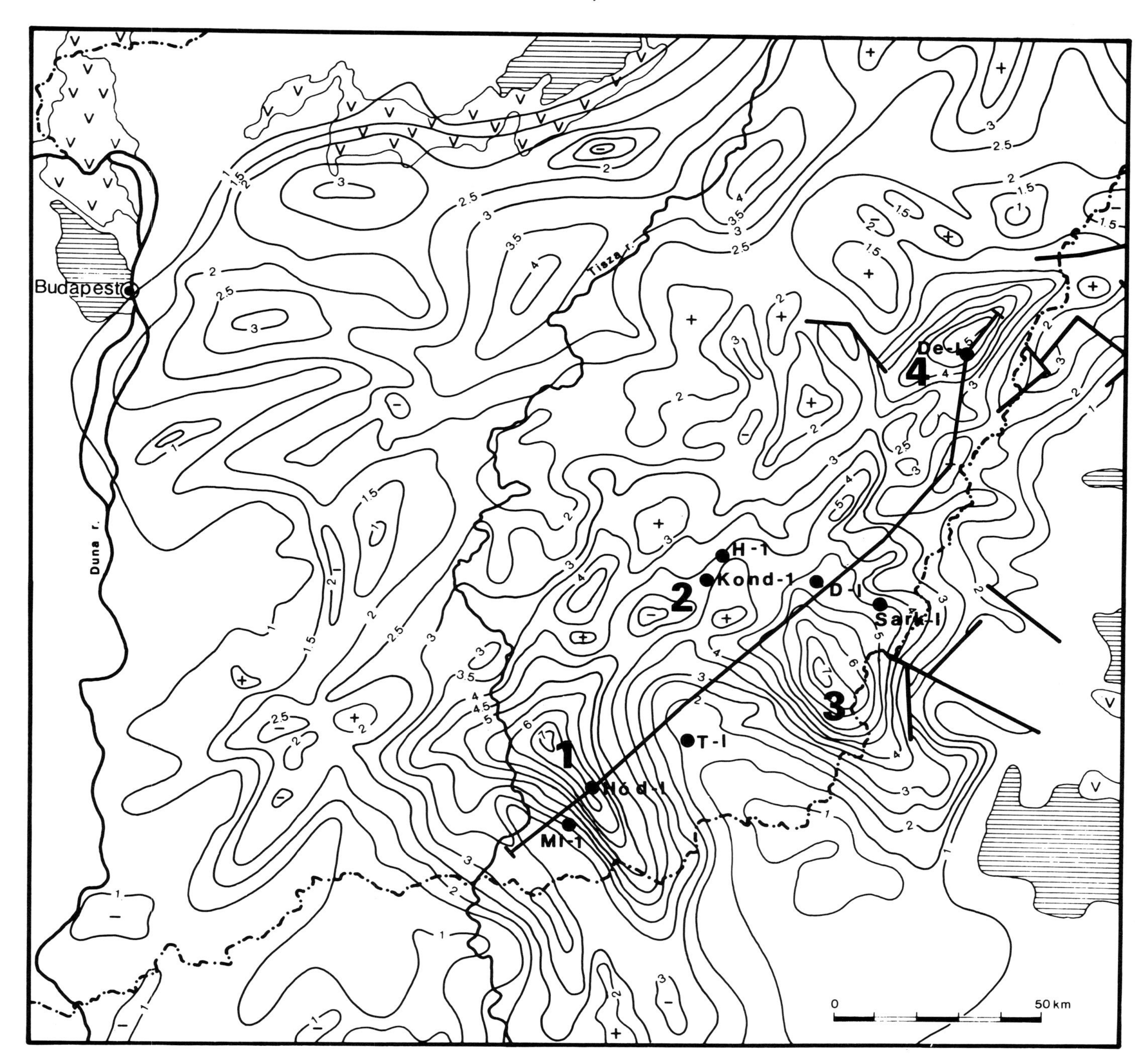

Figure 1. Isopach map of the Neogene–Quaternary sediments (in km) in the southeastern part of the Pannonian basin. Big numbers indicate the location of the Makó–Hódmezővásárhely (1), Nagykunság (2), Békés (3) and Derecske (4) subbasins. The position of the geological section in Figure 2 and boreholes are also shown.

Table 1 gives the pelite to psammite ratio and the average organic carbon content for lithogenetic units in the four subbasins. Where the lithogenetic units have a large pelite to psammite ratio (>5) the organic carbon content is relatively high. This suggests that they are potential source rocks, therefore they should be subject to further geochemical investigation.

The type of organic matter present in the Great Hungarian Plain is not well known because of insufficient analyses. The middle Miocene sediments ($M_{4,5}$) contain organic matter mainly of marine origin, while the Pannonian (*s. l.*) sediments contain organic matter mainly of freshwater origin (Holczhacker et al., 1981).

Table 2 shows the highest values of vitrinite reflectance (R_o) that were measured or calculated in the pelitic lithogenetic units. We defined the different zones of hydrocarbon generation as follows: oil generation as $0.6\% \leq R_o \leq 1.3\%$, wet gas generation as $1.3\% \leq R_o \leq 2.0\%$, and dry gas generation as $R_o > 2\%$. The most favorable conditions for hydrocarbon generation

Table 1. Pelite to Psammite Ratio and Average Organic Carbon Content (‰)

	Lithogenetic Unit									
	$M_{4,5}$		Pa_1^{1a}		Pa_1^{1b}		Pa_1^2		Pa_2	
Depression	p/ps	C_{org}	p/ps	C_{org}	p/ps	C_{org}	p/ps	C_{org}	p/ps	C_{org}
Makó–Hódmezővásárhely	>5.0	8.9	>5.0	5.9	<1.0	5.3	1.0–5.0	5.0	<1.0	3.4
Nagykunság	?	?	>5.0	5.8	<1.0	4.5	>5.0	5.5	<1.0	3.2
Békés	>5.0	8.7	>5.0	5.7	1.0–5.0	4.5	>5.0	4.8	<1.0	3.4
Derecske	>5.0	10.1	>5.0	8.6	1.0–5.0	8.0	>5.0	8.6	<1.0	4.1

Table 2. Highest Values of Virtrinite Reflectance (R_o in %) Measured in the Pelitic Lithogenetic Units

Depression	$M_{4,5}$	Pa_1^{1a}	Pa_1^2
Mako-Hódmezővásarhely	>2.0	1.8	0.6
Nagykunság	1.4	1.0	0.6
Békés	>2.0	1.3	1.0
Derecske	>2.0	>2.0	0.8

exist in the Derecske depression where the unit $M_{4,5}$ (Badenian, Sarmatian) and part of the unit Pa_1^{1a} (lowermost Pannonian) are now within the zone of dry gas generation. Much of unit Pa_1^2 has already reached the zone of oil generation (Figure 2).

In general $M_{4,5}$ source rocks along the axis of the deep troughs are within the zone of dry gas generation, source rocks of unit Pa_1^{1a} are within the zone of wet gas generation, and some of the source rocks of unit Pa_1^2 are within the zone of oil generation (Figure 2). Figure 3 shows the extent, thickness, and geochemical data of the lithogenetic units $M_{4,5}$ and Pa_1^{1a} together. The best prospects for hydrocarbon generation are given by the thick pelitic sequences along the axes of the depression because they have already passed through the main phases of oil generation.

Geological and geochemical data of the younger potential source rocks of unit Pa_1^2 are shown in Figure 4. Three main facies were distinguished within this lithogenetic unit: (1) pelite/psammite ratio greater than 5, (2) pelite/psammite ratio between 1 and 5, and (3) pelite/psammite ratio less than 1. Figure 4 shows that the unit Pa_1^2 has nowhere passed out of the zone of oil generation. The most favorable area of hydrocarbon generation appears to be the eastern part of the Nagykunság and the Derecske subbasins where the pelite/psammite ratio of Pa_1^2 is greater than 5, the organic carbon content is high, and the degree of maturation is adequate.

TIMING OF HYDROCARBON MATURATION AND FORMATION OF PRESSURE SEALS

The reconstruction of subsidence, compaction and thermal history of the sediments allows us to estimate the time of hydrocarbon generation and the coeval pressure conditions (Horváth et al., this volume).

In a pile of sedimentary rocks, an increase in overburden due to progressive burial during sedimentation reduces the porosity of the rocks by compaction. Chemical changes (e.g., precipitation or mineral transformation) may also occur but were not taken into consideration in this study. When fluid flow is not restricted during compaction, the pore pressure should be hydrostatic and the porosity should decrease more or less exponentially with depth. If fluid flow is restricted, the pore pressure becomes greater than the hydrostatic pressure. Such zones of anomalously high pore pressure (overpressured zones) are also zones of abnormally high porosity. Our analysis of the compaction history is based upon observed porosity versus depth relationships (Figures 5 and 6). Porosities were determined from continuous well logs for each borehole. The normal compaction curve was defined by plotting the data on a semilog plot, and calculating the best-fitting straight line for porosity data from normally compacted beds.

By combining the burial depth of the potential source rocks through time with the subsidence and thermal history of the basin, the timing of hydrocarbon generation can be determined. This varies in the individual troughs because of the combined effect of different burial histories and paleogeothermal conditions. Table 3 summarizes the times when the potential source rock units $M_{4,5}$ + Pa_1^{1a} and Pa_1^2 in each subbasin passed into the different hydrocarbon generation zones. There is significant variation in the timing of hydrocarbon generation in the different subbasins.

Paleo-pore pressure can be estimated from the subsidence history and from porosity versus depth data. Porosity versus depth was calculated for pelitic rocks at different times in the past. We have assumed that: (1) the normal porosity-depth relationship observed at present has remained unchanged through time; (2) anomalously high porosities formed as the rocks first followed the normal compaction curve and later became isolated and subsided without further compaction (Plumpley, 1980); and (3) pressure increase below this isolation depth is caused by an increase in the lithostatic load.

Figure 7 shows the calculated pore pressure versus depth history for well Hód-I. Shown are the paleo-pore pressures at the beginning of deposition of units Q, Pa_2, Pa_1^2, Pa_1^{1b}, Pa_1^{1a}, and $M_{4,5}$. In this well, pelitic rocks are currently normally compacted from the surface down to 2.8 km depth. Below 2.8 km, the pelites are overpressured and have porosities in excess of the normal porosity depth values. Figure 7 also shows the position of the present pressure seals and their development through time. The present pressure seals (the first between 4.9 and 5.1 km depth, the second between 4.1 and 4.2 km depth, and the third between 3.9 and 4.0 km depth) already existed at 2.4 Ma. However, 8 million years ago only the lower one was developed at a depth of 3.2 km. Ten million years ago the pressure of the pore fluid was hydrostatic all along the section.

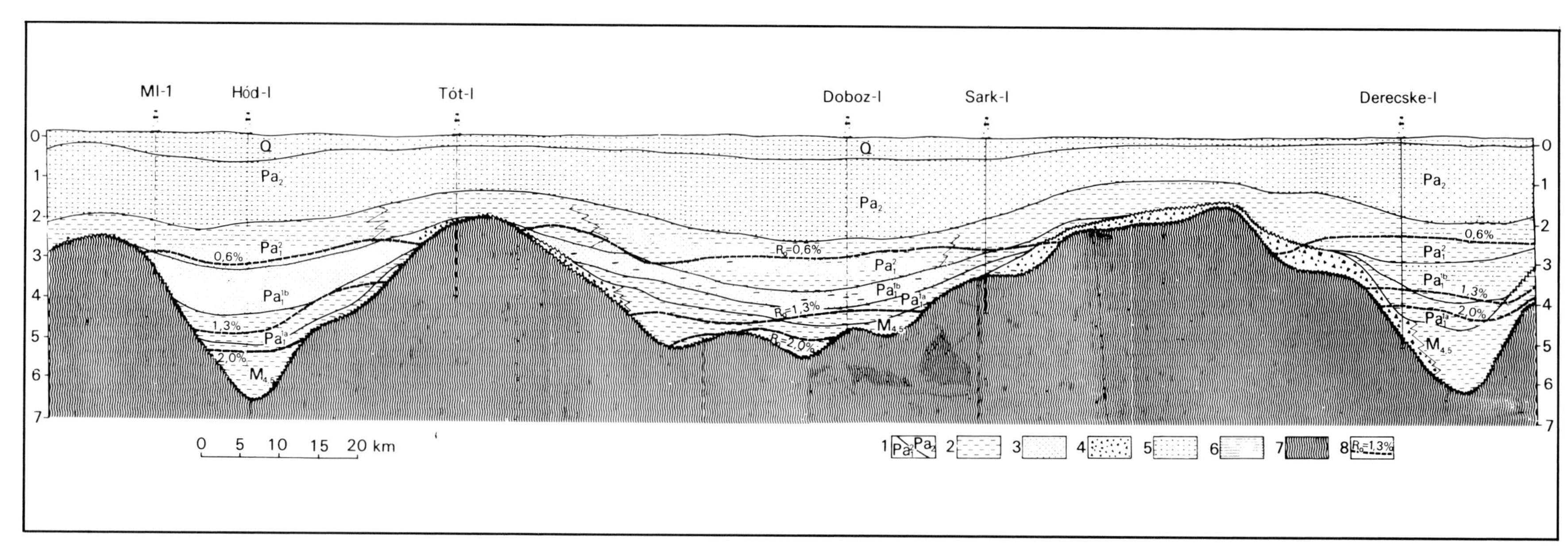

Figure 2. Geological section in southeastern Hungary to show the lithogenetic units and the maturity conditions. 1 = lithogenetic units and their boundaries determined by seismic correlation between boreholes; 2 = predominantly pelitic rocks; 3 = predominantly psammitic rocks; 4 = conglomerates and breccias; 5 = fluvio-lacustrine development made up from thin and alternating pelitic and psammitic beds; 6 = water; 7 = Mesozoic and/or Paleozoic basement; and 8 = lines of constant vitrinite reflectance determined by modeling of the temperature history (Horváth et al., this volume) and by measurements in Hód-I and Derecske-I wells.

Table 3. Time (Ma) When Potential Source Rock Units Passed Into Hydrocarbon Generation Zones

	Oil Generation Window $0.6\% \leq R_o \leq 1.3\%$		Wet Gas Zone $1.3\% \leq R_o \leq 2.0\%$		Dry Gas Zone $2\% < R_o$	
Drill hole	$M_{4,5} + Pa_1^{1a}$	Pa_1^2	$M_{4,5} + Pa_1^{1a}$	Pa_1^2	$M_{4,5} + Pa_1^{1a}$	Pa_1^2
Hód-I	9.5	—	7.1	—	5.5	—
Derecske-I	8.6	3.9	6.0	—	4.1	—
Hunya-1	6.9	2.6	0.3	—	—	—
Kondoros-1	6.5	2.7	—	—	—	—

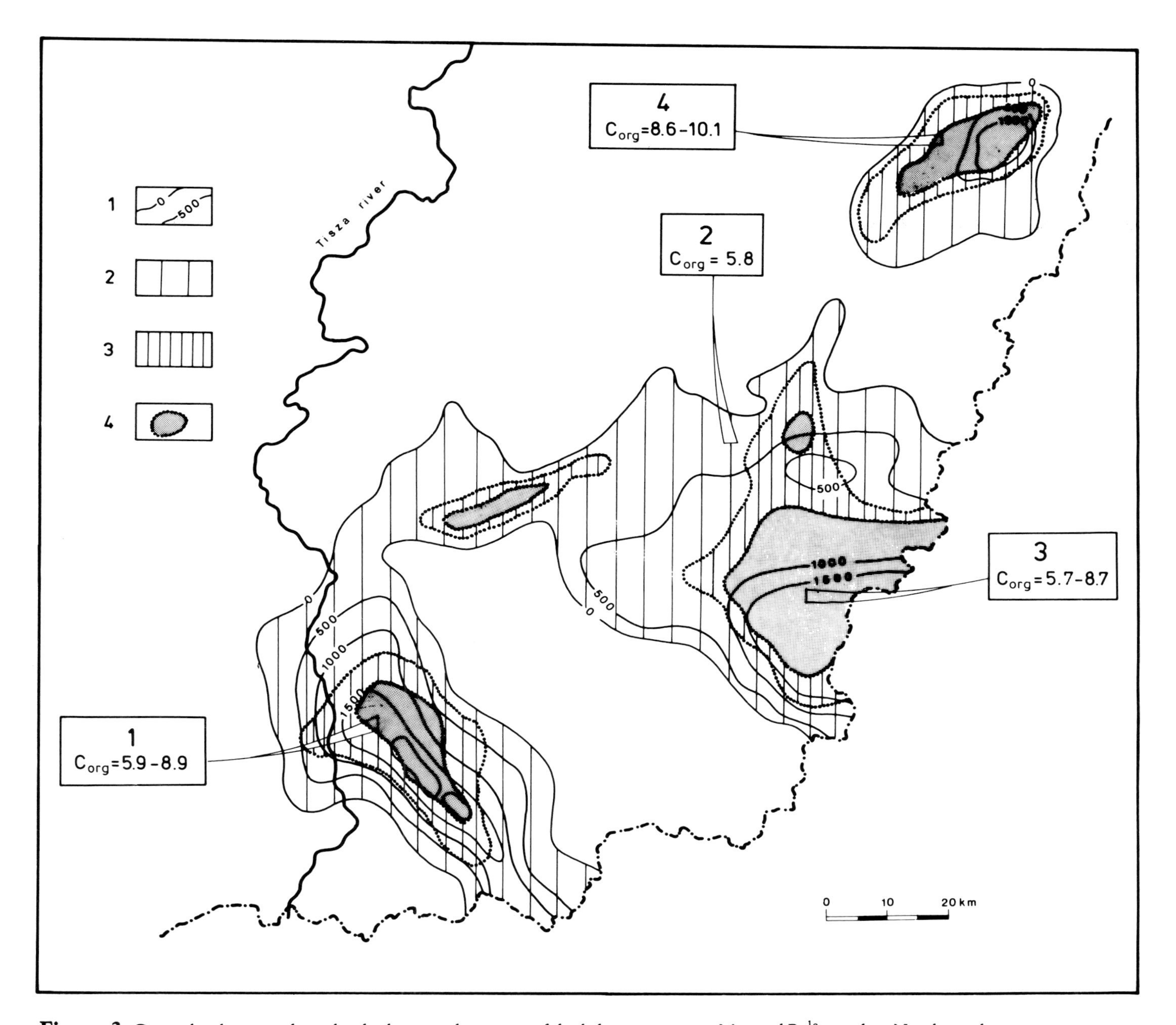

Figure 3. Generalized map to show the thickness and maturity of the lithogenetic units $M_{4,5}$ and Pa_1^{1a} together. Number in box gives an average value of organic carbon content of the two units (in ‰) for the four subbasins considered. 1 = thickness isoline of the $M_{4,5} + Pa_1^{1a}$ sediments, in meters; 2 = extent of rocks characterized by vitrinite reflectances of $0.6\% \leq R_o \leq 1.3\%$; 3 = extent of rocks with $1.3\% \leq R_o \leq 2\%$; and 4 = extent of rocks with $R_o > 2\%$.

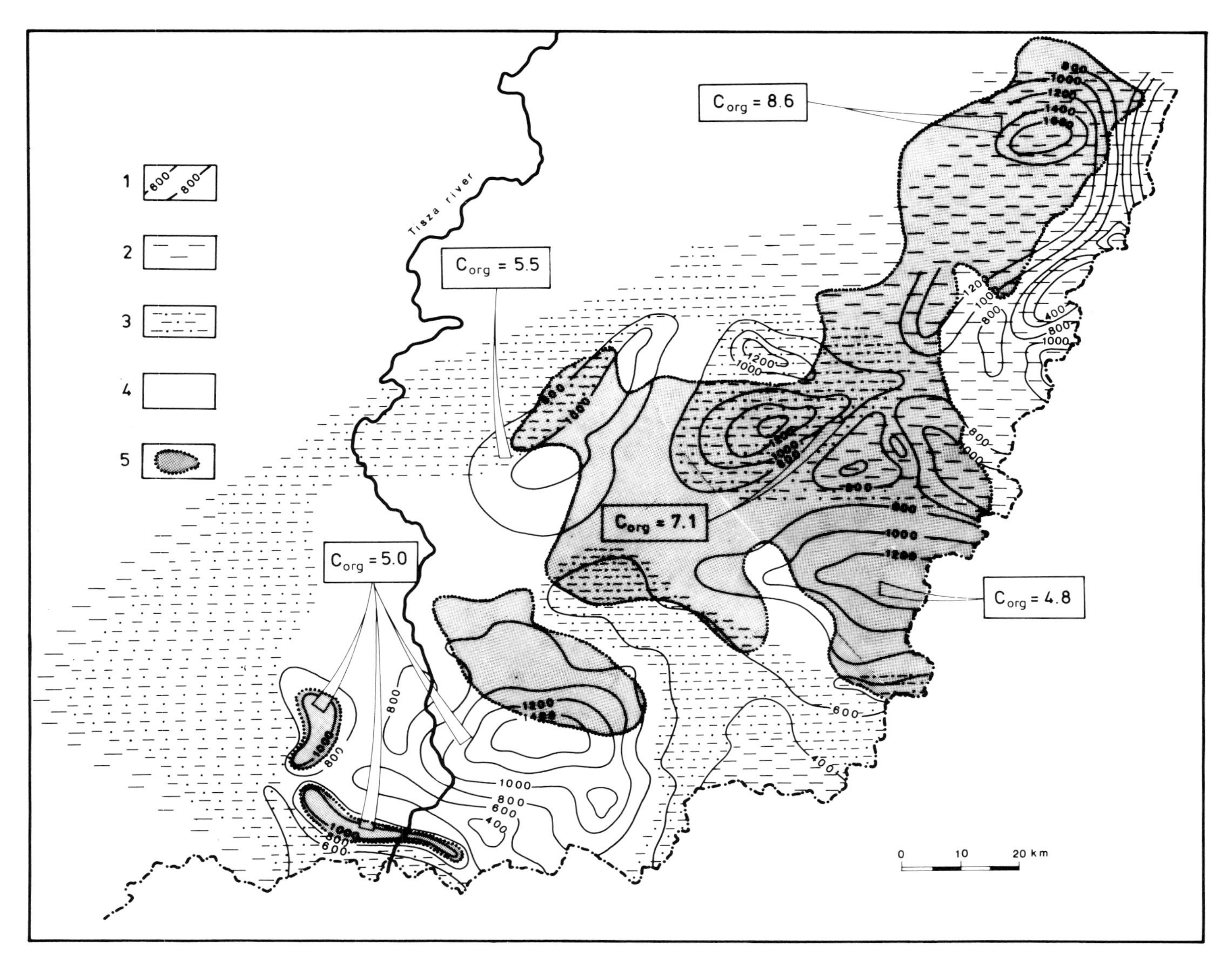

Figure 4. Generalized map to show the thickness, lithology, and maturity of the Pa_1^2 lithogenetic unit. Number in box gives an average value of organic carbon content of the unit (in ‰) for the four subbasins considered. 1 = thickness isolines of the Pa_1^2 lithogenetic unit; 2 = areal extent of highly pelitic beds (p/ps > 5); 3 = areal extent of moderately pelitic beds ($1 \leq p/ps \leq 5$); 4 = areal extent of predominantly psammitic beds (p/ps < 1); and 5 = extent of rocks characterized by vitrinite reflectances of $0.6\% \leq R_o \leq 1.3\%$.

The timing of the formation of such pressure seals influences the migration of hydrocarbons (Magara, 1978). If the pressure seal forms before maturation of the hydrocarbons in the underlying rocks, migration can only occur downward or laterally.

TEMPORAL RELATIONSHIP OF GENETIC ZONES AND PRESSURE SEALS

Figure 8 illustrates the temporal relationships between the potential source rocks and the lower pressure seal. It also shows the migration possibilities of two source rock units ($M_{4,5}$, Pa_1^{1a}) for well Hód-I.

The lowermost part of unit M_4 passed into the oil generation window at 9.5 Ma. The uppermost part of this unit passed out of oil generation window at 2.3 Ma. According to our basin reconstruction, however, a pressure seal currently at about 5 km depth was formed at a depth of 2.7 km at about 9.3 Ma. This pressure seal hampered upward migration during the time of oil generation in $M_{4,5}$ source rocks. Accordingly, migration could occur only horizontally and downwards. Horizontal migration in thick pelitic formation is, however, less efficient (Magara, 1976). Downward migration followed by lateral migration was probably promoted by the permeable zones at the unconformity along the basement or at the top of the $M_{4,5}$ unit. The migration from the thick, pelitic sequences down toward the basement may be favorable for hydrocarbon accumulation in deeper parts of depressions.

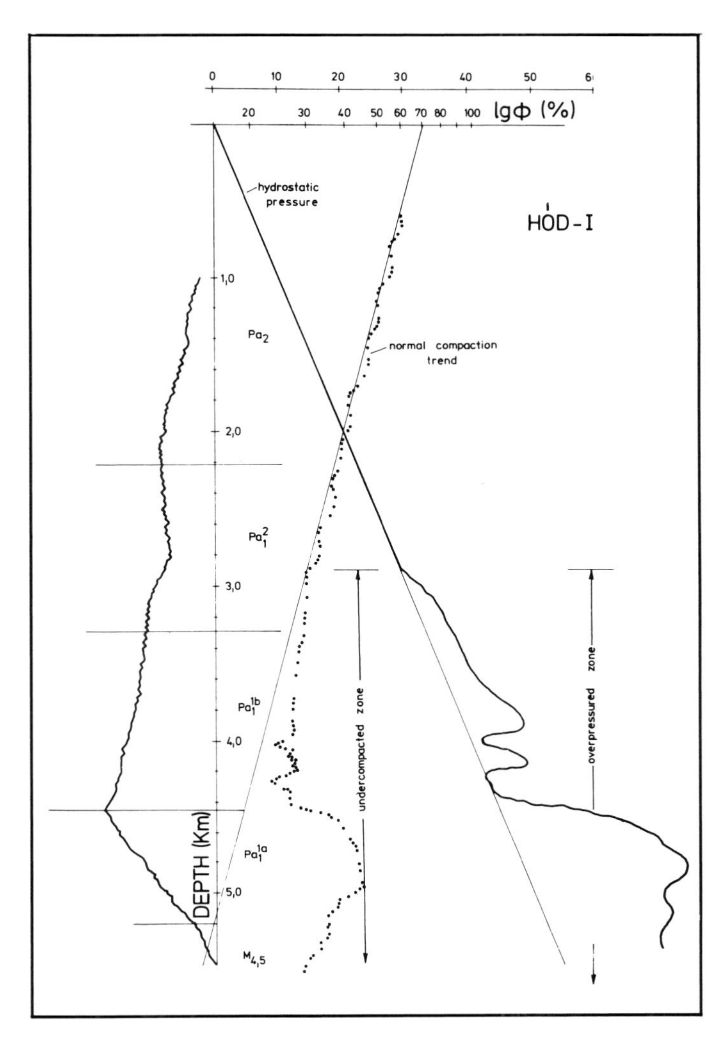

Figure 5. Lithogenetic trend diagram (left), porosity (middle), and pore pressure (right) as function of depth for the Hód-I well. Note that the predominantly shaley Pa_1^{1a} and $M_{4,5}$ units are remarkably overpressured and undercompacted. Pore pressure exceeds the hydrostatic pressure by a maximum of about 30 MPa at a depth of 4900 m.

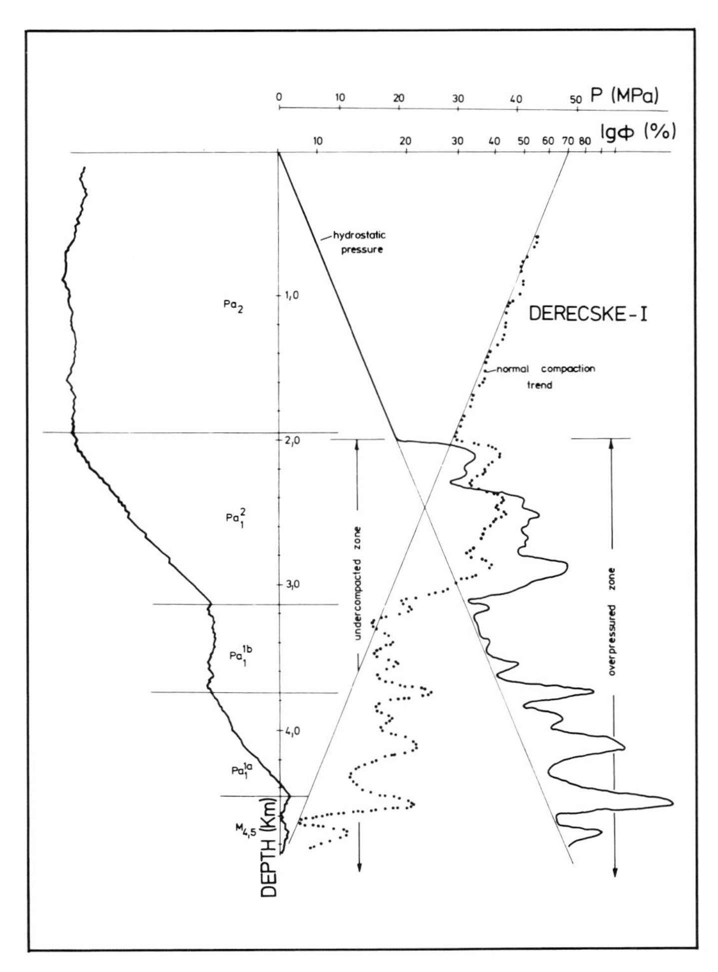

Figure 6. Lithogenetic trend diagram (left), porosity (middle), and pore pressure (right) as function of depth for the Derecske-I well. Pa_1^2 and Pa_1^{1a} units consist mainly of shale and are overpressured and undercompacted. Note that the Pa_1^{1b} and $M_{4,5}$ units contain frequent interbeds of permeable psammitic rocks and the pore pressure is close to hydrostatic.

A lesser volume of potential source rocks (the upper part of unit Pa_1^{1a}) has always remained above this pressure seal. Migration out of these rocks may have been upward into the dominantly psammitic and therefore permeable Pa_1^{1b} unit, and then laterally updip within Pa_1^{1b}. These hydrocarbons may have accumulated in structural and lithological traps adjacent to basement highs. Traps of this type contain most of the oil found to date in Hungary (Dank, this volume).

CONCLUSIONS

The dominantly pelitic units of the Great Hungarian Plain ($M_{4,5}$ + Pa_1^{1a} and Pa_1^2) are potential source rocks for hydrocarbon generation. Maturity data show that in the deep troughs, the unit $M_{4,5}$ and part of unit Pa_1^{1a} have passed through the oil generation window. Generally, unit Pa_1^2 has just reached the oil generation window. Reconstruction of subsidence and maturation histories, together with paleo-pore pressure conditions for the Hód-I well, gives an idea about the possible migration paths of the hydrocarbons. It shows that an overpressured zone was formed within the upper part of the potential source rocks (Pa_1^{1a}) at about the same time as they entered the oil generation window. Downward and lateral migration should have occurred. Accordingly, hydrocarbons could have accumulated in relatively deep traps associated with unconformities above the basement and/or above middle Miocene ($M_{4,5}$) strata. Hydrocarbons generated within unit Pa_1^{1a} and above the pressure seal should have migrated into the psammitic Pa_1^{1b} unit above. Where the unit above Pa_1^{1b} (Pa_1^2) was pelitic and therefore relatively impermeable, these hydrocarbons likely would have been trapped within Pa_1^{1b}, and sealed by Pa_1^2.

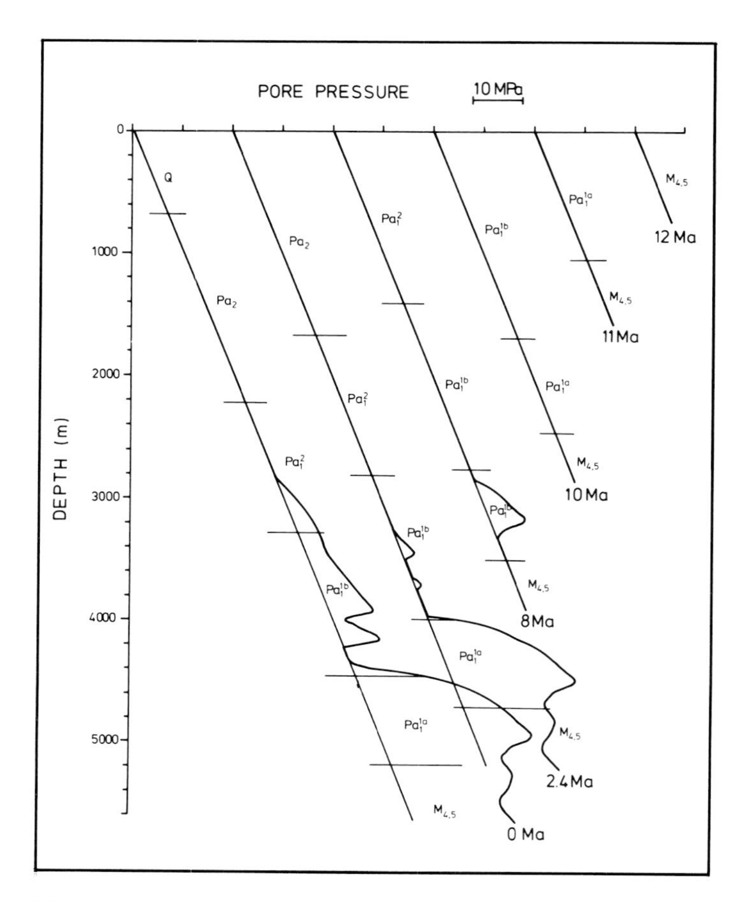

Figure 7. Pore pressure as a function of depth in the Hód-I borehole at the present (0 Ma), and its reconstruction at the beginning of Quaternary (2.4 Ma), Pa_2 (8 Ma), Pa_1^2 (10 Ma), Pa_1^{1b} (11 Ma) and Pa_1^{1a} (12 Ma). Overpressuring in the pelitic Pa_1^{1a} unit began to develop during the deposition of the Pa_1^2 unit (between 10 and 8 Ma).

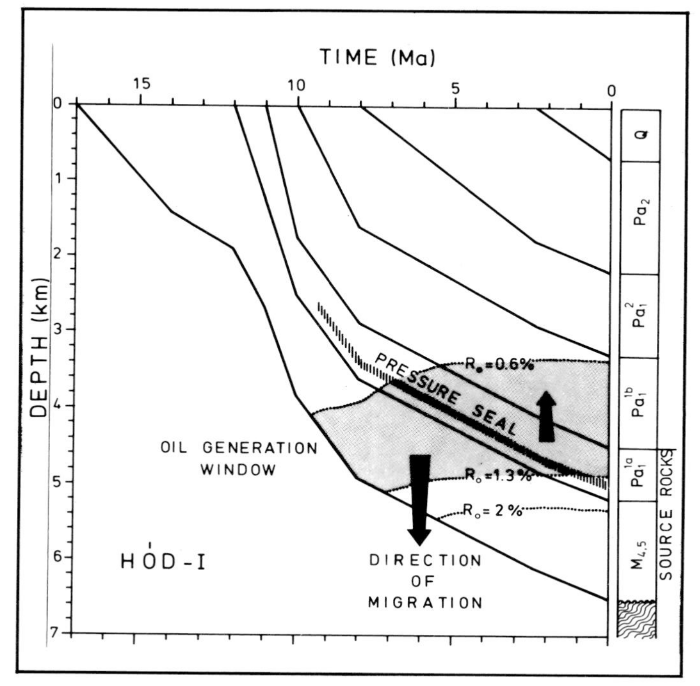

Figure 8. Subsidence and maturation history of the Hód-I well (after Horváth et al., this volume). The depth of the maximum overpressure in unit Pa_1^{1a} is shown through time. This pressure seal has controlled the direction of fluid migration. Hydrocarbons that have been generated in the upper part of unit Pa_1^{1a} should have had a tendency to move upwards. However, most of the hydrocarbons generated in the $M_{4,5}$ unit and in the lower part of Pa_1^{1a} should have been driven downwards towards the basement, which may be fractured, and thus permeable.

Analysis of other subbasins in the eastern part of the Great Hungarian Plain suggests that migration paths were similar to those described in this chapter for the Makó–Hódmezővásárhely trough. This suggests that downward migration of hydrocarbons out of deep source rocks, followed by lateral migration within more permeable rocks below, may be a common phenomenon within the Great Hungarian Plain. Therefore, future prospects for hydrocarbon accumulations at greater depths within the Pannonian basin are promising.

REFERENCES

Holczhacker, K., I. Koncz, and I. Fisch, 1981, Stabil szénizotóparány adatok felhasználási lehetőségei [Application possibilities of stable carbon isotope ratios]: Kőolaj és Földgáz, Budapest, v. 6, p. 178–187.

Magara, K., 1976, Water expulsion from clastic sediments during compaction: Direction and volumes: Amer. Assoc. Petr. Geol. Bull., v. 64, p. 414–430.

Magara, K., 1978, Compaction and fluid migration: Amsterdam/New York, Elsevier, p. 1–349.

Plumpley, W. J., 1980, Abnormally high fluid pressure: Survey of some basic principles: Amer. Assoc. Petr. Geol. Bull., v. 64, p. 414–430.

Szalay, Á. and K. Szentgyörgyi, 1979, Contribution to the knowledge of lithologic subdivisions of Pannonian basin formations explored by hydrocarbon drilling: Reconstructions based on trend analysis: MTA X. Osztály Közleményei, Budapest, v. 12, n. 4, p. 401–423.

CHAPTER 26

Subsidence, Thermal, and Maturation History of the Great Hungarian Plain

F. Horváth
P. Dövényi
Geophysical Department, Eötvös University,
H-1083 Budapest, Kun Bela tér 2, Hungary

Á. Szalay
Petroleum Exploration Company
H-5001 Szolnok, Munkásőr út 43, Hungary

L. H. Royden
Department of Earth, Atmospheric and Planetary Sciences
MIT, Cambridge, Massachusetts 02139

Recently available vitrinite reflectance data, temperature-depth profiles, and heat flow values have made it possible for us to develop an improved computer model to simulate the subsidence, thermal, and maturation history of the Pannonian basin. This model takes into consideration different rates of extension in the lower and upper lithosphere, heat generation in the crust, the thermal blanketing effect of fast sedimentation, the change in porosity and thermal conductivity in space and time, and both normal and abnormal compaction of sediments. Model parameters are constrained by comparison of predicted and observed subsidence histories, present crustal thicknesses, temperature versus depth profiles, and heat flow values. We conclude that lithospheric stretching, combined with major additional thinning of the subcrustal lithosphere, is adequate to explain the formation and evolution of the Pannonian basin.

Thermal maturation of organic matter was calculated by the use of the Lopatin method. Model results were compared with measured vitrinite reflectance (R_o) in eight master holes that range in depth between 1792 m and 5865 m. This provides an improved relationship between the time-temperature index (TTI) and vitrinite reflectance for $0.25\% \leq R_o \leq 3\%$, and for temperatures up to 230 °C. This relationship was then used to reconstruct the maturation history of potential source rocks in the basin. It shows that hydrocarbon generation started at 10 Ma and progressed so rapidly that sedimentary rocks currently below a depth of 4–5 km have completely passed through the oil generation window. The upper 2 to 3 km of the sedimentary rocks are, however, immature over the whole Pannonian basin.

Comparison of the known oil and gas fields with the present maturity conditions in the Great Hungarian Plain shows that 49 out of 53 fields are located outside of, and at some distance from, the areas of hydrocarbon generation. Traps are associated mostly with compactional anticlines above basement highs and with positive structures along strike-slip faults. Future prospects may be predicted in deeper zones where hydrocarbons could have accumulated in stratigraphic and unconformity traps and in the fractured basement.

INTRODUCTION

Plate tectonics has had important implications for the formation and evolution of passive continental margins and inland basins. Since the early work of Bott (1976), Lachenbruch (1978), and McKenzie (1978), much attention has been paid to quantitative evaluation of the evolution of continental extensional basins.

The Pannonian basin has had a long tradition of geological and geophysical exploration and contains an unusually large number of boreholes, drilled principally for hydrocarbon prospection. Bally and Snelson (1980) have classified the Pannonian basin as the type example of a back arc basin associated with continental collision and situated on the concave side of an A-type subduction arc.

The first plate tectonic model for formation of the Pannonian basin was suggested by Stegena et al. (1975). They showed that thin crust, elevated asthenosphere, high temperatures and heat flow, and intensive volcanism characterized the region. They argued correctly that these features were genetically related to the formation of the basin. Their model of active mantle diapirism, however, underestimated the role of crustal extension, which was not well documented at that time.

Sclater et al. (1980) applied a lithospheric stretching model to the intra-Carpathian basins. They suggested that the peripheral basins (e.g., the Vienna and Transcarpathian depressions) were the result of uniform stretching of the lithosphere by about of a factor of two ($\beta = 2$). They showed that in the Pannonian basin, however, unrealistically excessive stretching ($\beta = 3$ to 5) would be required to explain the observed thermal subsidence and high heat flow. A tenable alternative was moderate lithospheric stretching with additional subcrustal thinning. Subsequent work has supported this conclusion. Horváth and Royden (1981) and Royden et al. (1982) have suggested about 100 km of east-west extension across the Pannonian basin during the middle and late Miocene. Analysis of subsidence and thermal data for 23 wells in the Pannonian basin led to a modification of the stretching model (Royden et al., 1983 a,b). They combined finite crustal extension with large attenuation of the mantle lithosphere resulting in significant heating of the crust and the overlying sediments.

Vitrinite reflectance data were used first by Stegena et al. (1981) to reconstruct paleogeothermal conditions. They suggested that heat flow was normal in the Miocene and rapidly increased during the last 5 million years to its present high value. This initial work, however, did not model satisfactorily the temperature history of the accumulating sediments, and used a generalized relationship (Waples, 1980) to convert a time–temperature index (TTI) to vitrinite reflectance.

In the last two years much new data have become available for the Pannonian basin, including new measurements and interpretations, most of which are reviewed in this volume. This higher level of knowledge made it possible to improve our earlier model calculations (Royden et al., 1983 a,b), and to reconstruct the maturation history of organic matter in Neogene sedimentary rocks. In this paper we first review and evaluate the most important input data used in our model calculations. Model results are then compared with actual data in a number of "master wells" on the Great Hungarian Plain. Finally, the maturation history of Neogene sediments in the Great Hungarian Plain is predicted and its implication for hydrocarbon prospection discussed.

CONCEPT OF MODEL CALCULATIONS

A simplified flow diagram of the computer program which was used to model basin evolution is shown in Figure 1. Like many other problems in geophysics, modeling of basin evolution is a mathematical inversion. By supposing simple and geologically reasonable processes for basin formation, a set of mathematical equations (mathematical model in Figure 1) can be derived to solve the direct problem. It one assumes some stretching parameters for stretching of the lithosphere, the amount of subsidence, thermal features, and so on, can thus be calculated as a function of geological time. In reality it is the inverse problem which has to be solved. Geological information is available from boreholes and cross sections. This includes the stratigraphy of the basin fill, structural data, porosity, permeability and other rock physical parameters, and geothermal and geochemical data. Taken together, the data comprise the geological input (geological model in Figure 1). A satisfactory solution consists of stretching parameters that produce results in good agreement with the observed data. This is done by trial and error.

We have developed a one-dimensional mathematical model for extensional basins. Unknown model parameters are as follow: (1) stretching factors for the crust and the mantle lithosphere; and (2) radiogenic heat production of the crust. Constraints on these parameters can be obtained by comparison of the observed and predicted present temperature-depth profile, the present heat flow, the present crustal thickness, and the subsidence history. These constraints are usually adequate and lead to reasonably unique and stable solutions.

MATHEMATICAL MODEL

The mathematic model is based on the differential equation of heat conduction:

$$\nabla^2 T = \frac{1 \delta T}{\kappa \, \delta t}$$

where the temperature (T) is a function of position and time (t), and κ is the thermal diffusivity. A one dimensional solution for an infinite slab is given by Carslaw and Jaeger (1959). Using this solution our computation method is seminumerical, similar to that of Royden et al. (1983 a,b). The main differences are:

1. The thickness of the slab (l_i') is not constant but

$$l_i' = 1 - h_{wi}$$

where h_{wi} is the water depth above the sediments in the ith time interval.

2. The sediments were decompacted numerically using Szalay's normal compaction trends or measured porosity-depth profiles [$\phi(z)$] for pelites and psammites (Szalay and Szentgyörgyi, this volume). The thickness of the jth sediment layer (s_j) had to satisfy the equation

$$\int_{s_j} [100 - \phi(z)]\, dz = \text{constant}$$

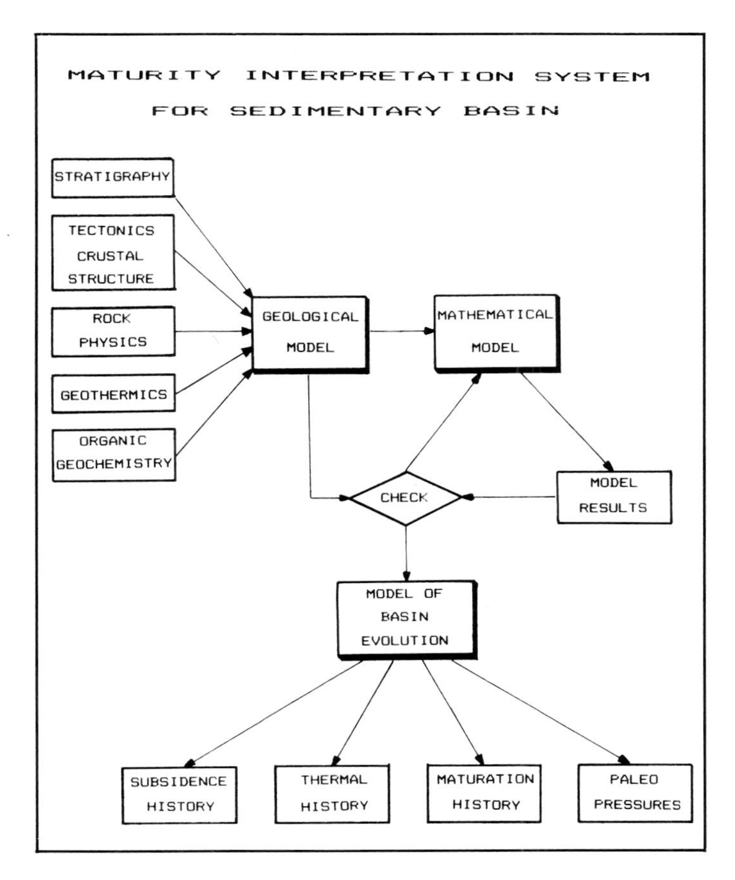

Figure 1. Simplified flow diagram to illustrate the concept of model calculations.

3. The thermal conductivity of sediments was taken to be a function of depth and time. We used Airy-type isostatic compensation at the bottom of the slab to calculate the subsidence caused by stretching, by thermal expansion or contraction of lithosphere, and by sediment and water loading.

Prediction of the maturity of organic matter was made by slightly modifying the Lopatin method (Royden et al., 1983b), except that TTI versus vitrinite reflectance relationship was recalibrated. The parameters used in our study are mostly from Parsons and Sclater (1977) and are listed in Table 1.

GEOLOGICAL MODEL

Stratigraphic Units

The age of sedimentary rocks in the Pannonian basin ranges from early Miocene to Pleistocene. The central Paratethys regional stage system is used for stratigraphic division of the early and middle Miocene strata (Steininger et al., this volume; Nagymarosy and Müller, this volume). Dating of these sediments was achieved by paleontological examination of core samples from exploration wells, in combination with well log markers and seismic correlation between drill holes. In the Great Hungarian Plain, commonly thin, early Miocene beds are overlain by thicker middle Miocene (M_4 = Badenian and M_5 = Sarmatian) strata.

Table 1. Parameters Used in the Model Calculation

Symbol	Parameter	Value
l	Initial lithosphere thickness	126 km
l_c	Initial crustal thickness	36 km
d	Initial depth of radiogenic plane source	8 km
T_a	Temperature of asthenosphere	1350 °C
T_o	Surface temperature	10 °C
ρ_{0c}	Crustal density at temperature of 0 °C	2800 kg/m^3
ρ_{01}	Density of the subscrustal material at temperature of 0 °C	3310 kg/m^3
ρ_w	Density of water	1000 kg/m^3
ρ_m	Matrix density of sedimentary rocks	2700 kg/m^3
K_1	Thermal conductivity of the lithosphere	3.14 mW/m °K
κ	Thermal diffusivity	8.122 10^{-7}m^2/s
α	Coefficient of thermal expansion	3.1 10^{-5} °C^{-1}

The late Miocene through Pliocene sedimentary complex is called Pannonian (*s. l.*). This is the thickest and most widely distributed stratigraphic unit in the Pannonian basin (Figure 2). Its subdivision into units Pa_1^{1a}, Pa_1^{1b}, Pa_1^2 and Pa_2 has been done using lithological trend diagrams and seismic correlations (Szalay and Szentgyörgyi, this volume). The Pannonian/Quaternary boundary is determined on the basis of lithological logs. The chronostratigraphic significance of these units is discussed by Horváth and Pogácsás (this volume). In making the model calculations we used the following dates for the beginning of the different units: 17 Ma Badenian, 14 Ma Sarmatian, 12 Ma Pa_1^{1a}, 11 Ma Pa_1^{1b}, 10 Ma Pa_1^2, 8 Ma Pa_2, and 2.4 Ma Quaternary.

Lithology

The sedimentary rocks of the Pannonian basin consist mainly of pelites (claystone, shale, siltstone) and psammite (gravel, sandstone, conglomerate). Locally thick volcanoclastics and occasional lava flows occur. Lithology is determined from continuous well logs which are usually available from the surface down to the bottom in each exploration and production borehole.

Porosity

Apart from direct measurements on drill cores, porosity data come from the interpretation of geophysical well logs (see the example in Szalay, this volume). Porosity versus depth profiles, combined with pore-pressure measurements, can define the normal compaction trend, and zones characterized by undercompaction and overpressure. In making the model calculations, decompaction of the sedimentary rocks was calculated by assuming that the sediments followed the normal compaction trend during progressive burial. We also assumed that the undercompacted rocks followed first the normal trend, then, after reaching an "isolation depth," underwent no further compaction (Magara, 1978; Szalay, this volume).

Depositional Depth

Subsidence histories can be determined from sedimentation histories if there are data available to constrain the water depth

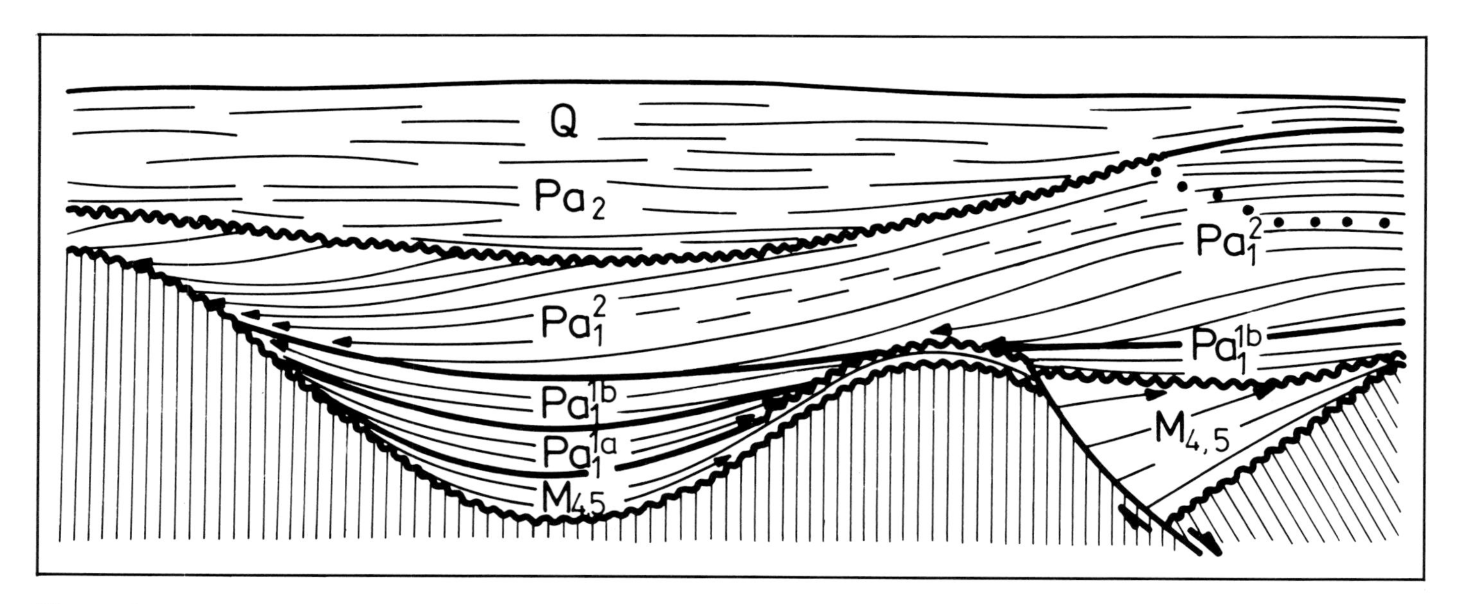

Figure 2. Schematic cross section (not to scale) showing the main stratigraphic units and the characteristic geometry of bedding planes in the Great Hungarian Plain. Thick, smooth, and rippled lines are conformable and unconformable boundaries of the stratigraphic units, respectively. $M_{4,5}$ denotes middle Miocene sedimentary rocks. Late Miocene and Pliocene rocks are called Pannonian (*s. l.*), and can be divided into units Pa_1^{1a}, Pa_1^{1b}, Pa_1^2 and Pa_2. The base of the Quaternary is an unconformity elsewhere in the Pannonian basin, and may be correlated with a lithological boundary in the interior of the Great Hungarian Plain.

in which deposition of the sediments occurred. We did not consider the possible influence of eustatic sea level changes on the surface of the Pannonian lake. We assumed that the water depths were controlled by the relationship between sedimentation rate and subsidence rate. Traditionally, the water depths in the Pannonian basin were thought to have been always shallow (0 m to 100 m) during the deposition of the basin sediments. Recent stratigraphic analysis of seismic sections and drill core studies have led to a better understanding of facies conditions and paleodepositional surfaces (Mattick et al., this volume; Bérczi and Phillips, this volume). We inferred from these studies a generalized water depth trend (Figure 3) for troughs currently deeper than 3 km in the Great Hungarian Plain. It is thought that areas with thinner sediment cover were first exposed and then covered by a shallow lake.

Temperature and Maturity Conditions

Temperature versus depth profiles and heat flow are important constraints on the unfixed model parameters. These two quantities are related through the thermal conductivity of rocks. Analysis of geothermal data has shown that purely conductive heat transfer in the basin fill of the Great Hungarian Plain is a good assumption (Dövényi and Horváth, this volume). Thermal conductivities depend on lithology, which is simple in the Pannonian basin. Conductivity, however, is also strongly dependent on porosity, because pore water is a much poorer conductor than the rock matrix. Analysis of measured conductivities in the Pannonian basin shows that a simple relationship can be used to describe the dependence of thermal conductivity on porosity in pelitic and psammitic rocks (Dövényi and Horváth, this volume). This relationship is an important part of our geological model because the thickness and porosity profile of the sedimentary column is reconstructed for past times by backstripping and decompaction. Accordingly, thermal conductivity of the sedimentary complex can be calculated as a function of depth and time. This helps to construct more reliably the temperature and maturation history of the sediments. Vitrinite reflectance (R_o) data from organic matter, mostly of terrestrial origin, was used as the measure of the observed maturity.

Synsedimentary Tectonics

Interpretation of reflection seismic profiles and drillhole data suggests that the Pannonian basin formed by crustal extension during the middle Miocene (Horváth and Rumpler, 1984; Rumper and Horváth, this volume). Normal faults are connected to a system of coeval strike-slip faults. Typical structures are half grabens and pull-apart basins. Flower structures related to strike-slip faults indicate that minor displacement occurred also during the late Miocene. In the model calculations the time interval of active lithospheric extension (initial phase) was usually fixed to the time interval between 17 Ma to 11 Ma, but in a few cases it was allowed to continue until 10 Ma. The 11 Ma (10 Ma) to recent period is the postextensional phase of basin evolution.

Structure of the Lithosphere

Refraction and reflection seismic data indicate that the crust of the Pannonian basin is continental, and that the depth to Moho varies from 24 km to 28 km under the Great Hungarian Plain (Dövényi and Horváth, this volume). When the sedimentary fill of the basin is taken into account the minimum thickness of the continental crust is about 18 km. The preextensional

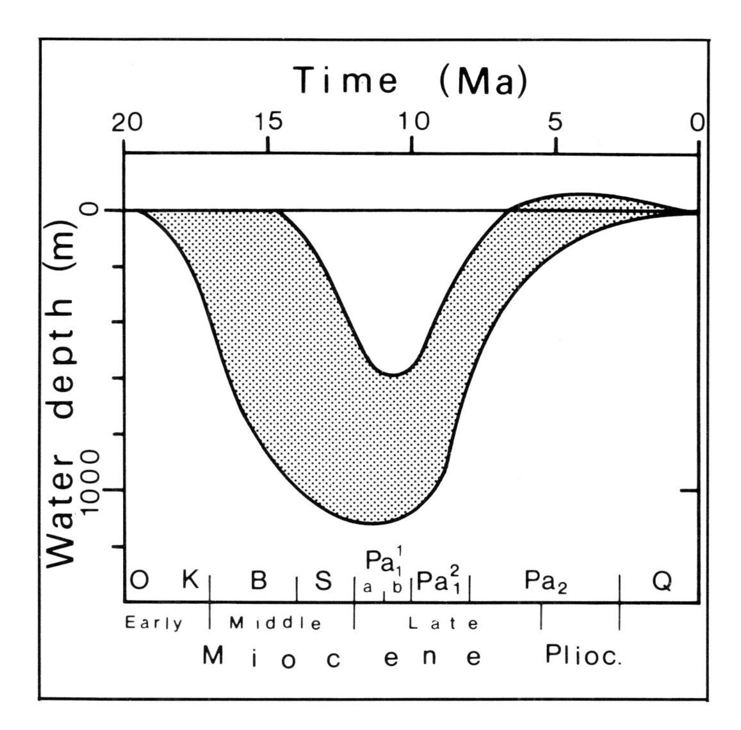

Figure 3. Generalized water depth trend for sedimentary troughs currently deeper than 3 km in the Great Hungarian Plain. The boundary lines of the dotted area give upper and lower limit of water depths during the deposition of early Miocene through Quaternary sediments. Actual water depth curve for a given locality should plot within the dotted area. Water depth less than 0 m indicates uplift. Fast basement subsidence during the phase of lithospheric stretching (middle Miocene) resulted in deep water basins (with up to 1000 m water depth). Slower subsidence rates after the end of the stretching phase and rapid inflow of clastic material led to shallow water environments towards the end of the Miocene.

crustal thickness is, of course, not known, but it is reasonable to suppose that it was similar to the present thickness of the internal Eastern Alps and Western Carpathians (35 to 45 km). This constrains crustal extension to be no more than 150% (β_c = 2.5).

The present thickness of the lithosphere is markedly less than normal. Deep seismic soundings (Posgay et al., 1981), magnetotellurics (Ádám and Wallner, 1981), and travel-time residuals (Hovland and Husebye, 1982; Babuska et al., 1984) indicate that a 40 to 80 km thick lithosphere beneath the Pannonian basin is underlain by a marked asthenosphere dome.

Lithospheric Evolution, Subsidence Rates, and Sedimentation

A schematic summary of the lithospheric structure present during formation of the Great Hungarian Plain, and the related subsidence and sedimentation is shown in Figure 4. This summary is based partly on general geological observations, and partly on the results of model calculations, a description of which follows.

In early Miocene time, the basement was above sea level, presumably with remarkable topographic relief. Basin formation began with local block faulting and intense volcanism. Crustal extension and lithospheric thinning culminated during the middle Miocene and were accompanied by a second phase of volcanism. The sedimentation rate did not keep pace with the fast basement subsidence, resulting in deep water basins in most of the troughs. Previous morphological highs or areas characterized by little or no extension remained above sea level in places. Sedimentation was dominated by fine-grained clastic material deposited from suspension. Turbiditic flows commonly occurred down the flanks of local basement highs. Tectonic breccias were formed along strike-slip faults.

During the first period of thermal phase (late Miocene), water flood became general and large volumes of clastic material were transported towards the basin centers by rivers. A prograding delta system mostly filled the Pannonian basin by the end of the Miocene. Additional thermal subsidence of the basement was roughly equal to the rate of sedimentation during the Pliocene and Pleistocene, giving rise to shallow lacustrine and marshy depositional environments. A few hundred meters of uplift occurred at about the end of the Pliocene in Transdanubia, but did not occur in the Great Hungarian Plain, which is still subsiding today.

RESULTS OF MODEL CALCULATIONS IN MASTER WELLS

Calculations were performed for 17 hydrocarbon exploration wells located in the Great Hungarian Plain (Figure 5). We selected these wells because they were characterized by relatively complete and reliable data as needed for thermal modeling.

Figure 6 shows an example of the results of these calculations. The Derecske-I well is located on the flank of a deep trough, which probably formed as a pull-apart basin. The drill penetrated the Paleozoic basement at a depth of 4988 m. At the Pannonian/Quaternary boundary, a few hundred meters of the stratigraphic section may be missing, but the rest of the section appears to be continuous. Figure 6A shows the time-space evolution of this well. The line starting from the surface at 17 Ma and reaching nearly 5 km depth at the present shows the subsidence history of the basement. The other five lines from left to right show the subsidence history of the top of the $M_{4,5}$, Pa_1^{1a}, Pa_1^{1b}, Pa_1^2 and Pa_2 units, respectively. Broken lines indicate the depth to the 40 °C, 80 °C, 120 °C, 160 °C, and 200 °C isotherms through time. The temperature versus depth profile that was calculated for the present can be compared with the measured temperature-depth profile (Figure 6B). One can see that the agreement is good.

The water depth (h_w in meters), determined by comparing the calculated basement subsidence with the coeval sediment thickness, is shown at the bottom of figure 6A. Deep water conditions prevailed from 12 Ma to 8 Ma (Pa_1^{1a}, Pa_1^{1b}, and Pa_1^2 time). Model parameters that were used for the calculation are also indicated in the figure. It can be seen that in the first 6 m.y. of basin formation (i.e., during $M_{4,5}$ and Pa_1^{1a} time), the crust stretched by a factor of 1.9, the mantle lithosphere was effectively thinned by a factor of 100. Only thermal subsidence occurred during the last 11 m.y.

Results for the Jászladány-I borehole are shown in Figure 7. This borehole is situated in the Jászság basin (Figure 5), and penetrated the Paleozoic basement at a depth of 3637 m. Litho-

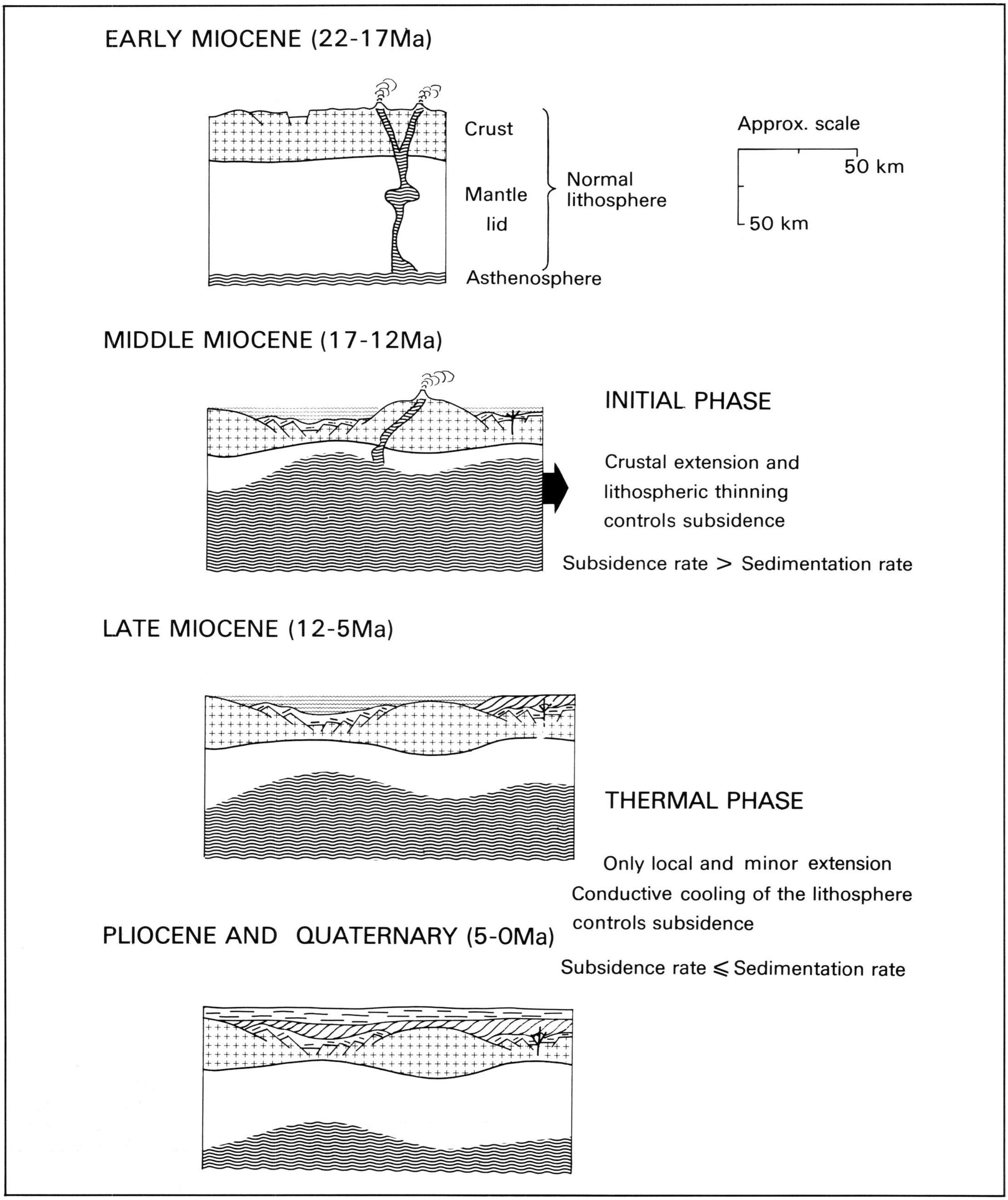

EARLY MIOCENE (22-17Ma)

Crust
Mantle
lid
Normal
lithosphere
Asthenosphere

Approx. scale
50 km
50 km

MIDDLE MIOCENE (17-12Ma)

INITIAL PHASE

Crustal extension and
lithospheric thinning
controls subsidence

Subsidence rate > Sedimentation rate

LATE MIOCENE (12-5Ma)

THERMAL PHASE

Only local and minor extension
Conductive cooling of the lithosphere
controls subsidence

Subsidence rate ⩽ Sedimentation rate

PLIOCENE AND QUATERNARY (5-0Ma)

Figure 4. Illustration shows the change of the structure of the lithosphere during basin formation, and the related subsidence and sedimentation processes in the Great Hungarian Plain.

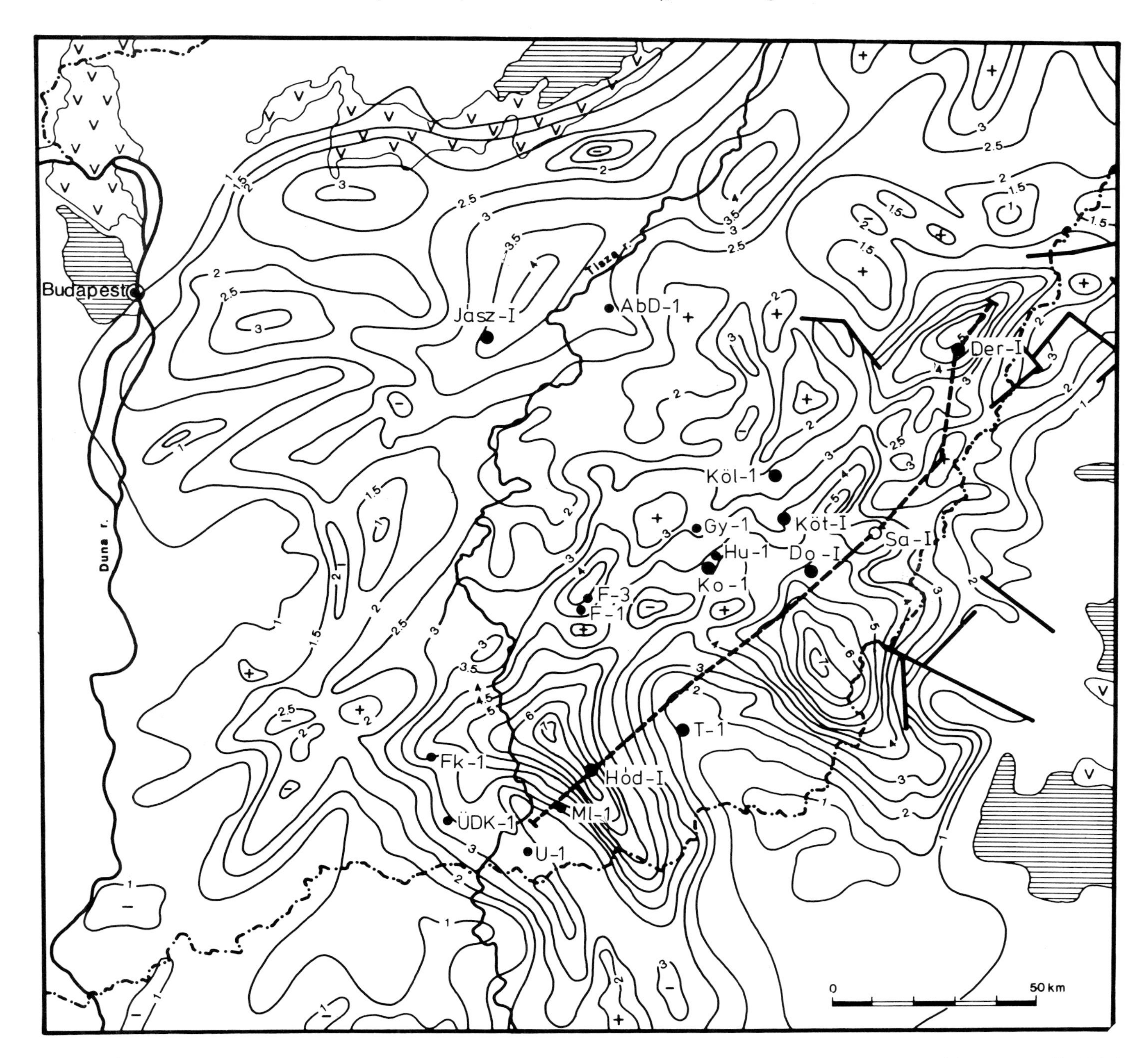

Figure 5. Isopach map (in km) of the Neogene–Quaternary basin fill in the Great Hungarian Plain and the location of the master wells. Solid black circles indicate wells that we used to derive the new TTI-R_o conversion: Hód-I, T-I (Tót-I), Ko-1 (Kondoros-1), Do-I (Doboz-I), Köt-I (Köröstarcsa-I), Köl-1 (Körösladány-1), Der-I (Derecske-I), and Jász-I (Jászladány-I). Small full circles indicate the rest of the master wells: U-1 (Uszi-1), Ml-1 (Maroslele-1), ÜDK-1 (Üllés-DK-1), Fk-1 (Fkút-1), F-1 (Fáb-1), F-3 (Fáb-3), Hu-1 (Hunya-1), Gy-1 (Gyoma-1), and AbD-1 (Abádszalók-D-1). Broken line shows the location of cross section in Figure 15A. The open circle gives the location of Sa-I (Sarkad-I) borehole shown in Figure 15A.

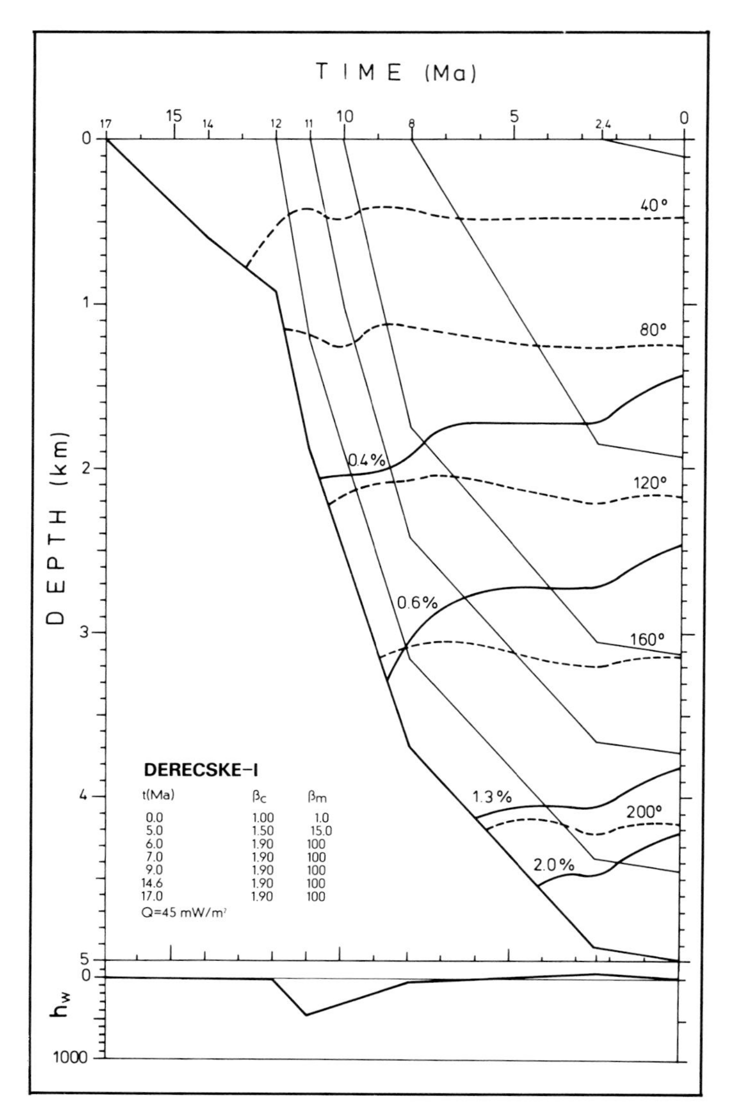

Figure 6A. Time-space evolution of the Derecske-I well. Subsidence history of the basement and of the boundary of the main stratigraphic units are shown together with the depth to 40 °C, 80 °C, 120 °C, 160 °C, and 200 °C isotherms, and to surfaces of constant vitrinite reflectance of 0.4%, 0.6%, 1.3%, and 2.0%. Water depth during deposition (h_w in meters) is given at the bottom of the figure. Parameters used in making the calculations are also shown. t (Ma) is the time elapsed since the beginning of the structural changes. β_c and β_m are the stretching parameters for the crust and mantle lithosphere, respectively. Q is the radiogenic heat production of the crust before stretching.

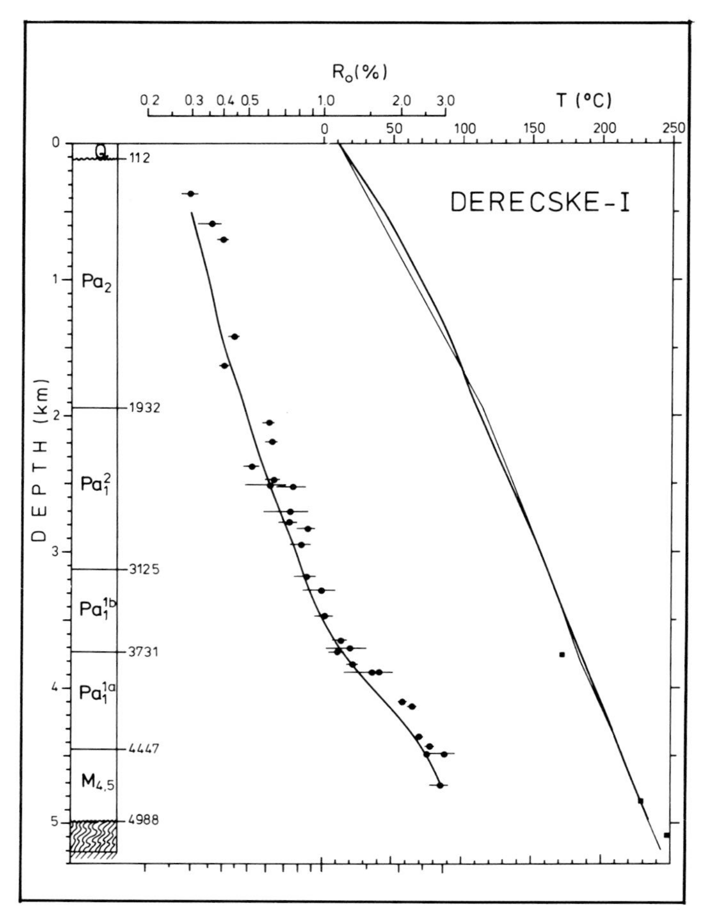

Figure 6B. Thickness of stratigraphic units in the Derecske-I well, measured and calculated vitrinite reflectances (R_o) and temperatures (*T*). Horizontal bars across the solid circles give the standard deviation of the measured vitrinite reflectances. The continuous curve is the vitrinite reflectance versus depth profile from the model calculations. Full squares are measured temperatures (drill stem test or corrected bottom hole data), and the thick continuous curve is the temperature versus depth profile from the model calculations. Thin line shows the temperature versus depth function determined in the course of heat flow evaluation.

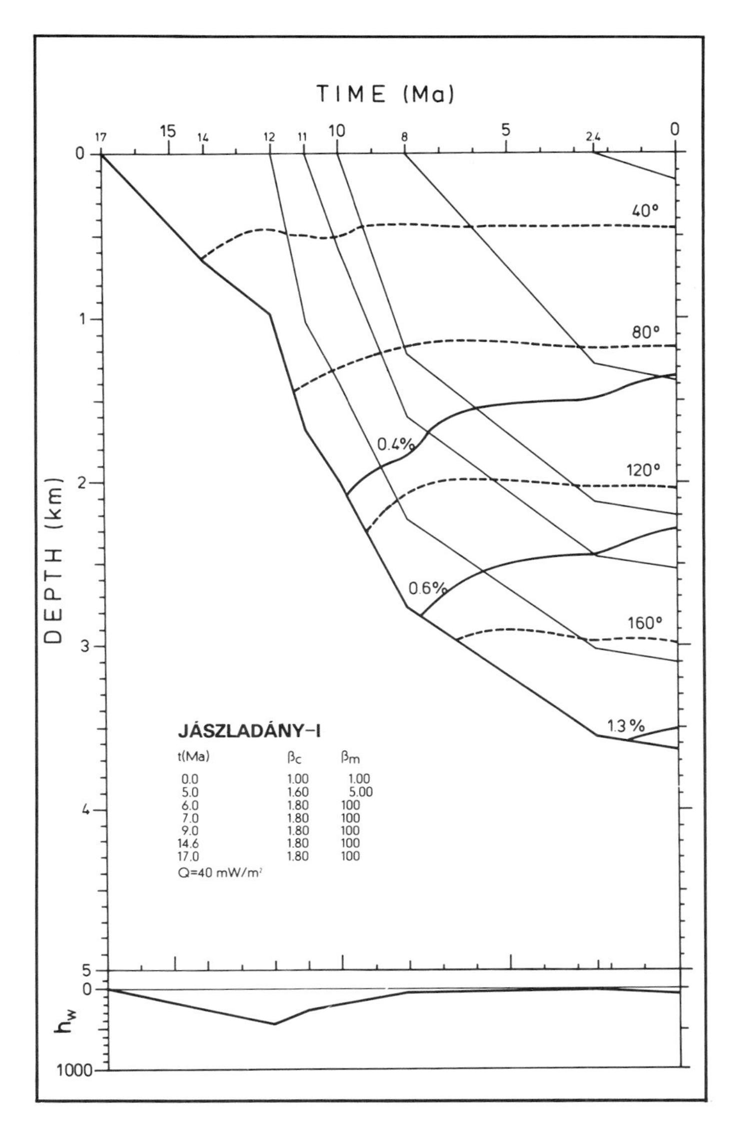

Figure 7A. Time-space evolution of the Jászládány-I well. (See also the explanation in the caption to Figure 6A and in the text.)

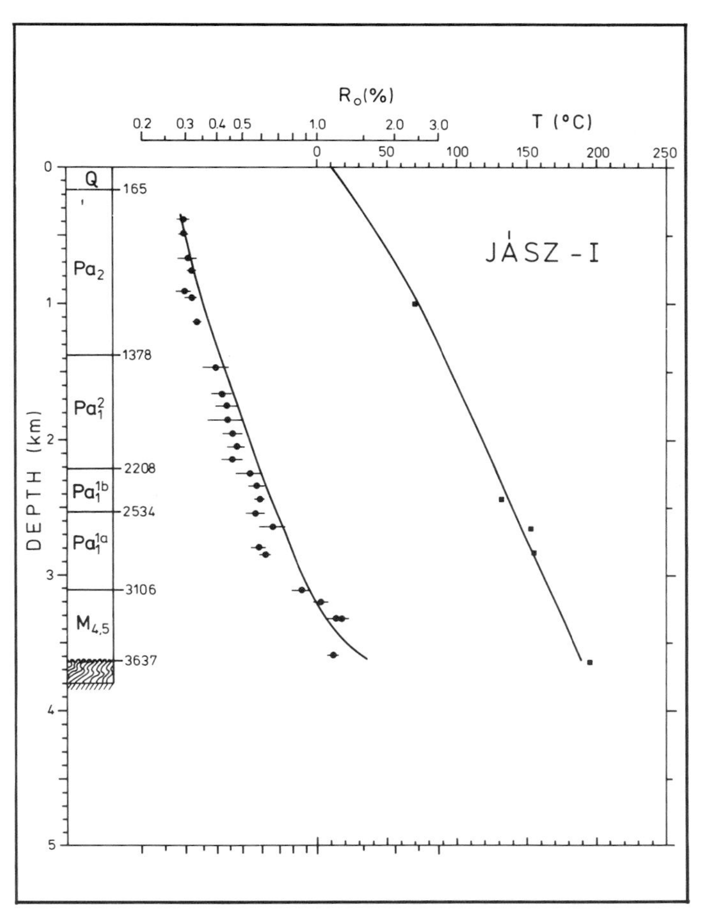

Figure 7B. Thickness of stratigraphic units in the Jászládány-I well, measured and calculated vitrinite reflectances (R_o) and temperatures (T). (See also the explanation in the caption to Figure 6B and in the text.)

spheric stretching occurred during the first 6 m.y. with parameters $\beta_c = 1.8$ and $\beta_m = 100$. The subsidence rate appears to have been higher than the sedimentation rate during this initial phase, and resulted in a relatively deep water basin. The basin filled up progressively during the postextensional phase of subsidence.

Figure 8 shows the results for the Körösladány-1 borehole. This hole is located in the shallower part of the basin (Figure 5), where the upper part of the $M_{4,5}$ and Pa_1^{1a} and Pa_1^{1b} strata are missing. This can be explained by marked uplift after the first 3 m.y. of stretching, as shown by the water depth curve. The computed stretching parameters are $\beta_c = 1.65$ and $\beta_m = 100$.

Similar results were obtained for the Hód-I, Doboz-I, Tót-I, Köröstarcsa-I and Kondoros-1 boreholes (Figure 5). In these eight boreholes we calculated the current TTI for each sedimentary horizon where there were measurements of vitrinite reflectance. These TTI values were then plotted against vitrinite reflectance (Figure 9). The scatter about a best-fit line through these points is small, and an empirical relationship is given in Table 2. Figure 10 shows the range of measured vitrinite reflectance (up to 3%) and temperatures (up to 230 °C) in the eight master wells, which were used to derive the new TTI-R_o relationship. This relationship between TTI and R_o can therefore be used for vitrinite reflectance and temperatures up to 3% and 230 °C, respectively.

Maturity calculations and observations in the other master wells (Uszi-1, Maroslele-1, Fáb-1, Fáb-3, Hunya-1, Gyoma-1, Fkút-1, Abádszalók-D-1, Üllés-DK-I) were used to check the applicability of this relationship. TTI values calculated for the present in each of these wells were converted to vitrinite reflectances according to this relationship and compared with measured data. Good agreement was obtained for all wells as illustrated by the two examples in Figures 11 and 12. Finally, our TTI-R_o relationship derived for the Pannonian basin was com-

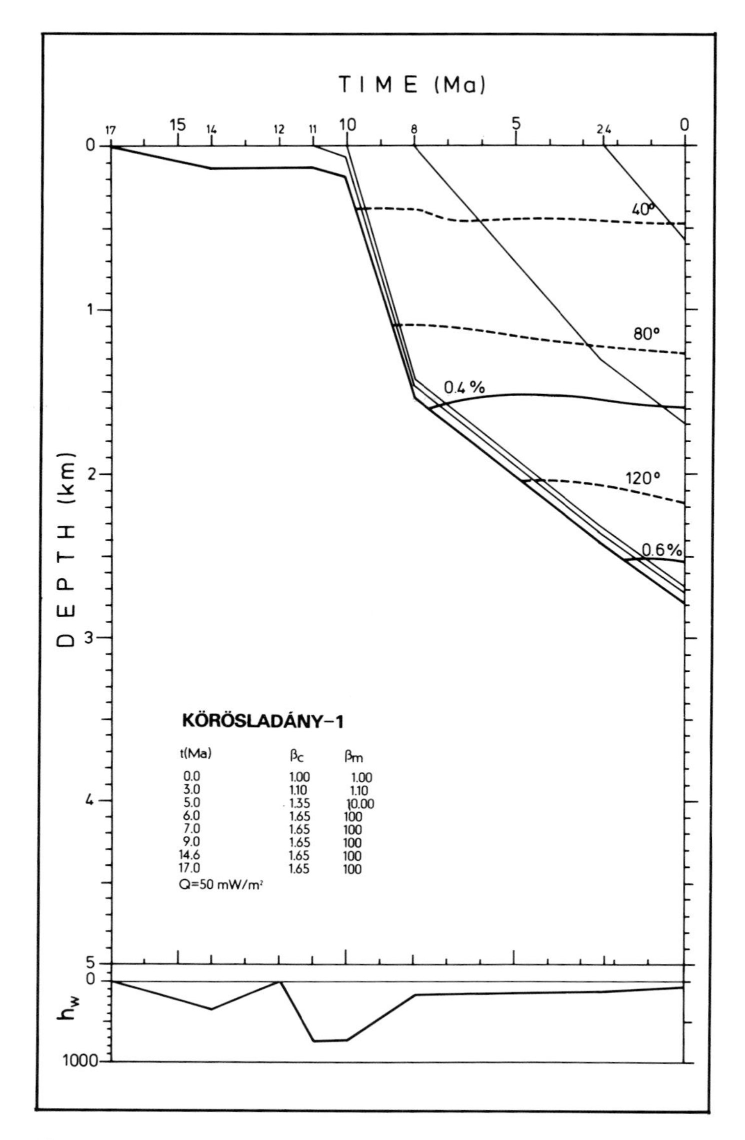

Figure 8A. Time-space evolution of the Körösladány-I well. (See also the explanation in the caption to Figure 6A and in the text.)

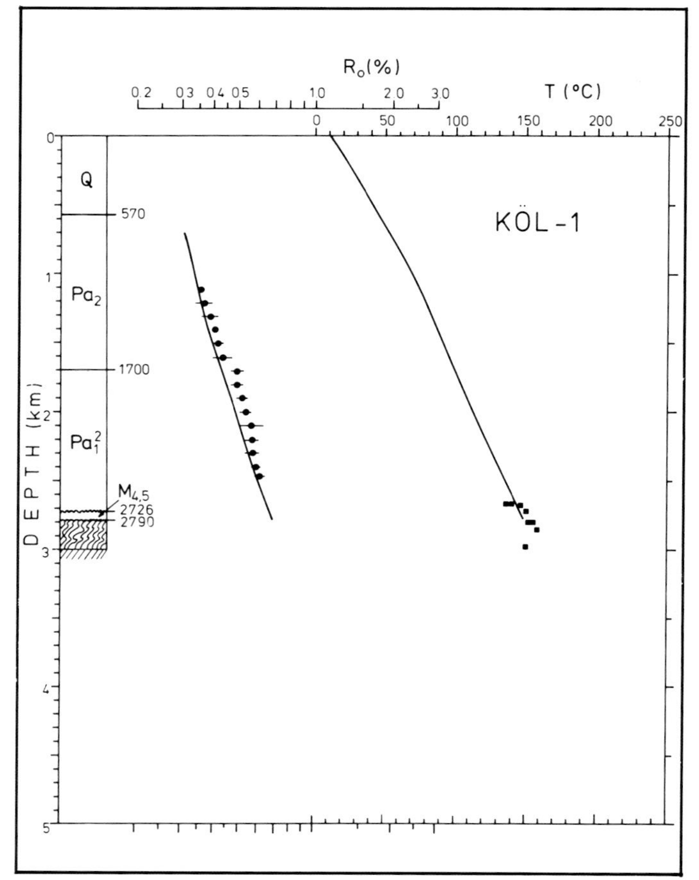

Figure 8B. Thickness of stratigraphic units in the Körösladány-I well, measured and calculated vitrinite reflectances (R_o) and temperatures (T). (See also the explanation in the caption to Figure 6B and in the text.)

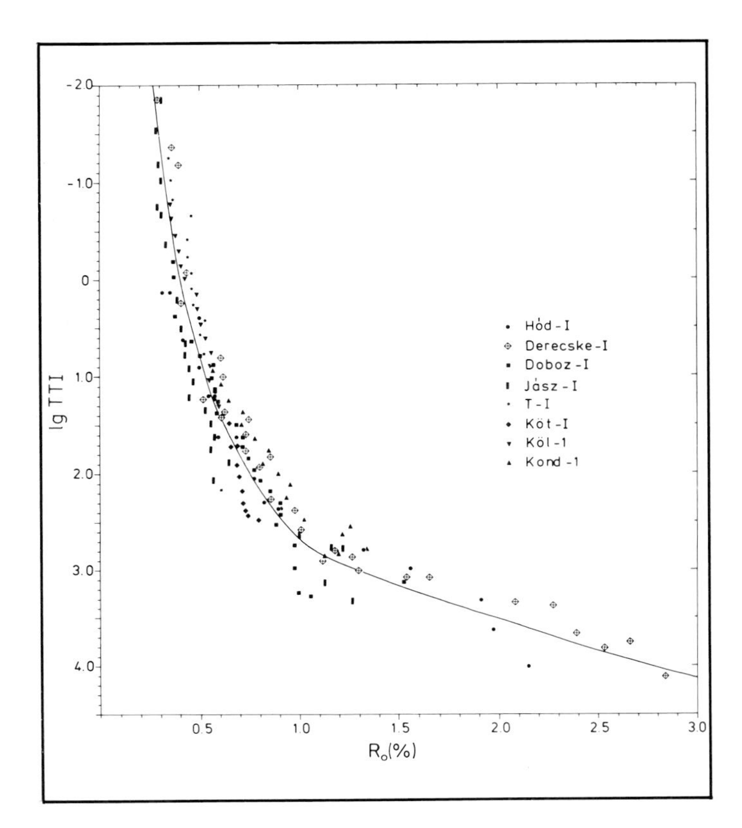

Figure 9. Logarithm (base 10) of the calculated Time-Temperature Index (TTI) plotted against the measured vitrinite reflectance (R_o) in eight master wells in the Great Hungarian Plain. The best-fit line defines a new relationship (Table 2) for TTI-R_o conversion in young and hot sedimentary basins.

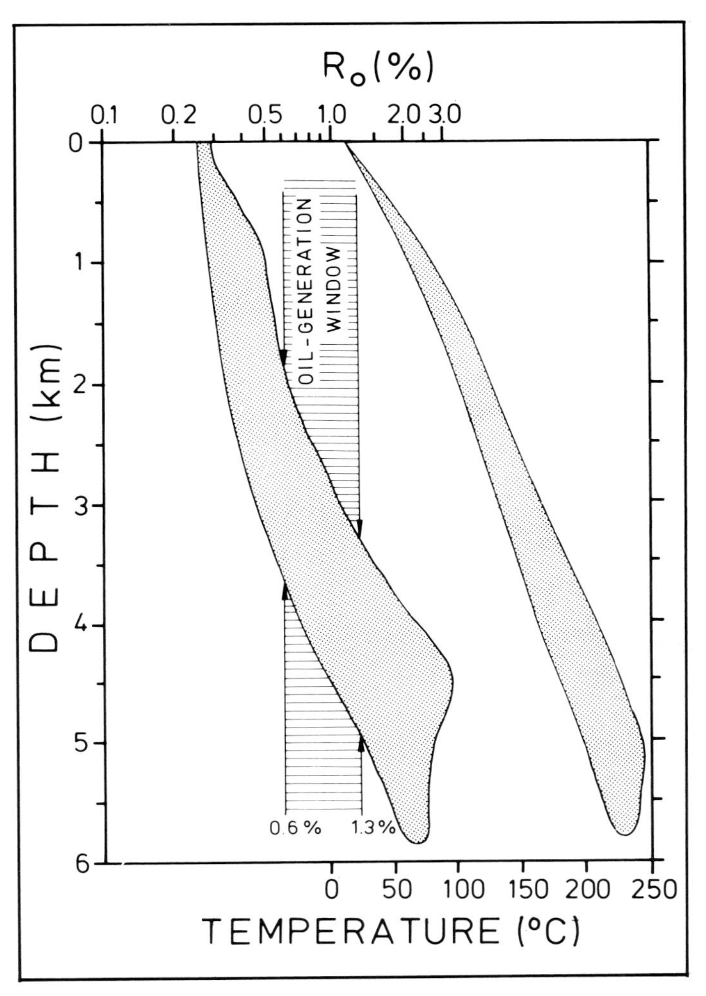

Figure 10. The range of vitrinite reflectances (dotted area on the left) and temperatures (dotted area on the right) as a function of depth in the eight master wells in the Great Hungarian Plain. The maturity and temperatures vary from R_o = 0.26% to 3% and T = 10°C to 250°C.

pared with published relationships (Figure 13). Our relationship is significantly different from that of Waples (1980) for $R_o > 0.6\%$ interval. (See also comments by Cohen, 1981, and Katz et al., 1982). Issler (1984) recently derived another relationship by studying temperature and maturation on the Atlantic continental margin of Canada. His results are in excellent accord with our TTI-R_o relationship. We are confident that the data in Table 2 provide a good conversion for young and hot sedimentary basins. The evolution diagram of the master wells (Figures 7, 8, 9, 11, and 12) also shows the maturation history of sedimentary rocks, as calculated by our TTI-R_o conversion.

Note that for all of the master wells (Figures 6, 7, 8, 11, and 12), crustal stretching parameter (β_c) is always below 2.2, while the stretching parameter for the mantle lithosphere (β_m) is larger by at least one order of magnitude. Royden and Dövényi (this volume) point out that β_m does not necessarily reflect actual extension of the mantle lithosphere. Rather, it is an expression of the amount of heat put into the mantle during extension. Therefore, our model results can be interpreted as moderate crustal extension during the initial phase of basin evolution, combined with dramatic attenuation of the mantle lithosphere.

Implication for Hydrocarbon Prospection

A test area was selected in the center of the Great Hungarian Plain, between the Tisza River and the Hungarian/Romanian national boundary (Figure 5). This covers an area of about 17,000 km^2 and is well studied. The total length of multifold seismic sections in this area is about 10,000 km. More than 200 wells deeper than 2 km have been drilled for hydrocarbon exploration. Fifty-three oil and gas fields have been found, most of which are still productive.

The test area was covered by a rectangular grid system with a

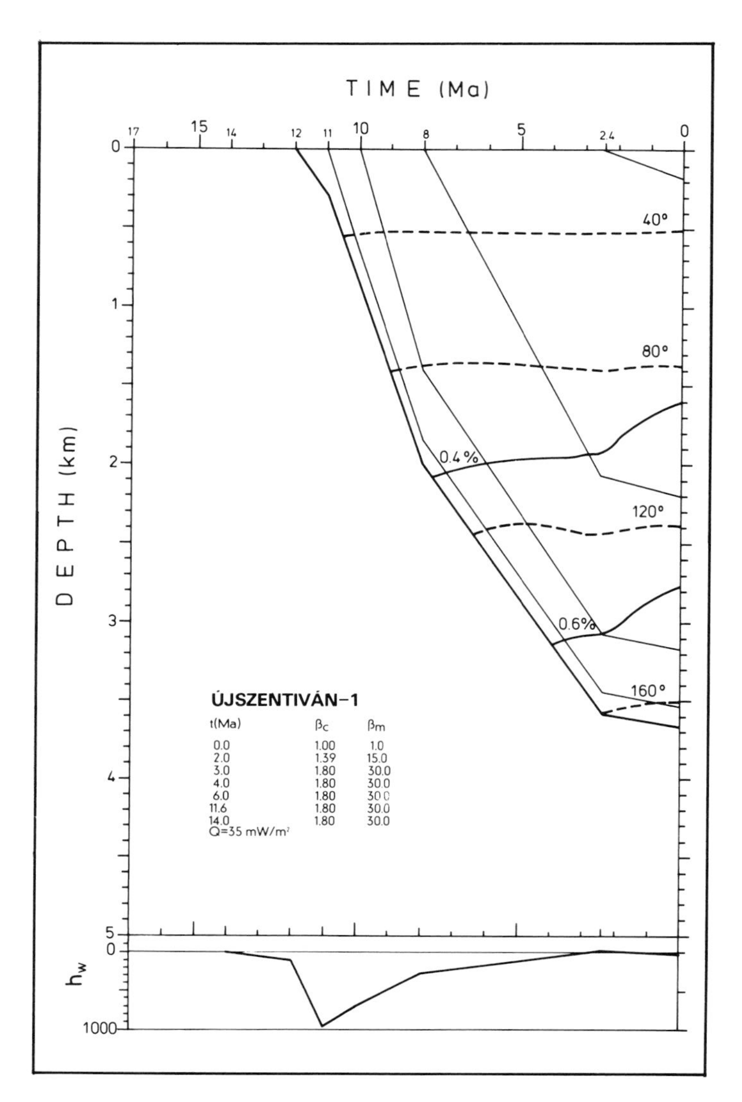

Figure 11A. Time-space evolution of the Uszi-1 well. (See also the explanation in the caption to Figure 6A and in the text.)

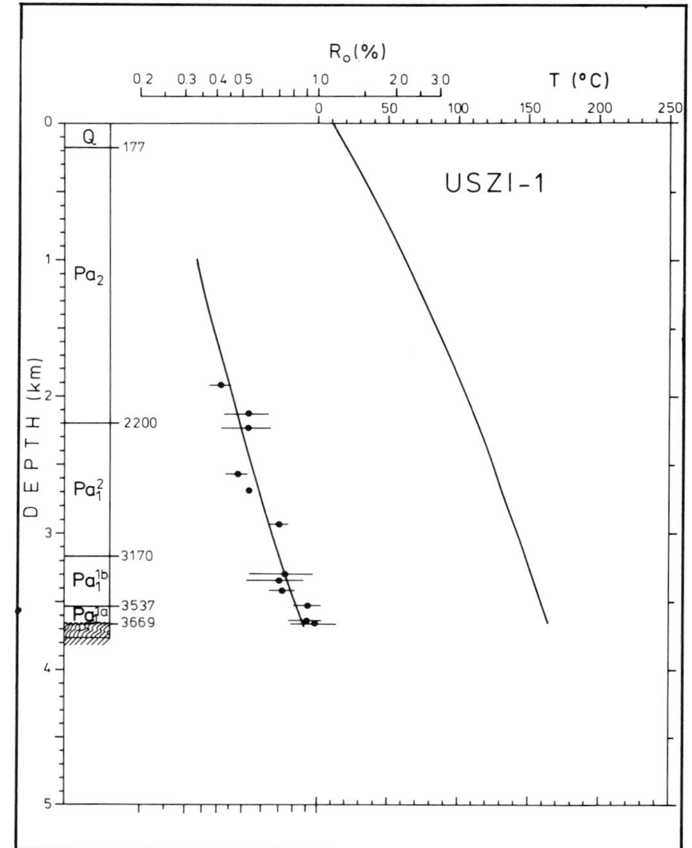

Figure 11B. Thickness of stratigraphic units in the Uszi-1 well, measured and calculated vitrinite reflectances (R_o) and temperatures (T). (See also the explanation in the caption to Figure 6B and in the text.)

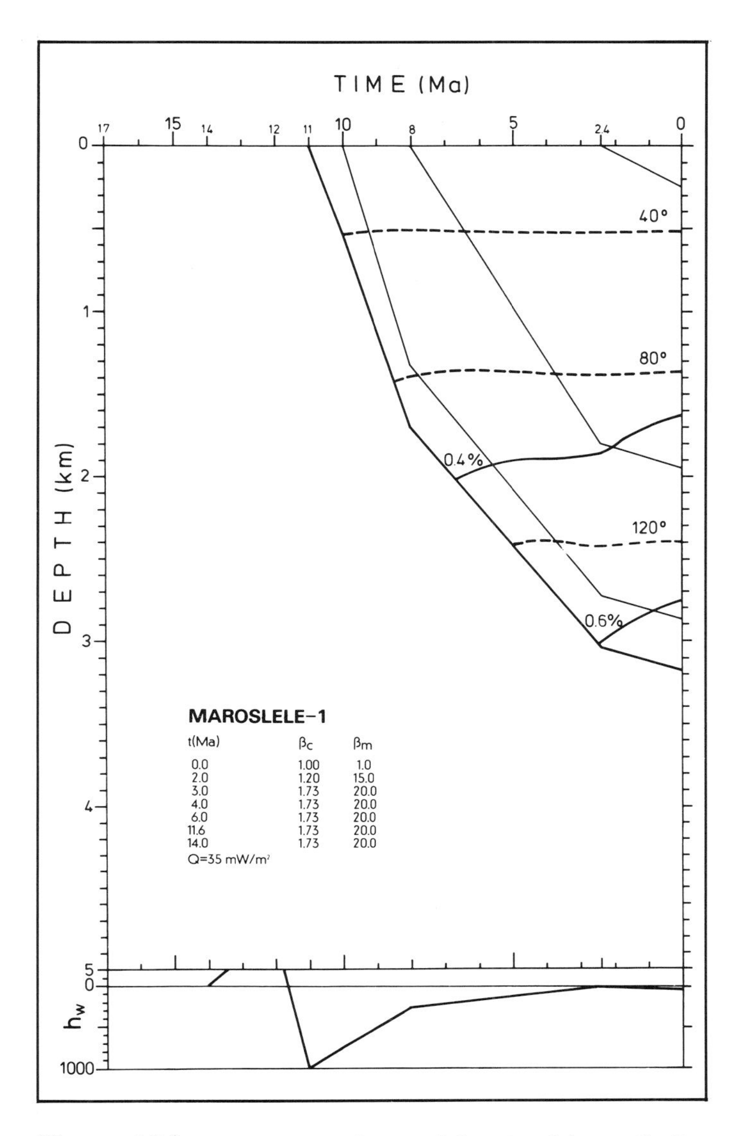

Figure 12A. Time-space evolution of the Maroslele-1 well. (See also the explanation in the caption to Figure 6A and in the text.)

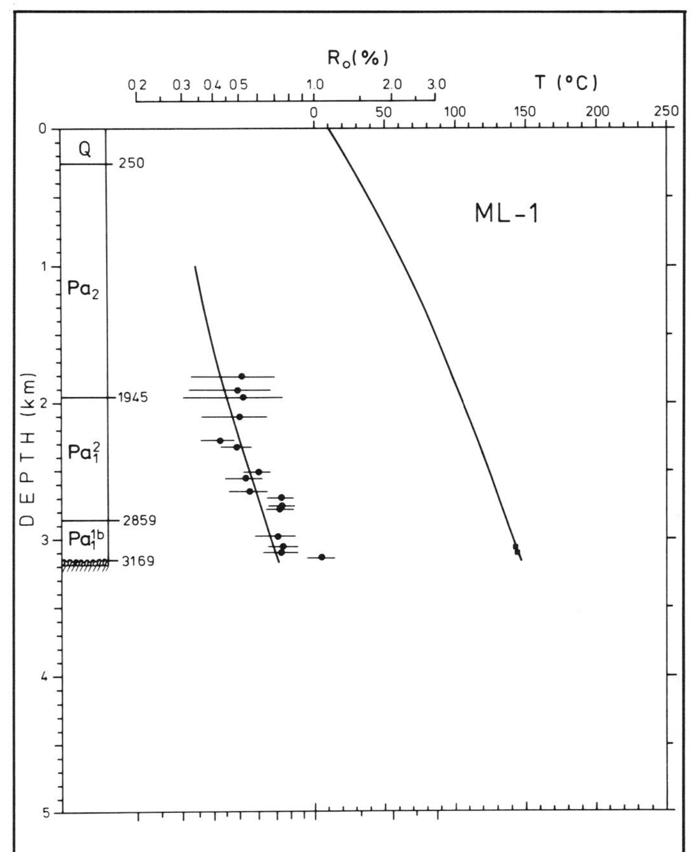

Figure 12B. Thickness of stratigraphic units in the Maroslele-1 well, measured and calculated vitrinite reflectances (R_o) and temperatures (T). (See also the explanation in the caption to Figure 6B and in the text.)

spacing of 5 km × 5 km, and calculations were performed at each grid point. We determined the present vitrinite reflectances as a function of depth and reconstructed the maturity for three different time slices (2.4 Ma, 8 Ma and 10 Ma), as well as the coeval temperatures and thicknesses of the stratigraphic units. Figure 14 is a map showing the present depth (below sea level) of the surface $R_o = 0.6\%$ in the basin fill. (This value was chosen because it is generally accepted as the beginning of oil generation.) It can be seen that this isoreflectance surface lies at greatest depth (4200 m) in the southwestern (lower left) corner of the area and locally reaches depths as shallow as 2200 m. Horizontal hatched lines indicate areas where the surface does not lie within the basin fill because, in these places, the basement is shallower than the $R_o = 0.6\%$ surface. Accordingly, such areas correspond to basement highs that are overlain by immature ($R_o < 0.6\%$) sedimentary rocks. It can be seen from Figure 14 that nearly all oil and gas fields (49 out of 53) have been found in such shallow areas of the Great Hungarian Plain. This map is therefore useful as it delimits areas where similar fields can be searched for in the future.

A section across the test area is shown in Figure 15A, and helps to understand the formation of the oil and gas fields in the Great Hungarian Plain. It can be seen that the oil generation window ($0.6\% < R_o < 1.3\%$) begins at a depth between 2 km to 3 km and terminates at 3.5 km to 5 km along the section. The shaley $M_{4,5}$ and Pa_1^{1a} units, which contain relatively high amounts of organic material (Szalay, this volume), have largely passed through this window. Other potential source rocks in unit Pa_1^2 have only just started to generate oil in some places. Sedimentary rocks above basement highs are usually immature.

Three hydrocarbon fields are crossed by the section in Figure 15A. The Algyő field on the SW is the biggest Hungarian oil

Table 2. Relationship Used to Convert Time-Temperature Index (TTI) to Vitrinite Reflectance (R_o)

TTI	$\log_{10}$TTI	R_o(%)
0.01	−2.0	0.27
0.03	−1.5	0.30
0.10	−1.0	0.33
0.32	−0.5	0.36
1.00	0.0	0.40
2.51	0.40	0.45
6.03	0.78	0.50
13.2	1.12	0.55
23.4	1.37	0.60
39.8	1.60	0.65
61.7	1.79	0.70
93.3	1.97	0.75
134.9	2.13	0.80
190.6	2.28	0.85
275.4	2.44	0.90
363	2.56	0.95
457	2.66	1.00
575	2.76	1.05
660	2.82	1.10
741	2.87	1.15
832	2.92	1.20
1000	3.00	1.30
1202	3.08	1.40
1445	3.16	1.50
2138	3.33	1.75
3162	3.50	2.00
4677	3.67	2.25
7413	3.84	2.50
9550	3.98	2.75
13183	4.12	3.00

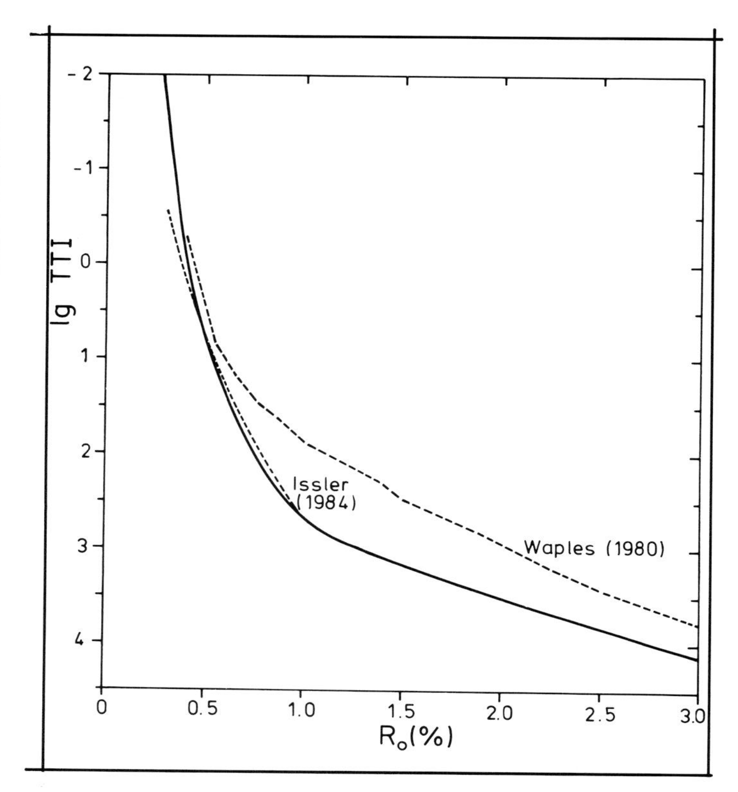

Figure 13. Comparison of our new TTI-R_o relationship (thick line) with that of Waples (1980) and Issler (1984).

field and is situated in a compaction anticline of Pa_1^2 and Pa_2 strata above the elevated basement. The Pusztaföldvár field was found in a similar structural/lithological trap right above the elevated basement. Körösszegapáti is a small field along the section and is particularly interesting, because most of the oil and gas were accumulated in the Paleozoic basement. The basement here exhibits high permeability due to fractures and fissures. It seems clear that the hydrocarbons in these fields were generated in deep troughs that contain mature and overmature source rocks ($M_{4,5}$ and Pa_1^{1a} units). Migration occurred updip over a horizontal distance of about 20 to 30 km, so that hydrocarbons were trapped in compactional anticlines and lithological traps above basement highs and in places within the basement itself. Elsewhere in the Great Hungarian Plain, hydrocarbon fields have been found in positive structures associated with strike-slip faults. Locations of known oil and gas fields suggest that this has been a common pattern in the Pannonian basin. We think, however, that different migration paths may have resulted in accumulation of hydrocarbons in other types of traps. This is suggested by studying the maturation history of the source rocks in the test area, together with the contemporaneous pore pressures.

Results are illustrated in Figures 15B–D, which show the reconstructions of the cross section at 2.4 Ma, 8 Ma, and 10 Ma. Ten million years ago, the initial phase of basin evolution was already finished and all of the fault-controlled structures were essentially complete. Deep sedimentary troughs (down to 4 km) existed, but part of the basement was still above sea level. Apart from the very bottom of the Derecske basin, source rocks had not reached the oil generation window. Further rapid subsidence and deposition of the thick Pa_1^2 unit occurred over the next two million years. It can be seen in Figure 15C that hydrocarbon generation started in each deep trough in this time interval. During this short time, however, expulsion of pore fluid from the pelitic $M_{4,5}$ and Pa_1^{1a} stratigraphic units did not occur. Instead, large overpressured zones formed with maximum overpressuring in the Pa_1^{1a} unit (Szalay, this volume). This pressure seal existed for the next 8 million years, during which additional subsidence and maturation occurred (Figures 15A and B). This suggests that significant amounts of hydrocarbons, generated in source rocks below the pressure seal, should have migrated downward and/or laterally toward the flanks of the depressions. Reservoir rocks in this domain can be found (1) in association with the unconformity between $M_{4,5}$ and younger Pannonian strata, (2) at the bottom of $M_{4,5}$ strata (e.g., basal conglomerates), and (3) in the fractured basement.

We suggest that traps inside the fractured basement and subtle traps at the flanks of basement highs may contain exploitable amounts of hydrocarbons. Reliable migration models and a good understanding of stratigraphic and tectonic conditions at depth are thought to be most powerful tools in searching for these subtle traps in the Pannonian basin.

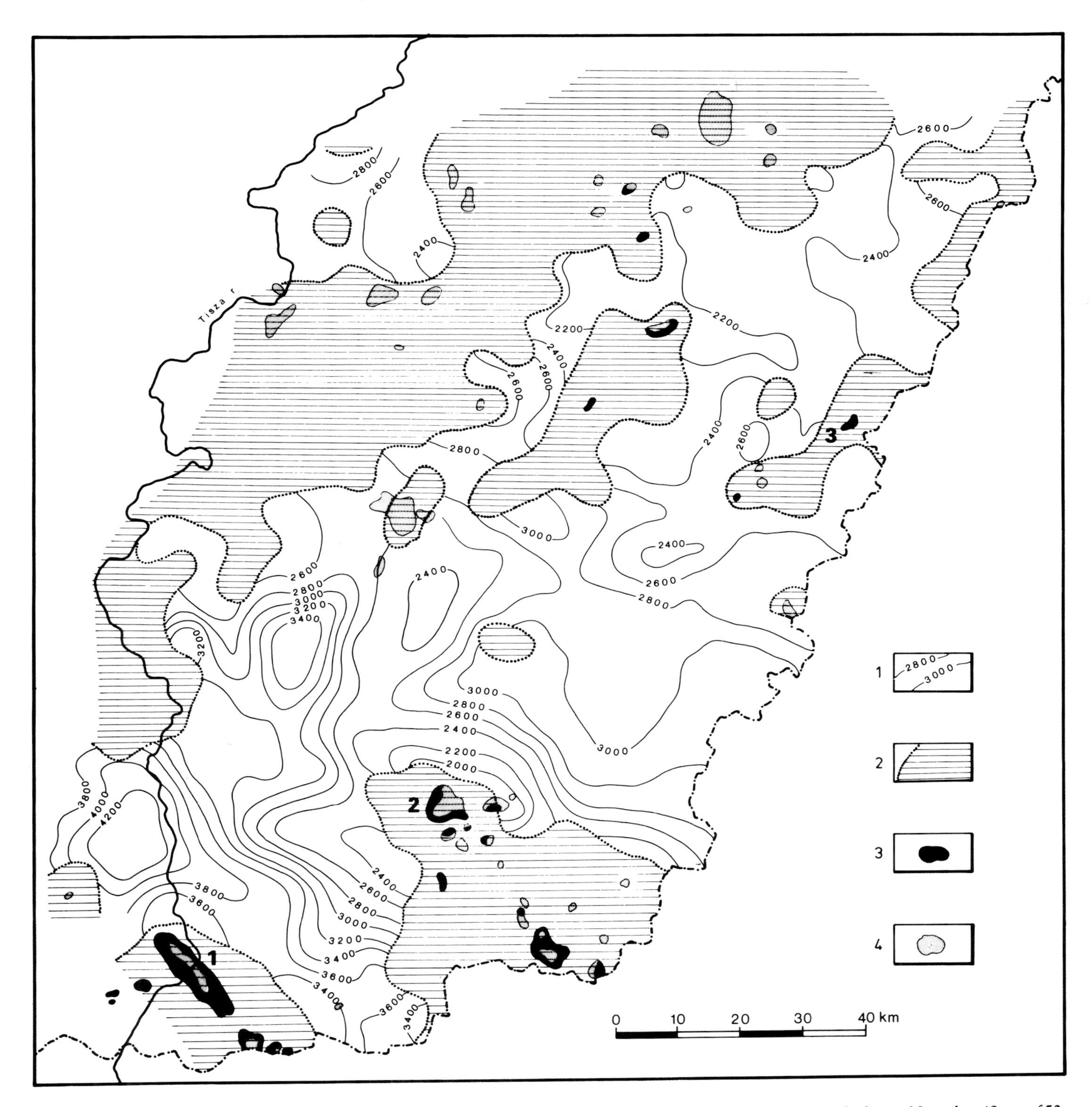

Figure 14. Oil and gas fields in the Great Hungarian Plain and the depth to R_o = 0.6% isoreflectance surface in the basin. Note that 49 out of 53 fields are located in immature reservoir rocks above basement highs (compare Figure 5). The numbers 1, 2, and 3 indicate the Algyő, Pusztaföldvár, and Körösszegapáti fields, respectively, which are shown on the cross section in Figure 15A. Keys: 1 = depth (in meters) below sea level of the R_o = 0.6% isoreflectance surface; 2 = areas where the elevated basement is directly overlain by immature (R_o < 0.6%) sedimentary rocks; 3 = oil fields; and 4 = gas fields.

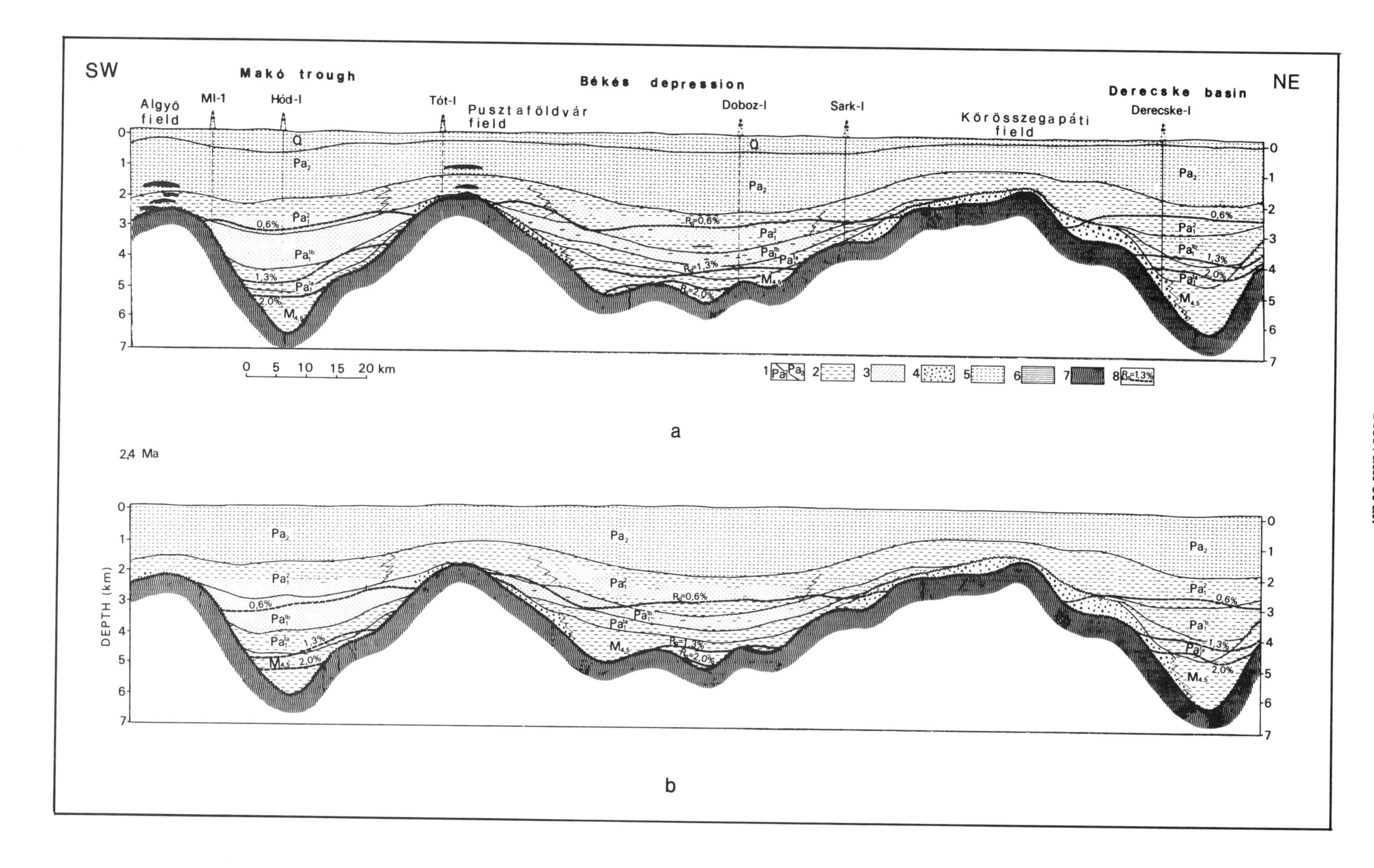
SW
Makó trough
Békés depression
Derecske basin
NE
Algyö field
Ml-1
Hód-I
Tót-I
Pusztaföldvár field
Doboz-I
Sark-I
Körösszegapáti field
Derecske-I
0 5 10 15 20 km
a
2,4 Ma
DEPTH (km)
b

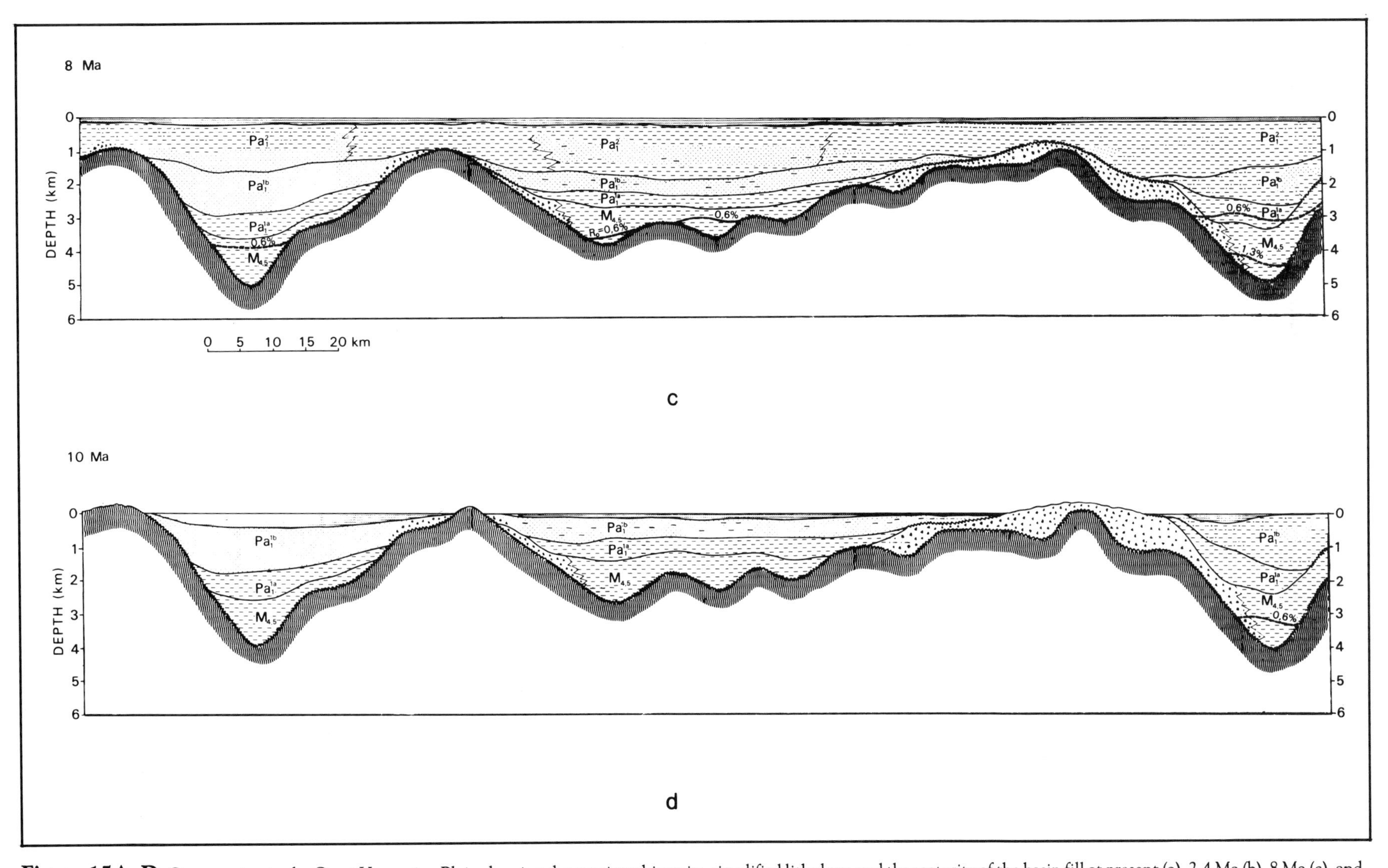

Figure 15A–D. Cross section in the Great Hungarian Plain showing the stratigraphic units, simplified lithology, and the maturity of the basin fill at present (a), 2.4 Ma (b), 8 Ma (c), and 10 Ma (d). The three hydrocarbon fields in Figure 15A are simplified and vertically exaggerated. The location of the section is shown in Figure 5. Key: 1 = stratigraphic units; 2 = marls and shales; 3 = sandstone; 4 = conglomerate and tectonic breccia; 5 = mixed development consisting of alternating sand and marl beds; 6 = water; 7 = basement; and 8 = isoreflectance lines (R_o = 0.6%, 1.3%, and 2%).

CONCLUSIONS

The main results of this work can be summarized as follows:

1. An improved computer model has been developed to simulate the subsidence, thermal, and maturation history of the Pannonian basin.
2. Computer simulation in the Great Hungarian Plain shows that moderate crustal extension, combined with major thinning (or heating) of the subcrustal lithosphere, has controlled the formation of the basin.
3. The simple Lopatin–Waples method works well for prediction of the maturity level of organic matter, provided that it is calibrated properly for use in a young and hot sedimentary basin.
4. Oil and gas fields in the Great Hungarian Plain have been found in immature reservoir rocks at relatively shallow depths (1.5 to 2.5 km) above basement highs, or locally within the basement. These hydrocarbons are derived from mature source rocks in nearby deep troughs ("hydrocarbon kitchens") and migrated updip a few tens of kilometers.
5. Model calculations suggest that a large volume of hydrocarbons should have migrated downward and/or laterally out of the source rocks, and could have been trapped in the basement and at the flanks of deep troughs at moderate depths (2 to 4 km).

ACKNOWLEDGMENTS

The Hungarian authors are indebted to the Petroleum Exploration Company, Szolnok (Hungary) for support in making a large part of this research. We thank them for their permission to publish.

REFERENCES

Ádám, A. and Á. Wallner, 1981, Information from electromagnetic induction data on Carpatho–Pannonian geodynamics: Earth Evol. Sci., v. 1, n. 3–4, p. 280–284.

Babuska, V., J. Plomerova, and J. Sileny, 1984, Large-scale oriented structures in the subcrustal lithosphere of Central Europe: Annales Geophys. v. 2, n. 6, p. 649–662.

Bally, A. W. and S. Snelson, 1980, Realms of subsidence, *in* A. D. Miall, ed., Facts and principles of world petroleum occurrence: Can. Soc. Petrol. Geol. Memoir 6, p. 9–94.

Bott, M. H. P., 1976, Formation of sedimentary basins of graben type by extension of the continental crust: Tectonophysics, v. 36, p. 77–86.

Carslaw, H. S. and J. C. Jaeger, 1959, Conduction of heat in solids: Oxford, Clarendon Press, p. 1-510.

Cohen, C. R., 1981, Time and temperature in petroleum formation: Application of Lopatin's method to petroleum exploration: Discussion. Amer. Ass. Petrol. Geol. Bull., v. 65, n. 9, p. 1647–1648.

Horváth, F. and L. Royden, 1981, Mechanism for the formation of the intra-Carpathian basins: A review: Earth Evol. Sci., v. 1, n. 3–4, p. 307–316.

Horváth, F. and J. Rumpler, 1984, The Pannonian basement: Extension and subsidence of an Alpine orogene: Acta Geol. Hung., v. 27, n. 3–4, p. 229–235.

Hovland, J. and E. S. Husebye, 1982: Upper mantle heterogeneities beneath Eastern Europe: Tectonophysics, v. 90, p. 137–151.

Issler, D. R., 1984, Calculation of organic maturation levels for offshore eastern Canada: Implications for general application of Lopatin's method: Canadian Journ. Earth Sci., v. 21, n. 4, p. 477–488.

Katz, B. J., L. M. Liro, J. E. Lacey, and H. W. White, 1982, Time and temperature in petroleum formation: Application of Lopatin's method to petroleum exploration: Discussion. Amer. Ass. Petrol. Geol. Bull., v. 66, n. 8, p. 1150–1152.

Lachenbruch, A. H., 1978, Heat flow in the basin and range province and thermal effects of tectonic extension: Pageoph, v. 117, p. 34–50.

Magara, K., 1978, Compaction and fluid migration: Amsterdam/New York, Elsevier, p. 1–319.

McKenzie, D., 1978, Some remarks on the development of sedimentary basins: Earth Planet. Sci. Let., v. 40, p. 25–32.

Parsons, B. and J. G. Sclater, 1977, An analysis of the variation of ocean floor bathymetry: J. Geophys. Res., v. 82, n. 5, p. 803–827.

Posgay, K., I. Albu, I. Petrovics, and G. Ráner, 1981, Character of the earth's crust and upper mantle on the basis of seismic reflection measurements in Hungary: Earth Evol. Sci., v. 1, n. 3–4, p. 272–279.

Royden, L., F. Horváth and B. C. Burchfiel, 1982, Transform faulting, extension and subduction in the Carpathian–Pannonian region: Geol. Soc. Amer. Bull., v. 73, p. 717–725.

Royden, L., F. Horváth, and J. Rumpler, 1983a, Evolution of the Pannonian basin system: 1. Tectonics: Tectonics, v. 2, n. 1, p. 63–90.

Royden, L., F. Horváth, A. Nagymarosy and L. Stegena, 1983b, Evolution of Pannonian basin system: 2. Subsidence and thermal history: Tectonics, v. 2, n. 1, p. 91–137.

Sclater, J. G., L. Royden, F. Horváth, B. C. Burchfiel, S. Semken, and L. Stegena, 1980, The formation of the intra-Carpathian basins as determined from subsidence data, Earth Planet. Sci. Lett., v. 51, p. 139–162.

Stegena, K., B. Géczy and F. Horváth, 1975, Late Cenozoic evolution of the Pannonian basin: Tectonophysics, v. 26, p. 71–90.

Stegena, K., F. Horváth, J. G. Sclater, and L. Royden, 1981, Determination of paleotemperature by vitrinite reflectance data: Earth Evol. Sci., v. 1, n. 3–4, p. 292–300.

Waples, D. W., 1980, Time and temperature in petroleum formation: Application of Lopatin's method to petroleum exploration: AAPG Bull., v. 64, p. 916–926.

Afterword: A General Approach to Basin Analysis

B. C. Burchfiel
L. H. Royden
Department of Earth, Atmospheric and Planetary Sciences
Massachusetts Institute of Technology
Cambridge, Massachusetts 02139

INTRODUCTION

The aim of this volume has been to present a study of basin evolution within the Pannonian basin, which is more accurately called the Pannonian basin system because it consists of several individual basins. The papers contained in this volume were designed to present basic data from a wide variety of geological disciplines, and to integrate these different data sets into a comprehensive study of the evolution of the Pannonian basin. These papers have focused not only on evolution of the basins themselves, but also on attendant processes within the crust and upper mantle that controlled the development of the basin in some fundamental ways.

The Pannonian basin system is a particularly good candidate for basin analysis, partly because the evolution of this young basin system is relatively simple. In addition, the basin system had one brief period of extension, a simple sedimentary history and a clearly defined regional tectonic setting. The active processes that formed the basin system were short-lived and recent, are essentially finished, and have not been overprinted by subsequent tectonic events. Events within the basin system itself can be spatially and temporally related to regional tectonic events outside of the basin area. The young age of the basin system ensures that many geological and geophysical data such as heat flow, seismic velocities, earthquakes, and so on, provide useful constraints on the processes of basin formation. Because dating is more accurate in young rocks than in older ones, events that are diachronous by as little as one or two million years can be documented in this young basin system, whereas in older basins these events would appear to be contemporaneous everywhere.

Through study and improved understanding of young, simple basin systems, many basic general processes of basin evolution can be recognized in areas where they can be documented most simply and unambiguously. One can then apply the insights gained from study of simple basins towards evaluation of older and more complex basins where basic processes may be much harder to identify. The papers contained in this volume provide an analysis of some of the basic basin processes that controlled the development of one young, simple extensional basin system that developed on continental crust. The results of these studies should have direct and obvious application to other basins of similar type. This volume also serves a broader purpose of illustrating one type of approach to basin analysis that could prove useful in studies of other basins in a variety of tectonic settings. A few of the basic concepts inherent in this approach are outlined as follows.

BASIN CLASSIFICATION IN BASIN ANALYSIS

Several classifications of basins have been published, such as those by Bally and Snelson (1980) or Kingston et al. (1983). Such basin classifications are useful because they provide a framework for grouping basins of generally similar origin and suggest a range of features to be expected within similar basin types. They offer a starting point for basin study and evaluation. As more detailed study proceeds, however, each basin or basin system must be analyzed from its own data sets. There are no type basins which can be used as a complete model for any other basin. Basins, like most other geological features, never develop from exactly the same set of circumstances, and any basin will belong to a spectrum of possible basin types, rather than mimicking a well-defined type basin.

The data and interpretations presented in this volume indicate that the Pannonian basin system has developed within a combined back arc and escape tectonic setting. Although the Pannonian basin system fits easily into a specific category in a basin classification system (back arc extensional on continental crust), it is composed of isolated but structurally related basins that display large individual variation. In fact, this system is an excellent example of the complexity of basin development that can be produced within a single broad tectonic setting. The differences between the Vienna, Transcarpathian, Great Hungarian and Transylvanian basins are extreme, yet they all are part of the Pannonian basin system. Detailed analysis of their tectonic setting within the basin system in this volume offers an interpretation of this variability and, to a first approximation, explains much of the data from these basins. It is at this scale of analysis that the broad basin classifications become less useful, and analysis must proceed based on specific data from the basin being analyzed. Here is where the classifications must be set aside and the analysis must proceed on an individual and unique course.

DIFFERENT SCALES IN BASIN ANALYSIS

A broadly based understanding of the Pannonian basin system only can be achieved by examining data and processes concurrently at several different scales. Interpretation of processes operating at one physical scale support or enhance interpretations of processes operating at another scale. Such a multiscale approach can often explain phenomena that could not be explained by examining the problem at only one particular scale. For example, it is clear from the studies presented in this volume that study of only one basin within the Pannonian basin system would be insufficient to understand the overall lithospheric and mantle processes operating in the area, or to relate the formation of that basin to events in adjacent basins and in the thrust belt. Only by studying the whole Carpathian–Pannonian system can one understand how the local development of an individual basin relates to the system as a whole.

A good understanding of the tectonic development of the Pannonian basin system requires tectonic analysis on at least three different scales: (1) the Cenozoic regional tectonic framework of western and eastern Europe; (2) the Carpathian-Pannonian system within this regional tectonic framework; and (3) individual basins within the Pannonian system and individual structures within the Carpathian thrust belt.

At the first and largest scale, examination of the Alpine belts of both western and eastern Europe in Cenozoic time provides a useful tectonic framework for the Pannonian basin system. For example within the Alpine chain, the small Pannonian lithospheric fragment east of the Austrian Alps is best interpreted as the product of Cenozoic escape of a continental fragment eastward away from the zone of diachronous continental collision and crustal shortening in the Austrian Alps. At this scale, the Pannonian area can be viewed as a roughly coherent fragment bounded by strike-slip faults systems on its northwest and southwest sides, and by a subduction zone (convergent boundary along the Carpathian mountains) on its northern and eastern sides. Within this tectonic escape setting, the boundary conditions for the Pannonian fragment are established.

At the second scale, the internal development of the Pannonian fragment can be analyzed in light of the boundary conditions imposed by the larger regional tectonic setting. Within the upper crust, the general relations between strike-slip, thrust and extensional faults can be developed and related to deformation along the Pannonian fragment boundaries. It is no coincidence that formation of the Pannonian basin system was coeval with shortening in the Carpathian thrust belt. For the Pannonian basin system, such analyses show that within the upper crust, areas of contemporaneous thrusting and extension are connected by a conjugate system of strike-slip faults, and that the overall direction of extension within the basin area is consistent with the direction and distribution of crustal shortening within the Carpathian thrust belt.

From the kinematics and timing of upper crustal deformation and displacements within the Pannonian fragment and deformation along the fragment boundaries, a qualitative interpretation of deep lithospheric and asthenospheric processes for the entire tectonic system can be made. Such deep processes control many of the fundamental aspects of basin development at the surface, such as subsidence and sedimentation rates, heat flow, temperatures, and organic maturation. Conversely, accurate measurements of these and other features can be used to further refine and test mantle processes related to basin formation. Often even quantitative analyses may not be sufficient to distinguish amongst several viable hypotheses for lithospheric behavior, but usually they can eliminate at least some of the alternatives.

At the third, even smaller scale, individual basins within the Pannonian system can be examined for their similarities, differences and relationships to one another and to the Carpathian thrust belt. From the studies presented in this volume, there are clear and significant differences between individual basins. For example, initiation of subsidence, extension and sediment accumulation within the basins varies by a few million years from one part of the basin system to another. These temporal shifts in the locus of extension within the basin system can be related to diachronous shortening within the Carpathian thrust belt and clearly indicate that, at the scale of individual basins, basin formation is linked to detailed changes in thrust belt activity. Other important differences between basins of the Pannonian system, such as their heat flow and thermal subsidence, and fundamental and systematic variations in upper mantle structure beneath the basin system, can be linked to the geometry of the subduction zone beneath the Carpathian thrust belt.

A similar type of multiscale analysis is probably necessary for all major basins. Most, if not all, large basins are formed as a result of lithospheric and deeper mantle processes. Each basin forms part of a broader dynamic system and must be understood within this broader dynamic system. The scale at which this broader dynamic system operates is never clear from restricted studies of data from within a basin. The Pannonian basin system requires an understanding of part of the eastern and western European Alpine system, but other basins may require analysis of tectonic systems that extend over a much larger region. It is clear, however, that basins must be studied in the context of a tectonic setting that extends well beyond the basin itself.

INTEGRATED DATA SETS IN BASIN ANALYSIS

One of the main aims of this volume, and of the cooperative project behind it, was to develop an integrated approach to basin analysis, and provide a geologic framework for evolution of the Pannonian basin system. The chapters in this volume show how many different aspects of basin evolution are genetically related, and how such different aspects of basin development can be studied and combined to understand the processes that controlled the basin history.

In these particular studies, integration was achieved by bringing together results from many different working groups, rather than by assembly of a large working group with specialists in many disciplines. Through such cooperation, a broadly based and multifaceted understanding of the Pannonian basin has been produced. A representative sample of this material has been presented in this volume. Other types of data could, and should, be used to focus on specific problems or processes, and this volume cannot be considered complete in this regard. Thus, the studies put forth here provide one level of understanding of the evolution of the Pannonian basin system, and should provide a broadly based framework on which to hang more detailed future work. Moreover, the level of integration achieved is clearly only a first pass at integrating huge existing data sets of very different types. Much room exists for better correlations of existing data.

One example of such integration of different data sets is the study of extensional styles within the different basins of the Pannonian system. Extension in some of the basins involved only the upper crust, while in others it involved the entire lithosphere. This conclusion was reached only by analyzing a combination of data sets, including thermal studies, subsidence history, sedimentological and depositional environments, as well as local structures and regional tectonics. No single data set, taken by itself, could have yielded the same result.

A few of the types of data integration used implicitly or explicitly in the Pannonian Basin studies can be outlined as follows:

Local tectonics: Depends on correlation with regional events; local structures (as determined from surface mapping, drilling and seismic data); basin filling history (sedimentology, paleoecology); dating of sedimentary rocks (paleontology, paleomagnetic and isotopic dating); active tectonics and seismicity.

History of basin filling: Dating of sedimentary fill (see above); facies reconstruction (lithofacies, seismic facies, relationship to paleontologic, radiometric and paleomagnetic dates); identification of sedimentary processes and environments; relationships to coeval tectonic processes and local structures.

Deep crustal and upper mantle processes: Implications of surface mapping and its downward projection by seismic reflection and refraction studies, drilling data, construction of balanced cross-sections, gravity and magnetic studies, crustal and mantle velocities, thermal studies, and subsidence studies.

Regional correlations: Coeval regional events, local tectonics, deep crustal and upper mantle processes, kinematics, palinspastic reconstructions, and possible dynamics.

Maturation studies: Measurements (vitrinite, T_{max}); mapping of organic facies (kerogen type, and so on); mapping of lithofacies; thermal reconstructions (heat flow history, burial history, thermal conductivity, etc.); and timing of maturation.

Hydrocarbon prospects: Local structures and potential traps, location of potential source rocks, probable migration directions, and permeable and impermeable zones.

The data sets and types of analysis given above are not intended as a complete listing to be used in basin analysis. They are simply meant to illustrate the kinds of data integration that must be achieved in basin analysis, and to emphasize that even a single aspect of basin analysis is dependent on many different data sets. The types of data available and the types of analyses used vary from basin to basin, but a successful study of any basin requires broadly based analyses of integrated data sets.

CONCLUSIONS

The chapters in this volume, and the cooperative work behind them, make up a geological investigation of the Pannonian basin system, and illustrate one approach to basin analysis. Three fundamental aspects of this basin study, which we feel are necessary in any successful basin study, are: (1) integration of data sets; (2) interdisciplinary cooperation and analyses, including a blending of qualitative and quantitative data and methods; and (3) synthesis of observations at many different scales, including both shallow crustal and deeper mantle processes. The fruitful exchange and cooperation between a wide variety of organizations and individuals has been important in this study of the Pannonian basin system. Major contributions to this study have come from academic institutions, government agencies and industrial groups working together, and the cooperation and exchange of data and ideas has been free and open. In this regard, the present volume can serve as an example of a successful international scientific cooperative effort.

REFERENCES

Bally, A. W. and S. Snelson, 1980, Realms of subsidence, *in* A. D. Miall, ed., Facts and principles of world petroleum occurrence: Canadian Soc. Petrol. Geol. Mem., v. 6, p. 9–75.

Kingston, D. R., C. P. Dishroon and P. A. Williams, 1983, Global basin classification system: Amer. Assoc. Petrol. Geol. Bull., v. 67, p. 2175–2193.

Explanatory Notes to Maps

MAP 1

The Carpathian–Pannonian region is shown and names and locations are given in the basins. The areas where the Neogene–Quaternary rocks in the Pannonian basin exceed 3 km in thickness are circled by a green line. Scale 1:2,000,000. Compiled by F. Horváth.

MAP 2

The Carpathian–Pannonian region is shown along with outer Carpathian units. Flysch and molasse units within the outer Carpathian thrust belt are shown. The total thickness of Neogene–Quaternary rocks within the Pannonian region is also shown, with contour lines every 500 meters. The sources for these contour lines are given in the explanation to Map 8. Contour lines showing the total thickness of Neogene–Quaternary rocks in the Carpathian foredeep are from Mahel (1973) and Paraschiv (1979). Scale 1:2,000,000. Compiled by L. Royden and M. Sandulescu.

REFERENCES

Mahel, M., ed., 1973, Tectonic map of the Carpathian-Balkan Mountain system and adjacent areas, scale 1:1,000,000: D. Stúr Geological Institute, Bratislava and UNESCO, Vienna.

Paraschiv, D., 1979, Oil and gas in Romania: Institutul de Geologie si Geofizica, Studii Tehnice si Economice, series A, Bucurest, 382 pages.

MAP 3

This map was compiled from Landsat imagery (MSS 7) by J. Lerner. Scale 1:2,000,000.

MAP 4

Locations of oil and gas fields in the Carpathian–Pannonian region are numbered on the map (compiled from Paraschiv, 1979; Wessely, this volume, and the Map of Petroleum-Bearing Basins in European Scoialist Countries). The numbers correspond to the names of the fields as listed below. The total thickness of Neogene–Quaternary rocks in the Pannonian region is also shown (from Map 8). Oil and gas fields in Hungary are discussed in greater detail in Dank (this volume). Fields in the Austrian part of the Vienna basin are shown in more detail in Wessely (this volume). Map compiled by V. Dank. Scale 1:2,000,000.

REFERENCES

Dank, V., this volume, Petroleum geology of the Pannonian basin, Hungary: An overview.

Paraschiv, D., 1979, Oil and Gas in Romania: Institutul de Geologie si Geofizica, Studii Tehnice si Economice, series A, Bucurest, 382 pages.

Wessely, G., this volume, Structure and development of the Vienna basin in Austria.

Map of petroleum-bearing basins in European Socialist Countries (in Russian), 1979, Moscow, scale 1:2,000,000.

KEY

P_z, Paleozoic or earlier; T, Triassic; J, Jurassic; K, Cretaceous; Pg_1, Paleocene; Pg_2, Eocene; Pg_3, Oligocene; N_1, lower and middle Miocene; N_2, upper Miocene and Pliocene.

LIST OF FIELD NAMES

Capital letters after each field name give the age of the reservoir rock(s) in which oil is trapped. Small letters show the age of reservoir rocks in gas and/or gas condensate fields.

West Carpathian Foredeep

1. Nikolcice j_3
2. Menin–Zhalchani N_1
3. Nizkovice n_1 p_z
4. Lubna P_z n_1 p_{g3} p_z
5. Príbor–Klokocov n_1 p_z
6. Staric p_z
7. Kozlovice (Frydek) n_1 p_z
8. Paskov n_1 p_z
9. Lískovec–Bruzovice p_z
10. Dolni Zukov n_1 p_z
11. Stonava n_1 p_z
12. Marklowice p_z
13. Debowiec n_1
14. Poguz n_1
15. Racibórske n_1
16. Nieznanowice–Grabira n_1
17. Bochnia–Gerzice n_1
18. Plavovice K_2 J_3
19. Grobla K_2 J_3
20. Lonkta n_1 k_2 j_3
21. Sufzin n_1
22. Tarnów n_1 j_3
23. Ladna n_1
24. Swazów n_1 k_2
25. Dabrowa Tarnowska J_3 k_2
26. Smogorzów J_3 k_2

27. Medrzechów n_1
28. Partynia N_1 J_3
29. Smoczka n_1
30. Wojstaw n_1 j_3
31. Kozhenuv N_1 J_3
32. Debica–Brzezówka N_1 J_3
33. Sedziszów n_1
34. Czarna–Sedziszowska n_1
35. Niwiska k_1 t_3 p_z
36. Czesniki n_1
37. Wola Ranizewska n_1
38. Lipnica n_1
39. Kamién n_1
40. Ezhove n_1
41. Sarzyna n_1
42. Zokynia–Lezajsk n_1
43. Albigowa–Krasne–Gluchow n_1
44. Uskovce n_1 j_3 j_2 p_z
45. Ryszkowa Wola n_1
46. Lubaczow–Kochanowka–Swidnica N_1 J_3 n_1 j_3
47. Cetynia n_1 p_z
48. Mirocin n_1
49. Jarostaw n_1
50. Kanczuga n_1
51. Rudolovice n_1
52. Radymno n_1
53. Zalaze n_1
54. Svente n_1
55. Zadabrowie n_1
56. Przemysl–Jaksmanice–Hodnoviczi n_1
57. Sadkowice n_1
58. Pinyani n_1
59. Rudki n_1 j_3
60. Nowosiótki n_1
61. Zaluzhani n_1
62. Malo–Gorozhanka n_1 k_2
63. Medinichi n_1 k_2
64. Gruszów n_1
65. North Bilche–Wolica n_1 k_2
66. Opari n_1
67. Bilche-Wolica n_1 k_2
68. Kavsko n_1
69. Ugersko n_1 k_2
70. Dasava n_1
71. Kadobno n_1
72. Kalus n_1
73. Grinovka n_1
74. Bogorodchani n_1
75. Jablonov n_1
76. Kosov n_1
77. Kovalevka–Tscheresenka n_1
78. Staryava Pg_2 K_2
79. Rossoha Pg_2 K_2
80. Strelbichi Pg_1
81. Staryj Sambor Pg_1
82. Naguevichi N_1 $Pg_{2\text{-}3}$
83. Popeli $Pg_{2\text{-}3}$
84. Borislav N_1 Pg
85. Opaka Pg K_2
86. St. Kropivnik $Pg_{2\text{-}3}$
87. Ribnik $Pg_{2\text{-}3}$
88. "MEP" Pg_1
89. "Miriam" K_2
90. Shodnica $Pg_{1\text{-}2}$ K_2
91. Urich Pg K_2
92. Orov Pg_1
93. Vapnyarka K_2
94. Ivaniki N_1 $Pg_{2\text{-}3}$ $pg_{2\text{-}3}$
95. Orov–Ulichno $Pg_{2\text{-}3}$
96. Stinava $Pg_{2\text{-}3}$
97. Tanyava Pg_3
98. Yuzhn. Tanyava Pg_1
99. Sev. Dolina N_1 $Pg_{2\text{-}3}$
100. Dolina N_1 Pg
101. Vitvica (deep) Pg_3
102. Vigoda k_2
103. Vitvica K_2
104. Strutin $Pg_{2\text{-}3}$
105. Spas Pg_3
106. Ripne Pg_3
107. Nebulov Pg_3
108. Maidan Pg_2
109. Rossolnaya pg_2
110. Kosmach $pg_{2\text{-}3}$
111. Gvizd $Pg_{2\text{-}3}$
112. Pniv Pg_3
113. Pasechnaya Pg K_2
114. Bitkov–Babche N_1 $Pg_{2\text{-}3}$ $pg_{2\text{-}3}$ k_2
115. Sloboda Rungurskaya K_2
116. Kosmach Pokutsky $Pg_{2\text{-}3}$ K_2

West Carpathian Flysch Zone

117. Gluk Pg k_2
118. Klenchani Pg_3 K_2
119. Librantowa–Stara Wies Pg_1 K_2
120. Stopnice pg_3
121. Rzepiennik K_2
122. Lichana K_2 $Pg_{1\text{-}2}$
123. Salova pg
124. Simbark Pg_2 K_2
125. Syari K_2
126. Senkova–Ropica Pg_1 K_2
127. Magdalena Pg K_2
128. Kryg–Lipinki $Pg_{1\text{-}2}$ K_1
129. Mecina–Wielka Pg_1 K
130. Ganka–Felneruvka Pg_3 K_2
131. Harklowa $Pg_{2\text{-}3}$
132. Bech $Pg_{1\text{-}2}$ K_2
133. Osobnica $Pg_{1\text{-}2}$ K_2
134. Folusv–Pielgrzymaka $Pg_{2\text{-}3}$
135. Mrukova $Pg_{2\text{-}3}$
136. Sverhova Pg_1 K_2
137. Rostoki Pg K_2 pg k_2
138. Yasev Pg_2 K_2
139. Potok $Pg_{1\text{-}2}$
140. Turaszówka $Pg_{1\text{-}2}$
141. Kobylyani Pg_2 K_2
142. Draganowa Pg_3
143. Ropienka Pg_1 K_2
144. Weglówka K_1

145. Wola Jasienicka Pg_1 K_2
146. Wola Komborska Pg_1 K_2
147. Kroscienko Pg_{1-2}
148. Targoviska Pg_3
149. Bubrka–Rogi–Ruvne Pg_{1-2} K_2
150. Zbojska Pg_3
151. Iwonicz–Wieś–Klimkuvka Pg K_2
152. Iwonicz–Zdruy Pg K_2
153. Rudawka Rymanovska Pg_{2-3} K_2
154. Stara Wieś K_2
155. Tuzhe Pole Pg_{1-2}
156. Tshesnuv Pg_{2-3}
157. Grabownica Pg_2 K_1
158. Grabownica Ves pg k_2
159. Strahocina pg k_2
160. Tokarnia Pg_3 K_2
161. Zabtotce pg
162. Zagórz Pg_3
163. Tarnawa Velopole Pg_{2-3}
164. Mokre Pg_3
165. Slonne Pg_2 K
166. Vara K_1
167. Vitriluv Pg_{1-2} K_2
168. Hlomcha Pg_1
169. Tirava Solna Pg_{2-3}
170. Uherce Pg_3
171. Vankova Pg_{2-3}
172. Ustyanova Pg_3
173. Raiske Pg_3
174. Zatvarnica Pg_3
175. Lodina Pg_{2-3}
176. Brzegi Dolne Pg_{2-3}
177. Stebnik Pg_3
178. Bandruv Pg_3
179. Hosuv Pg_3
180. Bistre Pg_3
181. Czarna–Lipe Pg_3
182. Polana Pg_3
183. Gronzeva Pg_{2-3} pg_{2-3}
184. Golovecko Pg_3 pg_1
185. Maly Volosyanka Pg_3
186. Lopusanka Pg_3
187. Lugi Pg_3
188. Bitlya Pg_3
189. Pogar Pg_3
190. Kozeva Pg_{2-3} K_2
191. Yasinya Pg_3
192. Dihtinec Pg_{2-3}

Outer Zones of the East and South Carpathians

193. Frasnin N_1 n_1
194. Valea Seaca n_1
195. Malini n_1
196. Cuejd N_1
197. Tazlaul Mare Pg_3
198. Frumoasa Pg_3
199. Geamana Pg_{2-3}
200. Gropile lui Zaharache Pg_3 pg_3
201. Toporu–Chilii Pg_3
202. Arsita Pg_3
203. Mihoc Pg_3 pg_3
204. Cucueti Pg_3
205. Chilii W Pg_3 pg_3
206. Tasbuga Pg_3
207. Zemes–Chilioaia Pg_3
208. Foale–Tazlaul Pg_3
209. Uture–Moinesti Pg_3
210. Comanesti N_1 Pg_{2-3}
211. Vasiesti Pg_{2-3}
212. Darmanesti Pg_3
213. Doftenita Pg_3
214. Dofteana–Bogata Pg_{2-3}
215. Pacurita Pg_2
216. Slanic–Bai Pg_3
217. Slanic–Ferastrau Pg_3 pg_3
218. Cerdac Pg_3 pg_3
219. Lepsa Pg_3
220. Ojdula Pg_3 pg_3
221. Ghelinta Pg_3 pg_3
222. Catiasu Pg_2
223. Popesti N_1
224. Cosminele N_1 Pg_3 pg_2
225. Virfuri–Visinesti Pg_2 pg_2
226. Cimpeni N_1
227. Tescani N_1 Pg_3 pg_3
228. Casin N_1
229. Cimpuri N_1
230. Bisoca N_1
231. Plopeasa N_1 n_{1-2}
232. Arbanasti–Berca N_1 n_2
233. Barbuncesti N_1 n_{1-2}
234. Grajdana N_1 n_{1-2}
235. Sarata–Monteoru N_1 n_2
236. Surani–Carbunesti N_1 Pg_3
237. Copaceni–Opariti N_1 Pg_3 pg_3
238. Matita–Podenii Noi N_{1-2}
239. Apostolache N_1
240. Tataru N_1
241. Malu Rosu n_1
242. Urlati–Ceptura N_1 n_{1-2}
243. Podenii Vechi N_1 n_{1-2}
244. Pacureti N_{1-2} n_{1-2}
245. Magurele n_{1-2}
246. Scaiosi N_1
247. Vilcanesti N_2
248. Boldesti N_1 n_{1-2}
249. Cutorani N_1
250. Aricesti N_1 n_{1-2}
251. Baicoi–Tintea N_{1-2} n_1
252. Cimpina–Gura Draganesii N_1 Pg_3
253. Bustenari–Runcu N_1 Pg_3
254. Silistea N_1
255. Ochiuri N_{1-2} Pg_3
256. Colibasi–Ocnitei N_1
257. Valea Resca N_1
258. Glodeni N_1 n_1
259. Aninoasa N_1 n_3
260. Teis–Viforita N_1 n_{1-2}
261. Gura Ocnitei W–Razvad N_{1-2}
262. Moreni–Gura Ocnitei N_{1-2} n_{1-2}

263. Margineni N_1 n_{1-2}
264. Manesti–Vladeni n_{1-2}
265. Frasin–Brazi n_{1-2}
266. Finta–Gheboaia n_{1-2}
267. Gura Sutii n_1
268. Bucsani N_{1-2} n_{1-2}
269. Bratesti n_2
270. Sotinga N_1
271. Dragaesti N_1 n_1
272. Dragomiresti N_1 n_1
273. Dobresti n_1
274. Ludesti N_1 n_1
275. Botesti n_1 pg_3
276. Cobia N_1
277. Suta Seaca N_1 n_1
278. Leordeni N_1 n_1
279. Glimbocelu N_1 n_1
280. Bogati N_1 n_1
281. Folesti n_1
282. Vilcele N_1 Pg_3 n_1
283. Sapunari Pg_3 pg_{2-3}
284. Merisana N_1 Pg_3 n_1
285. Colibasi N_1 n_1
286. Slatioarele N_{1-2}
287. Calinesti–Oarja N_1 n_1
288. Silistea-Ciresu N_1 n_1
289. Vata N_1 n_{1-2}
290. Otesti N_1 n_1
291. Cazanesti n_1
292. Babeni N_1 n_1
293. Galicea n_{1-2}
294. Folesti Pg_3
295. Gradiste n_1
296. Romanesti N_1 n_1
297. Zatreni n_1
298. Hurezani n_1
299. Bibesti–Sardanesti n_1
300. Alunu N_1 n_1
301. Bustuchini N_1 n_1
302. Colibasi N_1 n_{1-2}
303. Tg. Jiu N_1 n_1
304. Tamasesti n_1
305. Strimba–Rogojelu n_1
306. Bala n_1
307. Bilteni N_1 n_1
308. Ticleni N_1 n_1
309. Socu n_1
310. Roman–Secueni n_1
311. Bacau n_1
312. Gaiceana n_1
313. Glavanesti N_1 n_1
314. Huruesti N_1 n_1
315. Negulesti N_1 n_1
316. Adjud n_1
317. Tepu N_1 n_1
318. Matca n_1
319. Frumusita N_1 n_{1-2}
320. Independenta N_{1-2} n_{1-2}
321. Suraia N_1 n_1
322. Boldu n_2
323. Balta Alba n_2
324. Bobocu n_2
325. Rosioru n_2
326. Ghergheasa n_2
327. Oprisenesti N_1 n_1
328. Perisoru N_2 n_2
329. Bordei Verde N_2 n_2
330. Filiu N_2
331. Jugureanu N_1 K_1 n_1
332. Victoria n_2
333. Padina N_1 K_1 n_1
334. Amara
335. Bragareasa K_1 n_1
336. Lipanesti K_1
337. Girbovi n_1
338. Sinaia n_1
339. Baraitaru n_1
340. Urziceni N_1 K_1 n_1
341. Colelia N K_1
342. Colelia S N_1
343. Orezu n_2
344. Ileana N_1 n_1
345. Pasarea n_2
346. Cozieni n_1
347. Catelu N_1
348. Balaceanca N_1 n_1
349. Popesti n_2
350. Fierbinti N_1 n_1
351. Moara Vlasiei n_2
352. Peris K_1 n_1
353. Bilciuresti n_2
354. Serdanu K_1
355. Brincoveanu K_1 n_1
356. Titu K_1
357. Vultureanca N_1 K_1
358. Dumbrava K_1
359. Draghineasa N_1 K_1
360. Brosteni K_1 n_1
361. Petresti N_1 K_1 n_1 k_1
362. Stoenesti N_1 n_1
363. Gradinari N_1
364. Bragadiru N_1 n_1
365. Novaci N_1 n_1
366. Balaria N_1 K_1
367. Talpa N_1 K_1
368. Hirlesti East K_1 k_1
369. Videle–Blejesti N_1 K_1 n_1 k_1
370. Sopirlesti N_1 K_1
371. Hirlesti K_1 k_1
372. Cartojani N_1 n_1
373. Preajba N_1
374. Glavacioc N_1 k_1
375. Glogoveanu N_1 K_1
376. Stefan cel Mare N_1 K_1 k_1
377. Dumbreveni K_1
378. Tataresti n_2
379. Ciolanesti K_1 k_1
380. Caldararu k_1
381. Gliganu K_1 n_1
382. Recea K_{1-2} k_1
383. Surdulesti N_1 K
384. Birla N_1 K_1 n_1 k_1

385. Tufeni K_1 J_2
386. Ciuresti South K_1 J_2 T_3
387. Bacea N_2 K_1
388. Ciuresti North N_1 K_1 J_2 n_1 j_2
389. Spineni J_2
390. Oporelu J_2 T j_2 t
391. Strejesti n_2
392. Malu Mare J_2 n_2 j_2
393. Simnic J_2 n_2 j_2
394. Ghercesti J_2 n_2 j_2
395. Iancu-Jianu J_2 T_3 n_1 j_2
396. Fauresti J_2 j_2
397. Bradesti T t
398. Melinesti T
399. Virteju n_{1-2}
400. Bibesti N_1 T Pz n_1 t
401. Gorni Dibnik T_2
402. Dolni Dibnik T_2 t_3
403. Devetaki t_2
404. Dolni Lukovit J_1
405. Birdarski Geran J_1
406. Ciren j_1 t_1

Little Plain (Hungarian and Slovakian Part)

407. Nizná n_1
408. Madunice n_1
409. Krupá n_1
410. Spacince n_1
411. Trakovice n_1
412. Ivanka n_1
413. Mihályi n_1
414. Kecskemét n_1
415. Vése n_1
416. Ikervár n_1

Zala Basin and Dráva Trough

417. Nagylengyel N_1 K_2 T
418. Barabásszeg N_1 K_2
419. Szilvágy N_1 K_2
420. Ederics n_2
421. Hahót N_1 T n_2
422. Kilimán n_2
423. Filovci N_2 n_2
424. Dolina (Lendava) N_1 n_2
425. Petisovci N_{1-2} n_{1-2}
426. Lovászi N_2 n_1
427. Budafa N_{1-2} n_2
428. Selnica N_2
429. Peklenica N_2
430. Veliki Otak n_{1-2}
431. Legrad N_1 n_{1-2}
432. Belezna N_1 n_2
433. Bajcsa n_2
434. Inke n_{1-2}
435. Buzsák N_1
436. Őrszentmiklós Pg_3
437. Füzesgyarmat
438. Fedémes Pg_3 pg_3
439. Bükkszék Pg_3
440. Demjén Pg_3
441. Mezőkeresztes Pg_{2-3} T
442. Farmos n_2
443. Tarany N_1
444. Lepavina N_1 Pz
445. Jagnedovac N_{1-2}
446. Mosti N_2 n_2
447. Ferdinandovac N_{1-2} n_{1-2}
448. Heresznye N_2 n_2
449. Görgeteg-Babocsa N_2 n_2
450. Hampovica N_1 n_{1-2}
451. Cepelovac N_2 n_2
452. Pitomaca n_{1-2}
453. Sandrovac N_{1-2}
454. Gakovo N_2 n_2
455. Pepelana n_2
456. Cabuna N_1
457. Boksic pz
458. Benicanci N_1
459. Obod N_1
460. Vizovav N_1

Sava Trough

461. Dugo Selo N_2 n_2
462. Jezevo N_1
463. Klostar N_{1-2} Pz
464. Sumetani N_1 Pz
465. Ivanic-grad N_1
466. Prkos N_{1-2}
467. Zutica N_{1-2}
468. Bunjani N_1 Pz
469. Okoli n_{1-2}
470. Voloder N_1 n_2
471. Struzec N_2 n_2
472. Mramor Brdo N_{1-2} n_2
473. Goilo N_{1-2}
474. Janja Lipa n_2
475. Jamarica N_{1-2}
476. Bujavica n_2
477. Lipovljani N_{1-2} n_2

Great Hungarian Plain and Its Southern Extension

478. Moftinu n_2
479. Madaras N_1
480. Piscolt n_{1-2}
481. Curtuiuseni N_1 n_1
482. Hajdúszoboszló n_{1-2} pg
483. Kaba North n_2
484. Ebes n_2
485. Nádudvar n_2
486. Kaba n_2
487. Sacueni n_2
488. Ciocaia n_{1-2} pz
489. Abramut N_{1-2}

490. Suplacu de Barcau N_2
491. Derna N_2
492. Siniob N_1
493. Mihai Bravu N_1 n_1
494. Tatárüllés–Kunmadaras n_2
495. Karcag–Bucsa n_2
496. Püspökladány n_2
497. Fegyvernek n_2
498. Nagykörü n_2
499. Kisujszállás n_{1-2}
500. Turgony n_2
501. Turkeve n_2
502. Biharnagybajom N_{1-2} n_2
503. Furta n_{1-2}
504. Körösszegapáti n_{1-2}
505. Bors N_1 Pg_3
506. Tiszapüspöki n_2
507. Szandaszöllos n_2
508. Rákóczifalva n_2
509. Jászkarajenő n_2
510. Nagykörös n_{1-2}
511. Nagykörös S–Kecskemét N_1
512. Szarvas n_2
513. Szank N_1 n_1
514. Soltvadkert n_2
515. Tázlár N_{1-2} Pz n_2
516. Pusztaföldvár N_2 n_2
517. Pusztaszöllös N_2 n_2
518. Tótkomlós N_2 n_2
519. Mezőhegyes N_2 n_2
520. Battonya N_2 n_2
521. Turnu N_2 n_2 pz
522. Rém n_2
523. Öttömös N_2
524. Üllés N_{1-2} n_{1-2}
525. Dorozsma N_2
526. Algyő N_{1-2} Pz n_{1-2}
527. Ferencszállás N_2 n_2
528. Tompa n_2
529. Kelebia N_2
530. Ásotthalom N_2
531. Mezőcsokonya
532. Horgos n_{1-2}
533. Backi Vinogradi n_1
534. Palic N_1 Pz-T
535. Cherestur N_2 n_2 pz
536. Teremia N_2 n_2 pz
537. Majdan N_2
538. Novi Knezevac t
539. Velebit N_{1-2}
540. Cantavir n_1 t
541. Gorni Breg n_{1-2}
542. Mokrin N_{1-2} n_{1-2}
543. Tomnatek n_2
544. Varjas N_1 n_1 pz
545. Sandra–Satkinez N_{1-2} Pz n_{1-2} pz
546. Kalacea N_{1-2} Pz n_{1-2} pz
547. Ada n_1 pz
548. Kikinda-Varas N_{1-2} Pz n_2
549. Kikinda-Pole N_2 Pz n_2
550. Milosevo n_2
551. Srpska Cpna n_2
552. Becei n_2
553. Padicevicevo n_2
554. Ruski Krstur n_{1-2}
555. Srbobran n_1 pz
556. Melenci n_2 k
557. Karadzordzevo K n_{1-2}
558. Begeici n_2
559. Media n_2 k
560. Sinmartin n_2
561. Elemir N_{1-2} J
562. Gospodinci n_1 k
563. Boka N_1
564. Velika Greda n_1
565. Plandiste n_1
566. Janosik N_1
567. Ermenovci N_1 Pz
568. Lokve N_1 Pz
569. Alibunar n_1
570. Nikolinci n_1 pz
571. Tilva n_1 pz
572. Mramorak n_1 pz
573. Mramorak Selo n_1
574. Darány N_2
575. Barcs N_1
576. Vízvár N_1 N_2
577. Pusztaapáti T
578. Sávoly T N_1
579. Oltárc
580. Ortaháza
581. Pusztamagyaród
582. Kiskunhalas Pz
583. Zsana
584. Kiskunmajsa
585. Forráskút
586. Szeged
587. Endrőd N_1 N_2
588. Erosztő
589. Sarkadkeresztúr
590. Szeghalom
591. Kismarja
592. Komádi
593. Dévaványa

Transcarpathian Depression

594. Pozdisovce N_1
595. Trgoviste n_1
596. Banovce n_1
597. Lastomir n_1
598. Stretava n_1
599. Trebisov n_1
600. Ptruska n_2

Vienna Basin

601. Lednice pg
602. V. Bilovice–Podivin N_1 n_1
603. Poddvorov n_1

604. Mutenice n_1
605. Vacenovice N_1 Pg n_1
606. Josefov N_1
607. Luzice N_1 n_1
608. Godovin N_1
609. Breclav N_1 n_1
610. Gruski N_1 n_1
611. Tynec N_1 Pg
612. Cunin N_1 n_1
613. Kostice N_1
614. Gbely N_1
615. Petrova Ves N_1 n_1
616. Stefanov N_1 n_1
617. Lanzhot N_1 n_1
618. Brodské N_1
619. Kúty n_2
620. Zavod N_1 n_1
621. Studenka N_1 n_1
622. Jakubov n_1
623. Suhograd n_{1-2}
624. Malacky n_1
625. Láb n_{1-2}
626. Visoka n_{1-2}
627. Zwerndorf
628. Schönkirchen
629. Mühlberg
630. St. Ulrich Hauskirchen
631. Pirawarth, Hochleithen
632. Hirschstett
634. Fischamend, Orth
635. Wienerherberg

Transylvanian Basin (n_1 for All Fields)

636. Bendin
637. Strugureni
638. Encin
639. Fintinele
640. Puini
641. Buza
642. Lunca
643. Sarmasel
644. Craesti–Ercea
645. Voivodeni
646. Ibanesti
647. Ulies
648. Sinmartin
649. Bozed
650. Paingeni
651. Feleac
652. Damieni
653. Sincai
654. Dumbraviora
655. Zaul
656. Saulia
657. Grebenis
648. Dobra
659. Ernei
660. Miercurea Nirajulni
661. Magherani
662. Ludus
663. Singer
664. Iclanzel
665. Vaidei
666. Sause
667. Tirgu Mures
668. Corunca
669. Ghinesti–Trei Sate
670. Cusmed
671. Acatari
672. Bogata
673. Lechinta–lernut
674. Cucerdea
675. Suveica
676. Galateni
677. Singeorgiu de Padure
678. Firtusu
679. Tarcesti
680. Bentid
681. Bradesti
682. Laslau Mare
683. Filitelnic
684. Soimus
685. Telina
686. Delenii
687. Prod-Seleus
688. Nades
689. Eliseni
690. Cristuru
691. Cetalea de Balta
692. Bazna
693. Noul Sasesc
694. Taureni
695. Copsa Mica
696. Birghis
697. Daia Telina
698. Bunesti Crit
699. Beia
700. Petis
701. Rusi
702. Ilimbav

MAP 5

The sources and method used to compile this heat flow map are summarized in Dövényi and Horváth (this volume). Scale 1:2,000,000. Compiled by P. Dövényi and F. Horváth.

REFERENCES

Dövényi, P. and F. Horváth, this volume, A review of temperature, thermal conductivity and heat flow data for the Pannonian basin.

MAP 6

The volcanic rocks and the volcanoclastic sediments in the Pannonian basin are in large part covered by younger sedimentary rocks. The rhyolite tuff horizons and the ignimbrites are mainly known from borehole data. A discussion of major element chemistry and ages of the volcanic rocks is given by Póka (this volume). Some trace element geochemical data is presented by Salters et al. (this volume). Sources used to compile this map include: Gheorghieva (1978), Mahel (1978), and Ravasz (1981). Scale 1:2,000,000. Compiled by T. Póka.

REFERENCES

Dimitrova, E., 1978, Magmatic associations in the Carparthian–Balkan area: Commission on Magmatism-Metamorphism of the Carpatho–Balkan Geological Association, Geological Institute of the Bulgarian Academy of Sciences.

Mahel, M., 1978, Geotectonic position of magmatites in the Carpathians, Balkans and Dinarides: Zapadné Karpaty, Séria geológia 4, Bratislava, 173 pages.

Póka, T., this volume, Neogene and Quaternary volcanism of the Carpathian–Pannonian region: Changes in chemical composition and its relationship to basin formation.

Ravasz, Cs., 1981, Magyarország medencebeli miocén vulkanitjainak elterjedése (Spatial distribution of Miocene volcanic rocks in the basins of Hungary): unpublished map, Document Department of the Hungarian Geological Survey (MÁFI), scale 1:500,000.

Salters, V., S. R. Hart and Gy. Pantó, this volume, Origin of Late Cenozoic volcanic rocks of the Carpathian Arc, Hungary.

MAP 7

Epicenters for earthquakes in the East Alpine, Dinaric and Carpathian regions. Faults shown on the map have been taken from Mahel, 1973 and the epicentral data from Csomor and Kiss (1959), Schenkova et al (1979), Karnick et al (1980), Drimmel (1980), and Ribaric (1982). Circles indicate epicentral locations for earthquakes with a focal depth less than 40 km, squares indicate earthquakes with focal depths between 40 km and 100 km, and triangles indicate earthquakes with focal depths greater than 100 km. Seismic data is discussed in greater detail in Gutdeutsch and Aric (this volume). Scale 1:2,000,000. Compiled by R. Gutdeutsch and K. Aric.

REFERENCES

Csomor, D. and C. Kiss, 1959, Die Seismizitat von Ungarn: Studia Geophysica et Geodetica, p. 33–42.

Drimmel, J., 1980, Rezente Seismizitat und Seismotektonik des Ostalpenraumes: Der Geologische Aufbau Österreichs, Springer-Verlag, Wien, New York, p. 507–527.

Gutdeutsch, R. and K. Aric, this volume, Seismicity and neotectonics of the East Alpine, Carpathian and Pannonian Region.

Karnick, V., D. Prochazkova and Z. Schenkova, 1980, Atlas of seismological maps, Central and Eastern Europe: Geophys. Inst. Czech. Akad. Sci., Prague, 11 maps.

Mahel, M., 1973, Tectonic map of the Carpathian–Balkan mountain system and adjacent areas: Published by D. Stúr geological Institute in Bratislava/Unesco, Paris, 1 map.

Ribaric, V., 1982, Seizmicnost Slovenije, Katalog Potresov 792 n.e.: Catalogue of Earthquakes (792 A.D.–1981), Ljubljana, 649 pages.

Schenkova, Z., V. Karnik, and V. Schenk, 1979, Earthquake epicenters in Czechoslovakia and adjacent area (map): Geophysical Institute of the Academie of Sciences, Prague.

MAP 8

Thicknesses of the Neogene–Quaternary sedimentary rocks in the Pannonian basin are shown by contour lines every 500 meters. In the Transylvanian basin, the contour lines show the thickness of sediments above the base of the Badenian (Tortonian). Sources include Jiricek and Tomek (1981), Fusán et al. (1972), Rudinec et al. (1981), Mahel (1973), Kőrössy (1970), Pogácsás (1980), Dank (1963), Filjak et al. (1969), Ali-Mehmed et al. (1978), and Visarion and Veliciu (1981). The areas covered by each source are shown in Horváth and Royden (1981). Depth indicated in kilometers. Scale 1:1,000,000. Compiled by F. Horváth.

REFERENCES

Ali-Mehmed, E., T. Bandrabur, P. Craciu, C. Ghenea, P. Polonic, and M. Visarion, 1978, Contributions to the knowledge of structures with thermal waters in the eastern part of the Pannonian depression (Romania): Proceedings of the Conference on the Hydrogeology of Great Sedimentary Basins, Budapest, May/June 1976, Annales Inst. Geol. Publ. Hung., v. 59, n. 1–4, p. 431–447.

Dank, V., 1963, Stratigraphy of the Neogene basins of southern Alföld and their relation to the areas of south Baranya and Yugoslavia: Földt. Közl., v. 93, n. 3, p. 304–324.

Filjak, R., Z. Pletikapić, D. Nikolić, and V. Askin, 1969, Geology of petroleum and natural gas from the Neocene complex and its basement in the southern part of the Pannonian basin, Yugoslavia, *in* P. Hepple, ed., The Exploration of Petroleum in Europe and North Africa, Institute of Petroleum, London, p. 113–130.

Fusán, O., J. Slavik, J. Plancar, and J. Ibrmajer, 1972, Geological map of the substratum of the covered areas in the southern part of the inner West Carpathians: Geol. Inst. D. Stúr, Bratislava, scale 1:500,000.

Horváth, F., and L. Royden, 1981, Mechanism for formation of the intra-Carpathian basins: A review: Earth Evol. Sci., v. 1, n. 3–4, p. 307–316.

Jiricek, R., and C. Tomek, 1981, Sedimentary and structural evolution of the Vienna basin: Earth Evol. Sci., v. 1, n. 3–4, p. 307–316.

Korössy, L., 1970: Entwicklungsgeshichte der neogenen becken in Ungarn: Acta Geol. Acad. Sci. Hung., v. 14, p. 421–429.

Mahel, M., ed., 1973, Tectonic map of the Carpathian–Balkan mountain system and adjacent areas: D. Stúr Geological Institute, Bratislava and UNESCO, Vienna, scale 1:1,000,000.

Pogácsás, Gy., 1980, Evolution of Hungary's Neogene depressions in the light of geophysical surface measurements: Földt. Közl, v. 110, n. 3–4, p. 485–497.

Rudinec, R., C. Tomek, and R. Jiricek, 1981, Sedimentary and structural evolution of the Transcarpathian depression: Earth Evol. Sci., v. 1, n. 3–4, p. 205–217.

Visarion, M., and S. Veliciu, 1981, Some geological and geophysical characteristics of the Transylvanian Basin: Earth Evol. Sci., v. 1, no. 3–4, p. 212–217.

Index

A reference is indexed according to its important, or "key" words.

Three columns are to the left of a keyword entry. The first column, a letter entry, represents the AAPG book series from which the reference originated. In this case, ME stands for AAPG Memoir Series. Every five years, AAPG will merge all its indexes together, and the letters ME will differentiate this reference from those of the Studies in Geology Series (ST) or from the AAPG Bulletin (B).

The following number is the series number. In this case, 45 represents a reference from AAPG Memoir 45. The third column lists the page number of this volume on which the reference can be found. The fourth column represents the type of entry: K = keyword, A - author, and T = title.

Index entries without page numbers represent the title or compilation editors of this volume.